Osvaldo Gervasi · Beniamino Murgante
Sanjay Misra · Marina L. Gavrilova
Ana Maria Alves Coutinho Rocha · Carmelo Torre
David Taniar · Bernady O. Apduhan (Eds.)

Computational Science and Its Applications – ICCSA 2015

15th International Conference
Banff, AB, Canada, June 22–25, 2015
Proceedings, Part III

 Springer

Editors
Osvaldo Gervasi
University of Perugia
Perugia
Italy

Ana Maria Alves Coutinho Rocha
University of Minho
Braga
Portugal

Beniamino Murgante
University of Basilicata
Potenza
Italy

Carmelo Torre
Polytechnic University
Bari
Italy

Sanjay Misra
Covenant University
Canaanland
Nigeria

David Taniar
Monash University
Clayton, VIC
Australia

Marina L. Gavrilova
University of Calgary
Calgary, AB
Canada

Bernady O. Apduhan
Kyushu Sangyo University
Fukuoka
Japan

ISSN 0302-9743   ISSN 1611-3349 (electronic)
Lecture Notes in Computer Science
ISBN 978-3-319-21469-6   ISBN 978-3-319-21470-2 (eBook)
DOI 10.1007/978-3-319-21470-2

Library of Congress Control Number: 2015943360

LNCS Sublibrary: SL1 – Theoretical Computer Science and General Issues

Springer Cham Heidelberg New York Dordrecht London

Printed on acid-free paper

Springer International Publishing AG Switzerland is part of Springer Science+Business Media
(www.springer.com)

Preface

The year 2015 is a memorable year for the International Conference on Computational Science and Its Applications. In 2003, the First International Conference on Computational Science and Its Applications (chaired by C.J.K. Tan and M. Gavrilova) took place in Montreal, Canada (2003), and the following year it was hosted by A. Laganà and O. Gervasi in Assisi, Italy (2004). It then moved to Singapore (2005), Glasgow, UK (2006), Kuala-Lumpur, Malaysia (2007), Perugia, Italy (2008), Seoul, Korea (2009), Fukuoka, Japan (2010), Santander, Spain (2011), Salvador de Bahia, Brazil (2012), Ho Chi Minh City, Vietnam (2013), and Guimarães, Portugal (2014). The current installment of ICCSA 2015 took place in majestic Banff National Park, Banff, Alberta, Canada, during June 22–25, 2015.

The event received approximately 780 submissions from over 45 countries, evaluated by over 600 reviewers worldwide.

Its main track acceptance rate was approximately 29.7 % for full papers. In addition to full papers, published by Springer, the event accepted short papers, poster papers, and PhD student showcase works that are published in the IEEE CPS proceedings.

It also runs a number of parallel workshops, some for over 10 years, with new ones appearing for the first time this year. The success of ICCSA is largely contributed to the continuous support of the computational sciences community as well as researchers working in the applied relevant fields, such as graphics, image processing, biometrics, optimization, computer modeling, information systems, geographical sciences, physics, biology, astronomy, biometrics, virtual reality, and robotics, to name a few.

Over the past decade, the vibrant and promising area focusing on performance-driven computing and big data has became one of the key points of research enhancing the performance of information systems and supported processes. In addition to high-quality research at the frontier of these fields, consistently presented at ICCSA, a number of special journal issues are being planned following ICCSA 2015, including TCS Springer (*Transactions on Computational Sciences,* LNCS).

The contribution of the International Steering Committee and the International Program Committee are invaluable in the conference success. The dedication of members of these committees, the majority of whom have fulfilled this difficult role for the last 10 years, is astounding. Our warm appreciation also goes to the invited speakers, all event sponsors, supporting organizations, and volunteers. Finally, we thank all the authors for their submissions making the ICCSA conference series a well recognized and a highly successful event year after year.

June 2015

Marina L. Gavrilova
Osvaldo Gervasi
Bernady O. Apduhan

Organization

ICCSA 2015 was organized by the University of Calgary (Canada), the University of Perugia (Italy), the University of Basilicata (Italy), Monash University (Australia), Kyushu Sangyo University (Japan), and the University of Minho, (Portugal)

Honorary General Chairs

Antonio Laganà	University of Perugia, Italy
Norio Shiratori	Tohoku University, Japan
Kenneth C.J. Tan	Sardina Systems, Estonia

General Chairs

Marina L. Gavrilova	University of Calgary, Canada
Osvaldo Gervasi	University of Perugia, Italy
Bernady O. Apduhan	Kyushu Sangyo University, Japan

Program Committee Chairs

Beniamino Murgante	University of Basilicata, Italy
Ana Maria A.C. Rocha	University of Minho, Portugal
David Taniar	Monash University, Australia

International Advisory Committee

Jemal Abawajy	Deakin University, Australia
Dharma P. Agrawal	University of Cincinnati, USA
Claudia Bauzer Medeiros	University of Campinas, Brazil
Manfred M. Fisher	Vienna University of Economics and Business, Austria
Yee Leung	Chinese University of Hong Kong, SAR China

International Liaison Chairs

Ana Carla P. Bitencourt	Universidade Federal do Reconcavo da Bahia, Brazil
Alfredo Cuzzocrea	ICAR-CNR and University of Calabria, Italy
Maria Irene Falcão	University of Minho, Portugal
Marina L. Gavrilova	University of Calgary, Canada
Robert C.H. Hsu	Chung Hua University, Taiwan
Andrés Iglesias	University of Cantabria, Spain
Tai-Hoon Kim	Hannam University, Korea
Sanjay Misra	University of Minna, Nigeria
Takashi Naka	Kyushu Sangyo University, Japan

Rafael D.C. Santos   Brazilian National Institute for Space Research, Brazil
Maribel Yasmina Santos   University of Minho, Portugal

Workshop and Session Organizing Chairs

Beniamino Murgante   University of Basilicata, Italy
Jorge Gustavo Rocha   University of Minho, Portugal

Local Arrangement Chairs

Marina Gavrilova   University of Calgary, Canada (Chair)
Madeena Sultana   University of Calgary, Canada
Padma Polash Paul   University of Calgary, Canada
Faisal Ahmed   University of Calgary, Canada
Hossein Talebi   University of Calgary, Canada
Camille Sinanan   University of Calgary, Canada

Venue

ICCSA 2015 took place in the Banff Park Lodge Conference Center, Alberta (Canada).

Workshop Organizers

Agricultural and Environment Information and Decision Support Systems (AEIDSS 2015)

Sandro Bimonte   IRSTEA, France
André Miralles   IRSTEA, France
Frederic Hubert   University of Laval, Canada
François Pinet   IRSTEA, France

Approaches or Methods of Security Engineering (AMSE 2015)

TaiHoon Kim   Sungshin W. University, Korea

Advances in Information Systems and Technologies for Emergency Preparedness and Risk Assessment (ASTER 2015)

Maurizio Pollino   ENEA, Italy
Marco Vona   University of Basilicata, Italy
Beniamino Murgante   University of Basilicata, Italy

Advances in Web-Based Learning (AWBL 2015)

Mustafa Murat Inceoglu   Ege University, Turkey

Bio-inspired Computing and Applications (BIOCA 2015)

Nadia Nedjah	State University of Rio de Janeiro, Brazil
Luiza de Macedo Mourell	State University of Rio de Janeiro, Brazil

Computer-Aided Modeling, Simulation, and Analysis (CAMSA 2015)

Jie Shen	University of Michigan, USA, and Jilin University, China
Hao Chen	Shanghai University of Engineering Science, China
Xiaoqiang Liun	Donghua University, China
Weichun Shi	Shanghai Maritime University, China

Computational and Applied Statistics (CAS 2015)

Ana Cristina Braga	University of Minho, Portugal
Ana Paula Costa Conceicao Amorim	University of Minho, Portugal

Computational Geometry and Security Applications (CGSA 2015)

Marina L. Gavrilova	University of Calgary, Canada

Computational Algorithms and Sustainable Assessment (CLASS 2015)

Antonino Marvuglia	Public Research Centre Henri Tudor, Luxembourg
Beniamino Murgante	University of Basilicata, Italy

Chemistry and Materials Sciences and Technologies (CMST 2015)

Antonio Laganà	University of Perugia, Italy
Alessandro Costantini	INFN, Italy
Noelia Faginas Lago	University of Perugia, Italy
Leonardo Pacifici	University of Perugia, Italy

Computational Optimization and Applications (COA 2015)

Ana Maria Rocha	University of Minho, Portugal
Humberto Rocha	University of Coimbra, Portugal

Cities, Technologies and Planning (CTP 2015)

Giuseppe Borruso	University of Trieste, Italy
Beniamino Murgante	University of Basilicata, Italy

Econometrics and Multidimensional Evaluation in the Urban Environment (EMEUE 2015)

Carmelo M. Torre	Polytechnic of Bari, Italy
Maria Cerreta	University of Naples Federico II, Italy
Paola Perchinunno	University of Bari, Italy

Simona Panaro   University of Naples Federico II, Italy
Raffaele Attardi   University of Naples Federico II, Italy
Claudia Ceppi   Polytechnic of Bari, Italy

Future Computing Systems, Technologies, and Applications (FISTA 2015)

Bernady O. Apduhan   Kyushu Sangyo University, Japan
Rafael Santos   Brazilian National Institute for Space Research, Brazil
Jianhua Ma   Hosei University, Japan
Qun Jin   Waseda University, Japan

Geographical Analysis, Urban Modeling, Spatial Statistics (GEOGAN-MOD 2015)

Giuseppe Borruso   University of Trieste, Italy
Beniamino Murgante   University of Basilicata, Italy
Hartmut Asche   University of Potsdam, Germany

Land Use Monitoring for Soil Consumption Reduction (LUMS 2015)

Carmelo M. Torre   Polytechnic of Bari, Italy
Alessandro Bonifazi   Polytechnic of Bari, Italy
Valentina Sannicandro   University Federico II of Naples, Italy
Massimiliano   University of Salerno, Italy
   Bencardino
Gianluca di Cugno   Polytechnic of Bari, Italy
Beniamino Murgante   University of Basilicata, Italy

Mobile Communications (MC 2015)

Hyunseung Choo   Sungkyunkwan University, Korea

Mobile Computing, Sensing, and Actuation for Cyber Physical Systems (MSA4CPS 2015)

Saad Qaisar   NUST School of Electrical Engineering and Computer
                               Science, Pakistan
Moonseong Kim   Korean Intellectual Property Office, Korea

Quantum Mechanics: Computational Strategies and Applications (QMCSA 2015)

Mirco Ragni   Universidad Federal de Bahia, Brazil
Ana Carla Peixoto   Universidade Estadual de Feira de Santana, Brazil
   Bitencourt
Roger Anderson   University of California, USA
Vincenzo Aquilanti   University of Perugia, Italy
Frederico Vasconcellos   Universidad Federal de Bahia, Brazil
   Prudente

Remote Sensing Data Analysis, Modeling, Interpretation and Applications: From a Global View to a Local Analysis (RS2015)

Rosa Lasaponara	Institute of Methodologies for Environmental Analysis, National Research Council, Italy

Scientific Computing Infrastructure (SCI 2015)

Alexander Bodganov	St. Petersburg State University, Russia
Elena Stankova	St. Petersburg State University, Russia

Software Engineering Processes and Applications (SEPA 2015)

Sanjay Misra	Covenant University, Nigeria

Software Quality (SQ 2015)

Sanjay Misra	Covenant University, Nigeria

Advances in Spatio-Temporal Analytics (ST-Analytics 2015)

Joao Moura Pires	New University of Lisbon, Portugal
Maribel Yasmina Santos	New University of Lisbon, Portugal

Tools and Techniques in Software Development Processes (TTSDP 2015)

Sanjay Misra	Covenant University, Nigeria

Virtual Reality and Its Applications (VRA 2015)

Osvaldo Gervasi	University of Perugia, Italy
Lucio Depaolis	University of Salento, Italy

Program Committee

Jemal Abawajy	Deakin University, Australia
Kenny Adamson	University of Ulster, UK
Filipe Alvelos	University of Minho, Portugal
Paula Amaral	Universidade Nova de Lisboa, Portugal
Hartmut Asche	University of Potsdam, Germany
Md. Abul Kalam Azad	University of Minho, Portugal
Michela Bertolotto	University College Dublin, Ireland
Sandro Bimonte	CEMAGREF, TSCF, France
Rod Blais	University of Calgary, Canada
Ivan Blecic	University of Sassari, Italy
Giuseppe Borruso	University of Trieste, Italy
Yves Caniou	Lyon University, France
José A. Cardoso e Cunha	Universidade Nova de Lisboa, Portugal
Leocadio G. Casado	University of Almeria, Spain

Carlo Cattani	University of Salerno, Italy
Mete Celik	Erciyes University, Turkey
Alexander Chemeris	National Technical University of Ukraine KPI, Ukraine
Min Young Chung	Sungkyunkwan University, Korea
Gilberto Corso Pereira	Federal University of Bahia, Brazil
M. Fernanda Costa	University of Minho, Portugal
Gaspar Cunha	University of Minho, Portugal
Alfredo Cuzzocrea	ICAR-CNR and University of Calabria, Italy
Carla Dal Sasso Freitas	Universidade Federal do Rio Grande do Sul, Brazil
Pradesh Debba	The Council for Scientific and Industrial Research (CSIR), South Africa
Hendrik Decker	Instituto Tecnológico de Informática, Spain
Frank Devai	London South Bank University, UK
Rodolphe Devillers	Memorial University of Newfoundland, Canada
Prabu Dorairaj	NetApp, India/USA
M. Irene Falcao	University of Minho, Portugal
Cherry Liu Fang	U.S. DOE Ames Laboratory, USA
Edite M.G.P. Fernandes	University of Minho, Portugal
Jose-Jesus Fernandez	National Centre for Biotechnology, CSIS, Spain
Maria Antonia Forjaz	University of Minho, Portugal
Maria Celia Furtado Rocha	PRODEB/UFBA, Brazil
Akemi Galvez	University of Cantabria, Spain
Paulino Jose Garcia Nieto	University of Oviedo, Spain
Marina Gavrilova	University of Calgary, Canada
Jerome Gensel	LSR-IMAG, France
Maria Giaoutzi	National Technical University, Athens, Greece
Andrzej M. Goscinski	Deakin University, Australia
Alex Hagen-Zanker	University of Cambridge, UK
Malgorzata Hanzl	Technical University of Lodz, Poland
Shanmugasundaram Hariharan	B.S. Abdur Rahman University, India
Eligius M.T. Hendrix	University of Malaga/Wageningen University, Spain/The Netherlands
Tutut Herawan	Universitas Teknologi Yogyakarta, Indonesia
Hisamoto Hiyoshi	Gunma University, Japan
Fermin Huarte	University of Barcelona, Spain
Andres Iglesias	University of Cantabria, Spain
Mustafa Inceoglu	EGE University, Turkey
Peter Jimack	University of Leeds, UK
Qun Jin	Waseda University, Japan
Farid Karimipour	Vienna University of Technology, Austria
Baris Kazar	Oracle Corp., USA
DongSeong Kim	University of Canterbury, New Zealand
Taihoon Kim	Hannam University, Korea

Ivana Kolingerova	University of West Bohemia, Czech Republic
Dieter Kranzlmueller	LMU and LRZ Munich, Germany
Antonio Laganà	University of Perugia, Italy
Rosa Lasaponara	National Research Council, Italy
Maurizio Lazzari	National Research Council, Italy
Cheng Siong Lee	Monash University, Australia
Sangyoun Lee	Yonsei University, Korea
Jongchan Lee	Kunsan National University, Korea
Clement Leung	Hong Kong Baptist University, Hong Kong, SAR China
Chendong Li	University of Connecticut, USA
Gang Li	Deakin University, Australia
Ming Li	East China Normal University, China
Fang Liu	AMES Laboratories, USA
Xin Liu	University of Calgary, Canada
Savino Longo	University of Bari, Italy
Tinghuai Ma	NanJing University of Information Science and Technology, China
Sergio Maffioletti	University of Zurich, Switzerland
Ernesto Marcheggiani	Katholieke Universiteit Leuven, Belgium
Antonino Marvuglia	Research Centre Henri Tudor, Luxembourg
Nicola Masini	National Research Council, Italy
Nirvana Meratnia	University of Twente, The Netherlands
Alfredo Milani	University of Perugia, Italy
Sanjay Misra	Federal University of Technology Minna, Nigeria
Giuseppe Modica	University of Reggio Calabria, Italy
José Luis Montaña	University of Cantabria, Spain
Beniamino Murgante	University of Basilicata, Italy
Jiri Nedoma	Academy of Sciences of the Czech Republic, Czech Republic
Laszlo Neumann	University of Girona, Spain
Kok-Leong Ong	Deakin University, Australia
Belen Palop	Universidad de Valladolid, Spain
Marcin Paprzycki	Polish Academy of Sciences, Poland
Eric Pardede	La Trobe University, Australia
Kwangjin Park	Wonkwang University, Korea
Ana Isabel Pereira	Polytechnic Institute of Braganca, Portugal
Maurizio Pollino	Italian National Agency for New Technologies, Energy and Sustainable Economic Development, Italy
Alenka Poplin	University of Hamburg, Germany
Vidyasagar Potdar	Curtin University of Technology, Australia
David C. Prosperi	Florida Atlantic University, USA
Wenny Rahayu	La Trobe University, Australia
Jerzy Respondek	Silesian University of Technology Poland
Ana Maria A.C. Rocha	University of Minho, Portugal

Humberto Rocha	INESC-Coimbra, Portugal
Alexey Rodionov	Institute of Computational Mathematics and Mathematical Geophysics, Russia
Cristina S. Rodrigues	University of Minho, Portugal
Octavio Roncero	CSIC, Spain
Maytham Safar	Kuwait University, Kuwait
Chiara Saracino	A.O. Ospedale Niguarda Ca' Granda - Milano, Italy
Haiduke Sarafian	The Pennsylvania State University, USA
Jie Shen	University of Michigan, USA
Qi Shi	Liverpool John Moores University, UK
Dale Shires	U.S. Army Research Laboratory, USA
Takuo Suganuma	Tohoku University, Japan
Sergio Tasso	University of Perugia, Italy
Ana Paula Teixeira	University of Tras-os-Montes and Alto Douro, Portugal
Senhorinha Teixeira	University of Minho, Portugal
Parimala Thulasiraman	University of Manitoba, Canada
Carmelo Torre	Polytechnic of Bari, Italy
Javier Martinez Torres	Centro Universitario de la Defensa Zaragoza, Spain
Giuseppe A. Trunfio	University of Sassari, Italy
Unal Ufuktepe	Izmir University of Economics, Turkey
Toshihiro Uchibayashi	Kyushu Sangyo University, Japan
Mario Valle	Swiss National Supercomputing Centre, Switzerland
Pablo Vanegas	University of Cuenca, Equador
Piero Giorgio Verdini	INFN Pisa and CERN, Italy
Marco Vizzari	University of Perugia, Italy
Koichi Wada	University of Tsukuba, Japan
Krzysztof Walkowiak	Wroclaw University of Technology, Poland
Robert Weibel	University of Zurich, Switzerland
Roland Wismüller	Universität Siegen, Germany
Mudasser Wyne	SOET National University, USA
Chung-Huang Yang	National Kaohsiung Normal University, Taiwan
Xin-She Yang	National Physical Laboratory, UK
Salim Zabir	France Telecom Japan Co., Japan
Haifeng Zhao	University of California, Davis, USA
Kewen Zhao	University of Qiongzhou, China
Albert Y. Zomaya	University of Sydney, Australia

Reviewers

Abawajy Jemal	Deakin University, Australia
Abdi Samane	University College Cork, Ireland
Aceto Lidia	University of Pisa, Italy
Acharjee Shukla	Dibrugarh University, India
Adriano Elias	Universidade Nova de Lisboa, Portugal
Afreixo Vera	University of Aveiro, Portugal
Aguiar Ademar	Universidade do Porto, Portugal

Aguilar Antonio	University of Barcelona, Spain
Aguilar José Alfonso	Universidad Autónoma de Sinaloa, Mexico
Ahmed Faisal	University of Calgary, Canada
Aktas Mehmet	Yildiz Technical University, Turkey
Al-Juboori AliAlwan	International Islamic University Malaysia, Malaysia
Alarcon Vladimir	Universidad Diego Portales, Chile
Alberti Margarita	University of Barcelona, Spain
Ali Salman	NUST, Pakistan
Alkazemi Basem Qassim	University, Saudi Arabia
Alvanides Seraphim	Northumbria University, UK
Alvelos Filipe	University of Minho, Portugal
Alves Cláudio	University of Minho, Portugal
Alves José Luis	University of Minho, Portugal
Alves Maria Joo	Universidade de Coimbra, Portugal
Amin Benatia Mohamed	Groupe Cesi, France
Amorim Ana Paula	University of Minho, Portugal
Amorim Paulo	Federal University of Rio de Janeiro, Brazil
Andrade Wilkerson	Federal University of Campina Grande, Brazil
Andrianov Serge	Yandex, Russia
Aniche Mauricio	University of São Paulo, Brazil
Andrienko Gennady	Fraunhofer Institute for Intelligent Analysis and Informations Systems, Germany
Apduhan Bernady	Kyushu Sangyo University, Japan
Aquilanti Vincenzo	University of Perugia, Italy
Aquino Gibeon	UFRN, Brazil
Argiolas Michele	University of Cagliari, Italy
Asche Hartmut	Potsdam University, Germany
Athayde Maria Emilia Feijão Queiroz	University of Minho, Portugal
Attardi Raffaele	University of Napoli Federico II, Italy
Azad Md. Abdul	Indian Institute of Technology Kanpur, India
Azad Md. Abul Kalam	University of Minho, Portugal
Bao Fernando	Universidade Nova de Lisboa, Portugal
Badard Thierry	Laval University, Canada
Bae Ihn-Han	Catholic University of Daegu, South Korea
Baioletti Marco	University of Perugia, Italy
Balena Pasquale	Polytechnic of Bari, Italy
Banerjee Mahua	Xavier Institute of Social Sciences, India
Barroca Filho Itamir	UFRN, Brazil
Bartoli Daniele	University of Perugia, Italy
Bastanfard Azam	Islamic Azad University, Iran
Belanzoni Paola	University of Perugia, Italy
Bencardino Massimiliano	University of Salerno, Italy
Benigni Gladys	University of Oriente, Venezuela

Bertolotto Michela	University College Dublin, Ireland
Bilancia Massimo	Università di Bari, Italy
Blanquer Ignacio	Universitat Politècnica de València, Spain
Bodini Olivier	Université Pierre et Marie Curie Paris and CNRS, France
Bogdanov Alexander	Saint-Petersburg State University, Russia
Bollini Letizia	University of Milano, Italy
Bonifazi Alessandro	Polytechnic of Bari, Italy
Borruso Giuseppe	University of Trieste, Italy
Bostenaru Maria	"Ion Mincu" University of Architecture and Urbanism, Romania
Boucelma Omar	University of Marseille, France
Braga Ana Cristina	University of Minho, Portugal
Branquinho Amilcar	University of Coimbra, Portugal
Brás Carmo	Universidade Nova de Lisboa, Portugal
Cacao Isabel	University of Aveiro, Portugal
Cadarso-Suárez Carmen	University of Santiago de Compostela, Spain
Caiaffa Emanuela	ENEA, Italy
Calamita Giuseppe	National Research Council, Italy
Campagna Michele	University of Cagliari, Italy
Campobasso Francesco	University of Bari, Italy
Campos José	University of Minho, Portugal
Caniato Renhe Marcelo	Universidade Federal de Juiz de Fora, Brazil
Cannatella Daniele	University of Napoli Federico II, Italy
Canora Filomena	University of Basilicata, Italy
Cannatella Daniele	University of Napoli Federico II, Italy
Canora Filomena	University of Basilicata, Italy
Carbonara Sebastiano	University of Chieti, Italy
Carlini Maurizio	University of Tuscia, Italy
Carneiro Claudio	École Polytechnique Fédérale de Lausanne, Switzerland
Ceppi Claudia	Polytechnic of Bari, Italy
Cerreta Maria	University Federico II of Naples, Italy
Chen Hao	Shanghai University of Engineering Science, China
Choi Joonsoo	Kookmin University, South Korea
Choo Hyunseung	Sungkyunkwan University, South Korea
Chung Min Young	Sungkyunkwan University, South Korea
Chung Myoungbeom	Sungkyunkwan University, South Korea
Chung Tai-Myoung	Sungkyunkwan University, South Korea
Cirrincione Maurizio	Université de Technologie Belfort-Montbeliard, France
Clementini Eliseo	University of L'Aquila, Italy
Coelho Leandro dos Santos	PUC-PR, Brazil
Coletti Cecilia	University of Chieti, Italy
Conceicao Ana	Universidade do Algarve, Portugal
Correia Elisete	University of Trás-Os-Montes e Alto Douro, Portugal
Correia Filipe	FEUP, Portugal

Correia Florbela Maria da Cruz Domingues	Instituto Politécnico de Viana do Castelo, Portugal
Corso Pereira Gilberto	UFPA, Brazil
Cortés Ana	Universitat Autònoma de Barcelona, Spain
Cosido Oscar	Ayuntamiento de Santander, Spain
Costa Carlos	Faculdade Engenharia U. Porto, Portugal
Costa Fernanda	University of Minho, Portugal
Costantini Alessandro	INFN, Italy
Crasso Marco	National Scientific and Technical Research Council, Argentina
Crawford Broderick	Universidad Catolica de Valparaiso, Chile
Crestaz Ezio	GiScience, Italia
Cristia Maximiliano	CIFASIS and UNR, Argentina
Cunha Gaspar	University of Minho, Portugal
Cutini Valerio	University of Pisa, Italy
Danese Maria	IBAM, CNR, Italy
Daneshpajouh Shervin	University of Western Ontario, Canada
De Almeida Regina	University of Trás-os-Montes e Alto Douro, Portugal
de Doncker Elise	University of Michgan, USA
De Fino Mariella	Polytechnic of Bari, Italy
De Paolis Lucio Tommaso	University of Salento, Italy
de Rezende Pedro J.	Universidade Estadual de Campinas, Brazil
De Rosa Fortuna	University of Napoli Federico II, Italy
De Toro Pasquale	University of Napoli Federico II, Italy
Decker Hendrik	Instituto Tecnológico de Informática, Spain
Degtyarev Alexander	Saint-Petersburg State University, Russia
Deiana Andrea	Geoinfolab, Italia
Deniz Berkhan	Aselsan Electronics Inc., Turkey
Desjardin Eric	University of Reims, France
Devai Frank	London South Bank University, UK
Dwivedi Sanjay Kumar	Babasaheb Bhimrao Ambedkar University, India
Dhawale Chitra	PR Pote College, Amravati, India
Di Cugno Gianluca	Polytechnic of Bari, Italy
Di Gangi Massimo	University of Messina, Italy
Di Leo Margherita	JRC, European Commission, Belgium
Dias Joana	University of Coimbra, Portugal
Dias d'Almeida Filomena	University of Porto, Portugal
Diez Teresa	Universidad de Alcalá, Spain
Dilo Arta	University of Twente, The Netherlands
Dixit Veersain	Delhi University, India
Doan Anh Vu	Université Libre de Bruxelles, Belgium
Durrieu Sylvie	Maison de la Teledetection Montpellier, France
Dutra Inês	University of Porto, Portugal
Dyskin Arcady	The University of Western Australia, Australia

Eichelberger Hanno	University of Tübingen, Germany
El-Zawawy Mohamed A.	Cairo University, Egypt
Escalona Maria-Jose	University of Seville, Spain
Falcão M. Irene	University of Minho, Portugal
Farantos Stavros	University of Crete and FORTH, Greece
Faria Susana	University of Minho, Portugal
Fernandes Edite	University of Minho, Portugal
Fernandes Rosário	University of Minho, Portugal
Fernandez Joao P.	Universidade da Beira Interior, Portugal
Ferrão Maria	University of Beira Interior and CEMAPRE, Portugal
Ferreira Fátima	University of Trás-Os-Montes e Alto Douro, Portugal
Figueiredo Manuel Carlos	University of Minho, Portugal
Filipe Ana	University of Minho, Portugal
Flouvat Frederic	University New Caledonia, New Caledonia
Forjaz Maria Antónia	University of Minho, Portugal
Formosa Saviour	University of Malta, Malta
Fort Marta	University of Girona, Spain
Franciosa Alfredo	University of Napoli Federico II, Italy
Freitas Adelaide de Fátima Baptista Valente	University of Aveiro, Portugal
Frydman Claudia	Laboratoire des Sciences de l'Information et des Systèmes, France
Fusco Giovanni	CNRS - UMR ESPACE, France
Gabrani Goldie	University of Delhi, India Galleguillos Cristian, Pontificia Universidad Catlica de Valparaso, Chile
Gao Shang	Zhongnan University of Economics and Law, China
Garau Chiara	University of Cagliari, Italy
Garcia Ernesto	University of the Basque Country, Spain
Garca Omar Vicente	Universidad Autònoma de Sinaloa, Mexico
Garcia Tobio Javier	Centro de Supercomputación de Galicia, CESGA, Spain
Gavrilova Marina	University of Calgary, Canada
Gazzea Nicoletta	ISPRA, Italy
Gensel Jerome	IMAG, France
Geraldi Edoardo	National Research Council, Italy
Gervasi Osvaldo	University of Perugia, Italy
Giaoutzi Maria	National Technical University Athens, Greece
Gil Artur	University of the Azores, Portugal
Gizzi Fabrizio	National Research Council, Italy
Gomes Abel	Universidad de Beira Interior, Portugal
Gomes Maria Cecilia	Universidade Nova de Lisboa, Portugal
Gomes dos Anjos Eudisley	Federal University of Paraba, Brazil
Gonçalves Alexandre	Instituto Superior Tecnico Lisboa, Portugal

Gonçalves Arminda Manuela	University of Minho, Portugal
Gonzaga de Oliveira Sanderson Lincohn	Universidade Do Estado De Santa Catarina, Brazil
Gonzalez-Aguilera Diego	Universidad de Salamanca, Spain
Gorbachev Yuriy	Geolink Technologies, Russia
Govani Kishan	Darshan Institute of Engineering Technology, India
Grandison Tyrone	Proficiency Labs International, USA
Gravagnuolo Antonia	University of Napoli Federico II, Italy
Grilli Luca	University of Perugia, Italy
Guerra Eduardo	National Institute for Space Research, Brazil
Guo Hua	Carleton University, Canada
Hanazumi Simone	University of São Paulo, Brazil
Hanif Mohammad Abu	Chonbuk National University, South Korea
Hansen Henning Sten	Aalborg University, Denmark
Hanzl Malgorzata	University of Lodz, Poland
Hegedus Peter	University of Szeged, Hungary
Heijungs Reinout	VU University Amsterdam, The Netherlands
Hendrix Eligius M.T.	University of Malaga/Wageningen University, Spain/The Netherlands
Henriques Carla	Escola Superior de Tecnologia e Gestão, Portugal
Herawan Tutut	University of Malaya, Malaysia
Hiyoshi Hisamoto	Gunma University, Japan
Hodorog Madalina	Austria Academy of Science, Austria
Hong Choong Seon	Kyung Hee University, South Korea
Hsu Ching-Hsien	Chung Hua University, Taiwan
Hsu Hui-Huang	Tamkang University, Taiwan
Hu Hong	The Honk Kong Polytechnic University, China
Huang Jen-Fa	National Cheng Kung University, Taiwan
Hubert Frederic	Université Laval, Canada
Iglesias Andres	University of Cantabria, Spain
Jamal Amna	National University of Singapore, Singapore
Jank Gerhard	Aachen University, Germany
Jeong Jongpil	Sungkyunkwan University, South Korea
Jiang Bin	University of Gävle, Sweden
Johnson Franklin	Universidad de Playa Ancha, Chile
Kalogirou Stamatis	Harokopio University of Athens, Greece
Kamoun Farouk	Université de la Manouba, Tunisia
Kanchi Saroja	Kettering University, USA
Kanevski Mikhail	University of Lausanne, Switzerland
Kang Myoung-Ah	ISIMA Blaise Pascal University, France
Karandikar Varsha	Devi Ahilya University, Indore, India
Karimipour Farid	Vienna University of Technology, Austria
Kavouras Marinos	University of Lausanne, Switzerland
Kazar Baris	Oracle Corp., USA

Keramat Alireza	Jundi-Shapur Univ. of Technology, Iran
Khan Murtaza	NUST, Pakistan
Khattak Asad Masood	Kyung Hee University, Korea
Khazaei Hamzeh	Ryerson University, Canada
Khurshid Khawar	NUST, Pakistan
Kim Dongsoo	Indiana University-Purdue University Indianapolis, USA
Kim Mihui	Hankyong National University, South Korea
Koo Bonhyun	Samsung, South Korea
Korkhov Vladimir	St. Petersburg State University, Russia
Kotzinos Dimitrios	Université de Cergy-Pontoise, France
Kumar Dileep	SR Engineering College, India
Kurdia Anastasia	Buknell University, USA
Lachance-Bernard Nicolas	École Polytechnique Fédérale de Lausanne, Switzerland
Laganà Antonio	University of Perugia, Italy
Lai Sabrina	University of Cagliari, Italy
Lanorte Antonio	CNR-IMAA, Italy
Lanza Viviana	Lombardy Regional Institute for Research, Italy
Lasaponara Rosa	National Research Council, Italy
Lassoued Yassine	University College Cork, Ireland
Lazzari Maurizio	CNR IBAM, Italy
Le Duc Tai	Sungkyunkwan University, South Korea
Le Duc Thang	Sungkyunkwan University, South Korea
Le-Thi Kim-Tuyen	Sungkyunkwan University, South Korea
Ledoux Hugo	Delft University of Technology, The Netherlands
Lee Dong-Wook	INHA University, South Korea
Lee Hongseok	Sungkyunkwan University, South Korea
Lee Ickjai	James Cook University, Australia
Lee Junghoon	Jeju National University, South Korea
Lee KangWoo	Sungkyunkwan University, South Korea
Legatiuk Dmitrii	Bauhaus University Weimar, Germany
Lendvay Gyorgy	Hungarian Academy of Science, Hungary
Leonard Kathryn	California State University, USA
Li Ming	East China Normal University, China
Libourel Thrse	LIRMM, France
Lin Calvin	University of Texas at Austin, USA
Liu Xin	University of Calgary, Canada
Loconte Pierangela	Technical University of Bari, Italy
Lombardi Andrea	University of Perugia, Italy
Longo Savino	University of Bari, Italy
Lopes Cristina	University of California Irvine, USA
Lopez Cabido Ignacio	Centro de Supercomputación de Galicia, CESGA
Lourenço Vanda Marisa	University Nova de Lisboa, Portugal
Luaces Miguel	University of A Coruña, Spain
Lucertini Giulia	IUAV, Italy
Luna Esteban Robles	Universidad Nacional de la Plata, Argentina

M.M.H. Gregori Rodrigo	Universidade Tecnológica Federal do Paraná, Brazil
Machado Gaspar	University of Minho, Portugal
Machado Jose	University of Minho, Portugal
Mahinderjit Singh Manmeet	University Sains Malaysia, Malaysia
Malonek Helmuth	University of Aveiro, Portugal
Manfreda Salvatore	University of Basilicata, Italy
Manns Mary Lynn	University of North Carolina Asheville, USA
Manso Callejo Miguel Angel	Universidad Politécnica de Madrid, Spain
Marechal Bernard	Universidade Federal de Rio de Janeiro, Brazil
Marechal Franois	École Polytechnique Fédérale de Lausanne, Switzerland
Margalef Tomas	Universitat Autònoma de Barcelona, Spain
Marghany Maged	Universiti Teknologi Malaysia, Malaysia
Marsal-Llacuna Maria-Llusa	Universitat de Girona, Spain
Marsh Steven	University of Ontario, Canada
Martins Ana Mafalda	Universidade de Aveiro, Portugal
Martins Pedro	Universidade do Minho, Portugal
Marvuglia Antonino	Public Research Centre Henri Tudor, Luxembourg
Mateos Cristian	Universidad Nacional del Centro, Argentina
Matos Inés	Universidade de Aveiro, Portugal
Matos Jose	Instituto Politecnico do Porto, Portugal
Matos João	ISEP, Portugal
Mauro Giovanni	University of Trieste, Italy
Mauw Sjouke	University of Luxembourg, Luxembourg
Medeiros Pedro	Universidade Nova de Lisboa, Portugal
Melle Franco Manuel	University of Minho, Portugal
Melo Ana	Universidade de São Paulo, Brazil
Michikawa Takashi	University of Tokio, Japan
Milani Alfredo	University of Perugia, Italy
Millo Giovanni	Generali Assicurazioni, Italy
Min-Woo Park	SungKyunKwan University, South Korea
Miranda Fernando	University of Minho, Portugal
Misra Sanjay	Covenant University, Nigeria
Mo Otilia	Universidad Autonoma de Madrid, Spain
Modica Giuseppe	Università Mediterranea di Reggio Calabria, Italy
Mohd Nawi Nazri	Universiti Tun Hussein Onn Malaysia, Malaysia
Morais João	University of Aveiro, Portugal
Moreira Adriano	University of Minho, Portugal
Moerig Marc	University of Magdeburg, Germany
Morzy Mikolaj	University of Poznan, Poland
Mota Alexandre	Universidade Federal de Pernambuco, Brazil
Moura Pires João	Universidade Nova de Lisboa - FCT, Portugal
Mourão Maria	Polytechnic Institute of Viana do Castelo, Portugal

Mourelle Luiza de Macedo	UERJ, Brazil
Mukhopadhyay Asish	University of Windsor, Canada
Mulay Preeti	Bharti Vidyapeeth University, India
Murgante Beniamino	University of Basilicata, Italy
Naghizadeh Majid Reza	Qazvin Islamic Azad University, Iran
Nagy Csaba	University of Szeged, Hungary
Nandy Subhas	Indian Statistical Institute, India
Nash Andrew	Vienna Transport Strategies, Austria
Natário Isabel Cristina Maciel	University Nova de Lisboa, Portugal
Navarrete Gutierrez Tomas	Luxembourg Institute of Science and Technology, Luxembourg
Nedjah Nadia	State University of Rio de Janeiro, Brazil
Nguyen Hong-Quang	Ho Chi Minh City University, Vietnam
Nguyen Tien Dzung	Sungkyunkwan University, South Korea
Nickerson Bradford	University of New Brunswick, Canada
Nielsen Frank	Université Paris Saclay CNRS, France
NM Tuan	Ho Chi Minh City University of Technology, Vietnam
Nogueira Fernando	University of Coimbra, Portugal
Nole Gabriele	IRMAA National Research Council, Italy
Nourollah Ali	Amirkabir University of Technology, Iran
Olivares Rodrigo	UCV, Chile
Oliveira Irene	University of Trás-Os-Montes e Alto Douro, Portugal
Oliveira José A.	University of Minho, Portugal
Oliveira e Silva Luis	University of Lisboa, Portugal
Osaragi Toshihiro	Tokyo Institute of Technology, Japan
Ottomanelli Michele	Polytechnic of Bari, Italy
Ozturk Savas	TUBITAK, Turkey
Pagliara Francesca	University of Naples, Italy
Painho Marco	New University of Lisbon, Portugal
Pantazis Dimos	Technological Educational Institute of Athens, Greece
Paolotti Luisa	University of Perugia, Italy
Papa Enrica	University of Amsterdam, The Netherlands
Papathanasiou Jason	University of Macedonia, Greece
Pardede Eric	La Trobe University, Australia
Parissis Ioannis	Grenoble INP - LCIS, France
Park Gyung-Leen	Jeju National University, South Korea
Park Sooyeon	Korea Polytechnic University, South Korea
Pascale Stefania	University of Basilicata, Italy
Parker Gregory	University of Oklahoma, USA
Parvin Hamid	Iran University of Science and Technology, Iran
Passaro Pierluigi	University of Bari Aldo Moro, Italy
Pathan Al-Sakib Khan	International Islamic University Malaysia, Malaysia
Paul Padma Polash	University of Calgary, Canada

Peixoto Bitencourt Ana Carla	Universidade Estadual de Feira de Santana, Brazil
Peraza Juan Francisco	Autonomous University of Sinaloa, Mexico
Perchinunno Paola	University of Bari, Italy
Pereira Ana	Polytechnic Institute of Bragança, Portugal
Pereira Francisco	Instituto Superior de Engenharia, Portugal
Pereira Paulo	University of Minho, Portugal
Pereira Javier	Diego Portales University, Chile
Pereira Oscar	Universidade de Aveiro, Portugal
Pereira Ricardo	Portugal Telecom Inovacao, Portugal
Perez Gregorio	Universidad de Murcia, Spain
Pesantes Mery	CIMAT, Mexico
Pham Quoc Trung	HCMC University of Technology, Vietnam
Pietrantuono Roberto	University of Napoli "Federico II", Italy
Pimentel Carina	University of Aveiro, Portugal
Pina Antonio	University of Minho, Portugal
Piñar Miguel	Universidad de Granada, Spain
Pinciu Val	Southern Connecticut State University, USA
Pinet Francois	IRSTEA, France
Piscitelli Claudia	Polytechnic University of Bari, Italy
Pollino Maurizio	ENEA, Italy
Poplin Alenka	University of Hamburg, Germany
Porschen Stefan	University of Köln, Germany
Potena Pasqualina	University of Bergamo, Italy
Prata Paula	University of Beira Interior, Portugal
Previtali Mattia	Polytechnic of Milan, Italy
Prosperi David	Florida Atlantic University, USA
Protheroe Dave	London South Bank University, UK
Pusatli Tolga	Cankaya University, Turkey
Qaisar Saad	NURST, Pakistan
Qi Yu	Mesh Capital LLC, USA
Quan Tho	Ho Chi Minh City University of Technology, Vietnam
Raffaeta Alessandra	University of Venice, Italy
Ragni Mirco	Universidade Estadual de Feira de Santana, Brazil
Rahayu Wenny	La Trobe University, Australia
Rautenberg Carlos	University of Graz, Austria
Ravat Franck	IRIT, France
Raza Syed Muhammad	Sungkyunkwan University, South Korea
Rinaldi Antonio	DIETI - UNINA, Italy
Rinzivillo Salvatore	University of Pisa, Italy
Rios Gordon	University College Dublin, Ireland
Riva Sanseverino Eleonora	University of Palermo, Italy
Roanes-Lozano Eugenio	Universidad Complutense de Madrid, Spain
Rocca Lorena	University of Padova, Italy
Roccatello Eduard	3DGIS, Italy

Rocha Ana Maria	University of Minho, Portugal
Rocha Humberto	University of Coimbra, Portugal
Rocha Jorge	University of Minho, Portugal
Rocha Maria Clara	ESTES Coimbra, Portugal
Rocha Miguel	University of Minho, Portugal
Rodrigues Armanda	Universidade Nova de Lisboa, Portugal
Rodrigues Cristina	DPS, University of Minho, Portugal
Rodrigues Joel	University of Minho, Portugal
Rodriguez Daniel	University of Alcala, Spain
Rodrguez Gonzlez Alejandro	Universidad Carlos III Madrid, Spain
Roh Yongwan	Korean IP, South Korea
Romano Bernardino	University of l'Aquila, Italy
Roncaratti Luiz	Instituto de Física, University of Brasilia, Brazil
Roshannejad Ali	University of Calgary, Canada
Rosi Marzio	University of Perugia, Italy
Rossi Gianfranco	University of Parma, Italy
Rotondo Francesco	Polytechnic of Bari, Italy
Roussey Catherine	IRSTEA, France
Ruj Sushmita	Indian Statistical Institute, India
S. Esteves Jorge	University of Aveiro, Portugal
Saeed Husnain	NUST, Pakistan
Sahore Mani	Lovely Professional University, India
Saini Jatinder Singh	Baba Banda Singh Bahadur Engineering College, India
Salzer Reiner	Technical University Dresden, Germany
Sameh Ahmed	The American University in Cairo, Egypt
Sampaio Alcinia Zita	Instituto Superior Tecnico Lisboa, Portugal
Sannicandro Valentina	Polytechnic of Bari, Italy
Santiago Jnior Valdivino	Instituto Nacional de Pesquisas Espaciais, Brazil
Santos Josué	UFABC, Brazil
Santos Rafael	INPE, Brazil
Santos Viviane	Universidade de São Paulo, Brazil
Santucci Valentino	University of Perugia, Italy
Saracino Gloria	University of Milano-Bicocca, Italy
Sarafian Haiduke	Pennsylvania State University, USA
Saraiva João	University of Minho, Portugal
Sarrazin Renaud	Université Libre de Bruxelles, Belgium
Schirone Dario Antonio	University of Bari, Italy
Schneider Michel	ISIMA, France
Schoier Gabriella	University of Trieste, Italy
Schuhmacher Marta	Universitat Rovira i Virgili, Spain
Scorza Francesco	University of Basilicata, Italy
Seara Carlos	Universitat Politècnica de Catalunya, Spain
Sellares J. Antoni	Universitat de Girona, Spain
Selmaoui Nazha	University of New Caledonia, New Caledonia
Severino Ricardo Jose	University of Minho, Portugal

Shaik Mahaboob Hussain	JNTUK Vizianagaram, A.P., India
Sheikho Kamel	KACST, Saudi Arabia
Shen Jie	University of Michigan, USA
Shi Xuefei	University of Science Technology Beijing, China
Shin Dong Hee	Sungkyunkwan University, South Korea
Shojaeipour Shahed	Universiti Kebangsaan Malaysia, Malaysia
Shon Minhan	Sungkyunkwan University, South Korea
Shukla Ruchi	University of Johannesburg, South Africa
Silva Carlos	University of Minho, Portugal
Silva J.C.	IPCA, Portugal
Silva de Souza Laudson	Federal University of Rio Grande do Norte, Brazil
Silva-Fortes Carina	ESTeSL-IPL, Portugal
Simão Adenilso	Universidade de São Paulo, Brazil
Singh R.K.	Delhi University, India
Singh V.B.	University of Delhi, India
Singhal Shweta	GGSIPU, India
Sipos Gergely	European Grid Infrastructure, The Netherlands
Smolik Michal	University of West Bohemia, Czech Republic
Soares Inês	INESC Porto, Portugal
Soares Michel	Federal University of Sergipe, Brazil
Sobral Joao	University of Minho, Portugal
Son Changhwan	Sungkyunkwan University, South Korea
Song Kexing	Henan University of Science and Technology, China
Sosnin Petr	Ulyanovsk State Technical University, Russia
Souza Eric	Universidade Nova de Lisboa, Portugal
Sproessig Wolfgang	Technical University Bergakademie Freiberg, Germany
Sreenan Cormac	University College Cork, Ireland
Stankova Elena	Saint-Petersburg State University, Russia
Starczewski Janusz	Institute of Computational Intelligence, Poland
Stehn Fabian	University of Bayreuth, Germany
Sultana Madeena	University of Calgary, Canada
Swarup Das	Ananda Kalinga Institute of Industrial Technology, India
Tahar Sofiène	Concordia University, Canada
Takato Setsuo	Toho University, Japan
Talebi Hossein	University of Calgary, Canada
Tanaka Kazuaki	Kyushu Institute of Technology, Japan
Taniar David	Monash University, Australia
Taramelli Andrea	Columbia University, USA
Tarantino Eufemia	Polytechnic of Bari, Italy
Tariq Haroon	Connekt Lab, Pakistan
Tasso Sergio	University of Perugia, Italy
Teixeira Ana Paula	University of Trás-Os-Montes e Alto Douro, Portugal
Tesseire Maguelonne	IRSTEA, France
Thi Thanh Huyen Phan	Japan Advanced Institute of Science and Technology, Japan

Thorat Pankaj	Sungkyunkwan University, South Korea
Tilio Lucia	University of Basilicata, Italy
Tiwari Rupa	University of Minnesota, USA
Toma Cristian	Polytechnic University of Bucarest, Romania
Tomaz Graça	Polytechnic Institute of Guarda, Portugal
Tortosa Leandro	University of Alicante, Spain
Tran Nguyen	Kyung Hee University, South Korea
Tripp Barba, Carolina	Universidad Autnoma de Sinaloa, Mexico
Trunfio Giuseppe A.	University of Sassari, Italy
Uchibayashi Toshihiro	Kyushu Sangyo University, Japan
Ugalde Jesus	Universidad del Pais Vasco, Spain
Urbano Joana	LIACC University of Porto, Portugal
Van de Weghe Nico	Ghent University, Belgium
Varella Evangelia	Aristotle University of Thessaloniki, Greece
Vasconcelos Paulo	University of Porto, Portugal
Vella Flavio	University of Rome La Sapienza, Italy
Velloso Pedro	Universidade Federal Fluminense, Brazil
Viana Ana	INESC Porto, Portugal
Vidacs Laszlo	MTA-SZTE, Hungary
Vieira Ramadas Gisela	Polytechnic of Porto, Portugal
Vijay NLankalapalli	National Institute for Space Research, Brazil
Vijaykumar Nandamudi	INPE, Brazil
Viqueira José R.R.	University of Santiago de Compostela, Spain
Vitellio Ilaria	University of Naples, Italy
Vizzari Marco	University of Perugia, Italy
Wachowicz Monica	University of New Brunswick, Canada
Walentynski Ryszard	Silesian University of Technology, Poland
Walkowiak Krzysztof	Wroclav University of Technology, Poland
Wallace Richard J.	University College Cork, Ireland
Waluyo Agustinus Borgy	Monash University, Australia
Wanderley Fernando	FCT/UNL, Portugal
Wang Chao	University of Science and Technology of China, China
Wang Yanghui	Beijing Jiaotong University, China
Wei Hoo Chong	Motorola, USA
Won Dongho	Sungkyunkwan University, South Korea
Wu Jian-Da	National Changhua University of Education, Taiwan
Xin Liu	École Polytechnique Fédérale de Lausanne, Switzerland
Yadav Nikita	Delhi Universty, India
Yamauchi Toshihiro	Okayama University, Japan
Yao Fenghui	Tennessee State University, USA
Yatskevich Mikalai	Assioma, Italy
Yeoum Sanggil	Sungkyunkwan University, South Korea
Yoder Joseph	Refactory Inc., USA
Zalyubovskiy Vyacheslav	Russian Academy of Sciences, Russia

Zeile Peter	Technische Universität Kaiserslautern, Germany
Zemek Michael	University of West Bohemia, Czech Republic
Zemlika Michal	Charles University, Czech Republic
Zolotarev Valeriy	Saint-Petersburg State University, Russia
Zunino Alejandro	Universidad Nacional del Centro, Argentina
Zurita Cruz Carlos Eduardo	Autonomous University of Sinaloa, Mexico

Sponsoring Organizations

ICCSA 2015 would not have been possible without the tremendous support of many organizations and institutions, for which all organizers and participants of ICCSA 2015 express their sincere gratitude:

University of Calgary, Canada (http://www.ucalgary.ca)

University of Perugia, Italy (http://www.unipg.it)

University of Basilicata, Italy (http://www.unibas.it)

Monash University, Australia (http://monash.edu)

Kyushu Sangyo University, Japan (www.kyusan-u.ac.jp)

Universidade do Minho, Portugal (http://www.uminho.pt)

Contents – Part III

Workshop on Econometrics and multidimensional evaluation in the urban environment (EMEUE 2015)

Multicriteria Prioritization for Multistage Implementation of Complex Urban Renewal Projects

Teresa Cilona and Maria Fiorella Granata[✉]

Department of Architecture, University of Palermo, Palermo, Italy
{teresa.cilona,maria.granata}@unipa.it

Abstract. Renewal projects on neighborhood scale often include several sub-projects aiming at fulfill the different functions of urban areas, while public funds or activated public-private partnership can be insufficient to face the implementation of subprojects simultaneously. Therefore single parts of wider projects are frequently realized at different period of time, according to the availability of public funds or the interest of private partners in the most profitable projects. Although a major organizational effort of funding could be necessary, a more rational approach is required, which will be able to identify the most suitable sequence of realization ensuring the greatest advantage if non-completion of the entire urban renewal project occurs or during the time of its fulfillment which is usually long. This paper proposes a multi-criteria and multistage evaluation model supporting prioritization of sub-projects belonging to a wider renewal plan. The model is applied to a real world decision problem.

Keywords: Multiple-criteria and multistage prioritization · Dynamic decision-making · Promethee · Urban renewal

1   Introduction

Urban regeneration is a main issue today in Europe [1], where the need of avoiding a further shift of natural and agricultural land in artificial areas suggests improving the performance of already existing urban areas instead of building new ones [2]. Urban regeneration is a multi faced issue, involving social, economic, physical, environmental and financial aspects and multi-criteria decision analysis quite fits the relative decision-making [3, 4].

While the need for urban regeneration interventions is increasing, available financial resources are shrinking [5, 6]. Local governments have to resort to financing forms other than traditional national subsidy, as super-national subsidy funds, private funding and special domestic financial policies [7, 8]. Financial difficulties of current recession times may sometimes result in the implementation over time of subsequent parts of complex and expensive urban projects. The choice of priority renewal subprojects is in general imposed by the amount of available funds or by the interest of

M.F. Granata—This work results from an interdisciplinary collaboration of the authors, with the following contribution of T. Cilona (Chapter 4) and M.F. Granata (Chapters 1, 2, 3, 5 and 6).

O. Gervasi et al. (Eds.): ICCSA 2015, Part III, LNCS 9157, pp. 3–19, 2015.
DOI: 10.1007/978-3-319-21470-2_1

private partners in the most profitable projects. Although a major organizational effort of funding could be necessary, it may be appropriate realizing the parts of renewal projects according to a more rational sequence, in order to achieve the maximum benefit over time for the community during the frequently long time of completion of the entire project or in case of non-completion of the entire project.

The purpose of this paper is to present a multiple criteria and multistage model for handling the decision problem of prioritization in multistage implementation of complex urban renewal projects. After the description of the methodology, the model is illustrated with a real decision problem case. The special decision problem is approached with a procedure inspired by the basic idea of ri-conducting the dynamic assessment to a static one using classical multicriteria models, as it is in [9] but unlike the quoted approach we consider prefixed alternatives and criteria.

The present paper is organized as follows. After the introduction, identifying the decision problem, in the second and third sections the theoretical background of the proposed model and the same model are respectively described. Then the methodology is applied to a real world case, which is depicted in the fourth section. After the problem structuring, the assessment procedure is implemented and the relative recommendations are obtained (section 5), in the light of a sensitivity analysis. Finally, some conclusive remarks are provided.

2   Theoretical Background of the Proposed Model

The fulfillment of a renewal urban project through the implementation of its single parts over time induces economic, environmental and social changes that may affect the preferences of decision-makers about the performance of remaining parts of the same renewal project against criteria and on the relative importance of criteria. The choice of preferable sequence in the realization of complex urban renewal projects is somehow a dynamic decision problem, as time dimension cannot be neglected.

Typical features of dynamic decision making are the necessity of making real time succession of decisions, the dependence of single decisions on previous ones and the external and internal changes of decision environment [10]. In order to deal with time dependent decision making, in which the elements of evaluation (alternatives, criteria, values and weights) may suddenly change over time owing to unexpected events, special procedures have been generated. They are based on various sophisticated algorithm and approaches, as rule-based models, heuristic architecture, Bayesian inference [11], dynamic programming [12], Markov decision processes, dynamic decision networks [13], computer simulation microworld tools [14], multi-attribute utility theory, stochastic optimal control theory, game theory, decision trees, influence diagrams, reinforcement learning models [15]. Specific versions of AHP and ANP, the Dynamic Hierarchy Process (DHP) and the Dynamic Network Process (DNP) respectively, have been also provided [16]. Dynamic multicriteria approach, in a great variety of specific proposals, is widely used in industrial production and business management, in clinical decision making, in air traffic control, in emergency management, in politics management and in psychological and cognitive sciences [11].

In order to deal with dynamic decision problem with a finite number of alternatives, some approaches involving a lower computational burden have been proposed, as the multi-period approach in which current and past performances of alternatives are integrated in the evaluation with the aim of making a reliable choice [17] or specific dynamic operators are defined to deal with information collected in different period [18]. Besides, T.L. Saaty proposes to consider various scenarios to integrate uncertainties owing to the effects of time into the evaluation [16], while G. Campanella and R.A. Ribeiro suggest to lead back multicriteria decision making in a framework based on static model and taking the final decision at the end of an exploratory process aimed at reveal new criteria and alternatives not considered at the beginning of the decisional path or at excluding unnecessary criteria and alternatives [9].

Urban or territorial renewal decision problem are widely faced through multicriteria procedures [19, 20, 21, 22, 23] or by financial analysis used alone or in conjunction with multicriteria tools [8], [24, 25]. Multicriteria dynamic approach appears suitable as well in treating particular decision making within urban planning, although a small interest has been expressed so far [26].

3   The Proposed Assessment Model

In the present work, the problem of identification of the better succession in the realization of the parts in which a wider urban renewal project can be separated is dealt with. The decision problem under consideration has some typical features of dynamic decision problem, as the conditions of an urban area evolve over time while the subsequent stages of a renewal urban project are carried out. Decision-maker's preferences may then change when during the subsequent implementation of some parts of the project new states different from the original one are attained.

Urban renewal projects are often realized in a relatively restricted time interval, if compared to the time horizon of other planning instrument, as the town planning scheme. Therefore, in the present model we assume that no significant alteration of economic, social and environmental local conditions happens and then alternatives and criteria can be considered the same during all the interested time. Besides, we assume that changes results only from the order of implementation of the separate parts in which the overall urban project is divided, neglecting changes in inner preferences of decision maker over time [27], and changes over time of valuation parameters due to new elements influencing the decision, like external phenomena which can modify the same alternatives or criteria [27; 28]. This simplification of the reality can be considered realistic, if the fulfillment of all the sub-projects does not take a very long time. While considered alternative and criteria are unchanged along the interested interval of time, we admit the criteria weights and the evaluation of alternatives on criteria may vary when subsequent sub-projects are realized.

This model does not fully fit the definition of dynamic decision problem, as it does not requires immediate decisions and the simplified but realistic supposition is assumed according which only changes in decision environment caused by previous stage of decision making will occur and that decisions are all made before the implementation of the single parts of the project. Facing the decision problem under consideration by means of typical tools of dynamic decision making would result in an onerous computational task that in general is not easily understandable by decision makers. As the particular decision making under consideration has only a few charac- teristics of typical dynamic decision problems, a more simple approach is adopted.

We propose to deal the decision problem through the multiple application of a suit- able multicriteria algorithm for the aggregation of decision-maker's preferences in an interdependent and multistage decision making framework. The special procedure re- conduces the dynamic assessment to a static application of traditional multicriteria tools, as Campanella and Ribeiro do in [9], but we consider prefixed alternatives and criteria.

In the proposed assessment model, the overall urban renewal project is divided into sub-projects that are assumed as alternatives in the decision model. The preferable succession in the realization of the subprojects can be achieved through a certain number of multicriteria assessment, according to the scheme represented in figure 1.

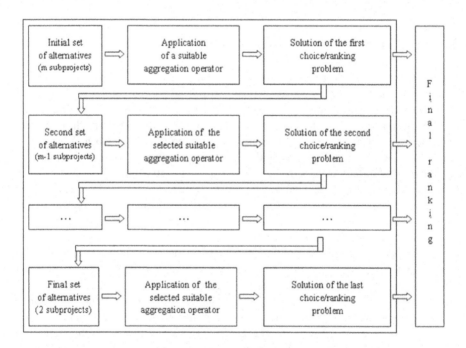

Fig. 1. Scheme of the multicriteria and multistage assessment model

In the first iteration, all the m subprojects in which the complete urban renewal project has been divided constitute the set of alternatives. The application of a suitable procedure aggregating decision-maker's preferences provides the solution of a choice or a ranking problem, permitting the identifying of the best subproject. In the second assessment step, the selected algorithm is applied to the set of remaining alternatives, and so on until only one subproject remains. The final ranking, representing the preferable succession in the realization of the parts of overall urban renewal project, is achieved after the iteration for m-1 times of the basic decision-making procedure.

The application of a suitable procedure aggregating decision-maker's preferences permits the transit from a decision-making step to another one. A great number of traditional multicriteria decision-making methods could be applied. The requested features for the aggregation operator will guide the choice among the available tool solving choice or ranking problems.

4   A Real World Case: Urban Regeneration in the Sicilian South West Coast (Italy)

4.1   The Littoral of St Leone

The Sicilian coastal heritage, as the Italian one, has undergone deep changes besides the irreversible lost of hundreds of thousands square meter of beach. The erosive phenomenon of the coast will carry on and without appropriate interventions the environmental heritage could be threatened for ever [29]. Moreover, the environmental decay of many urban spaces and the pollution of sea water have made the coasts less accessible for both citizens and tourists. This penalized also the economic sector. Today in Italy a lot of attention is paid to the renewal and the requalification of the sea coast thanks to the European rules about coast areas (COM 95/511/CE; COM 00/547/CE; RAC 02/413/CE) and to the international project as INTERREG Adriatic Action Plan 2020 [30]. In many Italian regions and also in Sicily a set of rules and decrees have been issued which oblige the use of specific urban instruments like for example the Beach Plan.[1] The following study is about the Sicilian South-West coast, in particular St Leone which is a little marine village, situated a few kilometers from the Archeological Park of the Valley of the Temples, UNESCO site (Fig. 2).

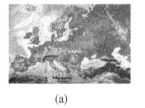   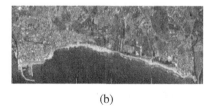

(a)   (b)

Fig. 2. St Leone – Agrigento – Italy (*a*), Part of the coast of Agrigento - St Leone beach (*b*)

---

[1] The Beach Plan is a fundamental global action which imposes rules to a specific territory from an ecologic point of view.

4.2   Urban Analysis and Renewal Interventions

St Leone suburb, whose name derives from the Sicilian Pope Leone II (682-683), develops along the coast area of Agrigento for 3.8 km and it is famous for its beachfront. It borders on the Valley of the Temples to the North, Cannatello suburb to the East, Akragas river to the West, and beautiful sandy hills to the South. From the '60s St Leone transformed from a small fisher suburb into a lively seaside place that then developed in a wild way and nowadays is overcrowded in the summer. In fact the harbor and the waterfront were expanded and part of the coast was cemented. The actual urban structure of St Leone is not only the result of a lack of an adequate urban planning and its planning instruments, but also a lack of respect towards the urban legislation and the environment by the citizens [29], [31]. It means lack of infrastructures and services. Therefore some actions are necessary like for example: 1- Making the area useful for touristic purposes; 2- Assuring reasonable buildings demands; 3- Bringing the dunes to its former state; 4- Creating efficient public means of transport from and to St Leone connected to outdoor parking; 5- Making the seaview possible; 6- Realizing small infrastructures addressed to nautical science, sailing and tourism; 7- Assuring entrances to the beaches to disabled; 8- Enlarging the harbor; 9- Inserting new sporting structures among the trees of a small wood; 10- Readapting open spaces used for cultural, touristic and recreational purposes; 11- Adopting a colour plan which establishes not only the colour but also the buildings, the materials and the building particularities.[2] Those actions can be attained by the following interventions. They are taken from the requalification project of this area which was presented at the International competition of IDEAS to Evaluate St Leone waterfront in Agrigento[3] [31].

Sub-project 1 - Road Conditions: Accessibility, Mobility, Usability. The study of the road conditions is a real Urban Plan of the Traffic (Fig. 3a) and it aims at a functional connection with other areas of the archeological town and the old town centre, moreover it aims also at the creation of parking zones, environmental sustainability through bike lane and walkways. At present from June to September there is a limited use of private cars for non-residents. This situation will be balanced out through the creation of car and bus parking called T1 and T2 (Fig. 3b) in the North of St Leone for about 8,000 car parks, in combination with not polluting, specific and organized public means of transport. The parking zones are provided with photovoltaic panels well integrated in the environment. The construction cost is about € 11,000,000.

Sub-Project 2 - Eco-sustainability - Environmental Renewal - Dunes System - Use of the Territory. The progressive process of decay of the coast will be stopped

---

[2] The adoption of such an instrument, which is complementary to the urban plans, will constitute a valid system of control and a vademecum for designers and owners as concern the maintenance, also because the building, even if it is not very recent, shows all the characteristics of a premature decay. This plan will take into consideration all the actions which aim at reducing the environmental impact of the private constructions which are 6 or 7 levels high.

[3] Ideas Competition for the development of the waterfront in St Leone-Agrigento. Designers: T. Cilona, coordinator; F. Caruso, A. Cimino, G. Riccobene. Freelancer P. Pontei, F. Maccari, D. Vinti.

by a model of environmental system based on naturalistic protection [32]. It is essential to restore the beach areas, today ruined, thanks to interventions of naturalistic engineering that will be able to protect and to build again the dunes (Fig. 4).

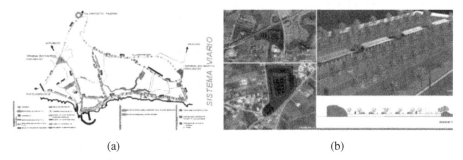

(a)   (b)

Fig. 3. Road system (*a*); Terminal T1 - T2 and Parking (*b*)

This kind of interventions is also addressed to the Akragas banks' river bringing them back to their native state and the creation of equipped naturalistic paths from which it is possible to get to the Valley of the Temples. The creation of naturalistic paths is also addressed to Maddalusa wood giving impulse to the waterfront. The construction cost is approximately € 6,000,000.

Fig. 4. Reconstruction and stabilization of the dune through consolidation constituted by wicker and windbreak guards

Sub-Project 3 - Port Infrastructures and Marine Works. To protect the coasts, the beaches and the structures which are on the promenade it has been necessary to create a system of reef (Fig. 5*c*) in order to soften the effects of the waves, reduce the erosion phenomenon and the risk of marine immersion [33].

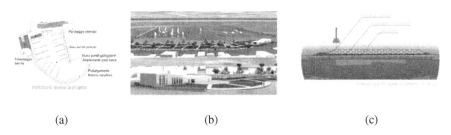

(a)   (b)   (c)

Fig. 5. Port Project hypothesis (*a*), garaging zone and polifunctional building (*b*), breackwater (*c*)

The reef will allow the recolonization of organisms which belong to the marine flora and fauna. Moreover it is thought to extend the west branch to protect from sirocco wind (Fig. 5a), and also to rise up the number of the boats places of about 250. It will also be necessary to extend the shipyard area and build a polifunctional building (Fig. 5b) which will guest the offices of the port authority, a café, a restaurant, shops and other touristic services, information centre. To go to this polifunctional building you can use the underground parking. The construction cost is about € 14,500,000.

Sub-project 4 - Tourist Equipments - Fishing and Sailing Sports. As concerns sailing sport the coast of St Leone is one of the most appreciated place in the island but the lack of specific structures has never allowed it to have success [34]. The solutions presented in this study aim at using and enjoying the coast enhancing new launching pads and the creation of windsurfing schools kite surfing, paraflying etc. The enhancement of the touristic port will be also guaranteed by other complementary fishing and touristic activities and underwater archaeology. Precise hypothesis about underwater activities have been formulated in order to give value to the archaeologic and environmental heritage [35]. The marine evidences pave the way to fixed underwater archaeological paths which today are just devolved to some enthusiasts; this new procedure which is also based on guided excursions, will allow the best preservation and promotion of the important sites. The construction cost is about € 2,000,000.

5   Application of the Assessment Model

5.1   The Set of Alternatives and the Family of Criteria

The considered alternatives are the four above described subprojects of the urban renewal project of the littoral of St Leone.

Using the proposed methodology, urban renewal projects can be easily provided with a recommendation of the same planner regarding the preferable order in carrying out the single parts of the overall project in order to better achieve urban quality. We consider the same planners the best decision-makers in this question since they have only to elicit the values underlying the urban renewal project.

The criteria have been fixed through an interactive process. A preliminary study on specific literature [3, 4] enabled the analyst to suggest an initial set of criteria, in order to help the task of the decision-maker (DM), who was a representative of the planners in this first application. On the base of a critical examination of the original set, the decision-maker, under the guide of the analyst, has finally attained the family of criteria (Tab. 1), expressing all the relevant goals in the decision problem avoiding redundancies [28]. All the criteria have to be maximized and the performances of the alternatives against them are measured on a value scale ranging from 0 to 10, where 0 represents the worst performance and 10 the best one. As specified in table 1, these–values, all assigned in the present test by the decision maker, are based on quantitative measures of performances when they were at planners' disposal (for example, the number of parking place for criterion C6). Otherwise, the qualitative judgments of DM on the expected performance are considered, like the awaited proportional reduction of the issued greenhouse gas in the case of the criterion C1 or the aesthetical quality (C4) reached by means of the implementation of an alternative.

Table 1. The family of criteria

Goal	Code	Criteria	Description	Type of measure
URBAN QUALITY OF THE NEIGHBORHOOD	C1	Air quality	It measures the impact of the considered alternative in terms of greenhouse gas emissions on high season and on rush-hours.	Qualitative
	C2	Acoustic quality	It measures the acoustic impact of the considered alternative on high season and on rush-hours.	Qualitative
	C3	Preservation of the environment	It expresses to what degree the preservation of the environment is attained by the considered alternative.	Qualitative
	C4	Aesthetical quality	It regards the quality of the urban landscape generated by the implementation of the intervention, including the view of the coastline from the sea.	Qualitative
	C5	Development of places for entertainment, social gathering and sports activities	It concerns the creation of places for entertainment, social gathering and sports activities, including bathing establishments.	Quantitative
	C6	Quality of motor traffic and parking	It regards the flow of the traffic and the enabled parking on high season and on rush-hours.	Qualitative and quantitative
	C7	Connection with the rest of the city	It expresses the level of physical and functional link between the neighborhood and the rest of territory.	Qualitative
	C8	Attraction capacity of the tourist port	It expresses the attractiveness of the tourist port. It concerns the functionality of the tourist port, measured by specific technical features, like the number of wharfs and boat-places, and general facilities and services	Quantitative and qualitative
	C9	Development of economic activities	It expresses the contribution to the development of existing or new craft and trade activities.	Qualitative

5.2   The Aggregation Operator of the Multicriteria Preferences

The proposed assessment model for prioritizing the identified parts of an overall urban renewal project requires in each stage of evaluation the use of a suitable multicriteria procedure aggregating the decision-maker's preferences.

Any tool dealing with choice or ranking decision problem may fit to the considered decision-making, agreeing specific characteristics of each procedure. In the present test we use the PROMETHEE (Preference Ranking Organization METHod for Enriched Evaluation) II method, based on pairwise comparison of actions. It applies to a finite set of alternatives and, given the decision maker's preferences, provides a ranking on the set of alternatives in the form of a complete preorder. The procedure permits modeling decision maker performances in a simple way, through concepts and parameters whose interpretation is easily understandable by non-skilled decision-makers [36]. Furthermore, the procedure shows robust ranking for small changes of the values of parameters [37] and is partially compensatory [38].

Promethee methods have aroused lively interest among scholars and a great number of applications are retraceable in literature concerning environment management, business and financial management, water management, chemistry, logistic and trans-portation, Energy management manufacturing and social topics [39]. There are also a plenty of more recent applications on management of natural resources [40], on com-prehensive evaluation of global cities [41] and on decision problem related to the building scale [42], while on urban renewal, Y.-K. Juan et al [43] combine the Prome-thee method with a fuzzy set theory in order to evaluate urban projects.

In multicriteria decision-aid problems, given a set of alternatives $A = \{A_1, A_2, ..., A_m\}$ and a family of criteria $C = \{c_1, c_2, ..., c_n\}$, the evaluation of alternatives on the criteria are aggregated in order to make a choice among the alternatives, to sort them or to rank the actions [38]. Using PROMETHEE II, the requested information on decision maker's preferences consists of weights (w_j), expressing the relative impor-tance of criteria, and preference functions related to criteria, $P_j (A_i, A_k)$, that are de-fined by a suitable functional form and the corresponding necessary parameters. Pre-ference functions assigns to differences between the performance of two alternatives on a criterion, $d_j(A_i, A_K) = c_j(A_i) - c_j(A_k)$, a preference degree ranging from 0 to 1. The pair $\{c_j(\cdot), P_j(A_i, A_k)\}$ is called "generalized criterion" and is defined in six types [44]. The intensity of preference of an alternative A_i over an alternative A_k is given by the "aggregated preference index", a number in the interval [0,1]. The higher is the score and the greater is the preference of an action over the other one on the considered criterion. Given a couple of alternatives A_i and A_k, the "aggregated preference index" are defined as follows:

$$\pi(A_i, A_k) = \sum_{j=1}^{n} P_j(A_i, A_k) w_j ; \qquad \pi(A_k, A_i) = \sum_{j=1}^{n} P_j(A_k, A_i) w_j . \qquad (1)$$

The aggregated preference index $\pi(A_i, A_k)$ expresses the preference of A_i over A_k according all the criteria, while $\pi(A_k, A_i)$ represents the degree of the preference of A_k over A_i [44]. Finally, the complete preorder on the set of alternatives is obtained according to the net outranking flow $\varphi(A_i)$ for each alternative, given by:

$$\varphi(A_i) = \varphi^+(A_i) - \varphi^-(A_i) \qquad (2)$$

where $\varphi^+(A_i)$ is the positive flow and $\varphi^-(A_i)$ is the negative flow, whose expressions are the following ones:

$$\varphi^+(A_i) = \frac{1}{m-1} \sum_{k \neq i} \pi(A_i, A_k), \text{ with } i = 1, 2, ..., m \qquad (3)$$

$$\varphi^-(A_i)=\frac{1}{m-1}\sum_{k\neq i}\pi(A_k,A_i), \text{ with } i = 1, 2, \ldots, m. \quad (4)$$

They respectively represent how the alternative A_i is outranking the other ones and how A_i is outranked by the other alternatives [44].

5.3 Assignment of Values, Weights and Other Assessment Parameters

The requested information on decision maker's preferences consist of the evaluations of alternatives with respect to the criteria, the weights of criteria, and the preference functions for the enrichment of criteria.

Table 2 gives the evaluation of alternatives on each criterion and the weights of the criteria in the first assessment stage. Performances of alternatives are expressed on a 10-point scale where 0 stands for a very insufficient level of the performance, while 10 for an optimal level of the performance. The weights have been assigned by a coplanner of the urban project, according his direct knowledge of the place. With the help of the analyst, the direct rating method [45] has been applied, by ordering the criteria according their importance and rating them on a 10-point scale, where 0 stands for the insignificance and 10 for the extremely importance of a criterion.

Table 2. Evaluation table and weights at the first stage of the assessment model application

Projects		Criteria								
		C1	C2	C3	C4	C5	C6	C7	C8	C9
	Weights	2	3,75	10	7	8	9	6,75	6	5
Alternative 1		5	7.5	5	2.5	8.75	10	9	8.75	6.25
Alternative 2		6.25	6.25	10	8	5	4	3.75	5	5.75
Alternative 3		2.5	4	8.75	3.75	6.5	7.5	6.25	10	5
Alternative 4		6.25	3	5	3.75	8.75	7.5	8.75	7.5	10

On the base of information gathered from the decision maker about his preference on pairwise comparison, the generalized criteria have been identified along with the relative parameters (Tab. 3).

Table 3. Type and parameters of generalized criteria

Criterion	C1	C2	C3	C4	C5	C6	C7	C8	C9
Max or min	max	max	max	max	max	max	max	max	max
Type of gen. criterion	5	5	1	5	5	1	5	4	5
Indifference threshold	1	1	---	1	0.5	---	1	2	1
Preference threshold	3	3	---	2	1	---	4	5	1.5

The indifference and preference thresholds fixing has been supported by the analyst through a number of examples of differences between performance values in order to catch when such differences were perceived as unimportant or instead significant of a strong preference between the two alternatives. The identified generalized

criteria belong to the type 1 (usual criterion) and 5 (V-shape with indifference criterion), whose preference functions are formulated as follows:

$$P_j(A_i,A_k)=\begin{cases}0 & if \ \ d_j(A_i,A_k)\leq 0 \\ 1 & if \ \ d_j(A_i,A_k)>0\end{cases} \qquad (type\ 1) \qquad (5)$$

$$P_j(A_i,A_k)=\begin{cases}0 & if \ \ d_j(A_i,A_k)\leq q_j \\ \dfrac{d_j(A_i,A_k)-q_j}{p_j-q_j} & if \ \ q_j<d_j(A_i,A_k)<p_j \\ 1 & if \ \ d_j(A_i,A_k)\geq p_j\end{cases} \qquad (type\ 5) \qquad (6)$$

being q_j and p_j respectively the indifference and preference thresholds [44].

5.4   Option Analysis, Sensitivity Analysis and Discussion of Results

Given the assessment parameters, Promethee II procedure develops into the following
main steps: 1. computation of preference functions related to criteria; 2. computation
of aggregated preference index for each ordered couple of alternatives; 3. computation of net outranking flow for each alternative [36]. In the considered decision problem, being the alternatives four in number, the preferred order of implementation of
the single subprojects is obtained in three assessment stages. The proposed evaluation
model has been implemented through a special spreadsheet. In the first assessment
stage, the evaluation table and the weights are given in table 2. The application of
the aggregation algorithm provides the first complete preorder according to the net
outranking flows (Tab. 4).

Table 4. Net outranking flows in the first assessment stage

φ(A1)	0.159	φ(A2)	-0.178	φ(A3)	-0.164	φ(A4)	0.182

The alternative A4, that results the best one, is the first subproject to be realized.
In the second assessment stage, the set of alternatives is made by A1, A2 and A3.
The evaluation table and the weights are given in table 5.

Table 5. Evaluation table and weights in the second stage of assessment model application

Projects		C1	C2	C3	C4	C5	C6	C7	C8	C9
	Weights	4.25	6	7.5	7	7.5	9	8	8	9
Alternative 1		6	7.5	6.25	5	6.5	10	9	8	7.5
Alternative 2		6.5	6.5	10	6.5	5	6.5	5	6.5	6
Alternative 3		5	5	5	5	7	6	7	10	8

As expected, decision-maker's preferences about performances of the single
parts of the overall project change during the evolution of the original state of the

neighborhood caused by the implementation of sub-project A4. Also the weights changes from the first to the second stage of evaluation. The application of the algorithm provides the second complete preorder (Tab. 6).

According to the net flows obtained in the second assessment stage, alternative A1 results the best one, and then it is the second subproject to be realized. In the last assessment stage, the set of alternatives is made by the remaining sub-projects. The evaluation table and the weights are given in table 7. The application of the algorithm provides the last complete preorder, according the net outranking flows (Tab. 8).

Table 6. Net outranking flows in the second assessment stage

$\varphi(A1)$	0,236	$\varphi(A2)$	-0,324	$\varphi(A3)$	0,088

Table 7. Evaluation table and weights in the third stage of assessment model application

Projects		Criteria								
		C1	C2	C3	C4	C5	C6	C7	C8	C9
	Weights	7	8	10	9	9	6	6	10	9
Alternative 2		6	7	10	8	8	8	6	8	8
Alternative 3		6	6	5	7	8	6	6	10	10

Table 8. Net outranking flows in the third assessment stage

$\varphi(A2)$	-0,122	$\varphi(A3)$	0,122

The final ranking is then A4, A1, A3, A2. It represents the preferable order in implementing the subprojects in which the complete renewal urban projects has been divided. The constancy of the rankings resulting from the three assessment stages, in which values of outcomes and weights have been changed, can be interpreted as the result of a sensitivity analysis showing the great robustness of the used procedure. The same final ranking is also obtained from a sensitivity analysis by using the wider difference between the thresholds of preference functions given in table 9. These thresholds are defined in order to take into consideration of all the differences in the performances of alternatives on criteria, even if small. Therefore, the decision maker can be quite confident about the stability of the obtained solution. However, the small difference in the values of the net flows of the alternatives A4 and A1, in conjunction with the great difference of their construction costs (being the A4 about six time cheaper than A1) may suggest to realize the A1 alternative before the A4 one.

Table 9. A wider range of the thresholds of the preference functions

Criterion	C1	C2	C3	C4	C5	C6	C7	C8	C9
Indifference-Preference threshold	0.5-5	0.5-5	---	0.25-8	0.25-8	---	0.25-7	1-5	0.5-8

It is significant to highlight that solving the same decision problem through a much more compensatory multicriteria procedure than Promethee method, as the weighted sum model [44], gives a different final ranking (A1, A4, A2, A3), in which the priority of A2 and A3 are inverted. This result confirms that the choice of the procedure is itself a non-neutral part of the preference modeling, and that Promethee II rewards the more balanced alternatives.

6   Conclusions

The present paper proposes a pragmatic approach to deal with a particular decision problem on urban renewal. Complex urban regeneration projects, even if only at a neighborhood scale, require in general high financing efforts, in lack of sufficient funds to fulfill in a short time the overall planned interventions. This frequent situation gives rise to the need to identify the preferable sequence in the fulfillment of the sub-projects, in order to ensure the greatest advantage in case of non-completion of the project and during the often long time of completion of the entire project.

The decision about the preferable succession in the fulfillment of the sub-projects is a sort of dynamic decision. Considering the simpler nature of the particular decision making problem compared with the real-time dynamic decision-making, a special multicriteria and multistage approach has been used, characterized by simplicity and transparence for the decision maker. In the proposed approach, alternatives and criteria are fixed and only endogenous changes induced by the same fulfillment of the single sub-projects have been considered by means of a multi stage updating of the criteria weights and of performance values. The novelty of the approach consists on the multiple application of an adequate assessment procedure, considering an initial multidimensional condition of the urban area that is not a static one, but is modified by previous stages of implementation.

The application of the proposed decision making model to a real case shows its effectiveness and the crucial importance of the choice of the aggregation operator which contributes to a better definition of decision-maker' preferences.

The proposed multistage approach, here used with the purpose of allowing the planner to suggest a rational path in the fulfillment of the project, could be also used in a participative process aimed at eliciting perceived values of local community rather than the more technical values of designers.

If the realization of the overall project could take a large time, then, exogenous changes that do not derive from previous choices of decision-maker should be considered, as it is in the case of variations to town-planning schemes, forecasting of substantial demographic changes, etc., that can influence the definition of the same alternatives and criteria. In that case, the decision-making model should be improved using typical dynamic algorithms that are able to deal with the more general situation in which the considered assumptions are not plausible and more general changes in the conditions of the decision environment should be taken into account.

Acknowledgements. Authors are grateful to Giuseppe Riccobene for the evaluation of alternatives on criteria and the assignment of weights and preference functions parameters.

References

1. European Sustainable Cities and Towns Conference. http://ec.europa.eu
2. European Environment agency: The European Environment. State and Outlook 2010. Land use. EEA, Copenhagen (2010)
3. Brandon, P.S., Lombardi, P.: Evaluating sustainable development. Blackwell Publishing, Oxford (2005)
4. Rizzo, F.: Il capitale sociale della città. FrancoAngeli, Milano (2003)
5. Moore-Cherry, N., Vinci, I.: Urban regeneration and the economic crisis: past development and future challenges in Dublin, Ireland. Planum. The Journal of Urbanism **25**(2), 1–16 (2012)
6. Plöger, J., Kohlhaas-Weber, I.: Shock-proof cities? The impact of and responses to the recent financial and economic crisis in older industrial cities. Journal of Urban Regeneration and Renewal **7**(2), 1–14 (2013)
7. Kool, H.: Preface. In: Wassenberg, F., van Dijken, K.: A practitioner's view on neighbourhood regeneration, The Haugue, The Netherlands, pp. 3–4 (2011)
8. Stanghellini, S., Copiello, S.: Urban models in italy: partnership forms, territorial contexts, tools, results. In: Dalla Longa, R. (ed.) Urban Models and Public-Private Partnership, pp. 47–130. Springer, Heidelberg (2011)
9. Campanella, G., Ribeiro, R.A.: A framework for dynamic multiple-criteria decision making. Decision Support Systems **52**, 52–60 (2011)
10. Edwards, W.: Dynamyc decision theory and probabilistic information processing. Human Factors **4**, 59–73 (1962)
11. Fox, J., Cooper, R.P., Glasspool, D.W.: A canonical theory of dynamic decision-making. Frontiers in Psycology **4**, 1–19 (2013). art. 150
12. Bellman, R.: Applied Dynamic Programming. Princeton University Press, Princeton (1957)
13. Da Costa, C.G., Buede, D.M.: Dynamic Decision Making: A Comparison of Approaches. Journal of Multi-Criteria Decision Analysis **9**, 243–262 (2000)
14. Gonzales, C., Vanyukov, P., Martin, M.K.: The use of micro worlds to study dynamic decision making. Computers in Human Behavior **21**(2), 273–286 (2005)
15. Busemeyer, J.R., Pleshoc, T.J.: Theoretical tools for understanding and aiding dynamic decision making. Journal of Mathematical Psycology **53**, 126–138 (2009)
16. Saaty, T.L.: Time dependent decision-making; dynamic priorities in the AHP/ANP: generalizing from points to function and from real to complex variables. Mathematical and Computer Modelling **46**, 860–891 (2007)
17. Lin, Y.-H., Lee, P.-C., Ting, H.-I.: Dynamic multi-attribute decision making model with gray number evaluation. Expert Systems with Applications **35**, 1638–1644 (2008)
18. Xu, Z.: On multi-period multi-attribute decision making. Knowledge-Based Systems **21**, 164–171 (2008)
19. Cerreta, M., De Toro, P.: Assessing urban transformations: a SDSS for the master plan of castel capuano, naples. In: Murgante, B., Gervasi, O., Misra, S., Nedjah, N., Rocha, A.M.A., Taniar, D., Apduhan, B.O. (eds.) ICCSA 2012, Part II. LNCS, vol. 7334, pp. 168–180. Springer, Heidelberg (2012)

20. Cilona, T., Granata, M.F.: A choquet integral based assessment model of projects of urban neglected areas: a case of study. In: Murgante, B., Misra, S., Rocha, A.M.A., Torre, C., Rocha, J.G., Falcão, M.I., Taniar, D., Apduhan, B.O., Gervasi, O. (eds.) ICCSA 2014, Part III. LNCS, vol. 8581, pp. 90–105. Springer, Heidelberg (2014)
21. Ferretti, V., Bottero, M., Mondini, G.: An integrated approach for exploring opportunities and vulnerabilities of complex territorial systems. In: Murgante, B., Misra, S., Rocha, A.M.A., Torre, C., Rocha, J.G., Falcão, M.I., Taniar, D., Apduhan, B.O., Gervasi, O. (eds.) ICCSA 2014, Part III. LNCS, vol. 8581, pp. 667–681. Springer, Heidelberg (2014)
22. Girard, L.F., Torre, C.M.: The use of ahp in a multiactor evaluation for urban development programs: a case study. In: Murgante, B., Gervasi, O., Misra, S., Nedjah, N., Rocha, A.M.A., Taniar, D., Apduhan, B.O. (eds.) ICCSA 2012, Part II. LNCS, vol. 7334, pp. 157–167. Springer, Heidelberg (2012)
23. Giove, S., Rosato, P., Breil, M.: An application of multicriteria decision making to built heritage. The redevelopment of Venice Arsenale. Journal of Multi-Criteria Decision Analysis 17, 85–99 (2011)
24. Nesticò, A., De Mare, G.: Government tools for urban regeneration: the cities plan in italy. a critical analysis of the results and the proposed alternative. In: Murgante, B., Misra, S., Rocha, A.M.A., Torre, C., Rocha, J.G., Falcão, M.I., Taniar, D., Apduhan, B.O., Gervasi, O. (eds.) ICCSA 2014, Part II. LNCS, vol. 8580, pp. 547–562. Springer, Heidelberg (2014)
25. Morano, P., Tajani, F.: Break Even Analysis for the financial verification of urban regeneration projects. Applied Mechanics and Materials 439, 1830–1835 (2013)
26. Sabri, S., Majid, M.R., Ludin, A.N.M.: Integrated modeling of multicriteria evaluation and planning support systems in inner city redevelopment appraisal. In: Proceedings of the 2nd International Postgraduate Conference on Infrastructure and Environment. IPCIE 2010, vol. 2, pp. 16–24 (2010)
27. Yu, P.L., Chen, Y.-C.: Dynamic MCDM, habitual domains and competence. In: Ehrgott, M., Figueira, J.R., Greco, S. (eds.) Trends in Multiple Criteria Decision Analysis, pp. 1–35. Springer, USA (2010)
28. Keeney, R.L., Raiffa, H.: Decisions with multiple Objectives: Preferences and Value Tradeoffs. John Wiley & Sons, New York (1976)
29. Cilona, T.: Agrigento: la città e la costa - dal waterfront al piano spiaggia. Aa 22, pp. 75–79. Ordine degli architetti di Agrigento, Agrigento (2007)
30. Cilona, T.: Il litorale agrigentino. Strumenti di pianificazione e gestione delle coste. Cepasa Editore, Agrigento (2008)
31. Cilona, T.: Waterfront of Agrigento. Studies and researches of St Leone concerning its development. Cepasa Editore, Agrigento (2010)
32. Boccaloro, F.: Difesa delle coste e ingegneria naturalistica. Manuale di ripristino degli habitat lagunari, dunari, litoranei e marini. Palermo, Dario Flaccovio Editore (2012)
33. Ministero dell'Ambiente e della Tutela del Territorio e del Mare - Direzione Generale per la Salvaguardia Ambientale. Assetto della tutela naturalistica. http://www.va.min-am-bien-te.it
34. Federazione Italiana Vela. http://www.federvela.it
35. Pietro, G.: Akragas-Agrigento. La storia, la topografia, i monumenti, gli scavi. Legam-biente, Roma (1995)
36. Vincke, P., Brans, J.P.: A preference ranking organization method. The PROMETHEE method for MCDM. Management Science 31, 641–656 (1985)
37. Brans, J.P., Vincke, P., Mareschal, B.: How to select and how to rank projects: The PROMETHEE method. European Journal of Operational Research 24, 228–238 (1986)

38. Vincke, P.: Multicriteria Decision-aid. John Wiley & Sons, Chichester (1992)
39. Behzadian, M., Kazemzadeh, E.B., Albadvi, A., Aghdasi, M.: PROMETHEE: A comprehensive literature review on methodologies and applications. European Journal of Operational Research **200**, 198–215 (2010)
40. Kuang, H., Kilgour, D.M., Hipel, K.W.: Grey-based PROMETHEE II with application to evaluation of source water protection strategies. Information Sciences **294**, 376–389 (2015)
41. Kourtit, K., Macharis, C., Nijkamp, P.: A multi-actor multi-criteria analysis of the performance of global cities. Applied Geography **49**, 24–36 (2014)
42. Le Téno, J.F., Mareschal, B.: An interval version of PROMETHEE for the comparison of building products' design with ill-defined data on environmental quality. European Journal of Operational Research **109**, 522–529 (1998)
43. Juan, Y.-K., Roper, K.O., Castro-Lacouture, D., Kim, J.H.: Optimal decision making on urban renewal projects. Management Decision **48**(2), 207–224 (2010)
44. Brans, J.P., Mareschal, B.: Promethee methods. In: Figeira, J., Greco, S., Ehrgott, M. (eds.) Multiple Criteria Decision Analysis: State of the Art Surveys, pp. 163–195. Springer, New York (2005)
45. Von Winterfeldt, D., Edwards, W.: Decision Analysis and Behavioral Research. Cambridge University Press, Cambridge (1986)

Calculating Composite Indicators for Sustainability

Marta Bottero[✉], Valentina Ferretti, and Giulio Mondini

Department of Regional and Urban Studies and Planning, Politecnico di Torino, Turin, Italy
{marta.bottero,valentina.ferretti,giulio.mondini}@polito.it

Abstract. Sustainability indicators are gaining more and more attention as a powerful tool for supporting policy making, providing information on different fields, such as environment, economy, society, technology etc. A very important aspect when dealing with sustainability indicators is related to the procedure for the construction of composite indices, that combine the information coming from several indicators and that are easier to be managed and communicated. The paper investigates the use of indicators in sustainability assessment of projects, plans and programmes and proposes an application of the Multi Attribute Value Theory (MAVT) for the definition of synthetic indices of evaluation. With the aim of highlighting potentialities and limits of the proposed approach, the methodology is applied to a real case concerning a set of well-being indicators. The results of the study show that MAVT is suitable for dealing with the aggregation of indicators systems.

Keywords: Projects/plans/programmes · Indicators and indices · Multicriteria analysis · Swing method · Well-being indicators

1 Introduction

An indicator is an parameter associated with a phenomenon, which can provide information on the characteristics of the event in its global form [1]. Its purpose is to indicate the state, or the variation in the state, of a phenomenon which cannot be measured directly. In fact the data, even if suitably presented, does not constitute an indicator, and can only be used as such when linked to a phenomenon other than that measured. Alone, an indicator provides little information unless it is associated with a system of indicators, able to provide systematic information for the purpose of assessment. A system of indicators consists of several indicators correlated from a logical and functional point of view, able to describe and provide information on several phenomena associated with each other, or which need to be interpreted in a coordinated way. One consolidated instrument for the integrated analysis of the social-economic and environmental aspects in the field of sustainability assessment is the system of environmental indicators known as the DPSIR model (Driving forces, Pressures, State, Impacts and Responses).

Furthermore, one indicator alone cannot express the complexity of the system being observed, but it is possible to notice that a system of partial and extremely incoherent indicators can be an obstacle in the assessment procedure. Therefore, synthetic

© Springer International Publishing Switzerland 2015
O. Gervasi et al. (Eds.): ICCSA 2015, Part III, LNCS 9157, pp. 20–35, 2015.
DOI: 10.1007/978-3-319-21470-2_2

indices can be defined, based on a combination of the information with reference to a multitude of indicators, able to express a value which represents the phenomenon being studied. The importance of defining synthetic indices through the aggregation of several different indicators (even with the loss of information as a result of said aggregation), is clearly expressed by all experts in strategic assessment who must, due to the nature of these procedures, be able to make judgments on compatibility very quickly.

The objective of the present paper is to investigate the role of the Multi-Attribute Value Theory (MAVT, [2]) for the definition of synthetic indices for the evaluation. With the aim of highlighting potentialities and limits of the proposed approach, the methodology is applied to a real case [3]; in particular, the set of well-being indicators proposed by the BES report at the Italian level is used for the present investigation [4].

2   Sustainability Indicators and Indices

As it is well known, sustainability development has been introduced by the Bruntland Commission in 1987 as the development that meets the needs of the present without compromising the ability of future generations to meet their own needs. Since the first definition, many attempts have been made for measuring sustainability performances of policies and countries. Indicators and composite indicators are increasingly recognized as a useful tool for policy making and public communication in conveying information on countries' performance in fields such as environment, economy, society, or technological development [5].

When dealing with sustainability indicators, a crucial element is related to the construction of composite indices that aggregate the information provided by several indicators and sub-indicators. The construction of composite indices involves subsequent steps, including definition of policy goal for the index, data selection, data imputation, standardization, weighting and final aggregation [6].

The components and sub-components for synthetic indices need to be determined on the basis of theory, empirical analysis, pragmatism or intuitive appeal [7]. For composite indexing purposes, also statistical analysis can be employed in order to measure the correlation between pairs of variables. Moreover, scaling operations are of fundamental importance in the construction of indices, as they allow to combine non-commensurable items. Another important element in the procedure is related to the use of weighting systems for the aggregation. Normally implicit weights are introduced during scaling operations while explicit weights can be introduced in the aggregation rule. Finally, sensitivity analysis should be done in order to test the stability of the final numerical results.

Numerous systems are available that propose specific sustainability indices and rating, such as for example the Environmental Performance Index [8] in the environmental field, the Human Development Index [9] in the social development sector, or the Urban sustainability Index in the urban planning domain. Singht et al [6] surveyed the most relevant sustainability indices available in the scientific literature, highlighting the scaling/normalization procedure, the weighting system and the aggregation rule. Mention has to be made to the fact that no system has been surveyed proposing the use of the MAVT for the composite index construction.

3   Multi-attribute Value Theory

The theoretical foundations of Multi Attribute Value Theory (MAVT) were described by Fishburn [10], Raiffa [11] and Keeney and Raiffa [2].

MAVT can be used to address problems that involve a finite and discrete set of alternative options that have to be evaluated on the basis of conflicting objectives. For any given objective, one or more different attributes, which typically have different measurement scales, have to be identified in order to measure the performance in relation to that objective [2]. By being able to handle quantitative as well as qualitative data, MAVT plays a vital role in the field of environmental decision-making where many aspects are often intangible.

The intention of MAVT is to construct a means of associating a real number with each alternative, in order to produce a preference order of the alternatives consistent with the Decision Maker value judgments. To do this, MAVT assumes that in every decision problem a real value function exists that represents the preferences of the Decision Maker. This function is used to transform the evaluation of each alternative option on considered attributes into one single value. The alternative with the best value is then pointed out as the best [12].

In the case of an additive value function, MAVT aggregates the options' performance across all the attributes to form an overall assessment and is thus a compensatory technique. This means that the method does allow compensation of a weak performance of one attribute by a good performance of another attribute. It is interesting to notice that the compensatory approach is crucial in the field of sustainability assessment. Sustainability is often considered in terms of the three pillars of environment, society and economics and a critical issue is how to combine the different dimensions in the evaluation framework [13]. In this context, it is worth recalling that there are two competing theories about sustainability, one referring to the weak sustainability approach and the other to the strong sustainability approach [14]. The weak sustainability approach, in contrast to the strong sustainability one, assumes that there is substitutability between man-made capital (e.g. monetary capital, labour) and natural capital (e.g. natural resources and ecosystems services). Since it is a compensatory method, MAVT supports the evaluation process under weak sustainability assumptions.

From a methodological point of view, the process to be followed to build a MAVT model consists of the five fundamental steps presented in Table 1.

It is important to underline that different strategies are available for the development of a MAVT model. The holistic scaling and the decomposed scaling strategies are the most used in practice [16]. According to the former, an overall value judgment of multiattribute profiles has to be expressed. These can be either the real alternatives or artificially designed profiles. Weights and value functions are then estimated through optimal fitting techniques (e.g. regression analysis or linear optimisation) and

Table 1. The key steps for the development of a MAVT model

Fundamental steps	Description
1. Defining and structuring the fundamental objectives and related attributes	Objectives are "statements of something that one desires to achieve" [15] and they depend on the problem to be analysed, on the actors involved in the decision process, and on the environment in which the decision process takes place. The degree to which objectives are achieved is measured through a set of attributes, which may be natural (they follow directly from the definition of the objective), constructed (they specify a finite number of degrees to which objectives are met), and proxy (they are only indirectly linked to the definition of the objective) [15, 16].
2. Identifying and creating alternative options	The alternatives are the potential solutions to the decision problem. Methods and models such as visioning, problem structuring methods and scenario planning can help to promote creativity for the generation of good strategies and strategic options [17].
3. Assessing the scores for each alternative in terms of each attribute	The performances of each alternative specify for each attribute the outcome of the alternative. In some cases, the performances are readily available, in some other cases they have to be computed or estimated ad hoc for the problem at hand [16].
4. Modelling preferences and value trade-offs	Besides the multiattribute profiles under evaluation (i.e. the decision alternatives), the information items that play a role in a multiattribute value function model are the marginal value functions, the weights, and the multiattribute value function which associates an overall value with each alternative.
5. Ranking of the alternatives	MAVT includes different aggregation models, but the simplest and most used one is the additive model [18] as it is represented in equation (1): $$V(a) = \sum w_i \times v_i(a_i) \quad (1)$$ where $V(a)$ is the overall value of alternative a, $v_i(a_i)$ is the single attribute value function reflecting alternative a's performance on attribute i, and w_i is the trade-off ration among attributes.

are the best representation of the assessor's judgments. According instead to the decomposed scaling technique, the multiattribute value function is broken down into simpler sub-tasks (the marginal value functions and the weights) which are assessed separately. The aim of decomposed scaling is to construct the multiattribute model for evaluating decision alternatives while the aim of holistic scaling is to make an inference about the underlying value functions and weights [16]. Since the case study illustrated in the present paper will follow the decomposed scaling approach, para-graph 3.1 explains how its two sub-tasks (value functions and weights) can be assessed.

Finally, as shown in Table 1, the additive representation of the value functions is the simplest and most practical representation available. However, if dependence does not hold, the additive model is not appropriate [16], [19]. The key condition for the additive form in (1) is mutual preference independence. Preferences on attributes x_i and x_j are independent if trade-offs (substitution rates) between x_i and x_j are inde-pendent from all other attributes. Mutual preference independence requires that pref-erence independence holds for all pairs x_i and x_j.

In the context of sustainability assessments, which are characterized by an ever in-creasing need for participation and transparency, the added value of an additive value function is linked to the intuitiveness of the procedure and thus to its ability to facili-tate the communication of the results [16].

3.1   Modelling Preferences and Value Trade-offs

In the development of a MAVT model it is of crucial importance to express the per-ceived values on the impact that the options under consideration can have, measuring the relative worthiness of each impact. The way of modelling such preferences is via a value function [17]. Value functions are a mathematical representation of human judgments. They offer an analytical description of the value system of the individuals involved in the decision process and aim at capturing the parts of human judgments involved in the evaluation of alternatives. In particular, a value function translates the performances of the alternatives into a value score, which represents the degree to which a decision objective (or multiple decision objectives) is achieved. The value is a dimensionless score: a value of 1 indicates the best available performance and a high objective achievement, while a value of zero indicates the worst performance and a low objective achievement [16]. What characterises the use of value functions is the measure of "differences of preferences" using interval scales [20]. Since people do not naturally express preferences and values in this way, value functions have to be estimated through a specially designed interviewing process in which the relevant judgments for the decision are organized and represented analytically. In this sense value functions are at best an approximate representation of human judgments and are constructed or produced [16].

Different approaches are available to build a value function (e.g. direct rating tech-nique, curve fitting, bisection, standard differences, parameter estimation and seman-tic judgement). A detailed discussion of the characteristics of these techniques is beyond the scope of this paper but the reader is referred to [16] and [18] for a complete overview.

Dealing with value functions, mention has to be made to the fact that the applica-bility of value functions relies to a large extent upon the possibility of assessing value function models easily and reliably [16].

One of the greatest strengths of the decomposed scaling approach is that it can deal with a large number of alternatives without an increase of the elicitation effort compared to a study with a smaller number of alternatives. This is due to the fact that value functions are elicitated from the Decision Maker or stakeholder independently of the alternatives, based on his or her preferences about the fulfilment of the different objectives. The preference elicitation step is indeed based simply on the range of variation of each attribute over all alternatives, i.e., the best- and worst- possible level of each attribute. Once the value function has been constructed by means of one of the aforementioned approaches it becomes possible to evaluate the different alternatives. Therefore, the elicitation procedure is independent from the number of alternatives and any additional alternative can be introduced at a later stage as long as the extreme levels of its attributes stay within the ranges defined for preferences elicitation [24]. However, eliciting preference functions for complex decisions with many objectives is intellectually challenging and time consuming.

The subsequent step consists in the prioritization of the different objectives. Many different methods have been proposed in the literature for assessing criteria weights which are then used explicitly to aggregate criterion specific scores. It has been generally agreed that the meaning and the validity of these criteria weights are crucial in order to avoid improper use of Multicriteria Decision Making models and the procedures for deriving criteria weights should not be independent of the manner they are used (e.g. [25]). The weights in the additive model are scaling constants which allow marginal value functions to take on values in the same interval. The meaning of weights can be stated in terms of the end point of each attribute range: the value improvement obtained by switching an attribute from its worst to its best score corresponds to the attribute weight [16].

As for the elicitation of value functions, different techniques are available for the assessment of weights (e.g. swing weights, rating, pairwise comparison, trade-off, qualitative translation). Again, a detailed discussion of the characteristics of these techniques is beyond the scope of this paper and the reader is referred to [16] for a complete overview.

Among the aforementioned approaches, one of the most used methods for eliciting weights is the swing-weights procedure, which explicitly incorporates the attribute ranges in the elicitation question. In particular, the method asks to value each improvement from the lowest to the highest level of each attribute [17]. The Swing method uses a reference state in which all attributes are at their worst level and asks the interviewee to assign points to states in which one attribute at a time moves to the best state. The weights are then proportional to these points. One of the most important advantages of the Swing method is that it only requires to know the attribute ranges and is thus independent from the shape of the value functions of the objectives. On the other hand, the disadvantages are that the technique is based on direct rating, it does not include consistency checks, and the extreme outcomes to be compared may not correspond to a realistic alternative, which makes the questions difficult to answer [24].

4   Application

4.1   The BES Report

The present paper proposes an application of the MAVT for aggregating several indicators into a composite index. In particular, the analysis makes use of data and indicators available from a scientific report named BES which was developed by the Italian Institute of Statistics with the aim of assessing the national well-being [4].

The interest in well-being measurement has been highlighted for the first time by the Commission on the Measurement of Economic Performance and Social Progress set up by the French president in 2008. After that, a growing number of nations and international organizations have started activities for the measurement of well-being and progress of the society with the final aim of identifying an holistic measurement of the key features of nations' social, economic, and environmental progress and of providing this information to different users and decision-makers [26].

In this context, a very promising field of research refers to the use of Multicriteria Decision Analysis (MCDA) for the aggregation of heterogeneous sustainability indicators into synthetic indexes [29]. Among the different MCDA techniques, this study proposes the MAVT approach because, based on the authors' experience [25], this method is able to generate a learning effect throughout the development of the process, which is crucial in the field of sustainability assessments of territorial transformation process based on heterogeneous indicators.

Therefore, the central research question discussed in the paper is the following:

Can MAVT be an effective tool for indicators aggregation in sustainability assessments? As far as the Italian BES report is concerned, the indicators for the measurement of the national well-being have been organized according to 12 categories, namely 1) Health; 2) Education; 3) Work; 4) Economic well-being; 5) Social relationships; 6) Politics and institutions; 7) Safety; 8) Subjective well-being; 9) Landscape and cultural heritage; 10) Environment; 11) Research and innovation; 12) Quality of services. Mention has to be made to the fact the both quantitative and qualitative indicators have been considered in the aforementioned report. Moreover, the indicators have been evaluated at a regional scale.

Finally, in the BES report no aggregation methodology is provided. The scientific contribution of our work thus relates to the illustration of a formal and replicable methodology for the aggregation of heterogeneous indicators into a composite index useful for policy evaluation and proposal. In our study, the Swing Weights protocol for the elicitation of the trade-offs has been tested in order to investigate its applicability in the context of indicators' analysis and aggregation.

4.2   An Example of a Synthetic Index

In order to provide a detailed example of the applicability of the method and the type of results it can lead to, the following paragraphs will focus on only one of the abovementioned categories, i.e. the "quality of services" category, for which all the steps of the process will be shown and strengths and limits will be discussed.

The same procedure can be repeated for all the other categories and in other similar indicators' aggregation studies.

The reason why the "quality of services" category has been chosen is linked to the strong relation existing between the availability of services and the wellbeing of citizens.

In this sense, a low level of education, sanitary services accessibility, water services, participation to social and political life, among others, contributes to worsen poverty conditions and social exclusion.

According to the study reported in the BES report [4], the quality of the services has been evaluated with reference to four dimensions: accessibility, level of equipment, efficacy and satisfaction. Table 2 illustrates the indicators considered for this category. Mention has to be made to the fact that these indicators refer to three macro areas, i.e. social services (healthcare), public utilities (energy, water and waste) and mobility.

Table 2. The indicators belonging to the "quality of services" category

Indicator	Description
1. Hospital beds	Hospital beds for every 1000 inhabitants
2. Medical examination giving up	Percentage of people that gave up medical examinations due to the long waiting lists
3. Public child services	Percentage of children between 0 and 2 years old that used childhood services offered by the municipalities
4. Public aged services	Percentage of elderly people that benefited from public assistance
5. Electrical network interruption	Number of long interruption of the electrical service per user
6. Gas supply	Percentage of households connected to the methane gas network
7. Water network interruption	Percentage of households reporting irregularities in the water supply service
8. Waste dumping	Percentage of waste sent to landfill
9. Waste separate collection	Percentage of waste sent to selective waste collection
10. Prison crowding	Percentage of prisoners in the prisons over the total number of available places
11. Mobility time	Minutes spent in mobility in an average week day
12. Network density	Km of public transport network for every $100km^2$
13. Services accessibility	Percentage of households reporting difficulties in reaching three or more important services (e.g. pharmacy, postal office, police office, school, market, etc.)

4.3   Elicitation of Value Function

The first step for the application of the MAVT consists in structuring the decision problem according to the value tree approach (Figure 1).

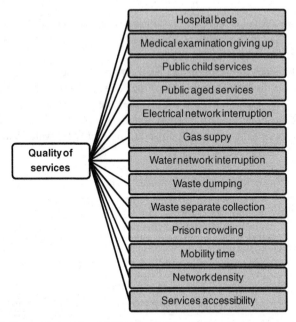

Fig. 1. Structuring of the decision problem according to the value tree approach

Once the structuring of the problem has been defined, the second steps for the application of the methodology considers the elicitation procedure, which consists in defining a value function for each indicator; this value function allows to scale the indicators between 0 and 1 in order to aggregate non-commensurable items.

4.4   Weighting and Aggregation

Once the alternatives have been evaluated, it is necessary to define the importance of the different attributes of the decision problem. In this case the Swing method has been used which explicitly incorporates the attribute ranges in the elicitation question. In particular, the method asks to value each improvement from the lowest to the highest level of each attribute [17] by using a reference state in which all attributes are at their worst level and asking the interviewee to assign points (e.g. in the range 0-100) to states in which one attribute at a time moves to the best state. The weights are then proportional to these values. In this study the evaluation has been performed by the project team (a panel of experts in the fields of environmental engineering, urban planning and economic evaluation led by the authors). Figure 2 shows the questionnaire they had to answer.

The overall set of weights resulting from the elicitation procedure is shown in Table 3.

Fig. 2. Questionnaire for the elicitation of the Swing Weights

Table 3. Normalized weights resulting from the Swing Weight elicitation procedure

Indicators	Weights (%)	Indicators	Weights (%)
1	11,72	8	3,91
2	7,81	9	3,91
3	10,42	10	8,46
4	6,51	11	11,07
5	5,21	12	7,55
6	5,21	13	13,02
7	5,21		

The single attribute value functions have then been aggregated using the obtained set of weights according to equation (1). Figure 3 shows the final ranking of the alternatives.

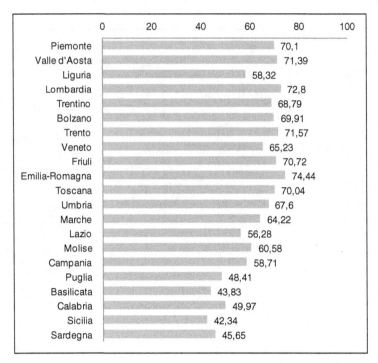

Fig. 3. Final ranking of the regions under analysis with reference to the "quality of services" index

The assumption concerning the difference independence between attributes was tested by asking the participants if they could think of preferences for several levels of attributes independently from the levels of other attributes and all participants stated they could.

4.5   Sensitivity Analysis

After obtaining a ranking of the alternatives, a sensitivity analysis has been performed on the final outcomes of the model in order to test its robustness. The sensitivity analysis is concerned with a "what if" kind of question to see if the final answer is stable when the inputs, whether judgments or priorities, are changed. As a matter of fact, it is of special interest to see whether these changes modify the order of the alternatives.

In the present paper two different sensitivity analysis have been undertaken in order to study the robustness of the model. In the first analysis, the stability of the results has been studied with reference to the variation of the weights of the criteria. In the second analysis, the research has explored the contribution of the standardization procedure in the definition of the final ranking.

With reference to the first sensitivity analysis, the One-at-a-Time (OAT) approach [27] has been used meaning that the weight of one attribute at a time has been increased to 0.40 while keeping all the others equal to 0.05 in order to observe the effects on the final results. As it is possible to see from Figure 4, 13 different scenarios have been obtained by having one indicator as predominant every time. The results show that there is variability in the ranking but the Regions of Emilia Romagna, Lombardy and Piedmont are overall ranked better than the others, thus confirming the validity of the obtained results.

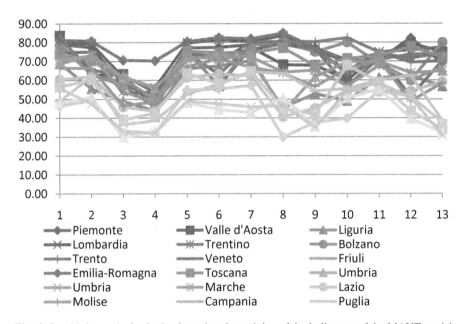

Fig. 4. Sensitivity analysis obtained varying the weights of the indicators of the MAVT model

In the second analysis, the value functions related to the three most important indicators have been modified in order to see if the changes affect the final results of the model. In particular, according to the set of weights resulting from the Swing procedure (Table 3), the value functions related to the indicators 1. Hospital beds, 11. Mobility time and 13. Services accessibility have been transformed according to different decision scenarios. Particularly, for the purpose of the sensitivity analysis, an optimistic scenario and a pessimistic scenario have been created. For the three aforementioned indicators, Figure 5 illustrates the original value functions and those related to the optimistic and pessimistic decision making scenarios.

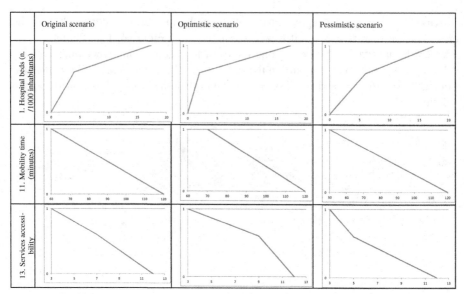

Fig. 5. Optimistic and pessimistic scenarios considered in the sensitivity analysis

Figure 6 represents the results of the second sensitivity analysis. As it is possible to see, the ranking is preserved and the Emilia Romagna and Lombardy regions are always ranked better than the others

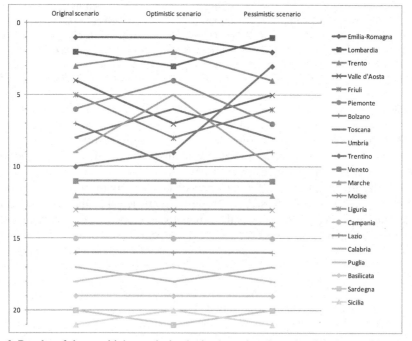

Fig. 6. Results of the sensitivity analysis obtained varying the value functions of the MAVT model

5   Conclusions

The results of the performed analysis show that MAVT represents a very useful tool for sustainability assessments [28] based on indicators and indices [29]. Particularly, it is possible to highlight that MAVT approach has many strengths. Firstly, MAVT enhances the understanding of the policy problem by forcing the Decision Makers to compose a value function that represents their preferences. Secondly, MAVT offers the possibility of reasoning about the problem by clarifying the strengths and weaknesses of the different alternative policies. Furthermore, MAVT strongly supports the decision-making process because it permits to clearly visualize and communicate the intermediate and final results. Finally, in the construction of the decision tree MAVT can incorporate the diverse views of stakeholder groups, considering the development of alternative options/solutions for the problem and the composition of the value functions. For these reasons, MAVT has been applied to many real-world decisions, in both the private and public sectors.

One of the most significant strengths of the MAVT methodology is represented by the fact that the Decision Makers and stakeholders gain more awareness of the elements at stake while structuring the model and thus learn about the problems while solving them. Moreover, MAVT can deal with a large number of alternatives without an increase of the elicitation effort compared to a study with a smaller number of alternatives. This feature is particularly important when dealing with regional and national sustainability ranking that could need to be updated periodically.

Moreover, the discussion oriented towards values rather than towards alternatives during both the value function construction and the level of trade-offs elicitation facilitated a better understanding of the problem and of the relationships among the considered aspects.

As a future development of the work, given the spatial nature of the decision problem under consideration, it would be of scientific interest to investigate the possibility of integrating the MAVT approach with Geographic Information Systems in order to develop a Multicriteria Spatial Decision Support System (MCSDSS) [30, 31] that will enable to investigate spatial correlations between the indicators under analysis.

Acknowledgments. The paper is the result of the joint work of the three authors. Despite the overall responsibility being equally shared, Marta Bottero is responsible for paragraphs 2, 4.3, and 4.4, Valentina Ferretti is responsible for paragraphs 3, 3.1 and 4.2, while paragraphs 4.1 and 4.5 are the result of the joint work of Marta Bottero and Valentina Ferretti. The abstract, the introduction and the conclusions have been jointly written by the three authors.

References

1. OECD - Organization for Economic Co-operation and Development: OECD Environmental Indicators. Development, measurement and use, Working paper (2003)
2. Keeney, R.L., Raiffa, H.: Decisions with Multiple Objectives: Preferences and Value Trade-offs. Wiley, New York (1976)

3. Bottero, M., Ferretti, V., Mondini, G.: From indicators to composite indexes: an application of the multi-attribute value theory for assessing sustainability. In: Bevilacqua, C., Calabrò, F., Della Spina, L. (eds.) New Metropolitan Perspectives, pp. 536–541. Trans Tech Publication Inc., Zurich (2014)
4. ISTAT: BES 2013. Il benessere equo e sostenibile in Italia (2013). http://www.istat.it/it/files/2013/03/bes_2013.pdf
5. KEI: Knowledge Economy Indicators, Work Package 7, Sate of the art report on simulation and indicators (2005)
6. Singh, R.K., Murty, H.R., Gupta, S.K., Dikshit, A.K.: An overview of sustainability assessment methodologies. Ecological Indicators 15, 281–299 (2012)
7. Booysen, F.: An overview and evaluation of composite indices of development. Social Indicators Research 59, 115–151 (2002)
8. Bohringer, C., Jochem, P.E.P.: Measuring the immeasurable – a survey of sustainability indices. Ecological Economics 63, 1–8 (2007)
9. United Nations: Human Development Report (2001). http://www.undp.org
10. Fishburn, P.C.: Methods of Estimating Additive Utilities. Management Science 13(7), 435–453 (1967)
11. Raiffa, H.: Preference for multi-attributed alternatives. RM-5868-DOT/RC. The RAND Corporation, Santa Monica (1969)
12. Herwijnen, M.V.: Spatial Decision Support for Environmental Management. Vrije Universiteit, Amsterdam (1999)
13. Munda, G.: Multicriteria Evaluation in a Fuzzy Environment. Theory and Applications in Ecological Economics. Physical–Verlag, Heidelberg (1995)
14. Daly, H.E., Cobb, J.: For the Common Good: Redirecting the Economy toward Community, the Environment, and a Sustainable Future. Beacon Press, Boston (1989)
15. Keeney, R.L.: Value focused thinking. Harvard University Press, Cambridge (1992)
16. Beinat, E.: Value functions for environmental management. Kluwer Academic Publishers, Dordrecht (1997)
17. Montibeller, G., Franco, A.: Decision and risk analysis for the evaluation of strategic options. In: O'brien, F.A., Dyson, R.G. (eds.) Supporting Strategy: Frameworks. Methods and Models. John Wiley & Sons, Chichester (2007)
18. Belton, V., Stewart, T.J.: Multiple Criteria Decision Analysis: An Integrated Approach. Kluwer Academic Press, Boston (2002)
19. Dryer, J.S.: MAUT – multiattribute utility theory. In: Figueira, J., Greco, S., Ehrgott, M. (eds.) Multiple Criteria Decision Analysis. State of the Art Survey, pp. 265–295. Springer, New York (2005)
20. Bouyssou, D., Marchant, Th, Pirlot, M., Tsoukiàs, A., Vincke, Ph: Evaluation and decision models with multiple criteria. Stepping stones for the analyst. Springer Verlag, Boston (2006)
21. Schuwirth, N., Reichert, P., Lienert, J.: Methodological aspects of multi-criteria decision analysis for policy support: A case study on pharmaceutical removal from hospital wastewater. European Journal of Operational Research 220, 472–483 (2012)
22. Roy, B., Mousseau, V.: A theoretical framework for analysing the notion of relative importance of criteria. Journal of Multi-Criteria Decision Analysis 5, 145–159 (1996)
23. Rinne, J., Lyytimaki, J., Kautto, P.: From sustainability to well-being: Lessons learned from the use of sustainable development indicators at national and EU level. Ecological Indicators 35, 35–42 (2013)
24. Daniel, C.: One-at-a-time-plans. Journal of the American Statistical Association 68, 353–360 (1973)

25. Ferretti, V., Bottero, M., Mondini, G.: Decision making and cultural heritage: an application of the Multi Attribute Value Theory for the reuse of historical buildings. Journal of Cultural Heritage **15**(6), 644–655 (2014). http://dx.doi.org/10.1016/j.culher.2013.12.007

26. Bottero, M.: Indicators assessment systems. In: Cassatella, C., Peano, A. (eds.) Landscape Indicators. Assessing and Monitoring the Landscape Quality. Springer, Berlin (2011)

27. Ferretti, V.: A Multicriteria- Spatial Decision Support System (MC-SDSS) development for siting a landfill in the Province of Torino (Italy). Journal of Multi-Criteria Decision Analysis **18**, 231–252 (2011)

28. Ferretti, V., Pomarico, S.: Ecological land suitability analysis through spatial indicators: an application of the Analytic Network Process technique and Ordered Weighted Average approach. Ecological Indicators **34**, 507–519 (2014)

29. FEEM, 2009. FEEM Sustainability Index. Methodological report. http://www.feemsi.org/pag/downloads.php (accessed on September 22, 2011)

Using Genetic Algorithms
in the Housing Market Analysis

Benedetto Manganelli[1(✉)], Gianluigi De Mare[2], and Antonio Nesticò[2]

[1] School of Engineering, University of Basilicata,
Viale dell'Ateneo Lucano, 10 85100, Potenza, Italy
benedetto.manganelli@unibas.it
[2] Department of Civil Engineering, University of Salerno,
Via Giovanni Paolo II, 132 84084, Fisciano (SA), Italy
{gdemare,anestico}@unisa.it

Abstract. This paper tests the use of Genetic Algorithms to interpret the relationship between real estate prices and the geographic locations of the properties. Issues of choosing algorithm parameters are discussed on the basis of applying data collected in the city of Potenza to 190 houses. The aim of the study is to show the potential and the limits of genetic algorithms in this field and how they can be effectively used in the analysis of the housing market.

Keywords: Genetic algorithms · Housing submarkets · Mass appraisal

1   Introduction

The need for accurate estimates of real estate, primarily to have control of the values and to keep their profitability updated, is strongly felt by the various operators in relation to several objectives [1,2]. For many of these, the observation of the housing market becomes critical, how it evolves and its dynamism [3,4]. This in turn leads to the need for tools and evaluation techniques that can interpret the phenomena typical of the real estate market.

A building is the result of a production process in which the land and what is built on it are involved. These are considered to be economic factors which are distinguishable even once the process is completed. The narrowness of the building is therefore closely related to the limited availability of land for building. This technical constraint depends on the location and the planning policies and therefore it cannot be overcome, at least not by factors that are internal to the market. If the urban land is, on the one hand, limited, on the other hand, it is of unlimited duration. Due to these features, in time, urban land is subject to rent. The fixed location inevitably makes real estate very sensitive to changes in economic conditions and social context; it therefore, on one hand accounts for the unique nature of each land, and on the other, it strongly conditions its usefulness and value [5].

The paragraphs 1, 2 and 3 are to be attributed in equal parts to the three authors; except subparagraphs 2.1 and 2.2, to be attributed only to Manganelli.

© Springer International Publishing Switzerland 2015
O. Gervasi et al. (Eds.): ICCSA 2015, Part III, LNCS 9157, pp. 36–45, 2015.
DOI: 10.1007/978-3-319-21470-2_3

The analysis of the housing market cannot be separated from its division into sub-markets which are distinguishable in relation to the factors that determine the demand and the supply, and the manner in which the two functions interact.

In urban house market studies, the urban housing market can be divided into a series of submarkets. The analysis of the real estate market becomes more accurate if performed within the same submarket. A key parameter for identifying different submarkets is certainly the geographic position. Housing submarkets are typically defined as geographic areas where the price per unit of housing quantity (defined using some index of housing characteristics) is constant [6]. In this study, a model that implements the Genetic Algorithms is developed and applied in order to specifically identify the marginal contribution to the value of the properties provided by their location. The analysis is carried out on a database of prices of properties in the city of Potenza.

2 Genetic Algorithms

Genetic Algorithms (GA) are complex, adaptive procedures, aimed at the resolution of optimization problems; they are procedures that seek the maximum (or minimum) point of a certain function, when this function is too complex to be maximized (minimized) quickly with analytical techniques and is unthinkable to a different process that randomly explores the solution space. Genetic Algorithms are a family of stochastic techniques, based on the metaphor of biological evolution, used in many fields [7,8].

In the analogy between a Genetic Algorithm and a biological system, the environment is constituted by a given problem to be solved and individuals of the population are potential solutions to this problem. The evolutionary pressure that the environment exerts on individuals is implemented through a fitness function, which provides a measure of fitness for each individual thus expressing the degree of accuracy of the answer given to the problem under consideration.

The main genetic operators of the individual transformations, borrowed from biology, are crossover and mutation. A completely new population of potential solutions is produced by selecting the best individuals in the current generation and coupling them to produce a new set of individuals. The subsequent generations in this way will consist, in a gradually increasing number of individuals who share the genetic characteristics of the best individuals that appeared in previous generations. The positive features are spread over the entire population, generation after generation, and are crossed and exchanged with other winning features. By favouring the coupling between individuals, the most suitable areas of the research space are explored until the optimal solution   is obtained.

The range of application of Genetic Algorithms could be divided into five large domains, although these do not have sharply defined boundaries, but in some cases, may overlap: 1) Planning, 2) Design, 3) Simulation and identification, 4) Control, 5) Classification, modelling and machine learning. The use of Genetic Algorithms in the latter domain is directed towards the construction of models able to interpret the underlying phenomenon: depending on the case, this model can allow for the simple classification of each observation to one of two or more classes, or as in this case, the model is built for forecasting purposes [9].

2.1   Structure of Genetic Algorithms

The first step provided for the construction of a Genetic Algorithm is the creation (random or using some heuristics) of an initial population; then an evolutionary cycle has to be generated, which, at each iteration (or generation), produces a new population by applying the genetic operators to the previous population.

The problem that genetic algorithms try to solve is substantially to seek the maximum or minimum point on a certain function (fitness). The idea at the basis of Genetic Algorithms is thus to select the best solutions and recombine in some way to each other in such a manner that they evolve towards an optimal point. A particular value of fitness is assigned to each individual that is generated from the initial population; this value depends on the quality of the solution as it is represented. The best individuals are recombined with each other in order to create a new generation, and so on until the average fitness of the species does not converge to the value of the best individual.

2.2   Assumptions of the Model

The model interprets the price of a property as the result of a multicriteria choice process. The selection criteria are represented by the features of properties that are decisive in the formation of value. In order to identify the contribution to the price of the geographic location of the property, the sample (initial population) has been selected by detecting purchase prices that refer to the recent past and to property for residential use that have a similar type of building in the urban area of the Potenza (n. 190).

In order to ensure that the model is able to better interpret the underlying phenomenon, other characteristics, besides the location, were recorded for each property of the sample. These other features are selected from those that best capture the non-homogenous features of property:

1) Size, square meters;

1) Age of construction, number of years going back in time from 2014;

2) Lift, it is a dichotomous variable that is valued at "0" if the property does not have a lift and in the opposite case at "1" (table 1).

With regards to the geographical position, this was measured by assigning to each building a sequence of characters capable of classifying the single unit as belonging to a certain urban area among those previously selected for a study that was already carried out in the city of Potenza. This study used the Geographically Weighted Regression (GWR) for housing market segmentation [10,11] and specifically classified the city of Potenza in 6 homogeneous areas as in figure 1 (A, B, C, D, E, F), so if the housing unit belongs to area D its sequence will resemble (0,0,0,1,0,0).

Table 1. Basic statistics characteristics of the population

Basic statistics	Size (sq.m)	Age (years)	Lift
Mean	99	19	
Standard deviation	40	12	
Max	212	52	1 =107
Min	30	0	0 = 83
Median	92	16	
Mode	59	41	
Range		190	

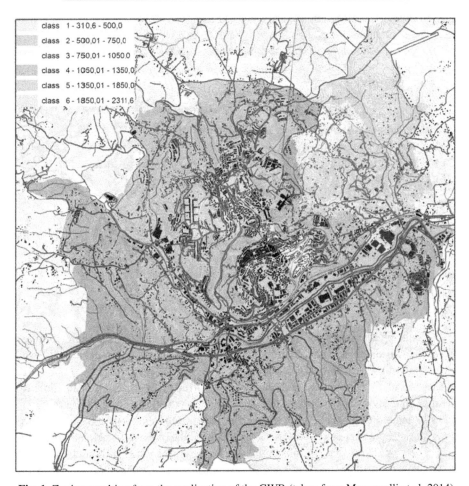

class 1 - 310,6 - 500,0
class 2 - 500,01 - 750,0
class 3 - 750,01 - 1050,0
class 4 - 1050,01 - 1350,0
class 5 - 1350,01 - 1850,0
class 6 - 1850,01 - 2311,6

Fig. 1. Zoning resulting from the application of the GWR (taken from Manganelli et al. 2014)

The fitness function is the Residual Sum of Squares (RSS), a residual being the difference between an observed price (y_i) and the fitted value provided by a model $(\bar{y}_i)$:

$$RSS = \sum_i \varepsilon_i^2 = \sum_i (y_i - \bar{y}_i)^2 = \sum_i (y_i - m_0 - m_i x_i)^2 \qquad for \; i = 1, 2, ..., n \quad (1)$$

Where:

x_i = measurements of the selected features;

m_0 and m_i are respectively the constant or statistical error and the coefficients of the individual features.

These represent the marginal contribution that the selected variables provide the property values.

The model was developed in MatLab ®, with the Genetic Algorithm tool.

The MatLab routine was performed numerous times by varying, one by one, all the optional parameters and constraints of the model, finalizing the choice to minimizing the sum of the squared residuals. This allowed to define the encoding able to generate the best solution.

The optimal solution produced the parameters set as follows:

Population type: {Double vector}

Population size: {190}

Fitness scaling: {Proportional}

Selection: {stochastic uniform}

Reproduction: {elite count=1; crossover fraction= 0.1}

Mutation: {Constraint dependent}

Crossover: {Heuristic}

Migration: {Both}

Stopping criteria:{Generations=5000}

Plot function: {best fitness; stopping}

The numerical values of the coefficients m are in table 2.

The coefficient vector allows the estimation of the marginal contributions of individual features. The price function is additive and its analytical form is:

$$y_i = f(C_j) = C_0 + \sum_{j=1}^{n} C_j = m_0 + \sum_{j=1}^{n} m_{ij} \cdot x_{ij} \qquad (2)$$

By replacing the parameters representative of each property in the function, the expected value is calculated and, therefore, the error between the latter and the observed price. The results are summarized in table 3.

These results are compared with those derived from the application of multiple regression analysis (MRA) on the same data. The comparison shows that the marginal prices of individual features are a little different to each other (even though in absolute value, the coefficients of the variables that identify the area of belonging are different). Even statistics show a substantial similarity between the two models, although the Genetic Algorithm shows a better ability to interpret the phenomenon given the lower value of the statistical error ($m_o = -16,181.12$).

Ultimately, the average market values for each homogeneous area were determined, by the following operation (3).

$$Average\ Market\ Value\ (€/sq.m) = \frac{\sum \bar{y}_i\ (€)}{\sum Size_i (sq.m)} \quad \forall\ i = unit \in same\ area \quad (3)$$

Table 2. Coefficients of geographic locations and housing structural characteristics

$m_0 = $ -2,874.95		m_{ij}
Characteristics / Area		
1	Size	1,919.48
2	Age	-216.04
3	Lift	3,101.23
4	A	-22,844.59
5	B	-10,783.20
6	C	5,846.50
7	D	3,451.50
8	E	12,779.26
9	F	-6,240.94

Table 3. Summary of the results

Meaningful statistics	
Max overestimation (%)	47
Max underestimation (%)	-60
Mean error (%)	17
No. Cases of overestimation (> 15%)	30
No. Cases of underestimation (< -15%)	55

The results obtained are in table 4.

The values obtained by the genetic algorithm were then compared with the values provided by the Revenue Agency (the Housing Market Observatory - OMI in Italian).

This Observatory has recently divided the territory of Potenza into ten homogenous areas (figure 2). It is not known if the Revenue Agency did the zoning. Some of them, even though not geographically contiguous, show similar prices in the residential market here investigated (table 5).

Given that the OMI classification is different for number of homogenous areas and extent of each area (10 areas), compared to the GWR classification (6 areas), it was necessary to merge some classes so that they could be interchangeable with a good level of accuracy to then carry out the comparison. In particular, classes A and D, of the GWR classification, correspond to area D1, of the OMI classification.

In order to carry out the comparison between the two classifications, it was therefore necessary to merge some homogenous OMI areas. This then produced the two classifications in table 6. The rankings and average values of the areas are not perfectly coherent especially with reference to the last two areas of table 6.

Table 4. Summary of the results

Classification GA – Average Market Value (€/sq. m)		ranking
A	1,566.00	6°
B	1,744.00	5°
C	1,925.00	2°
D	1,880.00	3°
E	2,001.00	1°
F	1,820.00	4°

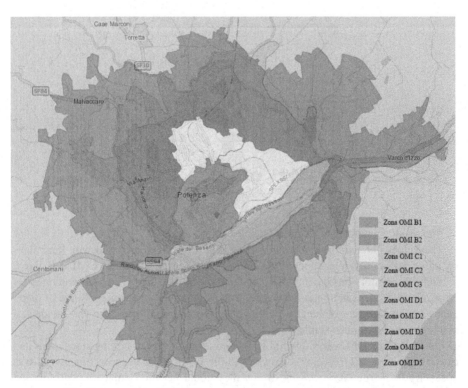

Fig. 2. Image from GEOPOI cartographic framework, Revenue Agency

Table 5. Average market values of the homogeneous areas (OMI) - Potenza

Classification OMI	Average market value (€/sq.m)
central areas	
B1	1,750.00
B2	2,000.00
semi central areas	
C1	1,750.00
C2	1,325.00
C3	1,700.00
peripheral areas	
D1	1,475.00
D2	1,700.00
D3	1,500.00
D4	1,700.00
D5	1,500.00

Table 6. Summary of the results

Area		Average market value and ranking				error	
GA	OMI	GA		OMI		Value	ranking
B	C2-C3	€ 1.744,00	4°	€ 1.510,00	4°	€ 234,00	
A-D	D1	€ 1.720,00	5°	€ 1.475,00	5°	€ 245,00	
C	B1-C3	€ 1.925,00	2°	€ 1.725,00	2°	€ 200,00	
E	B1-D2-D4	€ 2.001,00	1°	€ 1.715,00	3°	€ 286,00	
F	B2-C1	€ 1.820,00	3°	€ 1.875,00	1°	-€ 55,00	

3   Conclusions

The application of hedonic models is widespread in housing studies while less common is the use of other statistical techniques such as artificial intelligence (neural networks), structural equations, computational intelligence, linear programming [12,13,14,15,16]. There are very few examples of the use of Genetic Algorithms in this field [17,18]. Through the implementation of a Genetic Algorithm, this paper tries to measure the contribution that the geographic location provides to the formation of the housing price. This was done on the basis of a preliminary segmentation of urban land operated with the Geographically Weighted Regression on the same data. The results are good, although the Genetic Algorithms do not seem to greatly improve the performance obtained by the use of multiple regression analysis. This little improvement in results is probably due to the use of a linear function of fitness.

Hedonic models have been used to determine the relationship between characteristics of real estate and its selling price, using in most cases, as in this work, linear functions. This means that the price changes in direct proportion to each unit change of the dependent variable. However, other studies [19] have shown the importance of non-linear analysis, in particular on some variables. It is reasonable that the relationship between the price and the properties' characteristics is not linear and that the effect of the characteristics is different in relation to the level of the housing value or to a different location. The real estate market can be segmented according to the different perceptions that potential buyers have about the importance of the characteristics, determined by the existence of different preferences or budget. Choy et al. [20], studying transactions in the city of Hong Kong, show that the housing's characteristics receive a different level of appreciation by varying the price. It is difficult to map multi-attribute nonlinear relationships using regression analysis. The computing capabilities of Genetic Algorithms and the modelling ability of fitness functions able to treat the non-linearity of real phenomenon, might make to achieve even better results. Better results may be obtained in further developments, using more complex prediction and fitness functions, thereby exploiting the better adaptive capacity of Genetic Algorithms.

References

1. Del Giudice V., Torrieri F., De Paola P.: Property Value, Urban Quality and Maintenance Condition: A Hedonic Analysis in the City of Naples, Italy. Advanced Engineering Forum 11 (Economic-Estimative Dynamics and Valuation Tools), 560–565 (2014)
2. Manganelli B., Morano P.: Estimating the market value of the building sites for homogeneous areas. Advanced Materials Research, 869–870 (Sustainable Development of Industry and Economy), 14–19 (2014). doi: 10.4028/www.scientific.net/AMR.869-870.14
3. Salvo, F., Ciuna, M., De Ruggiero, M.: Property prices index numbers and derived indices. Property Management 32(2), 139–153 (2014)
4. De Mare, G., Manganelli, B., Nesticò, A.: Dynamic analysis of the property market in the city of avellino (Italy). In: Murgante, B., Misra, S., Carlini, M., Torre, C.M., Nguyen, H.-Q., Taniar, D., Apduhan, B.O., Gervasi, O. (eds.) ICCSA 2013, Part III. LNCS, vol. 7973, pp. 509–523. Springer, Heidelberg (2013)
5. Manganelli, B.: Real Estate Investing. Springer (2015). doi:10.1007/978-3-319-06397-3
6. Goodman, A.C., Thibodeau, T.G.: Housing Market Segmentation. Journal of Housing Economics 7, 121–143 (1998)
7. Michalewicz, Z.: Genetic Algorithms + Data Structures = Evolution Programs, 3rd edn. Springer (1995)
8. Mitchell, M.: An Introduction to Genetic Algorithms. MIT Press Cambridge, Massachusetts London (1999). Fifth printing
9. Nga, T., Skitmoreb, M., Wongc, K.F.: Using genetic algorithms and linear regression analysis for private housing demand forecast. Building and Environment 43(6), 1171–1184 (2008)
10. Manganelli, B., Pontrandolfi, P., Azzato, A., Murgante, B.: Urban residential land value analysis: the case of potenza. In: Murgante, B., Misra, S., Carlini, M., Torre, C.M., Nguyen, H.-Q., Taniar, D., Apduhan, B.O., Gervasi, O. (eds.) ICCSA 2013, Part IV. LNCS, vol. 7974, pp. 304–314. Springer, Heidelberg (2013)

11. Manganelli, B., et al.: Using geographically weighted regression for housing market segmentation. International Journal of Business Intelligence and Data Mining **9**(2), 161–177 (2014)
12. Del Giudice V., De Paola P.: Geoadditive Models for Property Market. Applied Mechanics and Materials, 584 – 586 (Project Management in Building and Construction Management), 2505–2509 (2014)
13. Narulaa, S.C., Wellingtonb, J.F., Lewisb, S.A.: Valuating residential real estate using parametric programming. European Journal of Operational Research **217**(1), 120–128 (2012)
14. Kontrimasa, V., Verikasb, A.: The mass appraisal of the real estate by computational intelligence. Applied Soft Computing **11**(1), 443–448 (2011)
15. Nguyen, N., Cripps, A.: Predicting Housing Value: A Comparison of Multiple Regression Analysis and Artificial Neural Networks. Journal of Real Estate Research **22**(3), 313–336 (2001)
16. Worzala, E., Lenk, M., Silva, A.: An Exploration of Neural Networks and Its Application to Real Estate Valuation. Journal of Real Estate Research **10**(2), 185–202 (1995)
17. Ahn, J.J., et al.: Using ridge regression with genetic algorithm to enhance real estate appraisal forecasting. Expert Systems with Applications **39**(9), 8369–8379 (2012)
18. De Mare, G., et al.: Economic Evaluations using Genetic Algorithms to Determine the Territorial Impact Caused by High Speed Railways. World Academy of Science, Engineering and Technology International Journal of Social, Education, Economics and Management Engineering **6**(11), 672–680 (2012)
19. Miles, W.: Boom-Bust Cycles and the Forecasting Performance of Linear and Non-Linear Models of House Prices. Journal of Real Estate Finance Economy **36**, 249–264 (2008)
20. Choy, L.H.T., Ho, W.K.O., Mak, S.W.K.: Housing attributes and Hong Kong real estate prices: a quantile regression analysis. Construction Management and Economics **30**(5), 359–366 (2012)

The Graduates' Satisfaction at Work Through a Generalization of the Fuzzy Least Square Regression Model

Francesco Campobasso[1(✉)] and Annarita Fanizzi[2]

[1] Department of Economics and Mathematics, University of Bari, Bari, Italy
francesco.campobasso@uniba.it
[2] Inter-University Department of Physics, University of Bari, Bari, Italy
annarita.fanizzi@uniba.it

Abstract. In previous works we provided some theoretical results on the estimates of a fuzzy linear regression model. In this paper we propose a generalization of such results to a polynomial model with multiplicative factors, which is actually more appropriate than the linear one. In fact, even in a fuzzy approach the growth rate of the dependent variable can vary depending on the values assumed by independent variables as well as on their interaction. In this application case, we regress the overall satisfaction for the working experience, expressed by the second cycle graduates in the 2008 of the University of Bari, on their satisfaction for specific aspects of job. Since the interviewed graduates express their own liking through scores which do not represent an objective measure of the personal opinions, but rather correspond to accumulation values on the submitted scale, the fuzzy approach is adequate to deal with such collected data.

Keywords: Fuzzy least square regression · Polynomial model · Multiplicative factors · Goodness of fit · Stepwise selection · Satisfaction at work · Graduates from University of Bari

1 Introduction

In previous works we provided some theoretical results on the estimates of a fuzzy linear regression model and we proposed a stepwise procedure to select the most appropriate independent variables [1].

It is known, however, that the effect on the dependent variable of a marginal change in one of the independent variables may vary depending on the particular value of the latter. A polynomial regression model can help us in addressing this circumstance, while remaining linear in the parameters.

It is also known that the introduction of multiplication factors between the independent variables allows us to measure the marginal effect of any interaction between them on the dependent variable; in fact this multiplier effect increases or decreases the additive effects of the independent variables individually considered.

Even such an introduction leaves the regression model linear in the parameters, that makes it possible to still estimate them by means of the least squares method.

© Springer International Publishing Switzerland 2015
O. Gervasi et al. (Eds.): ICCSA 2015, Part III, LNCS 9157, pp. 46–60, 2015.
DOI: 10.1007/978-3-319-21470-2_4

In this work, we propose a generalization of the Fuzzy Least Square Model and use it to analyse the overall satisfaction at work, expressed by the graduates from the University of Bari in 2008. In particular, we detect those factors which contribute more than others to such a satisfaction by a polynomial model, estimated on the data collected among the interviewees by the consortium *AlmaLaurea* five years after their graduation.

2   The Fuzzy Least Square Regression

In 1988 P. M. Diamond [2] introduced a metric into the space of triangular fuzzy numbers and derived the expression of the estimated coefficients in a simple fuzzy regression model.

A triangular fuzzy number $\tilde{X} = (x, x_L, x_R)_T$ for the variable X is characterized by a function $\mu_{\tilde{X}} : X \to [0,1]$, like the one represented in Fig. 1, that expresses the membership degree of any possible value of X to $\tilde{X}$ [3].

The accumulation value x is considered the core of the fuzzy number, while $\underline{\xi} = x_R - x$ and $\overline{\xi} = x - x_L$ are considered the left spread and the right spread respectively.

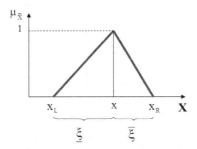

Fig. 1. Representation of a triangular fuzzy number

Note that x belongs to $\tilde{X}$ with the highest degree (equal to 1), while the other values included between the left extreme x_L and the right extreme x_R belong to $\tilde{X}$ with a gradually lower degree.

The set of triangular fuzzy numbers is closed under addition: given two triangular fuzzy numbers $\tilde{X} = (x, x_L, x_R)_T$ and $\tilde{Y} = (y, y_L, y_R)_T$, their sum $\tilde{Z}$ is still a triangular fuzzy number $\tilde{Z} = \tilde{X} + \tilde{Y} = (x + y, x_L + y_L, x_R + y_R)_T$. Moreover the opposite of a triangular fuzzy number $\tilde{X} = (x, x_L, x_R)_T$ is $-\tilde{X} = (-x, -x_R, -x_L)_T$.

It follows that, given n fuzzy numbers $\tilde{X}_i = (x_i, x_{Li}, x_{Ri})_T$, i = 1, 2, .., n, their average is

$$\overline{X} = \frac{\sum \tilde{X}_i}{n} = \left(\frac{\sum x_i}{n}, \frac{\sum x_{Li}}{n}, \frac{\sum x_{Ri}}{n} \right)^T.$$

Diamond introduced a metrics into the space of triangular fuzzy numbers; according to this metrics, the squared distance between $\widetilde{X}$ and $\widetilde{Y}$ is

$$d(\widetilde{X},\widetilde{Y})^2 = d\big((x,x_L,x_R)_T,(y,y_L,y_R)_T\big)^2 = (x-y)^2 + (x_L-y_L)^2 + (x_R-y_R)^2.$$

The same Author treated the fuzzy regression model of a dependent variable $\widetilde{Y}$ on a single independent variable $\widetilde{X}$, which can be written as

$$\widetilde{Y} = a + b\widetilde{X}, \quad a, b \in IR,$$

when the intercept a is non-fuzzy, as well as

$$\widetilde{Y} = \widetilde{A} + b\widetilde{X} \quad a, b \in IR,$$

when the intercept $\widetilde{A} = (a,a_L,a_R)_T$ is fuzzy, where it is $a_L = a - \underline{\gamma}$, $a_R = a - \overline{\gamma}$ and $\underline{\gamma}, \overline{\gamma}$ are non-negative.

The expression of the corresponding parameters a, $\underline{\gamma}$, $\overline{\gamma}$ and b is derived from minimizing, with respect to them, the sum of the squared Diamond distances between theoretical and empirical values of the fuzzy dependent variable $\widetilde{Y}$ in n observed units.

Such a sum takes different forms according to the signs of the regression coefficient b, as the product of a fuzzy number $\widetilde{X} = (x,x_L,x_R)_T$ and a real number depends on whether the latter is positive or negative.

Diamond demonstrated that the optimization problem has a unique solution under certain conditions.

In previous works [4] we provided some theoretical results about the estimates of the regression coefficients and about the decomposition of the sum of squares of the dependent variable in a multiple regression model. Recently [5] we generalized this estimation procedure to the case of a multiple regression model with a fuzzy asymmetric intercept, which seems more appropriate than the non-fuzzy one as it expresses the average value of the dependent variable (which is also fuzzy) when the independent variables equal zero.

Assuming to regress a dependent variable $\widetilde{Y}_i = (y_i, y_{Li}, y_{Ri})_T$ on k independent variables $\widetilde{X}_{ij} = (x_{ij}, x_{Lij}, x_{Rij})_T$ in a set of n units, the linear regression model with a fuzzy asymmetric intercept is given by $\widetilde{Y}_i^* = \widetilde{A} + \sum b_j \widetilde{X}_{ij}$, with i = 1, 2, ..., n, j = 1, 2, ..., k and a, $b_j \in IR$ (with * denoting the theoretical value) where $\widetilde{A} = (a,a_L,a_R)_T$, $a_L = a - \underline{\gamma}$, $a_R = a + \overline{\gamma}$ and $\underline{\gamma}, \overline{\gamma} > 0$.

The estimates of the fuzzy regression coefficients are determined by minimizing the sum of Diamond's squared distances between theoretical and empirical values of the dependent variable

$$\sum d(\widetilde{Y}_i, \widetilde{A} + \sum b_j \widetilde{X}_{ij})^2$$

respect to parameters a, b_1, ..., b_j, ..., b_k, $\underline{\gamma}$ and $\overline{\gamma}$. The function to minimize assumes different expressions according to the signs of the regression coefficients b_j.

In matricial terms, the estimates of the fuzzy regression coefficients are given by

$$\beta = (X'X + X_L'X_L + X_R'X_R)^{-1} (X'y + X_L'y_L + X_R'y_R)$$

where:

$y = [y_i]$ is the n-dimensional vector of cores of the dependent variable;

$y_L = [y_{Li}]$ and $y_R = [y_{Ri}]$ are the n-dimensional vectors of the lower extremes and the upper extremes respectively of the dependent variable;

X is the nx(k+3) matrix formed by vectors 1, k vectors of cores of the independent variables and two vectors 0;

X_L is the nx(k+3) matrix formed by vector 1, k vectors of lower bounds of the independent variables and vectors -1 and 0;

X_R is the nx(k+3) matrix formed by vector 1, k vectors of upper bounds of the independent variables and vectors 0 and 1;

β is the vector of the parameters (a, b_1, ..., b_j, ..., b_k, $\underline{\gamma}$, $\overline{\gamma}$)'.

We examined how the total sum of squares of the dependent variable can be decomposed according to Diamond's metrics. In particular we obtained the expressions of two components, the regression sum of squares and the residual one, like in the OLS estimation procedure for classical variables. Synthetically, denoting the fuzzy average of empirical values with

$$\overline{Y} = \left(\frac{\sum y_i}{n}, \frac{\sum y_{Li}}{n}, \frac{\sum y_{Ri}}{n} \right)_T = (\overline{y}, \overline{y}_L, \overline{y}_R)_T$$

and the fuzzy average of theoretical ones with

$$\overline{Y}^* = \left(\frac{\sum y_i^*}{n}, \frac{\sum y_{Li}^*}{n}, \frac{\sum y_{Ri}^*}{n} \right)_T = (\overline{y}^*, \overline{y}_L^*, \overline{y}_R^*)_T,$$

the expression of the total sum of squares

$$\text{Tot SS} = \sum d(\tilde{Y}_i, \overline{Y})^2 = \sum [(y_i - \overline{y})^2 + (y_{Li} - \overline{y}_L)^2 + (y_{Ri} - \overline{y}_R)^2]$$

can be written as $\text{Tot SS} = \text{Reg SS} + \text{Res SS}$, where

$$\text{Reg SS} = \sum d(\tilde{Y}_i^*, \overline{Y}^*)^2 = \sum [(y_i^* - \overline{y}^*)^2 + (y_{Li}^* - \overline{y}_L^*)^2 + (y_{Ri}^* - \overline{y}_R^*)^2]$$

represents the regression sum of squares,

$$\text{Res SS} = \sum d(\tilde{Y}_i, \tilde{Y}_i^*)^2 = \sum [(y_i - y_i^*)^2 + (y_{Li} - y_{Li}^*)^2 + (y_{Ri} - y_{Ri}^*)^2]$$

represents the residual sum of squares.

Conversely, in a model with a not asymmetric fuzzy intercept we obtained the expressions of two additional components, besides the regression sum of squares and the residual one, which arise from the diversity between theoretical and empirical values of the average fuzzy dependent variable (unlike in the OLS estimation procedure for classical variables). In such a case an increase in the regression sum of square does not necessarily imply a better fit to observed data: this is because the theoretical average value, from which the regression sum of squares is calculated, may be very different from the empirical one. Only a decrease in the residual sum of squares necessarily implies a better fit to observed data.

In order to assess the goodness of fit of the regression model, we proposed [6] the following index, for simplicity called Fuzzy Fit Index (FFI):

$$\text{FFI} = 1 - \frac{\text{Res SS}}{\text{Tot SS}} = 1 - \frac{\sum d(\tilde{Y}_i, \tilde{Y}_i^*)^2}{\sum d(\tilde{Y}_i, \overline{Y})^2}$$

where $\overline{Y}^* = (\overline{y}^*, \overline{y}_L^*, \overline{y}_R^*)_T$ and $\overline{Y} = (\overline{y}, \overline{y}_L, \overline{y}_R)_T$ denote the fuzzy theoretical average and the fuzzy empirical average of the dependent variable respectively.

The more this index is next to 1, the smaller the residual sum of squares is and the better the model fits the observed data.

In order to compare models that explain the same dependent variable by means of a different number of independent variables, it is appropriate to refer to an index that takes into account the corresponding degrees of freedom (closely linked at this number). As in the classic model, an increase in FFI does not necessarily mean that the new independent variable contributes significantly to explain $\tilde{Y}$; any excess in measuring the fit of the model can be corrected by deflating FFI for a term which increases with the number of independent variables included in the equation.

The proposed version of the adjusted FFI is

$$\overline{\overline{\text{FFI}}} = 1 - \left(\frac{\text{Res SS}}{\text{Tot SS}} \cdot \frac{n-1}{n-p-1} \right)$$

which increases only if the increase in FFI (i.e. in the regression sum of squares) exceeds the penalty induced by having one more independent variable in the model, and decreases otherwise.

3   The Fuzzy Polynomial Regression

In this paper we propose a polynomial regression model, which is actually more appropriate than the linear one, as the rate of increase or decrease of the dependent variable can vary depending on the values assumed by the independent variables.

For this purpose we first observe that the product $\tilde{Z}$ of two triangular fuzzy numbers $\tilde{X} = (x, x_L, x_R)_T$ and $\tilde{Y} = (y, y_L, y_R)_T$, based on Zadeh's extension principle [7], is still a triangular fuzzy number $\tilde{Z} = \tilde{X} * \tilde{Y} = (z, z_L, z_R)_T$, where

$$z = xy \quad z_L = \min(x_L y_L, x_L y_R, x_R y_L, x_R y_R,) \quad z_R = \max(x_L y_L, x_L y_R, x_R y_L, x_R y_R,).$$

Therefore, if x, x_L and x_R are non-negative, the square of a fuzzy number $\tilde{X} = (x, x_L, x_R)_T$ is still a fuzzy number $\tilde{X}^2 = (x^2, x_L^2, x_R^2)_T$.

In general terms, it is possible to define a fuzzy polynomial regression model of order r

$$\tilde{Y}_i^* = \tilde{A} + \sum_{j=1}^{k} b_j \tilde{X}_{ij}^m \quad m = 1, 2, .., r$$

so that the marginal effect of a single independent variable on the dependent variable is no more constant.

The matrix form of the estimator vector of the fuzzy regression coefficients does not change by adding independent variables of order r:

$$\beta = (X'X + X_L'X_L + X_R'X_R)^{-1} (X'y + X_L'y_L + X_R'y_R).$$

However, as $k \times r$ is the number of variables with order varying from 1 to r, the matrices of cores, lower bounds and upper bounds of the dependent variable take the following forms:

X is the $n \times (kr+3)$ matrix formed by vectors 1, $k \times r$ core vectors of the k independent variables of order varying from 1 to r and two vectors 0;

X_L is the $n \times (kr+3)$ matrix formed by vector 1, $k \times r$ lower bound vectors of the k independent variables of order varying from 1 to r and vectors -1 and 0;

X_R is the $n \times (kr+3)$ matrix formed by vector 1, $k \times r$ upper bound vectors of the k independent variables of order varying from 1 to r and vectors 0 and 1;

β is the vector of the parameters $(a, b_1, ..., b_j, ..., b_m, \underline{\gamma}, \overline{\gamma})'$, where $m = n \times (kr+3)$.

Since the marginal effect of one of the independent variables on the dependent variable also depends on the interaction among the former, we introduce multiplicative factors in the regression model. Considering for simplicity a quadratic order, such a model becomes:

$$\tilde{Y}_i^* = \tilde{A} + \sum_{j=1}^{k} b_j \tilde{X}_{ij}^2 + \sum_{j=1}^{k} \sum_{l=j+1}^{k-1} b_{j+(k*3)} \tilde{X}_{ij} \tilde{X}_{il}$$

and still remains linear in its coefficients.

Since the number of variables of first and second order is 2k, while the number of products between pairs of the k variables is $\binom{k}{2}$, the matrices of cores, lower bounds and upper bounds of the dependent variable take the following forms:

X is the $_{n\times\left(2k+\binom{k}{2}+3\right)}$ matrix formed by vectors 1, k×r core vectors of the k inde-

pendent variables of order varying from 1 to r, $\binom{k}{2}$ multiplicative factors e and two

vectors 0;

X_L is the $_{n\times\left(2k+\binom{k}{2}+3\right)}$ matrix formed by vector 1, k×r lower bound vectors of the

k independent variables of order varying from 1 to r, $\binom{k}{2}$ multiplicative factors e and

vectors -1 and 0;

X_R is the $_{n\times\left(2k+\binom{k}{2}+3\right)}$ matrix formed by vector 1, k×r upper bound vectors of

the k independent variables of order varying from 1 to r, $\binom{k}{2}$ multiplicative factors

and vectors 0 and 1;

β is the vector of the parameters $(a, b_1,...,b_j,...,b_m, \underline{\gamma}, \overline{\gamma})'$, where

$$m = \left(2k + \binom{k}{2} + 3\right).$$

4   An Analysis of the Graduates' Satisfaction at Work

Numerous studies on service delivery have made it possible to develop *ad hoc* statistical methods to measure performances.

In general, the concept of performance is identified in the meanings of efficacy and efficiency. With regard to university education, the first meaning becomes more relevant than the second, if you just consider the socio-economic context (the job market) in which such education is engaged.

In the job market the assessment of efficacy cannot be based only on objective data, but also on the subjective aspects related to satisfaction, which varies depending on each worker's preference scales and also on his cultural context belonging. The question we want to answer is: how the employed graduates appreciate their job and what are the issues that most affect their judgment? These aspects, in fact, are often overlooked and end up being ignored at the planning of the University training offer.

An accurate picture of the Italian human capital is provided by the consortium *AlmaLaurea* through its survey of graduates, which is a special reference for all those who look to our higher education system as a key sector for development. In particular *AlmaLaurea* supplies both a range of services to promote graduates' employment and also timely information on their characteristics, such as the type of contract, the time taken to find work, the branch of economic activity, the geographical and gender differences.

In this paper we examine the second cycle graduates in the 2008 summer session of the University of Bari, who declared to be employed five years after the end of their studies. The aim pursued is to detect the most influential factors in defining the level of the overall job satisfaction, separately for those who carried out at least one training period after graduation and for everyone else.

4.1   A General Profile of the Graduates in the 2008 Summer Session at the University of Bari

Among the 1626 graduates in the 2008 summer (May/July) session at the University of Bari, who were interviewed five years later by the consortium *AlmaLaurea*, 654 are the second cycle ones. In particular 400 of them declared to be employed, 159 declared not to be working and not to be looking for a job, while 95 declared not to be working and not to be looking for a job.

The 400 employed interviewees were asked to assign a score - on a graded series of natural numbers from 1 to 10 – both to the overall satisfaction with the working activities carried out (Y) and to the satisfaction with various related aspects, such as: acquisition of skills (X_1), prestige (X_2), correspondence to cultural interests (X_3), involvement in decision-making (X_4), future prospects of gain (X_5). As already mentioned, we intend to detect which of these aspects seem to be most influential in defining the interviewees' satisfaction at work.

It is likely that the latter is influenced by the result of a relevant choice which every student is faced with, after graduating: is it better a postgraduate education or an immediate entry into the world of work? The relevance of such a choice leads us to conduct our analyses after separating the 348 interviewees declaring they had carried out at least one training activity after the second cycle graduation by those others preferring to pursue a working career right away.

Sometimes, the intention to keep on studying after graduation is conditioned by the inability of university courses to prepare young people for the world of work; in fact the new reform of the education system is trying to fill this gap by introducing increasingly targeted and sector degree courses. Anyway, the satisfaction for the job in its entirety and for each of the related aspects seems to be on average higher among those who have not done any training since graduating.

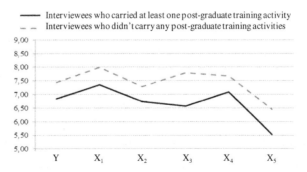

Fig. 2. Average scoring of the satisfaction for the job in its entirety (Y) and for each of the related aspects (X_1, X_2, ..., X_5)

The correlation between the scores of the satisfaction for the job in its entirety (Y) and for each of the related aspects (X_1, X_2, ..., X_5) is summarized both in Table 1, with reference to the interviewees declaring they had carried out at least one training activity after the second cycle graduation, and in Table 2, with reference to the interviewees preferring to pursue a working career right away.

Table 1. Spearman's correlation coefficients between the satisfaction for the job in its entirety (Y) and for each of the related aspects (X_1, X_2, ..., X_5), expressed by the graduates who have done at least one post-graduate activity

	Y	X_1	X_2	X_3	X_4	X_5
Y	-	0,52	0,60	0,48	0,47	0,49
X_1		-	0,56	0,55	0,42	0,42
X_2			-	0,54	0,50	0,52
X_3				-	0,49	0,28
X_4					-	0,42
X_5						-

Overall satisfaction (Y), acquisition of skills (X1), prestige (X2), correspondence to cultural interests (X3), involvement in decision-making (X4), future prospects of gain (X5)

Table 2. Spearman's correlation coefficients between the satisfaction for the job in its entirety (Y) and for each of the related aspects (X_1, X_2, ..., X_5), expressed by the graduates who have not carried out any post-graduate activity

	Y	X_1	X_2	X_3	X_4	X_5
Y	-	0,70	0,73	0,59	0,49	0,64
X_1		-	0,70	0,64	0,44	0,52
X_2			-	0,65	0,57	0,64
X_3				-	0,53	0,54
X_4					-	0,39
X_5						-

Overall satisfaction (Y), acquisition of skills (X_1), prestige (X_2), correspondence to cultural interests (X_3), involvement in decision-making (X_4), future prospects of gain (X_5)

Focusing on the first interviewees, we note that the satisfaction for the job in its entirety is more correlated with the satisfaction for the prestige, but also for the other aspects to some extent; going to focus on the second ones, we find a generally higher correlation between the satisfaction for the job in its entirety and the satisfaction for the issues related to professional and personal growth, such as acquisition of professionalism and prestige received, but also for prospects of gain (Table 2).

4.2   The Fuzzification of the Opinions on the Working Experience

In this section we restrict our attention on the opinions expressed by the 400 second cycle graduates who said they were busy at the time of the interview.

Such opinions have the status of scores ranging in a bipolar scale from 1 to 10, which is anchored to two descriptive adjectives ("unquestionably satisfied" and "unquestionably satisfied"). In particular, we can affirm that the latter ones function as extreme poles of a psychological continuum, along which the personal views are distributed.

Since each score represents a value of densification on the considered scale, then we associate a range of values centered around it in a fuzzy approach, whose width varies according to the intensity of the expressed opinion. In other words we assign a degree of representativeness to each value included in such range, which is inversely proportional to its distance from the center of the fuzzy number *ad hoc* identified.

In previous works Gola and Chiandotto [8] considered the categories "Unquestionably unsatisfied", "More unsatisfied than satisfied", "More satisfied than unsatisfied", "Unquestionably satisfied" in questionnaires on the graduates' preferences, and assigned respectively the scores 2, 5, 7 and 10 to them. Such a choice allowed to avoid that the responses were concentrated around a median category.

Therefore we use 2, 5, 7 and 10 as the centres of four fuzzy numbers, whose spreads are determined according to the intensity of the expressed opinions. In this way, it seems appropriate to match triangular asymmetrical numbers, characterized by quite extensive spreads, with interim assessments which describe a vague conviction, while a semi-trapezoidal fuzzy number with the lowest scores of the scale and a semi-triangular fuzzy number with the highest scores (because extreme assessments describe clear and unequivocal respondents' opinions).

On the basis of such heuristic evaluations, we express the graduates' responses in terms of the fuzzy numbers which are analytically represented in the following figure (Figure 3).

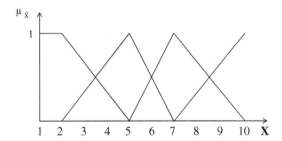

Fig. 3. Representation of the graduates' response categories by fuzzy numbers

Once transformed in fuzzy numbers, the satisfaction with the acquisition of skills (X_1), the prestige (X_2), the correspondence to cultural interests (X_3), the involvement in decision-making (X_4) and the future prospects of gain (X_5) can be used to explain the overall satisfaction with the working activities carried out (Y) in a fuzzy least square regression model.

In particular, we prefer to consider non-linear regression functions with multiplicative factors, according to which the marginal effect of the satisfaction with a single aspect on the overall satisfaction is not constant or depends on the value assumed by the (correlated) satisfaction with another aspect. Evidently, as the number of the parameters to be estimated increases further compared to a fuzzy linear regression model, the selection of the independent variables to be included becomes even more difficult from a computational point of view.

Recently we proposed a stepwise identification procedure which enables us to find the optimal combination of the independent variables by including at each iteration one of them according to two fundamental criteria: the significance of its contribution, measured by the relative increase of the total deviance in the dependent variable, and its originality, that is the ability to introduce information into the equation other variables have not already introduced (assessed in terms of correlation with the latter). Moreover, the procedure eliminates in each iteration the variable, whose explanatory contribution is subrogated by the combination of the other ones included after it was.

Let's examine how it works in detail, starting from the forward selection.

After identifying the simple regression model (in which $\tilde{X}_{(1)}$ presents the highest correlation with $\tilde{Y}$), in each successive iteration we select the variable less correlated with those already present, provided that it significantly explains the total sum of squares of the model. In other words, focusing on the q.th step (q = 2,3,...p), $\tilde{X}_{(q)}$ is candidate to be also included in the equation if its contribution is original with respect to the previous q-1 variables.

Such a contribution is evaluated by measuring the so called tolerance $T_q = 1 - \overline{FFI}_{q;1,2,\ldots,q-1}$, in which $\overline{FFI}_{q;1,2,\ldots,q-1}$ represents the share of variability of $\tilde{X}_{(q)}$ explained by $\tilde{X}_{(1)}$, $\tilde{X}_{(2)}$, ..., $\tilde{X}_{(q-1)}$. The tolerance ranges between 0 and 1, depending on the degree of linear correlation between $\tilde{X}_{(q)}$ and the other variables; therefore, only if T_q exceeds a threshold identified between 0 and 1, $\tilde{X}_{(q)}$ is candidate to become part of the model.

Note that a high value of the threshold allows to select very original variables, but it can also stop the process since from the initial steps; on the contrary, a low value allows to select a greater number of variables, although more correlated to each other. In any case, if none of the variables not yet included in the equation proves significantly its originality, the selection process would stop.

The opportunity of actually introducing the selected variable $\tilde{X}_{(q)}$ into the equation is now evaluated in terms of its explanatory contribution. In particular such a contribution is measured as the increase in the adjusted Fuzzy Fit Index of the model due to the entry submitted for consideration, i.e. as $\overline{FFI}_{y;1,2,\ldots,q} - \overline{FFI}_{y;1,2,\ldots,q-1}$ (where the two terms of the subtraction represent the proportion of the sum of squares of $\tilde{Y}$ explained by the model including and not including $\tilde{X}_{(q)}$ respectively).

The selected variable ends up being introduced into the equation if the increase in the adjusted Fuzzy Fit Index is higher than an arbitrary threshold value. The higher such an arbitrary value is, the easier the procedure inhibits the entry of new independent variables whose explanatory contribution is not that relevant.

If the explanatory contribution of $\tilde{X}_{(q)}$ is not significant, we pass to consider the inclusion of the remaining candidate variables on the basis of the same criterion.

In any case, the selection procedure stops when none of them contributes significantly to explain the sum of squares of $\tilde{Y}$.

The proposed procedure provides also the possibility of eliminating in each iteration one of the variables already included in the equation. For example, once $\tilde{X}_{(q)}$ is inserted, the explanatory contribution of every $\tilde{X}_{(i)}$ (i = 1, 2, ..q-1) is still valued as the reduction of the adjusted Fuzzy Fit Index caused by the elimination of $\tilde{X}_{(i)}$ from the model, i.e. as $\overline{FFI}_{y;1,2,...,q}$ - $\overline{FFI}_{y;1,2,...,q\,(-i)}$ (where the two terms of the subtraction represent the proportion of the sum of squares of $\tilde{Y}$ explained by the model excluding and not excluding $\tilde{X}_{(i)}$ respectively).

The variable which shows the smallest reduction is excluded from the equation if such a reduction does not exceed an arbitrary threshold value. This would happen if the explanatory contribution of the variable to be discarded was subrogated by the combination of the independent variables introduced after it was.

The just exposed procedure has been developed through the *MatLab Editor*.

4.3   The Estimates of the Polynomial Regression Model

In this application we use a polynomial regression model of the second order, which maintains a parsimonious number of independent variables and allows an easier interpretability of the obtained estimates. Moreover, the domain of the considered variables is too limited to use a model of a higher order.

The stepwise selection procedure leads us to affirm that, among the interviewees declaring to have carried out at least one post-graduate training activity before entering the job market (Table 3), the more influent aspects on the overall satisfaction with the working activities are the acquisition of skills (X_1), the prestige (X_2), the correspondence to cultural interests (X_3) and the involvement in decision-making (X_4).

In particular, the acquisition of skills and the prestige do not affect the overall satisfaction neither linearly nor quadratically; it is their iteration that helps to increase the latter one, replacing the linear or quadratic effects; that is to say, therefore, that when the satisfaction for one of the two factors increases, the marginal contribution of the second one on the overall satisfaction increases as well. Actually such factors are quite correlated between each other, as well as with the overall satisfaction (Table 1).

Also, while the involvement in decision-making contributes linearly to the overall satisfaction, the marginal effect of the correspondence to cultural interests is all the stronger, the higher the satisfaction for it.

Going to examine the interviewees declaring to have not carried any post-graduate training activity before entering the job market (Table 3), the interaction between the acquisition of skills and the prestige is still the most important component of overall satisfaction. Evidently the graduates' ambition does not depend on whether the professionalism was achieved through studies or it was materially acquired directly with the work experience.

The correspondence to cultural interests still affects, this time linearly, the overall satisfaction of those who have not carried any post-graduate training activity before entering the job market. This leads us to believe that they are sensitive to this factor, but less than their colleagues, because the latter ones probably perceive professional training such as the completion of their personal sphere.

The overall satisfaction seems to be less influenced by the involvement in decision-making than by the future prospects of gain. Therefore the interviewees who begin to work after their graduation, without any other training activities, seem to be more gratified by the economic aspects than by the relational ones of their job.

Table 3. The estimated regression coefficients of the fuzzy model

	At least one post-graduate activity carried out			No post-graduate activity carried out	
	Regression coefficients	Standardized regression coefficients		Regression coefficients	Standardized regression coefficients
Intercept	2,94	-	Intercept	2,41	-
Spread left	0,35	-	Spread left	0,28	-
Spread right	0,01	-	Spread right	0,01	-
X1*X2	0,04	0,48	X1*X2	0,04	0,50
X4	0,15	0,15	X5	0,16	0,16
X3^2	0,01	0,14	X3	0,20	0,20
FFI agg	0.52	0.52	FFI agg	0.65	0.65
FFI mod	0.53	0.53	FFI mod	0.68	0.68

Overall satisfaction (Y), acquisition of skills (X1), prestige (X2), correspondence to cultural interests (X3), involvement in decision-making (X4), future prospects of gain (X5)

Finally note that the two estimated regression models show adjusted Fuzzy Fit Indexes respectively equal to 0.52 and 0.65. Therefore both the models allow us to reproduce appropriately, albeit not exhaustively, the graduates' overall satisfaction at work.

5   Conclusions

In a job market characterized by a high rate of youth unemployment, the analysis of graduates' placement is extremely valuable. It is known, in fact, that a graduation increases the chances of finding a job, which however remains temporary and not corresponding to the acquired professional skills for young people.

This social and economic problem cannot be addressed by merely considering the young graduates' percentage of employment, but also by analysing their opinions on various aspects of job satisfaction. In this work, we intend to detect which aspects seem to be more influential in defining the job satisfaction expressed by the students from the University of Bari in 2008, interviewed five years after their graduation.

Specifically the employed interviewees were asked to assign a score - on a graded series of natural numbers from 1 ("unquestionably satisfied") to 10 ("unquestionably satisfied") - both to the overall satisfaction with the working activities carried out and to the satisfaction with various related aspects, such as: acquisition of skills, prestige, correspondence to cultural interests, involvement in decision-making, future prospects of gain. Since each score represents a value of densification on the considered scale, we associate a range of values centred around it in a fuzzy approach, whose width varies according to the intensity of the expressed opinion.

Once transformed in fuzzy numbers, the satisfaction with all the considered aspects can be used to regress the overall satisfaction with the working activities carried out. In particular, we propose a second order polynomial regression function, according to which the effect on the dependent variable of a marginal change in one of the independent variables may vary depending on the particular value of the latter. Moreover, the introduction of multiplication factors between the independent variables allows us to measure the marginal effect of any interaction between them on the dependent variable.

The above analysis, based on a generalization of the so called fuzzy least square regression model, is conducted after separating the interviewees who have carried out at least one training activity after the second cycle graduation by those others who pursued a working career right away. In general terms, the satisfaction both for the work in its entirety and for each of the related aspects seems to be on average higher among those who have not done any training after graduating than among the other ones.

In more detailed terms, after a selection process of the variables to be considered, we find that the interaction between the acquisition of skills and the prestige is the more significant factor in regressing the overall satisfaction in both the cohorts of interviewees. Also the correspondence to cultural interests is an important factor for both the cohorts, although its marginal effect on the overall satisfaction is quadratic among the graduates who have carried out at least one training activity, but linear among those who have not. At last the involvement in decision-making is relevant only in the first cohort, as well as the future prospects of gain in the second one.

Therefore all the interviewees are interested in increasing their professionalism; but those who begin to work after their graduation, without any other training activities, seem to be more gratified by the economic aspects than by the relational ones of their job.

Acknowledgements. Although the work is the result of joint reflections by the authors, chapters 3, 4.1, 4.2 and 5 are to be attributed to Francesco Campobasso, and chapters 1, 2, 4 and 4.3 to Annarita Fanizzi.

References

1. Campobasso, F., Fanizzi, A.: A stepwise procedure to select variables in a fuzzy least square regression model. In: International Conference on Fuzzy Computation Theory and Applications 2011, Paris, France (2011)
2. Diamond, P.: Fuzzy least squares. Information Sciences **46**, 141–157 (1988)
3. Zimmerman, P.R.: Fuzzy Set Theory. Kluwer Academic Publishers, Dordrecht (1991)
4. Campobasso, F., Fanizzi, A., Tarantini, M..: Some results on a multivariate generalization of the fuzzy least square regression. In: Proceedings of the International Joint Conference on Computation Intelligence, pp.75–78. SciTePress, Madeira (2009)

5. Montrone, S., Campobasso, F., Perchinunno, P., Fanizzi, A.: An analysis of poverty in italy through a fuzzy regression model. In: Murgante, B., Gervasi, O., Iglesias, A., Taniar, D., Apduhan, B.O. (eds.) ICCSA 2011, Part I. LNCS, vol. 6782, pp. 342–355. Springer, Heidelberg (2011)

6. Campobasso, F., Fanizzi, A.: Goodness of fit measures and model selection in a fuzzy least squares regression analysis. In: Madani, K., Dourado, A., Rosa, A., Filipe, J. (eds.) Computational Intelligence. SCI, vol. 465, pp. 241–257. Springer, Heidelberg (2013)

7. Zadeh, L.: The concept of a linguistic variable and its application to approximate reasoning, Part I. Inf. Sci. **8**, 199–249 (1975)

8. Chiandotto, B., Gola, M.: Rapporto Finale del Gruppo di Ricerca per la Valutazione della didattica da parte degli student. In: MURST-Osservatorio per la Valutazione del Sistema Universitario (1999)

Historic Buildings and Energetic Requalification A Model for the Selection of Technologically Advanced Interventions

Antonio Nesticò(✉), Maria Macchiaroli, and Ornella Pipolo

Department of Civil Engineering, University of Salerno,
Via Giovanni Paolo II 132 84084, Fisciano (SA), Italy
anestico@unisa.it, m.macchiaroli@live.it,
ornellapipolo@virgilio.it

Abstract. The recovery and enhancement of historic buildings require interventions aimed also at the energetic requalification. Indeed, on one side there is a growing attention to the reuse of these buildings not only for conservative purposes, but even for economic and environmental aims; by forcing to include in the project even the energy efficiency improvement. On the other side, the current legislation on environmental sustainability in construction gives clear guidelines in energetic improvement object regarding the entire real estate assets, including historic buildings.

The paper aims to define a technical-economic model for the selection of technologically advanced interventions useful to improve the buildings energy performance of historical and architectural interest. This can happens by respecting the needs and regulations on the protection of the valued property.

The software implementation for the energy characterization of the historic building and the use of cost-benefit analysis, allow to drawing up an analysis protocol to support investment decisions. The issues research about the low thermal and acoustic insulation of the building envelopment and inefficient plants leads to outline intervention strategies and corresponding technical solutions, listed in detail in the article.

The research appears important on the theoretic side because rationalizes the various phases necessary for the selection of the energetic requalification interventions, for which the cash flows analysis is essential to ensure the financial sustainability of the initiative. It's also true in operative terms since it provides an easy instrumentation to professionals and operators in this sector to lead towards technologically efficient interventions, profitable in management phase and respectful of the historical and architectural values of the built heritage.

Keywords: Economic evaluation · Historic buildings · Integrated Conservation · Technologies for energetic requalification

© Springer International Publishing Switzerland 2015
O. Gervasi et al. (Eds.): ICCSA 2015, Part III, LNCS 9157, pp. 61–76, 2015.
DOI: 10.1007/978-3-319-21470-2_5

1   Cultural Property and Historical-Artistic Interest Buildings. Notions and Regulatory Constraints

In the international arena, the first official recognition of "cultural property" dates back to the Convention for the Protection of Cultural Property in the Event of Armed Conflict, signed at The Hague in 1954 by forty states around the world. Participating Nations agreed to safeguard these assets in case of war, arguing that the attacks to the assets of any people were violence to the heritage of the entire international community.

For the purposes of the Convention, «the term "cultural property" shall cover, irrespective of origin or ownership (art. 1):

a) movable or immovable property of great importance to the cultural heritage of every people, such as monuments of architecture, art or history, whether religious or secular; archaeological sites; groups of buildings which, as a whole, are of historical or artistic interest; works of art; manuscripts, books and other objects of artistic, historical or archaeological interest; as well as scientific collections and important collections of books or archives or of reproductions of the property defined above;

b) buildings whose main and effective purpose is to preserve or exhibit the movable cultural property defined in sub-paragraph (a) such as museums, large libraries and depositories of archives, and refuges intended to shelter, in the event of armed conflict, the movable cultural property defined in subparagraph (a);

c) centres containing a large amount of cultural property as defined in sub-paragraphs (a) and (b), to be known as "centres containing monuments"».

In Rome, in 1964 the Franceschini Commission defines cultural property «the assets of archaeological, historical, artistic, environmental and landscape, archive and library interest and any property that constitutes material testimony having the value of civilization», by moving the legislature intervention in field of cultural heritage from the previous concept of simple guarantor of physical conservation to an active role in enhancing the cultural property.

The notion of cultural property has been subjected to a wide debate and has also interested the possible actions to its optimal use and evaluation techniques that allow to understand the different components of the complex social value[1] [1, 2, 3, 4, 5].

Legal regulations protect cultural heritage and report specific requirements pertaining to historic buildings for which the declaration of cultural interest has been notified[2].

---

[1]   The need to relate economic evaluations with the evaluations from other points of view, as the biological, ecological, social, induced to define the complex value, which for resources of public interest takes the name of complex social value. The multidimensional analysis are based on it, intending to consider in the process of evaluating the set of all values that coexist in a resource, which are many and varied [6]. The assessment, in this view, is no longer based on the only monetary indicator, but on a set of indicators, some of which are expressed in monetary form, others in terms of quality.

[2]   In Italy the real property (public or private) constructed from at least seventy years and work of an artist no longer living, "declared of cultural interest", are subjected to constraint (L. 106/2011 art.4).

In France, the Decree 178/04 Code du Patrimoine provides that the owner of the property has to preserve it, intervening with the necessary restoration work behind issuance of a specific prior authorization of the Ministry for Cultural Affairs [7, 8].

In England, the National Planning Policy Framework in 2012 in the "Conserving and enhancing the historic environment" orders that all demolition or alteration which may in any way undermine the historical or architectural character of the property, must be preceded by the release of special permissions by the Secretary of State supported by English Heritage.

In Italy, the Code of the Cultural and Landscape Heritage is in force and it was enacted by Legislative Decree 42/2004. According to art. 10 of this Code cultural property are in any case «the immovable and movable things belonging to the State, the Regions, other local governments, as well as any other public agencies and institution and non-profit private legal entities, including civilly recognized ecclesiastical agencies, which have artistic, historical, archaeological or ethno-anthropological interest».

The Cultural Heritage Code defines also the actions to be taken to protect and safeguard them. By affixing the constraint, the property passes from a normal regime to a special one, under which the owner cannot freely use the protected assets, but he must necessarily be subjected to specific rules concerning the actions of transformation and enhancement. This means that the execution of works of any kind on the cultural property is subject to authorization by the Superintendent, moreover these buildings cannot be used for purposes that are not compatible with their historic or artistic character, or such as to create injury to their conservation or integrity.

The owner of protected assets is obligated to keep it «through a coherent, coordinated and planned activity of study, prevention, maintenance and restoration» (art. 29). In particular, the constraints are referred to the conservation of the architectural and material and/or decorative features of building such as the maintenance of the geometry and shape of the building, windows and doors, flooring and ceilings, decorations (paintings, frames, shelves) of historical materials of walls, roofs [9].

Article 45 of the Code introduces the indirect protection aimed to what characterizes the immediate vicinity, suggesting to pay particular attention and sensitivity to interventions related to the context nearest to the monument, with the purpose to «not damage the prospect or the light or to not alter the conditions of environment and decor».

It is evident that the constraints to which cultural property are subject concern also interventions for energetic requalification. In fact, in Italy the properties subjected to landscape and cultural constraint, that are transferred with or without payment or rented, must be always provided of Energy Performance Attestation (APE) under the law 90/2013 (Energy performance in buildings. Urgent provisions for the transposition of Directive 2010/31/EU of the European Parliament and of the Council in May 19, 2010).

Article 3bis, introduced by this law in the Legislative Decree no. 192/2005, states that cultural properties are excluded from the obligation of energy attestation when compliance of requirements implies unacceptably alter their artistic-historical character. The competent authority to grant approval, judges the presence of "substantial alteration". So, the possibility of installing mechanical, thermal, electrical and/or electronic plants, useful to improve the functionality and energy performance in the building in operating conditions, cannot be excluded [10].

Ultimately, for protected assets, all the requirements of the law on the obligation to adopt energy performance attestation are worth. It means that the APE must be produced not only for each act of transfer or new lease, but even for the "major renovation" i.e. all those interventions of ordinary and extraordinary maintenance, renovation and rehabilitation conservative which affect more than 25% of the surface of the building envelope. As for the other buildings, even for the protected assets used by the public administration, opened to the public and with area greater than 500 m^2, there is the obligation to produce an APE.

In the presence of historic buildings, protected or not, it is necessary to pay attention to the risk of proposing lines of general action that, while representing good requalification techniques, cannot take into account the specific case, fundamental to the recovery of assets that have their characteristics by virtue of historical and cultural value.

2   Integrated Conservation and Energetic Requalification of Historic Buildings

In order to achieve the objectives promoted by the 2009 climate and energy package, the European Union stressed the need to intervene in the construction industry not only by increasing the energy performance of new buildings, but also redeveloping existing ones, since the reduction in consumption and the use of renewable energy sources are essential for environmental sustainability [11].

With specific regard to the historic buildings, it is necessary an "integrated conservation", defined as «the result of the combined use of the restoration technique and research of appropriate functions» (Congress of Amsterdam, 1975, European Architectural Heritage), i.e. of legal, administrative, financial and technical synergies within historical contexts, provided the environmental characteristics and used materials are satisfied [12].

An updated interpretation of the concept of integrated conservation, however, must encompass the possibility of making use of materials and technologies that, while respecting the historical and architectural significance of the building, allows also pursuing the energetic requalification objectives. On the other side, this is a theme that has often been discussed, as it is, among other things: in Europe, from the EU Rebuild Program "Renewable Energy and Historic Places" (2002) and the guide SECHURBA (2011); in Italy, from the Project "Shares of technology transfer to improve the energy and environmental performance of the historic buildings" (2010) and the ministerial guidelines, currently being drafted, for the efficient use of energy in the cultural heritage. The Green Building Council has also introduced the GBC Historic Building, protocol of voluntary certification of sustainability level of conservation, requalification, recovery and integration interventions of historic buildings with different uses.

The recent economic crisis has caused a considerable reduction in the available resources of the management agencies, particularly with regard to government grants. It must be found a solution that reduce the actual consumptions, through the use of renewable energy sources, ensuring in the same time the proper protection of artistic and cultural value of the buildings and the optimum use of space and art works

present in them [13, 14, 15]. So, the interventions of integrated conservation on historical and architectural building heritage allow to pursuing multiple objectives, namely:
- recovery, restoration or preservation of the building[3];
- assignment of different uses (schools, hotels, residences, museums), also useful to pursue the financial sustainability of management activities;
- optimization of indoor comfort [16];
- higher levels of energy performance;
- environmental sustainability.

The Table 1 summarizes the objectives and sub-objectives of initiatives for integrated conservation and for energy efficiency of historic buildings.

Table 1. Energy efficiency of historical buildings: objectives and sub-objectives

OBJECTIVES	SUB-OBJECTIVES
recovery, restoration or preservation of the building	- elimination of the deterioration - restoration of historical and architectural characters of the property - improvement of the urban context, environment and decorum - removal of architectural barriers - ability to attract the visitors flows
assignment of different uses	- new services for the community - increase of income flows - employment opportunities
indoor comfort optimization	- improving thermo-hygrometric levels - raising of the acoustic comfort
higher levels of energy performance	- thermal insulation - plant efficiency control - contraction of management costs
environmental sustainability	- lower consumption of resources - decrease of polluting emissions - reducing of noise pollution

---

[3] Recovery means a program of operations to rehabilitate a building in degradation so that it can "recover" functionality, independently from the previous one. It doesn't imply a previous recognition of historical and artistic value.

The restoration indicates an operation designed to ensure the preservation of architecture or retrieval to previous conditions for actions of agents of deterioration, vandalism acts, wear, sagging, etc. with essential historical knowledge and technical and scientific skills.

The conservative renewal regards prevention, preservation or maintenance works aimed mainly to the hygienic, static and functional recovery of buildings for which are necessary the consolidation and integration of structural elements and the planimetry modification [17, 18].

It be noted that the recovery, restoration or conservative renewal of the historic building have to be compared to the intended use for the building, since, varying the latter, change the parameters of required environmental comfort and therefore maximum energy derivable performance. Another factor related to the intended use is the number of users, which can determine a substantial increase of CO_2 in the air, variations in moisture and temperature, causing also significant changes in terms of comfort and even for the property preservation. Therefore it is necessary to carry out a preliminary assessment of compatibility between the new functions to be set up and the building casing.

The implementation of the interventions imposes always a methodological approach able to reconcile different needs, and sometimes conflicting, as the selection of advanced technologies, the respect for the specific characteristics and the property historicity, the reduction of investment costs [19].

3   A Model for the Economic Evaluation of Projects on the Protected Building Heritage

In order to optimize the energy performance of historic buildings, so to reduce the costs borne by private and/or public administrations, it is necessary to assess the interventions technical and economic feasibility [20, 21, 22, 23, 24]. This has to be done by following a protocol based on a sequence of logical-operating steps that begins with historical research on the property and with the collection and analysis of data for the measurement of energy audits of the building in the current state and for the simulation of its thermodynamic behaviour. It continues with the verification of energy requirements and model calibration. Next steps are the identification of the critical points, the selection of possible interventions, how to manage and search financial supports, before arriving to the economic evaluation and optimal investment choice [25].

Phase I implies, through inspections and surveys, information retrieval:
- on the property historical background
    - historical era
    - the district history and the building fabric evolution
- on the property specific characteristics
    - state of conservation
    - any renovations
    - architectural value
    - used materials
    - building type
    - building techniques
- climatic about places
    - climatic zone (°C)
    - degrees days DD
- geometric-dimensional
    - gross surface area in typical floor area (m^2) and heights (m)
    - net heated surface area S (m^2) and net heated volume V (m^3)
    - S/V coefficient (m^2/m^3)

- thermophysical of the casing
    - thermal resistance R (K/W)
    - transmittance of simple components $U = 1/R$ and stratigraphic $U = 1/\Sigma R_i$ (W/m^{2}K)
- performance data of the systems
    - type (heating, electrical, hot water)
    - fuel used (propane, methane, oil)
    - delivery systems (radiators, heaters, etc.)
    - distribution systems (vertical, horizontal)

Phase II processes the collected data to preparing an energetic statement on the thermodynamic behaviour of the building through the performance indicators proposed in the UNI TS 11300 and related regulations:

$$Ep_{wh} = \frac{\left(\frac{Q_h}{A_{floor}}\right)}{\eta_g} = \frac{\left[\frac{(Q_{h,tr} + Q_{h,ve}) - \eta_s \times (Q_{int} + Q_{sol})}{A_{floor}}\right]}{\eta_g}, \tag{1}$$

$$Ep_{sc} = \frac{(Q_{int} + Q_{sol}) - \eta_s \times (Q_{th,tr} + Q_{h,ve})}{A_{cool}}, \tag{2}$$

$$Ep_{dhw} = \frac{\left(\frac{Q_w}{A_{floor}}\right)}{\eta_r} = \frac{\frac{\rho_w \cdot c_w \cdot [V_w \cdot (\theta_s - \theta_0)] \cdot G}{A_{floor}}}{\eta_r}, \tag{3}$$

$$Ep_{gl} = Ep_{wh} + Ep_{sc} + Ep_{dhw} + Ep_l \tag{4}$$

where: Ep_{wh} = energy performance index for winter heating [kWh/m^{2}K], Q_h = thermal energy demand of the building [kWh], A_{floor} = useful floor area [m^2], η_g = average global seasonal performance coefficient, $Q_{h,tr}$ = transmission losses [W/K], $Q_{h,ve}$ = dispersions due to ventilation [W/K], η_s = coefficient of use of free inputs, generally assumed to be equal to 0.95, Q_{int} = free internal inputs [MJ], Q_{sol} = solar inputs [MJ]; Ep_{sc} = energy performance index for summer cooling of the building envelope [kWh/m^{2}K], A_{cool} = useful cooled surface [m^2]; Ep_{dhw} = primary energy for domestic hot water [kWh/m^{2}K], Q_w = energy demand for domestic hot water [kWh], ρ_w = volumetric mass density of water [1000 kg/m^3], c_w = specific heat of water [$1.162 * 10^{-3}$ kWh/(kg K)], V_w = daily volume of water required by activity or service [m3/day], θ_s = water supply temperature [40° C], θ_0 = entry temperature of cold water [15° C], G = number of days in the calculation period, η_g = global seasonal average performance coefficient; Ep_{gl} = global primary energy [kWh/m^{2}K]; Ep_l = energy performance index for artificial lighting [kWh/m^{2}K].

Phase III calibrates the model by comparing the real energy consumption $Q_{i, real}$ of historical building, taken from the bills, with the theoretical needs $Q_{i, th}$ obtained by the collected input data in Phase I and processed with software. It must be verified that the following relation is satisfied

$$\frac{Q_{i,real} - Q_{i,th}}{Q_{i,real}} \times 100 < 15\%, \tag{5}$$

with:

$Q_{i,th}$ = theoretical energy requirement for heating or dhw [kWh];

$Q_{i,real}$ = real energy demand for heating or dhw [kWh].

Phase IV identifies, through surveys, the critical points of housing. It is necessary identify the components of the building envelope and/or plants that cause the attribution of a low energy class. The diagnostic activity can happen through

- non-destructive techniques:
  - visual examination,
  - infrared thermography, detection technique of electromagnetic radiation emitted by bodies in the infrared range that have a temperature above absolute zero (T> 0 ° K i.e. - 275.15 °C), through radiometric measurements performed with special digital cameras,
  - ultrasonic analysis, based on the measurement and analysis of the propagation characteristics of the ultrasonic waves with a frequency between 50 KHz and 10 MHz, to determine the degree of homogeneity, the presence of voids, structural damages or discontinuity, the elastic module value, the concrete resistance,
  - blower door test and a smoke tracers, useful to measure the air tightness of a building subjected to pressure difference of 50 Pa through the relation

$$n\,(50) = \frac{\Phi}{V} \quad [1/h] \tag{6}$$

  with airflow Φ and volume V of the building,
  - flowmeter analysis, which allows to estimate with sufficient accuracy the opaque element transmittance U, after a period of continuous relief of the thermal resistance;
- weakly destructive techniques:
  - stratigraphic analysis, in order to know in detail the material of which the architectural work is made.

Phase V is the identification of possible actions to be implemented to optimize the energy consumption levels, so as to allow a significant reduction of global energy performance:

$$Ep_{gl,after\ intervention} < Ep_{gl,first\ intervention}\,. \tag{7}$$

The specification of possible actions to be taken for historic buildings is set out in following paragraph 4.

Phase VI examines the funding for the various types of intervention. This may be Europeans incentives (European Energy Efficiency Fund EEEF, European Local

Energy Assistance ELENA), national or regional facilitations. In Europe, the EEEF follows investments by 5mil / EUR to 25mil / EUR including projects carried out by energy service companies (ESCO), renewable energy services and energy efficiency on small scale. ELENA is instead funded a grant for 90%. The program covers the costs related to the support by technical and legal consultants for the preparation of a project and the publication of the contract notice for the award of works and services required for its realization. The minimum size for access to financing is 50 million euro. In Italy, the Code of cultural property provides for some time the possibility of access to government grants in the case of restoration or conservation work. The Ministry has, in fact, the right to contribute to the costs incurred by the owner for an amount not exceeding half. In the event that the interventions are of particular importance or concern buildings for public use, the subsidy may even reach the full amount. The work for energy efficiency can be managed directly, if conducted by the user or by the company that intends to improve the building energy performance, or indirectly, if they occur through the Energy Service Company (ESCo).

Phase VII evaluates the economic convenience of the investment comparing costs and revenues that are generated in the implementation and management of the works. In fact, once performed the energy audits in order to make feasible the project on the historical building, it is necessary verify that the envisaged works make convenient the initiative also in terms of cash flow. The interventions feasibility requires a measure of financial profitability through indicators such the Net Present Value (NPV), Internal Rate of Return (IRR), the Payback period (PB):

$$NPV = \sum_{i=0}^{n} \frac{CF_i}{(1+r)^i} - I_0, \tag{8}$$

$$IRR = r \rightarrow \sum_{i=0}^{n} \frac{CF_i}{(1+r)^i} - I_0 = 0, \tag{9}$$

$$PB = \frac{I_0 - A_f}{R}, \tag{10}$$

with CF_i cash flow to the i-th year; r discount rate, I_0 the initial investment cost. For the IRR and Payback period, the following relations are applied, where Af indicates any tax breaks and R the annual recovery.

Phase VIII recognizes greenhouse gases released from the building through parameters that measure – in CO_2 equivalent – the emissions contribution of various pollutants having different effects on climate.

The methodology for the calculation of the quantity E_i of annual emission CO_{2eq} (tCO_{2eq}) by part of the pollutant i, is based on the relationship

$$E_i = A_i \times FE_i \tag{11}$$

where:

A_i is the value of energy delivered from the heating, production of dhw and electricity (in MJ, kWh, Kg, m^3 o L);

FE_i is the emission factor, which depends on the fuel used.

Even the economic benefit obtainable by reducing CO_2 emissions can be achieved through cost-benefit analysis attributing a social cost to avoided carbon.

4   Possible Interventions for the Energetic Requalification of Historic Buildings

The protocol outlined in paragraph 3 provides, to the phase V, the selection of interventions aimed to improving the energy performance of the historic building. The aim is not to achieve energy savings by respecting the law parameters established for new construction, but rather to adopt technological solutions that allow a more rational use of resources, exploiting better the potential of the place and limiting consumptions [26].

As the Table 2 shows, the energetic issues can relate to the insufficient thermal and acoustic insulation, as well as the system equipment. Both cause, in turn, negative effects in terms of consumption and environmental sustainability. The corresponding action strategies must allow the increase in energy performance of the building-plants systems, but also the use of renewable sources and technologies for efficient lighting and home automation.

Table 2. Energetic issues and action strategies for historic buildings

ISSUES

- thermal insulation
- acoustic insulation
- absent, incomplete or inefficient plants
- excessive energetic consumption and environmental sustainability

INTERVENTION STRATEGIES

- thermal and acoustic performance rising of building envelope
- investment in new plants for the renewable energy production and exploitation
- increase of living comfort and energy savings through illuminating engineering and home automation

Intervention strategies are implemented through technical solutions to be determined in relation to the specific architectural, constructive and functional characteristics of the property and in compliance with regulatory restrictions which often undertake the historic building.

Given the extensive bibliography on the subject [27, 28], the Table 3 shows the set of possible technical solutions useful to pursuit of each strategy. The compatibility of each solution is evaluated on an ordinal scale (very high, high, medium, low and very low).

Table 3. Strategies, solutions and compatibility with historical buildings

INTERVENTION STRATEGIES AND TECHNICAL SOLUTIONS	COMPATIBILITY WITH HISTORIC BUILDINGS
THERMAL AND ACOUSTIC PERFORMANCE RISING OF BUILDING ENVELOPE	
opaque vertical closures	
- thermal plaster	very high, if with the use of specific plasters for the historical building
- insulation	low
- heat reflective materials	low
transparent vertical closures	
- fixtures replacement	average, however, to be evaluated in function of the building architectural and functional characteristics
- replacement of only glass inside the existing frames	very high, if permitted by the constructive characteristics of the fixtures
- apposition of additional fixtures on the inside of the opening compartment	average
horizontal and/or inclined closures	
- thermal and acoustic insulation of the floor	average, to be evaluated according to the floor characteristics
- thermal and acoustic insulation of the roof slab	high
- inner tube for natural ventilation of roof	high
- use of insulating materials or sheathing in roofs	high
- waterproof membranes and/or vapour barriers in the roof	high
ground horizontal closures	
- floor reconstruction aerated with crawl and layer of thermal insulation	low, to be evaluated in respect of the characteristics of the ground closure element
- construction of trenches for ventilation	low
- realization of ventilated cavity below the floor	low

Table 3. (*Continued*)

INVESTMENT IN NEW PLANTS FOR THE RENEWABLE ENERGY PRODUCTION AND EXPLOITATION	
- integrated photovoltaic system in the roof plane for the electric energy production	low
- small photovoltaic plants arranged in outdoor appurtenant areas	average, only in presence of suitable appurtenant areas
- traditional tiles replacement with photovoltaic tiles	high
- solar panels for heating and hot water	low
- energy production with superficial or deep geo-thermal probes	average, in relation to the site characteristics
- plant for the wind energy production (mini-wind)	low, in relation to the site characteristics
- plant for the hydraulic energy production (mini-hydraulic)	low, in relation to the site characteristics
- biomass use for energy production	media
- creation of a heating system with wood boiler or pellet	high
- cogeneration plant for the electricity and heat production	high
- trigeneration plant for the electricity, heat and dhw production	high
INCREASE OF LIVING COMFORT AND ENERGY SAVINGS THROUGH ILLUMINATING ENGINEERING AND HOME AUTOMATION	
- home automation to control the heating, cooling, lighting, security and safety in indoor rooms and to reduce pollutants emissions	very high
- existing lamps replacement with low energy light bulbs or LED	very high
- illuminating engineering for the exploitation of the indirect light (reflected from the ceiling, floor, walls)	high, to be evaluated according to architectural, constructive and functional characteristics of the individual rooms

Regarding to the thermal and acoustic performance rising of building envelope, it has be noted that the regulations often prevent changes or alteration of the facades. Moreover it is possible to achieve energy savings by improving the insulation of walls, windows, roofs and floors, often with considerable potential in terms of the energy losses reduction. For old fixtures, responsible for substantial heat loss, sometimes replacement is contrary to the dispositions on conservation. In such circumstances, an improvement can be achieved with the double glazing insertion inside the traditional frames and also with the increasing of the energy performance of the originating wooden frames. Another approach can be made by applying additional fixtures inside of the opening compartment, by preserving outside the original window [29, 30].

Among the investments in new plants for renewable energy, the use of photovoltaic tiles, able to guarantee high aesthetic integration, is worth mentioning. With this tile, maintenance operations, replacement of damaged panels or use of most technologically advanced panels can be easily performed. The intervention requires careful planning, shadows accurate study, correct positioning of the connection cables [31].

Illuminating engineering and home automation allow the increasing of living comfort and energy savings. The illuminating engineering allows to exploit the indirect light for the interior rooms optimum illumination and to offer to the public, in the case of museums, churches or libraries, the best vision of the works [32]. Home automation allows to control locally or remotely: the automation of motorized blinds; air conditioning for differentiating the desired temperature in zones; the lights with occupancy sensors and timed switch off; the light regulation in the various rooms. Despite the costs are higher than the traditional technologies, the actual energy savings and the revaluation of the property with the satisfaction of a better energy class, allow a fast return of investment [33, 34].

5   Conclusions

The environmental and energetic question, which has led to international agreements and regulations, aroused the interest in new construction; the economic crisis and ecological instances, which have slowed the construction of new structures, have instead shown the need to focus on the enhancement of the existing buildings.

Today, however, in many Nations with advanced economies, the supply is higher than the building demand, and this underlines how it is more productive to exploit the available resources, avoiding construction of other buildings used to increase the gap between supply and demand.

The decision to limit the consumption of new soil, the will to improve the welfare of the users and the need to decrease the cost of building management [35, 36], leads to increase furthermore the performance of the existing one, even for the historic buildings.

To optimize the energy efficiency of the historical heritage it is necessary to follow a techno-economic model that, already proposed for traditional buildings, is detailed in this paper for the valuable buildings. The rationalization of the phases implemented with the protocol is distinguished by the retrieval of the historical background and the property characteristics, the non-destructive or weakly destructive diagnosis to avoid damage, the selection of compatible interventions with the regulatory requirements and with the architectural, constructive and functional characteristics, the identification of specific incentives for historic renovations that can affect the results of the cash flow analysis. All these operations help to evaluate in financial terms the feasible works and in environmental terms the reduction of polluting emissions.

The specification of possible feasible interventions depends on the recognition of the problems related to historical buildings. In fact, the construction and materials characteristics of these buildings, manufactured to meet housing needs different from those of today, provide thermal-hygrometric performance far from those considered now essential to their enjoyment. This is caused from the envelope, not insulated, particularly permeable to the passage of heat and moisture, from thermal bridges and the not air tight fixtures, as well as from the absence of plants or from the presence of not sufficient heating systems to achieve a comfortable temperature in the indoor rooms. To act on such inefficiencies it is necessary to select the intervention strategy between the rise of thermal and acoustic performance of the building envelope, the investment in new plants for the renewable energy production and exploitation, increased living comfort and energy savings through illuminating engineering and home automation. Each of these strategies can be pursued through technical solutions specified in the paper.

The model has been applied to a real case in order to verify the efficiency of the methodological approach and the ease of application. The results are being analysed and will be the subject of an extended paper.

References

1. Williams, S.A.: The International and National Protection of Movable Cultural Property: A Comparative Study. Oceana Publications, NY (1978)
2. Gillman, D.: The idea of Cultural Heritage. Cambridge University Press, Cambridge (2010)
3. Genovese, R.A.: La politica dei Beni Culturali. Edizioni Scientifiche Italiane, Napoli (1995)
4. Bruno, I.: La nascita del Ministero per i Beni culturali e ambientali. Il dibattito sulla tutela. LED, Milano (2011)
5. Guarini, M.R., Battisti, F.: Benchmarking Multi-criteria Evaluation: A Proposed Method for the Definition of Benchmarks in Negotiation Public-Private Partnerships. In: Murgante, B., Misra, S., Rocha, A.M.A., Torre, C., Rocha, J.G., Falcao, M.I., Taniar, D., Apduhan, B.O., Gervasi, O. (eds.) ICCSA 2014, Part III. LNCS, vol. 8581, pp. 208–223. Springer, Heidelberg (2014)
6. Fusco Girard L.: Estimo ed economia ambientale: le nuove frontiere nel campo della valutazione. Studi in onore di Carlo Forte. Franco Angeli, Milano (1993)
7. Pontier J.M.: Codifications et évolution du droit du patrimoine. AJDA n. 25, Paris (2004)
8. Greffe, X.: La gestion du patrimoine culturel. Anthropos, Paris (1999)
9. Boriani, M., Giambruno M., Garzulino A.: Studio, sviluppo e definizione di schede tecniche di intervento per l'efficienza energetica negli edifici di pregio. Report 64, ENEA, Roma (2011)
10. Boarin, P., Zuppiroli, M.: Sostenibilità di processo nell'intervento per la salvaguardia del valore testimoniale dell'edilizia storica. De Lettera Editore, Milano (2014)
11. Bellomo M., Pone S.: Il retrofit tecnologico degli edifici esistenti: qualità dell'abitare, sostenibilità ambientale, rilancio economico. Techne. Journal of Technology for Architecture and Environment n. 1. Firenze University Press, Firenze (2011)
12. Powter, A., Ross, S.: Integrating Environmental and Cultural Sustainability for Heritage Properties. APT Bulletin, Springfield (2005)

13. World Business Council for Sustainable Development: Transforming the Market: Energy Efficiency in Buildings. WBCSD, Switzerland (2009)
14. From Culture and History to Sustainable Development. Securing the future, protecting the past. SECHURBA (2011)
15. Brooks, E., Law, A., Huang, L.: A comparative analysis of retrofitting historic buildings for energy efficiency in the UK and China. The Planning Review Published, Zurich, Swiss (2014)
16. Rohdin P., Dalewski M., Moshfegh B.: Indoor Environment and Energy Use in Historic Buildings - Comparing Survey Results with Measurements and Simulations. International Journal of Ventilation, Veetech Ltd., England (2012)
17. Musso, S.: Recupero e restauro degli edifici storici: guida pratica al rilievo e alla diagnostica. EPC Libri, Roma (2004)
18. Gasparoli, P.: Criteri, metodi e strategie per l'intervento sul costruito. Alinea, Firenze (2006)
19. Cessari, L., Di Marcello, S.: (a cura di): Techa 2008 Technologies exploitation for the cultural heritage advancement. Gangemi Editore, Roma (2009)
20. Pearce, D.W., Nash, C.A.: The social appraisal of project. A text in cost-benefit analysis. Macmillan, London (1981)
21. Towse, R., Khakee, A.: Cultural economics. Springer, Berlin (1992)
22. Nesticò, A., De Mare, G.: Government Tools for Urban Regeneration: The Cities Plan in Italy. A Critical Analysis of the Results and the Proposed Alternative. In: Murgante, B., Misra, S., Rocha, A.M.A., Torre, C., Rocha, J.G., Falcao, M.I., Taniar, D., Apduhan, B.O., Gervasi, O. (eds.) ICCSA 2014, Part II. LNCS, vol. 8580, pp. 547–562. Springer, Heidelberg (2014)
23. Calabrò F., Della Spina L.: The cultural and environmental resources for sustainable development of rural areas in economically disadvantaged contexts. Economic-appraisals issues of a model of management for the valorisation of public assets. In: ICEESD 2013, Advanced Materials Research, vols. 869–870, pp. 43–48. Trans Tech Publications, Switzerland (2014). doi:10.4028/www.scientific.net/AMR.869-870.43
24. De Mare, G., Manganelli, B., Nesticò, A.: The Economic Evaluation of Investments in the Energy Sector: A Model for the Optimization of the Scenario Analyses. In: Murgante, B., Misra, S., Carlini, M., Torre, C.M., Nguyen, H.-Q., Taniar, D., Apduhan, B.O., Gervasi, O. (eds.) ICCSA 2013, Part II. LNCS, vol. 7972, pp. 359–374. Springer, Heidelberg (2013)
25. Nesticò, A., De Mare, G., Fiore, P., Pipolo, O.: A Model for the Economic Evaluation of Energetic Requalification Projects in Buildings. A Real Case Application. In: Murgante, B., Misra, S., Rocha, A.M.A., Torre, C., Rocha, J.G., Falcao, M.I., Taniar, D., Apduhan, B.O., Gervasi, O. (eds.) ICCSA 2014, Part II. LNCS, vol. 8580, pp. 563–578. Springer, Heidelberg (2014)
26. Nuzzo E., Tomasinsig E.: Recupero e coefficiente del costruito. Confronto tra soluzioni migliorative di pareti, coperture e solai. Edicom, Udine (2008)
27. Morandotti M., et al.: Studio, sviluppo e definizione di linee guida per interventi di miglioramento per l'efficienza energetica negli edifici di pregio e per la gestione efficiente del sistema edificio-impianto. Report 63. ENEA, Roma (2011)
28. Lucchi, E., Pracchi, V.: Efficienza energetica e patrimonio costruito. Maggioli Editore, Milano (2013)
29. Montacchini, E.P., Tedesco, S.: Edilizia Sostenibile: requisiti, indicatori e scelte progettuali. Maggioli Editore, Milano (2009)
30. Latham, D.: Creative Re-Use of Buildings. Donhead Publishing Ltd., Dorset, United Kingdom (2000)

31. Losasso M.: Il progetto come prodotto di ricerca scientifica. Techne, n. 2. Journal of Technology for Architecture and Environment. Firenze University Press, Firenze (2011)
32. Bonomo, M.: Teoria e tecnica dell'illuminazione d'interni. Maggioli Editore, Milano (2009)
33. Piano, M.: Energie Rinnovabili e Domotica. Franco Angeli, Milano (2008)
34. Quaranta, G.G.: La domotica per l'efficienza energetica delle abitazioni. Maggioli Editore, Milano (2009)
35. Bencardino M., Greco I.: Smart Communities. Social Innovation at the service of the smart cities. TeMA SI. Journal of Land Use Mobility and Environment, 39–51. University of Naples,Federico II Print (2014). ISSN: 1970-9889. doi:10.6092/1970-9870/2533
36. Greco, I., Bencardino, M.: The Paradigm of the Modern City: SMART and SENSEable Cities for Smart, Inclusive and Sustainable Growth. In: Murgante, B., Misra, S., Rocha, A.M.A., Torre, C., Rocha, J.G., Falc\ {a}o, M.I., Taniar, D., Apduhan, B.O., Gervasi, O. (eds.) ICCSA 2014, Part II. LNCS, vol. 8580, pp. 579–597. Springer, Heidelberg (2014)

Investing in Sports Facilities:
The Italian Situation Toward an Olympic Perspective
Confidence Intervals for the Financial Analysis of Pools

Gianluigi De Mare[1(✉)], Maria Fiorella Granata[2], and Fabiana Forte[3]

[1] University of Salerno, Via Giovanni Paolo II, 132, Fisciano (SA), Italy
gdemare@unisa.it
[2] University of Palermo, Viale delle Scienze, 14, Palermo, Italy
maria.granata@unipa.it
[3] Second University of Naples, Via San Lorenzo, Aversa (CE), Italy
fabiana.forte@unina2.it

Abstract. Given the current economic crisis, there is a lack of investments also for public sport facilities, which are vital to the urban quality of the city and the quality of life of its citizens. The situation is serious in Italy, especially after the failure of public policies for funding of major events (World Cup, 1990; Winter Olympics, 2006). Some recent laws (147/2013) have revived the sector with particular attention to the medium size facilities but have neglected the small structures that represent the basic activities. In case of the latter, a convergence of objectives is needed between public administrations and private investors. Therefore, public authorities should think in terms of sustainability and investment performance, a perspective that involves the private sector. The research extends the knowledge framework for the financial outline of investments in the sector of swimming pools, starting with a technical and financial analysis of 18 case studies and building confidence intervals for the relevant variables.

Keywords: Sports facilities · Project financing · Financial evaluation

1   Introduction

The effects of the contingent economic crisis have led to a stagnation in all sectors of public and private investment associated with the construction sector. In particular network and one-off structures were affected by the macroeconomic framework. Among the latter, there are a few new achievements especially in sports facilities. This is due to the compression of per capita income that determines, as a first effect, the reduction of consumer spending for recreational activities and hobbies.

In the recent past the construction of sports facilities has been associated with the construction of residential and commercial buildings, in order to exploit possible derivable synergies [1,2,3,4]. This is the case for Italy. The law 147/2013 allows private individuals to propose the construction of sports facilities along with bars, restaurants, shopping malls, hotels, but prohibits the construction of residential spaces;

The work was developed equally among the three authors.

O. Gervasi et al. (Eds.): ICCSA 2015, Part III, LNCS 9157, pp. 77–87, 2015.
DOI: 10.1007/978-3-319-21470-2_6

however, the latter conjoint constructions are permitted in many European countries (see, among many, the Arsenal FC stadium in London). The Italian law pushes toward environmentally friendly solutions and above all financially and energy sustainable solutions [5].

The issues of financial sustainability of these specialized structures has an important role in the light of confidence for their concession (if publicly owned), given the need to quantify congruous royalties with the revenues that can be generated and, at the same time, the need to maintain the provided services socially accessible. In this case, the Italian law 221/12 has defined the terms of aligning national behaviour to the European reference standard for SGEI (services of general economic interest), imposing cost-Returns analysis based on the mechanisms of credit [6,7].

Evidently, in the project financing procedures, the two themes come together in a single issue, because the private lender will be the executor of the facility but also the concessionary of the same facility for a period until the return on investment.

This scenario corresponds to the present funds request. The aim is to highlight the elements of reference for the legislative discussion regarding economic evaluation of the matter, proposing the identification of the confidence intervals [8] for design parameters in the construction and management of swimming pools. In other words, the technical-Economic variables characterizing the swimming facilities in the study area are selected; a confidence interval for the average value is constructed for each of these variables (based on observations and data processing); the project performance indicator, namely the IRR (internal rate of return), has also a relative interval of confidence. This will provide a range of values attributable to technical and monetary indicators (independent variables). These values are useful in the previous phase to designing. The respect of the definition intervals for variables should ensure a project falling within the relevant range of yield (dependent variable).

The test is carried out on the base of development from scratch of three analytical designs of swimming pools under construction in the study area.

2   Sports Facilities as Economic Catalysts for the Development of the City

The sensitivity of the subject is immediately evident when observing that the last official census of sports facilities in Italy is dated 2003 [9].

More recent analysis concern only specific regions (Lombardy, Emilia Romagna, Marche, Sardinia, Piedmont and Friuli Venezia Giulia).

The data point out that the sports industry has a significant effect on GDP (1.6%) with nearly a billion equivalent value for Italian products sold abroad.

The great rise of the media has attributed to sporting events a previously unimaginable turnover capacity, therefore in many European and transoceanic countries the facilities have been adapted to the marketing needs dictated by globalization. Not all this happened in our country. Italy is trying to take the opportunity to revitalize the sector by the candidature to the Olympics that will be held in 2024. Previous experiences (1990 World Cup, the 2006 Winter Olympics, 2009 World Championships) have proved negative if not really disastrous in terms of posthumous sustainability of investment.

Evidently, the situation at a local level is equally degraded; a survey of the website sportindustry.com in 2014 revealed facilities whose construction has remained unfinished. Their value is over 100 million Euro.

Yet concession invitations to build and operate sports facilities continue to proliferate, demonstrating that the space for profit margins in the sector exists and it is well identified by private investors. It would seem that the solution to increase the efficiency of interventions resides in the forms of credit of the projects, passing from the use of public resources, lacking in financial viability checks, to private ones governed by rigid formulas to control the probability of return.

The transfer of entrepreneurial risks to the private sector opens perspectives in a field where the managerial know-How and the cultural innovation represent elements that determine the success or the failure of the initiative. This principle applies not only to the mega-structures but also at a local level, given the easy spread of behavioural patterns and trends due to the media.

The question of the scope of the provision of social services for the practice of sport remains complex, if interpreted as a means of raising the quality of life as well as of increasing the preventive health care. In this perspective, review procedures of the financial sustainability of investments are also vital for the public sector, in order to calibrate properly the economic terms of the relationship with the private investor, negotiating suitable rates for all social groups and compensation in terms of integrative services and infrastructure works.

The financial analysis conducted in this research helps also to provide preventive tools for the control of the bidding process in case of loans for the construction and management of energy efficient municipal pools [10].

3   The Law 147/2013 for Stadiums

The last frontier in the field of legislation regarding sports facilities is the law 147 of December 27, 2013, paragraphs 303-305. A careful reading [11] shows that the few paragraphs of interest concern the medium size sports facilities, as well as the recovery of already existing dilapidated facilities. Evidently, the norm dictates a tight schedule for the necessary permits for the intervention and it structures the path of the investment according to the criteria of a typical PPP (public private partnership). However, it was raised some controversy due to the difficulty of applying this law to small facilities, those that meet the needs of grassroots sport. The norm has been highly criticised, because it refers to exclusively private interventions for basic sport structures, with all the deriving difficulties from fiscal and managerial aspects. In fact, to access on the favourable credit terms from the Institute of Sports Credit, companies have to have special regimes that hinder a real business activity (according to the

Italian laws sports clubs and associations must be no profit). Precisely in this perspective, the present study helps to shed light on the economic and financial terms involved in swimming facilities investments, highlighting the typical constraints and expectations of private investors.

4   The Outsourcing of Sport Facility Management

A question of not secondary importance regards the assignment of the management of sports facilities. The law 221 of December 17, 2012 provides that the public body give evidence of compliance with European standards for the chosen forms of entrustment, as well as it require showing the financial management model of the project presented by the concessionary. These rules constitute an important breakthrough because they force the government to ensure transparent inspections of the monetary sustainability of credit lines, allowing also, if necessary, compensation arrangements for the Public administration. Therefore, the Public administration is called upon to undertake a proper analysis of costs and returns of the initiative thus guaranteeing the avoidance of rash projects that would lead to a termination of service with impairment of social rights to the use of the facilities. Additionally, this check also allows a conscious bargaining of tariffs applied to the service.

The proposed study helps to shed light on the imposed levels of price and on the applicable ones in sports facilities that, even though funded by private individuals (and thus projected on a return on invested capital), must also ensure access to the less wealthy population groups.

5   The Construction of the Confidence Intervals

The disclosed study takes into account the analysis of 18 swimming facilities in the province of Salerno

Figure 1 shows the location of the plants from whom the information was collected.

Fig. 1. The location of the swimming facilities in the territory

The main objective was to gather technical and economic information useful in the design phase, in order to select the variables that influence the financial viability of the investment, and to define eligibility intervals for the values of these variables. The interested contractors must take into account these intervals in order to achieve an acceptable return on the financial resources invested in the swimming facility sector.

Table 1 presents the significant variables and their definition intervals.

Table 1. Relevant variables and their definition intervals

Variables		Definition intervals		
	Olympic pool	length (m)	width (m)	depth (m)
	1	25 - 28	13 - 20	1.5 - 2
Technical characteristics of the facility	Complementary basins 0 - 2	10 - 16	6 - 8	0.5 - 1.5
	Energy recovery systems		Co-generation system	
			Photovoltaic system	
Delimitation of the catchment area	Population	50,000 - 250,000		
Annual per capita income		10,000 - 14,000		
Number of users		2,000 - 7,000		
Investment and management cost	Development cost (€)	1 - 4 mil.		
	Management cost (€/year)	0.2 - 0.4 mil.		
Types of provided service	Opening months	9 - 12		
	Opening hours	8 - 14		
	Swimming lessons	Free, with instructor		
	Additional services	No, bar, restaurant, solarium, sauna, gym		
Rates	With instructor (€/month)	60		
	Without instructor (€/month)	40		
Equity	Equity capital/total capital	0 - 50 %		
Value of money loan		5 - 7 %		
Discount rate of the project		5 % CE		
Pre-tax expected IRR (internal rate of return)		7 - 14 %		

5.1   Significant Variables and Confidence Intervals

For the selection of the relevant variables in the formation of the levels of return on equity, we proceeded with a monetary approach, reconstructing the usual route for the verification of financial sustainability of a project.

So the hypothesized steps can be summarized as follows:

1. recognition of the technical characteristics of the facility (size of the main swimming pool, characteristics of supplementary swimming pools, existence of energy recovery systems, etc.);
2. delimitation of the catchment area;
3. survey of demand levels;
4. investment and management cost analysis;
5. analysis of the types of the provided service (periods and hours of operation, types of provided courses, supplementary services like bar, restaurant, sauna, etc.);
6. analysis of rates of provided services;
7. financial structure of the capital;
8. investment results.

In the case studies, the variable 1 shows the prevalence of an Olympic basin, although the indications of Table 1 are oscillatory. This variability obviously depends on the historical moments of realization of the reservoirs and on different regulations. Some facilities are equipped exclusively with the main basin, while other ones have additional basins for children or for different activities such as therapeutic and rehabilitative activities. The presence (at least in the most modern facilities) of cogeneration systems and the production of energy from alternative sources (photovoltaic) is relevant. The impact on the investment costs of such systems fluctuates between 10 and 15% on average.

The variable 2 refers to the metropolitan conurbation where the facility is inserted. The urban area of Salerno (about 270,000 inhabitants), Nocera (about 280,000) and Cilento (about 50,000) have been substantially the surveyed ones.

The demand levels (variable 3) are determined by studying the actual users of the facilities but also the potential ones. In fact, one cannot exclude a portion of unmet demand due to inefficient business and management strategies employed in the surveyed facilities. The indicator is derived by the cost of travel method: the single facility is analysed as a part of the system of facilities that can compete with it; the costs that users have to incur to enjoy each considered facility is added; the spatial horizon of demand is constructed. The measured costs bring in account the cost of transport (private or public) and the cost of travel time (obtained in relation to the income level of the concerned populations).

The variable 4 is constructed from the recognition of the costs of construction and management. For the first, only the more recent systems (less than 10 years) were considered.

The variable 5 obviously depends on the specific location of the facility; in fact, it is near the sea, so it is little exploited in the summer (the opening period is limited to 9 months). The variable 5 also depends on commercial decisions, on opening hours (morning and afternoon), on swimming related services (free swimming and paid courses), on the provision of extra services like small and medium size restaurants as well as fitness and wellness.

Rates (variable 6) are identified by the case studies.

The reference financial structure (variable 7) reflects the matrix of most public facilities. The few private facilities were made with venture capital between 50 and 100%. The cost of financing varies between 6 and 7%.

The results obtained from these projects (variable 8, namely IRR) are affected by the need to bring a production process that is often not created in a perspective of private-Financial efficiency into a discount cash flow type structure. The latter perspective has slightly modern accounting and managerial practices that provide a greater transparency. The time alignment rate is 5% and is directly borrowed from the practice of the European Community in the definition of EU funding in similar works. The period of analysis is assumed 25 years.

6   Reliability Verification of the Confidence Intervals

Confidence intervals are constructed[1] at a 95% confidence level with the following formula:

$$\bar{x} - t_{n-1}\frac{s}{\sqrt{n}} \leq \mu_x \leq \bar{x} + t_{n-1}\frac{s}{\sqrt{n}}$$

with $n = 18$;

obviously $\bar{x}$ and s are computedin consideration of the average value and the standard deviation that the different variables take in the data sample.

The reliability verification of the identified intervals is conducted through the development from scratch of three case studies, each of which is implemented in one of the conurbations quoted in variable 2 paragraph 4.

Table 2 shows the design parameters for each considered facility. Based on these parameters the relative economic and financial plans were developed. In order to illustrate the conducted analysis, table 3 shows the economic and financial plan relevant to the Project 2. As you can see from the last two lines of Table 2, the analytically calculated IRR fall broadly into the related confidence interval, reaching the top of it.

---

[1] For those variables that allow to define them in base of the availability and accuracy of input data.

Table 2. Technical and financial parameters for the three projected facilities

Design variables	Project 1	Project 2	Project 3
Technical characteristics of the facility			
Olympic pool	1	1	1
length (m)	25.0	25.0	28.0
width (m)	17.0	13.0	20.0
depth (m)	1.8	2.0	1.6
Complementary basins	2	0	1
length (m)	17.0		11.0
width (m)	8.0		6.0
depth (m)	1.0		0.5
Energy recovery systems	1	1	1
Co-generation system		1	
Photovoltaic system	1		1
Delimitation of the catchment area			
Population	285,000	53,343	268,799
Annual per capita income	10,900	12,000	8,000
Number of users	7,263	1,908	1,964
Investment and management cost			
Development cost (€)	4 mil	1 mil	2.6 mil
Management cost (€/year)	262,090	213,400	266,860
Types of provided service			
Opening months	12	9	12
Opening hours	14	10	12
Swimming lessons	Free, with instructor	Free, with instructor	Free, with instructor
Additional services	bar, restaurant, solarium, sauna, gym	bar	Bar & restaurant
Rates			
With instructor (€/month)	60	50	60
Without instructor (€/month)	40	40	40
Equity			
Equity capital/total capital	40%		20%
Value of money loan	6%	7%	6%
Discount rate of the project	5 % CE	5 % CE	5 % CE
Pre-tax expected IRR (internal rate of return)	7 - 14 %	7 - 14 %	7 - 14 %
Pre-tax calculated IRR	12.93%	11.70%	14.00%

Table 3. Economical and financial plan of Project 2

Year	1	2	3	4	5	6	7	8	9...	20	21	22	25
Investment costs	560,000	500,000											
wages and salaries	0	0	69,290	69,290	69,290	69,290	69,290	69,290	69,290	69,290	69,290	69,290	69,290
miscellaneous expenditure	0	0	80,122	80,122	106,829	106,829	106,829	106,829	106,829	106,829	106,829	106,829	106,829
Taxes	0	0	6,300	6,300	6,300	6,300	6,300	6,300	6,300	6,300	6,300	6,300	6,300
Fees	0	0	31,000	31,000	31,000	31,000	31,000	31,000	31,000	31,000	31,000	31,000	31,000
Total Management costs	0	0	186,712	186,712	213,419	213,419	213,419	213,419	213,419	213,419	213,419	213,419	213,419
Residual value													1,060,000
Revenues	0	0	121,173	142,346	466,050	466,050	466,050	466,050	466,050	466,050	466,050	466,050	466,050
loan I	33,600	33,600	33,600	33,600	32,156	30,626	29,004	27,285	25,462				
loan II		30,000	28,711	27,345	25,897	24,362	22,734	21,010	19,181				
Tot. financial costs	33,600	63,600	62,311	60,944	58,053	54,987	51,738	48,294	44,643				
$CF_{before-Tax}$	-593,600	-563,600	-127,850	-5,311	194,578	197,643	200,892	204,336	207,987	252,631	252,631	252,631	252,631
IRES tax base				-5,311	194,578	197,643	200,892	204,336	207,987	252,631	252,631	252,631	252,631
IRES			-1,460		53,509	54,352	55,245	56,192	57,196	69,474	69,474	69,474	69,474
IRAP tax base	-627,200	-627,200	-155,516	-31,612	171,170	177,300	183,798	190,687	197,988	287,276	287,276	287,276	287,276
IRAP	-31,172	-31,172	-7,729	-1,571	8,507	8,812	9,135	9,477	9,840	14,278	14,278	14,278	14,278
$CF_{after-Tax}$	-562,428	-532,428	-120,121	-2,279	132,562	134,479	136,512	138,667	140,951	168,880	168,880	168,880	168,880

7 Conclusions

The investment industry for sports facilities is experiencing a severe crisis in Italy over the last twenty years. The problem does not regard only the facilities of national importance, but also the structures that are used for the various widespread and popular disciplines. The laws on recovery (221/12 and 147/13) pay for the current time of severe crisis for public finances, therefore the path for the development of industrial products resides in balanced Public-Private partnerships. In this context arises the present study that investigated the technical and management parameters of 18 swimming pools in the province of Salerno, deducing confidence intervals for the design and financial parameters. These intervals represent areas for the definition of indicators that can be used in a preventive and efficient way, whenever the negotiation between governments and investors is aimed at safeguarding adequate indexes of profitability and, at the same time, social issues regarding access to services.

The choice of the confidence intervals, instead of directly developing a statistical function of interpretation and possible prediction, depends on the strong inherent uncertainty in the process of developing a design proposal in project financing. This proposal has such a number of hypotheses that it makes uncertain any pressing future projection. In this context, the integration of additional case studies (8) will allow to experiment more accurate analysis tools than the actual ones [12,13], taking into account the qualitative aspects of the design.

Therefore, the strengths of model are the ease of implementation and a medium-high reliability; so, at the prior stage of planning or design, verifying the economic feasibility of investments becomes easier. On the other hand, the weaknesses are the narrow number of study cases in order to define the indicators of the model and the problems that these approaches can cause when it must decide quickly about the ranking of funds.

References

1. Feng X., Humphrey B.R.: Assessing the economic impact of sports facilities on residential property values: a spatial hedonic approach. In: IASE/NAASE Working Paper Series, No. 08–12 (2012)
2. Huang, H., Humphrey, B.R.: New sports facilities and residential housing markets. Journal of Regional Science **54**(4), 629–663 (2014)
3. De Mare, G., Nesticò, A., Tajani, F.: Building investments for the revitalization of the territory: a multisectoral model of economic analysis. In: Murgante, B., Misra, S., Carlini, M., Torre, C.M., Nguyen, H.-Q., Taniar, D., Apduhan, B.O., Gervasi, O. (eds.) ICCSA 2013, Part III. LNCS, vol. 7973, pp. 493–508. Springer, Heidelberg (2013)
4. Guarini, M.R., Battisti, F.: Social housing and redevelopment of building complexeson brownfield sites: the financial sustainability of residential projects for vulnerable social groups. In: Advanced Materials Research, vol. 869–870, pp. 3–13; Trans Tech Publications, Switzerland (2014). ISSN 1662-8985; doi: 10.4028/www.scientific.net/AMR.869-870.3
5. Zaretsky, A.M.: Should cities pay for sports facilities? In: The Regional Economist, Federal Reserve Bank of St. Louis (2001)

6. Pedicini, A., Tiani, T.: Lo sport e le risorse finanziarie nell'Unione europea – L'inquadramento della disciplina a favore della costruzione e della ristrutturazione degli impianti sportivi. In: Rivista di diritto ed economia dello sport, vol. VII (2012). ISSN 1825-6678, Fasc. 1
7. European commission, White Paper on Sport, English Full Version, Belgium (2007)
8. Huntsburger, D., Billingsley, P., Croft, D.J.: Statistical Inference for Management and Economics. Allyn and Bacon, Boston (1975)
9. CNEL: La situazione degli impianti sportivi in Italia al 2003 (2005)
10. Nesticò, A., De Mare, G., Fiore, P., Pipolo, O.: A model for the economic evaluation of energetic requalification projects in buildings. A real case application. In: Murgante, B., Misra, S., Rocha, A.M.A., Torre, C., Rocha, J.G., Falcão, M.I., Taniar, D., Apduhan, B.O., Gervasi, O. (eds.) ICCSA 2014, Part II. LNCS, vol. 8580, pp. 563–578. Springer, Heidelberg (2014)
11. CONI: Guida all'applicazione della legge per lo sviluppo dell'impiantistica sportiva. In: Spaziosport (2014). ISSN 1125-12568, numero 28, anno VIII
12. De Mare, G., Lenza, T., Conte, R.: Economic evaluations using genetic algorithms to determine the territorial impact caused by high speed railways. In: World Academy of Science, Engineering and Technology, (71) (2012). ser. ICUPRD 2012, ISSN: p 2010-376X e 2010-3778
13. Morano, P., Tajani, F.: Least median of squares regression and minimum volume ellipsoid estimator for outliers detection in housing appraisal. International Journal of Business Intelligence and Data Mining **9**(2), 91–111 (2014)

Urban Renewal:
Negotiation Procedures and Evaluation Models

Lucia Della Spina[✉], Raffaele Scrivo, Claudia Ventura, and Angela Viglianisi

Mediterranea University of Reggio Calabria, Reggio Calabria, Italy
{lucia.dellaspina,raffaele.scrivo,
claudia.ventura,angela.viglianisi}@unirc.it

Abstract. In complex programs of urban redevelopment recourse to negotiation procedures between PA and private developers emphasizes the role of evaluation, both in terms of the collective advantage and from the point of view of financial feasibility. The purpose of the contribution is twofold: the first is to investigate the role that can be exercised by the evaluation in order to confer efficiency, fairness, transparency and democratic participation in the processes of formation and implementation of strategic programs for urban and metropolitan development characterized by high levels of complexity, the second objective is to provide an integrated assessment model which can fulfill this role properly and adequately support the decision-makers and the public in decision-making ex-ante to an integrated assessment of the "public convenience" and at the same time check whether there are sufficient margins of feasibility and financial sustainability for the private developer to carry out the investment program.

Keywords: Integrated evaluation · Evaluation models · Multicriteria analysis · Financial feasibility · Strategic programs · Urban development · Public-private partnership

1   Introduction

The return of urban property to the center of the European agenda is not only an opportunity to redefine the profiles of competitiveness, sustainability and cohesion of nations overwhelmed by the tsunami of crisis, but also the opportunity to review the paradigms that guide both analysis and practices for the territorial government [13].

Since complex urban regeneration policies require huge investments in the face of limited resources or the increasingly stringent budgetary constraints arising from the credit crunch, and the domestic fiscal policy or the EU (European stability pacts and internal agreements) the state is increasingly urged to involve new actors to prosecute

---

The paper reflects the opinion and the serious commitment of its authors who all contributed to its writing. However, Della Spina L. wrote paragraphs 1, 2, 4.5, 4.6, 4.7 and Conclusions; Scrivo R. wrote paragraph 3; Ventura C. wrote paragraph: 4.1, 4.2 and 4.3; Viglianisi A. wrote paragraphs: 4.4.

© Springer International Publishing Switzerland 2015
O. Gervasi et al. (Eds.): ICCSA 2015, Part III, LNCS 9157, pp. 88–103, 2015.
DOI: 10.1007/978-3-319-21470-2_7

public works, to set priorities and resort to political incentives rather than coercion[13]. This implies a major change in the management of public works and it is in this context that we discuss the various forms of Public-Private Partnership (PPP) [4], [18].

With the definition of PPP as identified today, forms of public-private partnership have become ever more diverse and articulate [4], [7], [16]. Yet there is a very clear common goal: the involvement and use of private resources, financial, managerial and creative programs in complex urban regeneration (PUC), to build and / or manage equipment [5] or public service activities. In Italy, in the process of urban development, this involvement took place in the wake of the European experience, very gradually, especially during the nineties, and legislation has gradually introduced several sophisticated tools to cope with the many demands on the city, within a framework of increasingly limited availability of public resources.

These new instruments highlight the need to develop tools that enable flexibility in providing immediate and transparent assessments, whilst at the same time, prior assessment of the feasibility of such investments and a verification of the results and financial operations [4], [21]. The PUC in the PPP relationship is substantiated by the Public Administration (PA) in making the investment with a sufficient margin for the private developer, obtaining as much as possible by negotiation, in terms, for example, of works and additional services to the statutory minimum (standard).

There is obviously a model of optimal negotiation to be applied in all circumstances, but the explicit objectives of the actors should be quite clear: for the private developer: gain for the highest possible return commensurate to the risk, for the public to ensure that the program is realized and obtain the highest standards of construction for the city in terms of public works and additional services, thus ensuring competitiveness between developers [4], [5], [8].

The object of the exchange is the surplus or profit from the land defined by the PA through the definition of land use decisions, the areas to be developed, the works to be realized, indices of airspace, etc.

Bargaining can only take place based on the substance of a prior transparent assessment which allow the PA to measure the requests of the developer [4]. If this does not occur the consultation is likely to become an empty ideology making it an unlikely opaque process for the transformation of land. Here lies the risk.

This essay will therefore deepen the role that can be exercised by the assessment in negotiating procedures between public and private and then provide a useful methodological contribution for which it may perform properly. The aspiration is to help the parties involved in their negotiations - the PA on the one hand and private developers on the other - to arrange a PUC that reconciles the maximization of urban quality pursued by the PA with all the convenience of private investment in property. To this end, an integrated assessment model is proposed that combines two different methodologies: financial analysis (DCFA) with Multicriteria Decision Analysis (MCDA).

The verification of the model to the case study found that the model is able to overcome the weaknesses and exploit strategic synergy resulting from the integration of two different valuation methods to support transparent decision-making processes relative to complex programs of urban transformation brought by private developers.

The presentation of the model is preceded by an illustration of the problems that have stimulated the development and role of evaluation in the context of complex decision-making.

2   Methodology Used for "new rules" Research

In reference to the case study, Calabrian Region governmental legislation states that the municipalities must identify the areas of their territory to be covered by urban development and set the objectives of environmental, social and architectural quality that they intend to achieve. Under this law, municipalities should promote the involvement of public and private developers in the preparation and implementation of urban regeneration programs through public tender [18]. However, where privately owned properties are present in the areas to be covered by urban renewal it is stipulated that in place of a call for tender the Municipality activates negotiation procedures with developers which are aimed at defining the forms of their participation in the program [8], [9]. The outcome of such negotiation procedures may constitute a variation to the Master Plan (MP).

In such circumstances, this "transgression" of the rule refers to a specific asset or set of rules for which transformation is recognized as being in the public interest, as represented by the elimination of degraded conditions in the pursuit of an objective to improve urban quality. The "transgression" index of "buildability" is then commonly implemented in specific urban areas with new rules.

The argument then recognizes the important role of evaluation [10] which is essential for achieving the following goals:

• making the consultation and the final decision transparent;
• explicating public objectives and maximizing their level of achievement;
• finalizing the process of formation and appropriation of urban land.

The municipal administration would do well to officially recognize this reevaluation of the assessment procedure, further specifying that the assessment of a private project must be undertaken in an objective manner, with reference to data of a technical nature (town planning, environmental, financial, etc.) The "new rules" used in the evaluation procedure must be disclosed to all potential developers, i.e. to all owners of areas subject to urban regeneration. The contents of the procedure must be technically examined with the results verified at a political level.

The assessment model for consultation between the public administration and the owners/private developers is divided into two components, as shown in Figure 1:

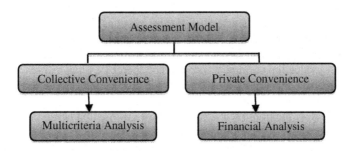

Fig. 1. Assessment model

- evaluation of the collective convenience (MCDA);
- evaluation of the   private convenience (DCFA).

The increased complexity of urban transformation requires a systematic approach and non-linear to complex problems, hence the need for an integrated assessment of valuation methods in order to overcome the weaknesses and exploit the synergies arising from the integration of strategic assessments MCDA and Discounted Cash Flow Analysis (DCFA), and support transparent decision-making processes relating to any plan for urban transformation proposed by private entities.

The integrated use of the two assessments allows the public entity on the one hand to check under ex-ante if the effects of the implementation of private investment are generally positive on the system environment, urban and social, and therefore acceptable to the community, and at the same time on the other side to check, through the DCFA, if there is sufficient room for feasibility and financial sustainability for the private party in making the investment.

The DCFA, here used in the public sector for the evaluation of private investment, of course, it is a technique of quantitative-monetary and optical private law does not take into account either the social effects of an intervention, or any opportunity costs of alternative projects, nor qualitative aspects not monetized: to measure these externalities recourse is to be used in conjunction with the AMC. The MCDA are a tool to support the public body in decision making ex-ante, and are designed to provide a rational basis to problems of choice but in reality are characterized by a multiplicity of objectives / criteria often in conflict with each other in order to identify possible alternative processing and the advantages and disadvantages that may result from their implementation [22, 23], [25, 26].

The private feasibility study provides an explanation of the pricing system, the costs for the private developer and also the value of the property to be transformed, what development will result in the area, the developer's profit and the interest on capital advances. This activity is intended to alleviate the lack of information between the public administration and the private developer: while the latter contains detailed information about the financial aspects of the project, the first does not [16], [18], [22].

However, this information is directed towards the public administration and is unavailable to the private developer, except to a partial extent and in a summarized form that can be deduced from Administration documents. The information on the need for goods and public services is part of the interests and skills of the public sphere. However, this information is often inaccessible to the public body to the extent that it can be profitably used in a contractual relationship, as witnessed in the case study, with the private developer. In addition, the PA - unlike the private developer - is interested in the effects produced by private development on the urban environment, i.e its externalities. An evaluation of public convenience therefore aims to increase the PA's level of information concerning the effects of the project and to state the desired public objectives [1, 2, 3].

As already noted, the improved accessibility of information is especially beneficial for the public sector. In a negotiating context such as that described, the combined use of the private convenience and the collective convenience evaluation is aimed at identifying a single solution, which satisfies both the private and public sectors [13].

As regards the private feasibility study, it is necessary that the proposal of urban transformation advanced by the private developer is accompanied by a pre-feasibility study, which can be prepared according to the technical standards defined by the public Administration. This must then also contain technical requirements of an economic-financial nature, such as the completed clarification market surveys, the costs outlined, all financial processing performed, and so on [21].

The feasibility study prepared by the private developer must justify the "buildability" and the proposed destinations. In turn, the PA, through its Technical Department, and possibly assisted by external experts, ensures the reliability of the feasibility study contents and the accuracy of its results.

In this regard, the evaluation technique is consolidated. Through the financial analysis of the investment project and then the calculation of the Net Present Value (NPV) and the Internal Rate of Return (IRR), the public administration can verify how reasonable the building costs and zoning requested from the private developer are.

As regards to the assessment of collective convenience, however, there is no single universally accepted procedure, unlike financial feasibility [21]. Consequently, when there is no monetization of problematic decisions and qualitative problems remain relevant, the preferred solution is sought is through the MCDA [25, 26]. The MCDA, which abandons the simplistic concept of optimization, incorporates a choice for the type of "justified" approach, in which subjective perspectives can be made explicit, that is, defined and justified in a way that can be subject to public debates [24]. For these reasons, its use is also beneficial in the decision-making context, search the "most satisfactory solution". Since the MCDA includes a combined set of models and methods [17], [22, 23], [25, 26] the evaluation procedure to be applied to private projects of urban regeneration must be specially prepared.

The assessment model proposed in this paper was used on an experimental basis, during the consultation procedure between the PA of Reggio Calabria and the owners of the properties within the subject area destined for urban regeneration. The urban and environmental degradation in a predominantly residential area, within a context of social degradation, led the PA to make it the subject of redevelopment and to invite private developers to present their plans on the general objectives of urban quality.

In response to the request, private parties (owners and property developers) have presented a project destined predominantly for a residential area. However, the proposed costs were significantly higher than the Master Plan had considered for areas of redevelopment. The assessment procedure therefore has become the technical framework for consultation between the PA and private entities, for the urban transformation project to be realized.

3   The Private Feasibility Evaluation

The evaluation of the private feasibility study requires that the value of the area be costed for the urban renewal program. These costs are compared with the current value of the real estate in order to verify the financial convenience for private developers to perform the transformation.

To estimate the value of the real estate in its current condition requires an estimation of the market value of the real estate under current market conditions in reference to the demand for real estate. In this respect, the information provided by real estate analysts and by sectorial real estate publications can give the first indications. More detailed information regarding the costs of sale which are similar to those estimated must be found through deeper investigations. Essentially the characteristics of the area being valued and the data collected mean the value of the building complex is identified with the aid of well known estimative methods [12], [14, 15].

The research into the value of real estate as a result of a transformative project is based on the Value of transformation estimate criteria (Vtr) [4, 5], [19]. The DCFA gives a more analytic estimate of the Vtr [21]. This technique estimates the transformation of the property by estimating the flow of costs and revenues related to the investment project carried out by a developer. Generally, the DCFA involves the resolution of the equation:

$$NPV = F_1 / (1+r) + F_2 / (1+r)^2 + ...+ F_n / (1+r)^n \qquad (1)$$

where *NPV* is the Net Present Value (NPV) of the investment project, *F* is the difference between revenues and costs for a period considered, *r* is the discount rate and *n* is the length of time considered for the implementation of the project.

The NPV is therefore the value that the good assumes due to the assumed conversion:

The use of DCFA requires that specific assumptions are made on the expectations of revenue and cost (Table 1), and are then identified:

- value (revenues) of the real estate under development;
- investment   costs;
- active and passive interest rates and the discount;
- profit.

In the application of the experimental method set out in this paper, the developer has made a summary estimate of the real estate in terms of construction costs and an estimate of the revenues from sales of real estate in order to explain the absolute profitability and percentage deriving from the building investment.

The PA, in turn, looks to gain from their assessments. The NPV of the property was estimated by a process of synthetic-comparative. This value was then compared with the value of transformation resulting from DCFA, mentioned in Table 2 and shows a pattern application.

Table 1. The assumptions used for the application of DCFA

Entries	Quantity	Unit value (Euro/m^2)	Percentage (%)
URBAN DATA			
Territorial area	30.000 m^2		
Building permission index	2,00 m^3/m^2		
Building area	16.100 m^2		
Building volume	60.000 m^3		
Total volume	59.000 m^3		
REVENUES			
Tourist accommodation areas	9640 m^2	180,00	
Tertiary-commercial areas	2.400 m^2	108,00	
Tertiary-business areas	4.088 m^2	143,00	
Total revenues		431,00	
COSTS			
Production costs		1.740,00	
License fees		n.a.	
Professional fees			10%
Total costs		1.740,00	
INTERESTS and PROFITS			
Passive Interest rates			8%
Active Interest rates			3,20%
Discount rate			5%
Profit			5,11%

However, rather than having only the Vtr, the PA can consider two temporal scenarios - a horizon of 6 years and one of 8 - and check the resulting value for the real estate property from a combination of the following variables for each of them:

- the index required by the building rights vs. a property index of 1.0 m^2/m^2 maximum considered by the PA in terms of the urban load permitted;
- the quantification of the costs of primary tabular urbanization vs cost estimate of urbanization that the new urban settlement would really require.

Therefore, the following disparity was verified:

$$Vtr > Vm \qquad (2)$$

where: *Vtr* is the Value of transformation obtained by the DCFA and *Vm* is the current market value of the real estate obtained through a synthetic comparative method.

The DCFA has highlighted how the transformation was financially sustainable with a lower urban load with an index equal to 1,0 m^2/m^2, but it has also made clear that by attributing the entire cost of urbanization to private developers, as deemed necessary by the PA, it would make intervention problematic and convenience questionable for the private developers.

In addition, in the model, combining traditional evaluations based on NPV with more flexible models was considered appropriate: tools such as Real Options Analysis (ROA) [9] allow a dynamic analysis of the investment, which makes the strategic dimension of urban regeneration programs explicit [6, 7]. This internalizes the estimate of the opportunity cost which can be deferred over time, pending administrative scenarios and a more favourable market value, until the uncertainty that characterizes the decision variables is resolved. It also allows the development of a comparison between several alternative scenarios on the density and the intended use [5], [9].

Table 2. A diagram of the DCFA application   (values €x1.000)

Entries	Total value	I year	... year	V year	VI year
REVENUES					
Tourist accommodation areas	€ 1.731,00	--		€ 1.731,00	€ 1.731,00
Tertiary-commercial areas	€ 258,00	--		€ 258,00	€ 258,00
Tertiary-business areas	€ 583,00	--		€ 583,00	€ 583,00
Residual value real estate	€ 25.450,00				€ 25.450,00
Total revenues	€ 2.572,00	--		€ 2.572,00	€ 27.872,00
COSTS					
Area market value	€ 3.600,00	€ 3.600,00		--	--
Production costs	€ 28.024,00	€ 4.203,00		--	--
Operating costs				€ 1.770,00	€ 1.770,00
Total of costs	€ 31.624,00	€ 7.803,00		€ 1.770,00	€ 1.770,00
BALANCE		-€ 7.803,00		€ 802,00	€ 26.102,00
Interests	€ 7.018,00	€ 1.145,00		--	--
Debt exposure	€ 2.133,00	--		€ 2.133,00	€ 2.133,00
PROFIT (NPV)	€ 8.411,00				

4   The Evaluation of the Collective Convenience

The evaluation of the convenience to the community resulting from the transformation of the area was achieved through MCDA, thus based on multiple evaluation criteria [17], [22, 23], [25, 26].

The objective of this evaluation is to determine whether the resulting effects from the implementation of the proposed project by the private developer compared to the current use of the area are generally positive in terms of the impact on the environment, urban and social system, and therefore acceptable to the community. In relation to this evaluation profile there must be a comparison between the zero option (the actual state or non-intervention) and option one (the transformation project presented by private developers) and option 2 (ideal project: with a lower urban load).

In this case, the MCDA evaluation process aims not only to measure the alternatives (i.e to indicate which of these is preferable) but also, to measure the impact of the various options on the urban system [18, 19, 20]. It is necessary in order to determine whether the difference, in environmental, social, economic and urban terms, between the realization of one alternative or another implies significant or negligible impact, made explicit in numerical terms, and therefore these differences are easily communicable [11].

The system evaluation alternatives are articulated into the following steps:

- description of the alternatives;
- choice of criteria for the evaluation of impact;
- construction of indicators for measuring the impact;
- analysis and survey of impact;
- standardization data and presentation of the weight of evaluation criteria;
- choice of the ranking technique for the alternatives.

However, in the experimental application of the proposed model, the multicriteria evaluation - as you will see - concludes with a measurement of impact. This is because the objective information given by the impact matrix was voted on by public decision makers, providing sufficient detail to start the negotiation process with individuals.

4.1   The Description Phase of the Alternatives

The first step is the description of the alternative projects, due to the zero option (status quo or non-intervention), option one (project submitted by private developers), option two requiring less urban infrastructure (ideal project). The different alternatives are represented (Table 3) as values that express the main urban characteristics, in terms of activities and public spaces laid down by law.

Table 3. Land use to design alternatives (m^2)

Land use	Option 0	Option 1	Option 2
Totale area	30.000	30.000	30.000
Green park, public spaces, walking and cycling paths	26.000	-	-
Parkings and roads	-	12.000	12.000
Public utilities (cultural, recreational, etc.).	-	4.088	4.088
Accommodation activity (hotel)	0	9.640	8.380
Accommodation activity (residence)	0	-	1.300
Skilled and administrative services	2.800	-	-
Business	200	2.400	2.400

4.2   The Definition of the Evaluation Criteria

For an assessment of urban projects a balanced evaluation characterized by three criteria was decided as the best strategy: the criterion environmental/cultural, the economic criterion and the ethics criterion (or social justice) [1,2], [13], [17], [23], [26]. In this case the criterion of environmental quality was also chosen as it allows a better assessment of "collective convenience" for the transformation intervention compared to the urban environment in which the area is located. Below is the valuation criteria used with respective measurement modes (Tables 4, 5 e 6).

- *Environmental sustainability* is measured by a lower consumption of natural resources, for consumption we intend the return on the resource in terms of "pollution" (e.g. polluted air, contaminated soil, etc.) and the increased provision through development that involves a reduction in the agents that cause a negative impact on the environment and humans (e.g. green barriers for the mitigation of noise pollution, energy and economic saving etc.).
- *Urban quality* is measured by the completeness of infrastructural facilities and punctual collective nature to the needs of the settled population.
- *Social solidarity* is measured by the degree of response to the potential demand. The higher the percentage of people served by the structures and spaces specially dedicated to the specific requirements, plus the criterion is satisfied.
- *Financial sustainability* is measured by evaluating the cash flows DCFA through the criterion of Net Present Value (NPV).

4.3   The Construction of Indicators for the Measurement of Impact

For each evaluation criterion sub-criteria were defined to enable a more specific general criterion. The next step was to define a set of indicators for the different sub-criteria, that could measure in quantitative terms to achieve the objective underlying each sub-criteria. The indicators were not only chosen according to their significance, but also by the availability of the data (or data necessary for the indicator construction), officialdom of the source and the possibility of an update [10]. As illustrated in the attached tables (Tables 4, 5 and 6), on each evaluation matrix given, the operational definition for each indicator is represented making the quantification feasible, allowing the expected trend with respect to the objective function of the corresponding sub-criterion (i.e. maximization, minimization or tending to zero) and giving values of the performance for each project option. Here, for lack of space, the operational definition of each indicator has been omitted. The indicators were obtained by processing the data of the project, sectoral plans (urban traffic plans, zoning acoustic plans, etc.) and detailed planning, from simulations conducted specifically in this sector.

4.4   The Analysis Phase and Detection of Impact

The analysis phase of the impact renews the techniques of impact assessment, from a definition of the effects to an analysis of the impact. Once the criteria is defined, it establishes the most appropriate indicators for measuring impact [17], [22, 23], [25, 26].

(e.g. for noise pollution, an indicator may be the daily number of cars in the input and output from peak hours, as a source of noise) [11]. The calculation of values was made using coded analysis techniques (traffic analysis for the estimation of cars, induced employment analysis, etc.). In a specific summary table the impact generated is illustrated for the project alternatives.

Table 4. The evaluation matrix for MCDA: Environmental sustainability

Evaluation criteria	Indicators category	Objective function	Option 0	Option 1	Option 2
Noise pollution	Permissible noise levels	tending to 0	0	0	0
	Presence of activities classified as sources of noise	min	0	0	0
	Noise levels as laid down by acoustic zoning classification	min	0	0	0
	Presence of air pollution from mobile sources	min	60	30	30
	Morphology of the settlement	h. min	6	21	18
	Noise impact mitigation measures	n. trees max	20	20	20
Air pollution	Road traffic entering and exiting from study area	n. cars min	0	320	246
	Presence of permanent sources of air pollution	m^2/m^2 min	250	600	450
Water consumption	Soil permeability	m^2 max	22.000	13.800	13.800
Energy saving and bioarchitecture	Sunlight and brightness	n. max	0	178	150
	Ventilation	n. max	0	0	20
Energy and economic saving	Cost of maintenance and management heating system (Convention Centre)	€/m^2 min	---	532,00	423,00

Table 5. The evaluation matrix for MCDA: Urban quality

Evaluation criteria	Indicators category	Expected developments (objective function)	Option 0	Option 1	Option 2
Standard facilities	Free public parking spaces	$m^2 \geq 0$	0	7.000	7.000
	Private parking spaces	$m^2 \geq 0$	....	5.000	5.000
	Urban green areas	$m^2 \geq 0$	....	1872	2100
Accessibility	Links to road network	ml max	0	900	900
	Area accesses by external users	n. max	1	2	2
Community facilities	Supra-municipal services	m^2 max	10	21	24
	Equipment of local interest	m^2 max	0	0	0
Functional complexity	Diversification of business in the area	m^2 max >10	2	6	8

Table 6. The evaluation matrix for the MCDA: Social solidarity

Evaluation criteria	Indicators category	Expected developments (objective function)	Option 0	Option 1	Option 2
Safety	Accessible, protected, Pedestrian and cycle paths towards the services and city center with rest areas	m^2 max	0	0	0
	Open spaces for meeting and socializing served by public services	m^2 max	0	250	250
	Presence of interventions included in the Program of Safety	m^2 max	750	750	750
Social integration	Green areas for play time and meeting	m^2/inhab $\leq 0{,}5$	0	0,3	0,4
	Commercial services in the Neighbourhood	m^2/inhab. > 0 and ≤ 4	0	2	3
Employment	Sustained employment	n. max	12	26	26

4.5 The Evaluation Phase: Standardization and Ponderation

The next steps are the standardization of values and the importance of weight assignment to different evaluation criteria.

If the data is used for quantitative analysis, it is necessary to conduct an operation of data homogenization, using a procedure of standardization. All values are mathematically processed so as to become dimensionless numbers between 0 and 1 and therefore comparable. Among the many methods of homogenization, generally one of the two most commonly used methods are chosen [26], obtained from the following formulas.

$$Eij = (eij - \min eij)/(\max ej - \min ej) \tag{3}$$

$$Eij = eij/\max ej \tag{4}$$

where *Eij* is the data corresponding to the criterion *j* and to alternative *i* standardized, *eij* is the data before standardization. and *max eij* e *min eij* respectively represented, the maximum value and the minimum value observed for the criterion j from all the other alternatives (i = 1, 2, ..., I). With the first method of standardization, the highest value is 1 and the lowest 0, while in the second method of standardization, the lowest value can be different from 0. For the estimation of the weights, in the trial illustrated, a method of rating was used, often applied in planning practice [23], [25, 26]. This type of method requires the interlocutor to assign a predetermined amount of points (e.g. 100) to the criteria identified in such a way that the number of points assigned to each criterion reflects its relative importance.

4.6   Ranking the Alternatives

Finally we calculate types of alternative by combining weights and indicators with respect to each alternative. The methods for doing so are many [11]. Among the qualitative methods, which rely on the retrieval of numerically measurable data, chosen for use - as part of the "Electre" methods processed within the French school [23]: the Analysis of concordance / dominance that accepts 'intransitivity' and lack of comparability in preference relations between the alternatives. The analysis of concordance / dominance is a relatively simplified analytical translation procedure of decision making, far from rigid constraints of a mathematical nature and, conversely, more suitable to perceived indications from decision-makers and opinion-leaders.

This analysis, beyond the structural differences [23], [25] expected:

- Allocation of weights to the criteria according to preferences;
- Calculation of the value of the coefficients Concordance $Con_{i,k}$ and Discordance $Dis_{i,k}$ between pairs of alternatives;
- Verification of the existence of a relationship between pairs of outranking alternatives   (on thresholds concordance and discordance thresholds);
- Ordering of alternatives;
- Sensitivity analysis.

The concordance (Con) measures the satisfaction of choosing the Alternative i (Ai) of the Alternative k (Ak) (compared to the criteria C1, C2, etc.) and is the sum of only the weights, for which the criteria explains the satisfaction of choosing the first alternative rather than the second (often, but not necessarily, normalized by the sum of the row):

$$Con_{i,k} = \Sigma_j w_j \tag{5}$$

Conversely, the discordance (Dis) measures the regret in discarding the Alternative i (Ai) of the Alternative k (Ak), and is measured (usually) as the maximum difference between the values of the indicators of those criteria for which regret is expressed at not choosing the second alternative rather than the first (usually, but not necessarily, normalized to the maximum difference in the column):

$$Dis_{i,k} = \max \left| a_{kj} - a_{ij} \right| \tag{6}$$

At this point it is possible to construct two matrices of pairwise comparisons (respectively that of discordance and that of concordance) with the values thus obtained. These values are aggregated respectively in two vectors, whose elements are the Indices concordance (Ic) and Indices discordance (Id), determined in accordance with the formulas:

$$Ic(i) = \Sigma_j Con_{i,j} - \Sigma_j Con_{j,i} \tag{7}$$

$$Id(i) = \Sigma_j Dis_{i,j} - \Sigma_j Dis_{j,i} \tag{8}$$

The alternatives are then sorted into two lists: an index of increasing rates for concordance and decreasing rates for discordance.

Finally, simultaneously taking into account the Test of concordance (Tc) and discordance (Td), an outranking report is built (S) (outranking.), on the basis of which the different alternatives are hierarchically arranged with all criteria taken together. The alternative to outperformance is that if *aj*, with reference to the pair (ai, aj), where both the tests of concordance and that of discordance are exceeded, according to the threshold values of concordance (C *) and discordance (D *) predetermined.

$$S(ai,aj) = \quad \{1 \text{ se } Tc\ (a_i,aj) \geq C^* \quad e \text{ se } Td\ (ai,aj) < D^*$$
$$\{0 \text{ if otherwise} \qquad\qquad\qquad (9)$$

Finally, the results of the model of decision support to determine which variations of the model can generate substantial differences in the performance of the alternatives are obtained through sensitivity analysis. Generally, these investigate the values of criteria and indicators which correspond to the "turning point", that is, when the ranking of alternatives is reversed [11], [17].

4.7   The Results of the Assessment

The successive methodological phases, including the standardization of data, the weighting of the criteria and the ranking of the alternatives, the application under examination did not take place. In fact, the PA considered that the matrix of impact provided a comprehensive information framework to start negotiating with individuals. Initial meetings between the PA and private developers have targeted a sharing of methodology whereby multicriteria analysis can be used to evaluate their project proposals including the development of the evaluation criteria and proposed indicators. As hoped, the sharing of criteria by private developers was immediate, and the choice of indicators has not been questioned. Private developers were then involved in gathering the data relating to the quantification of the indicators for two of the three options. In the next step the PA highlighted the impact resulting from the implementation of the various project proposals and together with the private developers explored the possibility of any intermediate design changes, between the proposed project and the one initially proposed as an alternative by the PA. Although there was an initial rejection of new solutions, the transparency of the evaluation procedure, designed to measure the negative and positive impacts associated with different alternatives, helped overcome the initial uncooperative attitude. Revision and project development then took place. The results of this evaluation led the PA to favour the alternative option 2 -project ideal that produces less negative impact and greater benefits for the community, while maintaining adequate profitability for private investment.

The outcome of the negotiations was positive: the PA was satisfied with the role played by the multicriteria evaluation as a support to the decision.

5   Conclusions

The trial adopted was accompanied by the emergence of various expectations. First, the "new rules" - that is, the evaluation procedures described - are capable of ensuring in the decision-making process the PA requirements of transparency and fairness to the citizens, thus obviating the serious lack of current experiences of concerted planning. Second, the quality of urban transformation occurred at the outset in several respects: not only aesthetic and functional, but also expressed by performance indicators of urban nature, environment, social and economic-financial.

Finally, the need to ensure the viability and sustainability of the redevelopment means that objectives, resources, the possible actions of the different actors can interact with each other and co-exist. Thus, there is the opportunity to assign evaluation as a process that accompanies every stage in the formation of the program, as a role of coordination and training through planning to satisfy divergent interests, that explores alternative hypotheses and, if necessary, the need to amend and / or supplement the initial proposals in the program in relation to both its structure and organization in its relationship with the social, economic, environmental and cultural context, all of which are selling points.

References

1. Aragona, S., et al.: The evaluation culture to build a network of competitive cities in the mediterranean. In: Bevilacqua, C., Calabrò, F., Della Spina, L. (eds.) New Metropolitan Perspectives. The Integrated Approach of Urban Sustainable Development through the Implementation of Horizon/Europe 2020. Advanced Engineering Forum, vol. 11, © Trans Tech Publications, Switzerland (2014)
2. Calabrò, F., et al.: Evaluating cultural routes for a network of competitive cities in the mediterranean sea: the eastern monasticism in western mediterranean area. In: Advanced Materials Research, vols. 1073–1076, © Trans Tech Publications, Switzerland (2015)
3. Calabrò, F., et al.: Quality Monitoring and Control Tools for the enhancement of the architectural heritage: the Code of Practice for Historic Centres Conservation. WSEAS Press (2015). ISBN: 978-1-61804-259-0
4. Calabrò, F., Della Spina, L.: The public-private partnerships in buildings regeneration: a model appraisal of the benefits and for land value capture. In: Advanced Materials Research, vols. 931–932, © Trans Tech Publications, Switzerland (2014)
5. Calabrò, F., Della Spina, L.: The cultural and environmental resources for sustainable development of rural areas in economically disadvantaged contexts. Economic-appraisals issues of a model of management for the valorisation of public assets. In: Advanced Materials Research, vols. 869–870, © Trans Tech Publications, Switzerland (2014)
6. Calabrò, F., Della Spina, L.: Innovative tools for the effectiveness and efficiency of administrative action of the metropolitan cities: the strategic operational programme. In: Bevilacqua, C., Calabrò, F., Della Spina, L. (eds.) New Metropolitan Perspectives. The Integrated Approach of Urban Sustainable Development through the Implementation of Horizon/Europe 2020. Advanced Engineering Forum, vol. 11, © Trans Tech Publications, Switzerland (2014)
7. Calabrò, F., et al.: Cultural planning: a model of governance of the landscape and cultural resources in development strategies in rural contexts. In: Proceedings of the XVII - Ipsapa Interdisciplinary Scientific Conference, vol. V (2013)
8. Calavita, N., et al.: Transfer of development rights as incentives for regeneration of illegal settlements. In: Bevilacqua, C., Calabrò, F., Della Spina, L. (eds.) New Metropolitan Perspectives. The Integrated Approach of Urban Sustainable Development through the Implementation of Horizon/Europe 2020. Advanced Engineering Forum, vol. 11, © Trans Tech Publications, Switzerland (2014)

9. Calavita, N., Mallac, A.: Inclusionary housing, incentives, and land value recapture. In: Land Lines, Lincoln institute of Land Policy (2009)
10. D'Alpaos, C., Marella, G.: Urbanistica consensuale e valori d'opzione. In: Valori e valutazioni. n. 12. Dei, Roma (2014)
11. Della Spina, L., et al.: The culture of evaluation to improve the airport competitiveness of the metropolitan city of the strait: analysis of alternative scenarios of connection. In: Advanced Materials Research, vols. 1065–1069, © Trans Tech Publications, Switzerland (2015)
12. Della Spina, L.: Procedure di valutazione della qualità abitativa. Gangemi, Roma (1999)
13. Forte, C., De Rossi, B.: Principi di economia ed estimo, ETAS, Milano (1992)
14. Khakee, A., et al. (eds.): New Principles in Planning Evaluation. Ashgate Publishing Ltd., England (2008)
15. Morano, P., Tajani, F.: Bare ownership evaluation. Hedonic price model vs. artificial neural network. International Journal of Business Intelligence and Data Mining **8**(4) (2013)
16. Morano, P., Tajani, F.: Least Median of Squares regression and Minimum Volume Ellipsoid estimator for outliers detection in housing appraisal. International Journal of business intelligence and data mining **9** (2014)
17. Morano, P., Tajani, F.: Concession and lease or sale? A model for the enhancement of public properties in disuse or underutilized. Wseas Transactions on Business and Economics **11** (2014)
18. Munda, G.: Social multi-criteria evaluation for a sustainable economy. Springer-Verlag, Berlin Heidelberg (2010)
19. Nesticò, A., De Mare, G.: Government tools for urban regeneration: the cities plan in italy. A critical analysis of the results and the proposed alternative. In: Murgante, B., et al. (eds.) ICCSA 2014, Part II, LNCS, vol. 8580. Springer, Heidelberg (2014)
20. Nesticò, A., De Mare, G.: Efficiency analysis for sustainable mobility. The design of a mechanical vector in Amalfi Coast (Italy). In: Advanced Materials Research, vols. 931–932. Trans Tech Publications, Switzerland (2014)
21. Nesticò, A., De Mare, G., Fiore, P., Pipolo, O.: A model for the economic evaluation of energetic requalification projects in buildings. A real case application. In: Murgante, B., et al. (eds.) ICCSA 2014, Part II, LNCS, vol. 8580. Springer, Heidelberg (2014)
22. Prizzon, F.: Gli investimenti immobiliari. Analisi di mercato e valutazione economica finanziaria degli interventi. Celid, Torino (2001)
23. Nijkamp, P., Delft, A.: Multicriteria Analysis and Regional Decision Making. Springer, New York (1977)
24. Roy, B.: Méthodologie multicritère d'aide à la decision. In: Economica-Collection Gestion, Parigi (1995)
25. Sen, A.K.: The Impossibility of a Paretian Liberal. Journal of Political Economy (78) (1970)
26. Van Delft, A., Nijkamp, P.: Multi-criteria analysis and regional decision-making. In: Martinus Nijhoff Social Sciences Division, Leiden (1977)
27. Voogd, H.: Multicriteria evaluation for urban and regional planning. Pion, London (1983)

Energy Production Through Roof-Top Wind Turbines A GIS-Based Decision Support Model for Planning Investments in the City of Bari (Italy)

Pierluigi Morano[1(✉)], Marco Locurcio[2], and Francesco Tajani[1]

[1] Department of Science of Civil Engineering and Architecture, Polytechnic of Bari, Bari, Italy
pierluigi.morano@poliba.it, francescotajani@yahoo.it
[2] Department of Architecture and Design, University "Sapienza", Rome, Italy
marco.locurcio@uniroma1.it

Abstract. In this paper the financial feasibility in the installation of roof-top wind turbines in the territory of the city of Bari (Italy) has been analyzed. The elaborations carried out have allowed to define "wind" maps, in terms of annual mean wind speed and annual energy production, and evaluative maps, in terms of total profit for the investor and land lease values of the flat roofs of existing buildings. The thematic maps generated constitute a primary support for the operators interested in taking advantage of European resources and/or the incentives offered by energy regulations for the installation of roof-top wind turbines and identifying the areas characterized by higher yields. Furthermore, the model obtained provides investors and flat roofs' owners motivated and contextualized economic values, referred to the local wind power capacities of the areas in which the installation of roof-top wind turbines is financially convenient.

Keywords: Renewable energy sources · Small-scale turbines · Discounted cash flow analysis · Land lease · Financial feasibility

1 Introduction

The United Nations Framework Convention on Climate Change and the Kyoto Protocol have set the energy dictates for the containment of greenhouse gas emissions in the major industrialized Countries. The European Program for climate change, that is the main instrument of the Community strategy for the implementation of the Kyoto Protocol, identifies several solutions in order to satisfy the commitments of the reduction of carbon dioxide production. Among these alternatives, a key role is assigned in the last decade to the generation of electricity from renewable energy sources (RES-E), through the use of feed-in tariff scheme, competitive bidding processes and tradable green certificates schemes.

The main forms of RES-E are photovoltaic, geothermal, small hydro, biomass and wind. In particular, Menanteau et al. [17] have highlighted that wind energy should be

---

The work must be attributed in equal parts to the authors.

O. Gervasi et al. (Eds.): ICCSA 2015, Part III, LNCS 9157, pp. 104–119, 2015.
DOI: 10.1007/978-3-319-21470-2_8

able to provide most of the extra renewable energy required to reach the objectives set by the European Commission.

Supported by the European Directive 2001/77/CE, the wind technology has been highly developed, so that in 2008 in Europe the wind energy power installed has exceeded every other electric production sources, including gas, coal and nuclear plants. In the same year, the European producers of wind turbines covered 60% of the global market, which has a value of 36 billion euro just for the production of wind turbines.

In Italy, also thanks to an effective system of feed-in tariff, the wind solution of RES-E has been widely spread, especially in the Regions of Puglia and Sicily, characterized by good geo-climatic conditions. In the last years, the need to limit the impacts on the landscape determined by the proliferation of wind farms and to reduce the soil sealing of agricultural lands [2, 13] has generated an increased attention to the implementation of energy policies of RES-E integrated with the existing urban properties. Small wind turbines installed on flat roofs of existing buildings for a period coinciding with the life of the plant constitute a type of "urban" RES-E. Compared to the classical wind systems, represented by plants which extend on large areas excluded from the original agricultural vocation, the roof-top wind turbines provide the best compromise between the need to produce an annual energy production (AEP) that justifies the investment in financial terms, and the landscape and environmental instances concerning the respect of the urban skyline and the soil sealing.

Favored in Italy by a simplified authorization system (L.D. 387/2003; L.D. 28/2011; art. 1122-bis of the Civil Code), the installation of roof-top wind turbines is particularly suitable in those areas characterized not only by a good wind speed distribution, but also by climatic conditions that have oriented to prefer the realization of flat roofs on the buildings, that represent a good support base for the structures of the micro-turbines. For these reasons, the analysis developed in this paper is contextualized to the city of Bari (Italy), which is characterized, in addition to the climatic and structural elements mentioned above, by an urban conformation that creates limited reductions in wind speed.

2   Aims

The placement of small-scale turbines on flat roofs of existing buildings allows to maximize the productivity of the wind energy plant in the area considered, because the wind speed is a function of the height of the turbine from the level of the sea.

In Italy, the incentive system for the installation of RES-E is regulated by the D.M. 07.06.2012, that for wind turbines on shore with a nominal power that ranges from 20 to 200 kW provides a feed-in tariff equal to 268 € for each MWh of energy produced.

With reference to this incentive system, the case carried out in this paper raises two evaluation issues of particular importance. The first relates to the assessment of the threshold of financial feasibility for an investor in the installation of a wind energy plant, that means the identification of the most productive areas for the location of the roof-top wind turbines. The second issue is the appraisal of the land lease value of the area (flat roof) on which the wind energy plant rests for a limited period of time (*concession period*). In ordinary terms, it is plausible to consider the split between the private/public entrepreneur that realizes the investment in small-scale turbines and the owner - usually a condominium - of the area that hosts the wind energy plant. Therefore, there is the need to develop an economic relationship between the owner (*lessor*) and the investor (*tenant*), concerning the land lease for a limited period of time.

Although the estimative doctrine proposes a logical procedure for assessing the land lease value [8], based on an income approach, the lack of comparables in terms of wind energy productivity makes it impossible to obtain reliable value judgements. In the majority of cases, the determination of the land lease value to be paid to the lessor is deduced from business plans related to national data, which are devoid of local references and do not provide to the lessor any instrument of negotiation, obliging him to passively accept the price offered by the tenant.

In this paper, the financial feasibility resulting from the realization of small wind energy plants in Bari, a city of Southern Italy of about 320,000 inhabitants and with a territorial extension of over 100 km^2, is analyzed [22]. The outputs obtained constitute a reference for the various - public and private - actors operating in the territory: first, for those that are engaged in the research of innovative and economically self-sustainable projects [1, 3, 4], able to meet the rewarding requirements provided by the European Commission in the allocation of Community funds [6, 7, 19]; secondly, for the entrepreneurs interested in exploiting the incentives provided by the energy sector legislation for investments in RES-E; thirdly, for the owners of the flat roofs, that require land lease values motivated, as support for the definition of the land lease price in the phase of negotiation with the investors.

The first phase of the research involves the study of the wind climatology of the urban areas of the city of Bari. Borrowing the methodology developed by the MET Office [18] and Heath [14], the division of the territory of the city of Bari in homogeneous areas in terms of potential wind energy resource has been performed. The use of the Regional Technical Map and the Map of Use of Soil for the Puglia Region[1] has allowed to implement the model developed in a GIS environment [15, 21] , generating two types of thematic maps for the city of Bari, regarding the estimation of the *annual mean wind speed* and the *AEP*. Subsequently, for the type of roof-top wind turbines considered, the revenues and the costs that concur to determine the financial feasibility of the "wind" investment, differentiated on the basis of the two thematic maps (*annual mean wind speed* and *AEP*) obtained in the previous phase, have been evaluated. The financial items determined have allowed to generate two evaluative maps: the first is a map of financial feasibility of the investment, that shows the geographical areas of the city of Bari in which the intervention ensures a reasonable profit to the investor; the second is a map of the unit land lease values per square meter of area covered by buildings.

The article is structured in the following parts. In paragraph 3 the urban wind model implemented is specified, and the phases and the parameters that allow the construction of the wind maps are outlined. Paragraph 4 presents the main legal aspects of the land lease, the areas of application and the assessment methodology used in this

---

[1] www.sit.puglia.it

research. In paragraph 5 the spatial modeling of the case study is delineated, the wind maps for the city of Bari are obtained, the financial model implemented for estimating the land lease value is defined and the evaluative maps generated are commented. Finally, paragraph 6 highlights the conclusions of the work.

3   Urban Wind Model

The methodology for estimating the total potential wind energy resource from small-scale turbines has been inferred by integrating the studies developed by the MET Office [18] with the results of several researches [12, 14, 16, 24, 25, 26]. Therefore, the operations provided are the following:

i) determination of the parameters for the estimation of the wind speed. These parameters represent the typical aerodynamic properties of homogenous zones in urban territory and are divided into two classes: the parameters for the characterization of the urban area (roughness length, height of the urban canopy, blending height, plan-area density, frontal-area density, displacement height); the parameters for the large-scale reference wind climatology (reference height, diffusion height, reference wind speed profile, blending wind speed profile). The model developed in this work concerns the installation of roof-top wind turbines in a highly urbanized area: therefore, the roughness length, the height of urban canopy, the blending height and the plan-area density are the parameters that play a key role in the estimation of the wind speed. The *roughness length* is a property of the surface connected to the friction generated by the passage of a fluid. The air masses in motion are slowed down at the earth's surface due to its roughness. In particular, the speed of the fluid particles to the ground is practically zero, whereas, moving away from the earth's surface, the speed of the air masses tends to be equal to the speed of the undisturbed current, creating a speed profile of logarithmic type. The *height of the urban canopy* represents the height of the substrate that is created by the passage of the wind in proximity of the buildings. This parameter can be assumed to be equal to the average height of the buildings in the area in analysis [18]. The *blending height* describes the portion of the atmosphere characterized by convective motions that make it significant the mixing of air masses. This parameter can be considered equal to two times the height of the urban canopy and in any case greater than 10 m [14, 18]. The *plan-area density* occurs in the description of the packing of the buildings, i.e. it characterizes the regime of wind flow in urban areas [12, 23]; this parameter is the ratio between the total plan-area of the roughness elements and the total plan-area;

ii) assessment of the wind speed at building-scale and construction of the wind maps. Estimated the basic parameters and known the altitude of placement of the wind turbine, the wind speed is determined by implementing the methodology of MET Office [18] and Heath [14]. In particular, since the method of Heath is more reliable in the proximity of the change of roughness, the Heath's formula is used where this condition occurs (perimeter zones), whereas the MET Office's formula is applied in central areas, where the effect related to the change of roughness is irrelevant. These assumptions have allowed to develop wind maps at different altitudes of installation of the roof-top wind turbines;

iii) choice of the wind turbines. The operating conditions must be considered in this phase, i.e. the wind speeds at different altitudes of placement and the respective wind speed frequency distributions. In general, in areas characterized by high mean wind speed, it is preferable to install a wind turbine with high nominal power and cut out[2] speed; vice versa, in areas characterized by low mean wind speed, it is better to use small wind turbines that require a low cut in speed. In any cases, the maximum wind speed must not exceed the maximum speed supported by the wind turbine (IEC 61400-2 standard);

iv) evaluation of the Annual Energy Production (AEP). The wind speed frequency distribution is combined with a turbine power curve to obtain an estimate of the AEP from a single turbine. The results obtained can be used to generate AEP maps. The estimation methodology of the AEP of a single wind turbine can be extended to a wind turbine plant, in which case the area of influence to be respected in order to ensure the maximum functionality of each wind turbine of the plant must be determined. According to the IEC 61400-1 standard, the distance (D) to be respected in order to minimize any interference between the wind turbines in urban areas is equal to three times the rotor diameter (d). Known this characteristic, it is possible to determine the area of influence of the wind turbine $\left(A_{inf}\right)$. Assuming that A_{inf} is wholly contained in the installation surface of the flat roof $\left(A_{bui}\right)$, the number of wind turbines of the plant (N) is determined by the Equation (1).

$$N = \frac{A_{bui}}{A_{inf}} = \frac{A_{bui}}{\left(D/2\right)^2 \cdot \pi} = \frac{A_{bui}}{\left(3d/2\right)^2 \cdot \pi} \qquad (1)$$

4   Outlines of Land Lease

The land lease provides a dual configuration: a) the owner of an area can constitute the right to realize, above (and/or below) the ground, a construction in favor of others (*ius ad aedificandum*); b) the owner of an existing building can alienate this property separately from the ownership of the area.

The land lease can be perpetual or temporary basis. In the second case, at the end of the period, the land lease expires and the "fullness" of the ownership of the building with the area on which it insists is consolidated. No compensation is due from the owner of the land for the acquisition of the building realized or for additions or improvements to the existing building, separately from the area.

Although in Italy the land lease is widely used for interventions on public areas - this is the case of kiosks, telephone booths, electrical cabins, bathing facilities, affordable housing, fuel stations, cemetery chapels - it has not yet found a widespread application in private negotiations. However, the failure of the process of alienation of public properties, caused by the worsening of the economic recession, the credit crunch, the current crisis in the housing market and the fear of a "fire sale" of public assets, has been leading many Governments to discuss alternative solutions to the

---

[2] The "cut in" speed is the minimum wind speed to activate the turbines. The "cut out" speed is the threshold speed, above which the turbine will not produce more energy.

transfer of ownership of public properties, including the land lease that provides the concession of the existing building for a limited period of time.

The literature illustrates the main procedures for estimating the land lease value. This methods are differentiated according to the concession period (perpetual or temporary) and the presence/absence of a building on the ground [5, 8].

In this research, the land lease value is a function of the profitability of the financial investment (small wind turbine plants) for an ordinary entrepreneur. Therefore, assuming that the land lease is purchased by paying a unique initial amount (V_s), expression of the sum of the discounted annual rents to be paid in the concession period, trough the development of a Discounted Flow Analysis (DCF) it is possible to determine V_s as the difference between the revenues generated by the sale of the energy produced by the wind turbines installed and the costs required to realize the intervention, including the expected profit of the entrepreneur. Equation (2) clarifies the financial terms involved in the determination of the land lease value.

$$V_s = \sum_{i=1}^{n} \frac{R_i}{(1+r)^i} - \sum_{i=1}^{n} \frac{(K_i + P_i)}{(1+r)^i} \tag{2}$$

where:

V_s = land lease value;

n = concession period;

R_i = revenues generated by the sale of the energy produced by the wind turbines at the i-th year;

K_i = costs of realization and management of the wind turbine plant at the i-th year;

P_i = expected profit of the entrepreneur at the i-th year;

r = actualization rate of the cash flows at the time of the assessment of the land lease value.

In summary, the formula (2) calculates V_s as Net Present Value (NPV) of the investment from the point of view of the lessor.

5   The Case Study: Application to the City of Bari

The defined model has been implemented to the city of Bari (Italy), through a GIS environment and the database of the Regional Technical Map (CTR) and the Map of Use of Soil for the Puglia Region (UDS). In particular, geometric and orographic information necessary for the model (height and surface of buildings, level curves) have been extrapolated by the CTR. The information obtained by the UDS have allowed the construction of the map of the roughness length [26].

Through a series of topological overlay and the methodology described in the previous paragraphs, the parameters that represent the characteristics of the territory have been derived. The spatial analysis has allowed to determine indispensable elements, as the distance from the change of roughness. Spatial queries have been used to define the wind field, the installed power and the financial results. Finally, through the geovisualization in maps the results obtained have been represented in an immediate and easy to read modality.

5.1   Discretization of the Territory of the City of Bari

The entire territory of the city of Bari has been divided into a grid of 12,288 square meshes, each with area equal to 10,000 m² (100 m x 100 m), for a total surface of 116.21 km². In proximity of the sea and the administrative limits, meshes assume different size and shape. This division allows to standardize the parameters to consider, which are then referred to the m-th mesh of the grid, assumed as the fundamental unit. The mesh size has been chosen with reference to the results of recent research [16, 26], in which meshes with side ranging between 80 and 120 meters have been employed.

In Fig. 1 thematic maps for the city of Bari, referred to the principal parameters for the assessment of the wind speed, are reported.

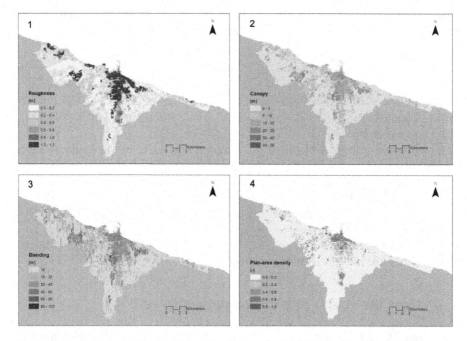

Fig. 1. Maps of roughness length (1), height of urban canopy (2), blending height (3) and plan-area density (4)

The analysis of the map of roughness length (Fig. 1.1) shows an area with a high roughness in the central part of the city of Bari, with values up to 1.2 m along the north-south direction from the coast and near the historic center, to reach 0.8 m in proximity of the southern border of the city. Sporadic significant values of roughness are also found in the west area. On the rest of the city rarely the roughness exceeds 0.7 m and reaches values around 0.3 m in rural areas near the inhabited zone.

The maps relating to the height of the urban canopy and the blending height (Fig. 1.2 and Fig. 1.3) show a distribution similar to the map of roughness length, which emphasizes that higher urbanized areas are characterized by higher roughness.

The analysis of the map of plan-area density (Fig. 1.4) highlights a trend similar to the distribution observed for the two previous parameters, but in this case the weight of the higher urbanization (together with the lower building heights) in proximity of the north-west coast on the parameter considered is more evident: densely built areas but with low heights do not have a high weight in determining the urban canopy, and their effect in terms of friction (i.e. the relative roughness length) is rather limited.

5.2   Definition of the Maps of Annual Mean Wind Speed and AEP

The analysis has been developed considering a standard roof-top wind turbine with a nominal power equal to 3 kW, horizontal axis (HAWT) and glass fiber reinforced blades, according to IEC 61400-2 standard. The characteristics of the roof-top wind turbine are summarized in Table 1.

Table 1. Technical characteristics of the turbine used in the application

Nominal power [kW]	3
Rotor diameter [m]	5
Swept area [m²]	19.6
Propeller blades [n]	3
Cut-in speed [m/s]	2.5
Nominal speed [m/s]	8
Cut-out speed [m/s]	20
Maximum speed tolerated [m/s]	40
Total weight [kg]	250
Height of the tower [m]	8
Influence area [m²]	177
Price of the support [€]	550
Price of the turbine [€]	1,554

Considering the division in meshes of the city of Bari, maps of annual mean wind speed and AEP are respectively represented in Fig. 2 and Fig. 3.

The map in Fig. 2 shows that the maximum value of the wind speed does not exceed 4 m/s, and this condition occurs in few cases, located in the west and in the south of the urban center. Wind speed values ranging between 3.0 m/s and 3.5 m/s are recorded in the north-west, in a wide peripheral band of the city and in proximity of the airport, because of low roughness and scarcity of buildings. The most common wind speed values are between 2.5 m/s and 3.0 m/s, that affect about 90% of the meshes considered. Mean wind speed values below the cut-in speed are mainly observed in the central area of the north-south trunk line.

The map in Fig. 3 shows that the AEP of a single roof-top wind turbine per mesh is mainly characterized by values ranging between 1,500 kWh and 2,500 kWh. Lower values occur in the trunk line characterized by low mean wind speed, whereas higher values are recorded for specific cases, prevailing in the north-west and in proximity of the airport, where the AEP values are between 4,500 kWh and 6,500 kWh.

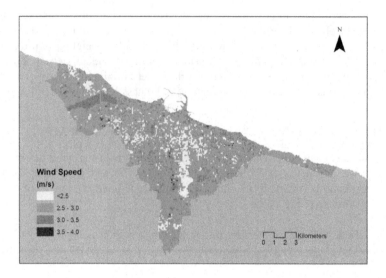

Fig. 2. Map of annual mean wind speed

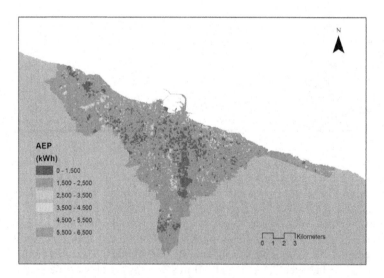

Fig. 3. Map of Annual Energy Production

5.3   The Appraisal of the Land Lease Value

In order to define evaluative maps as support to the operators interested in investing in HAWT plants (Table 1) to be placed on the flat roofs of existing buildings in the city of Bari, the assumptions used to determine revenues and costs of Equation (2) for the appraisal of V_s are the following: i) all covered surfaces in the meshes are equally useful for the installation of wind turbines. Therefore, any further "accommodation" costs of the supporting surfaces of the wind turbines are not considered; ii) for the areas that will be identified by the model as financially attractive for the "wind" investment $(V_s > 0)$, the intended destination of the covered surfaces (supporting surfaces of the wind turbines) coincides with the respective *highest and best use*; iii) the production capacity of the wind turbines is considered constant for the whole concession period. This means that the revenues R_i in Equation (2) are constant for all the concession period, and only differ according to the mesh considered (R_m).

On the basis of these assumptions, a DCF has been developed for each of the meshes considered. In particular: the analysis has been carried out assuming a constant prices system; the concession period (n) is equal to twenty years, corresponding to the ordinary period of useful life of the wind turbines; a residual value equal to zero is assumed for the wind turbines; the actualization rate (r) has been assumed to be equal to the average net return rate of twenty-year Italian Treasury Bonds (3.00%)[3].

Revenues (R_m)

Established the feed-in tariff according to the legislation (t_f), for each m-th mesh and for each i-th year revenues are a function of the annual energy production (AEP_m) and of the number of wind turbines installed (N_m), as reported in Equation (9) in Table 2.

Costs $(K_m + P_m)$

The construction costs consist of the purchase cost of the wind turbines (K_p) and the additional costs (K_{add}). The total purchase cost, estimated by Equation (11) in Table 2, is a function of the purchase cost of a single wind turbine (k_p) and the number of wind turbines placeable in the mesh (N_m), that is obtainable through Equation (1). The additional costs (K_{add}) include: the costs of installation and transportation; the costs for the connection to the electric network; the audit and management costs for the Energy Services Manager (GSE); the technical costs. The additional costs are determined as a percentage of 20% of the total purchase cost, as reported in Equation (12) in Table 2.

It is assumed that the full amount of the construction costs $(K_p + K_{add})$ is financed by a bank and repaid in ten years, with an annual fixed interest rate (r') equal to 5.50%. Equation (15) in Table 2 allows to determine the annual amortization charge (Q_{amm}) to be included in the first ten years of the development of the DCF.

Maintenance costs (K_{man}) and insurance costs (K_{ass}) of the wind turbine plants are estimated as a percentage equal, respectively, to 1.80% and 0.70% of the purchase costs of the wind turbines. The corresponding expressions are given in Equations (16) and (17) in Table 2.

---

[3] Menanteau et al. [17] state that for the renewable energy sources "the market risk is non-existent and the profitability of projects depends essentially on the ability of investors to control their costs".

An additional annual operating cost is the property tax (K_{IMU}) that weighs on the wind turbine plants to be realized. Circular No. 14/2007 of the Italian Revenue Agency has clarified that the wind turbine plants belong to the "D/1" ("factories") cadastral category. On the basis of the technical information contained in Circular No. 6/2012 of the Italian Revenue Agency, concerning the assessment of the cadastral income of non-ordinary property units (including units of D/1 category), and taking into account the IMU basic rate set by the City Offices of Bari for the year 2014, it is possible to estimate a plausible property tax by Equation (3):

$$K_{IMU,m} = \{[75\% \cdot K_{p,m} + 12.44\% \cdot (75\% \cdot K_{p,m})] \cdot 2\%\} \cdot 1.05 \cdot 65 \cdot 1.06\% \qquad (3)$$

where:

$75\% \cdot K_{p,m}$= mean value for a new plant, characterized by useful life of twenty years and residual value equal to zero (Table 2, Annex III, Circular No. 6/2012 of the Italian Revenue Agency);

$12.44\% \cdot (75\% \cdot K_{p,m})$= expected (cadastral) profit, whose percentage is assumed equal to the gross weighted average yield of a basket of Government Bonds in the biennium 1988-1989 (Annex II, Circular No. 6/2012 of the Italian Revenue Agency);

2% = return rate for the determination of the cadastral income of property units belonging to the "D" cadastral category (Circular No. 6/2012 of the Italian Revenue Agency);

$1.05 \cdot 65$= coefficient to be used for the determination of the cadastral value subject to the property tax;

1.06%= IMU basic rate (Resolution No. 39/2014 of the City Offices of Bari).

Finally, the expected profit of the entrepreneur (P_m) is determined as a percentage of revenues R_m, therefore it is considered constant in the concession period. On the basis of investigations conducted, a rate of 30%, gross of taxes pertaining to the specific tax regime of the investor, constitutes an appropriate amount for the financial intervention in analysis (Equation (19) in Table 2).

The development of the DCF based on the parameters defined above allows to estimate the land lease value for each mesh identified $(V_{s,m})$. In synthetic terms, taking into account the assumption of constant prices, it is possible to estimate the land lease value $V_{s,m}$ through Equation (4).

$$V_{s,m} = \left\{[R_m - (K_{man,m} + K_{ass,m} + K_{IMU,m} + P_m)]\frac{(1+r)^n-1}{r\cdot(1+r)^n}\right\} - Q_{amm,m}\frac{(1+r)^w-1}{r\cdot(1+r)^w} \qquad (4)$$

The meaning of each term of Equation (4) and the related estimating Equations are summarized in Table 2.

Table 2. Definition of the terms for the assessment of the land lease value $V_{s,m}$

$V_{s,m}$	land lease value for the mesh m	(5)
t_f	feed-in tariff established by legislation (DM 07/06/2012)	(6)
AEP_m	annual energy production for the mesh m	(7)
$N_m = \dfrac{A_{bui,m}}{A_{inf,m}}$	number of wind turbines to be installed in the mesh m	(8)
$R_m = t_f \cdot AEP_m \cdot N_m$	annual revenues for the mesh m	(9)
k_p	purchase cost of a single wind turbine	(10)
$K_{p,m} = k_p \cdot N_m$	total purchase cost of the wind energy plant in the mesh m	(11)
$K_{add,m} = 20\% \cdot K_{p,m}$	additional costs for the realization of the wind turbine plant in the mesh m	(12)
$r' = 5.50\%$	financing rate of the bank loan	(13)
$w = 10 years$	amortization period of the bank loan	(14)
$Q_{amm,m} = \left(K_{p,m} + K_{add,m}\right) \cdot \dfrac{r\prime \cdot (1+r\prime)^j}{(1+r\prime)^j - 1}$	annual amortization charge for the mesh m ($j = 1, ..., 10$)	(15)
$K_{man,m} = 1.80\% \cdot K_{p,m}$	annual maintenance cost of the wind turbine plant in the mesh m	(16)
$K_{ass,m} = 0.70\% \cdot K_{p,m}$	annual insurance cost of the wind turbine plant in the mesh m	(17)
$K_{IMU,m}$	annual property tax of the wind turbine plant in the mesh m	(18)
$P_m = 30\% \cdot R_m$	annual profit of the investitor for the mesh m	(19)

5.4   The Construction of the Evaluative Maps

The implementation of Equation (4) for all 12,288 meshes has allowed to obtain the maps shown in Fig. 4 and Fig. 5: the first represents the variation of the total actualized profit of the investor; the second describes the variation of the land lease value, parameterized per square meter of covered surface within each mesh.

The white areas of the maps represent the part of the territory of the city of Bari where the roof-top wind turbines cannot be installed, because of the absence of buildings where to locate the plants, or due to regulatory constraints (National-Regional Protected Areas, detectable from www.sit.puglia.it).

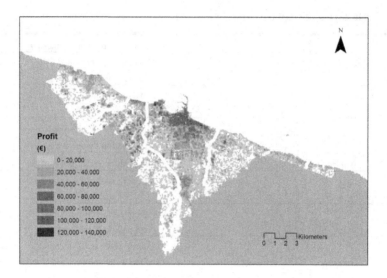

Fig. 4. Map of the total actualized profit of the investor

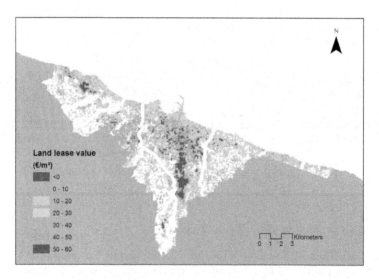

Fig. 5. Map of unit land lease value per m² of built-up area

The map in Fig. 4 shows that the most "financially hot" areas for a generic investor are the central and near the north coast, characterized not so much by a high AEP, but by a significant plan-area density, i.e. by a considerable urban concentration: in fact, in this part of the territory the total actualized profit varies between 80,000 € and 140,000 €, compared with a total investment cost amounting respectively to 65,000 € and 100,000 €. Moving away from the center, an appreciable convenience of the investment – with a total profit ranging between 40,000 € and 80,000 € - is still guaranteed where the plan-area density is consistent (west zone) or the wind speed values are high (north-west area in proximity of the airport). The peripheral territories are the less interesting areas, characterized by low values of the main parameters for estimating the wind speed profile.

The map in Fig. 5 shows a fairly good dispersion of unit land lease values on the whole territory of the city of Bari: the unique concentrations occur along the central north-south trunk line, characterized by negative land lease values or slightly above zero, and in the north-west, in proximity of the airport, where land lease values ranging from 30 €/m^2 and 50 €/m^2 are estimated and the plan-area density is low (<20%, Fig. 1.4). Land lease values between 10 €/m^2 and 30 €/m^2 are estimated in proximity of significant changes of roughness (coast and agricultural areas), where – especially in the central areas – the plan-area density values are also higher than 80% (Fig. 1.4). Land lease values above 50 €/m^2 are estimated in exceptional cases, related to the placement of roof-top wind turbines on higher buildings, and for meshes characterized by plan-area density of approximately 20%.

6   Conclusions

According to the forecasts of the European Wind Energy Association (EWEA), renewable energy generated by wind power systems is designed to be the largest contribution in order to satisfy the European objectives, arriving in 2020 to cover the 34% of the whole share of renewable energy.

In this paper the financial feasibility in the installation of a type of roof-top wind turbine in the territory of the city of Bari (Italy) has been analyzed. The elaborations carried out have allowed to define, first of all, "wind" maps of the city, in which the annual energy productions geo-referenced on the territory of Bari are represented; subsequently, evaluative maps have been obtained, in terms of total profit for the investor and land lease values. The results obtained highlight the financial feasibility of the investment in different areas of the city of Bari.

The thematic maps defined constitute an essential evaluative support, on the one hand, for the (public and/or private) operators, interested in taking advantage of European resources and/or the incentives offered by energy regulations for the installation of micro-turbines in a windy territory and identifying the areas characterized by higher yields; on the other hand, for the tenants (investors) and the lessors (condominiums), as a source of motivated and contextualized economic values, compared to the current estimation, that is, in practice, exclusively entrusted to the contractual capacities of the parties or to prices available from the web and referred to the entire national territory.

Finally, taking into account the current needs of the recent Italian Law on tax matters (L. 23/2014) in order to obtain a fairer tax system [11], the model and the evaluative maps described represent effective tools for defining a system of cadastral incomes equalized to the wind productivity of the areas ("wind" incomes). Through appropriate mass appraisal techniques [9, 10, 20], the results obtained can be used to define the "wind" microzones of the city of Bari, i.e. uniform territorial areas in terms of wind power capacity.

Acknowledgements. This study has been developed within research activities being carried out by Real Estate Valuation Center of the MITO-LAB (LABoratory of Multimedia Information for Territorial Objects) of the Polytechnic of Bari, Italy (web site: http://mitolab.poliba.it).

References

1. Attardi, R., Pastore, E., Torre, C.M.: ``Scrapping'' of quarters and urban renewal: a geostatistic-based evaluation. In: Murgante, B., Misra, S., Rocha, A.M.A., Torre, C., Rocha, J.G., Falcão, M.I., Taniar, D., Apduhan, B.O., Gervasi, O. (eds.) ICCSA 2014, Part III. LNCS, vol. 8581, pp. 430–445. Springer, Heidelberg (2014)
2. Balena, P., Sannicandro, V., Torre, C.M.: Spatial multicrierial evaluation of soil consumption as a tool for SEA. In: Murgante, B., Misra, S., Rocha, A.M.A., Torre, C., Rocha, J.G., Falcão, M.I., Taniar, D., Apduhan, B.O., Gervasi, O. (eds.) ICCSA 2014, Part III. LNCS, vol. 8581, pp. 446–458. Springer, Heidelberg (2014)
3. Calabrò, F., Della Spina, L.: The cultural and environmental resources for sustainable development of rural areas in economically disadvantaged contexts. Economic-appraisals issues of a model of management for the valorisation of public assets. Advanced Materials Research **869–870**, 43–48 (2014)
4. Calabrò, F., Della Spina, L.: The public-private partnerships in buildings regeneration: a model appraisal of the benefits and for land value capture. Advanced Materials Research **931–932**, 555–559 (2014)
5. Carrer, P.: Il diritto di superficie in Italia, relazione introduttiva al Convegno AREL. Istituto Bancario San Paolo di Torino, Milano (1992)
6. De Mare, G., Nesticò, A., Tajani, F.: Building investments for the revitalization of the territory: a multisectoral model of economic analysis. In: Murgante, B., Misra, S., Carlini, M., Torre, C.M., Nguyen, H.-Q., Taniar, D., Apduhan, B.O., Gervasi, O. (eds.) ICCSA 2013, Part III. LNCS, vol. 7973, pp. 493–508. Springer, Heidelberg (2013)
7. De Mare, G., Nesticò, A., Tajani, F.: The rational quantification of social housing. In: Murgante, B., Gervasi, O., Misra, S., Nedjah, N., Rocha, A.M.A., Taniar, D., Apduhan, B.O. (eds.) ICCSA 2012, Part II. LNCS, vol. 7334, pp. 27–43. Springer, Heidelberg (2012)
8. De Mare, G., Nesticò, A.: Il diritto di superficie nelle trasformazioni urbane: profili estimativi. Valori e Valutazioni 4/5 (2010)
9. Del Giudice, V., De Paola, P., Torrieri, F.: An integrated choice model for the evaluation of urban sustainable renewal scenarios. Advanced Materials Research **1030**, 2399–2406 (2014)
10. Del Giudice, V., De Paola, P.: The effects of noise pollution produced by road traffic of Naples Beltway on residential real estate values. Applied Mechanics and Materials **587–589**, 2176–2182 (2014)
11. Granata, M.F.: Rendita catastale e rendita solare degli impianti di produzione elettrica a fonti rinnovabili: perequazione fiscale energetico-immobiliare. Valori e Valutazioni **13**, 113–127 (2014)
12. Grimmond, C.S.B., Oke, T.R.: Aerodynamic properties of urban areas derived from analysis of surface form. Journal of Applied Meteorology **38**(9), 1262–1292 (1999)
13. Guarnaccia, C., Mastorakis, N.E.: Wind turbine noise: theoretical and experimental study. International Journal of Mechanics **5**(3), 129–137 (2011)
14. Heath, M.A., Walshe, J.D., Watson, S.J.: Estimating the potential yield of small building-mounted wind turbines. Wind Energy **10**(3), 271–287 (2007)

15. Las Casas, G., Lombardo, S., Murgante, B., Pontrandolfi, P., Scorza, F.: Open data for territorial specialization assessment territorial specialization in attracting local development funds: an assessment procedure based on open data and open tools. Tema. Journal of Land Use, Mobility and Environment (2014)
16. Macdonald, R.W.: Modelling the Mean Velocity Profile in the Urban Canopy Layer. Boundary-Layer Meteorology **97**(1), 25–45 (2000)
17. Menanteau, P., Finon, D., Lamy, M.L.: Prices versus quantities: choosing policies for promoting the development of renewable energy. Energy Policy **31**, 799–812 (2003)
18. MET Office: Small-scale wind energy. Technical Report, http://www.carbontrust.com
19. Morano, P., Locurcio, M., Tajani, F., Guarini, M.R.: Urban redevelopment: a multi-criteria valuation model optimized through the fuzzy logic. In: Murgante, B., Misra, S., Rocha, A.M.A., Torre, C., Rocha, J.G., Falc\ {a}o, M.I., Taniar, D., Apduhan, B.O., Gervasi, O. (eds.) ICCSA 2014, Part III. LNCS, vol. 8581, pp. 161–175. Springer, Heidelberg (2014)
20. Morano, P., Tajani, F.: Bare ownership evaluation. Hedonic price model vs. artificial neural network. International Journal of Business Intelligence and Data Mining **8**(4), 340–362 (2013)
21. Murgante, B., Tilio, L., Lanza, V., Scorza, F.: Using participative GIS and e-tools for involving citizens of Marmo Platano – Melandro area in European programming activities. Journal of Balkans and Near Eastern Studies **13**(1), 97–115 (2011)
22. Pagnini, L.C., Burlando, M., Repetto, M.P.: Experimental power curve of small-size wind turbines in turbulent urban environment. Applied Energy **154**(5), 112–121 (2015)
23. Raupach, M.R., Antonia, R.A., Rajagopalan, S.: Rough-wall turbulent boundary layers. Applied Mechanics Reviews **44**(1), 1–25 (1991)
24. Simiu, E.: Logarithmic profiles and design wind speeds. Journal of the Engineering Mechanics Division **99**(5), 1073–1083 (1973)
25. Solari, G.: Turbulence modeling for gust loading. Journal of Structural Engineering **113**(7), 1550–1569 (1987)
26. Università degli Studi di Genova: Studio inerente la modellistica numerica del vento nelle aree portuali di Genova, Savona e Vado Ligure, La Spezia, Livorno e Bastia, http://www.ventoeporti.net

"Flame": A Fuzzy Clustering Method to Detection Prototype in Socio-Economic Context

Silvestro Montrone[1,2], Paola Perchinunno[1,2(✉)], and Samuela L'Abbate[1,2]

[1] DISAG, University of Bari, Via C. Rosalba 53 70100, Bari, Italy
{silvestro.montrone,paola.perchinunno,samuela.labbate}@uniba.it
[2] Università Cattolica "Nostra Signora Del Buon Consiglio",
Rr. Dritan Hoxha, Tirane, Shqiperi

Abstract. Cluster analysis is highly advantageous as it provides "relatively distinct" (or heterogeneous) clusters, each consisting of units (families) with a high degree of "natural association". Different approaches to cluster analysis are characterized by the need to define a matrix of dissimilarity or distance between the n pairs of observations. The cluster analysis allows to identify the profiles families who meet certain descriptive characteristics, not defined a priori. Fuzzy clustering is useful to mine complex and multi-dimensional data sets, where the members have partial or fuzzy relations. Among the various developed techniques, fuzzy-C-means (FCM) algorithm is the most popular one, where a piece of data has partial membership with each of the pre-defined cluster centers. Fu and Medico [1] developed a clustering algorithm to capture dataset-specific structures at the beginning of DNA microarray analysis process, which is known as Fuzzy clustering by Local Approximation of Membership (FLAME). It worked by defining the neighborhood of each object and Identifying cluster supporting objects that have great importance in the field of market research in order to identify not only the average profiles (centroid) but the real prototypes and assigned to other units observed a degree of similarity.

Keywords: Fuzzy clustering · Hardship · Flame · Prototypes

1 Approaches and Methodologies for the Fuzzy Clustering

Clustering techniques are methods used to organize data into groups based on similarities among the individual data items. Most clustering algorithms do not rely on common assumptions to conventional statistical methods, such as the underlying statistical distribution of data, and therefore they are useful in situations where little prior knowledge exists. The potential of clustering algorithms to reveal the underlying structures in data can be exploited in a wide variety of applications, including classification, image processing, pattern recognition, modelling and identification.

The contribution is the result of joint reflections by the authors, with the following contributions attributed to S. Montrone (chapters 2), to P. Perchinunno (chapter 1 and paragraph 4.2), and to S. L'Abbate (chapter 3 and paragraph 4.1). The conclusions are the result of the common considerations of the authors.

© Springer International Publishing Switzerland 2015
O. Gervasi et al. (Eds.): ICCSA 2015, Part III, LNCS 9157, pp. 120–129, 2015.
DOI: 10.1007/978-3-319-21470-2_9

Fuzzy clustering methods allow the objects to belong to several clusters simultaneously, with a different degrees of membership. Objects on the boundaries between several classes are not forced to fully belong to one of the classes, but rather are assigned membership degrees between 0 and 1 indicating their partial membership. The discrete nature of the hard partitioning also causes difficulties with algorithms based on analytic functionals, since these functionals are not differentiable.

Cluster analysis is highly advantageous as it provides "relatively distinct" (or heterogeneous) clusters, each consisting of units (families) with a high degree of "natural association" [2]. Different approaches to cluster analysis are characterized by the need to define a matrix of dissimilarity or distance between the n pairs of observations. The cluster analysis allows identifying the profiles families who meet certain descriptive characteristics, not defined a priori. The cluster analysis is, in fact, a multivariate analysis technique through which you can group the statistical units in classes, so that the observations are as homogeneous as possible within the classes and the possible heterogeneous between the different classes [3,4,5]. This technique starts with the choice of an algorithm which defines the rules of how to group units into subgroups based on their similarity. Depending on of data, you have different sizes. For quantitative data have distance measures, for qualitative data have association measures. Once the choice of measurement to be used, there is the choice of method or algorithm of classification and the criterion of aggregation / subdivision.

Fuzzy clustering is used for complex data sets and multi-dimensional, where the members have fuzzy relationships. Among the various techniques developed, algorithm-Fuzzy C-means (FCM) is the most popular, in which a piece of data has partial membership with each of the cluster centers predefined. In FCM, the cluster centres and membership values of the data points with them are updated through some iterations [6].

Fu and Medico [1] developed a clustering algorithm to capture dataset-specific structures at the beginning of DNA microarray analysis process, which is known as Fuzzy clustering by Local Approximation of Membership (FLAME). It worked by defining the neighbourhood of each object and identifying cluster supporting objects. Fuzzy membership vector for each object was assigned by approximating the memberships of its neighbouring objects through an iterative converging process. Ma and Chan [7] proposed an Incremental Fuzzy Mining (IFM) technique to tackle complexities due to higher dimensional noisy data, as encountered in genetic engineering.

2   Fuzzy Clustering by Local Approximation of Membership (FLAME)

2.1   Introduction

The algorithm FLAME (Fuzzy clustering by Local Approximation of Membership) [1], worked by defining the neighbourhood of each object and identifying cluster supporting objects. Fuzzy membership vector for each object was assigned by approximating the memberships of its neighbouring objects through an iterative converging process.

The method involves three main steps:

1. definition of the information structure on which to base the classification;
2. application of a criterion of approximation to define the fuzzy membership of an object to one or more groups;
3. final classification.

1. The information structure must necessarily be based on a criterion of accumulation points. In particular it is necessary to define the particular points that are "poles of attraction" of a them neighborhood. This neighborhood, for a generic object i, is defined by k objects nearest or most similar to it. This k-nearest neighbors are denoted KNN *(i)* and points of attraction as *Cluster Supporting Object (CSO)*.

Considered the dataset X with n points and m variables, the density is calculated with the following expression:

$$d_i = \frac{S_{max}}{S_i} \tag{1}$$

where:

- $S_i = \frac{1}{k}\sum_{j \in KNN(i)} dist(X_i, X_j) \quad \forall i = 1, 2, \dots, n$

- $S_{max} = \max_i S_i$

- $dist(X_i, X_j)$ is the function of distance used between X_i and X_j profiles .

The density values can be divided into three classes:

- Defined a threshold level (percentile), all values below this threshold are considered outlier. In the original formulation of the algorithm [1], this set of points is considered as a cluster; in this work, however, outliers were excluded for not interfering in the process of local approximation and therefore the size of the data set is reduced to $n' = n -$ number of outliers.
- Identification of the *CSO*s consists of identifying those points with the maximum density. Number of *CSO*s, i.e. the number of clusters, is not defined a priori and depends on the fact that if a *CSO* belongs to neighborhood of another *CSO*, between the two will choose the one with the higher density. Furthermore another parameter that influences the number of *CSO*s is k: as k increases, the number of clusters decreases.
- The points that are not outliers or are not *CSO*s are the points that need to be classified.

2. The algorithm then provides the assignment of fuzzy membership by a method of local approximation. The number of clusters is defined by the number of *CSO*s, representative points of the dataset.

At the beginning, each object has the same degree of fuzzy membership at all clusters, with the exception of *CSO*s. In fact, to each *CSO* is assigned a full membership to himself as if it were a single cluster.

Therefore defined with $C = \#CSO$ (the number of *CSO*), for each profile (point) is associated a vector which indicates the membership degree to the cluster l:

$$p(i) = (p_{i1}, p_{i2}, \dots, p_{il}, \dots, p_{iC}) \qquad \forall \ i = 1, 2, \dots, n'$$

with $0 \leq p_{il} \leq 1$ and $\sum_{l=1}^{C} p_{il} = 1$

this vector is updated with an iterative process that minimizes an error function of local approximation.

$$E[p] = \sum_{i=1}^{n'} \sum_{l=1}^{C} \left\| p_{il} - \sum_{j \in KNN(i)} w_{ij} \, p_{jl} \right\|^2 \tag{2}$$

where w_{ij} are the coefficients to calculate the relative distances to the nearest neighbors obtained by

$$w_{i,j} = \frac{1/dist(X_i, X_j)}{\sum_{k \in KNN(i)} 1/dist(X_i, X_k)} \tag{3}$$

with $\sum_{j \in KNN(i)} w_{ij} = 1$.

The function $E[p]$ can be minimized by solving the following linear equations:

$$p_{il} - \sum_{j \in KNN(i)} w_{ij} \, p_{jl} = 0, \qquad i = 1, 2, \dots, n' \qquad \text{and} \qquad l = 1, 2, \dots, C \tag{4}$$

that have a unique solution found by the iterative procedure defined as

$$p_{il}^{t+1} = \sum_{j \in KNN(i)} w_{ij} \, p_{jl}^{t} \, . \tag{5}$$

3. The last step is the construction of the clusters that can be done in two ways. By assigning each object to the cluster which has a greater degree of membership in order to have a partition crisp. Or, considering a threshold value for the degree of membership, assigning each object to one or more clusters according to its degree of membership if greater than the threshold value set. In this work, we opted for the second mode in order to have the intermediate profiles.

3   The Construction of Sets of Indicators with a Fuzzy Method

3.1   Introduction

In this report the data source used in order to construct indicators of socio-economic hardship is that of the Family Lifestyles survey conducted by the University of Bari "A. Moro" (December 2012 - January 2013).

The *Family Lifestyles survey* collected significant information on income, spending behavior, and on the use of financial loans by families with children, resident in the metropolitan city of Bari. The objective of the survey, carried out by the University of Bari was that of analyzing issues associated with the measurement of socio-economic

124   S. Montrone et al.

hardship created by the difficulty of attributing a single and generally agreed definition. A methodology based on *objective variables* (those resources actually available
to families) was accompanied by *subjective measurements* based on the perception of
the family in terms of its social and economic condition.

In order to obtain a measurement of the level of socio-economic hardship of the
families interviewed, *sets of indicators* were constructed for the detection of the pos-
session or absence of functional goods, the ability to bear certain costs, the perception
of the evolution of the economic condition of the family etc.. Such sets of indicators
were used in order to obtain a fuzzy value corresponding to the level of hardship of
each family [2].

3.2   The Fuzzy Approach in the Analysis of Socio-Economic Hardship

The development of fuzzy theory initially stems from the work of Zadeh [8] and subsequently draws upon Dubois and Prade [9] and their definition of a methodological
basis. Fuzzy theory develops from the assumption that every unit is associated contemporarily to all categories identified and not univocally to only one, on the basis of
ties of differing intensity expressed by the concept of degrees of association. Fuzzy
methodology in the field of "poverty studies" in Italy has been recently employed in
the work of Cheli and Lemmi [10] who define their method "total fuzzy and relative"
(TFR) on the basis of the previous contribution from Cerioli and Zani [11].

Such a method consists in the construction of a function of membership to the
fuzzy totality of the poor which is continuous in nature, and able to provide a measurement of the degree of poverty present within each unit.

Given a set of **X** elements $x \in$ **X**, any fuzzy subset **A** of **X** is defined as follows:

$$A = \{X, f_A(x)\} \tag{6}$$

where $f_A(x): x \to [0,1]$ is defined as the membership function of the fuzzy subset
A and indicates the degree of membership of x to **A**. Therefore $f_A(x) = 0$ indicates
that x does not belong to **A**, while $f_A(x) = 1$ indicates that x belongs only to **A**. However, in the case of $0 < f_A(x) < 1$, x belongs partially to **A**, with a greater degree of
membership the closer $f_A(x)$ is to 1.

Supposing an observation of k poverty indicators for every family, the function of
membership of i_{th} family to the fuzzy subset of the poor may be defined thus:

$$f(x_i) = \frac{\sum_{j=1}^{k} g(x_{ij}).w_j}{\sum_{j=1}^{k} w_j} \qquad i = 1,....,n \tag{7}$$

The *Total Fuzzy and Relative* (TFR) model is used in order to summarize the values emerging from analysis in a single "blurred" fuzzy value which, as described

above, measures the degree of membership of an individual in the range between 0 (condition of well-being) and 1 (hardship).

The indices were chosen in order to identify levels of socio-economic hardship and were calculated so as to match the high values of the index with a high level of hardship and low values of the index with higher levels of well-being. The indices were grouped into several sets characterized by different situations:

- **Set 1: difficulty in paying debts/instalments or buying food staples** (mortgages, other debts and taxes, utility bills, food staples);
- **Set 2: difficulty in paying for education, health or unforeseen expenses** (costs of school meals and other subsidies for children; voucher for medical treatment in public hospitals, private medical care or other unexpected expenses);
- **Set 3: difficulty in purchasing other goods and services** (consumption of meat or fish at least once every two days, heating or air-conditioning in the home, purchase of clothing items when needed, going to the cinema/theatre at least once a month, going on holiday for one week a year);
- **Set 4: difficulty in participating in events** (social, religious, sporting, political, voluntary, or cultural) [2].

4   An Application with FLAME Fuzzy Clustering Algorithm

4.1   Identification of Cluster Supporting Object (CSO)

Primarily by the application method FLAME were identified *CSO* or those points with particular characteristics about which build the clusters. These points are the "prototype family" to which members of the group have similar profiles.

Table 1. CSO for different value of hardship

Cluster	Value of hardship Set 1	Value of hardship Set 2	Value of hardship Set 3	Value of hardship Set 4
CSO 1	0.82	0.90	0.96	1.00
CSO 2	0.70	0.84	0.95	0.60
CSO 3	0.13	0.63	0.61	0.39
CSO 4	0.36	0.71	0.54	0.49
CSO 5	0.21	0.58	0.53	0.64

From the five *CSO* identified emerges as each of them has a well-defined profile (Fig.1).

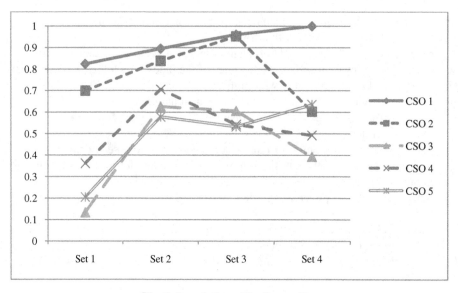

Fig. 1. Description of the five profiles

In particular, the *CSO* 1 presents high levels of hardship with respect to all 4 Set of indicators; then it will be a "prototype" of the family whit *hard economic and social conditions (high hardship)*. The *CSO* 2 also reflects a prototype of a family in conditions of *high hardship* as regards the first 3 Set of indicators; however they haven't difficult in Set 4 (difficulty in participating in events: social, religious, sporting, political, voluntary, or cultural). The *CSO* 3 concerns families who have low levels of hardship in all Set of indicators, thus showing a prototype of a *prosperous family*. The *CSO* 4, however, shows the average levels of hardship in all sets of indicators, except in Set 3 (difficulty in purchasing other goods and services) which doesn't show any difficulty. Finally, the *CSO* 5 also presents a *low profile of hardship* from the economic point of view (first 3 Sets of indicators) and slightly high regard on Set 4 (difficulty in participating in events: social, religious, sporting, political, voluntary, or cultural). It is therefore families in *good economic conditions* but with *some difficulty in social aspects*.

4.2   Identification of Fuzzy Clusters

After identifying the *CSO*, were generated clusters of two types: *Main clusters*, which have the characteristics similar to only one *CSO* and *Fuzzy Clusters*, with similar characteristics to one or more *CSO*, with a different level of membership (probability). The different families are classified in five "Main Clusters" and in nine "Fuzzy Clusters", characterized by profiles derived from a mixture of two or more characteristics of the main five *CSO*.

Table 2. Description of the clusters based on the number of households and the average value of hardship

Cluster	Number of families	%	Value of hardship Set 1	Value of hardship Set 2	Value of hardship Set 3	Value of hardship Set 4
Cluster 1	46	2.6%	0.84	0.87	0.94	0.98
Cluster 2	94	5.4%	0.84	0.88	0.95	0.63
Cluster 3	1	0.1%	0.13	0.63	0.61	0.39
Cluster 4	87	5.0%	0.69	0.78	0.82	0.73
Clusters 5	1	0.1%	0.21	0.58	0.53	0.64
Clusters 2,4	48	2.8%	0.79	0.78	0.89	0.48
Clusters 3,4	17	1.0%	0.12	0.67	0.63	0.33
Clusters 4,2	26	1.5%	0.81	0.74	0.75	0.48
Clusters 4,5	861	49.3%	0.39	0.67	0.65	0.62
Clusters 5,4	386	22.1%	0.21	0.51	0.46	0.80
Clusters 3,4,5	23	1.3%	0.11	0.63	0.59	0.36
Clusters 4,3,5	32	1.8%	0.16	0.65	0.60	0.28
Clusters 4,5,3	82	4.7%	0.22	0.50	0.53	0.28
Clusters 5,4,3	7	0.4%	0.12	0.52	0.60	0.50
Outliers	34	1.9%				
Total	1,745	100.0%				

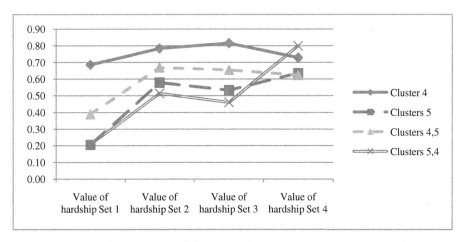

Fig. 2. Description of the profiles of two Fuzzy clusters and their Main clusters

In particular, we represent the profiles of two Main clusters (4 and 5) and two Fuzzy clusters (4.5 and 5.4). The Fuzzy Cluster 4.5 presents an intermediate profile than the two main clusters, with further similarities to the cluster 4, compared to cluster 5. However the fuzzy cluster 5.4 assumes a profile more similar to the cluster 5 compared to cluster 4.

The Table 3 shows the average probabilities membership to different clusters. In particular, the *Fuzzy cluster 3,4,5*, has a higher probability membership to the *CSO* 3 (0.34), then to the *CSO* 4 (0.31) and finally to the *CSO* 5 (0.27).

Table 3. Probability membership to the single clusters (values between 0 and 1)

Cluster	Number of families	Prob 1	Prob 2	Prob 3	Prob 4	Prob 5
1	46	0.32	0.18	0.11	0.21	0.18
2	94	0.12	0.34	0.12	0.23	0.18
2,4	48	0.09	0.31	0.14	0.26	0.20
3	1	0.00	0.00	1.00	0.00	0.00
3,4	17	0.02	0.05	0.37	0.31	0.25
3,4,5	23	0.02	0.05	0.34	0.31	0.27
4	87	0.12	0.20	0.15	0.30	0.23
4,2	26	0.08	0.26	0.15	0.29	0.22
4,5	861	0.04	0.09	0.20	0.36	0.30
4,3,5	32	0.02	0.06	0.31	0.34	0.27
4,5,3	82	0.03	0.06	0.26	0.35	0.30
5	1	0.00	0.00	0.00	0.00	1.00
5,4	386	0.03	0.07	0.20	0.33	0.37
5,4,3	7	0.02	0.06	0.27	0.32	0.33

5   Concluding Remarks

An increasing interest has, in recent years, developed in both the scientific and political fields towards the issue of poverty and, more generally, phenomena of social marginalization. Such studies are however often faced with the lack of specific statistical data. Analyses of poverty are based on surveys of a general nature and often at a "macro" level. It would be of particular interest to perform in Italy, as has already been the case for several years in the United States, surveys of "micro" areas.

The present analysis has attempted to quantify the influence of income and of family typology (number of members) in order to understand how family lifestyles may evolve. The risk of poverty estimates based on "objective" indicators, such as income or levels of debt are completely independent of the state of awareness of those directly involved. It is, however, also useful to observe the "subjective" perception of Italian people in relation to their standard of living and to the recurring causes of economic and social hardship.

The study presented seeks to overcome old classifications between the poor and non-poor by creating "blurred" profiles between those living in different circumstances. Through the two different applications carried out in this work it is possible to:

- Analyse situations of family hardship through the synthesis of multi-dimensional sets of indicators (*Total Fuzzy and Relative method*);
- Identify *Cluster Supporting Object (CSO)*, points with particular characteristics around which build Main clusters and Fuzzy Clusters.
- Create "fuzzy profiles" highlighting the specific peculiarities of small groups not strictly belonging to a defined profile but to a mix of different profiles, having different levels membership to the Main clusters.

It is hoped that the variations regarding the new family profiles emerging in general from analyses carried out with different criteria may provide a solid basis for not only a more accurate description and understanding of the phenomenon of economic hardship but also for developing indications for social policies that may contrast poverty [2].

References

1. Fu, L., Medico, E.: FLAME: a novel fuzzy clustering method for the analysis of dna microarray data. BMC Bioinformatics **8**(3), (2007). doi:10.1186/1471-2105-8-3
2. Montrone, S., Perchinunno, P., L`Abbate, S., Zitolo, M.R.: The lifestyles of families through fuzzy C-means clustering. In: Murgante, B., Misra, S., Rocha, A.M.A., Torre, C., Rocha, J.G., Falcão, M.I., Taniar, D., Apduhan, B.O., Gervasi, O. (eds.) ICCSA 2014, Part III. LNCS, vol. 8581, pp. 122–134. Springer, Heidelberg (2014)
3. Fabbris, L.: Analisi esplorativa di dati multidimensionali, Cleup editore (1990)
4. Green, P.E., Frank, R.E., Robinson, P.J.: Cluster Analysis in text market selection. Management science (1967)
5. Jardine, N., Sibson, R.: Mathematical Taxonomy. Wiley, London (1971)
6. Chattopadhyay, S., Pratihar, D.K., De Sarkar, S.C.: A comparative study of fuzzy c-means algorithm and entropy-based fuzzy clustering algorithms. Computing and Informatics **30**, 701–720 (2011)
7. Ma, P., Chan, K.: Incremental Fuzzy Mining of Gene Expression Data for Gene Function Prediction. IEEE Trans. on Biomedical Engineering (2010)
8. Zadeh, L.A.: Fuzzy sets. Information and Control **8**(3), 338–353 (1965)
9. Dubois, D., Prade, H.: Fuzzy sets and systems. Academic Press, Boston, New York London (1980)
10. Cheli, B., Lemmi, A.A.: Totally fuzzy and relative approach to the multidimensional analysis of poverty. Economic Notes **24**(1), 115–134 (1995)
11. Cerioli, A., Zani, S.: A fuzzy approach to the measurement of poverty. In: Dugum, C., Zenga, M. (eds.) Income and Wealth Distribution, Inequality and Poverty, pp. 272–284. Springer Verlag, Berlin (1980)

Green Marketing and Sustainable Development: A Statistical Survey on Ikea Customers' Perception

Pierlugi Passaro[1], Paola Perchinunno[1(✉)], and Dario Antonio Schirone[2]

[1] DISAG, University of Bari, Via C. Rosalba 53 70100, Bari, Italy
{pierluigi.passaro,paola.perchinunno}@uniba.it
[2] Law Department, University of Bari, P.zza Umberto I 70121, Bari, Italy
darioschirone@libero.it

Abstract. The definition and adoption of practices "sustainable" at economic, social and environmental level, is an important tool to respond to multiple pressures from different stakeholders operating at multiple levels (local, national and supranational). The idea of sustainability implies a crucial change in the relationship between the company and the community (understood in a more or less wide) belongs to. In order to assess whether the adoption of a socially responsible conduct has a positive impact on turnover of manufacturing enterprises, this works refers to the multinational IKEA, among those working in the field of furniture.

Keywords: Green marketing · Environmental sustainability · Statistical survey

1 Introduction

When we speak about Sustainability from a scientific-economic point of view, we refer to the condition of development, able to assure the satisfying of needs of present generation without compromising the possibility of future generations of doing the same. This concept, obviously, is not new, in fact it has been introduced by ONU during the first conference about environment dating back to 1972.

Recently, however, as we can notice by Fig. 1, the attention to eco-sustainable activities is increased exponentially, inciting the whole economy to change the point of view of its own strategic approaches.

Above all the economists have highlighted the environment's belonging to the economy, by diffusing the scarecrow, towards the public authorities and political movements, about the risks of the unrenewable resources' exhaustion, of environment's destruction and of rubbish' uncontrolled increase.

One of the main mistakes to put down to the past economy is just that of having underestimated the importance of human resources and of natural resources in the final production.,

---

The contribution is the result of joint reflections by the authors, with the following contributions attributed to P. Passaro (chapters 1,2), to P. Perchinunno (chapter 4), to D. A. Schirone (chapters 3, 5).

O. Gervasi et al. (Eds.): ICCSA 2015, Part III, LNCS 9157, pp. 130–145, 2015.
DOI: 10.1007/978-3-319-21470-2_10

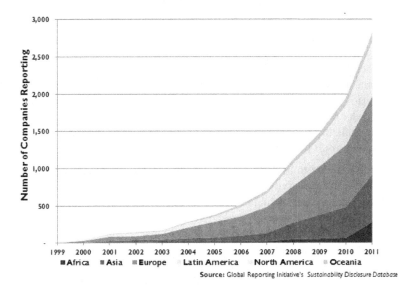

Fig. 1. Worldwide Growth in Business Sustainability Reporting, 1999-2001. Fonte: Global Reporting Initiative's Sustainability Disclosure Database.

The adoption of the objective of sustainable development radically changes the modus operandi towards the productive activities reflecting a new awareness of natural resources' lack and of inevitable repercussion on the whole system.

Since many years by then many firms have adopted strategies of "sustainable" Marketing (or Green Marketing) to be able to exploit at best a competitive advantage deriving from the increase of value proposition from the consumers who are more and more careful to the eco-sustainable politics.

But behind every Dr. Jekyll and Mr. Hyde always hides or, behind several serious and efficient initiatives, companies always hide which try to gain the advantages of this type of politics without actually investing on them, hiding behind false declarations and deceptive advertising: the Greenwashing [1].

2   What We Mean As "Sustainable Marketing"

2.1   Definition of Sustainable Marketing

We attribute the first definition of sustainable Marketing (or Green Marketing) to Donald Fuller who in 1999 defined it as " *The process of planning, implementing and controlling development,pricing, promotion, distribution of products so as to satisfy three criteria: satisfying of consumers' needs, achievement of business objectives and compatibilità with the eco-system"* [2].

The sensitizing to the environmental times is not, then, the only objective for the businesses. The Green Marketing is oriented also to the economic advantage and therefore represents a meeting point between environmental and commercial reality. A *win-win* between both the involved parts.

John Grant, universally known expert of Green Marketing, in his "Green Marketing - Manifesto" has theorized some basic characteristics and fundamental strategies to follow in order to be able to reach effective "sustainable Marketing" actions.

Grant thinks that, to be effective in the medium-long term, the GM action has to mirror what he himself has shown as the *5 Is of Green marketing*:

- **Intuitive**, able to be accessibile, understandable, usual for the consumers the ecological alternatives.
- **Integral,** has to combine the ecological aspect with the commercial, technological and social one. The business results must integrate with those environmental. The consumer has called to act.
- **Innovative**, has to produce a real innovation which creates new products and life styles.
- **Inviting**, has to underline the positive aspect and the attractivity of 'green' choices, releasing thus the environmentalism by the halo of uncomfortableness and sacrifice.
- **Informed**, has to stimulate the environmental education and the active participation of consumers

Always in his *masterpiece* he has identified a device to be able to evacuate which are, more than better or worse, the adequate strategies for each business profile: the matrix of Green marketing. This matrix decides to manage the trade-off among the exact objectives of GM and the levels on which marketing can generally act. On the horizontal axis of the matrix we can see reported the different objectives of GM:

- **Green:** Setting new standards for products, services, brands and responsible firms. It deals with the classical marketing applied to products and firms which pursue a major sustainability comparing to substitutes and competitors and which, for this reason, fix new standards by adopting specific politics, sensitive about control and verify. This type of activity only pursues the first of GM objectves, the commercial one. Communication must focus above all about the firm itself and the activities 'that are effectively coming about in an optics of great truthfulness and integrity'.
- **Greener:** sharing the responsibilities with consumers. It refers to the new marketing 2.0, where the traditional advertising leaves room to the new devices such as the events linked to the brand, the social networks, the communities and the word of mouth. We pass to a more cooperative approach with users and consumers in which the keyword becomes *"share with passion"* or creating participation and enthusiasm about the products. This strategy not only pursues the commercial objective but also the environmental one: reducing the barriers between clients and firm with the consequent possibility of being able to modify the products' directions for use.

- **Very green:** sustaining the innovation (proactive *green* marketing); new habits, new services, new business model. It deals with operating a real cultural revolution, starting from what people consume every day and arrivingS to introduce innovative products, services and habitus in an optics of normality. The marketing devices will have the task to comunicate new ways of living, radically different and surely better than the present ones, letting them perceive as accessible and familiar, as a new everyday life. This approach based on plannig and cultural creation answers all of the three GM objectives: commercial, cultural and environmental.

On the vertical axis, instead, the matrix sets three levels on which the society simultaneously acts:

- **Firms & markets:** nowadays it is the consumer to choose the firm, to look behind the brand to evaluate the reliability of its credentials, its political choices and its values. The figure of the consumer has changed, is more aware and has more knowledge of the product and of what exists behind the producer. We have passed from the brand power to the consumer power towards the brand.
- **Social Brand & identity':** People do not love to be easily labelled with regard to the purchase of a particular product and are more aware in comparison with the image communication, knowing how to separate the value of a success advertisement compared to the goodness of a particolar brand. The brand's meaning passes today from people, no more from advertising, but above all from the truthfulness of social meanings and of branding which are communicated.
- **Personal products & habits:** Most of the today's purchase behaviours start from the satisfying of practical needs, personal tastes and daily habitus and not from the research of status or visibility This type of purchase is mainly driver by habitual decisions and not by the will of changing. It is in this direction that we can try to act to insert new sustainable proposals [3].

Developing our matrix we will have a table 3x3 which will help us in the research of the more appropriate strategies to use during our campaign towards the sustainability.

2.2 Some Numbers

Only in the USA the investments in environmental sustainability and responsibility have increased to a 76% in the last two years, depending on the SIF Foundation (The Forum for Sustainable and Responsible Investment). This result is perhaps the best in the last ten years and helps us to understand how the sustainable investment strategies have nowadays become mainstream.

The foundation itself in its "Report on US Sustainable, Responsible and Impact Investing Trends 2014" has identified investments equal to $6.57 trillion at the beginning of 2014, compared to only $3.74 trillion measured at the beginning of 2012.

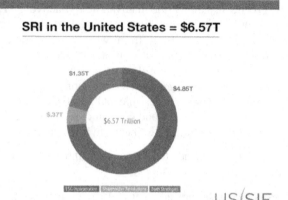

Fig. 2. Report on US Sustainable, Responsible and Impact Investing Trends 2014. Fonte: US – SIF Foundation.

In Italy, instead, from a recent research guided by professors Giovanni Maria Garegnani and Angelo Russo (University LUM Jean Monnet) and Emilia Merlotti (University Bocconi) on a sample of 248 Italian businesses quoted at MTA (Actionary Telematic Market), an increasing trend emerges as concerns the will of Italian companies towards investments in sustainability. Particularly the 60% of companies does its best in a continuous updating of sustainability devices, such as balance sheets and ethical codes.

Particular emphasis is noticed in the companies of energy and utility sector (83%) and of ICT sector (70%). At the same time, increasing attention is devoted to specific themes such as the integration of sustainability either at source or downstream of the production chain: respectively 89% and 83% of the sample declares particular attention to the use of devices and codes related to the management of relations with suppliers and consumers. Besides, the Italian companies go on supporting investments for workers' rights and conditions protection (87% of sample companies), as well as new elements compared to the past emerge with reference to the safeguard of women' rights and equal opportunities in the companies ranks (85%). Finally, great importance rises with reference to licence-to-operate compared to the local context, where 94% of companies declares new attention with regard to the past about themes of safeguard and improvement of the local context where it acts.

«Our yearly workshop has been the occasion to recap the subject of corporate sustainability (CS) in the light of impact of economic crisis on company system, trying to identify, even on the basis of quantitative data supplied by the presented research, the thematic trends of interest for the professional and entrepreneurial world. Our sensation is that, despite the crisis has made more and more delimited and shrewd the investments about corporate social responsibility, nonetheless they are by then perceived by top management of the great Italian companies as a structural element and then inalienable in the management of company itself» Massimo Boidi, President of Synergia Consulting Group [4].

2.3   Greenwashing

As well as it happens with the Free rider in bank sphere, even in the "Green" economy some companies have exploited, and go on exploiting, the green wave to be able to boast sustainable characteristics, which actually they do not have, and then to make the most of competitive advantages that they would not deserve. This is about the Greenwashing. The Greenwashing is a false ecological marketing and deals with a whole of procedures, above all communication activities, adopted by companies, interested to acquire an ecological reputation, without really having an effect on the environmental sustainability of adopted productive chains or of realized products, with the goal of acquiring more consumers and obtaining advantages in terms of image and then also of turnover. It is then generated by effective but not true communication strategies in the short term: the information contains a false or distorted environmental message which, in a first moment, allows the company to obtain advantages but, subsequently, damages its image and competitiveness on the market [5].

In the USA and in Canada a way has been tried to stop this problem, helping the consumers to protect themselves. Then two similar programmes have been devised which make some indexes available for consumers to avoid to be cheated by false environmental advertisements: one is *"The seven sins of greenwashing"* and the other one is *"GreenWashing Index Scoring Criteria"* [6].

Terrachoice, American company of environmental marketing which, in cooperation with *Underwriters laboratories,* has given rise to a useful programme to analyse the present products on the market, has identified "seven sins" to put down to companies which try to manipulate green. And they are:

1. **Sin of hidden trade off**, is committed suggesting that a product is "green" on the basis of a limited whole of characteristic, without paying attention to other important environmental problems. One type of paper, for example, is not necessarily preferable to another one from an environmental point of view, only because the trees from which it comes, are cultivated and cut according to ecologically correct methods. Other factors in the productive process of paper can be equally or more important, for example the energy consumptions, the the emissions of greenhouse gas, the air and water pollution.
2. **Sin of lack of proof** , is committed each time does some statements about the green nature of a product, which are not supported by data, information or evidences easily verifiable or by an independent certificaions. Common examples are fabrics which declare to contain percentages of recycled materials, without giving any measurable evidence.
3. **Sin of vagueness**, is committed when the statements are so generic and unprecise that their real meaning is not understandable by the consumer. The declaration that a product is 100% natural is an example of this type. Arsenic, uranium and formaldeyde are all 100% natural, but highly poisonous. To be 100% natural does not mean to be necessarily *green*.
4. **Sin of irrelevance**, is committed when the statements can be true, but are irrelevant or do not help the consumer in the selection of ecologically preferable products. The declaration *Cfc free*, for example, is often used to define a product green, despite the fact that *Cfc* (chlorofluorocarbon, believed one of the

responsible of the socalled ozone hole) is nowadays forbidden by law alreday since several years and then all products are *Cfc free*.

5. **Sin of less of two evils**, is committed when the statements can be true inside a specific products category, but tend to divert the consumer from the fact the consumption of that specific product has in its inside a great environmental impact. An example are the sports cars with a powerful engine which consume less than the average of the category.

6. **Sin of fibbing**, is committed when someone does simply false statements. It deals with a nowadays hardly frequent sin. A typical example is the false declaration that a product respects or is compatible with a determined standard, for example of energetic consumptions.

7. **Sin of worshipping of false labels**, is committed when though words, images or symbols, a product give the false impression of support or certification by ad independent subject, typically through false brands.

Il "**The Greenwashing index**" founded by EviroMedia, a company of Austin, Texas, which handles of social marketing, has promoted in cooperation with the University of Oregon School of Journalism and communication the "Greenwashinf index", an interesting website which deals with advertising that uses in its *claims* subjects which regard environment and nature. The goals of the site are to help the consumers to become more acute in evaluate the advertising *claims* reliability, that use nature as means to spread their own products, let the business responsible stick to what they declare in their *claims*.

The aim is of course that of making the consumers aware of advertising illusions in environmental sphere, stimulating their critical sense with regard to the growing mass of "ecological" messages of firms.

In the USA, already, one thinks about an appropriate prescription of the sector, above all because of the continuous grow of number of *borderline* communication campaigns about the theme of greenwashing, especially in America, the USA government, through the *federal trade commission* (principal consumers' protection body) has announced an intervention in the short term.

Even in Italy it begins to be a certain attention to the phenomenon, clear example is that or mineral water "San Benedetto" which will have to pay a fee of 70 thousands euro for having presented in its advertising messages the plastic bottles like "friend of environment" [7,8].

3   Sustainable Development: The Case of Ikea

3.1   Introduction

In order to assess whether the adoption of a socially responsible conduct has a positive impact on turnover of manufacturing enterprises, this works refers to the multinational IKEA, among those working in the field of furniture.

At the base of its strategic orientation, IKEA has as a mission the provision of furniture and furnishings that meet the needs of the users, who want to live in comfortable and functional environments. The offer, therefore, consists of a wide assortment of functional items of furniture and design, available at affordable prices to meet the demands of consumers who have different tastes and different levels of income. The application of competitive prices is the result of an attention to the quality not only in the production of goods, from the raw material supplying to the suppliers' selection, but also in retail.

The strategy, which aims at obtaining an economic result, is realized by minimizing the environmental and social impacts of the carried out activities. The strategic guidelines are defined by the head office and it is the duty of subsidiaries located in different countries to choose how to develop projects and initiatives in relation to the specificity of local contexts.

Within the strategy of the IKEA Corporate Social Responsibility (CSR) some areas are identified (IKEA Report, 2013); they can be divided into three groups:

- The suppliers: IKEA has working relationships with 2150 suppliers, to whom it passes on information and knowledge through training courses on quality, efficiency and environmental protection. IKEA, moreover, also performs directly part of the production, thanks to 35 companies, components of its industrial group Swedwood that are located in 11 countries. Employees of IKEA Purchasing Offices, that are present in 42 locations in 33 countries, regularly go to the supplying companies to follow the production, test new ideas, support partners in carrying out their activities and make controls and inspections [9].
- The environment: IKEA is focused on cost and on the efficient use of resources to avoid waste and harmful emissions; preferably it uses the wood to create its products because it is a recyclable, biodegradable and renewable material, and it certifies its origin from intact natural forests. In addition it forms and involves its employees to respect the environment.
- The social projects: numerous initiatives, concerning both the human resource management and the enhancement of the territories where it operates, have been started. IKEA considers employees fundamental to achieve its mission and therefore it has given way to a project Work-Life Balance to try to ensure them a balanced relationship between work and private life. A part of this project concerns the creation of a direct link between the company and the employees on maternity leave through training (distance-learning) and continuous information on the changes in the organization during their absence.

The evaluation by consumers-customers is very important, as the corporate social responsibility is perceived and measured through internal investigations. Every store, in fact, carries out a survey (for about a week) once every four month in Italy and the interviewees are approximately 1200 for each store.

Confidence in the brand, with an average of 87%, is certainly an indicator of how CSR is communicated and perceived. It is necessary to precise that every store records a different percentage because the difference also derives from the local individual initiatives conducted in socio-environmental context.

An enterprise strategic value has been identified in the sustainable development, so it allows IKEA to contribute to obtain, in the long term, both profit increases and a competitive advantage over competitors.

It is difficult to assess with precision the extent to which the strategies, adopted by the IKEA Group, IKEA Italy in terms of social and environmental protection, affect company turnover.

The operational aspects with which IKEA achieves its own socially responsible conduct are also important: it contributes to the development of the corporate strategy with a view to the sustainable development. The action fields can be identified in the following areas: raw materials, ethic code and IWAY audit, waste management, renewable resources, logistics, diversity management.

3.2   The Ethical Code and the IWAY Audit

The conduct code, called "The IKEA Way on Purchasing Home Furnishing Products" (IWAY), was adopted for the first time in 2000 and defines all the minimum requirements suppliers must meet regarding the legal aspects, the working conditions, the active prevention of the child labor[1] of the forced labour, the environmental protection and the forest management. Activity that is fundamental to ensure full effect to the IWAY requirements is the audit, the control activity which checks, in a systematic, documented, periodic and objective manner, the supplier compliance to the IWAY conduct codes. The checks are mostly programmed by mutual consent, but they can also be made "by surprise".

In order to exert a capillary and effective control, several subjects have been identified; they carry out checks on different levels. At a first level, the subjects responsible for carrying out audits are qualified personnel, within the same IKEA, specially formed to carry out this delicate task. At a second level, IKEA internal technical staff is involved but they belong to an international body, to calibrate the work of the IKEA auditor in different countries and in different areas. Finally, at a third level, external specialised company auditors act to give reliability and veracity to the system. After having inserted IWAY standards in all contracts, there is the operational phase of document verification and control, to ascertain the supplier compliance to the conduct code. During the last auditing activities carried out by IKEA were made 35 audits among environmental managers, pressmen, cleaning firms of and/or services, outsourcing and maintenance companies. Of these, 19 are those "approved by IWAY", 4 those "not approved", while the other 12 are still "in progress" or better during the phase of conformity assessment. All the suppliers who have been considered not suitable have been excluded from the professional relationship with IKEA [10].

---

[1] In order to comply with the regulations regarding the children's protection and against the child labour IKEA introduced in 2000 a special ethic code called: "The IKEA Way on Preventing Child Labour (conduct code on the prevention of child labour), to make sure that neither suppliers nor subcontractors employ minor workers).

3.3   Renewable Energy Source Supply and the Transports and Logistics

In order to reduce the direct emissions of carbon dioxide, starting from 2007 IKEA has formalized its commitment through the international project "IKEA goes renewable" which aims "at reducing energy consumption and at supplying energy from renewable sources" [2].

The renewable energy sources are the inexhaustible ones (unlike coal, fuel oil and methane), that is to say the solar energy, the waterpower, the wind power, the geothermal and biomass energy. The project aims at achieving the following long-term targets: -100% of energy acquired and/or produced from renewable sources; - 25% of consumption reduction on the basis of the indicator kWh/ m^3 of sold goods compared to 2005. The environmental targets are both to optimize the management and the requalification of already existing stores, and to plan new stores.

The cost is quantifiable in about 32 million Euros: between 2007 and 2010 about 12 million Euros were spent in order to qualify the existing buildings and the new ones from the energy efficiency point of view; and about 20 million Euros were allocated to provide 150,000 photovoltaic solar modules made of amorphous silicon in all the stores.

Even the logistics and transport modes affect CO_2 emissions in the environment, air pollution and congestion on the motorway network. To verify the effectiveness of logistical choices, IKEA Distribution uses various indicators, including the "average coefficient of filling the truck" for road transport; during the 2012 this coefficient had been decreasing slightly, mainly because of the increase in the direct deliveries from supplier to customer. On the other hand, direct deliveries avoid a passage of the supply chain (not passing from a warehouse), and this means less mileage, less emitted CO_2 and lower costs of stock of goods in the warehouse. The transport strategy adopted in a prevalent manner is the mixed-mode transport of the goods that, depending on the different cases, involves different means of transportation by rail and road, or even by sea [11].

The carriers are subject to continuous inspections by "IKEA Transport". In order to reduce the LGV environmental impact and to apply the due safety criteria, IKEA requires carriers some essential requirements which can be summed up in two main points:

1)   To use means that are less than 10 years old;
2)   To reach a minimum score in the test of the environmental profile. In other words, it verifies if drivers receive an appropriate training for a efficient driving; the fuel consumption of each driver is checked and its reduction is encouraged through bonus; the employed tires are low-rolling resistance; the tire pressure is monitored on a regular basis; the carrier controls and/or limits the time when the engine is on while the vehicle is stationary; the carrier communicates and verifies the actual prohibition of the use of alcohol and drugs during the working hours; the carrier has an environmental management system. If the carrier does not reach the minimum of necessary points it cannot begin to work with IKEA.

---

[2] Extract from "IKEA goes renewable" report published by IKEA in 2007.

The questionnaire should be updated annually and the carrier is subject to control, as for the IWAY audit.

4   Consumer Perception on Environmental Issues

4.1   Introduction

Environmental issues have become over the years a topic of great interest to public opinion, for engineers and scientists, for national and local governments. The scope is quite wide and requires approaches, strategies, different tools depending on the environmental issues discussed. In particular, the need to communicate the environmental risk is closely associated with the need to manage issues related to human activities that have an impact on the health and environmental quality.

4.2   The Istat Survey on Population and Environment

The survey conducted by ISTAT in 2012 on "Population and Environment: behaviors, evaluations and opinions" shows an overall framework negative. In particular, it is evident that only less than half of the population is interested in environmental issues and, therefore, environmentally sustainable behaviors are unable to emerge.

Than in previous years, in fact, at the local level, decreases the perception of environmental risk. The percentage of individuals expressing worries for its proximity to housing of installations potentially harmful is down slightly. Worries of the population are directed mainly to air pollution (indicated by 52% of citizens), to the production and disposal of waste and climate change (both 47%).

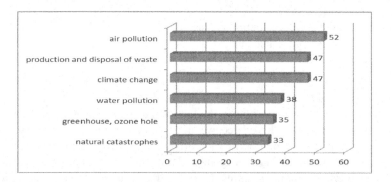

Fig. 3. The main concerns of the Italian on environmental issues

In 2012, fewer than half of the Italian population (46%) said to be interested "much" or "quite" to environmental issues. A restricted segment of the population expresses a high level of interest: people who are interested in "much" environmental issues are indeed about 9%, while the "quite" are 37%.

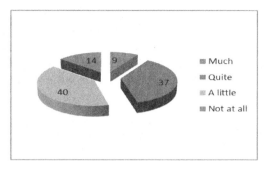

Fig. 4. Percentage of individuals who are interested in environmental issues

The percentage of people who declare be interested "much or quite" to environmental issues in 2012 results to be directly proportional to the level of education: from 26.5% of respondents with primary education, up to 70% for individuals graduates.

Regarding age, the youngest (14-24 years) and the over 75 are less interested (respectively 38% and 27%), compared to those who have between 45 and 54 years (51%) and 55-64 (53%).

As for the territorial aspect, the proportion of citizens who are interested in the environment is wider to the north: 51% in the North Est and 50% in the North Ovest. Despite the growth recorded over the years, the South is still little interest from this point of view (38%) and also the Center is placed slightly below the national standards (43%).

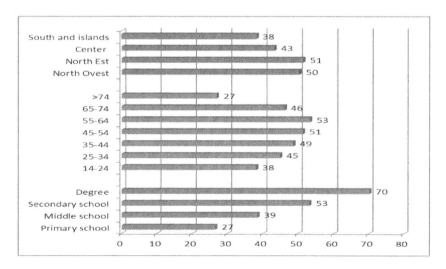

Fig. 5. Percentage of individuals who are interested in environmental issues classified by Geographical distribution, Age and level of Education

Citizens know the environmental issues through a wide range of behaviors, ranging by use of traditional mass media, to the activism and social participation (associations, participation in specialized events, financing of green initiatives). Prevails the use of television and radio programs (85%), reading newspapers and magazines (54%) or specialized books (12%). The percentages are much lower on the active participation, ranging from attending conferences on environmental issues (5% of respondents).

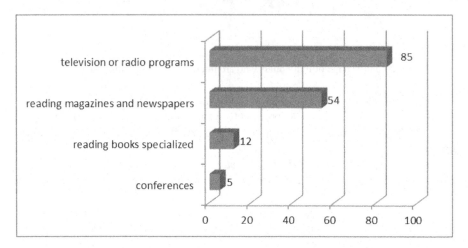

Fig. 6. Percentage of individuals who are interested in environmental issues based on the information source

In the complex emerges, then, as still a significant proportion of citizens do not have strong interests and definite opinions on environmental issues, on the importance of the role played by various actors in the improvement of the environmental situation and on the adequacy of the information. This share is stable over time. It suggests a low level of awareness of environmental issues among the population.

4.3   The Survey of Ikea on the Consumer's Perception

The statistical survey on consumer Ikea Bari was conducted in October 2014 on a sample of customers. The main objective of the survey is to measure what is the perception by the customer Ikea to the environment (both in terms of purchases of personal practice).

First we wanted to test if the customer perceives the attention paid by the company Ikea to environmental issues and if he can recognize it in the products and services offered by the company.

The sample has the following characteristics:

-    is mainly composed of customers aged between 31 and 49 years (60% of respondents);
-    54% of respondents have a high school diploma and 25% have a degree;
-    42% of respondents are represented by people who are married with children and 21% of singles.

People were subjected to a series of questions aimed at detect different aspects.

1st OBJECTIVE: Verify the customer's attention to the environment (both in terms of purchases of personal practice).

On the basis of the answers emerged as the customer Ikea is particularly sensitive to environmental issues. In fact, 82% of respondents said that it wouldn't buy a product from a company that doesn't respect the environment and 87% of respondents believe that the use of best practices and respect for the environment influences the quality of the product (much or quite).

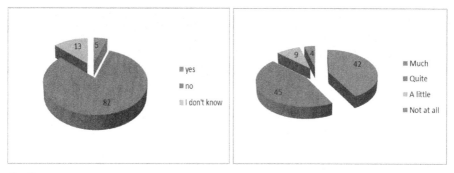

Fig. 7. a) Would you buy a product from a company not careful and respectful of the environment? b) The use of good practice and respect for the environment influences the quality of the product?

2nd OBJECTIVE: To assess customer perception on environmental policies of Ikea

As for the perception that the customer has on Ikea's environmental policies, emerges as the 72% of respondents believe that the commitment of Ikea in implementing sustainable environmental practices is high; less elevated is the perception about the creation of initiatives by Ikea in favor of the local community.

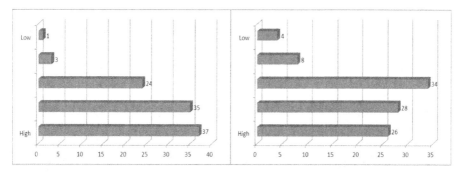

Fig. 8. a) How do you assess the commitment of Ikea in implementing sustainable environmental practices? b) How do you assess the commitment of Ikea in favor of the local community?

3rd OBJECTIVE: Verify the customer's ability to identify the attention paid to the environment by Ikea

The survey shows that 87% of respondents recognize the care taken from Ikea on the environment; however this is especially true for furniture (63%) and to a lesser extent for the accessories (13%) and Ikea-food (11%).

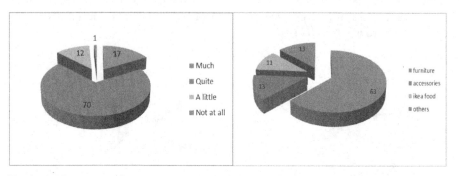

Fig. 9. a) When you are in a store Ikea fails to recognize the attention paid to the environment? b) To which products believes that there is, by Ikea, greater attention to the environment?

5   Concluding Remarks

In conclusion, the growing environmental sensitivity of consumers has necessarily persuaded the professionals to give more and more prominence to the planning of their own advertising campaigns, to the environmental compatibility' of offered products or services. The environmental or green claims, directed to suggest the less or reduced environmental impact of product or service, have, then, become a powerful marketing device able to have a significant effect on the purchase choices of consumers.

As we know that, by now, market is almost undifferentiated and companies try anyway to be able to boast a distinctive element inside their own brands, the claims that describe or evoke a quality worth to distinguish the offered product under a profile which is positively evaluated by consumers, a minimum duty by the professionals who want to use these merits in their own marketing politics, is that of presenting them in a clear, true, accurate way, neither ambiguous nor deceptive. This burden implies that the environmental claim must be reliable and verifiable and, then, not utilized in a generic, indemonstrable way, then lacking in precise scientific and documentable confirmations.

Anyway the interest shown about this subject must be positive, for example, according to WWF within 2030 we will need 3 planets, if we do not intervene in trends of use of resources, due to the today consumption and production styles. For this reason it is very important to teach people, to drive the planet to sustainable activities, not so much for our generation as for the future ones. "Business is the only mechanism on the planet today powerful enough to produce the changes necessary to reverse global environmental and social degradation".

At the base of its strategic orientation, IKEA has as a mission the provision of furniture and furnishings that meet the needs of the users, who want to live in comfortable and functional environments. The offer, therefore, consists of a wide assortment of functional items of furniture and design, available at affordable prices to meet the demands of consumers who have different tastes and different levels of income. The application of competitive prices is the result of an attention to the quality not only in the production of goods, from the raw material supplying to the suppliers' selection, but also in retail.

The strategy, which aims at obtaining an economic result, is realized by minimizing the environmental and social impacts of the carried out activities. The strategic guidelines are defined by the head office and it is the duty of subsidiaries located in different countries to choose how to develop projects and initiatives in relation to the specificity of local contexts.

By the survey conducted emerges as the customer IKEA is careful to environmental policies, believes that the quality of the product is influenced by attention to the environment, considers that the commitment of Ikea in implementing sustainable environmental practices is high, especially with regard to production of furniture.

Moreover, he knows, but not perfectly, what are the initiatives of IKEA in favor of the local community and not fully recognizes the attention paid by Ikea to the environment, especially with regard to accessories, food products, and more.

References

1. Grant J.: Green Marketing; Il Manifesto, p. 7. Brioschi Editore (2009)
2. Lambin, J.J.: Market-driven Management. Marketing Strategico e Operativo, McGraw Hill (2004)
3. Masi, D.: Go Green: Il nuovo trend della comunicazione, Logo Fausto Lupetti Editore (2010)
4. Citterio, A., Migliavacca, S., Pizzurno, E.: Impresa e ambiente: un'intesa sostenibile. Strategie, strumenti ed esperienze, Libri Scheiwiller (2009)
5. Koudate, A., Samaritani, G.: Eco – eco management. Sinergia tra ecologia ed economia nell'impresa, Franco Angeli (2004)
6. Iraldo, F., Michela, M.: Green Marketing. Come evitare il Greenwashing comunicando al mercato il valore della sostenibilità. Il Sole 24 Ore (2012)
7. Hawken, P., Lovins, A.B., Lovins, L.H.: Capitalismo naturale. Edizioni ambiente (2011)
8. Levinson, J.C., Horowitz, S.: Guerrilla Marketing diventa green, Brioschi Editore (2010)
9. De Felice, I.M., Schirone, D.A.: The Exchange of Knowledge: a Case Study. Advances in Management & Applied Economics 3(2), 67–82 (2013)
10. IKEA: Social and Environmental Responsibility, Report IKEA, 2013 (www.ikea.com)
11. Schirone, D.A., Torkan, G.: New transport organization by IKEA, An example of social responsibility in corporate strategy. Advances in Management and Applied Economics 2(3), 181–193 (2012)

GIS-Based Multi-Criteria Decision Analysis for the "Highway in the Sky"

Maria Rosaria Guarini[1(✉)], Marco Locurcio[2], and Fabrizio Battisti[1]

[1] Department of Architecture and Design (DIAP), Faculty of Architecture,
"Sapienza" University of Rome, Via A. Gramsci 53 00197, Rome (RM), Italy
{mariarosaria.guarini,fabrizio.battisti}@uniroma1.it
[2] Doctoral School in Architecture and Construction (DRACO),
Department of Architecture and Design (DIAP), Faculty of Architecture,
"Sapienza" University of Rome, Via A. Gramsci 53 00197, Rome (RM), Italy
marco.locurcio@uniroma1.it

Abstract. This paper describes a methodology for analyzing the feasibility of a system of minor airports designed to increase the use of air travel for short trips in Italy. The study starts with an analysis of Italian territory and the positioning of its minor airports; the database of the Geographical Information System (GIS) provided by Italy's National Institute of Statistics (ISTAT) is used to describe the country's main geographical and economic features. A multi-criteria decision analysis (MCDA) is developed for processing the dataset and creating a platform shared by all stakeholders. Thanks to the Spatial Multi-Criteria Analysis (SDSS) arrived at, it is possible to localize the best position for new minor airports and to formulate strategies for improving existing ones.

Keywords: Multi-Criteria decision analysis · Decision support system · Geographical information systems · Spatial decision system · Minor airports

1 Introduction

The National Transportation Plan (2012) states that the main network of the Italian airport system should consist (as of 2015) of 42 airports of national importance that can be divided, based on, among other factors, their volumes of traffic, two distinct categories of facilities: primary (24) and service (18) airports [1]. Also of note is the link with the framework identifying the TEN-T European network[1], whose Italian air transport sector includes 33 of these national airports, distributed, in accordance with the system of classification and organisation of the European Union airports, on two

---

The contribution is the result of the joint work of the authors. Although scientific responsibility is attributable to all three authors, paragraph nos. 1 and 2 were written by M.R. Guarini; the abstract and paragraphs nos. 4 and 5 were written by M. Locurcio, and paragraph no. 3 was written by F. Battisti.

[1] The National Transportation Plan (2012) refers to the European Union proposal COM (2011) 650, subsequently approved as EU Regulation 1315/2013 for the development of a trans-European transportation network.

© Springer International Publishing Switzerland 2015
O. Gervasi et al. (Eds.): ICCSA 2015, Part III, LNCS 9157, pp. 146–161, 2015.
DOI: 10.1007/978-3-319-21470-2_11

levels: 1) a comprehensive network, to be established by 2050; 2) a core network, to be established by 2030, meant to serve as the backbone of the trans-European air transportation network.

While these airports of national importance handle the bulk of the aircraft and passengers involved in public commercial air transport, the Italian air transportation system also includes the infrastructures used for working flights, civil defence, business aviation, air-taxi services, leisure and tourist activities etc. (54 airports not open to commercial traffic, plus 250 airfields, 460 helipads and 3 seaplane bases). If effective initiatives are taken, these infrastructures can play an increasingly significant role in local economic development.

The above scenario places particular importance on the use of aircraft as a widespread "individual" means of transportation [2], constituting an alternative to roadway and railway modes of travel for medium-range trips (roughly between 100 km and 500 km) made for a variety of reasons, including business and tourism-leisure activities. This type of travel should utilise the "minor" airports, meaning facilities that fall between simple airfields, on the one hand, and the airports of strategic or national interest, on the other, with this last group obviously handling the bulk of passenger flows. A well structured and distributed network of "minor" facilities (compared to the network of nationally important airports) throughout Italian territory could allow "limited" flows of passengers to arrive in relatively out-of-the-way parts of the country that are currently hard to reach, and not only by air travel, serving as an economic stimulus for such areas.

The use of aircraft for individual transportation has become increasingly widespread, especially in the United States, where 150,000 aircraft and 6,000 airports are in operation, though the trend can also be seen in many European countries, including Germany, France and England.

The sharp increase in many countries in the use of aircraft for individual transportation can be traced to the growing presence on the aircraft market of new models of "latest generation" planes, such as advanced ultra-light planes (UL) and very light jets (VLJ) whose "relatively" limited weight and cost put them within the possibilities of new segments of users who, in the eyes of industry operators, will continue to expand in the future.

The association I.LAN (Italian Light Airport Network), founded in 2006, is promoting the establishment of this network of minor airports in Italy, so that it can operate in parallel with, and as a complement to, the airports of national importance, allowing business and leisure fliers who utilise advanced ultra-light planes (UL) and very light jets (VLJ) to reach a noteworthy number of provincial destinations in Italy, as well as the rest of Europe, on a point-to-point basis [3].

The development of individual air transportation in Italy has been hindered by regulations and procedures tied to the characteristics and requirements of commercial air traffic and passengers, which call for structural facilities (arising from prerequisites indispensable to the operation of the aircraft involved: landing, taxiing, take-off, mobility on the ground) and services (connected with the resources needed to accommodate the users frequenting a given facility) that obviously differ to a significant extent from those hold to be necessary, or requested, by operators involved in air traffic and passenger flows at minor facilities.

Therefore, the development and expansion of "individual" air traffic within Italy is closely tied to the establishment of a network of second-level airport infrastructures for civil aviation (minor airport) both adequately distributed throughout the national territory and efficient, accessible and interconnected with other modes of transport, in addition to being easy to use.

With regard to the existing situation, as outlined above, research has been undertaken, at the request of, and in collaboration with, DEMETRA (Development of European Mediterranean Transportation)[2], in pursuit of the general objective of identifying elements that could be of use in supporting the decisions that must be made to redesign Italy's system of minor airports in such a way as to establish a network of second-level airport infrastructures (the destinations of the so-called "highway of the sky") that is integrated with the general airport and infrastructure system, all in support of the sustainable local development of the different areas of Italian territory.

The research is organized in successive phases:

• Phase 1: identify potentiality and problems of the minor airport system current configuration (2014); expected results (basic and reference to set the subsequent phases): i) identify the characteristics needed to qualify a minor airport's resources and consequently identify the reference characteristics for voicing a value judgement for each airport (existing or planned) regarding the infrastructures and services that meet the needs of users; ii) establish status quo related to 52 Italian minor airport's resources (structural and services) through the construction and implementation of a multi-criteria evaluation procedure;

• Phase 2: define the actions must be taken to built an interface for data processing and their geo-referenced restitution and, in particular, developing integration between the multi-criteria methodology formulated in the first part of the research and GIS instruments; expected results (necessary for implementing subsequent research phase): i) create an integrated database system that can be managed through a single geo-referenced geographical rendering database interface for processing and geo-referenced return of data; ii) geo-reference the results obtained in the phase 1 of the research; iii) identify Italian territorial areas "covered" and "uncovered" by minor airport infrastructure. In the "uncovered" areas must be provided new minor airport to give continuity to the network;

---

[2] DEMETRA (http://www.demetracentrostudi.it) is a Research Centre set up in 2006 with the aim of promoting, in Italy, educational, legal, administrative and economic initiatives in the transport sector, with specific reference to civil aviation. DEMETRA disburses a study grant on an annual basis for in-depth study of a specific issue. In 2013, the study grant was assigned to A. Chiovitti for drafting of a degree thesis on Appraisal and Valuation at the Faculty of Architecture of Sapienza University of Rome (Supervisors: Prof. L. di Paola, Prof. M. R. Guarini). The research's methodological organisation and thesis charts were presented during the first session on 15 December 2014 ("Enhancement of the airport network: from the National Plan to minor airports") of the Legal-Administrative Training Course (9th year), held in Rome at the offices of Italy's National Civil Aviation Authority (E.N.A.C.).

- Phase 3: identify the different configurational scenarios of the complete minor airport network; expected results: i) identify the best location for the individual nodes; ii) appraise the necessary investment and identify the socio-economic, financial and environmental repercussions arising at each individual potential node from construction of the network in the reference geographical areas; iii) define the structure of the network in relation to the time and investment necessary to make it fully operational.

The noteworthy number of factors to be considered in the different phases of the research, in order to resolve the decision-making issues being addressed, as well as the close correlation and interaction of those factors, calls for the use of a methodology of assessment based on Multi-Criteria Decision Analysis (MCDA) [7], an approach that the European Commission also deemed to be a useful evaluation tool when the objective is to express value judgments in complex situations that call for consideration of a diverse, heterogeneous set of factors, as well as attention to the objectives and points of view of one or more stakeholders [8].

The decision-making issues addressed by the research are purely spatial in nature [9]; as a result, and in light of the objectives, the write-ups and the evaluations to be developed in the different phases of the research, it became necessary to draw on the support of a Geographical Information Systems (GIS) with the capacity to acquire, register, analyse, visualise ad provide feedback on the information generated by the geographical (geo-referenced) data [10]. The combined application of a multi-criteria analysis (MCA), together with a territorial analysis utilising geographic information systems (GIS), takes on the features of a multi-criteria spatial analysis (MCDA-SDSS) [11, 12].

The scope of the research and application of the multi-criteria analysis is distinct from that of the GIS tools, but in terms of the international literature, and especially over the last ten years, a number of different articles have been published in Scopus illustrating a very real, and fully functional, integration of these two different to modes for carrying out research and supporting decision-making [13, 14].

All the steps implemented in the first phase of the research prove functional, and can serve as part of the reference framework, when it comes time to develop the subsequent phases, which follow as a consequence of the first.

In the following portion of the present paper, phase 2 of the research, shall be addressed by illustrating how integration was developed between the multi-criteria methodology prepared during phase 1 of the research and the GIS tools, together with a preview of the procedures used to integrate the two systems in the subsequent phase of the research as well; the illustration will specifically cover, in paragraphs: 2) the objectives and motives for the integration of the multi-criteria procedure with the geographic information systems; 3) a summary of the multi-criteria methodology prepared for the research and of the data utilised; 4) a description of the model for the GIS-MCDA integration; 5) the results from phase 2 of the research.

2   Objectives and Motivations for Integrating a Multi-criteria Procedure with Geographic Information Systems

As already noted, the objective of this text is to describe, from a methodological perspective, the operating procedures used to develop the integration between the multi-criteria methodology established in the first phase of a wider-ranging research effort on minor airports and the GIS tools, so as to be able to construct an integrated database system that can be operated with a single interface for the geo-referenced feedback of data, something that will also prove useful in the course of the implementation of the subsequent phases of the research.

Scientific articles having to do with the joint application of a multi-criteria approach (Multi Criteria Decision Analysis - MCDA) and territorial analysis by means of Geographical Information Systems (GIS), or the so-called Multi-Criteria Spatial Analysis (MCDA-SDSS) [11], show that this is a field of research that has come into its own, featuring wide-ranging and extremely significant work, as well as a trend towards growth, both internationally and, to an even greater extent, in Italy, where it has yet to become truly widespread. The number and the variety of specialised reviews in which articles related to MC-SDSS have been published show that this methodology can be used to address different types of decision-making problems in may different spheres of application, with the related experiments demonstrating the benefits that can be obtained through joint use of the two tools [13, 14, 15, 16].

Joint, integrated use of the two methodologies (MCDA and GIS) makes it possible to overcome their respective limitations while improving the effectiveness and reliability of the assessment/decision-making process.

With GIS, geospatial data able to play a fundamental role in the evaluation/decision-making process can be obtained, recorded, filed, analysed, processed and visualised. As for the MCDA, it provides a wide range of procedures and techniques that can be used as support with assessment/decision-making problems in which the multiple, diverse factors to be considered must be structure and estimated in order to examine all the possible alternative solutions.

In short, MCDA-GIS can be described as a process that transforms and combines geographical data and value judgments to obtain information that can prove to be of use in decision-making processes [9, 10, 11,13].

Joint use of MCDA and GIS takes the form of a Spatial Decision Support Systems (SDSS). SDSS are defined as "interactive computer systems designed to support a user or a group of users in achieving a higher effectiveness of decision making while solving a semi-structured spatial decision problem" [16]. Use of SDSS is called for when: explicit geographical components must be dealt with; decision-makers need support to reach their conclusions; data, models and structured tools of evaluation must be used in the decision-making process.

The use of SDSS during the different phases of the research underway on minor airports thus proved appropriate, seeing that both the general and specific objectives called for: the solution of spatial type problems, the formulation of complex decision-making problems, the use of multi-criteria models to structure the assessment/decision-making process, interaction between the decision-making users and supervisors, employment of support methods for spatial decisions, the construction of scenarios pertinent to the spatial decisions to be made, and which are typical of SDSS.

A review of the literature on SDSS points to three levels of interaction, in practical terms, between MCDA and GIS [13]:

1. loose coupling: the two tools can reciprocally use as input data the information found in, as well as the processing operations carried out by, the other; in such cases, the databases and the interfaces of the two tools remain separate;
2. tight coupling: MCDA utilises a separate database, although it is inserted as an integral component of the GIS; in such cases, the two systems share the GIS data interface;
3. full integration: MCDA and GIS share both database and interface; in fact, the existing set of commands and routines in the GIS package is supplemented, through the use of generic programming languages, with specific routines for processing the data pertaining to both systems.

Efficient application of the two systems is arrived at only when they are fully integrated [13].

For the purposes of this work, a "tight" coupling was established, with the possibility, at a later point in time, of creating a user-friendly interface for operation of the MCDA module within the GIS software.

Development of the multi-criteria methodology prepared during phase 1 of the research was implemented in the GIS environment through use of the ArcGIS program. ArcGIS-ESRI offers advanced functions for the management, visualisation and modelling of geographical data, making it possible to perform complex geospatial analyses by interfacing with the primary database management software.

3   The Multi-criteria Evaluation Procedure

An analysis of the context that preceded the development of the multi-criteria evaluation process, carried out during the first phase of the research, highlights all the elements of use in its construction (and implementation); specifically, the analysis of the context, which included interviews with experts[3] in the sector, led to identification: i) structural resources (criteria) essential for the functioning of VL and ULJ aircrafts; services level, in the minor airport, to be included in the network; ii) relevant key variables (sub-criteria) of system of structural and services resources; iii) mutual relations between key variables to establish: hierarchical order, indicators and classification parameters that must be adopted in the evaluation procedure.

---

[3] For the implementation of the contextual analysis and of Stakeholders Analysis has been interviewed a representative sample of 43 persons exponents of the business sector and institutional engaged in legal, administrative, technical and economic sector of civil aviation: i) Pilots, Ground staff and flight crew in domestic airlines and international airlines; ii) Pilots, Ground staff and flight crew in general aviation; iii) Airport directors; iv) Air traffic controllers; v) Executives and Professionals in technical and administrative sector of *Ente Nazionale per l'aviazione civile (ENAC)* and *Società Nazionale per l'assistenza al Volo (ENAV S.p.a)*; vi) Executives and Professionals in the legal and economic sector of *ENAC* and *ENAV S.p.a.*

With regard to the two criteria established, the resources that qualify a minor airport were identified[4]: i) structural: runways, taxiway, aprons (structural sub-criteria); ii) service-related: airport services; air traffic support; passenger services; inter-modal transport connections (service sub-criteria).

Working on the expected results and the analysis of the context, the evaluation procedure, for each of 52 minor airport, has been structured to assign: "structural classification", level of judgment A (JLA): suitable (S), transformable (T), not suitable (NS); "services classification", level of judgment B (JLB): very high (VH), high (H), average (A), low (L), very low (VL); "overall resources classification", level of judgment C (JLC), resulting from the sum of the results obtained in the two previous judgement levels (LGA+LGB): SVH, SH, SA, SL, SVL, TVH, TH, TA, TL, TVL, NSVH, NSH, NSA, NSL, NSVL.

Table 1 and Table 2 provides a succinct description of the procedure followed in classifying airports according respectively to their structural and services resources (JLA, JLB), overall resources (JLC).

With reference to the subdivisions of the Nomenclature of Territorial Units for Statistics (NUTS[5]) adopted at a European level, the results of the evaluation (for each judgment level) allowed: i) rank the airports according to the configuration class obtained; ii) calculate the number of airports that are part of the different classes related to: structural resources, services resources and synthetic judgement configurations; ii) identify subdivision and distribution of minor airports by geographical area (NUTS 1) according to: structural, services, overall resources.

4   Description of the Model for GIS-MCDA Integration

The different phases of the research, as summarised in the introductory paragraph, are all geared towards reconstructing an overall, fully developed framework of those elements that can result, following a multi-level evaluation process, in identification of the best possible locations for the destinations comprising the "highway in the sky" of minor airports[6]. The bulk of the information that must be collected, processed and examined in order to come up with responses during the different phases of the research prove to be highly correlated with data of a spatial and territorial type.

---

[4] It should be noted that because the research is "in progress", following a further opportunity for a new discussion with experts, the sub-criteria identified could be partially different. It is believed however that this possibility is not going to change the model of multi-criteria analysis proposed.

[5] NUTS (Nomenclature of Territorial Units for Statistics), introduced for statistical purposes by the European Union identify 3 levels of territorial subdivision of Member States. As regards Italy, NUTS 1 comprises 5 geographical subdivisions (North-West, North-East, Centre, South and Islands); NUTS 2 comprises 21 units: 19 regions and the two autonomous provinces of Trento and Bolzano; NUTS 3 comprises 107 provinces.

[6] It is important to note the centrality of economic valuation techniques for the selection of projects for sustainable mobility [17].

Table 1. Procedure for structural and classification: Judgement level A, B (JLA, JLB)

Phase A: Judgement level A (JLA)	Creation of the JLA matrix	Identification, for each alternative A.n (minor airport), of the sub-criteria *SA.n* and relative indicators *IA.n* needed to measure the specific resource (structural) parameter (impact) *i(IA.n;A.n)*
		Identification, for each individual indicator, of specific classification indicators (threshold values) pcsp.n(IA.n)
	Attribution of relations among sub-criteria and ind.	Two levels (JLA1 and JLA2) of data aggregation to be entered in the JLA assessment matrix have been identified
	Entry of data in JLA matrix	Collection of data related to each indicator, or of specific resource (structural) parameters i(IA.n;A.n) for each minor airport
		Entry of these data i(IA.n;A.n) in the assessment matrix
	Processing of data entered in the assessment matrix for the "structural resources" classification	Attribution of a specific classification to each airport csp.n(IA.n;A.n),referring to each indicator included in the two aggregation levels, JLA1 and JLA2
		Aggregation of specific classification data csp.n(IA.n;A.n) attributed to each minor airport, with reference to sub-criteria indicators (Tab. 3) for each judgement level JLA1 and JLA2
		Attribution of concise classification to each airport which makes it possible to classify each airport in relation to the judgement level: JLA1: suitable, csnt.a(JLA1); transformable, csnt.b(JLA1); not suitable csnt.c (JLA1); JLA2: suitable csnt.a(JLA2); transformable csnt.b(JLA2)
Structural classification: results judgement level A (JLA)	General structural classification JLA	Combining the data related to the two concise judgement levels JLA1 and JLA2 for each airport, functions (logics) leads to the classifications for each airport with regard to structural characteristics: "suitable", "transformable" and "not suitable"
	Ranking of minor airports according to struct. res.	The number and territorial distribution of minor airports, classified according to the suitability of structural resources: "suitable", "transformable" and not suitable" can be obtained on the basis of the general classification
Phase B: Judgement level B (JLB)	Creation of the JLB matrix	Identification, for each alternative A.n (minor airport), of the sub-criteria SB.n and relative indicators IB.n needed to measure the specific resource (services) parameter (impact) i(IB.n;A.n)
		Identification for each individual indicator IB.n, of the potential resource parameters ppd.n(IB.n)
	Entry of data in the JLB matrix	Collection of data related to the specific service parameters i(IB.n;A.n) of each indicator, for each minor airport
		Entry of these data i(IB.n;A.n) in the assessment matrix
	Stakeholder Analysis (SA)	Creation of two interview forms (Questionnaire 1 and Questionnaire 2) to be submitted to each Stakeholder St.n. interviewed, in order to obtain the level of: • importance attributed to the four (4) sub-criteria p(SB.n;St.n) and 15 indica-tors p(IB.n;St.n) related to the services criterion (JLB) - Questionnaire 1; • satisfaction with regard to the specific service parameters JLB - Questionnaire 2
		Selection of sizeable sample of "Experts" in the aeronautical and airport sectors St.n to be interviewed
		Administration of questionnaires
		Processing of data obtained from interviews
	Processing of data entered in assessment JLB matrix	Attribution of a value classification to each airport csp.n(IB.n;A.n) for each indi-
		Weighting and standardisation cdmn(IB.n;A.n) of value classifications cdmn.n(IB.n;A.n) attributed to each indicator
Classification of services: results judgement level B (JLB)	General services classification JLB	The values related to weighted and standardised value classification cdmn.n(IB.n;A.n) (mathematical average) are aggregated using a general classification function f(cgnr.n). Attribution of one of the five general satisfaction levels corresponds to the concise judgement and JLB result
	Ranking of minor airports according to services resources	Organisation of the minor airports in accordance with JLB services in relation to the attributed satisfaction level, which expresses a concise satisfaction indicator that can be "Very High", "High", "Average", "Low" and "Very Low"

Table 2. Procedure for overall classification of resources. Judgement level C (JLC)

			Results JLB (services classification)				
			Very High	High	Average	Low	Very Low
			(VH)	(H)	(A)	(L)	(VL)
Results JLA (structural classification)	Suitable	(S)	SVH	SH	SA	SL	SVL
	Transformable	(T)	TVH	TH	TA	TL	TVL
	Not suitable	(NS)	NSVH	NSH	NSA	NSL	NSVL
			JLC results (general equipment classification)				

Integration of procedures of multi-criteria evaluation and spatial analysis makes it possible to collect, structure analyse, process, evaluate, provide feedback and visualise both the input and output data, with the option of utilising different layers of geo-referenced maps to compare the performance rankings identified for the various alternatives on the different levels of judgment taken into consideration.

It follows that even the resolution of the types of problems of assessment raised in the first part of the research calls for implementation of the MCDA (as described in paragraph 3) in the GIS environment; in practical terms, this results in the formulation of a Spatial Decision Support System – SDSS [16] in which the alternatives identified also present a geographical significance, meaning that the ranking takes on a spatial characterisation relevant to all of Italian territory.

Implementation of the model in the GIS environment was made possible by the maps and data provided by the Italian national statistics institute (ISTAT)[7]. The maps in question are geo-referenced according to a system of representation consisting of the WGS 1984 datum and the UTM (Universal Transverse Mercator) reference system, zone 32N.

The primary operations leading to the construction of a territorial information system (phase 2 of the research) are summarised below, with respect to the other phases of the research, as illustrated in paragraph 1:

- With regard to phase 1 of the research:

1. analysis of the airport geographical database made available by ISTAT, with information characterising the different airports;
2. integration of ISTAT's airport shape file, which localises a number of minor airports not found in the database, geo-referencing them for the purposes of the present topic;
3. addition of a number of elements meant to describe the airports more accurately (services provided, runway dimensions etc.), draw from sector studies [18, 19] and collected for multi-criteria assessment;
4. addition of a number of elements meant to implement the identified parameters in order to ensure a minimum operational capacity for minor airports, meaning minimum basic characteristics that the airports must possess (characteristics of size, services, infrastructures etc.);
5. identification of activities to be undertaken in order to upgrade to the level referred to under point 4 those airports that lack the minimum operational capacity;
6. localisation of territories that lack the airports needed to ensure the continuity of the highway of the sky, in relation to journey data of aircrafts flight;
7. formulation of a new dedicated map for the GIS-MCDA application [20], based on the parameters identified above.

---

[7] www.istat.it

- With regard to phase 3 of the research:

8. identification, within the territories referred to under point 6, of areas whose geo-morphological characteristics make possible the potential localisation of new airports;
9. determination from the ISTAT database of parameters that can be of use in studying the national territory as a whole (resident population subdivided by age group, altitude, production sites present, infrastructure system etc.);
10. determination of an index of correlation based on geo-statistical analyses [18] and capable of cross-evaluating the parameters found under point 9 with those of point 6 and point 8;
11. a comparative analysis of the index of correlation (point 10) with the areas referred to under point 6 and point 8, in order to determine which of the available localisations is the best;
12. confirmation of the continuity of the highway of the sky, based on the data on the distances flown by the aircraft.

The correlation index introduced under point 10 serves as a function that brings together a series of parameters held to be significant (identified during phase 1), so as to guarantee that the minor airports, as well as the new ones, are actually used. By obtaining dimension-based and economic characterisations of the zones where the projects are to be undertaken, the potential financial sustainability of the initiative (analysed in detail in phase 3 of the research) is guaranteed, limiting the risk of planning and designing pointless "cathedrals in the desert".

The MCDA of phase 1 was implemented by constructing a database and processing the data with Microsoft Excel; the Excel spreadsheet was then connected to ArcGIS using ArcCatalog. This operation is carried out through a three-stage procedure: i) creation of an Open Database Connectivity (ODBC) data source in Windows; ii) reformatting of the data in Excel; iii) connection to the Excel file in ArcCatalog.

Analysing the operations carried out with ArcGIS in detail, it can be noted that the geo-processing made it possible to plan the activities and automate the work flows by assembling ordered sequences of operations, in this was providing for the preparation of the basic information needed for the subsequent processing.

With reference to the evaluation procedure elaborated in the phase 1 of the research, a series of topological overlays were used to generate parameters able to describe the characteristics of the territory (point 4). Spatial analysis made it possible to extrapolate elements indispensable to our models, such as the distances between the infrastructures found within the territory (point 4). The spatial queries and the buffering were used to process the analyses referred to under point 6.

Finally, geo-visualisation with maps made it possible to illustrate the results obtained in an immediate and easily understandable form through a MCDA and GIS joint implementation.

Specifically, the geo-referenced cartographic feedback data has obtained; this allows, for each airport, to visualise: the localisation and distribution of the airports within Italian territory (Fig. 1); the classification related to resources: structural - JLA (Fig. 2); services - JLB (Fig. 3); overall - JLC (Fig. 4).

Fig. 1. Airports localisation within Italian territory in geographical areas (NUT 1), regions (NUT 2) and provinces (NUT3)

Through geo-referenced data elaboration, from overview of the levels of resources found at the airports that currently (2014) constitute the system of minor airports (JLC), the following can be identified:

- numerosity and geographical distribution of existing airports that could potentially become destinations of the so-called "highway of the sky" (JLA minimum classification = transformable);

- areas within italian territory "covered" from minor airport classified transformable (from JLA). These airports, depending on distance radius (200 km) between network nodes considered as optimal by stakeholders, can give continuity to so-called "highway of the sky".

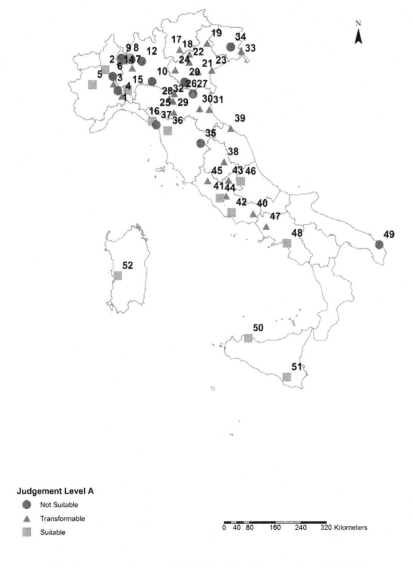

Fig. 2. Results of structural classification: Judgement level A (JLA)

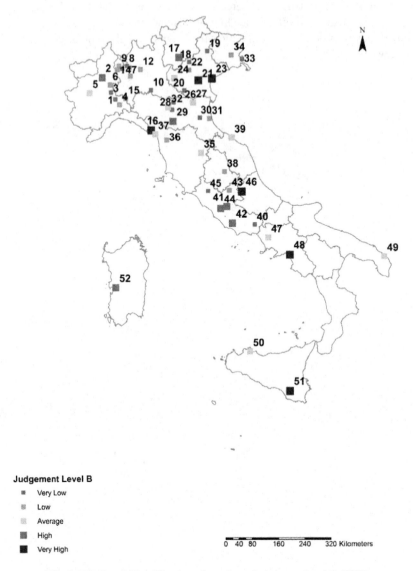

Fig. 3. Results of Classification of services: Judgement level B (JLB)

In particular, the geo-references data elaboration, shows as in the main axes of Italian peninsula: longitudinal along Tyrrhenian and Adriatic coasts and latitudinal along Appennines and major islands; must be identified areas where new minor airports can be built for the low presence/absence of "adequate" minor airports; along Po Valley, where many airport have enough resources, must be identified the nodes which should be adjusted to minimum standards in the research detected.

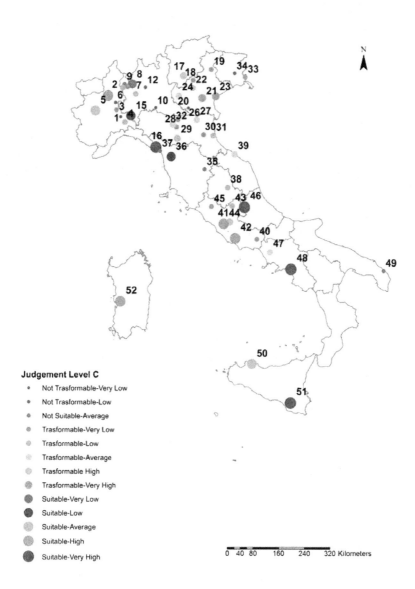

Fig. 4. Results Overall classification of resources. Judgement level C (JLC)

5   Conclusions

The research underway moves along the same lines as the initiatives pursued since 2006 by the Italian Light Airport Network (I.Lan) to develop a widespread and efficient network of minor airports in Italy. In fact, I.Lan was founded with the goal of supporting the creation of a network of airports meant for civil traffic, to be operated in parallel with the network of hubs of national importance (meant for commercial traffic), so as to provide point-to-point connections at a limited cost and in a manner tailored to meet the needs of business and tourism users. As is pointed out in the National Transportation Plan (2012) [1], and more recently in a study carried out by the Italian Union of Chambers of Commerce [21], another reason to preserve and reinforce minor airports is to keep the existing gap between Italy's more economically developed regions and areas and those that are less advanced economically from growing. A further motive for focussing particular attention on the optimisation of minor airports, together with the potential creation of new ones, so as to establish a widespread, integrated network, especially in the wake of the recent, albeit timid signs of economic recovery, is the new need to dispose of rapid connections with the different zones of Italian territory, as well as with the rest of Europe, in this way limiting the costs, together with the consumption of land [22], tied to the creation of fixed, linear infrastructure systems (railway networks, highways etc.) that require significant flows of passengers. The present work formulates a vector data model that integrates the criteria designed to qualify minor airports with specific spatial criteria meant to be used to characterise the territory being examined. The development of the various phases of the research effort is aimed at coming up with a web-GIS capable of monitoring the current state of minor airports, in this way allowing its users to enter information and suggestions of use in the development of the facilities.

The result is a shared platform [9], [20] with which the various stakeholders taking part in the decision-making process, as well as the community affected in various ways by the transformations within the territory, can remain in contact with the decision-makers as the strategies of development are determined.

The process of evaluation established through the integrated use of a methodology of multi-criteria analysis and GIS makes it possible to address the different levels of analysis to be taken into consideration in the course of arriving at a configuration that proves flexible, but also well structured throughout the territory, for the minor airport destinations that must comprise the network of the highway of the sky.

References

1. MIT (Ministero delle Infrastutture e dei trasporti), ENAC (Ente nazionale per l'aviazione civile): Piano nazionale degli aeroporti, febbraio 2012 (2012)
2. Monti, P.: Le Infrastrutture in Italia: dotazione, realizzazione, programmazione. Università Bocconi, Banca d'Italia Eurosistema, Milano (2012)
3. Criscuolo, C.: Aeroporti leggeri: scelta o necessità?. In: Italian Light Airport Network (i.LAN), Roma (2007)
4. Tajani, F., Morano, P.: Concession and lease or sale? A model for the enhancement of public properties in disuse or underutilized. Wseas Transactions on Business and Economics 11, 787–800 (2014)
5. Morano, P., Tajani, F.: Estimative analysis of a segment of the bare ownership market of residential property. In: Murgante, B., Misra, S., Carlini, M., Torre, C.M., Nguyen, H.-Q., Taniar, D., Apduhan, B.O., Gervasi, O. (eds.) ICCSA 2013, Part IV. LNCS, vol. 7974, pp. 433–443. Springer, Heidelberg (2013)

6. Morano, P., Tajani, F.: Bare ownership evaluation. Hedonic price model vs. artificial neural network. International Journal of Business Intelligence and Data Mining **8**(4), 340–362 (2013)
7. Morano, P., Locurcio, M., Tajani, F., Guarini, M.R.: Urban redevelopment: a multicriteria valuation model optimized through the fuzzy logic. In: Murgante, B., Misra, S., Rocha, A.M.A., Torre, C., Rocha, J.G., Falcão, M.I., Taniar, D., Apduhan, B.O., Gervasi, O. (eds.) ICCSA 2014, Part III. LNCS, vol. 8581, pp. 161–175. Springer, Heidelberg (2014)
8. European Commission, Europe Aid Cooperation Office, Evaluation, Methodology, Evaluation tools, Multi-criteria analysis
http://ec.europa.eu/europeaid/evaluation/methodology/tools/too_cri_def_en.htm
9. Malczewski, J.: A GIS based approach to multiple criteria group decision making. Int. J. of Geogr. Syst. **10**(8), 955–971 (1996)
10. Malczewski, J.: A note on integrating Geographic Information Systems and MCDA. Newsletter of the European Working Group Multiple Criteria Decision Aiding. **3**(21), 1–3 (2010)
11. Malczewski, J.: GIS and Multicriteria Decision Analysis. John Wiley and Sons, New York (1999)
12. Attardi, R., Cerreta, M., Franciosa, A., Gravagnuolo, A.: Valuing cultural landscape services: a multidimensional and multi-group sdss for scenario simulations. In: Murgante, B., Misra, S., Rocha, A.M.A., Torre, C., Rocha, J.G., Falcão, M.I., Taniar, D., Apduhan, B.O., Gervasi, O. (eds.) ICCSA 2014, Part III. LNCS, vol. 8581, pp. 398–413. Springer, Heidelberg (2014)
13. Malczewski, J.: GIS-based multicriteria decision analysis: a survey of the literature. International Journal of Geographical Information Science **20**(7), 703–726 (2006)
14. Greco, S., Ehrgott, M., Figueira, J.R.: Trends in multiple criteria decision analysis. Springer Science & Business Media, Germany (2010)
15. Chakhar, S., Martel, J.M.: Enhancing Geographical Information Systems Capabilities with multi-criteria evaluation functions. Journal of Geographical Information and Decision analysis **7**(2), 47–71 (2003)
16. Sugumaran, R., Meyer, J.C., Davis, J.: A web-based environmental decision support system for environmental planning and watershed management. In: Handbook of Applied Spatial Analysis, pp. 703–718. Springer, Berlin Heidelberg (2010)
17. De Mare, G., Nesticò, A.: Efficiency analysis for sustainable mobility. The design of a mechanical vector in Amalfi Coast (Italy). Advanced Materials Research pp. 931–932, 808–812 (2014)
18. Chiles, J.P., Delfiner, P.: Geostatistics modeling spatial uncertainty. John Wiley & Sons, New York, Chichester (1999)
19. Crossland, M.D., Wynne, B.E., Perkins, W.C.: Spatial decision support systems: An overview of technology and a test of efficacy. Decision Support Systems **14**(3), 219–235 (1995)
20. Prévil, C., St-Onge, B., Waaub, J.P.: Analyse multicritère et SIG pour faciliter la concertation en aménagement du territoire: vers une amélioration du processus décisionnel? Cahiers de géographie du Québec **48**(134), 209–238 (2004)
21. UNIONCAMERE – Camere di Commercio d'Italia: Dal Piano degli aeroporti alle scelte di ruolo: strategie del sistema camerale, Gennaio (2014).
http://www.unioncamere.gov.it/download/3144.html
22. Balena, P., Sannicandro, V., Torre, C.M.: Spatial Multicrierial Evaluation of Soil Consumption as a Tool for SEA. In: Murgante, B., Misra, S., Rocha, A.M.A., Torre, C., Rocha, J.G., Falcão, M.I., Taniar, D., Apduhan, B.O., Gervasi, O. (eds.) ICCSA 2014, Part III. LNCS, vol. 8581, pp. 446–458. Springer, Heidelberg (2014)

A Model of Multi-Criteria Analysis to Develop Italy's Minor Airport System

Maria Rosaria Guarini[1]([✉]), Fabrizio Battisti[1], Claudia Buccarini[2], and Anthea Chiovitti[3]

[1] Department of Architecture and Design (DIAP), Faculty of Architecture, "Sapienza" University of Rome, Via A. Gramsci 53 00197, Rome (RM), Italy
{mariarosaria.guarini,fabrizio.battisti}@uniroma1.it
[2] Doctoral School in Architecture and Construction (DRACO),
Department of Architecture and Design (DIAP), Faculty of Architecture,
"Sapienza" University of Rome, Via A. Gramsci 53 00197, Rome (RM), Italy
claudia.buccarini@uniroma1.it
[3] Faculty of Architecture, "Sapienza" University of Rome,
Via A. Gramsci 53 00197, Rome (RM), Italy
antheachiovitti@gmail.com

Abstract. In the forecasts related to air traffic in the period 2015-2030, as worldwide as in Italy, is expected a substantial increase in air traffic individual. In Italy, the development and increase of "individual" air traffic is closely linked to the construction of a network of second-level airport infrastructures for civil aviation (minor airports), suitably distributed throughout the national territory, efficient, accessible, interconnected with other means of transport and easy-to-use. A network structured in this way can also contribute to the improvement of the conditions of the context of peripheral areas.

The model of multi-criteria analysis proposed is aimed to identify useful elements for supporting the choices needed to redefine Italy's minor airport system so as to construct a second-level airport infrastructure network (nodes of a so-called "highway in the sky") integrated with the general airport and infrastructure system and aimed at supporting local sustainable development of the different areas of Italy.

Keywords: Appraisal · Multi-criteria analysis · Stakeholders · Minor airports · Geographic information system

1 Introduction

The "National Airport Plan" (2012), lists possible doubling of passenger traffic (approximately 300 million) for Italy by 2030 [1]. Part of this increase can be attributed

The contribution is the result of the joint work of the four authors. Although scientific responsibility is attributable equally to M.R. Guarini, F. Battisti and C. Buccarini, the abstract and the paragraph nn. 2, 5.1, 5.2, 7 were written by M.R. Guarini; the paragraphs nn. 1, 3, 5.3, were written by F. Battisti and the paragraphs nn. 4, 5.4, 6 were written by C. Buccarini. A. Chiovitti collected and processed the data used in the evaluation procedure.

O. Gervasi et al. (Eds.): ICCSA 2015, Part III, LNCS 9157, pp. 162–177, 2015.
DOI: 10.1007/978-3-319-21470-2_12

to the widespread use of airplanes as an "individual" means of transport, an alternative to road and rail transport for medium-range travel (indicatively between 100 and 500 km), for a number of purposes including business and sports tourism. This type of travel should involve "minor" airports, medium-size airports positioned between airfields and strategic airports of national interest, where most collective air transport (commercial) vehicles and passengers are directed [2].

Affirmation of the airplane for individual transport is in major expansion in many states thanks to the presence on the market of latest-generation, advanced ultra-light aircraft (UL) and very light jets (VLJ) boasting a "relatively" low cost and weight. The production and supply of ever-new models of UL aircraft and VLJs, with increasingly lower costs and consumption, offers proof of the focus placed by manufacturers on new groups of uses that could be much more numerous in the future [3].

The development of this segment of air transport in Italy is made difficult by legislation and procedures that are still linked to the characteristics of mass collective air traffic (commercial).

Therefore, the development and increase of "individual" air traffic within Italy is closely linked to the construction of a network of second-level airport infrastructures for civil aviation (minor airports). Said network must be suitably distributed throughout the national territory, efficient, accessible, interconnected with other means of transport and easy-to-use.

The association I.LAN (Italian Light Airport Network), founded in 2006, is promoting the establishment of this network of minor airports in Italy, so that it can operate in parallel with, and as a complement to, the airports of national importance, allowing business and leisure fliers who utilise advanced ultra-light planes (UL) and very light jets (VLJ) to reach a noteworthy number of provincial destinations in Italy, as well as the rest of Europe, on a point-to-point basis [3].

Minor airports must have: i) fundamental requisites for operations (landing, runway manoeuvring, take-off, ground mobility) of air carriers (UL aircraft and VLJs); ii) service resources considered valuable by users in order to make use of the airport (airport attractiveness) [4, 5].

Consequently, development of the minor airport system must be envisaged in accordance with a "tendential" network scenario, focused on "diffusion and specialisation", comprising airports that:

- have suitable infrastructures and services for the aircraft and users using them;
- are positioned in relation to the airport network of national standing;
- have the possibility of acting as "boosts" for local development.

Therefore, there is the need to plan the development of minor airports within a network logic that guarantees the possibility of connections between all areas of the country where, at the present time, some marginal areas are completely lacking in suitable facilities, and that takes into account the local vocations of the areas where the individual airports are already located or will be located[1].

---

[1] Generally in careful planning, it is very important the support of appropriate socio-economic and financial evaluation tools, to highlight the growth of an area not only in purely financial terms, but also social, cultural and environmental [6, 7, 8].

In order to understand how to develop and optimise the potential related to the increase of this segment of civil aviation, first, the following need to be examined and assessed with regard to minor airports that are currently operational (2014) in Italy:

- distribution throughout the national territory;
- services and structural resources of each airport;
- improvement/streamlining needs of the network in its current state (2014) in relation to each airport's resources and current distribution throughout the national territory;
- level of accessibility from surrounding area and interconnection with other means of transport.

2   General Aims and Research Phases

The paper presents the procedures selected and implemented for the context analyses and assessment procedure (multi-criteria) adopted to develop the first part of a wider research in progress, launched at the request of and together with Research Centre DEMETRA (Development of European Mediterranean Transportation)[2]. The main aim of the research is to identify useful elements for supporting the choices needed to redefine Italy's minor airport system so as to construct a second-level airport infrastructure network (nodes of a so-called "highway in the sky") integrated with the general airport and infrastructure system and aimed at supporting local sustainable development of the different areas of Italy.

To this end, the research has been organised in the following and different phases:

- Phase 1: identify potentiality and problems of the minor airport system current configuration (2014); expected results (basic and reference to set the subsequent phases): i) identify the characteristics needed to qualify a minor airport's resources and consequently identify the reference characteristics for voicing a value judgement for each airport (existing or planned) regarding the infrastructures and services that meet the needs of users; ii) establish status quo related to 52 Italian minor airport's resources (structural and services) through the construction and implementation of a multi-criteria evaluation procedure;
- Phase 2: define the actions must be taken to built an interface for data processing and their geo-referenced restitution and, in particular, developing integration

---

[2] DEMETRA (http://www.demetracentrostudi.it) is a "Research Centre set up in 2006 with the aim of promoting, in Italy, educational, legal, administrative and economic initiatives in the transport sector, with specific reference to civil aviation". DEMETRA disburses a study grant on an annual basis for in-depth study of a specific issue. In 2013, the study grant was assigned to A. Chiovitti for drafting of a degree thesis on Appraisal and Valuation at the Faculty of Architecture of Sapienza University of Rome (Supervisors: Prof. L. di Paola, Prof. M. R. Guarini). The research's methodological organisation and thesis charts were presented during the first session on 15 December 2014 ("Enhancement of the airport network: from the National Plan to minor airports") of the Legal-Administrative Training Course (9th year), held in Rome at the offices of Italy's National Civil Aviation Authority (E.N.A.C.).

between the multi-criteria [9, 10] methodology formulated in the first part of the research and GIS instruments [11, 12]; expected results (necessary for implementing subsequent research phase): i) create an integrated database system that can be managed through a single geo-referenced geographical rendering database interface for processing and geo-referenced return of data; ii) geo-reference the results obtained in the phase 1 of the research; iii) identify Italian territorial areas "covered" and "uncovered". In the "uncovered" areas must be provided new minor airport to give continuity to the network;

- Phase 3: identify the different configurational scenarios of the complete minor airport network; expected results: i) identify the best location for the individual nodes; ii) appraise the necessary investment and identify the socio-economic, financial and environmental repercussions arising at each individual potential node from construction of the network in the reference geographical areas [13]; iii) define the structure of the network in relation to the time and investment necessary to make it fully operational.

Therefore, the contextual analysis and multi-criteria assessment methodology formulated and implemented in the phase 1 of the research to identify the "potentiality and problems of the minor airport system's current configuration (2014) will be illustrated later in this paper, with regard to the first first part of the research. Specifically, the following will be illustrated: Section (Sec.) 3 - "Specific aims, structure and analysis and assessment methodology"; Sec. 4 - "Contextual Analysis "; Sec. 5 - "Organisation of the multi-criteria assessment procedure" split into "Structural classification. Judgement Level A (JLA)" (5.1), "Classification of services. Judgement Level B (JLB)" (5.2), "Overall classification of resources. Judgement Level C (JLC)" (5.3); Sec. 6 – "Implementation of procedure and first results"; Sec. 7 - "Conclusions".

3   Specific Aims, Structure and Analysis and Assessment Methodology for Determining Potentiality and Problems of Italy's Minor Airport System's Current Configuration (2014)

The first phase of the research referred to herein is aimed at:

- identifying the equipment needed to qualify a minor airport as regards structural resources (necessary requisites for efficiency of the following vehicles: UL aircraft and VLJs) and services (requisites considered valuable by users in order to make use of the airport, hence that make an airport "attractive");
- assessing the level of "suitability" of each of the minor airports in relation to (reference criteria): i) "structural" resources by identifying the level of correspondence to necessary requisites for the accessibility and efficiency of the following vehicles: UL aircraft and VLJs; ii) services resources, identifying the level of correspondence to requisites considered valuable by users in order to make use of the airport (airport attractiveness);
- classifying the minor airports in relation to the level of satisfaction of all the requisites considered essential for each type of resource;

- identifying the minor airports that could be included in the "highway in the sky", considering the combined level of satisfaction achieved with regard to the resources examined.

4   Contextual Analysis

The contextual analysis was allowed to highlighting the main factors and elements, both within and outside the minor civil aviation segment, to be taken into account in order to assess, and possibly plan development strategies and actions. Specifically, the following were examined:

- current scenarios and future growth of air traffic at a national and European level. The analyses were performed in order to check sir traffic trends at both a European and national level [1], [3];
- civil aviation aircraft categories. Specifically, the analysis was aimed at highlighting the technical operating aspects of UL and VLJ aircrafts and hence at understanding what structural characteristics an airport (minor) must have in order to allow for landing and take-off and ground management of these aircraft [2], [5];
- Italy's airport system and minor airports located throughout Italy. In particular, general and specific data (related to resources) of the 52 Italian minor airports have been collected and analyzed [5], [14];

An analysis of the context that preceded the development of the multi-criteria evaluation process, carried out during the first phase of the research, highlights all the elements of use in its construction (and implementation); specifically, the analysis of the context, which included interviews with experts[3] in the sector, led to identification: i) structural resources (criteria) essential for the functioning of VL and ULJ aircrafts; services level, in the minor airport, to be included in the network; ii) relevant key variables (sub-criteria) of system of structural and services resources; iii) mutual relations between key variables to establish: hierarchical order, indicators and classification parameters that must be adopted in the evaluation procedure.

With regard to the two criteria established, the resources that qualify a minor airport facility were identified[4]:

---

[3] For the implementation of the contextual analysis and of Stakeholders Analysis has been interviewed a representative sample of 43 persons exponents of the business sector and institutional engaged in legal, administrative, technical and economic sector of civil aviation: i) Pilots, Ground staff and flight crew in domestic airlines and iternational airlines; ii) Pilots, Ground staff and flight crew in general aviation; iii) Airport directors; iv) Air traffic controllers; v) Executives and Professionals in technical and administrative sector of *Ente Nazionale per l'aviazione civile (ENAC)* and *Società Nazionale per l'assistenza al Volo (ENAV S.p.A)*; vi) Executives and Professionals in the legal and economic sector of *ENAC* and *ENAV S.p.A.*

[4] It should be noted that because the research is "in progress", following a further opportunity for discussion with experts, the sub-criteria identified could be partially different. It is believed however that this possibility is not going to change the model of multi-criteria analysis proposed.

- structural: runways, taxiway, aprons (structural sub-criteria) selected in relation to the ability to express the fundamental requisites related to landing, runway manoeuvring, take-off and management of aircraft within the minor airport;
- service-related: airport services; air traffic support; passenger services; inter-modal transport connections (service sub-criteria).

As regards the large number of aspects to be considered and their close correlation and interaction, it is necessary to adopt an assessment methodology based on Multi-Criteria Analysis (MCA) [9] for the various research phases. The European Commission acknowledges the MCA as a useful assessment instrument to be used in complex situations when diverse and heterogeneous aspects and aims and viewpoints of one or more stakeholders needed to be taken into consideration [10].

Consequently, the procedure has been structured to express judgements with regard to the level of satisfaction of requisites taken as essential (by Stakeholders) in relation to the categories of equipments looked at.

5   The Model of Multi-criteria Analysis

5.1   Structuring of the Evaluation Procedure

Depending on the forecast results and contextual analysis, the assessment procedure has been organised as follows to obtain a:

- Classification of airports according to resources:
  − structural, judgement level A (JLA): "structural classification";
  − services, judgement level B (JLB): "classification of services";
  − overall, judgement level C (JLC): "overall resource classification", resulting from the sum of the results obtained in the two previous judgement levels (JLA+JLB) and expression of the quality of resources found at each airport.
- Ranking of the airports according to the classification obtained.

5.2   Structural Classification: Judgement Level A (JLA)

The assessment procedure for the "structural" classification of airports is as follows:

- Creation of the JLA matrix:
  − imputation, for each alternative $A.n$ (minor airport), of the sub-criteria $SA.n$ and relative indicators $IA.n$ needed to measure the specific resource (structural) parameter (impact) $i(IA.n;A.n)$. In total, 7 indicators were selected depending on the ability to express the connotative requisites of the structural characteristics of minor airports (Tab. 1);

— imputation for each individual indicator, of specific classification indicators (threshold values) *pcsp.n(IA.n)*, obtained from (and depending on) sector regulations as well as from technical data for the categories of aircraft looked at. The specific classification parameters represent the threshold values to be used to check the performance of each airport (alternatives) for each indicator. In this way, specific classification (or value) can be attributed to each of the alternatives *csp.n(IA.n;A.n)* with regard to structural resources, in accordance with three categories: suitable, can be adapted, not suitable (Tab. 1).

- Attribution of relations among sub-criteria and indicators for each alternative (Tab. 1). According with Stakeholders indications, two levels (JLA1 and JLA2) of data aggregation to be entered in the JLA assessment matrix have been identified, distinguishing the characteristics linked to vehicle efficiency in strictly flight-related landing or take-off manoeuvres (JLA1) and hence to the category of aircraft, from ground manoeuvring/stationing characteristics (JLA2).

Table 1. JLA matrix, specific classification (parameters)

Criteria (CA.n)	Sub-criteria (SA.n)	Indicators (IA.n)	Specific resource parameter i(IA.n) for alternative (A.n)	Specific classification indicator (pcsp.n)	Specific classification [csp.n(A.n)]		Levels
Structural requisites (JLA)	SA.1 Tracks	IA.1 Lenght (m)	i(IA.1;A.n)	pcsp.a(IA.1)	>900	csp.a(IA.1;A.n) = suitable	JLA1
				pcsp.b(IA.1)	800≤JA.1≤900	csp.b(IA.1;A.n) = adaptable	
				pcsp.c(IA.1)	<800	csp.c(IA.1;A.n) = not suitable	
		IA.2 Width (m)	i(IA.2;A.n)	pcsp.a(IA.2)	>18	csp.a(IA.2;A.n) = suitable	
				pcsp.b(IA.2)	15≤JA2≤18	csp.b(IA.2;A.n) = adaptable	
				pcsp.c(IA.2)	<18	csp.c(IA.2;A.n) = not suitable	
		IA.3 Accessibility (corner obstacle)	i(IA.3;A.n)	pcsp.a(IA.3)	α	csp.a(IA.3;A.n) = suitable	
				pcsp.b(IA.3)	β	csp.b(IA.3;A.n) = adaptable	
				pcsp.c(IA.3)	γ	csp.c(IA.3;A.n) = not suitable	
		IA.4 Paving (mat.)	i(IA.4;A.n)	pcsp.a(IA.4)	asphalt/tarmac	csp.a(IA.4;A.n) = suitable	
				pcsp.b(IA.4)	grass	csp.b(IA.4;A.n) = adaptable	
	SA.2 Taxiway	IA.5 Width (m)	i(IA.5;A.n)	pcsp.a(IA.5)	>10,5	csp.a(IA.5;A.n) = suitable	JLA2
				pcsp.b(IA.5)	≤10,5	csp.b(IA.5;A.n) = adaptable	
		IA.6 Paving (mat.)	i(IA.6;A.n)	pcsp.a(IA.6)	asphalt/tarmac	csp.a(IA.6;A.n) = suitable	
				pcsp.b(IA.6)	grass	csp.b(IA.6;A.n) = adaptable	
	SA.3 Apron	IA.7 Paving (mat.)	i(IA.7;A.n)	pcsp.a(IA.7)	asphalt/tarmac	csp.a(IA.7;A.n) = suitable	
				pcsp.a(IA.7)	grass	csp.b(IA.7;A.n) = adaptable	

- Entry of data in JLA matrix: imputation of data collected referred to specific resource (structural) parameters (one for each sub-criterion) *i(IA.n;A.n)* for each minor airport *A.n*;
- Processing of data entered in the assessment matrix for the "structural resources" classification:
  - attribution of a specific classification to each airport *csp.n(IA.n;A.n)*, referring to each indicator included in the two aggregation levels, JLA1 and JLA2. Specific classification is performed by applying a specific classification function *f(csp.n)*; comparison between specific resource (structural) parameters *i(IA.n;A.n)* and specific classification parameters *pcsp.n(IA.n)* makes it possible to assign specific classification to each indicator *csp.n(Ia.n;A.n)*, or to classify it as: suitable, can be adapted, not suitable (Tab. 2).

— Aggregation of specific classification data *csp.n(IA.n;A.n)* attributed to each minor airport, with reference to sub-criteria indicators (Tab. 3) for each judgement level JLA1 and JLA2, by applying specific concise classification functions (logics) *f(csnt.n)* for each of the two judgement levels: *csnt.n(JLA1)*, *csnt.n(JLA2)*. In this way a concise classification value is attributed to each airport which can be: for JLA1: suitable, *csnt.a(JLA1)*; transformable, *csnt.b(JLA1)*; not suitable *csnt.c (JLA1)*; for JLA2: suitable *csnt.a(JLA2)*; transformable *csnt.b(JLA2)*

Table 2. JLA matrix, specific classification (function)

Criteria (CA.n)	Sub-criteria (SA.n)	Sotto-criteri (SA.n)	Indicators (IA.n)	Specific resource parameter i(IA.n) for	Specific classification indicator (pcsp.n)	Specific classification function [f(csp.n)]	Specific classification (csp.n)
Structural requisites (JLA)	SA.1 Tracks	SA.1 Piste	IA.1 Lenght (m)	i(IA.1;A.n)	pcsp.a(IA.1)	if i(IA.1;A.n)>pcsp.a(IA.1) =	csp.a(IA.1;A.n) = suitable
					pcsp.b(IA.1)	if pcsp.a(IA.1)≤i(IA.1;A.n) = ≤pcsp.b(IA.1)	csp.b(IA.1;A.n) = adaptable
					pcsp.c(IA.1)	if i(IA.1;A.n)<pcsp.c(IA.1) =	csp.c(IA.1;A.n) = not suitable
			IA.2 Width (m)	i(IA.2;A.n)	pcsp.a(IA.2)	if i(IA.2;A.n)>pcsp.a(IA.2) =	csp.a(IA.2;A.n) = suitable
					pcsp.b(IA.2)	if pcsp.a(IA.2)≤i(IA.2;A.n) = ≤pcsp.b(IA.2)	csp.b(IA.2;A.n) = adaptable
					pcsp.c(IA.2)	if i(IA.2;A.n)<pcsp.c(IA.2) =	csp.c(IA.2;A.n) = not suitable
			IA.3 Accessibility	i(IA.3;A.n)	pcsp.a(IA.3)	if i(IA.3;A.n)>pcsp.a(IA.3) =	csp.a(IA.3;A.n) = suitable
					pcsp.b(IA.3)	if pcsp.a(IA.3)≤i(IA.3;A.n) = ≤pcsp.b(IA.3)	csp.b(IA.3;A.n) = adaptable
			(corner obstacle)		pcsp.c(IA.3)	if i(IA.3;A.n)<pcsp.c(IA.3) =	csp.c(IA.3;A.n) = not suitable
			IA.4 Paving (mat.)	i(IA.4;A.n)	pcsp.a(IA.4)	if i(IA.4;A.n)=pcsp.a(IA.4) =	csp.a(IA.4;A.n) = suitable
					pcsp.b(IA.4)	if i(IA.4;A.n)=pcsp.b(IA.4) =	csp.b(IA.4;A.n) = adaptable
	SA.2 Taxiway	SA.2 Taxiway	IA.5 Width (m)	i(IA.5;A.n)	pcsp.a(IA.5)	if i(IA.5;A.n)>pcsp.a(IA.5) =	csp.a(IA.5;A.n) = suitable
					pcsp.b(IA.5)	if i(IA.5;A.n)<pcsp.b(IA.5) =	csp.b(IA.5;A.n) = adaptable
			IA.6 Paving (mat.)	i(IA.6;A.n)	pcsp.a(IA.6)	if i(IA.6;A.n)=pcsp.a(IA.6) =	csp.a(IA.6;A.n) = suitable
					pcsp.b(IA.6)	if i(IA.6;A.n)=pcsp.b(IA.6) =	csp.b(IA.6;A.n) = adaptable
	SA.3 Apron	SA.3 Apron	IA.7 Paving (mat.)	i(IA.7;A.n)	pcsp.a(IA.7)	if i(IA.7;A.n)=pcsp.a(IA.7) =	csp.a(IA.7;A.n) = suitable
					pcsp.a(IA.7)	if i(IA.7;A.n)=pcsp.b(IA.7) =	csp.b(IA.7;A.n) = adaptable

- General structural classification JLA, *cgnr.n(LGA)*, of minor airports which is obtained through structural classification functions (logics) *f(cgnr.n)*, by combining the datum related to the two concise judgement levels JLA1 and JLA2 for each airport. In this way each airport can be classified with regard to structural characteristics as: suitable [*if csnt.a(JLA1; JLA2)*], transformable [*if csnt.a(JLA1), csnt.b(JLA2); or if csnt.b(JLA1), csnt.a(JLA2); or if csnt.b(JLA1), csnt.b(JLA2)*] and not suitable [*if csnt.c(JLA1)*];
- Ranking of minor airports according to structural resources, JLA results; the number and territorial distribution of minor airports can be obtained on the basis of general structural classification: suitable (S), transformable (T) not suitable (NS).

5.3   Classification of Services: Judgement Level B (JLB)

The assessment procedure for classification of "services" is as follows:

- Creation of the JLB matrix:
  - imputation, for each alternative *A.n* (minor airport), of the sub-criteria *SB.n* and relative indicators *IB.n* needed to measure the specific resource (services) parameter (impact) *i(IB.n;A.n)*. In total, 15 indicators were selected depending on their ability to express the connotative requisites of the quality of services for users (security, recreational, logistic) (Tab. 4);

- imputation for each individual indicator *IB.n*, of the potential resource parameters *ppd.n(IB.n)*, in relation to the quality of the services resources of minor airport services (Tab. 4);
- Entry of data in the JLB matrix: imputation of data collected referred to specific resource (services) parameters (one for each sub-criterion) *i(IB.n;A.n)* for each minor airport *A.n*;

Table 3. JLA matrix, concise classification (function)

Criteria (CA.n)	Sub-criteria (SA.n)	Indicators (IA.n)	Specific resource parameter i(IA.n) for alternative (A.n)	Specific classification (csp.n)	Concise classification function [f(csnt.n)]	Concise classification (csnt.n)
Structural requisites (JLA)	SA.1 Tracks	IA.1 Lenght (m)	i(IA.1;A.n)	csp.a(IA.1;A.n) = suitable		
				csp.b(IA.1;A.n) = adaptable		
				csp.c(IA.1;A.n) = not suitable	if cspc.a(IA.1; IA.2; IA.3; IA.4) =	csnt.a(JLA1) (suitable)
		IA.2 Width (m)	i(IA.2;A.n)	csp.a(IA.2;A.n) = suitable		
				csp.b(IA.2;A.n) = adaptable	if cspc.b(IA.1); cspc.b(IA.2); cspc.b(IA.3); cspc.b(IA.4) =	csnt.b(JLA1) (transformable)
				csp.c(IA.2;A.n) = not suitable		
		IA.3 Accessibility (comer	i(IA.3;A.n)	csp.a(IA.3;A.n) = suitable	if cspc.c(IA.1); cspc.c(IA.2); cspc.c(IA.3) =	csnt.c(JLA1) (not suitable)
				csp.b(IA.3;A.n) = adaptable		
				csp.c(IA.3;A.n) = not suitable		
		IA.4 Paving (mat.)	i(IA.4;A.n)	csp.a(IA.4;A.n) = suitable		
				csp.b(IA.4;A.n) = adaptable		
	SA.2 Taxiway	IA.5 Width (m)	i(IA.5;A.n)	csp.a(IA.5;A.n) = suitable		
				csp.b(IA.5;A.n) = adaptable	if cspc.a(IA.5; IA.6; IA.7) =	csnt.a(JLA2) (suitable)
		IA.6 Paving (mat.)	i(IA.6;A.n)	csp.a(IA.6;A.n) = suitable		
				csp.b(IA.6;A.n) = adaptable	if cspc.b(IA.5); cspc.b(IA.6); cspc.b(IA.7) =	csnt.b(JLA2) (transformable)
	SA.3 Apron	IA.7 Paving (mat.)	i(IA.7;A.n)	csp.a(IA.7;A.n) = suitable		
				csp.b(IA.7;A.n) = adaptable		

- Stakeholder Analysis (SA) to obtain the opinion of "Experts" (Stakeholders)[5], through interviews, as regards the type and level of services that should be found at a minor airport in order to meet users' needs.

The Stakeholders Analysis is split into the following phases:

- Creation of two interview forms (Questionnaire 1 and Questionnaire 2) to be submitted to each Stakeholder *St.n.* interviewed, in order to obtain the level of:
  - importance attributed to the four (4) sub-criteria *p(SB.n;St.n)* and 15 indicators *p(IB.n;St.n)* related to the services criterion (JLB) - Questionnaire 1;
  - satisfaction with regard to the specific service parameters JLB - Questionnaire 2;
- Administration of questionnaires; each expert interviewed is asked to assign:
  - a score (weight) *p(IB.n;St.n)* for each indicator *IB.n*; the score expressed as a percentage determines the level of importance of said indicators and sub-criteria (base of 100 per total indicators and sub-criteria); the total scores of the indicators related to a sub-criterion *SB.n* represent the sub-criterion score *p(SB.n;St.n)*.– completion of Questionnaire 1;

---

[5] See note n. 3.

    o a satisfaction threshold *ssd(IB.n;St.n)* obtained for each specific potential resource parameter *ppd.n(IB.n)*, using the following scale of judgement: very high (VH), high (H), average (A), low (L), very low (VL) – completion of Questionnaire 2;

— Processing of data obtained from interviews; the data obtained from administration of each questionnaire are totalled and the average of the judgements voiced by the Experts for each interview area is calculated; in this way the concise value, to be used in the assessment, is obtained with regard to:

    o the weight to be assigned to each sub-criterion and relative indicators (interview results - Questionnaire 1);

    o the average satisfaction threshold *ssd(IB.n;St.n)* referring to potential services parameters (interview results – Questionnaire 2).

Table 4. JLB matrix, potential resource (parameters)

Criteria (*CB.n*)	Sub-criteria (*SB.n*)	Indicators (*IB.n*)		Specific resource parameter *i(IB.n)* for alternative (*A.n*)	Potential resource parameter (*ppd.n*)	
Services requisites (JLB)	SB.1 Airport services	IB.1	VDS "advanced" (access)	i(IB.1;An)	ppd.a(IB.1;An)	open
					ppd.b(IB.1;An)	closed
		IB.2	Fire service (ICAO category)	i(IB.2;An)	ppd.a(IB.2;An)	1
					ppd.b(IB.2;An)	2
					ppd.c(IB.2;An)	3
					ppd.d(IB.2;An)	4
		IB.3	ATS (time slots activities)	i(IB.3;An)	ppd.a(IB.3;An)	0
					ppd.b(IB.3;An)	8
					ppd.c(IB.3;An)	12
					ppd.d(IB.3;An)	24
		IB.4	Refueling (time slots activities)	i(IB.4;An)	ppd.a(IB.4;An)	0
					ppd.b(IB.4;An)	8
					ppd.c(IB.4;An)	12
					ppd.d(IB.4;An)	24
		IB.5	Handling (time slots activities)	i(IB.5;An)	ppd.a(IB.5;An)	0
					ppd.b(IB.5;An)	8
					ppd.c(IB.5;An)	12
					ppd.d(IB.5;An)	24
		IB.6	Lighting signs for RWY and TWY (presence)	i(IB.6;An)	ppd.a(IB.6;An)	yes
					ppd.b(IB.6;An)	no
	SB.2 Air traffic support services	IB.7	Hangar for aircraft in transit (disponibility)	i(IB.7;An)	ppd.a(IB.7;An)	yes
					ppd.b(IB.7;An)	no
		IB.8	Service repairs (disponibility)	i(IB.8;An)	ppd.a(IB.8;An)	yes
					ppd.b(IB.8;An)	no
	SB.3 Services to passenger	IB.9	Dining (presence)	i(IB.9;An)	ppd.a(IB.9;An)	yes
					ppd.b(IB.9;An)	no
		IB.10	Hotel (km from airoport)	i(IB.10;An)	ppd.a(IB.10;An)	<1
					ppd.b(IB.10;An)	$1 \leq A.10 \leq 5$
					ppd.c(IB.10;An)	<5
	SB.4 Intermodal connectivity	IB.11	City (min. from airport by car)	i(IB.11;An)	ppd.a(IB.11;An)	<10
					ppd.b(IB.11;An)	$10 \leq A.11 \leq 20$
					ppd.c(IB.11;An)	<20
		IB.12	Chief town (min from airoport by car)	i(IB.12;An)	ppd.a(IB.12;An)	<20
					ppd.b(IB.12;An)	$20 \leq A.11 < 40$
					ppd.c(IB.12;An)	<40
		IB.13	Railway station (min. from airport by foot)	i(IB.13;An)	ppd.a(IB.13;An)	<10
					ppd.b(IB.13;An)	$10 \leq A.13 \leq 20$
					ppd.c(IB.13;An)	<20
		IB.14	Taxy (presence)	i(IB.14;An)	ppd.a(IB.14;An)	yes
					ppd.b(IB.14;An)	no
		IB.15	Car rental (presence)	i(IB.15;An)	ppd.a(IB.15;An)	yes
					ppd.b(IB.15;An)	no

- Processing of data entered in assessment matrix for "services" classification:
  - attribution of a value classification to each airport $csp.n(IB.n;A.n)$ for each indicator $IB.n$. The specific resource parameter $i(IB.n;A.n)$ is compared with the potential resource parameter $ppd.n(IB.n;St.n)$ and relative average satisfaction threshold $ssd(IB.n;St.n)$ expressed (by processing of Questionnaire 2 SA) for each indicator $IB.n$, using a specific value classification function (logic) $f(cdm.n)$. In this way, a value classification is obtained for each airport referring to each indicator, expressed through a value that represents the satisfaction indicator on the specific resource parameter according to the following scale: very high (VH), high (H), average (A), low (L), very low (VL) which the following values correspond to VH = 1, H = 0,75, A = 0,5, L = 0,25, VL = 0; (Tab. 5);
  - weighting and standardisation $cdmn(IB.n;A.n)$ of value classifications $cdm.n(IB.n;A.n)$ attributed to each indicator; the following are obtained;
  - weighted value classification for each individual indicator by multiplying the value classification score $cdm.n(IB.n;A.n)$ by the indicator weight $p(IB.n;St.n)$ obtained from Questionnaire 1 of SA;
  - weighted and standardised value classification through "vertical" standardisation taking into account all the values obtained with the weighted value classification for each alternative $A.n$; "vertical" standardisation is performed by placing the highest score obtained with weighting of 1 followed by linear interpolation of all the other indicator scores (Tab. 6). In this way a weighted and standardised value classification $cdmn(IB.n;A.n)$ expressed by a reference value of between 0 and 1 is attributed to each specific resource parameter $i(IB.n;A.n)$;

Table 5. JLB matrix, value classification (function) and related score

Criteria Sub- (CB.n)	Sub-criteria (SB.n)	Indica tors (IB.n)	Specific resource parameter i(IB.n) for alternative (A.n)	Potential resource parameter (ppd.n)	Thresholds satisfaction (ssd.n)	Value classification function [f(cdm.n)]	Value classification and related score (cdm.n)
Services requisites (JLB)	SB.1	IB.1	i(IB.1; An)	ppd.a(IB.1;A.n)	ssd(IB.1;St.n)MA	if ppd.a(IB.1;A.n)>ssd(IB.1;St.n)VH	= cdm.a(IB.1;A.n)VH=1
					ssd(IB.1;St.n)A	if ssd(IB.1;St.n)VH>ppd.a(IB.1;A.n)>ssd(IB.1;St.n)H	= cdm.a(IB.1;A.n)H=0,75
					ssd(IB.1;St.n)M	if ssd(IB.1;St.n)H>ppd.a(IB.1;A.n)>ssd(IB.1;St.n)A	= cdm.a(IB.1;A.n)A=0,5
					ssd(IB.1;St.n)B	if ssd(IB.1;St.n)A>ppd.a(IB.1;A.n)>ssd(IB.1;St.n)L	= cdm.a(IB.1;A.n)L=0,25
					ssd(IB.1;St.n)MB	if ppd.a(IB.1;A.n)<ssd(IB.1;St.n)L	= cdm.a(IB.1;A.n)VL=0
				ppd.b(IB.1;A.n)	ssd(IB.1;St.n)MA	if ppd.n(IB.1;A.n)>ssd(IB.1;St.n)VH	= cdmn(IB.1;A.n)VH=1
					ssd(IB.1;St.n)A	if ssd(IB.1;St.n)VH>ppd.n(IB.1;A.n)>ssd(IB.1;St.n)H	= cdmn(IB.1;A.n)H=0,75
					ssd(IB.1;St.n)M	if ssd(IB.1;St.n)H>ppd.n(IB.1;A.n)>ssd(IB.1;St.n)A	= cdmn(IB.1;A.n)A=0,5
					ssd(IB.1;St.n)B	if ssd(IB.1;St.n)A>ppd.n(IB.1;A.n)>ssd(IB.1;St.n)L	= cdmn(IB.1;A.n)L=0,25
					ssd(IB.1;St.n)MB	if ppd.n(IB.1;A.n)<ssd(IB.1;St.n)L	= cdmn(IB.1;A.n)VL=0
				...	...	...	...
				ppd.n(IB.n;A.n)	ssd(IB.n;St.n)MA	if ppd.n(IB.n;A.n)>ssd(IB.n;St.n)VH	= cdmn(IB.n;A.n)VH=1
					ssd(IB.n;St.n)A	if ssd(IB.n;St.n)VH>ppd.n(IB.n;A.n)>ssd(IB.n;St.n)H	= cdmn(IB.n;A.n)H=0,75
					ssd(IB.n;St.n)M	if ssd(IB.n;St.n)H>ppd.n(IB.n;A.n)>ssd(IB.n;St.n)A	= cdmn(IB.n;A.n)A=0,5
					ssd(IB.n;St.n)B	if ssd(IB.n;St.n)A>ppd.n(IB.1;A.n)>ssd(IB.n;St.n)L	= cdmn(IB.n;A.n)L=0,25
					ssd(IB.n;St.n)MB	if ppd.n(IB.n;A.n)<ssd(IB.n;St.n)L	= cdmn(IB.n;A.n)VL=0

- Attribution of "General services classification" (JLB results). The values related to weighted and standardised value classification $cdmn.n(IB.n;A.n)$ (mathematical average) are aggregated using a general classification function $f(cgnr.n)$ thus obtaining an appraisal score between 0 and 1. The 0 and 1 interval is split into five satisfaction levels: interval 1 - 0.8 = very high (VH); interval 0.8 - 0.6 = high (H), interval 0.8 - 0.6 = average (A) interval 0.6 - 0.4 = low (L), interval 0.4 - 0.2 = very low (VL). The general classification value *cgnr.n(JLB)* of each airport is obtained by examining in which of the five intervals the appraisal score lies and attributing the relative level of satisfaction;

- Ranking of the minor airports in accordance with JLB services resource; the number and territorial distribution of minor airports can be obtained on the basis of the general services classification: "VH", "H", "A", "L" and "VL".

Table 6. JLB matrix, weighted and standardised value (function)

Criteria Sub-criteria (CB.n) (SB.n)	Indicators (IB.n)	Specific resource parameter i(IB.n) for alternative (A.n)	Potential resource parameter (ppd.n)	Value classification and related score (cdm.n)	Weighting and standardisation of value classification function [fnl(cdm.n)]	Weighted and standardised value classification (cdmn.n)
Services requisites (JLB) SB.1	IB.1	i(IB.1;An)	ppd.a(IB.1;A.n)	cdm.a(IB.1;A.n)VH=1 or	fnl[cdm.a(IB.1;A.n)VH*p(IB.1;St.n)]	cdmn.a(IB.1;A.n)
				cdm.a(IB.1;A.n)H=0,75 or	fnl[cdm.a(IB.1;A.n)H*p(IB.1;St.n)]	
				cdm.a(IB.1;A.n)A=0,5 or	fnl[cdm.a(IB.1;A.n)A*p(IB.1;St.n)]	
				cdm.a(IB.1;A.n)L=0,25 or	fnl[cdm.a(IB.1;A.n)L*p(IB.1;St.n)]	
				cdm.a(IB.1;A.n)VL=0	fnl[cdm.a(IB.1;A.n)VL*p(IB.1;St.n)]	
			ppd.n(IB.1;A.n)	cdm.n(IB.1;A.n)VH=1 or	fnl[cdm.n(IB.1;A.n)VH*p(IB.1;St.n)]	cdmn.n(IB.1;A.n)
				cdm.n(IB.1;A.n)H=0,75 or	fnl[cdm.n(IB.1;A.n)H*p(IB.1;St.n)]	
				cdm.n(IB.1;A.n)A=0,5 or	fnl[cdm.n(IB.1;A.n)A*p(IB.1;St.n)]	
				cdm.n(IB.1;A.n)L=0,25 or	fnl[cdm.n(IB.1;A.n)L*p(IB.1;St.n)]	
				cdm.n(IB.1;A.n)VL=0	fnl[cdm.n(IB.1;A.n)VL*p(IB.1;St.n)]	
			...	...		...
			ppd.n(IB.n;A.n)	cdm.n(IB.n;A.n)VH=1 or	fnl[cdm.n(IB.n;A.n)VH*p(IB.1;St.n)]	cdmn.n(IB.n;A.n)
				cdm.n(IB.n;A.n)H=0,75 or	fnl[cdm.n(IB.n;A.n)H*p(IB.1;St.n)]	
				cdm.n(IB.n;A.n)A=0,5 or	fnl[cdm.n(IB.n;A.n)A*p(IB.1;St.n)]	
				cdm.n(IB.n;A.n)L=0,25 or	fnl[cdm.n(IB.n;A.n)L*p(IB.1;St.n)]	
				cdm.n(IB.n;A.n)VL=0	fnl[cdm.n(IB.n;A.n)VL*p(IB.1;St.n)]	

5.4 Overall Classification of Resources. Judgement Level C (JLC)

A concise value *ct.n(LGC;An)* is obtained by combining the results obtained during the phases JLA and JLB, which expresses the connotation in relation to the actual level of resources found (both structural and of services) at each minor airport in Italy. Juxtaposition of the results obtained from the two levels of assessment generate 15 possible configurations (Tab. 7). So we can have the ranking of the minor airports in accordance with JLC.

Table 7. Concise value (evaluation results)

			Results JLB (services classification)				
			Very High (VH)	High (H)	Average (A)	Low (L)	Very Low (VL)
	Suitable	(S)	SVH = ct.1(LGC;A.n) [if cgnr.a(LGA)SU; cgnr.a(LGB)VH]	SH = ct.2(LGC;A.n) [if cgnr.a(LGA)SU; cgnr.a(LGB)H]	SA = ct.3(LGC;A.n) [if cgnr.a(LGA)SU; cgnr.a(LGB)A]	SL = ct.4(LGC;A.n) [if cgnr.a(LGA)SU; cgnr.a(LGB)B]	SVL = ct.5(LGC;A.n) [if cgnr.a(LGA)SU; cgnr.a(LGB)MB]
Results JLA (structural classification)	Transformable	(T)	TVH = ct.6(LGC;A.n) [if cgnr.a(LGA)TR; cgnr.a(LGB)VH]	TH = ct.7(LGC;A.n) [if cgnr.a(LGA)TR; cgnr.a(LGB)H]	TA = ct.8(LGC;A.n) [if cgnr.a(LGA)TR; cgnr.a(LGB)A]	TL = ct.9(LGC;A.n) [if cgnr.a(LGA)TR; cgnr.a(LGB)B]	TVL = ct.10(LGC;A.n) [if cgnr.a(LGA)TR; cgnr.a(LGB)MB]
	Not suitable	(NS)	NSVH = ct.11(LGC;A.n) [if cgnr.a(LGA)NSU; cgnr.a(LGB)VH]	NSH = ct.12(LGC;A.n) [if cgnr.a(LGA)NSU; cgnr.a(LGB)H]	NSA = ct.13(LGC;A.n) [if cgnr.a(LGA)NSU; cgnr.a(LGB)A]	NSL = ct.14(LGC;A.n) [if cgnr.a(LGA)NSU; cgnr.a(LGB)B]	NSVL = ct.15(LGC;A.n) [if cgnr.a(LGA)NSU; cgnr.a(LGB)MB]
			JLC results (general equipment classification)				

6   Implementation of the Procedure and First Results

The evaluation procedure has been implemented to check the adequacy of 52 Italian minor airport's resources.

Table 8 shows - as the subdivisions of the Nomenclature of Territorial Units for Statistics (NUTS[6]) adopted at a European level – the results of the evaluation; for each minor airport is stated the classification relating to:

- structural resources (JLA): suitable (S), transformable (T), not suitable (NS);
- services resources (JLB): satisfaction very high (VH), high (H), average (A), low (L), very low (VL);
- synthetic judgement about resources, combination between JLA and JLB results (JLC).

With reference to JLA, JLB and JLC, processed data obtained through the evaluation procedure, it has been possible to:

- rank the airports according to the configuration class obtained;
- calculate the number of airports that are part of the different classes related to: structural resources, services resources and synthetic judgement configurations;
- identify subdivision and distribution of minor airports by geographical area (NUTS 1) according to: structural, services, overall resources (tab. 9).

It has been also possible to identify which existing airports:

- could potentially be included as airports of the so-called "highway in the sky" with resources adjustment:
  - nil: 10 minor airports (JLC results: SHV, SH, SA);
  - very low: 3 minor airports (JLC results: SL, SVL);
  - low: 10 minor airports (JLC results: TVH, TH, TA);
  - high: 17 minor airports (JLC results: TL, TVL).
- could not be included as airports of so colled "highway in the sky" because they require a substantial adjustment of their structural and services resources: 12 minor airport (JLC results: NSVH, NSH, NSA, NSL, NSVL).

---

[6] NUTS (Nomenclature of Territorial Units for Statistics), introduced for statistical purposes by the European Union identify 3 levels of territorial subdivision of Member States. As regards Italy, NUTS 1 comprises 5 geographical subdivisions (North-West, North-East, Centre, South and Islands); NUTS 2 comprises 21 units: 19 regions and the two autonomous provinces of Trento and Bolzano; NUTS 3 comprises 107 provinces.

Table 8. Evaluation procedure results

NUTS 1: Geographical areas	NUTS 2: Regions	N°	Minor airport	NUTS 3: Provinces	Geographical coordinate	Structural Judg. Level A (JLA)	Services Judg. Level B (JLB)	Overall Judg. Level C (JLC)
					Geographical informations		Classification resources	
North-West (ITC)	Piemonte	1	Alessandria	AL	44°55'30"N 008°37'31"E	NS	VL	NSVL
		2	Biella Cerrione	BL	45°29'45"N 008°06'09"E	S	H	SH
		3	Casale Monferrato	AL	45°06'40"N 008°27'22"E *	T	VL	TVL
		4	Novi Ligure	AL	44°46'48"N 008°47'11"E *	T	L	TL
		5	Torino Aeritalia	TO	45°05'04"N 007°36'11"E	S	A	SA
		6	Vercelli	VC	45°18'40"N 008°25'03"E *	NS	L	NSL
	Lombardia	7	Alzate Brianza	CO	45°46'12"N 009°09'39"E *	NS	VL	NSVL
		8	Como Idroscalo	CO	45°48'53"N 009°04'11"E *	S	VL	SVL
		9	Calcinate del Pesce	VA	45°48'35"N 008°46'05"E *	NS	A	NSA
		10	Cremona Migliaro	CR	45°10'02"N 010°00'07"E *	NS	VL	NSVL
		11	Milano Bresso	MI	45°32'29"N 009°12'08"E *	T	L	TL
		12	Valbrembo	BG	45°43'14"N 009°35'37"E *	NS	L	NSL
		13	Varese Venegono	VA	45°44'29"N 008°53'12"E *	T	VL	TVL
		14	Vergiate	VA	45°42'52"N 008°41'59"E	T	L	TL
		15	Voghera Rivanazzano	PV	44°57'37"N 009°00'35"E *	S	L	SL
	Liguria	16	Sarzana Luni	SP	44°05'20"N 009°59'20"E *	S	VH	SVH
	Val D'Aosta		Region without minor airports					
North-East (ITD)	Trentino Alto-Adige	17	Trento Mattarello	TN	46°01'24"N 011°07'30"E	T	H	TH
	Veneto	18	Asiago	VI	45°53'16"N 011°31'00"E *	T	VL	TVL
		19	Belluno	BL	46°10'02"N 012°14'52"E *	T	VL	TVL
		20	Legnago	VR	45°07'59"N 011°17'32"E *	NS	VL	NSVL
		21	Padova	PD	45°23'46"N 011°50'53"E *	T	VH	TVH
		22	Thiene	VI	45°40'32"N 011°29'47"E *	T	L	TL
		23	Venezia	VE	45°25'44"N 012°23'16"E	T	VH	TVH
		24	Verona Boscomantico	VR	45°28'23"N 010°55'37"E *	T	A	TA
	Emilia Romagna	25	Carpi Budrione	MO	44°50'06"N 010°52'18"E *	T	VL	TVL
		26	Ferrara Aguscello	FE	44°47'24"N 011°40'22"E	NS	L	NSL
		27	Ferrara	FE	44°48'57"N 011°36'48"E *	T	A	TA
		28	Modena	MO	44°38'05"N 010°48'37"E *	T	VL	TVL
		29	Pavullo nel Frignano	MO	44°19'20"N 010°49'54"E *	T	H	TH
		30	Lugo di Romagna	RA	44°23'53"N 011°51'17"E *	T	VL	TVL
		31	Ravenna	RA	44°21'52"N 012°13'29"E *	T	L	TL
		32	Reggio Emilia	RE	44°41'56"N 010°39'45"E	T	A	TA
	Friuli V Giulia	33	Gorizia	GO	45°54'24"N 013°35'57"E *	T	VL	TVL
		34	Udine Campoformido	UD	46°01'55"N 013°11'12"E *	NS	L	NSL
Centre (ITE)	Toscana	35	Arezzo	AR	43°27'21"N 011°50'49"E *	NS	A	NSA
		36	Lucca Tassignano	LU	43°49'47"N 010°34'44"E *	S	L	SL
		37	Massa Cinquale	MS	43°59'09"N 010°08'34"E *	NS	A	NSA
	Umbria	38	Foligno	PG	42°55'58"N 012°42'36"E *	T	L	TL
	Marche	39	Fano	PU	43°49'33"N 013°01'39"E *	T	A	TA
	Lazio	40	Aquino	FR	41°29'10"N 013°43'07"E *	T	VL	TVL
		41	Guidonia	RM	41°59'46"N 012°44'05"E	T	H	TH
		42	Latina	LT	41°32'49"N 012°54'30"E	S	H	SH
		43	Rieti	RI	42°25'36"N 012°51'00"E *	T	L	TL
		44	Roma Urbe	RM	41°57'07"N 012°30'03"E *	S	H	SH
		45	Viterbo	VT	42°25'49"N 012°03'51"E	T	VL	TVL
South (ITF)	Abruzzo	46	L'Aquila Preturo	AQ	42°22'46"N 013°18'34"E	S	VH	SVH
	Molise		Region without minor airports					
	Campania	47	Capua	CE	41°06'57"N 014°10'41"E *	T	A	TA
		48	Salerno Pontecagnano	SA	40°37'12"N 014°54'45"E	S	VH	SVH
	Puglia	49	Lecce Lepore	LE	40°21'27"N 018°17'38"E *	NS	A	NSA
	Basilicata		Region without minor airports					
	Calabria		Region without minor airports					
Islands (ITG)	Sicilia	50	Palermo Boccadifalco	PA	38°06'39"N 013°18'48"E	S	A	SA
		51	Comiso	RA	36°59'45"N 014°36'32"E	S	VH	SVH
	Sardegna	52	Oristano Fenosu	OR	39°53'42"N 008°38'35"E	S	H	SH

Table 9. JLA, JLB and JLC results by geographical areas (NUTS 1)

	Resources classification	North-West (ITC) Va	V%	North-East (ITD) Va	V%	Centre (ITE) Va	V%	South (ITF) Va	V%	Islands (ITG) Va	V%	Italy Va	V%
JLA	S = "Suitable"	5	38%	0	0%	3	23%	2	15%	3	23%	13	25%
	T = "Transformable"	5	19%	15	56%	6	22%	1	4%	0	0%	27	52%
	NS = "Not Suitable"	6	50%	3	25%	2	17%	1	8%	0	0%	12	23%
	TOTAL											52	100%
JLB	VH = "Very High"	1	17%	2	33%	0	0%	2	33%	1	17%	6	12%
	H = "High"	1	14%	2	29%	3	43%	0	0%	1	14%	7	13%
	A = "Average"	2	18%	3	27%	3	27%	2	18%	1	9%	11	21%
	L = "Low"	6	46%	4	31%	3	23%	0	0%	0	0%	13	25%
	VL = "Very Low"	6	40%	7	47%	2	13%	0	0%	0	0%	15	29%
	TOTAL											52	100%
JLC	SVH = "Suitable-Very High"	1	25%	0	0%	0	0%	2	50%	1	25%	4	8%
	SH = "Suitable-High"	1	25%	0	0%	2	50%	0	0%	1	25%	4	8%
	SA = "Suitable-Average"	1	50%	0	0%	0	0%	0	0%	1	50%	2	4%
	SL = "Suitable-Low"	1	50%	0	0%	1	50%	0	0%	0	0%	2	4%
	SVL = "Suitable-Very Low"	1	100%	0	0%	0	0%	0	0%	0	0%	1	2%
	TVH = "Trasformable-Very High"	0	0%	2	100%	0	0%	0	0%	0	0%	2	4%
	TH = "Trasformable High"	0	0%	2	67%	1	33%	0	0%	0	0%	3	6%
	TA = "Trasformable-Average"	0	0%	3	60%	1	20%	1	20%	0	0%	5	10%
	TL = "Trasformable-Low"	3	43%	2	29%	2	29%	0	0%	0	0%	7	13%
	TVL = "Trasformable-Very Low"	2	20%	6	60%	2	20%	0	0%	0	0%	10	19%
	NSVH = "Not Suitable-Very High"	0	0%	0	0%	0	0%	0	0%	0	0%	0	0%
	NSH = "Not Suitable-High"	0	0%	0	0%	0	0%	0	0%	0	0%	0	0%
	NSA = "Not Suitable-Average"	1	25%	0	0%	2	50%	1	25%	0	0%	4	8%
	NSL = "Not Trasformable-Low"	2	50%	2	50%	0	0%	0	0%	0	0%	4	8%
	NSVL = "Not Trasformable-Very Low"	3	75%	1	25%	0	0%	0	0%	0	0%	4	8%
	TOTAL											52	100%

7   Conclusions

The research underway moves along the same lines as the initiatives pursued since 2006 by the Italian Light Airport Network (I.Lan) and by Research Centre DEMETRA (Development of European Mediterranean Transportation) to develop a widespread and efficient network of minor airports in Italy. Analysis and evaluation results illustrated in the paper (phase 1 of a wider research) are the starting necessary input to redefine the Italian minor airport system (second-level of civil aviation) in order to built an airport infrastructure network (node of a so- called "highway in the sky"): efficient, integrated with aeroportual and general infrastructural systems, aimed at support local sustainable development of the different Italian territories. Through elaboration and implementation of a multi-criteria evaluation procedure how 52 Italian minor airport's (alternative in the evaluation) structural and services resources (criteria in the evaluation) comply with suitability parameters has been determinated; suitability parameters have been established through the data collected and processed in the context and stakeholders analysis.

Consequently the operating minor airport that can be included in the so-called "highway in the sky", with or without resources adjustment, have been identified.

The data obtained by applying the procedure form the base of the route marked out to achieve results, with systematic approach, which can be especially useful for planning actions to develop second-level civil aviation in Italy.

References

1. MIT (Ministero delle Infrastrutture e dei trasporti), ENAC (Ente nazionale per l'aviazione civile): Piano nazionale degli aeroporti, febbraio 2012 (2012)
2. Criscuolo, C.: Aeroporti leggeri: scelta o necessità?, presentazione all'incontro, Italian Light Airport Network (i.LAN), Presentazione di uno studio finalizzato alla codifica di una nuova tipologia di infrastruttura di volo intermedia fra "Aviosuperficie" e "Aeroporto". giugno 2007, Roma http://www.filas.it/Page.aspx?IDPage=207 (2007)
3. AIRBUS sas: Global Market Forecast. Flying on demand 2014–2033, Art & Caractère (France) http://www.airbus.com/company/market/forecast/ (2014)
4. Ente Nazionale per l'Aviazione Civile (ENAC): Regolamento per la costruzione e l'esercizio degli aeroporti (ed. 2), Roma http://www.enac.gov.it/La_Normativa/Normativa _Enac/Regolamenti/Regolamenti_ad_hoc/info-1548018725.html (2014)
5. Italian Light Airport Network (i.LAN): Studio finalizzato alla codifica di una nuova tipologia di infrastruttura di volo intermedia fra "aviosuperficie" e "aeroporto", giugno 2007, Roma http://www.filas.it/Page.aspx?IDPage=207 (2007)
6. Nesticò, A., De Mare, G.: Government tools for urban regeneration: the cities plan in italy. a critical analysis of the results and the proposed alternative. In: Murgante, B., Misra, S., Rocha, A.M.A., Torre, C., Rocha, J.G., Falcão, M.I., Taniar, D., Apduhan, B.O., Gervasi, O. (eds.) ICCSA 2014, Part II. LNCS, vol. 8580, pp. 547–562. Springer, Heidelberg (2014)
7. Nesticò, A., De Mare, G., Fiore, P., Pipolo, O.: A model for the economic evaluation of energetic requalification projects in buildings. a real case application. In: Murgante, B., Misra, S., Rocha, A.M.A., Torre, C., Rocha, J.G., Falcão, M.I., Taniar, D., Apduhan, B.O., Gervasi, O. (eds.) ICCSA 2014, Part II. LNCS, vol. 8580, pp. 563–578. Springer, Heidelberg (2014)
8. Morano, P., Tajani, F.: Bare ownership evaluation. Hedonic price model vs. artificial neural network. International Journal of Business Intelligence and Data Mining **8**(4), 340–362 (2013)
9. Nijikamp, P., Rietveld, P., Voogd, H.: Multicriteria evaluation in physical planning. North Holland Publ, Amsterdam/New York (1990)
10. European Commission, Europe Aid Cooperation Office: Evaluation, Methodology, Evaluation tools, Multi-criteria analysis http://ec.europa.eu/europeaid/evaluation/methodology/ tools/too_cri_def_en.htm (2005)
11. Malczewski, J.: GIS and Multicriteria Decision Analysis. John Wiley and Sons, New York (1999)
12. Malczewski, J.: GIS-based multicriteria decision analysis: a survey of the literature. International Journal of Geographical Information Science **20**, 703–726 (2006)
13. Tajani, F., Morano, P.: Concession and lease or sale? A model for the enhancement of public properties in disuse or underutilized. Wseas Transactions on Business and Economics **11**(2014), 787–800 (2014)
14. Ente Nazionale per l'Aviazione Civile (ENAC), Ministero delle infrastrutture e dei Trasporti (MIT) OneWorks, KPMG, Nomisma: Atlante degli aeroporti Italiani. Studio per lo sviluppo futuro della rete aeroportuale nazionale, Roma http://www.enac.gov.it/La_ Comunicazione/Pubblicazioni/info464245000.html (2010)
15. Fusco Girard, L., Nijkamp, P.: Le valutazioni per lo sviluppo sostenibile della città e del territorio. Franco Angeli, Milano (2000)

Financial Sustainability and Morphogenesis of Urban Transformation Project

Grazia Napoli[✉]

Department of Architecture, Viale Delle Scienze, University of Palermo,
Edificio 14 90128, Palermo, Italy
grazia.napoli@unipa.it

Abstract. The urban transformation projects are very complex and have to be examined from several points of view (socio-cultural, environmental, infrastructural, administrative, and economic-financial) to determine their sustainability. This study aims to test the financial analysis as a tool for outlining the morphogenesis of the project's characteristics and exploring the frontiers of the financial feasibility especially when the urban projects, according to Italian laws, involve Public Private Partnerships (PPPs). A financial model is applied to a case study (the transformation of an abandoned railway area) in which the absence of an adequate returns on investment, because of the crisis of the real estate market, requires one to iteratively modify the project's characteristics and define various alternative scenarios for obtaining the project's form that achieves the financial feasibility.

Keywords: Financial analysis · Urban planning · Morphogenesis · PPP

1 Introduction

The urban transformations are complex projects that activate an anabolic process of creation of urban form for reversing the catabolic drift that brings the decline or dysfunction of urban areas, and managing the trans(in)formation of the capital/energy/information inputs into social fixed capital and private capital, by involving private and public actors.

To reach the aim of the physical and functional renewal of an urban area, it is necessary to delineate a balanced solution between public city and private city, among infrastructures, public equipments and services as well as private real estate and economic activities, so that the new wealth produced by the transformation can be distributed among social use value, income and profit.

The projects' realization depends, obviously, on their complex sustainability (that is socio-cultural, environmental, infrastructural, technical-administrative, and economic-financial) and can be valued by using multicriteria models. If the public financial funds are not sufficient and the private investors search for profits able to balance the high risk caused by the economic crisis, the financial sustainability constitutes a decisive constraint to realize a project.

© Springer International Publishing Switzerland 2015
O. Gervasi et al. (Eds.): ICCSA 2015, Part III, LNCS 9157, pp. 178–193, 2015.
DOI: 10.1007/978-3-319-21470-2_13

The financial sustainability can be carried out in different phases of the decisional process (strategic or operational phase) according to different reasons: it is a tool for choosing and describing alternatives with fixed characteristics if the aim is summarizing or certificating the alternatives; otherwise, it becomes a tool for defining the characteristics that make a project sustainable if the aim is morphogenetic. The application of the financial analysis according to the morphogenetic aim is full of future prospects because it activates a creative process supporting the promoter and/or decision maker, even fixing undeniable goals, for the exploration of the financial feasibility frontiers and the building of a good sustainable solution.

2   Feasibility Analysis of the Projects and the PPPs

To achieve the allocation of resources, public authorities have to select these projects that maximise their positive impact on the social-economic development. The Italian law "Regolamento attuativo del Codice dei contratti pubblici" (Implementing regulation of the code of the public contracts) establishes that the public investment projects have to be assessed by the Feasibility Analysis (FA). Despite some differences, the FA's rules proposed by several official documents, e.g. "Notes" by UVAL and "Linee Guida" by ITACA [1,2,3], require the verification of the following:

- Environmental sustainability;
- Administrative-procedural sustainability;
- Financial feasibility (financial analysis);
- Economic-social feasibility (cost-benefit analysis).

The FA's structure reflects the complexity of the public decision process, even if it does not establish the confluence of each verification in a multicriteria evaluation model as, instead, it has already been applied in various case studies [4,5,6,7]. Moreover, according to the law, the FA's positive verification is binding for accessing to public financing sources, above all, where the municipality intends to use them to realize a project also involving private capitals and applying Public Private Partnerships (PPPs).

PPPs combine public resources with those of private agents in order to deliver societal goals [8] and can be defined as a long term agreement between the government and one or more private partners where the service delivery objectives of the government are aligned with the profit objectives of the private partner and there is a sufficient transfer of risk to the private partners [9,10,11]. According to Hodge and Greve [12], there are five families of arrangements encompassed in the PPP concept: institutional cooperative for joint production and risk sharing, long-term infrastructure contracts, public policy networks, civil society and community development, and urban renewal and downtown economic development.

The agreement on PPPs is not concordant, in fact «It is not a simple matter to judge whether PPPs are the next chapter in the privatization story; another promise in our ongoing attempts to better define and measure public sector service performance; a renewed support scheme for boosting business in difficult times; or a language game camouflaging the next frontier of conquering transaction merchants, legal advisors, and merchant bankers pursuing fat commissions» [12]. Despite the scepticism about the attractiveness and effectiveness of PPPs, most people consider them an important approach to designing and implementing economic development strategies [13].

The European Commission distinguishes between "purely contractual PPPs" and "institutionalised PPPs" [14], whereas in Italy, where the PPP projects have been applied since the 1990s to many urban transformations, it is possible to group them in the following three categories [15]:

- Complex Programs Partnership (PRU, PRUSST, Urban, "Contratto di quartiere", "Harbours and Stations Innovative Program", etc.);
- Institutional Partnership (Società di trasformazione Urbana - STU);
- Contractual Partnership (Società di progetto private).

The main characteristics that are common to the three PPP categories are the following: the source of financing that is both public and private, the cooperative interaction among public and private actors during all the project's phases (planning, financing, realization, and management), and the sharing of the risks. The different characteristics are numerous and the most meaningful characteristics concern the role assumed by the public and private stakeholders and the composition of the project investment cost. The STUs are characterized by the formation of a new project promoter that can be the municipality up to a plurality of public and private actors (private bank, real estate society, etc.) and that provides to the project's realization and management. The investment capital is differentiated for qualitative (land, building, grant and loan, private and public capital) and quantitative composition (prevalence/balance among the capital categories) as well as the corresponding profitability (Fig. 1). Instead, in the Contractual Partnership, the project's realization and management are charged to the PPP private partner and the public administration maintains the role of supervisor of the process (Fig. 2) [16,17,18].

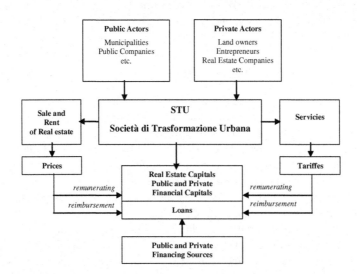

Fig. 1. PPP Institutional partnership: STU - Società di Trasformazione Urbana

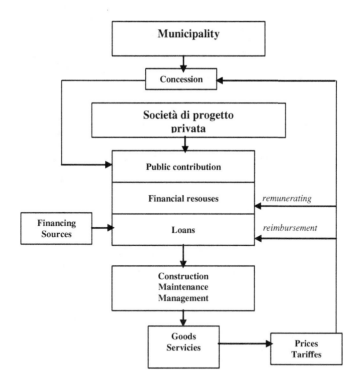

Fig. 2. PPP Contractual partnership: Società di progetto privata

3   Morphogenetic Functions of the Financial Analysis

If the city is the result of the synergistic interaction among the forecasts of the planning and the effective realization made by public and private actors, every trans-formation process implies a different vision of the city that has to be shared by the most of the citizens. «The city is culture, language, de-sign. The city is market, capital, money, ..., project, plan. The city is project, hope, desire, desire of desire» [19].

In the urban transformation projects with PPPs, the public administration has the goal to prefigure, promote, and realize the renewal of urban areas verifying the project's complex sustainability and feasibility as well as the convenience for all the actors. The economic and financial performances are, therefore, crucial in involving the private capitals and can be applied with morphogenetic purpose to support the project's development in order to explore the frontiers of the financial sustainability [20].

The economic and financial analyses are largely being discussed in the economic literature [21,22,23,24,25,26] and in some European documents, as "Guides to cost-benefit analysis of investment projects" (European Commission) [27], offering theoretical and operational tools. In the Financial Analysis, the cash flow represents the temporal and monetary distribution of the outflows (investment and operating costs) and inflows (revenues for sale and operating revenues) (Table 1). The main indicators

Table 1. Revenues and costs

Investment costs	Outflows	Cash flow	FNPV
Operating costs			FRR
Operating revenues (tariff)	Inflows		
Revenues for sale (price)			

for testing the project's financial performance are the Financial Net Present Value of the project (FNPV) and the Financial Internal Rate of Return (FRR) (Formulas 1 and 2). More specifically, the FNPV(C) measures the performance of the investment independent of the sources or methods of financing.

$$FNPV = \sum_{t=1}^{n} \frac{I_{kt}}{(1+r_k)^t} - \sum_{t=1}^{n} \frac{O_{kt}}{(1+r_k)^t} \tag{1}$$

$$\sum_{t=1}^{n} \frac{I_t}{(1+FRR)^t} = \sum_{t=1}^{n} \frac{O_t}{(1+FRR)^t} \tag{2}$$

where: FNPV = Financial Net Present Value; I = inflows; O = outflows; r = discount factor; k = capital; t = time; FRR = Financial Internal Rate of Return

This work proposes a model of financial analysis of urban transformation project with the PPP, considering it as a part of the FA. The application of the model enables choosing the undeniable characteristics/goals of a "baseline scenario" (invariants) and delineating the characteristics useful for the project's morphogenesis (*morphogenetic variables*). Subsequently, the parameters of these variables can be modified for defining the limit conditions of the financial feasibility.

The morphogenetic variables of an urban transformation project can be grouped in:

- *Technical morphogenetic* variables
    - *GA, general area* of the project, eventually articulated in small areas;
    - *AB, area of buildings.* It expresses the square meters of public and private buildings and depends on the urban density;
    - *FM, functional mix.* It represents the different mix of public and private use of land;
    - *TC, technological and architectural characteristics.* It points out the architectural and technical characteristics of the works.
- *Operational-financial morphogenetic* variables
    - *PC, PPP's category.* It expresses the PPP's categories (qualitative), STU, or contractual partnership;
    - *CR, capital ratio.* It represents the ratio (quantitative) among the public and private invested capitals;
    - *CC, capital composition.* It distinguishes among the categories (quality) of invested capitals (and their corresponding profitability), as monetary capital, real estate, equity, and loans.

The steps of the model are:

1. Analysing the "baseline project" and choosing the invariants and *morphogenetic variables* (*technical and operational-financial*);
2. Defining the range of the *technical morphogenetic variables*;
3. Defining I level scenarios;
4. Calculating I level scenarios' FNPV and FRR;
5. Choosing the scenarios that verify the following conditions FNPV≥0 and FRR≥s (s= minimum value of the discount rate);
6. Defining the range of the operational-financial variables;
7. Defining II level scenarios (combination of each *technical and operational-financial morphogenetic variable*);
8. Verifying the conditions FNPV≥0 and the profitable FFR of each invested capital;
9. Choosing the best scenarios and defining the project's thresholds that verify the financial feasibility;
10. Whether the FNPVs in steps 5 and 8 are negative, it is possible to go back to step 1 and modify the project's characteristics (invariants and variables).

As the model has the aim of evaluating the urban transformation projects, the discount rate applied in the FNPV's formula is regarded as:

- the public financing cost, e.g. the discount rate applied to the loans from the "Cassa Depositi e Prestiti" (CDP) to the municipality;
- the opportunity cost of low risk investments (Buoni Pluriennali del Tesoro BTP);
- the investments return of the real estate.

However, the heterogeneity of the invested capitals requires the calculation of the wacc (weighted average cost of capital).

4   Application of a Morphogenetic Model of Financial Analysis to the Renewal Project of Abandoned Railway Areas

A morphogenetic model of financial analysis is applied to a case study, the renewal project of abandoned railway areas in Palermo (Italy), with the aim of:

- verifying whether the baseline scenario defined in 2008 respects the conditions of the financial feasibility also in 2014;
- individualizing the thresholds that define the extreme conditions of the financial feasibility;
- choosing the alternative scenarios that better verify the financial feasibility.

4.1   The ATI-2 Project

The ATI-2 project (Areas of Integrated Transformation) has been elaborated by the Municipality of Palermo with the Harbours Authority, the Italian Railway Net (RFI), and the Railroads Real Estate (FRE) as an application of the "Programma Innovativo in Ambito Urbano" (PIAU) called "Harbours and Stations" and it has the aim to realize "new urban centralities" in the abandoned railway areas (183.000 m^2) of the Notarbartolo station and the ex Lolli station [28].

For the application of this PIAU, a contract between several public actors has been stipulated and the FA has been elaborated to select a private promoter (PPP) for the provision of public assets, services, and private buildings (Fig. 3, Table. 2).

The FA has analyzed the project for verifying the complex sustainability, i.e. environmental, administrative and institutional, technical and economic-financial.

From the point of view of the environmental sustainability, the whole area is completely urbanized and the project does not modify any natural habitat or existing green areas. The technical sustainability involves the respect of the Master Plan's guidelines. The social sustainability is verified by the realization of public services and infrastructures and Social Housing. The technical-administrative sustainability can be reached by stipulating Partnership with RFI and Protocols with the Regional administration, launching International Planning competition, modifying the Master Plan, and searching a private partner for the PPP. In 2008, the financial feasibility of the baseline scenario (the time horizon was 30 years and the discount rate was 5,5%) calculated the FNPV equal to 36 million of Euros and the FRR equal to 7%.

Fig. 3. The ATI-2 project's area (baseline scenario)

Table 2. Land use of the project's areas

Area N°	Land use
1	Museum and cultural facilities in the ex Lolli station
14	Cultural facilities near the Notarbartolo station
23	Sports equipment
2, 3, 5, 19	Green areas
4, 10, 20	Local road system
22	Parking areas
6, 12, 16	Underground parking garages
7	Social housing
7, 8, 13, 18, 23B	Residential and commercial buildings
9, 11, 17, 21	Shopping centres

4.2   Application of the Model

The model is built by defining the *morphogenetic variables*, the range of each variable, and the alternative scenarios that are tested to verify their financial performances after having updated the market prices and the discount rate of the cash flow (Table 3).

Updating the market prices. The costs of construction and maintenance have increased by 10%, while the land tax concession is supposed to be lower. The greater variation in prices involves the real estate market prices that are the main revenue for sale. In I semester 2008, the prices were high because there was a point of *displuvium* of the "real estate basin" [12] [20], but, since the bursting of the speculative bubble caused by subprime loans in USA, a consistent contraction of prices has been occurring and it is equal to -17% (weighed average) in the period 2008-2014 (Table 4). The descending real estate prices are also reflected in a lower value of the land, appraised by the "value of transformation". The operating costs and revenues and the time horizon of the project are not changed.

Updating the discount rate. The choice of the discount rate has referred to the mixed nature of the investment (public assets and private real estate); furthermore, its value can be at least the cost of the CDP's loan to the municipality (time horizon 29 years) that is equal to 2,3%, and at maximum the opportunity cost of a lower risk investment (BTP 30 years) that is equal to 3,25% (this value is also lined up to the real estate investments return in the Lolli and Notarbartolo areas that is 3,15%).

Table 3. Project's cash flow

Cash flow	Outflows	Investment costs *(years 1-7)*	Land acquisition
			Land tax concession
			Construction
			Maintenance
		Operating costs *(years 8-30)*	Public infrastructures and equipments (local road system, green area, cultural facilities business centres, sports equipment and underground parking garages)
	Inflows	Operating revenues *(years 8-30)*	Cultural facilities, business centres, sports equipment and underground parking garages
		Revenues for sale *(years 4-8)*	Residential and commercial buildings, offices, underground parking garages

Table 4. Real estate market prices in the project's area (2008 and 2014)

Area	Land use	Market prices FA 2008*	Market prices 2014**	Var % 2008-14
Ex Lolli station	Residential	2.887 €/m^2	1.995 €/m^2	-30,9%
	Residential for Social housing	1.500 €/m^2	1.365 €/m^2	-13,3%
	Commercial and offices	3.060 €/m^2	2.205 €/m^2	-24,5%
	Garage	34.000 €	24.000 €	-29,4%
	Garages for Social Housing	23.000 €	16.000 €	-30,4%
Notarbartolo station	Residential	3.050 €/m^2	3.045 €/m^2	+0,0%
	Commercial and offices	4.136 €/m^2	3.080 €/m^2	-25,5%
	Garage***	36.000 €	41.000 €	+13,9%

*Source: OMI. ** Source: market data directly collected and OMI. *** Garage of 20 m^2.

Outline of the model's variables. The model's *technical morphogenetic variables* are:

- *AB, area* (square meters) *of buildings* (range: "baseline area", +10%, +20%, +25%);
- *FM, functional mix* (range: residential 38%, office 35% and commercial 27%; residential 73% and commercial 27%; residential 38% and office 62%)

The *operational-financial morphogenetic variables* are:

- *PC, PPPs category* (range: STU, contractual partnership);
- *CC, capital composition* of the financial resources and the loan (range: 50-50%, 25-75%, 75-25%) (Fig. 4).

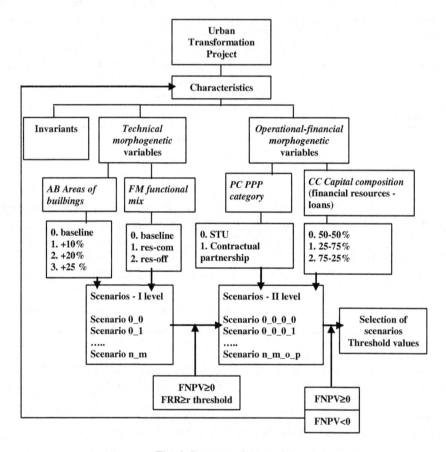

Fig. 4. Structure of the model

Afterwards, I level scenarios are obtained by the combination of the *technical morphogenetic variables* and their FNPVs and FRRs are calculated. Observing the results illustrated in Table 5, we can see where the discount rate has previous values (2,3-3,25%):

- the FNPV is positive only for 5 scenarios (1_0, 2_0, 3_0, 2_1, and 3_1):
- the mix scenarios of residential-office are never profitable;
- the other scenarios are not profitable whether they maintain the "baseline area" even if the functional mix is modified;
- where the square meters of buildings increase (+10%, +20%, and +25%), the scenarios' FNPVs become positive and are greater, especially for the baseline mix (Figs. 5-7).

According to these results, the threshold values of the *technical morphogenetic variables* are *FM* = baseline mix; *AB*≥10%. The scenarios 1_0, 2_0, and 3_0 are the best ones and are selected for the next step.

In this step, the selected scenarios are analysed on the base of the *operational-financial morphogenetic* characteristic and, aiming to notice how each project's element affects the FNPV, the elements of the cash flow are grouped in three categories (Table 6):

- *A*. public assets having investment and operating costs and no revenue;
- *B*. public assets having investment and operating costs and operating revenues;
- *C*. public assets having investment and operating costs and revenues for sale.

Consequently, we can establish that category *A* has always a negative FNPV because of the absence of any revenues (an aspect that characterizes many public works), for the categories *B* and *C*, the FNPV can be both positive and negative, depending on the combination (and corresponding returns) of public/private capital and own resources/loan capital.

Table 5. I level scenarios

	FNPV (r=2,3%)	FNPV (r=3,25%)	FRR
Scenario 0_0	€ 23.652.707	-€ 11.355.871	2,92%
Scenario 1_0	**€ 44.540.329**	**€ 8.163.145**	**3,49%**
Scenario 2_0	**€ 65.427.951**	**€ 27.682.161**	**4,10%**
Scenario 3_0	**€ 75.871.762**	**€ 37.441.670**	**4,41%**
Scenario 0_1	€ 12.349.759	-€ 22.003.067	2,62%
Scenario 1_1	€ 32.107.086	-€ 3.548.770	3,15%
Scenario 2_1	**€ 51.864.413**	**€ 14.905.527**	**3,70%**
Scenario 3_1	**€ 61.743.077**	**€ 24.132.676**	**3,98%**
Scenario 0_2	-€ 37.047.885	-€ 68.534.854	1,41%
Scenario 1_2	-€ 22.230.322	-€ 54.733.736	1,76%
Scenario 2_2	-€ 7.412.760	-€ 40.932.618	2,12%
Scenario 3_2	-€ 3.978	-€ 34.032.059	2,30%

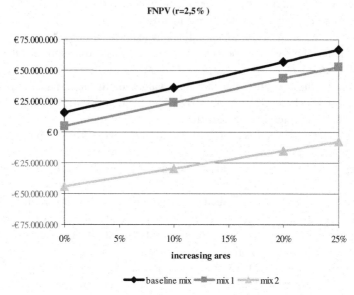

Fig. 5. FNPV of the I level scenarios (r=2,5%)

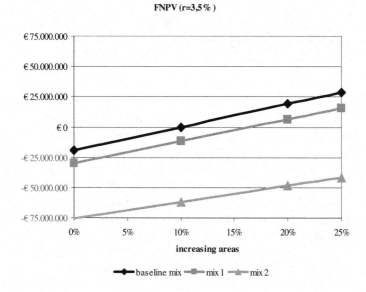

Fig. 6. FNPV of the I level scenarios (r=3,5%)

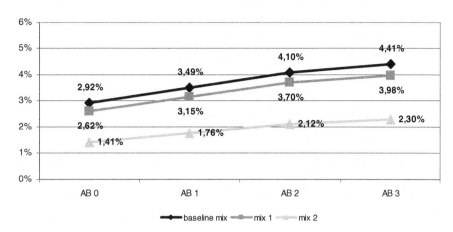

Fig. 7. FRR of the I level scenarios

Table 6. Categories of assets of the project's areas

Asset	Category	Area n°	Land use
Public assets	A. Investment costs and Operating costs (No inflows)	2, 3, 5, 19	Green areas
		4, 10, 20	Local road system
	B. Investment costs, Operating costs and Operating revenues	1	Museum and cultural facilities in the ex Lolli station
		14	Cultural facilities near the Notarbartolo station
		23	Sports equipment
		22	Parking areas
		6, 12, 16	Underground parking garages
		17	Shopping centre
Private assets	C. Investment costs and Revenues for sale	7	Social housing
		7, 8, 13, 18, 23B	Residential and commercial buildings
		9, 11, 21, 17B	Shopping centres

The cash flow of each area is modified according to the variable *CC* (capital composition financial resources/loan) by including the reimbursement loans and the calculation of the wacc (Table 7), and according to the variable *PC* (PPP category) by adding outflows for taxes, management expenses, and concession fees. The FNPV of each area is calculated and, by way of example, the results of the scenario 2_0_0_0 (*AB*= +20%, *FM*= baseline, *PC*= STU, *CC*= 50-50%) are summarized in Table 8:

- the total FNPV of the category *A* is negative and partially compensated by the positive FNPV of the category *B*, while the FNPV of the category *C* is strongly negative;
- the overall FNPV of the scenario 2_0_0_0 is -47.810.576 million of euros.

Instead, if the capital composition *CC* varies and is 25% own resources and 75% loan, as in the scenario 2_0_0_1 (*AB*= +20%, *FM*= baseline, *PC*= STU, *CC*= 25-75%), the FNPV becomes less negative and is -2.765.302 million of euros (Table 9).

Where the main revenues of the cash flows are derived from the sale of the real estate, a strongly negative FNPV (of the category C) reveals that the project, even in the case of the concession of a surplus of square meters of buildings (+20%), is notable to produce neither profits (for the remuneration of the private capital), nor supra-normal profits that could be converted in public equipments or infrastructures (category A). The negative FNPV could depend on exogenous elements and/or on the project's endogenous characteristics that can be modified, for instance, going back to step 1 and renouncing to the shopping centres in the areas n° 9 and 17. The new results are synthesized in Table 9: the FNPV is less negative than the previous scenario where the ratio own resources/loans is 50-50%; instead, it becomes positive where the ratio is 25-75% because the investment risk is postponed and transferred to the loans. The threshold value of the variable CC is <40%.

Similar results could be obtained modifying the variable *PC* (PPP Category) in Contractual Partner and varying the outflows for taxes, management expenses, and concession fees.

Where the conditions FNPV≥0 and FRR≥s are not verified, introducing a switch in the early characteristics, the model's application can start again fixing progressively some parameters but, at the same time, exploring the frontiers of the financial feasibility as a consequence of his flexibility.

In this case study, the project's difficulties in reaching a stable financial feasibility mainly depend on exogenous causes referred to the real estate market. The dynamics of the urban real estate market is the result of the combination of macroeconomic, microeconomic, macro-territorial, and micro-territorial factors, and it can be analyzed in terms of "real estate basins" delimited by points of *dipluvium*, the beginning of a raising phase, and points of *compluvium*, the beginning of a descending phase in which the market prices could be too low to make a project profitable.

Table 7. Discount rates

Capital		Discount rate	
Public and private real estate	risk	Real estate return in the project areas	$r_{im} = 3,15\%$
Public financial	risk	Opportunity cost (BOT return 30 years)	$r_{op} = 3,25\%$
	loan	CDP's loan cost	$r_{fp} = 2,30\%$
Private financial	risk	Opportunity cost (BOT return 30 years)	$r_{op} = 3,25\%$
	loan	Private loan cost	$r_{fr} = 4,65\%$

Table 8. FNPV of the Scenario 2_0_0_0

Category of assets					
Area n°	Category A FNPV	Area n°	Category B FNPV	Area n°	Category C FNPV
2	-€ 1.513.072	1	-€ 13.660.166	7	-€ 15.407.549
3	-€ 336.716	14	-€ 3.831.697	8	-€ 6.657.131
5	-€ 190.958	23	-€ 13.981.217	13	€ 4.270.756
19	-€ 1.388.222	22	€ 10.805.091	18	€ 5.098.977
4	-€ 3.349.751	6	€ 16.025.798	17B	-€ 12.296.659
10	-€ 20.888.916	12	€ 27.683.104	9	-€ 23.733.805
20	-€ 22.462.976	16	€ 25.542.568	11	€ 16.417.467
		17	-€ 19.750.537	21	€ 3.029.705
				23B	€ 2.765.331
tot	*-€ 50.130.611*	*tot*	*€ 28.832.944*	*tot*	*-€ 26.512.909*
Overall FNPV -€ 47.810.576					

Table 9. Comparison between scenarios' FNPV

FNPV	Scenario 2_0_0_0 CC 50-50%	Scenario 2_0_0_1 CC 25-75%
Overall FNPV	-€ 47.810.576	-€ 2.765.302
Overall FNPV without areas 9 and 17	-€ 4.326.234	**€ 40.719.040**

Despite these facts, it is important to remark that the real estate market is character-ized by uncertainties, instability, speculative forecasts, and liquidity transmutation that feed a continuous and asymmetric process of plus-minus-evaluation [29,30,31]. According to Rizzo's theory of the capital [30], the expected value of a capital is a function of the investor's expected plus-minus evaluation (Formula 3) that can coin-cide or diverge from the market trends, and so each economic operator can be willing to participate even in an investment with a negative FNPV if he expects a future plus-evaluation of the real estate (a>0) that can reach to compensate, after being dis-counted, the FNPV's negative value [32].

$$V = k \pm ak \tag{3}$$

where: V= expected value of the capital, K= current price of the capital, a= coefficient of plus-minus evaluation.

5   Conclusions

The realization of the urban transformation projects with PPP is bound to the evalua-tion of their sustainability by the FA, in which financial analysis can be applied with morphogenetic purpose. In this work, a model, supporting the exploration of the fi-nancial sustainability frontiers, has been proposed and applied to a case study. The application has allowed verifying the flexibility of the model in comparison with the peculiarities of the case study in order to outline the characteristics that make the project sustainable; furthermore, it has emphasized the necessity to deepen the relationships among some operational-evaluation elements of the financial analysis (as the discount rate and the FNPV) and theoretical-economic elements (as the theory of the capital and the dynamics of the real estate markets) to understand the causes of low profitability of private and public capitals better.

References

1. Uval: Note per la redazione degli studi di fattibilità. Ministero dell'Economia e delle finanze, Dipartimento per le politiche di sviluppo e di Coesione, Roma (2000). http://db.formez.it/fontinor.nsf/021efd2fc2123c86c1256cc200435aff/76BB5B1856FC4E9 FC1256E6700474410/$file/Note_per_redazione_studi_fattibilita_maggio_2000.pdf
2. Autorità per la vigilanza sui contratti pubblici: Linee guida per la redazione dello studio di fattibilità. Allegato alla Determinazione n. 1/2009 Linee guida sulla finanza di progetto. http://www.lavoripubblici.it/merloni/autorita/2009/autorita_1_2009.pdf
3. ITACA: Linee guida per la redazione di Studi di fattibilità. www.itaca.org/documenti/news/LG%20ITACA%20SDF_Completo_240113.pdf
4. Figueira, J., Greco, S., Ehrgott, M. (eds.): Multiple Criteria Decision Analysis. State of the Art Survey. Springer, New York (2005)
5. Bottero, M., Lami, I.M., Lombardi, P.: Analytic Network Process. La valutazione di scenari di trasformazione urbana e territoriale. Alinea, Firenze (2008)
6. Napoli, G., Schilleci, F.: An application of analytic network process in the planning process: the case of an urban transformation in palermo (Italy). In: Murgante, B., Misra, S., Rocha, A.M.A., Torre, C., Rocha, J.G., Falcão, M.I., Taniar, D., Apduhan, B.O., Gervasi, O. (eds.) ICCSA 2014, Part III. LNCS, vol. 8581, pp. 300–314. Springer, Heidelberg (2014)
7. Trovato, M.R., Giuffrida, S.: A DSS to assess and manage the urban performances in the regeneration plan: the case study of pachino. In: Murgante, B., Misra, S., Rocha, A.M.A., Torre, C., Rocha, J.G., Falcão, M.I., Taniar, D., Apduhan, B.O., Gervasi, O. (eds.) ICCSA 2014, Part III. LNCS, vol. 8581, pp. 224–239. Springer, Heidelberg (2014)
8. Skelcher, C.: Public-private partnerships and hybridity. In: Ferlie, E., Lynn, I.J., Pollitt, C. (eds.) The Oxford Handbook of Public Management. Oxford University Press, Oxford (2005)
9. OECD: Public-Private Partnerships: In Pursuit of Risk Sharing and Value for Money. OECD Publishing (2008)
10. OECD: Dedicated Public-Private Partnerships Units: A Survey of Institutional and Governance Structures. OECD Publishing (2010)
11. Rossi, M., Civitillo, R.: Public Private Partnerships: a general overview in Italy. Social and Behavioral Science **109**, 140–149 (2014)
12. Hodge, G.A., Greve, C.: Public-Private Partnerships: An International Performance Review. Public Administration Review **67**(3), 545–558 (2007)
13. Mullin, S.P.: Public-Private Partnerships and State and Local Economic Development, Leveraging Private Investment. Reviews of Economic Development Literature and Practice **16**, 1–38 (2002)
14. European Commission: Green Paper on Public-Private Partnerships and Community Law on Public Contracts and Concessions. Brussels (2004)

15. Copiello, S.: Progetti urbani in partenariato, Studi di Fattibilità e Piano economico finanziario. Alinea, Firenze (2011)
16. Unità Tecnica Finanza di Progetto.
    http://www.utfp.it/docs/documenti/project_financing/introduzione.pdf
17. Unità Tecnica Finanza di Progetto.
    http://www.utfp.it/docs/documenti/eco_fin/valutazione_eco-fin.PDF
18. Public Investiment Evaluation Unit,
    http://www.dps.gov.it/opencms/export/sites/dps/it/documentazione/servizi/materiali_uval/
    Metodi/MUVAL_30_Guida_Eng.pdf
19. Rizzo, F.: Dalla rivoluzione keynesiana alla nuova economia. In: Dis-equilibrio, tras-informazione e coefficiente di capitalizzazione. FrancoAngeli, Milano (2002)
20. Patassini, D.: Valutazione e politiche territoriali: le dimensioni di un concetto pervasivo. In: Stanghellini, S. (ed.) Valutazione e processo di piano, pp. 23–55. Alinea, Firenze (1996)
21. Pearce, D.W., Nash, C.A.: A text in Cost-Benefit Analysis. The Macmillan Press LTD, London (1981)
22. Pearce, D.W., Atkinson, G., Mourato, S.: Cost-benefit analysis and environment: recent developments. OECD, Paris (2006)
23. Shofield, J.A.: Cost-benefit analysis in urban and regional planning. Allen & Unwin, London (1989)
24. Nuti, F.: L'analisi costi-benefici. Il Mulino, Bologna (1987)
25. Pennisi, G., Ruta, P., Scandizzo, P.: Tecniche di valutazione degli investimenti pubblici. Istituto Poligrafico e Tecnica dello Stato, Roma (1991)
26. Pennisi, G.: Valutare l'incertezza. L'Analisi costi benefici nel XXI secolo. Giappichelli, Torino (2003)
27. European Commission: Guide to Cost-Benefit Analysis of investment projects. http://ec.europa.eu/regional_policy/sources/docgener/guides/cost/guide2008_en.pdf
28. ANCE: Riqualificare le città per arginare il declino. L'area Lolli Notarbartolo. Quaderni Opus Concretum, Palermo (2011)
29. Rizzo, F.: Valori e valutazioni. La scienza dell'economia o l'economia della scienza. FrancoAngeli, Milano (1999)
30. Rizzo, F.: La dinamica dei capitali. FrancoAngeli, Milano (2006)
31. Napoli, G.: Teoria e pratica dei capitali urbani. La forma temporale e monetaria della città. FrancoAngeli, Milano (2007)
32. Giuffrida, S., Ferluga, G., Valenti, A.: Clustering analysis in a complex real estate market: the case of ortigia (Italy). In: Murgante, B., Misra, S., Rocha, A.M.A., Torre, C., Rocha, J.G., Falcão, M.I., Taniar, D., Apduhan, B.O., Gervasi, O. (eds.) ICCSA 2014, Part III. LNCS, vol. 8581, pp. 106–121. Springer, Heidelberg (2014)

Property Valuations in Times of Crisis. Artificial Neural Networks and Evolutionary Algorithms in Comparison

Francesco Tajani[1], Pierluigi Morano[1(✉)], Marco Locurcio[2], and Nicola D'Addabbo[2]

[1] Department of Science of Civil Engineering and Architecture, Polytechnic of Bari, Bari, Italy
francescotajani@yahoo.it, pierluigi.morano@poliba.it
[2] Department of Architecture and Design, University "Sapienza", Rome, Italy
marco.locurcio@uniroma1.it, nda.archt@gmail.com

Abstract. In the current economic situation, characterized by a high uncertainty in the appraisal of property values, the need of "slender" models able to operate even on limited data, to automatically capture the causal relations between explanatory variables and selling prices and to predict property values in the short term, is increasingly widespread. In addition to Artificial Neural Networks (ANN), that satisfy these prerogatives, recently, in some fields of Civil Engineering an hybrid data-driven technique has been implemented, called Evolutionary Polynomial Regression (EPR), that combines the effectiveness of Genetic Programming with the advantage of classical numerical regression. In the present paper, ANN methods and the EPR procedure are compared for the construction of estimation models of real estate market values. With reference to a sample of residential apartments recently sold in a district of the city of Bari (Italy), two estimation models of market value are implemented, one based on ANN and another using EPR, in order to test the respective performance. The analysis has highlighted the preferability of the EPR model in terms of statistical accuracy, empirical verification of results obtained and reduction of the complexity of the mathematical expression.

Keywords: Property valuations · Artificial neural networks · Evolutionary polynomial regression · Genetic algorithms · Estimative analysis · Market value

1 Introduction

In current economic situation, the use of tools for the evaluation of real estate values has become essential for sector operators (buyers, sellers, institutions, insurance companies, banks, etc.) [6, 17]. The continuous change of the boundary conditions [4, 31, 32, 33] causes that it is necessary to use, rather than models characterized by a strong theoretical and methodological basis, "slender" models, able to operate even on limited data and to automatically capture the causal relations between explanatory variables and prices, as well as to predict property values in the short term [30].

The work must be attributed in equal parts to the authors.

O. Gervasi et al. (Eds.): ICCSA 2015, Part III, LNCS 9157, pp. 194–209, 2015.
DOI: 10.1007/978-3-319-21470-2_14

The artificial neural networks (ANN) satisfy these prerogatives. Many studies [3, 12, 27, 28, 29, 34, 37] have highlighted that: ANN provide very good performance in forecasting market values, even when the data are limited; they avoid the econometric problems linked to the multicollinearity, the heteroskedasticity and the spatial autocorrelation, that are typical of other models (e.g. hedonic prices); they are more robust to model misspecification regarding how explanatory variables are measured.

However, ANN models have several weaknesses [9, 14, 21]. First of all, they provide that the structure of the neural network (e.g. model inputs, transfer functions, number of hidden layers, etc.) is exogenously defined. Furthermore, over-fitting problems are frequent in parameter estimation. Another disadvantage is the inability to incorporate known economic laws into the learning process.

On these aspects, some steps have been made through the Genetic Programming (GP), that is an approach of artificial intelligence capable of generating a structured representation of the system model. The most common method of GP, named symbolic regression, allows to obtain mathematical expressions to fit a set of data points using operations analogous to the evolutionary processes that occur in nature [23].

Several authors have borrowed the logic of genetic algorithms to improve the application of mathematical procedures to the real estate market. Among them, Kròl et al. [24] have developed a fuzzy rule-based system to assist the real estate appraisal, employing an evolutionary algorithm to generate the rule base. Wang [36] has elaborated a decision support system that, through a data envelopment analysis, converts numerical data into information that can be used to evaluate real estate investments. Dzeng e Lee [10] have proposed a model to optimize the development schedule of resort projects using a polyploidy genetic algorithm.

Recently, in some fields of Civil Engineering an hybrid data-driven technique has been implemented, called Evolutionary Polynomial Regression (EPR), that combines the effectiveness of Genetic Programming with the advantage of classical numerical regression [2, 22, 25]. An advanced version of this method, called EPR-MOGA, uses Multi-Objective Genetic Algorithms to search those model expressions that simultaneously maximize accuracy of data and parsimony of mathematical functions. This approach generates a set of explicit expressions with different accuracy to experimental data and different degree of complexity of the structural models. The analysis of the expressions generated allows to select the solution which corresponds to the best compromise in terms of accuracy and complexity and better suited for specific applications.

In the present paper, ANN methods and the evolutionary approach based on EPR-MOGA are compared for the construction of estimation models of real estate market values. On the same database and with reference to the same explanatory variables of the unit price of residential apartments, two estimation models are implemented, one based on ANN and another using EPR, in order to test the respective performance.

The paper is structured as follows. In section 2, notes on ANN and on EPR-MOGA theory are reported. In section 3 the case study is introduced: it is relative to a sample of residential apartments recently sold in a segment of the real estate market of the city of Bari (Italy). In section 4 the ANN model and the EPR-MOGA model are specified, the calculations are carried out and the results are compared and then illustrated. In section 5 the conclusions of the work are discussed.

2   Outline of ANN Models and EPR Models

ANN are complex systems [26], formed by a set of elementary unit, the neurons, combined in an opportune way in a netting structure made of *layers* presenting an elevated degree of interconnection.

The complexity of the structure of a neural network depends on the number of neurons and the number of existing connections. Neurons can be classified on the basis of the level they occupy in the network. The first level, called the *input layer*, is formed by neurons which contain the exogenous information, translated in terms of the pulse for the neurons of the upper level. Opposite to the input layer is the *output layer*, whose neurons return the result generated by the operation of the network. Between these two levels it is possible to predict one or more intermediate levels of hidden neurons or *hidden layers*, which are entrusted with the task of developing the information coming from the input layer and translating them in the output.

In ANN the training is the responsibility of the only variable part of the structure, i.e. the weights of the connections. By altering the weights of the connections through a learning rule, the neural network is able to learn a distinctive function from couples of examples input/output (*training set*) that are repeatedly presented to it.

The transmission of information between the neurons takes place through an activation function, associated to each neuron and almost always common to all the nodes of the model.

For connections between the neurons is generally adopted the hierarchical structure, in which the connections are present only between neurons of two successive levels, and the pulses of the neurons are direct (one way) from the input layer to the output layer. In this model, called *feed-forward*, the input of the neurons that are located at a level higher than the first is given by the set of all the signals from the neurons of the lower level. Indicating with w_{ij} the weight associated to the connection between the neuron i and the neuron j, the input I_j of the neuron j can be defined as follows:

$$I_j = \sum_{i=1}^{N} O_i \cdot w_{ij} \tag{1}$$

where N is the number of neurons present in the level lower than the level in which the neuron j is located, O_i is the output of neuron i, which, through the activation function f, can be expressed as:

$$O_i = f(I_i) \tag{2}$$

So that the neural network is able to learn a particular task assigned, it is necessary to transfer to the system a learning technique, that is a rule by which it is possible to appropriately update the weights of the connections of the network.

There are many learning techniques, each of them suited to achieving the objective. The learning rule more frequently implemented is the *back-propagation* method. This rule allows to modify iteratively the weights of the connections of the network in function of the error, i.e. the difference from time to time found between the actual output of the network and the target value.

The method of EPR can be considered as a generalization of the original stepwise regression, that is linear with respect to regression parameters, but it is non-linear in the model structures. The following equation summarizes a generic non-linear model structure that can be implemented in EPR:

$$Y = a_0 + \sum_{i=1}^{n} [a_i \cdot (X_1)^{(i,1)} \cdot \ldots \cdot (X_j)^{(i,j)} \cdot f((X_1)^{(i,j+1)} \cdot \ldots \cdot (X_j)^{(i,2j)})] \qquad (3)$$

where n is the number of additive terms, a_i are numerical parameters to be valued, X_i are candidate explanatory variables, (i, l) - with $l = (1, \ldots, 2j)$ - is the exponent of the l-th input within the i-th term in Eq. (3), f is a function selected by the user among a set of possible mathematical expressions. The exponents (i, l) are also selected by the user from a set of candidate values (real numbers).

The iterative investigation of model mathematical structures, implemented by exploring the combinations of exponents to be attributed to each candidate input of Eq. (3), is performed through a population based strategy that employs a Genetic Algorithm, whose individuals are constituted by the sets of exponents in Eq. (3) and chosen by the user.

The algorithm underlying EPR does not require the exogenous definition of the mathematical expression and the number of parameters that fit better the data collected, since it is the iterative process of the genetic algorithm that returns the best solution.

The accuracy of each equation returned by EPR is checked through its Coefficient of Determination (COD), defined as:

$$COD = 1 - \frac{N-1}{N} \cdot \frac{\sum_{N}(y_{EPR} - y_{detected})^2}{\sum_{N}(y_{detected} - mean(y_{detected}))^2}, \qquad (4)$$

where y_{EPR} are the values of the dependent variable estimated by the EPR model, $y_{detected}$ are the collected values of the dependent variable, N is the sample size in analysis [15]. The fitting of a model is greater when the COD is close to the unit value.

A recent version of EPR, called EPR-MOGA [16], reproduces an evolutionary multi-objective genetic algorithm, as optimization strategy based on the Pareto dominance criterion. These objectives are conflictual, and aim at i) the maximization of model accuracy, through the satisfaction of appropriate statistical criteria of verification of the equation; ii) the maximization of model's parsimony, through the minimization of the number of terms (a_i) of the equation; iii) the reduction of the complexity of the model, through the minimization of the number of the explanatory variables (X_i) of the final equation. Through the use of a Microsoft Office Excel add-in function (EPR MOGA-XL v.1, freely available at site: www.hydroinformatics.it), the optimization strategy defined above, based on the Pareto dominance criteria, allows to obtain, at the end of the modeling phase, a set of model solutions (i.e. the Pareto front of optimal models) for the three objectives considered. In this way, a range of solutions is offered to the operator, among which it is possible to select the most appropriate solution according to the specific needs, the knowledge of the phenomenon in analysis and the type of experimental data used.

3   The Case Study

With the help of estate agents operating on site, an estimative sample of 90 residential properties sold in 2013-2014 in the Madonnella district of the city of Bari (Italy) has been collected. The boundary of the district has been defined so as to coincide with the relative "Microzone". The Italian real estate agents consider a geographical segmentation of the market in Microzones, defined according to the Presidential Decree 138/1998 and ensuing Regulation issued by the Ministry of Finance. For the Italian regulation, the "Microzone" is a part of the urban area that must be urbanistically homogeneous and at the same time must constitute a homogeneous real estate market segment. A Microzone, in other words, is an area of the real estate market in which extrinsic factors (accessibility, presence of services, building characteristics, green areas, pedestrian zones, etc.), involved in the formation of real estate values, evolve in a substantially uniform manner [11].

The Madonnella district is a central area of Bari, characterized by numerous Liberty style buildings realized in the late nineteenth century and several public buildings of cultural value made during the Fascist era. It is predominantly a residential district, with a population of about 18,000 inhabitants. The district is next to the Nazario Sauro waterfront and near the historical centre of the city (San Nicola district). It is quite accessible thanks to several bus lines.

With the help of estate agents operating in the district, the following information have been obtained for each housing unit: the *unit selling price* (*PRZ*), in euro per square meter of floor area of the property; the *floor* (*F*) the apartment is on; the *number of bathrooms* (*B*) of the property; the *panoramic view* (*P*) of the apartment, taken as a qualitative variable and differentiated, with a synthetic evaluation, by the categories "none", "enough" and "good". In the model, for this explanatory variable two dummies have been considered, respectively for the state "enough" (P_e) and "good" (P_g); the presence of *independent heating* (*H*), expressed with a dichotomous criterion, with 1 if the heating is independent in the apartment and 0 if the heating is centralized; the distance from the *center* (*C*) of the city, expressed in minutes it takes to walk to it; the *rental situation* of the apartment (*R*), expressed through a dichotomous criterion, with 1 if the apartment is rented and 0 otherwise; the *surface* (*S*) of the apartment, expressed in square meters of floor area of the property.

Other common intrinsic characteristics have been excluded from the analysis due to the relative mechanism of appreciation of the market value in the segment investigated, identifying equal conditions between the building units of the sample and not contributing to the explanation of the market value.

Table 1 provides, for the sample collected in Madonnella district, statistics for quantitative variables: continuous (unit selling price, distance from center, surface), discrete (floor the apartment is on, number of bathrooms) and dummies (panoramic view, presence of independent heating, rental situation).

Both in the ANN model as in the EPR-MOGA developed in this work, the logarithm of the unit selling price (*PRZ*) identifies the dependent variable, whereas the other parameters collected (*F*, *B*, P_e, P_g, *H*, *C*, *R*, *S*) are the explanatory variables.

3.1   The ANN Model

To specify the ANN model it is necessary to define the network topology (number of hidden layers and the neurons in each of them), the propagation rule, the activation function and the learning rule.

Here the Multi-Layer Perceptron network is employed, with one hidden layer which includes thirteen nodes, an input layer with the eight exogenous variables defined, and an output layer with the natural logarithmic of real unit prices.

A fully connected feed forward network is assumed, which means activation travels in a direction from the input layer to the output layer, and the units in one layer are connected to every other unit in the next layer up.

In the input layer and in the output layer an activation function of sigmoidal type is associated to each neuron, which is modeled continuously between 0 and 1 using the following analytical expression:

$$f(x) = \frac{1}{1 + e^{-kx}} \tag{5}$$

where *x* is the input and *k* is the slope of the tangent to the curve at the inflection point.

Several alternative topologies have been tried, with two and three hidden layers, and different number of neurons and activation functions. However the best results have been obtained with the ANN model that will be presented.

The model is implemented through the software "BKP – Neural Network Simulator" [1], that employs the algorithm of Back-Propagation (BKP) to adjust iteratively the connection weights.

In accordance to standard analytical practice, the estimative sample has been divided in a random basis into two sets, the "training set" and the "test set". The training set includes 80% of the sample, corresponding to 72 transactions, leaving the remaining 18 cases as the test set.

In the software employed, a random starting point has been used, with the following values for the main training parameters: slope *k* term = 1; learning rate = 0.65; momentum term = 0.1; maximum number of iteration = 25,000.

The Root Mean Squared Error (RMSE) has been the error function selected. A value of RMSE equal to 1.2623 is related to the model defined. The determination index (R^2) is equal to 0.9932. The Mean Absolute Percentage Errors (MAPE), that is the average percentage error between the prices of the original sample and the values estimated with ANN, is equal to 3.9155. The Maximum Absolute Percentage Errors (MaxAPE), that is the maximum percentage error between the prices of the original sample and the values estimated with the ANN model, is 10.0744.

Table 1. Sample descriptive statistics

Variable	Mean	Standard Deviation	Levels/Intervals	Frequency
Unit selling price [€/m²]	2,764.30	433.80		
Floor [n.]	2.97	1.23		
			0	0.02
			1	0.12
			2	0.14
			3	0.39
			4	0.23
			5	0.10
Number of bathrooms [n.]	1.66	0.56		
			1	0.38
			2	0.57
			3	0.05
Panoramic view			none	0.22
			enough	0.23
			good	0.55
Presence of independent heating (1-independent, 0-centralized)	0.42	0.51		
			0	0.58
			1	0.42
Distance from the center [min. walking]	9.36	4.15		
			<5	0.11
			6-10	0.55
			11-15	0.07
			16-20	0.14
Rental situation (1-rented, 0-available)	0.16	0.36		
			0	0.85
			1	0.15
Floor surface [m²]	88.92	34.87		
			<50	0.11
			51-70	0.23
			71-90	0.29
			91-110	0.08
			>110	0.29

Sensitivity analysis (Table 2) allows the evaluation of the influence of each ex-ogenous variable using its error ratio, obtained as the RMSE of the model without the explanatory variables in analysis compared to the RMSE of the model including all the variables.

Table 2. Sensitivity analysis for the ANN model

VARIABLE	RATIO	ORDER OF IMPORTANCE
C	2.016161	1
H	1.596292	2
F	1.509742	3
P_g	1.487650	4
R	1.354432	5
S	1.349758	6
P_e	1.298589	7
B	1.222877	8

It is noted that with ANN the importance of the variable, in descending order, is as follows: the *distance from the center (C)*, the presence of the *independent heating (H)*, the *floor (F)*, a "good" *panoramic view* (P_g), a "sufficient" *panoramic view* (P_e), the *rental situation (R)*, the *surface (S)* of the apartment and finally the *number of bath-rooms (B)* of the property.

3.2   The EPR Model

The base model structure reported in Eq. (3) is used with no function f selected, whe-reas each additive monomial term is assumed to be a combination of the inputs (i.e. the explanatory variables) raised to the proper exponents., Candidate exponents be-long to the set (0; -1; 1), in order to facilitate the interpretation of the results generated and to represent the direct/inverse relationship between candidate inputs and the de-pendent variable. The maximum number n of additive terms in final expressions is assumed to be 8, that is equal to the number of explanatory variables considered.

With these conditions, the run of the EPR-MOGA software has returned the twelve models (M_i) reported in Table 3, characterized by different number of additive terms and explanatory variables, as well as different accuracy in terms of COD (Figure 1).

The examination of the models shows that starting from the model *M4*, the mathe-matical expressions generated by the software are complicated, since the same expla-natory variables appear in more terms of the model, also combined with other variables. This circumstance, if it leads to functions that can reproduce better prices, on the other hand generates complicated expressions that make it difficult to interpret the phenomenon.

Table 3. Equations obtained by the implementation of EPR-MOGA

M1	$ln(PRZ) = 8.2664 - 0.0379\,C$	(6)
M2	$ln(PRZ) = 8.2046 - 0.0339C + 0.0574H$	(7)
M3	$ln(PRZ) = 8.1536 - 0.03159C + 0.0563H + 0.01\,F$	(8)
M4	$ln(PRZ) = 8.1905 - 0.0325C + 0.3561\dfrac{H}{C}$	(9)
M5	$ln(PRZ) = 8.1641 - 0.0295\,C + 0.0711\,H + \\ + 0.4403\,R - 0.0284\,C\,R$	(10)
M6	$ln(PRZ) = 8.1139 - 0.0243\,C + 0.4705\,R + \\ - 0.0322\,C\,R + 0.5132\,\dfrac{H}{C}$	(11)
M7	$ln(PRZ) = 8.0293 - 0.0258\,C + 0.3784\,R + 0.0344\,B + \\ - 0.0247\,C\,R + 0.5147\,\dfrac{H}{C} + 0.0022\,\dfrac{F\,C}{B}$	(12)
M8	$ln(PRZ) = 8.1398 - 0.0266\,C + 0.5637\,R + 0.1025\,\dfrac{1}{B} + \\ - 0.0314\,C\,R + 0.545\,\dfrac{H}{C} + 0.0034\,\dfrac{F\,C}{B} - 0.0508\,F\,R$	(13)
M9	$ln(PRZ) = 8.096 - 0.0299\,C - 0.3860\,R + 0.0232\,B + \\ + 0.5324\,\dfrac{H}{C} + 0.0039\,\dfrac{F\,C}{B} + 5.9505\,\dfrac{R}{C} - 0.02\,\dfrac{F}{B}$	(14)
M10	$ln(PRZ) = 8.0916 - 0.0234\,C + 0.5066\,R + 0.0456\,H + \\ - 0.0341\,C\,R + 0.2252\,\dfrac{B\,H}{C} - 3.7734\,\dfrac{P_g\,H}{S} + 1.2658\,\dfrac{F\,P_g}{B\,S}$	(15)
M11	$ln(PRZ) = 8.1135 - 0.0249\,C + 0.4646\,R - 0.0312\,C\,R + \\ + 0.2671\,\dfrac{B\,H}{C} - 2.4002\,\dfrac{P_g\,H}{S} + 1.1754\,\dfrac{F\,P_g}{B\,S}$	(16)
M12	$ln(PRZ) = 8.1041 - 0.0244\,C + 0.475\,R - 0.0318\,C\,R + \\ + 0.3073\,\dfrac{B\,H}{C} + 1.5295\,\dfrac{F\,P_g}{B\,S} - 0.9643\,\dfrac{F\,P_g\,H}{S}$	(17)

The COD relative to the models obtained through EPR-MOGA is always close to unity, ranging from the minimum value of 91.43%, corresponding to the model *M1*, to the maximum value of 96.29%, which is obtained for the model *M10* (Fig. 1). However, after the model *M3*, characterized by a COD equal to 93.69%, the increase of accuracy becomes relatively significant. In particular, for the models *M4*, *M9* and *M11*, in addition to the reduction in COD, the evident increase of the complexity of the mathematical function, compared to the model that immediately precedes them, should be highlighted.

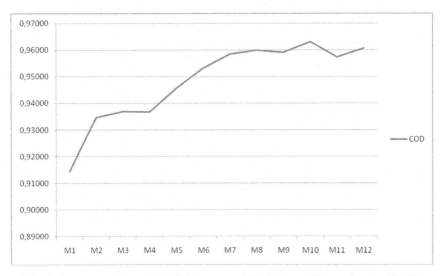

Fig. 1. Accuracy of the models (M_i) in terms of Coefficient of Determination (COD)

These considerations suggest to focus only on the first three models generated by the software.

The evolution of the mathematical equations obtained from the model *M1* to the model *M3*, provides some important information: the first concerns the importance of the main exogenous variables in the explanation of the observed prices, which can be deduced by the sequence in the appearance of the variables in the first three models: the *distance from the center* (*C*), the presence of the *independent heating* (*H*), the *floor* (*F*); the second information is related to the empirical coherence of the signs of the coefficients of the independent variables, which make explicit the inverse proportionality between the unit price (*PRZ*) and the variable *distance from the center* (*C*) and, vice versa, the direct proportionality with the variables presence of the *independent heating* (*H*) and *floor* (*F*); the third information is that the combination of the variables *distance from the center* (*C*), presence of the *independent heating* (*H*) and *floor* (*F*) already provides a good explanation of the prices. The inclusion of the other variables in the equations of the models subsequent to *M3* would generate too complex expressions, unusable for the interpretation of the phenomenon.

Therefore, *M3* is the model that is effectively able to explain and reproduce the mechanism of formation of the prices of the sample analyzed. For this model, a value of RMSE equal to 1.0723 is obtained; the determination index (R^2) is equal to 0.9998; the MAPE is equal to 3.3172; the MaxAPE is 8.8689. It should be added that the model *M3* allows to calculate the *hedonic prices* of the explanatory variables (Table 4). In fact, in a log-linear model for a non-dichotomous variable the coefficient multiplied by 100 directly provides the percentage change of the dependent variable caused by the change in the explanatory variable; whereas, for dichotomous variables, the percentage effect on the dependent variable can be obtained by the relation $\Delta p_{boj} = 100 \cdot (e^{\beta j} - 1)$, where βj is the coefficient of the *j*-th dummy variable [13, 20].

Table 4. Percentage effects of the explanatory variables selected by the model *M3* on the unit prices (*PRZ*)

Variable	Coefficient	Percentage effect [%]
distance from the center (C)	-0.03159	-3.159
indipendent heating (H)	0.0563	5.792
floor (F)	0.01	1.00

Table 4 shows the following results: each additional minute that it takes to walk to the *center* determines a decrease in the unit price equal to -3.159%. The incidence of this reduction ranges from 12.636% to 63.18%, respectively for travel time walk in about 4 minutes and about 20 minutes; the presence of the *independent heating* produces an increase of the unit price equal to 5.792%; every additional *floor* level originates an increase of the price equal to 1.00%. The incidence of this raise ranges from 0% to 5.00%, respectively for the ground floor and the fifth floor level.

3.3   Comparison Between the ANN and the EPR-MOGA Models and Interpretation of Results

The performance of ANN and EPR-MOGA models, computed in terms of R^2, RMSE, MAPE and MaxAPE, are summarized in Table 5 and Fig. 2. The indicators selected and the Graph in Figure 3 show the best accuracy of EPR-MOGA *M3* model for the evaluation of property prices relative to Madonnella district of Bari.

The comparison between the 90 detected prices and the corresponding values determined by the ANN model and the EPR-MOGA *M3* model is graphically shown in Fig. 3. The correspondence is excellent in both cases.

It is interesting to note that the EPR-MOGA model selected (*M3*) confirms the order returned by the ANN model related to the importance of the explanatory variables, at least for the positions of the "podium" (Table 2). The analysis confirms that in the Madonnella district the location (variable *C*) is the feature with the highest influence on the appreciation of a residential property. In this regard, it should be noted that the importance of this variable is enhanced by the absence of fast connections in the Madonnella district (e.g. subway or light railway) with the center of Bari.

The presence of the independent heating (*H*) obtains the second place. The relevance of this variable expresses, on the one hand, the appreciation of buyers for more home comfort determined by the independent heating; on the other hand, the possibility to avoid disagreements that often arise in apartment blocks about the heating: indeed, although the Italian Condominium Act (L. 220/2012) has simplified the procedures for the separation from the central heating system - provided that the intervention does not entail any additional expenses and/or inconvenience to the operation of the central heating system - other regulations (D.P.R. 59/2009; D.L. 192/2005) require that the owner of the apartment who wants to perform the separation from the central heating system must verify the energy saving capacity of the independent heating to be realized. The legal disputes that may arise from the failure to comply these regulations are often a disincentive to the conversion of the central heating system in an independent heating of an apartment.

The significant impact of the floor (F) the apartment is on, reflects the appreciation that is normally attributed to this feature when the building in which the apartment is located has a lift, which happens to all the apartments of the sample collected.

The other variables of the initial database (P_g, R, S, P_e, B) participate in the mechanism of price formation in the ANN model, but they are not included in the EPR-MOGA *M3* model.

Furthermore, both models allow to overcome the effects of collinearity that might occur with other procedures (e.g. hedonic pricing method), that, for example, can involve the variables *floor* and *panoramic view*. In particular, the EPR-MOGA recognizes the *floor* level as a "proxy" variable of the *panoramic view* in all models generated: indeed, only the variable "good" *panoramic view* (P_g) appears in the last three models (*M10, M11* and *M12*), characterized by mathematical expressions considerably complex, in which the variables F and P_g end up to being combined in a unique additive term.

Table 5. Comparison of the performance of ANN and EPR-MOGA M_3 models

Performance measures	ANN model	EPR model
R^2	0.9932	0.9998
RMSE	1.2623	1.0723
MAPE	3.9155	3.3172
MaxAPE	10.0744	8.8689

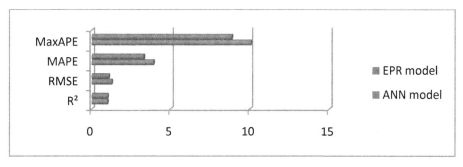

Fig. 2. Graphical comparison of the performance of ANN and EPR-MOGA M_3 models

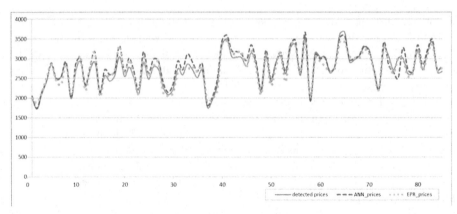

Fig. 3. Comparison between the detected unit prices (continuous line), the unit prices estimated with the ANN model (broken line) and the EPR-MOGA M_3 model (dashed line)

4   Conclusions

In the phase of uncertainty that is characterizing the Italian real estate sector, the use of innovative assessment tools may allow market operators to formulate more reliable estimates, as well as to effectively monitor the evolution of property values [5, 7, 8, 18, 19, 35].

In this paper, two methods have been implemented on the same database and with reference to the same explanatory variables of market prices of residential apartments recently sold in a district of the city of Bari (Italy): one based on ANN theory, the other one using EPR-MOGA procedure and selecting the most appropriate equation in terms of accuracy and interpretability of the results.

Both the models tested have shown excellent performance, but the EPR-MOGA M_3 model allows to obtain simultaneously the best statistical accuracy in the prediction of the market prices, a quick check of the empirical consistency of results obtained as well as the overcoming of the main limitations of ANN models. In fact, ANN are "black boxes", i.e. it does not allow to generate a straightforward functional relationship between the input and the output values nor punctually investigate and reproduce the mechanisms of the prices formation. Furthermore, it can happen that the results obtained through ANN could not be stable, but could improve with increasing sample size, as well as results from models prepared with the same data but generated by different software packages could be different. The EPR model, instead, overcomes these shortcomings: the transparency of the mathematical expression obtained allows to verify (and quantify) the significance of the explanatory variables in the formation of property prices.

The EPR-MOGA *M3* model selected in this work is configured as a semi-logarithmic equation obtainable from the application of hedonic pricing method, considering as independent variables the only three characteristics identified by the EPR-MOGA *M3* model as the most representative of the property prices. The advantage of the implementation of the genetic algorithm underlying the logic of

EPR-MOGA is the ability to automatically and quickly select the optimal regressive functions in terms of accuracy and understanding. This result can be hardly obtained through the use of traditional hedonic pricing method. Therefore, this is an important added value in the real estate appraisal, in which it is essential to have forecasting functions characterized by high performance and at the same time easy to interpret.

The EPR-MOGA model identified in this work could have interesting applications: indeed, expressing property prices as a function of only three explanatory variables (the distance from the center, the presence of the independent heating and the floor level) and of a constant additive term, it is likely to be a simple and effective tool in the reform of Italian Cadastre (L. 23/2014) for the assignment of a mass appraisal function to properties located in the same territorial microzone and whose, in this case, the location and two intrinsic characteristics, easily acquirable, should be taken over.

Acknowledgements. This study has been developed within research activities being carried out by Real Estate Valuation Center of the MITO-LAB (LABoratory of Multimedia Information for Territorial Objects) of the Polytechnic of Bari, Italy (web site: http://mitolab.poliba.it).

References

1. Barile, S., Magna, L., Marsella, M., Miranda, S.: A marketing decision problem solved by application of neural networks. In: International Conference on Computational Intelligence and Multimedia Applications (1999)
2. Berardi, L., Kapelan, Z., Giustolisi, O., Savic, D.: Development of pipe deterioration models for water distribution systems using EPR. Journal of Hydroinformatics **10**(2), 113–126 (2008)
3. Brunson, A.L., Buttimer, R.J., Rutherford, R.C.: Neural Networks, Nonlinear Specification and Industrial Property Values. Working Paper Series, pp. 94–102. University of Texas at Arlington (1994)
4. Calabrò, F., Della Spina, L.: The cultural and environmental resources for sustainable development of rural areas in economically disadvantaged contexts. Economic-appraisals issues of a model of management for the valorisation of public assets. Advanced Materials Research **869–870**, 43–48 (2014)
5. Calabrò, F., Della Spina, L.D.: The public-private partnerships in buildings regeneration: A model appraisal of the benefits and for land value capture. Advanced Material Research **931–932**, 555–559 (2014)
6. D'Alpaos, C., Canesi, R.: Risks assessment in real estate investments in times of global crisis. WSEAS Transactions on Business and Economics **11**, 369–379 (2014)
7. Del Giudice, V., De Paola, P.: Geoadditive models for property market. Applied Mechanics and Materials **584**, 2505–2509 (2014)
8. Del Giudice, V., De Paola, P.: The effects of noise pollution produced by road traffic of Naples Beltway on residential real estate values. Applied Mechanics and Materials **587**, 2176–2182 (2014)
9. Do, A.Q., Grudnitski, G.: A neural network analysis of the effect of age on housing values. Journal of Real Estate Research, American Real Estate Society **8**(2), 253–264 (1993)

10. Dzeng, R.J., Lee, H.Y.: Optimizing the development schedule of resort projects by integrating simulation and genetic algorithm. International Journal of Project Management **25**(5), 506–516 (2007)
11. Forte, F.: Costs of noise and Italian urban policies. In: 36th International Congress and Exhibition on Noise Control Engineering, vol. 6, pp. 3991–3999. Istanbul, Turkey (2007)
12. Gallego, J.: La inteligencia artificial aplicada a la valoraciòn de inmuebles. Un ejemplo para valorar Madrid. Revista CT/Catastro **50**, 51–67 (2004)
13. Giles, D.E.: Interpreting Dummy Variables in Semi-logarithmic Regression Models: Exact Distributional Results. Econometrics Working Paper EWP1101, University of Victoria, Canada (2011)
14. Giustolisi, O., Laucelli, D.: Increasing generalisation of input-output artificial neural networks in rainfall-runoff modelling. Hydrological Sciences Journal **3**(50), 439–457 (2005)
15. Giustolisi, O., Savic, D.: A symbolic data-driven technique based on evolutionary polynomial regression. Journal of Hydroinformatics **8**(3), 207–222 (2006)
16. Giustolisi, O., Savic, D.: Advances in data-driven analyses and modelling using EPR-MOGA. Journal of Hydroinformatics **11**(3–4), 225–236 (2009)
17. Guarini, M.R., Battisti, F.: Social Housing and Redevelopment of Building Complexes on Brownfield Sites: The Financial Sustainability of Residential Projects for Vulnerable Social Groups. Advanced Materials Research **869–870**, 3–13 (2014)
18. Guarnaccia, C., Quartieri, J., Mastorakis, N.E., Tepedino, C.: Development and Application of a Time Series Predictive Model to Acoustical Noise Levels. WSEAS Transactions on Systems **13**, 745–756 (2014)
19. Guarnaccia, C.: Advanced Tools for Traffic Noise Modelling and Prediction. WSEAS Transactions on Systems **12**(2), 121–130 (2013)
20. Halvorsen, R., Palmquist, R.: The interpretation of dummy variables in semilogarithmic regressions. American Economic Review **70**, 474–475 (1980)
21. Islam, K.S., Asam, Y.: Housing market segmentation: a review. Review of Urban & Regional Development Studies **21**(2–3), 93–109 (2009)
22. Javadi, A.A., Rezania, M.: Applications of artificial intelligence and data mining techniques in soil modeling. Geomechanics and Engineering **1**(1), 53–74 (2009)
23. Koza, J.R.: Genetic programming: on the programming of computers by natural selection. MIT Press, Cambridge (1992)
24. Król, D., Lasota, T., Trawiński, B., Trawiński, K.: Investigation of evolutionary optimization methods of TSK fuzzy model for real estate appraisal. International Journal of Hybrid Intelligent Systems **5**(3), 111–128 (2008)
25. Laucelli, D., Giustolisi, O.: Scour depth modelling by a multi-objective evolutionary paradigm. Environmental Modelling & Software **26**(4), 498–509 (2011)
26. Limsombunchai, V., Gan, C., Lee, M.: House price prediction: hedonic price model vs. artificial neural network. American Journal of Applied Sciences **1**(3), 193–201 (2004)
27. Liu, J.-G., Zhang, X.-L., Wu, W.-P.: Application of fuzzy neural network for real estate prediction. In: Wang, J., Yi, Z., Żurada, J.M., Lu, B.-L., Yin, H. (eds.) ISNN 2006. LNCS, vol. 3973, pp. 1187–1191. Springer, Heidelberg (2006)
28. McCluskey, W.J., Dyson, K., McFall, D., Anand, S.: The mass appraisal of residential property in Northern Ireland. In: Computer Assisted Mass Appraisal: An International review, pp. 59–77. Ashgate Publishing Limited, England (1997)
29. Morano, P., Tajani, F.: Bare ownership evaluation. Hedonic price model vs. artificial neural network. International Journal of Business Intelligence and Data Mining **8**(4), 340–362 (2013)

30. Morano, P., Tajani, F.: Least median of squares regression and minimum volume ellipsoid estimator for outliers detection in housing appraisal. International Journal of Business Intelligence and Data Mining **9**(2), 91–111 (2014)
31. Oppio, A., Corsi, S., Mattia, S., Tosini, A.: Exploring the relationship among local conflicts and territorial vulnerability: The case study of Lombardy Region. Land Use Policy **43**, 239–247 (2015)
32. Scorza, F., Casas, G.L., Murgante, B.: Overcoming interoperability weaknesses in e-government processes: organizing and sharing knowledge in regional development programs using ontologies. In: Lytras, M.D., Ordonez de Pablos, P., Ziderman, A., Roulstone, A., Maurer, H., Imber, J.B. (eds.) WSKS 2010. CCIS, vol. 112, pp. 243–253. Springer, Heidelberg (2010)
33. Scorza, F., Casas, G.B., Murgante, B.: That's ReDO: ontologies and regional development planning. In: Murgante, B., Gervasi, O., Misra, S., Nedjah, N., Rocha, A.M.A., Taniar, D., Apduhan, B.O. (eds.) ICCSA 2012, Part II. LNCS, vol. 7334, pp. 640–652. Springer, Heidelberg (2012)
34. Selim, H.: Determinants of house prices in Turkey: hedonic regression versus artificial neural network. Expert Systems with Applications **36**(2), 2843–2852 (2009)
35. Torre, C.M., Mariano, C.: Analysis of fuzzyness in spatial variation of real estate market: Some Italian case studies. Smart Innovation, System and Technologies **4**, 269–277 (2010)
36. Wang, W.K.: A knowledge-based decision support system for measuring the performance of government real estate investment. Expert Systems with Applications **29**(4), 901–912 (2005)
37. Wong, K.C., Albert, P.S., Hung, Y.C.: Neural network vs. hedonic price model: appraisal of high-density condominiums. Real Estate Valuation Theory **8**(2), 181–198 (2002)

Spline Smoothing for Estimating Hedonic Housing Price Models

Vincenzo Del Giudice[1], Benedetto Manganelli[2(✉)], and Pierfrancesco De Paola[1]

[1] University of Naples, Piazzale V. Tecchio 80125, Naples, Italy
vincenzo.delgiudice@unina.it, pfdepaola@libero.it
[2] University of Basilicata, Viale dell'Ateneo Lucano 85100, Potenza, Italy
benedetto.manganelli@unibas.it

Abstract. The exact prediction of housing selling prices is a relevant issue for real estate market, also to evaluate alternative forms of financial investment. In this paper a hedonic price function built through a semiparametric additive model is implemented. This model use penalized spline functions and aims to achieve a significant improvement in the prediction of the market price of the properties.

Keywords: Penalized spline · Semiparametric regression · Additive models · Property investment · Real estate valuation

1 Introduction

The complex evolution of the real estate market is influenced by quantitative and qualitative characteristics, as well as by differentiation and change of the appreciation mode of real estate goods. Then these aspects suggest the development of new and advanced models for quantitative analysis of property prices, able to recognize the different kinds of appreciation, based on the detection and analysis of statistical market data [1][2].

In this paper, we proposed as aims the formulation of a semi-parametric statistical models that can improve the performance, in terms of estimation, respect to the usual predictive multiparametric models.

In international context, many recent studies have applied some special nonparametric or semiparametric additive regressions for formulate hedonic price models for the analysis of the housing market. Mainly, in these studies were used the Generalized Additive Models, among the most common techniques regressive nonparametric multivariate, and the "backfitting algorithm" [3] that represents the main method for resolution of additive models in base to available statistics data.

As regards the Generalized Additive Models, they are based on the sum of q nonparametric functions, relating to q variables of T, plus a constant term (α); also providing for the use of a link function, note and parametric [$G(\cdot)$], such as to connect the different functions that bind the dependent variable (Y) for each predictor [4]:

This paper is to be attributed in equal parts to the three authors.

© Springer International Publishing Switzerland 2015
O. Gervasi et al. (Eds.): ICCSA 2015, Part III, LNCS 9157, pp. 210–219, 2015.
DOI: 10.1007/978-3-319-21470-2_15

$$E[Y|T] = G\left[\alpha + \sum_{j=1}^{q} f_j(T_j)\right]$$

where T is a generic vector of numerical explanatory variable: $T = (T_1, \ldots, T_q)^T$.

On the other hand, the backfitting algorithm represents a very flexible tool, which allows to build models with components very articulate and defined by an iterative procedure which suggests the algorithm that can be used for the estimation of individual functions in an additive model, removing the effects of all other functions (other of *i-th* considered) towards the dependent variable. This iteration of the algorithm is then performed until the individual function of the model does not suffer more variations between one iteration and the other [5].

An alternative approach to the studies since carried out here, with limited computational difficulties in estimating the individual functions that define an additive model, consists to place and match to each of these functions some specific smoothing spline on which was later given account.

The application of smoothing spline functions currently interest many fields of scientific research like chemistry, natural and physical sciences, medicine, economy (albeit limited to issues relating to the costs).

Early implementations of the techniques that use smoothing spline are attributable to some applications carried out in the field of physical sciences [6], with the first significant studies in the economic [7][8][9], but very limited are currently the applications in real estate field.

It should be also reported that studies relating to the property market and to urban economy, focused on the use of smoothing spline functions, are mainly due to Anderson [10][11], which estimate forecast models of urban residential density; Sunderman et al. [12] formulate models to check the iniquity of the U.S. taxation systems on the real estate; Speyrer and Ragas [13] highlight the impact of the risk of flooding on property values; Zheng [14] applies density functions to examine the spatial structure of the metropolitan area of Tokyo; Coulson [15] formulates a model for estimating hedonic price based on smoothing spline for a properties sample in Pennsylvania; and, finally, Bao and Wan [16] analyse the residential property market in Hong Kong based on a sample of about 170,000 transactions relating to real estate units.

In several recent contexts in the field of statistics, the international scientific research has also further implemented the semiparametric techniques regressive employing functions smoothing spline, combining them with mixed models characterized by a partial systematic component. This has given rise to applications of semi-parametric regression models with mixed effects that use specific penalized spline functions, greatly limiting the random effects [17][18][19][20][21][22].

The application of semiparametric models to the field of real estate appraisal is currently the subject of specialized literature, and particularly it concerns the collection and processing of the prices and the features of the properties [23][24][25][26].

2   Model Specification

About the theoretical aspects of the proposed model, it should be noted that the relationship between the selling housing price and the explanatory variables is measured with a semi-parametric additive model, characterized by the combination of a generalized additive model which expresses the relationship between the non-linear response and the explanatory variables, and a linear mixed effects with which you express the spatial correlation of observed values[27][28]:

$$prezzo = \beta_0 + \beta_1 serv + \beta_2 affacci + \beta_3 man + \beta_4 liv + f_1(epoca) + $$
$$+ f_2(cvend) + f_3(\text{sup int}) + f_4(\text{sup } balc) + f_5(\text{sup } coll) + \varepsilon_i \tag{1}$$

More precisely, in the expression (1) the additive component, the mixed effects and the erratic component (ε), are independent. Furthermore, in order to obtain a function estimated using the procedures relating to models mixed effects, it is considered a version of low rank both for the additive component both for the mixed effects[25].

The proposed semiparametric model can then be briefly defined by the following general formula:

$$y = X\beta + Zu + \varepsilon \tag{2}$$

where:

- $y = (y_1, \ldots, y_N)^T$
- $X = [\,1\ x_i\,]_{1 \le i \le N}$
- Z contains $T \le N$ truncated power basis functions of p-degree for the approximation of nonlinear structure in f functions:

$$Z = \begin{bmatrix} (x_1 - \kappa_1)^p_+ & \ldots & (x_1 - \kappa_k)^p_+ \\ \ldots & \ldots & \ldots \\ (x_n - \kappa_1)^p_+ & \ldots & (x_n - \kappa_k)^p_+ \end{bmatrix}$$

alternatively, in reduced form:

$$Z = \left[(x_i - \kappa_k)^p_+ \right]_{\substack{1 \le k \le K \\ }}{}_{1 \le i \le n}$$

- $u = (u_1, \ldots, u_k)^T$ is the vector of random effects with:

$$E(u) = 0, \quad Cov(u) = \sigma^2_u I, \quad Cov(\varepsilon) = \sigma^2_\varepsilon I,$$

considering the coefficients (u_k) of knots (κ_k) as random effects independent of ε term [29].

Note that the formulation (2) is a particular case of linear mixed-effects model of Gaussian type. For non-linear components of the model are used Penalized spline functions qualified by the following general expression:

$$f(x) = \alpha_0 + \alpha_1 x + \dots + \alpha_p x^p + \sum_{k=1}^{K} \alpha_{pk} (x - \kappa_k)^p_+ \tag{3}$$

in which the base of the generic function (3) is represented by the following terms:

$$1, x, \dots, x^p, (x - \kappa_1)^p_+, \dots, (x - \kappa_k)^p_+$$

Where the generic function $(x - \kappa_k)^p_+$ has (p – 1) continuous derivatives.

For *p > 0* the expression that is used to determine the fitted values is as follows:

$$\hat{y} = X(X^T X + \lambda^{2p} D)^{-1} X^T y \tag{4}$$

Where:

$$X = \begin{bmatrix} 1 & x_1 & .. & x_1^p & (x_1 - \kappa_1)^p_+ & \dots & (x_1 - \kappa_k)^p_+ \\ .. & .. & .. & \dots & \dots & \dots & \\ .. & .. & .. & \dots & \dots & \dots & \\ .. & .. & .. & \dots & \dots & \dots & \\ 1 & x_p & .. & x_n^p & (x_n - \kappa_1)^p_+ & \dots & (x_1 - \kappa_k)^p_+ \end{bmatrix}$$

$$D = diag\left(0_{p+1}, 1_K\right) \tag{5}$$

Simplifying, the relation (4) becomes:

$$\hat{y} = S_\lambda \cdot y \tag{6}$$

The smoother matrix is defined as follows:

$$S_\lambda = X(X^T X + \lambda^{2p} D)^{-1} X^T \tag{7}$$

The λ term is usually referred to as smoothing parameter.

The smoothing parameter intervenes in the determination of the degrees of freedom for nonlinear component of the model and allows also to control the trade-off between fitting model to the observed values (smoothing parameter near to zero value) and the smoothness of the same (high values of smoothing parameters).

The selection of the smoothing parameter, for a spline function of *p-degree*, occurs by the Restricted Maximum Likelihood condition:

$$\hat{\lambda}_{REML} = \left(\hat{\sigma}^2_{\varepsilon,REML} \Big/ \hat{\sigma}^2_{u,REML} \right)^{1/2p} \tag{8}$$

In the formula (8) values are determined by minimizing the following condition:

$$\ell_R(V) = -\frac{1}{2}\left[n\log(2\pi) + \log|V| + \log\left|X^T V^{-1} X\right| + y^T V^{-1}\left\{I - X\left(X^T V^{-1} X\right)^{-1} X^T V^{-1}\right\}y\right]$$

with:

$$V = \hat{\sigma}^2{}_u ZZ^T + \hat{\sigma}^2{}_\varepsilon \tag{9}$$

Following the definition of the parameter λ obtained through the condition of restricted maximum likelihood, the estimates of the β coefficients and the predictions of the u variables may be obtained as follows:

$$\hat{y} = X\hat{\beta} + Z\hat{u} \tag{10}$$

The degrees of freedom for smoother function (df) are finally defined as follows:

$$df_{fit} = tr(S_\lambda) \tag{11}$$

$$df_{fit} = tr\{(X^T X + \lambda^2 D)^{-1} X^T X\} \tag{12}$$

Note, finally, that the representation of low rank considered in the model allows that the number of bases functions used does not increase with the sample size.

3   Application

The market price analysis of the property carried out with the use of an additive semi-parametric model provides for the adoption of statistical tools (significance test, measures of residues, etc.) able to select both the sample data and the endogenous variables [30]; these tools also allow to verify the reliability and the quality of results.

The algebraic structure of proposed model has been specified on the basis of real estate data of the sample, as well with the help of statistical and empirical-argumentative tests, by implementing the following semiparametric additive model:

$$prezzo = \beta_0 + \beta_1 serv + \beta_2 affacci + \beta_3 man + \beta_4 liv + f_1(epoca) +$$
$$+ f_2(cvend) + f_3(\text{sup int}) + f_4(\text{sup } balc) + f_5(\text{sup } coll) + \varepsilon_i$$

The data sample refers to a defined real estate market segment of a medium size urban center (Campania region, Italy) and, specifically, no. 148 sales of residential property units located in an urban central area during eight years (tables 1 and 2).

The sampled properties have the same build type and quality (residential units located in used multi-storey buildings recently built), and they are included in a homogeneous urban area in terms of qualification and distribution of main services.

In the absence of multicollinearity phenomena, given the low correlation between the explanatory variables, the main verification indexes of the model are shown for completeness in tables and graphs.

The amounts related to the percentage error (7,50%) and absolute percentage error (6,47%) appear congruent, because the forecast values obtained using the proposed model show a trend compliant to observed data, also even residue analysis shows no abnormalities (figure 1).

From the statistical point of view, significant is the determination index, equal to 0,941 (corrected index equal to 0,938), as well as the F test is significant for a 95% confidence level.

Table 1. Variable description

Variable	Description
Real estate price (prz)	expressed in thousands of Euros
Property's age (epoca)	expressed retrospectively in no. of years
Sale date (cvend)	expressed retrospectively in no. of months
Internal area (supint)	expressed in sqm
Balconies area (supbalc)	expressed in sqm
Connected area (supcoll)	expressed in sqm (lofts, cellars, etc.)
Number of services (serv)	no. of services in residential unit
Number of views (affacci)	no. of views on the street
Maintenance (man)	expressed with dichotomous scale (1 or 0, respectively, for the presence or absence of optimal maintenance state)
Floor level (liv)	no. of floor level of residential unit

Table 2. Statistical description of variables

Variable	Std. Dev.	Median	Mean	Min	Max
prz	32.134,11	114.585	113.502	41.316	250.000
epoca	7,40	24	23,51	10	35
cvend	23,77	33	37,30	2	96
supint	25,88	117	118,80	48	210
supbalc	10,19	16	17,56	0	59
supcoll	15,78	0	9,53	0	62
serv	0,52	2	1,61	1	3
affacci	0,60	2	2,27	1	4
man	0,33	0	0,12	0	1
liv	1,63	3	3,16	1	7

The fixed effects of model's linear component that result statistically significant coincide with two variables: maintenance (*man*) and services number (*serv*); the intercept and the variables *affacci* (no. of views on the street) and *liv* (floor level) do not exceed the *t* test, even though they are not excluded from the analysis for the importance that have in the formation of real estate prices.

With regard to the nonlinear part of model, there are no significant abnormalities encountered in the values assumed by smoothing parameters (*spar*) or freedom degrees (*df*), (table 3).

In the model's linear component, the variables' coefficients directly express the implicit marginal prices; for the nonlinear component, marginal prices for each variable are obtained by processing and examination of estimated functions (figure 2).

For brevity, for each nonlinear variable of model, the marginal prices are only graphically shown, being a primary objective of this paper the experimentation of proposed model (figure 2).

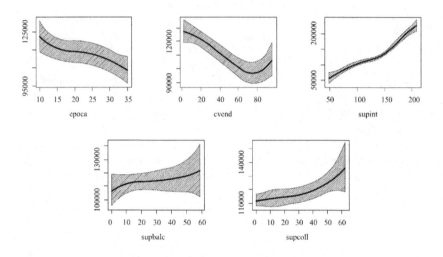

Fig. 1. Effects of nonlinear components on selling prices with representation of 95% confidence level

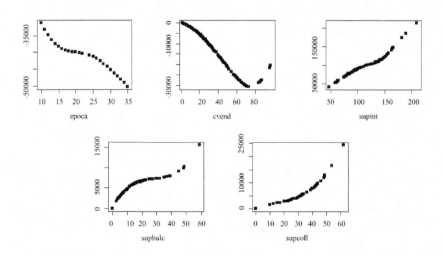

Fig. 2. Predicted values for each nonlinear variable of model

In conclusion, this work leads to results which, for their consistency with buying and selling prices detected, can be considered representative of the validity of the methodology used.

The tool used for analyse the real estate data is the *R-project* software.

Table 3. Main results of semiparametric model

No. of variables	9
Freedom degrees	138
Mean for PRZ variable (€)	113.502
Standard Error (€)	8.561
Percentage Error (SE/Mean for PRZ variable) (%)	7,50
Determination index (R^2)	0,941
Corrected determination index (R^2_c)	0,938
Absolute percentage error (%)	6,47
F-test	284,88 ($> F_{(\alpha, n, m\text{-}n\text{-}1)}$)

Estimation of fixed effects for linear and nonlinear components of the model:

Variable	coef	SE	Ratio	P-value	t-test	
intercept	54.200	70.280	0,77	0,44	0,77	
serv	3.798	2.203	1,72	0,08	1,72	
affacci	878,4	1.554	0,56	0,57	0,56	
man		6.831	2.711	2,52	0,01	2,52
liv	-461,9	544,8	-0,85	0,40	0,85	

Variable	df	spar	knots
f (epoca)	3,101	8,69	5
f (cvend)	3,540	33,28	16
f (supint)	5,139	26,24	15
f (supbalc)	3,000	302,20	8
f (supcoll)	3,000	251,70	6

4   Conclusions

The excellent results obtained by the application of the proposed model suggest that semi-parametric models can be successfully used for prediction of the residential property selling price.

In this case, the error committed in the prediction of selling prices is in fact lower by about 18%, compared to conventional multiple regression models, showing among other things the high potential of proposed model.

More generally, as amply confirmed by the international literature, the application of semi-parametric models leads to an improved estimated between 10 and 20% in the forecast of the properties market price, compared to conventional multiparametric methods, in line with the experimental results obtained in this contribution.

In the analysis of the housing market, the proposed model fits to the knowledge of real estate (price trends, housing demand, construction's cost indexes, etc.), determining a more precise measurement of the examined phenomenon, quantifying the uncertainty related to such measuring and being able to identify any "alarm state" about the data used such as, for example, in the evaluation of potential real estate speculation leading up to a "bubble" real estate.

Then, semiparametric econometric models used for forecasting purposes are of interest in the economy for every productive enterprise that devote resources to the planning of its future activities and, therefore, fundamental is the usefulness of having reliable scenarios on the future value of the variables that influence the trends of costs and income.

The aims pursued with the theoretical model proposed are many and varied, as the study of different segments in local real estate market, or even the prediction and interpretation of phenomena related to the genesis of building income, with particular reference to the problems of transformation of urban areas affected from projects or plans of action, and in order to optimize the choices of use of public or private goods and resources.

Looking ahead, further developments of semi-parametric model proposed will depend to a large extent by the results of research on methods for detection and processing of the data property, through appropriate analysis tools that can also orient the choice of the functional form of the model best suited to the phenomenon examined.

References

1. Morano, P., Tajani, F.: Bare ownership evaluation. Hedonic price model vs. artificial neural network. International Journal of Business Intelligence and Data Mining **8**(4), 340–362 (2014)
2. Manganelli, B., Pontrandolfi, P., Azzato, A., Murgante, B.: Using geographically weighted regression for housing market segmentation. International Journal of Business Intelligence and Data Mining **9**(2), 161–177 (2014)
3. Friedman, J.H., Stuetzle, W.: Projection pursuit regression. Journal of the American Statistical Association **76**, 817–823 (1981)
4. Ruppert, D., Wand, M.P., Carroll, R.J.: Semiparametris regressions. Cambridge University Press (2003)
5. Hastie, T., Tibshirani, R.: Generalized Additive Models. Chapman & Hall, New York (1990)
6. Whittaker, E.T.: On a new method of graduation. Proceedings of the Edinburgh Mathematical Society **41**, 63–75 (1923)
7. Engle, R.F., Granger, C.W.J., Rice, J., Weiss, A.: Semiparametric estimates of the relation between weather and electricity sales. Journal of the American Statistical Association **81**, 310–320 (1986)
8. Koenker, R., Ng, P.T., Portnoy, S.: Quantile smoothing splines. Biometrika **81**(4), 673–680 (1994)
9. Craig, S.G., Ng, P.T.: Using quantile smoothing splines to identify employment subcenters in a multicentric urban area. Journal of Urban Economics **49**(1), 100–120 (2001)
10. Anderson, J.E.: Cubic-spline urban density functions. Journal of Urban Economics **12**(2), 155–167 (1982)
11. Anderson, J.E.: The changing structure of a city: temporal changes in cubic spline urban density patterns. Journal of Regional Science **25**(3), 413–426 (1985)
12. Sunderman, M.A., Birch, J.W., Cannaday, R.E., Hamilton, T.W.: Testing for vertical inequity in property tax systems. Journal of Real Estate Research **5**(3), 319–334 (1990)

13. Speyrer, J.F., Ragas, W.R.: Housing prices and flood risk: an examination using spline regression. Journal of Real Estate Finance and Economics **4**(4), 395–407 (1991)
14. Zheng, X.P.: Metropolitan spatial structure and its determinants: a case study of Tokyo. Urban Studies **28**(1), 87–104 (1991)
15. Coulson, N.E.: Semiparametric estimates of marginal price of floorspace. Journal of Real Estate Finance and Economics **5**(1), 73–83 (1992)
16. Bao, H., Wan, A.: On the use of spline smoothing in estimating hedonic housing price models: empirical evidence using Hong Kong data. Real Estate Economics **32**(3), 487–507 (2004)
17. Parise, H., Wand, M.P., Ruppert, D., Ryan, L.: Incorporation of historical controls using semiparametric mixed models. Applied Statistics **50**(1), 31–42 (2001)
18. Coull, B.A., Ruppert, D., Wand, M.P.: Simple incorporation of interactions into additive models. Biometrics **57**(2), 539–545 (2001)
19. Coull, B.A., Schwartz, J., Wand, M.P.: Respiratory health and air pollution: additive mixed model analyses. Biostatistics **2**, 337–349 (2001)
20. Zheng, H., Little, R.J.A.: Penalized spline nonparametric mixed models for inference about a finite population mean from two-stage samples. Survey Methodology **30**, 209–218 (2004)
21. Opsomer, J.D., Claeskens, G., Ranalli, M.G., Kauermann, G., Breidt, F.J.: Nonparametric small area estimation using penalized spline regression. Journal of the Royal Statistical Society, Series B **70**(1), 265–286 (2008)
22. Montanari, G.E., Ranalli, M.G.: A mixed model-assisted regression estimator that uses variables employed at the design stage. Statistical methods and applications **15**(2), 139–149 (2006)
23. Bin, O.: A prediction comparison of housing sales prices by parametric versus semiparametric regressions. Journal of Housing Economics **13**(1), 68–84 (2004)
24. Clapp, J.: A semiparametric method for estimating local house price indices. Real Estate Economics **32**(1), 127–160 (2004)
25. Gencay, R., Yang, X.: A forecast comparison of residential housing prices by parametric versus semi-parametric conditional mean estimators. Economics Letters **52**(2), 129–135 (1996)
26. Pace, K.: Appraisal using generalized additive models. Journal of Real Estate Research **15**(1), 77–99 (1998)
27. Del Giudice, V., De Paola, P.: Geoadditive Models for Property Market. Applied Mechanics and Materials **584–586**, 2505–2509 (2014)
28. Del Giudice, V., De Paola, P.: Undivided real estate shares: appraisal and interactions with capital markets. Applied Mechanics and Materials **584–586**, 2522–2527 (2014)
29. Wand, M.P.: Smoothing and mixed models. Computational Statistics **18**, 223–249 (2003)
30. Morano, P., Tajani, F.: Least median of squares regression and minimum volume ellipsoid estimator for outliers detection in housing appraisal. International Journal of Business Intelligence and Data Mining **9**(2), 91–111 (2014)

Urban Heritage Regeneration in the Old Town of Ragusa (Italy): An Architecture-Centred Analysis-Valuation-Plan Pattern

Salvatore Giuffrida$^{(\boxtimes)}$, Filippo Gagliano, and Vittoria Ventura

Department of Civil Engineering and Architecture, University of Catania, Catania, Italy
sgiuffrida@dica.unict.it,
{fmgagliano,vittoriaventura01}@gmail.com

Abstract. The old town of Ragusa is undertaking a regeneration process in some different directions. Its differently valued areas need to be comprised in a whole valorisation strategy, connecting the decision making process with functional and symbolic values. In such a multi-layered urban sub-system, the architectural unit can play the role of one of the main information/value units which the valuation/decision making pattern starts from. The pattern works as an impact assessment tool, that generates a great number of intervention strategies in order to maximize different and conflicting objectives, and provides for each of them an axiological profile aimed at defining a trade-off scheme between them. Some GIS tools and a WebGIS interface allow the real-time control of the different scenarios envisaged in the decision making process.

Keywords: Old town valorisation · Architectural values · Equalization · GIS spatial analysis · Semiotic approach

1 Introduction

The dense and structured historic centres constitute urban units mainly characterized by the urban fabric. The latter is the form of diffuse capital in which the biggest quota of the accumulated social value, measurable in terms of global cost value and patrimonial value as well, is immobilized. This circumstance justifies the interest convergence of several architectural disciplines, and legitimizes the integration of the valuation programmatic themes in the formation of programming and management tools of the valorisation processes [18].

The value and valuation issues precede the design ones, which are not justified outside the constant reference to the produced differential value.

In the current political and cultural context – nowadays more and more demanding regarding the need for validation of the processes concerning modifications of the collective and personal wealth allocation model – this differential is requested to be represented in an effective way for the purpose of social communication that means choice sharing and participation. The valuation science tackles a great challenge in

© Springer International Publishing Switzerland 2015
O. Gervasi et al. (Eds.): ICCSA 2015, Part III, LNCS 9157, pp. 220–236, 2015.
DOI: 10.1007/978-3-319-21470-2_16

this field, having to adjust the structural and deep value matrix to the many languages crossing the surface of the valuations, and to individuate from time to time the most suitable valorising substance for this purpose.

This methodology address is currently widely established in the planning practice, and it has induced the abandonment of the restraint approach and the zoning; as a consequence, wide planning perspectives "from the bottom", and for projects, have opened up. The project evaluation privileges these types of operative field, in which its commitment for the recovery of a global vision inspired by values and founded in the argumentation on the correct combination of personal and collective interest is manifest.

The "deconstruction" of the plan structure is coherent with a "digital" or numerical approach to the city, based on the abstraction and the management of information, therefore on a representation by record and fields, that on the one hand extracts the significant characteristics of the diverse goods, on the other allows to manage great quantities of data reducing their informative content *to the source,* in order to improve their chances to be transmitted to the *recipient.*

This study is inspired by the above mentioned type of approach; it is aimed at the formation of a support model for the "generation" of valorisation strategies of the Ragusa's historic centre built heritage. A cognitive, evaluative and planning approach based on the architectural unity, individuated and identified as minimum informative entity, which is entrusted the capital value maximum attribute, on the urban and human points of view [4].

It is an analytic approach to the city, in which the fraction of urban information concentrated in the architectural unit is represented in abstract forms of classification inside a coherent comparison system, and returned to the urban context in the form of punctual choices, derived from the former classifications through a system of rules.

The adherence of the choice system to the "built realm" depends on the resolution of the descriptive model that means on the articulation of the classes of value attributes, and on the coherence of the correspondences between valuations and decisions.

The objective of this study is the formation of a model capable to assist planners and decision makers in the definition of multiple conservation/transformation strategies, controlling their global effects on several points of view, and referring to potentially conflicting goals. The model is applied to a consistent database deduced from the masterplan for Ragusa's historic centre, and extended to a portion of 1788 architectural units (records) characterized by 140 "original fields" and by 800 "other fields" in 30 sheets; the contents of the latter are mostly derived from the process of translation of the data in information units, and subsequently in attributes, valuations and classifications in intervention categories.

2   Materials: The Old Town of Ragusa and the Database

2.1   Characteristics of the Old Town an Definition of the Sample

Ragusa is the second most important town (after Syracuse) of the south-eastern part of Sicily, well known as Val di Noto. Its historic centre has been included in the Unesco World Heritage list since 2002.

The reconstruction process begun after the earthquake that in 1693 destroyed Ragusa and all the main cities of south-eastern Sicily originated the current separation of its historic centre in two distinct entities which reproduced the class distinction [6] [23]: Ragusa Ibla, the area reconstructed by the feudal aristocracy, that inserted late baroque architectural models inside the pre-existing medieval urban fabric; Ragusa Superiore, where the emerging middle class realized from scratch a square grid urban centre structured by two central axes, with several sizes of block in order to adapt to the topography of the site [1] [3] (Fig.1).

Fig. 1. Ragusa: the ancient and the modern old towns

The studied context is located inside the historic centre of Ragusa Superiore; it is part of the "first period" urban systems, as indicated by the "Norme Tecniche di Attu-azione" of the "P.P.E." (technical regulations of the masterplan). The area comprises 118 blocks and 1788 architectural units (Fig. 2). "An architectural unit is meant as that built organism generally formed by one or more real estate units, and that shows a certain level of autonomy on the functional, constructive and aesthetic points of view. It is characterized by an entrance from a road or another public use space, and by a system – even fragmentary – of vertical connection (stairs, walkways, etc.) through which the properties are accessible" [7].

Table 1 synthesizes in quantitative data the allocation of the built heritage in the diverse quality classes for each characteristic taken into account by the PPE.

Table 1. Exsemple of the Architectural Units database

			1	2	3	4	5	6
identificazione	unità ed		1	2	3	4	5	6
	id unità ed		1298	1299	1300	1301	1302	1303
localizzazione	settore		4	4	4	4	4	4
	isolato		112	112	112	112	112	112
	id via		128	125	125	125	125	125
	nome via		Via Scale	Corso Mazzini	Corso Mazzini	Corso Mazzini	Corso Mazzini	Corso Mazzini
superfici	sedime	mq	32	74	65	20	41	41
	perim	mq	27	35	35	18	28	28
	commerciale	sf mq	64	223	131	60	123	82
		pr mq						
aratteri dimensiona	H med. parete		3,1	3,1	3,1	3,1	3,1	3,1
	elevazioni		2 elevazione intera	3 elevazione intera	2 elevazione intera	3 elevazione intera	3 elevazione intera	2 elevazione intera
	N. elev. F.t	sf n	2	3	2	3	3	2
		pr n						
	volume	sf mc	197	691	405	187	380	254
		pr mc						
	largh. fronte princ	ml	9	4	12	5	6	7
	larghezza altri fronti	ml	11	7	2		13	
	superficie fronti	sf mq	121	106	89	50	178	41
		pr mq						
	sup vetrata	mq	19	17	14	8	28	6
tipologia	cod		T1	T1	T1	T1	T1	T1
	descr		Edilizia di base	Edilizia di base	Edilizia di base	Edilizia di base	Edilizia di base	Edilizia di base
epoca stimata	cod		B03	B01	B01	B01	B01	B01
	descr		Epoche sovrapposte	Antecedente 1950	Antecedente 1950	Antecedente 1950	Antecedente 1950	Antecedente 1950
grado di utilizzo	cod		E01	E03	E02	E01	E01	E01
	descr		Utilizzato	Parzialmente utilizzato	Non utilizzato	Utilizzato	Utilizzato	Utilizzato
destinazione p.t	cod		D01	D01	D01	D01	D01	D01
	descr		Abitazione	Abitazione	Abitazione	Abitazione	Abitazione	Abitazione
caratteri strutturali	cod		G03	G01	G01	G01	G01	G01
	descr		Misto	Muratura	Muratura	Muratura	Muratura	Muratura
caratt prosp tipici	cod		H04	H03	H04	H03	H04	H05
	descr		Presenza in parti residue	Prevalenza elementi tipici	Presenza in parti residue	Prevalenza elementi tipici	Presenza in parti residue	Assenza
caratt copertura	cod		I04	I02	I04	I02	I02	I02
	descr		Misto tetto-terrazza	Tetto tegolato tipico	Misto tetto-terrazza	Tetto tegolato tipico	Tetto tegolato tipico	Tetto tegolato tipico
caratt cromatici	cod		L20	L19	L06	L07	L07	L01
	descr		Ocra gialla	Colori diversi per p.	Salmone	Grigio	Gngio	Bianco
caratt infissi	cod		N06	N01	N07	N01	N06	N06
	descr		Misto Legno - metal.	Legno	Assenti	Legno	Misto Legno - metal.	Misto Legno - metal.
caratt parapetto	cod		O01	O01	O01	O01	O01	O09
	descr		Ferro a disegno semplice	Ferro a disegno semplice	Ferro a disegno semplice	Ferro a disegno semplice	Ferro a disegno semplice	Assenti
grado conservazione	cod		F05	F05	F9	F01	F02	F02
	descr		Pessimo	Pessimo	Lavori in corso	Buono	Normale	Normale
	interventi		Da recuperare	Da recuperare	Da manutenzionare	Da manutenzionare	Da manutenzionare	Da manutenzionare
superf evidenti	cod		M10	M10	M09	M09	M09	M09
	descr		Superf. evidenti	Superf. evidenti	Assenti	Assenti	Assenti	Assenti
CME	tip CME		c CME	h CME	c CME	g CME	h CME	f CME
	cod		3	8	3	7	8	6
	cat. di interv.		MO1	MO3	MO2	MS1	MS3	MO1
	cod		1	3	2	4	6	1

Fig. 2. Ragusa: Usage, Preservation, Typologies and Age of the AU in the studied area

3   Methods and Procedures

3.1   Concepts: A Semiotic Interpretation

As we have observed so far, the database of the analysis, valuation and project model here presented comes from the GIS of Ragusa's historic centre, that represents in the discrete the building heritage as a multilayer entity informed by the logic of the database; it selects only some forms of the intelligence of the planning process, and some areas of the rationality of the latter, and goes through some of the itineraries of the reasoning of the decisional process only, and only in the circumstances in which: the system may be discretized; the actions on its parts may mostly be considered independent, the choices base on sequences of logic conditions and recursive operations.

The circular sequence Analysis, Valuation and Project represented in Fig. 3 shows the articulation of the knowledge and action process whose heuristic and logic coherence guarantee consequentiality and reproducibility of the decisional process. This coherence may be consolidated basing on a semiotic interpretation of the above mentioned process, where the AU assume the function of signs forming a semantic chain.

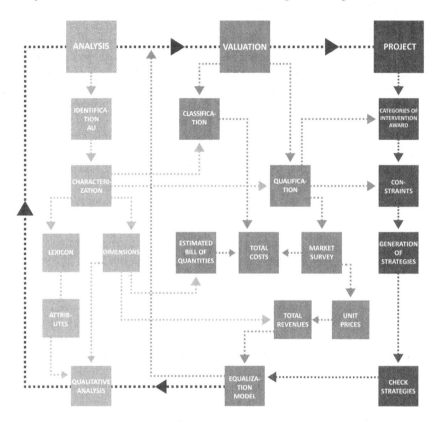

Fig. 3. Conceptual map of the general pattern

In the **analysis** phase the architectural units (AU) are identified as *referents,* as they constitute the occurrences mainly described concerning their geometric and materic component; the further qualities are selected and represented basing on a system of codes (the lexicon) foreseeing the characterization of the AU as *signifiers* liable to attributions and classifications for the formalization of the qualitative analysis.

The **valuation** phase concerns all the above sense attributes addressing the observations and the classifications; in this phase the semantic link between the referent and the correspondent terms of qualitative and quantitative-monetary value is defined; the latter allow to appreciate in what extent the modifications of significance (value) of the single AU can influence the overall configuration of the context - the semantic chain - represented in terms of value.

The **project** phase is conducted through the activation of the syntactic mechanisms, the rules binding motivations (values) to merits (decisions), to adjust progressively the semantic chain in order to achieve the configuration that realizes the highest differential of value, with the chance to verify whether the general syntactic necessities do not overwhelm the punctual or specific conditions of signification.

Moreover, this interpretation outlines the possibility to classify the different architectural units referring to the "sign-symbol-icon" status they may assume in an axiological crescendo: the AU that are valued as a consequence of their relationship to the other ones, due to the sense attributes consolidated in the semiosis process characterizing the semantic field, belong to the first class: these AU have a main contextual value [9]; the AU that are valued themselves and are capable to irradiate their value to their own environment generating positive externalities, belong to the last class; as icons they don't need any relationship with the semantic field they belong to, and in some cases they may also oppose the latter or assume a critical relation (in constructive sense) to it.

Another aspect of this interpretation concerns the distinction between context (linguistic or general) and co-text (situational – or communicative – and particular), inside which the signs (the AU) differently assume, modify and consolidate their significance (value):

— The *context* regards the structural characteristics and the general causes of value the architectural sciences more specifically deal with, involving the syntactic aspects of an "urban grammar";
— The *co-text* more specifically involves the specific aims of acting, undertaken when the urban grammar loses its grip on signs, and the semantic links between the characteristics of the AU and their importance in the localization model are relaxed. The processes of physical and social abandonment of the historic centres are phenomena activating new necessities of communication between the social system and the built environment [19, 20]; in these circumstances new tools – not considered or not allowed by the *context*, that means the architectural sciences – are required, as, for example, the use of equalization [5] and the introduction of transformative intervention categories, whose argumentation involves the sciences of value an of valuations, and, unaivoidably question of capital funding [24].

3.2   The Pattern

In its complex, the model is aimed at the assignment to each AU of a specific Intervention Category [11, 12, 13] [22] IC_{hk} – where $h = (1, 2, ..., 5)$ is the number of groups of IC (1. Ordinary Maintenance, 2. Extraordinary Maintenance, 3. Renovation, 4. Demolition and Reconstruction with or without extension, 5. Restoration), while $h = (1, ..., 3)$ is the grade of conservation status of the AU (according to which a soft, medium or consistent intervention is envisaged) – basing on a set of restraints v representing a set of values.

To assign the IC to each AU the model uses the set of restraints which have a dual nature and function. For each IC a certain number of qualification restraints $\vec{v}_g$, $g = (1, 2, ..., m)$ and an access restraint $\bar{v}$ are envisaged:

- $\vec{v}_g$ indicates the conditions of conformity of the AU to the set of values expressed by the specific *IC* so that it can be part of it: to select the AU it combines their diverse characteristics – scored from 1 (low grade) to 5 (high grade) according to a set of valuation functions: therefore, each AU is described by a vector of valuations which are filtered by the set of conditions of conformity associated to the *IC*.
- $\bar{v}$ indicates the minimum number of conditions of conformity which must be satisfied so that the AU can have access to the *IC*.

The AU has access to the *IC* if the number of filtered scores is bigger or equal to the access threshold $\bar{v}$, that is, if the following generic condition is true:

$$IC \leftrightarrow \exists i_1, ..., i_k (k \leq t, t \leq 1) \mid b_{i_j} \geq \bar{v}_{i_j} \forall j = 1, ..., k \qquad (1)$$

where $\bar{v}_{i_j}$ represents the threshold of the *j*th criterion, k is the number of the criteria by means of which $\bar{v}_{i_j} \geq t_{i_j}$ is verified, and t_{i_j} is the minimum threshold;

furthermore: if more than one IC are selected, the highest one in the sorted G_m list prevails. Sorting of G_m can be chosen according to the type of global strategy (conservative or transformative). As a consequence, a strategy is defined by the combination of the set of restraints which, progressively relaxed, admit the AU to the diverse intervention categories. This combination can filter the AU addressing a bigger number towards the conservative or transformative IC. This assortment can be variously composed, creating multiple intermediate strategies (Fig. 10).

Each strategy is verified at real time from the point of view of the reference values. A MAVT model has been implemeded in order to evaluate each strategy by aggregating the different quantitative and qualitative characters in standard measurements of value. The functions of value are constructed filtering the distribution of values of each variable from the minimum to the maximum calculated values, and basing on the arrangement of the cases inside this range.

The impacts on the **landscape value matrix** [16, 17] are valued basing on the *building density*, referring to which the extent of the transformative IC envisaging the increment of volume is controlled.

The impacts on the **identity value matrix** are valued on the grounds of the *presence of typical facade elements,* and at the level of the *building volume* which an overall conservative IC (Maintenance or Restoration) is associated to.

The impacts on the **functional value matrix** are valued basing on the total *dimension* of the UA and on the *technologic level* that depends on the combination of the transformation grade and the maintenance status.

The impacts on the **economic value matrix** of the urban context due to the implementation of each strategy are valued basing on the overall *revenue-cost margin*, envisaged as result of the implementation of the strategy. Each IC includes works whose cost might or might not be compensated by the increment of market value due to the improvement of the technologic and architectural characteristics and to the increase of volume in the envisaged cases.

The *cost* of the works associated to the specific *IC* is extracted from a matrix (Table 2) that establishes a relation between each of the 22 building types identified in the studied area and each 15 IC_{hk}, providing a total 330 typical unitary costs; each cost has been calculated by means of a general Bill of Quantities including all the main voices relating to the envisaged works, and selecting those inherent to each IC_h, whose intensity is graduated according to the observed conservation status kth.

Fig. 5 shows the composition of the costs aggregated per classes (masonries, roofs, finishes, plants, window frames, etc.) for each IC and for each conservation grade for one of the 22 typologies.

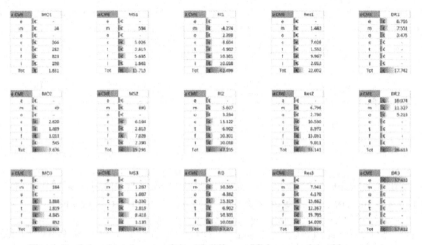

Fig. 4. Partial and total costs of the 15 different IC for one of the 22 typologies

Table 2. Unit costs for each building type and for the each of the 15 different IC

Building types							Intervention categories								
	MO1	MO2	MO3	MS1	MS2	MS3	RI1	RI2	RI3	Res1	Res2	Res3	DR1	DR2	DR3
1	€ 39	€ 91	€ 120	€ 225	€ 276	€ 356	€ 405	€ 563	€ 682	€ 524	€ 696	€ 1.056	€ 423	€ 634	€ 806
2	€ 32	€ 74	€ 96	€ 184	€ 235	€ 315	€ 357	€ 502	€ 622	€ 406	€ 571	€ 875	€ 423	€ 634	€ 806
3	€ 32	€ 56	€ 77	€ 150	€ 194	€ 260	€ 342	€ 492	€ 622	€ 369	€ 587	€ 891	€ 423	€ 634	€ 806
4	€ 26	€ 42	€ 57	€ 117	€ 160	€ 226	€ 304	€ 443	€ 574	€ 274	€ 488	€ 749	€ 423	€ 634	€ 806
5	€ 33	€ 69	€ 91	€ 173	€ 218	€ 291	€ 362	€ 517	€ 651	€ 403	€ 620	€ 941	€ 423	€ 634	€ 806
6	€ 27	€ 55	€ 72	€ 140	€ 185	€ 258	€ 323	€ 469	€ 602	€ 308	€ 520	€ 797	€ 423	€ 634	€ 806
7	€ 26	€ 50	€ 66	€ 132	€ 176	€ 244	€ 319	€ 463	€ 594	€ 296	€ 515	€ 787	€ 423	€ 634	€ 806
8	€ 31	€ 63	€ 84	€ 162	€ 207	€ 276	€ 354	€ 506	€ 637	€ 380	€ 604	€ 916	€ 423	€ 634	€ 806
9	€ 32	€ 60	€ 81	€ 159	€ 205	€ 271	€ 356	€ 509	€ 639	€ 386	€ 609	€ 922	€ 423	€ 634	€ 806
10	€ 26	€ 45	€ 60	€ 123	€ 168	€ 234	€ 315	€ 457	€ 587	€ 283	€ 504	€ 771	€ 423	€ 634	€ 806
11	€ 26	€ 49	€ 65	€ 130	€ 175	€ 239	€ 322	€ 467	€ 598	€ 301	€ 521	€ 794	€ 423	€ 634	€ 806
12	€ 31	€ 61	€ 82	€ 161	€ 206	€ 273	€ 357	€ 510	€ 640	€ 383	€ 608	€ 921	€ 423	€ 634	€ 806
13	€ 26	€ 41	€ 56	€ 114	€ 158	€ 225	€ 301	€ 439	€ 567	€ 266	€ 480	€ 739	€ 417	€ 625	€ 804
14	€ 29	€ 50	€ 68	€ 135	€ 180	€ 248	€ 326	€ 470	€ 599	€ 324	€ 543	€ 830	€ 417	€ 625	€ 804
15	€ 26	€ 48	€ 64	€ 127	€ 171	€ 238	€ 313	€ 454	€ 584	€ 283	€ 499	€ 764	€ 417	€ 625	€ 804
16	€ 26	€ 49	€ 65	€ 136	€ 185	€ 256	€ 338	€ 485	€ 612	€ 312	€ 541	€ 831	€ 417	€ 625	€ 804
17	€ 28	€ 47	€ 64	€ 132	€ 178	€ 247	€ 328	€ 473	€ 600	€ 313	€ 537	€ 823	€ 417	€ 625	€ 804
18	€ 26	€ 54	€ 71	€ 140	€ 180	€ 251	€ 308	€ 453	€ 587	€ 323	€ 525	€ 800	€ 429	€ 643	€ 809
19	€ 25	€ 51	€ 68	€ 135	€ 175	€ 243	€ 308	€ 451	€ 582	€ 318	€ 527	€ 800	€ 429	€ 643	€ 809
20	€ 21	€ 40	€ 52	€ 108	€ 149	€ 217	€ 280	€ 416	€ 547	€ 237	€ 450	€ 691	€ 429	€ 643	€ 809
21	€ 22	€ 45	€ 59	€ 126	€ 172	€ 250	€ 334	€ 448	€ 578	€ 372	€ 545	€ 838	€ 446	€ 669	€ 818
22	€ 22	€ 43	€ 56	€ 123	€ 169	€ 243	€ 323	€ 447	€ 577	€ 343	€ 534	€ 818	€ 446	€ 669	€ 818

The revenue has been calculated as difference between the present market value and the current market value, basing on a market survey extended to 80 study cases in the historic centre of Ragusa Superiore. Having verified the quality of the sample (whose analysis is here omitted as it will be subject to a subsequent specific study) the coefficients of the single regressors of the synthetic pluriparametric valuation function $V_m = a_1 k_e + a_2 k_i + a_3 k_t + a_4 k_a + b$ – where k_e indicates the extrinsic characteristics or of localization, k_i the intrinsic ones or of position in the building, k_t the technologic ones, k_a the architectural-environmental ones [10] – have been extracted, and the function has been implemented to determine the market value of each 1788 AU of the sample, in the existing status and in the project status as well. The diagrams in Fig. 6 show a synthesis of the relationship between the characteristics (aggregated in a single weighed average variable) and the surveyed prices per room and sq.m.

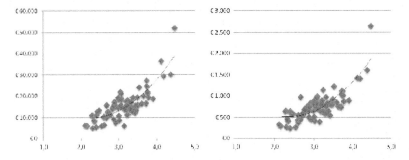

Fig. 5. Relationship between price (per room and per sq.m) and overall quality variable

4   Applications and Results

In this chapter the main arguments of the valuation functions necessary for the implementation of the so far described model will be synthetically illustrated.

4.1   Notes about Some Geographical Information System Applications

The informative base utilized for the characterization and the qualification of the studied sample is constituted by a database in (Postgresql) with PostGIS extension structured in records and fields. The records are made of objects topologically characterized as points, lines and polygons; the fields represent a set of alphanumeric type attributes. On this base some elaborations of spatial analysis aimed at characterizing single units on the points of view of the positional extrinsic characteristics (localization) and the intrinsic ones (proximity and altimetry relationship with the other AU) have been carried out.

The positional extrinsic characteristic has been attributed through the calculation of the distances from the single architectural unit to the central locations and the main arterial roads, specifically classified in order of importance. The normalization of these distances in one standard scale has provided the relative score of each unity from this point of view.

The positional intrinsic characteristic has been calculated by means of a spatial function, measuring for each AU the distances of each exposed facade, from the facades of the closest AU, and for each elevation above ground level, and calculating the exposure basing on the rotation angle from the West (or the East) and/or the North (or the South). This way an index of medium weighted distance measuring the volume of the free space around the AU and an exposure index have been calculated. The exposure index does not take into account distances longer than 50 m, limit assumed as threshold of indifference of the utility function. The utilized algorithm is proposed as follows. The spatial function implemented in the geo-database analyzes the polygonal feature representing the shapes of the AU (Fig. 6) [8] [21].

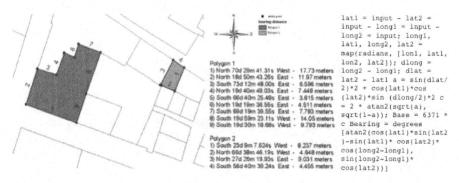

Fig. 6. Sample of the spatial calculation of the exposure/distance index, and the code

4.2   Contents

The model transforms concrete characteristics of the AU turning them in attributes of value useful for the assignment of the IC to each AU. Fig. 7 shows the plan allocation of the attributes relating to the k_e, k_i, k_t, k_a. The implementation of the different 15 strategies (Table 3), impacts on the urban fabric with relevant changes in volume and value. Fig. 9 shows the allocation of the changes in volume envisaged by 4 of the 15.

Table 3. Generation oft he 15 strategies by relaxing the restraints

strategy	demolition and reconstruction								cubage increase						ordinary				extraordinary				renovation				restoration					
	age	typol	facade	roof	sup.ev.	preserv	floors	yes/no	age	typol	facade	roof	sup.ev.	si/no	structure	car.inf	preserv	yes/no	structure	windows	preserv	yes/no	structure	windows	preserv	yes/no	age	typol	facade	roof	sup.ev.	yes/no
1	3	4	3	3	3	2	3	7	3	4	3	3	3	5	3	3	2	3	3	3	1	3	3	3	1	2	5	2	3	2	3	5
2	3	4	3	3	3	2	3	7	3	4	3	3	3	5	3	3	2	3	3	3	1	3	3	3	1	2	5	2	3	3	3	5
3	3	4	3	3	3	2	3	7	3	4	3	3	3	5	3	3	2	3	3	3	1	3	3	3	1	2	5	3	2	2	2	5
4	3	4	3	3	3	2	3	7	3	4	3	3	3	5	2	2	3	3	3	3	2	3	5	5	1	2	5	3	3	3	3	5
5	3	4	3	3	3	2	3	6	3	4	3	3	3	3	3	3	3	3	3	3	2	3	5	5	1	2	5	3	3	3	3	5
6	3	4	3	3	3	2	3	6	3	4	3	4	3	3	2	2	4	3	2	2	3	3	5	5	2	2	5	3	3	3	3	5
7	3	4	3	4	3	2	3	6	5	4	3	4	3	3	3	2	4	3	2	2	3	3	5	5	2	2	5	3	3	3	3	5
8	3	4	3	4	3	2	3	6	5	4	4	5	5	3	3	2	4	3	2	2	3	3	5	5	2	2	5	3	3	3	3	5
9	3	4	3	3	3	2	3	5	5	4	5	5	5	3	3	2	4	3	2	2	3	3	5	5	2	2	5	3	4	3	3	5
10	3	4	4	3	3	2	3	5	5	4	5	5	5	3	3	2	4	3	2	2	3	3	5	5	2	2	5	3	4	4	4	5
11	3	4	4	4	3	2	3	5	5	4	5	5	5	3	3	2	4	3	2	2	3	3	5	5	2	2	5	3	4	4	4	5
12	3	4	4	4	3	2	3	5	5	5	5	5	5	3	3	2	4	3	2	2	3	3	5	5	2	2	5	3	4	4	4	5
13	3	4	4	4	4	2	2	5	5	5	5	5	5	3	3	2	4	3	2	2	3	3	5	5	2	2	5	3	4	4	4	5
14	3	4	4	4	5	3	2	5	5	5	5	5	5	3	3	2	4	3	2	2	3	3	5	5	2	2	5	3	4	4	4	5
15	5	4	4	4	4	3	2	5	5	5	5	5	5	3	3	2	4	3	2	2	3	3	5	5	2	2	5	3	4	4	4	5

Fig. 7. Spatial allocation of the main characteristics of the sample

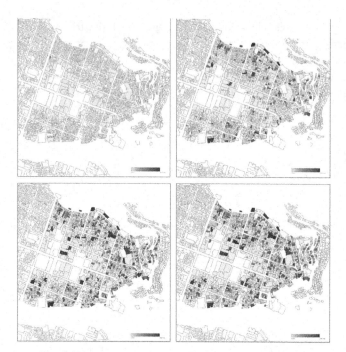

Fig. 8. Volume increase supposed by 4 of the 15 strategis

4.3   Axiology and Generation of Strategies

The 15 strategies are implemented by relaxing progressively the restraints as shown in table 3. The trade off functions between the diverse axiological matrixes have been studied basing on the model. The first diagram of Fig. 10 shows the change in value

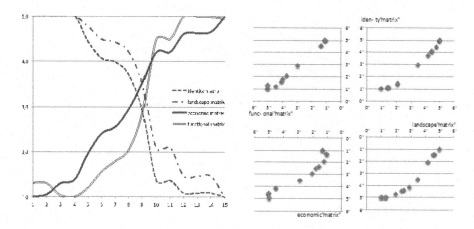

Fig. 9. Multidimensional evaluation of the 15 strategies and trade off between the different axiologic matrixes

of the whole urban context from the perspectives of the four axiological matrixes. The second diagram shows the trade-off relations between the diverse axiological matrixes. The formation of a multidimensional decision model, by means of which it is possible to select the optimal choice among the possible ones, is postponed to a subsequent research study, as we believe that the proposed analytic approach has thoroughly laid its premises.

Fig. 10. Identification of urban signs, symbols and icons

5   Discussions. Semiotics in the Equalization Perspective

On the grounds of the presented semiotic framework, the proposed model allows to map the AU according to their "sign density", classifying them as signs, symbols [14] and icons. This operation is carried out:

— inside the *contextual* sphere, basing on the *signification* process, referring to the assigned values, k_e, k_i, k_t, k_a, and to their aggregated value, by means of which it is possible to distinguish the AU as signs, symbols and icons;
— inside the *co-textual* sphere, referring to the IC associated to each AU: the progressive distribution of more transformative strategies allows to individuate the most "resistant" elements (the icons), that means those keeping longer the conservative IC; vice versa, the less resistant elements, the "signs", once the strategic objectives

have been modified, may be associated to interventions that are very different from those hypothesized at the beginning.

The classification of the urban objects in signs, symbols and icons, allows to individuate the AU – and therefore, according to their localization, the diverse homogeneous urban areas as well – from which it is possible to insert a programme to irradiate their intrinsic qualities; in fact, while the significance (value) of the icons is not questioned, the significance of the symbols, and successively of the signs, allows for modifications requiring a step by step less consistent argumentative support.

At the opposite, the acknowledgment of the conditions of iconicity selects the AU – or possibly their characteristics – which are not negotiable, then not compensable in a monetary form or in any other contingent forms.

The "pragmatic" approach of the contemporary town planning discipline has therefore its own positive function when it doesn't affect iconic values that in certain cases are not manifest; further researches envisage the extension of this approach, from the AU to the adjacent urban context, as in many cases the value of the single AU is of typically complementary nature. Fig. 11 shows the identification of these three categories of architectural-urban signs.

6   Conclusions

The proposed model comprises a consistent network of logic functions connecting a set of economic-evaluative calculations carried out on observed quantities and qualities of the architectural-urban heritage of the historic centre of Ragusa Superiore. It is meant as a constructive and propositional verification of the valorisation program carried out by the municipality of Ragusa since the approval of the historic centre's PPE (masterplan) in 2012. This town planning tool has allowed to form Intervention Categories basing only on the observations, and without any verification in terms of value and evaluative procedures. The here proposed axiological approach has instead reversed this approach, integrating in a coherent global system the recursive sequence of the *judgements of fact*, the *judgements of value*, and the *judgements of merit*. In this integration, in which the evaluative tools have been proved useful for the project purpose, it has been possible to verify the inverse relation, as the project itself has proved to be a useful evaluation tool. Therefore, the possibility to iteratively form strategies with different conservation grades has allowed to distinguish the AU basing on their "resistance" or "condescendence", and to pass from a more conservative IC to a more transformative one.

This has allowed experiencing the importance of valuations as project tools, and, at the opposite, that the project is an important valuation tool, that means of exploration of the possibilities of an object to generate new values.

In this direction, this study has fostered the conceptual reflection on the distinction between *contextual* valuations and *co-textual* explorations – a similar relation connects *conventional* and *conversational* implicature [15] – that allow associating, in extreme synthesis, each AU to the notions of sign, symbol and icon, in a crescendo of strategic importance in the urban fabric. This approach practically bases on the complementarily between the categories of knowledge (signification in the context) and those of action (transformation basing on the co-text). Something similar has been addressed in cognitive and non cognitive view opposition, within the contemporary environmental aesthetic [2]

On the grounds of this classification it is possible to articulate contingent compensative and equalization approaches, as: icons and symbols will be allowed to be objects of public interest projects, concentrating on them one more or less high quota of the overall margin produced inside the whole area, and utilizing the latter to sustain these projects with funding in capital interest and mechanisms of mainly public partnership involving tools of project funding; the signs, instead, will be allowed to be objects of those forms of transfer of real rights, by means of which the margin between revenues and costs deriving from the application of more advantageous intervention categories is meant to finance the application of the less advantageous ones.

Acknowledgements. Salvatore Giuffrida edited paragraphs 1, 3, 4.3, 5 and 6, drew up the valuation model; Filippo Gagliano edited paragraph 4.1 and pictures 2, 8, 9, 11; Vittoria Ventura edited paragraphs 2, 4.2. Special thanks to the Municipality of Ragusa for providing us the database of the old town of Ragusa.

References

1. A.A. V.V.: Il centro antico di Ragusa superiore – decadenza e rinascita di una città, Italia nostra, Ragusa (1991)
2. Allen, C., Lintott, S. (eds.): Nature, Aesthetics, and Environmentalism: From Beauty to Duty. Columbia University Press, New York (2007)
3. Bentivegna, V., Miccoli, S.: Valutazione Progettazione Urbanistica. DEI, Roma (2010)
4. Boscarino, S., et al.: Petralia Soprana. Ipotesi di restauro urbano e studi di analisi multi-criteriale. Medina, Palermo (1994)
5. Carbonara, S., Torre, C. (eds.): Urbanistica e perequazione: dai principi all'attuazione. FrancoAngeli, Milano (2008)
6. Caruso, M., Perra, E.: Ragusa - La città e il suo disegno. Genesi di un organismo urbano tra '600 e '700. Tarquinia, Gangemi (1994)
7. Comune di Ragusa: Norme Tecniche di Attuazione of the Ragusa's historic centre "Piano Particolareggiato" – masterplan, Ragusa (2012)
8. Dale, P.: Introduction to Mathematical Techniques used in GIS. CRC Press, US (2004)
9. Drdacky, T., Teller, J.: Strategic assessment of urban cultural heritage: the case of Old Town Square in Prague. http://orbi.ulg.ac.be/bitstream/2268/29235/1/Arcchip.pdf
10. Forte, C.: Elementi di estimo urbano. Etas Kompass, Milano (1968)
11. Giuffrida, S., Ferluga, G.: Renewal and conservation of the historic water front. Analysis, evaluation and project in the grand harbor area of Syracuse. BDC. Bollettino del Dipartimento di Conservazione dei Beni Architettonici ed Ambientali dell'Università degli Studi di Napoli, vol. 12, pp. 735–754 (2012). ISSN: 1121-2918
12. Giuffrida, S., Ferluga, G., Gagliano, F.: Social Housing nei quartieri portuali storici di Siracusa. Valori e Valutazioni **11**, 121–154 (2013). ISSN: 2036-2404
13. Giuffrida, S., Gagliano, F.: Sketching smart and fair cities WebGIS and spread sheets in a code. In: Murgante, B., Misra, S., Rocha, A.M.A., Torre, C., Rocha, J.G., Falcão, M.I., Taniar, D., Apduhan, B.O., Gervasi, O. (eds.) ICCSA 2014, Part III. LNCS, vol. 8581, pp. 284–299. Springer, Heidelberg (2014)

14. Gottdiener, M., Lagopoulos, A.: The City and the Sign: An Introduction to Urban Semiotics. Columbia University Press, New York (1986)
15. Grice, P.: Logic and conversation. In: Cole, P., Morgan, J. (eds.) Syntax and Semantics. New York Academic Press (1975)
16. Lynch, K.: The Image of the City. Cambridge, MA (1960)
17. Lynch, J.A., Gimblett, R.H.: Perceptual Values in the Cultural landscape: a computer model for assessing and mapping perceived mystery in rural environments. Computer, Environment and Urban Systems 16, 453–471 (1992)
18. Patassini, D.: Paradigmi e strategie di valutazione di piani, programmi e politiche. Urbanistica 105, INU edizioni, Roma (1995)
19. Rizzo, F.: Economia del patrimonio architettonico-ambientale. FrancoAngeli, Milano (1989)
20. Rizzo, F.: Valore e Valutazioni. La scienza dell'economia o l'economia della scienza. FrancoAngeli, Milano (1999)
21. Sinnot, R.W.: Virtues of the Haversine. Sky and Telescope 68(2), 159 (1984)
22. Trovato, M.R., Giuffrida, S.: The choice problem of the urban performances to support the Pachino's redevelopment plan. Int. J. Business Intelligence and Data Mining 9(4), 330–355 (2015)
23. Valente, L.: Ragusa Ibla: memoria, calamità e piano. Consiglio Scientifico Internazionale Sviluppo Sostenibile, Milano (2001)
24. Vernieres, M. (ed.): Methods for the Economic Valuation of Urban Heritage: A Sustainability-based Approach. Agence Francaise de Developpment (2012). http://recherche.afd.fr

Decision Trees Analysis in a Low Tension Real Estate Market: The Case of Troina (Italy)

Alberto Valenti[1(✉)], Salvatore Giuffrida[1], and Fabio Linguanti[2]

[1] Department of Civil Engineering and Architecture, University of Catania, Catania, Italy
albvlt79@gmail.com, sgiuffrida@dica.unict.it
[2] Special Educational Department of Architecture, University of Catania, Catania, Italy
fabio.linguanti@virgilio.it

Abstract. Troina is a town in the central mountainous area of Sicily, in the Province of Enna, and well represents the general social and economic profile of this territory. Its real estate market is assumed in this study as one of the most significant ones for the description of this profile, because of its characteristics that, especially during the current economic-financial crisis, are particularly evident. The study of this market has been carried out as a basis for a possible redevelopment capital-centered policy, so that both urban/architectural and real estate characteristics have been considered within the proposed pattern. This pattern is based on the decision trees technique, a data mining procedure that allows defining the different submarkets under some specified hypotheses. The different aggregations we have figured out express different ways of assuming the real estate market profile and the directions of any policy that could boost the preservation of the historical urban context instead of promoting the outward urban spreading with further land consumption.

Keywords: Minor inner old towns · Real estate market segmentation · Decision trees · Mass appraisal

1 Introduction

The real estate market can be assumed as one of the most important information sources about both the structural and the contingent economic profiles of a land/urban community. The real estate capital assets can be considered, in each of their different articulations, the phenomena of the accumulation process of the social economic surplus, and in two couples of ways: 1a. in constructive terms; 1b. in destructive (or deconstructive) terms; 2a. in private terms; 2b. in public terms.

1a. A bullish trend increases the expectancies and encourages the accumulation of further capital asset, increasing its implicit liquidity, that means the distance between yield and price: the cap rate drops.

1b. A bearish trend decreases the expectancies and encourages the disinvestments increasing the real estate explicit liquidity: cap rate increases [13, 17].

2a. Despite the general trends and the consequent expected behaviours, implicit liquidity arises during the bearish trend as well, so that prices do not always drop as

© Springer International Publishing Switzerland 2015
O. Gervasi et al. (Eds.): ICCSA 2015, Part III, LNCS 9157, pp. 237–252, 2015.
DOI: 10.1007/978-3-319-21470-2_17

expected when demand falls down. In this case ask prices prevail on bid prices. Symmetrically, in case of pathologic bearish trend, low prices don't persuade the purchasers that show, as a consequence, a precise speculative demand of money, that in this case is much more liquid than real estate: bid prices prevail on ask prices.

2b. The real estate liquidity can be considered as the general perspective of a local community that assumes its location as the general context of its identity, so that the real estate market can be considered the social-economic place in which the preferences about identity and functional values are revealed.

This study, by means of the identification of the different market segments [18], assumes such perspective in order to "measure and map" the willingness to "invest in itself" expressed by the community of Troina, especially during the current structural bearish trend, a situation of economic crisis more general than the one started in 2008.

2   Materials. Troina and Its Real Estate Market Survey

Troina is a medium size town in the province of Enna. Its inner location makes it recognizable as a typical mountainous centre strongly characterized by local usages and traditions still alive and vital. Its urban structure is articulated in three parts: the old town dating since Muslim-Norman age, the first expansion dating since the earthquake in 1693, that gave rise to the Borg neighbourhood, the final expansion after the Second World War, and the recent sprawl toward the countryside.

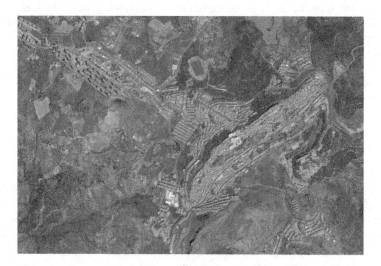

Fig. 1. Aerial view of Troina

The old town (Fig. 1) is characterized by smaller architectonic units composed by one to three multi-storey properties, located alongside narrow and curvy lanes, except for the area of Via Conte Ruggero where larger buildings are located. The intermediate area, the Borgo, is characterized by two different settlement layouts: the eastern one, the more ancient, with traditional buildings and streets following the ground morphology, and the newer one, with standardized buildings located alongside parallel streets; the recent urban expansion mostly comprises larger building complexes and multi-storey buildings located alongside larger and straight streets.

The general economic crisis affects the real estate market of Troina mainly because of the difficulty of funding, that discourages the young couples from buying the houses or apartments recently built in the newer and peripheral areas. These dwellings are larger, more expensive and more exposed than in the past. They usually buy the older dwellings, located in the central areas (cheaper and generally not exposed to the cold winds from NW), in order not to have to pay for high heating costs as well.

The real estate market of Troina has been surveyed analysing properties belonging to the whole urban area. The sample includes 51 properties described by $h = e, i, t, a$ features k_h, including 15 characteristics k_{hi} (i varies for each group): k_e, *location*, (infrastructures k_{e1}, accessibility k_{e2}, parking k_{e3}, environmental quality k_{e4}, amenity k_{e5}); k_i, *intrinsic* (floor k_{i1}, brightness k_{i2}, view k_{i3}); k_t, *technological* (building structure, k_{t1}, finishes, k_{t2}, maintenance, k_{t3}, equipment, k_{t4}); k_a, *architectural* (size and functional adequacy k_{a1}, accessories k_{a2}, decorum k_{a3}) [8]. Each k_h is the weighed average score of the correspondent k_{hi} ($k_h = \sum_l k_{hl} c_{hi}$; $\sum_i c_{hi} = 1$), ranging from 1 (low quality degree) to 5 (high quality degree); the weight c_{hk} measures the relative importance of each k_{hi} within each of each group.

Table 1 synthesizes the main quantitative and qualitative characteristics of the properties, ask prices, unit prices and the overall quality score, the weighed average score of k_h ($k^* = \sum_h k_h c_h$; $\sum_h c_h = 1$); k^* has been uses in order to compare the results of the further analyses as a synthetic qualitative independent variable.

Because of its typological heterogeneity both the surface area (sq. m) and the number of rooms (r) have been considered as quantitative characters: in fact, ancient buildings might have oversized rooms or some rooms might not be regular, so that the actual utility does not always correspond to the property surface area.

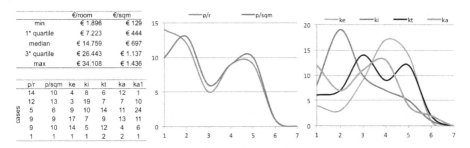

Fig. 2. Unit ask prices and quality distribution within the sample

The range of unit price and qualitative features are synthesized in Fig. 2: the first table shows the main prices (minimum, 1st quartile, median, 3rd quartile, maximum); the second shows the distribution of the properties within six classes of prices and k_h, as displayed in the two graphs: both regarding prices and qualities the sample presents some anomalies due, on the one hand, to the great number of low price properties and, on the other hand, to the qualitative heterogeneity (especially technological and architectural features); therefore a deepening of the analysis has been carried out in order to describe the market segmentation that can explain this singular articulation.

Table 1. Real estate market survey synthesis

neighborhood	location	id	ask price	rooms	unit price (€/room)	surface	unit price (€/sqm)	ke	ki	kt	ka	k*
a	Castellano, 4	1	€ 30.000	6,5	€ 5.192	153	€ 227	3,1	2,8	2,2	3,0	2,8
a	Scialfa, 25	2	€ 40.000	3,5	€ 8.963	50	€ 697	3,9	2,1	2,0	2,6	2,7
a	San Domenico, 44-46	3	€ 80.000	3,0	€ 23.694	60	€ 1.243	4,1	2,3	3,0	4,1	3,6
b	Mercato, 5-7	4	€ 80.000	5,8	€ 13.729	120	€ 654	2,6	1,9	1,3	2,7	2,2
b	Mercato, 19-21	5	€ 70.000	6,8	€ 11.201	53,3	€ 1.214	2,9	2,6	2,0	2,8	2,6
b	Liccardi, 15	6	€ 130.000	7,5	€ 18.924	150	€ 893	3,7	2,4	3,2	3,3	3,2
b	Umberto 30	7	€ 150.000	4,5	€ 31.882	110	€ 1.338	4,7	4,0	4,1	4,6	4,4
b	San Pietro, 54	8	€ 90.000	3,5	€ 23.248	90	€ 949	3,7	2,3	3,9	4,1	3,6
a	Garibaldi, 54	9	€ 20.000	3,5	€ 3.248	29	€ 560	4,6	3,1	2,7	1,5	2,8
a	Cavallotti, 88	10	€ 25.000	2,6	€ 6.170	35	€ 592	3,0	1,8	1,2	1,2	1,7
b	San Silvestro, 49	11	€ 100.000	12,8	€ 14.759	285	€ 552	3,9	3,0	2,9	3,2	3,3
b	San Silvestro, 51	12	€ 150.000	10,3	€ 19.014	200	€ 841	3,9	3,0	3,2	3,1	3,3
b	San Silvestro, 18	13	€ 150.000	6,5	€ 23.654	150	€ 1.027	3,9	2,8	4,0	3,1	3,4
b	San Silvestro, 52	14	€ 50.000	3,9	€ 10.818	120	€ 404	3,9	2,6	1,3	1,9	2,4
b	Oreto, 1	15	€ 30.000	3,5	€ 6.106	55	€ 449	3,1	2,6	2,8	1,4	2,3
b	Nicosia, 19	16	€ 180.000	8,3	€ 24.170	180	€ 1.066	4,0	3,1	4,0	3,6	3,7
b	Mazzini, 52	17	€ 120.000	4,5	€ 25.215	100	€ 1.162	3,8	3,3	3,6	4,2	3,8
b	Cavallotti, 51	18	€ 160.000	4,5	€ 34.104	110	€ 1.429	3,6	3,0	2,4	3,8	3,3
b	Cavallotti, 51	19	€ 120.000	4,5	€ 25.215	90	€ 1.282	3,5	2,5	2,9	3,7	3,2
a	Roma, 14	20	€ 70.000	8,8	€ 10.859	207	€ 439	3,7	2,1	2,8	2,1	2,6
a	Reliquia di San Silvestro, 37	21	€ 25.000	5,3	€ 4.071	75	€ 263	1,4	2,1	2,5	1,3	1,7
b	Umberto, 96	22	€ 60.000	9,0	€ 9.779	120	€ 488	4,4	3,7	3,5	2,7	3,5
a	nuova del Carmine, 27	23	€ 25.000	3,5	€ 4.677	92,5	€ 223	2,3	2,8	3,8	2,3	2,7
a	Sollima, 88	24	€ 130.000	4,5	€ 27.437	120	€ 1.071	4,4	3,0	3,5	3,5	3,7
b	Sollima, 21	25	€ 130.000	4,5	€ 27.437	120	€ 1.071	4,0	3,4	3,6	4,3	3,9
a	nuova del Carmine, 27	26	€ 25.000	6,9	€ 5.321	150	€ 227	1,7	2,4	2,6	1,5	1,9
b	Umberto, 154	27	€ 180.000	13,0	€ 21.015	360	€ 799	4,1	4,4	3,8	4,5	4,2
c	Vittorio E. Orlando, 3	28	€ 180.000	6,5	€ 28.269	170	€ 1.111	4,3	3,5	4,0	4,0	4,0
c	Vittorio E. Orlando, 3	29	€ 180.000	6,5	€ 28.269	170	€ 1.111	4,3	3,6	3,9	4,0	4,0
a	Via Papa Urbano II, 5	30	€ 60.000	10,4	€ 10.290	250	€ 396	2,8	3,7	2,7	2,5	2,8
a	Cagnone, 8	31	€ 30.000	5,6	€ 5.023	107	€ 251	1,5	2,4	1,8	1,2	1,6
a	Abbate Romano, 66	32	€ 30.000	5,9	€ 5.049	120	€ 238	1,3	2,6	2,4	1,3	1,7
b	piazza A. Majorana, 2	33	€ 15.000	3,3	€ 1.896	75	€ 129	3,3	2,6	2,1	1,4	2,2
b	via della Resistenza, 16	34	€ 80.000	6,8	€ 12.682	220	€ 481	4,3	2,8	2,4	2,3	2,9
b	Schinocca, 14	35	€ 45.000	4,8	€ 8.276	90	€ 449	2,8	1,9	2,0	1,8	2,1
b	Nazionale, 117	36	€ 80.000	4,5	€ 16.326	110	€ 702	4,1	2,8	2,7	2,9	3,1
b	Regalbuto, 27	37	€ 120.000	4,3	€ 26.530	80	€ 1.436	4,2	2,8	3,8	4,0	3,8
c	Nazionale, 181	38	€ 180.000	5,5	€ 32.290	130	€ 1.385	3,9	3,9	3,6	4,1	3,9
c	Nazionale, 182	39	€ 180.000	6,5	€ 28.269	150	€ 1.227	3,4	3,9	1,8	2,9	3,0
b	Croce, 26	40	€ 20.000	5,5	€ 3.199	130	€ 155	3,4	2,1	2,5	1,9	2,4
b	XVIII Ottobre, 19	41	€ 60.000	6,5	€ 9.807	120	€ 488	3,4	3,1	4,0	3,1	3,4
b	Aldo Moro	42	€ 180.000	5,5	€ 32.290	180	€ 1.066	3,9	3,1	2,5	3,5	3,3
c	Mustica , 2	43	€ 30.000	5,5	€ 5.017	70	€ 352	3,1	2,1	2,5	1,9	2,3
a	Bonanno, 64	44	€ 120.000	6,8	€ 18.608	229	€ 653	3,1	4,3	2,5	3,8	3,4
c	Pietro Nenni, 11	45	€ 140.000	4,5	€ 29.659	110	€ 1.248	4,3	2,9	4,3	3,9	3,9
a	Sotera, 40	46	€ 60.000	6,8	€ 9.719	105	€ 540	3,6	2,8	1,3	2,3	2,5
b	piazzale Lazio, 14	47	€ 150.000	5,6	€ 26.356	110	€ 1.338	4,6	2,8	4,0	4,4	4,1
a	Santa Lucia, 5	48	€ 30.000	3,5	€ 6.106	70	€ 352	2,6	3,7	3,6	2,5	2,9
a	nuova del Carmine, 38	49	€ 70.000	9,9	€ 11.088	120	€ 571	3,2	2,8	3,2	2,9	3,0
c	Generale dalla Chiesa, 5	50	€ 190.000	5,5	€ 34.108	150	€ 1.293	4,3	4,3	4,4	5,0	4,5
c	Generale dalla Chiesa, 5	51	€ 180.000	5,5	€ 32.290	150	€ 1.227	4,3	3,8	4,4	5,0	4,4

3   Methods and Procedures

3.1   Clustering Analysis: The Decision Trees

Decision trees algorithms belongs to the tools of data mining techniques, generally used in decision making process as well as in big data sorting and analyses. They look like top-down flowcharts that split a sample into sub-groups in order to provide more consistent aggregations. Each node of this flowchart represents a "test" on an attribute, and each branch represents the outcome of the test; the result of the disaggregation can be the response to the specific decision problem [7].

Unlike the bottom-up clustering methods [2] [9, 10] [15, 16], the decision trees algorithms [1] [12] aggregate elements from the initial sample into sets increasingly smaller and increasingly internally homogeneous with respect to a specific variable, called driving-variable [5, 11].

In real market analysis this type of procedure, called "supervised" method, is quite suitable because of the specific kind phenomenon that is typically social and economic, and requires the previous awareness of the analyst who is requested to make some hypotheses about the most influent variables [14]. The disaggregation is carried out according to a criterion, dependent from the chosen algorithm. So a final partition of the initial sample, constituted by disjoint and exhaustive subsets, is created. To visualize better the result of the division of the initial sample, it is possible to use a graphical tool, called tree; the latter is a typical concept of mathematics which has some applications in the context of data mining.

In the context of data mining, trees are used as graphical support to visualize top-down algorithms.

The disaggregation is carried out starting from the initial sample and choosing subsets in which it is possible to realize the subdivision (partition) based on a specific split rule applied recursively.

Starting from a statistical sample, the construction phases of a tree associated to a top-down algorithm are:

1. Generation
2. Stop
3. Pruning

The tree associated to a top-down algorithm is constructed associating to every subset a node of the tree itself.

The first node (root) corresponds to the entire statistical sample. The successors (sons) correspond to the subsets in which every node is subdivided. Every division of a generic node (Fig. 3) is carried out basing on a split rule applied recursively. The split rule depends on the chosen algorithm. The last possible split provides nodes not further divisible (leaves). Generally the stop criterion is based on the achievement of a maximum level of depth, or on the achievement of a minimum number of elements for every terminal node.

In the phase of pruning, once stopped the generation process, the tree is reduced, or, using the jargon, is "pruned". In Fig. 3 the tree has been pruned at the third level.

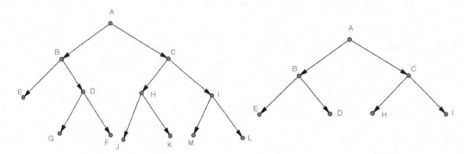

Fig. 3. Exemplification of the sequence of nods and the tree pruning

Since real estate market is a social-economic phenomenon, and because of the heterogeneity of this particular real estate, a *supervised* classification has been carried out, so that one classification for each feature k_h, has been drew up in order to identify the most significant ones.

In particular a *CART algorithm* has been carried out, a binary algorithm that divides each node into two further successors nodes. *CART* algorithms split every node by minimizing a measure of "impurity", or internal heterogeneity.

This measure changes when the type of variable changes, as follows:

Type of variables	Impurity measure
Nominal	Gini index
Ordinal	Entropy
Scaling	Standard Deviation

Formally, the decrease of impurity measure depends on the node t and the split s. It is defined as follows:

$$\Delta_{iY}(t,s) = \{i_Y(t) - [i_Y(t_l)p(t_l) + i_Y(t_r)p(t_r)]\} \tag{1}$$

where:

Y is the driving variable;

t is the father node;

t_l is the son node on the left;

t_r is the son node on the right;

$i_Y(t), i_Y(t_l), i_Y(t_r)$ are respectively the impurity measures of the node t, t_l, t_r;

$p(t_l)$ is the ratio between the number of elements in the node t_l and the number of elements in the node t;

$p(t_r)$ is the ratio between the number of elements in the node t_r and the number of elements in the node t.

Therefore, $i_Y(t_l)p(t_l) + i_Y(t_r)p(t_r)$ is the weighed sum of the impurities of the left and right nodes with weights $p(t_l)$ and $p(t_r)$. $\Delta_{iY}(t,s)$ is the decrease between the impurity measure at the node t and the impurity measures at the successor nodes t_l and t_r, corresponding to the split s. The algorithm identifies the split s which maximizes this decrease $\Delta_{iY}(t, s)$ and goes on to the division.

In case of scaling variable the split s is based on a numerical threshold of one variable describing the sample, so that the elements are divided into the following two sets: the ones under this threshold and the ones over this threshold. Therefore, the possible splits are identified as pairs (k_h, x_{hj}) $h = e, i, t, a$, $j = 1, 2, ..., n_h$ where k_h is the generic feature and x_{hj} is its possible value.

4 Application and Results

4.1 Numeric Decision Tree Application

The exposed procedure has been applied using the four driving variables k_h but the most relevant results has been obtained by using location (k_e) and intrinsic (k_i) ones.

Driving variable k_e. The decision tree obtained fron the perspective of the location feature is displayed in Fig. 4, where the number of elements that form the set and the value of the standard deviation of the aggregate quality within the node are indicated beside each node. A_{ij} is the set at level i settled in the position j starting from the left. Each A_{ij} corresponds to a group of properties, whose scatter in synthetically represented by a graph displaying the relation between prices per sq. m. *vs* overall quality", as displayed in Fig 5.

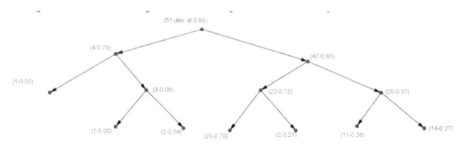

Fig. 4. Decision tree by driving variable k_e

At the beginning the sample has a standard deviation of 0.83. At the second level two groups have been formed: A_{21} and A_{22} respectively composed by 4 and 47 elements. A_{22}, including almost the whole sample, is a heterogeneous group of properties apparently to be divided into two further groups, the first located in the lower-left part of the graph, the second located in the upper-right one. In fact, the regression function shows negative values for both unit prices (Fig. 5).

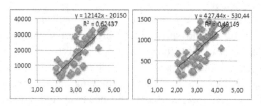

Fig. 5. Price *vs* k related to A_{22}

At the third level, the split of A_{22} gives A_{33} and A_{34} situated on the right of the tree, and are formed by 22 and 25 elements. A_{33} shows a contained prices and quality range so that it can be considered a sub-market itself.

Similarly A_{34} could be considered a submarket because of the low standard deviation (0.37) of the overall quality; in particular it could be considered a high-value segment, because, mainly from the point of view of the prices/room the scatter is adapted by the trend line having a very high intercept, and includes the most worthy properties (Fig. 6).

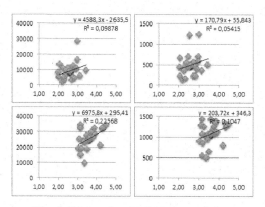

Fig. 6. Price *vs* k related to A_{33} and A_{34}

The later split gives rise to A_{46} (11 elements) and A_{47} (14 elements). The first one can be considered a consistent cluster only taking into account the low range of the overall quality, but not the prices, whose range is very high; therefore it does not provide useful information to allow us considering these properties as representative elements of a real submarket; the second one shows a clear relation price/room *vs* overall quality, so that it can be considered a submarket in a worthy segment formed by the finest properties in the A_{22} group (25 properties) (Fig. 7).

By concluding, two sub-markets can be clearly outlined: the first one is well represented by A_{33}, formed by properties with a low value in terms of overall quality; the second one (A_{47}) is formed by high quality properties that, for this reason, are mostly placed in the higher segment of the market.

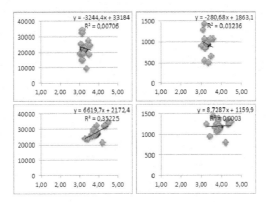

Fig. 7. Price *vs* k related to A_{46} and A_{47}

The intermediate band of the sample comprises a group of properties (A_{46}) whose prices range significantly. This group might contain more sub-markets hardly to be outlined on the basis of the available information. Moreover, in this segment the overall scores, ranging from 2.9 to 3.5 are the weight average of very diverse scores of the different k_h whose differences compensate each other, so that the properties cannot be considered strictly similar. These circumstances give rise to some uncertainties about the fair relationship price *vs* quality.

As a consequence, from the point of view of the driving variable k_e, the real estate market can be considered well structured in the low and high-price bands and quite immature in the intermediate one: in that band, the scarcity of the information doesn't support supply that hardly manages to find out the fair ask price.

About the other nodes, it is possible observe that almost all of them have few elements except the node A_{43}, that however differs from the node from which it derives just for two elements and, therefore, it is not convenient to analyse it, since it has the same characteristics of the group from which it descends, while other nodes have very few elements and might represent latent submarkets. In particular, the group A_{32} has a very low standard deviation of the overall quality (0.03). Although these groups are small in number, it is appropriate to consider them as possible sub-markets that would arise if information was more complete and the sample was larger.

Driving variable k_i. Assuming k_i as driving variable, the results of the clustering process mostly confirm the previous ones, outlining three sub-markets.

Differently from k_e, the tree shows eight leaves rather then seven, and quite different standard deviations (Fig. 8).

The first sub-market, outlined by group A_{21}, shows a quite low R^2 coefficient, but, since the scatter is concentrated in the low-left area of the graph it can be considered a significantly consistent segment. It comprises 13 properties and can be considered a low-band market segment, whose prices just overcome 500 € per sq. m (Fig. 9).

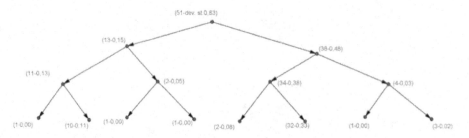

Fig. 8. Decision tree by driving variable k_i

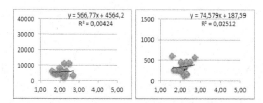

Fig. 9. Price *vs* k related to A_{21}

The second and the third segments both belong to group A_{22}, showing a higher R^2 coefficient. It looks like A_{22} group outlined in the previous analysis (driving variable k_e). It is composed of 38 elements and, although it doesn't significantly split at the further levels, in reality it is reasonable to distinguish two different sub-scatters: the first one, a low-band sub-market, ranging from the minimum up to 750 € per sq. m; the second one, the high-band sub-market, ranging from 750 € per sq. m up to the maximum price (Fig. 10).

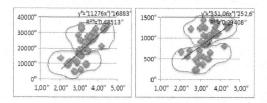

Fig. 10. Price *vs* k related to A_{22}

About the low-band group, however, we observe a certain dispersion in terms of quality and a range of prices wider than in the higher band. This effect had already been noticed in the previous k_e driving variable subdivision as well, where an intermediate band had a high price fluctuation.

Two difference can be observed: in k_e driving variable analysis, the algorithm outlined a non significant cluster; in k_i driving variable, the two significant clusters, A_{22a} and A_{22b}, weren't been split by the algorithm.

The successive nodes do not comprise a significant number of elements, so that they cannot be considered actual sub-market or they might be latent sub-markets.

At last, considering the node A_{34}, although it comprises four elements, the position of them suggests it can be considered a latent sub-market.

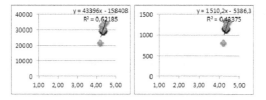

Fig. 11. Price *vs* k related to A_{34}

5   Discussions

5.1   Spatial Display

The analysis carried out so far shows significant results regarding the way the clusters are formed and the extent they are indicative of the relation between the aggregated characteristics and the prices of the sample. In particular, it is possible to recognize the tendency towards the main segmentation between ancient centre and external areas, with further differentiations concerning the direct accessibility (relation road/house) and the typology, connected to the significant difficulties of use of the small real estate units arranged on several stories. Further considerations may be done basing on the visualization of the clusters on the map.

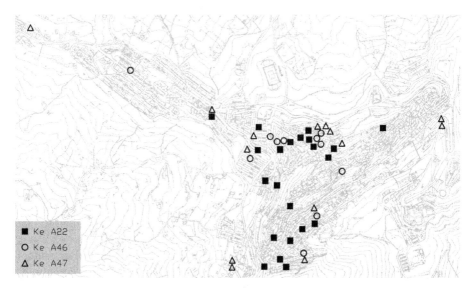

Fig. 12. Driving variable k_e, level 3

Fig. 12 represents the disaggregation regarding the variable k_e and shows the localization of the low-band market within the old town and the intermediate area; the middle-band market is located mostly within the middle area and less in the ancient and new; the high-band market is spread mostly in the intermediate and in the outer areas.

Fig. 13 represents the disaggregation regarding the variable k_i and shows the localization of the low-band properties mostly within the old town, the middle-band ones, in the old town and in the intermediate area, and the high-band ones, in the intermediate and outer areas hosting more comfortable and wide apartments.

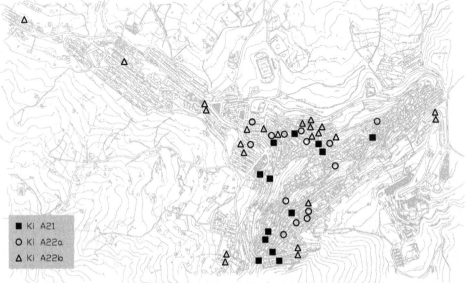

Fig. 13. Driving variable k_i, level 2

5.2   Nominal Decision Tree Application

A successive in-depth analysis aimed at researching the correlation of the single variables with the prices, has however highlighted some negative correlations, like for example the technologic characteristic one. As a consequence, above all in consideration of the particular conformation of the historic centre, and to accentuate its character discriminating the localization choices, a topographic variable k_l concerning the assignment of the properties to each of the three urban sectors has been introduced, and a further in-depth analysis has been carried out adopting a nominal approach.

The values of the variables have been turned in three classes of belonging, so that the resolution of the observation has been reduced.

To represent the nominal decision tree, a compromise between the level of internal homogeneity and the minimum number of elements has been made for every cluster; the best partition is forth level (nodes marked in red) which includes a cluster with 24 elements and Gini index 0.50, and a cluster with 11 items with index 0.30. These clusters are more numerous while the others cannot be considered significant because of the small number of elements.

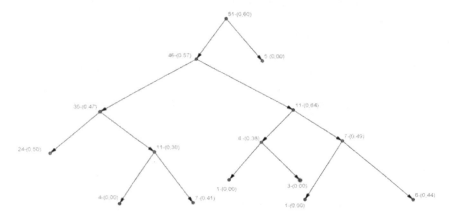

Fig. 14. Nominal decision tree related to k_l

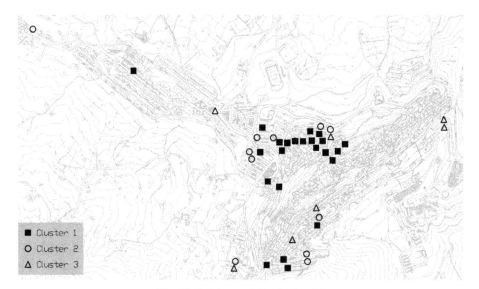

Fig. 15. Driving variable k_i, level 2

Fig. 15 shows the location of the three clusters for the nominal analysis. Also in this case, the qualitative clustering approximately corresponds to the market appreciation, even if the clusters are a little bit more compact, especially the low-band one

Table 2 shows the nominal representation of the sample and the comparison between numeric and nominal clusters.

The first block of columns indicates the nominal class of each property for each variable; the second one the comparison by position of each property from the perspectives of the three analyses; the third block shows the comparisons in pairs between the different analyses: the dark square indicates if the property belongs to both the clusters; the percentage on the top indicates the rate of overlapping between the two clusters in that analysis.

Table 2. Nominal approach and comparison with the numerical approach

id	Okl	1ke	2ki	3kt	4ka	24 11 7 nom	22 11 14 numke	13 14 24 numki	73% 36% 86% nom numke	69% 9% 86% nom numki	69% 27% 100% numke numki
1	a	b	b	a	b						
2	a	c	a	a	b						
3	a	c	a	b	c						
4	b	b	a	a	b						
5	b	b	a	a	b						
6	b	c	a	b	b						
7	b	c	c	c	c						
8	b	c	a	c	c						
9	a	c	b	b	a						
10	a	b	a	a	a						
11	b	c	b	b	b						
12	b	c	b	b	b						
13	b	c	b	c	b						
14	b	c	a	a	a						
15	b	b	a	b	a						
16	b	c	b	c	b						
17	b	c	b	c	c						
18	b	c	b	b	c						
19	b	b	a	b	c						
20	a	c	a	b	a						
21	a	a	a	b	a						
22	b	c	c	c	b						
23	a	a	b	c	a						
24	b	c	b	c	b						
25	b	c	b	c	c						
26	a	a	a	b	a						
27	b	c	c	c	c						
28	c	c	b	c	c						
29	c	c	c	c	c						
30	a	b	c	b	b						
31	a	a	a	a	a						
32	a	a	a	b	a						
33	b	b	a	a	a						
34	b	c	b	b	a						
35	b	b	a	a	a						
36	b	c	b	b	b						
37	b	c	b	c	c						
38	c	c	c	c	c						
39	c	b	c	a	b						
40	b	b	a	b	a						
41	b	b	b	c	b						
42	b	c	b	b	b						
43	c	b	a	b	a						
44	a	b	c	b	c						
45	c	c	b	c	c						
46	a	c	b	a	a						
47	b	c	b	c	c						
48	a	b	c	c	b						
49	a	b	b	b	b						
50	c	c	c	c	c						
51	c	c	c	c	c						

6   Conclusions

The real estate market in Troina is characterized by a wide oscillation of prices with a very reduced minimum extreme. The variegated characterization of the real estate asset suggests in-depth analyses to understand the motivations of such a wide variation, and then which segments may be individuated to define aimed strategies of valorisation, through the differentiation of the supports to the punctual and diffused redevelopments as well.

The real estate market analysis has been carried out by applying the decision tree approach that, unlike the hierarchical grouping techniques [3, 4], [6], sorts the elements of the sample by using a decisional process based on a top-down classification. In general it is possible to observe the position in the urban centre and the accessibility of the dwellings and the building typology, that accumulate disadvantaged situations in the internal areas of the historic centre.

The extensive analysis carried out to delimit the sub-markets through the application of the "decision trees" method has been conducted referring to four guide variables; its most significant results, concerning the k_e and the k_i, have been commented. Moreover, a further verification has been conducted introducing a new topographic variable, which has been subject to a nominal clusterization by classes of belonging. By comparing the results obtained by means of the two above mentioned techniques, it has been possible to observe that a certain amount of properties are included in the same clusters; therefore, the real estate-urban structure of Troina may be significantly represented by means of a numerical or nominal top-down type data mining approach.

In the two mentioned approaches the progressive refinement of the sample has generated diverse clusters, concerning in particular the numeric analysis, but considerations on their dimension have led to acknowledge as significant only three clusters.

When the localization variables have been assumed as guide variables, the analysis has shown a significant correspondence between the tripartition of the characteristics and the tripartition by price categories: the biggest cluster comprises in both cases low price properties, while the smaller clusters comprise worthier properties.

The segmentation by means of the driving variable k_i instead, provides a partition including a very consistent cluster even in relation to the successive cuts (at the third one it passes from 38 to 34 elements), and contains in equal proportions properties of both price categories; for the purpose of the comparisons, two parts of this cluster have been considered, according to the good correlation between price and overall quality: one low-price part and one-high price part.

The comparison between the three clusterizations has been carried out through the verification of the overlapping between the three clustering modalities, concerning the number of properties belonging to the same cluster in the three approaches. Table 2 shows an average overlapping higher than 60%.

The analysis presented so far confirms the first instance observation (Fig. 2) concerning the big amount of properties belonging to very low price categories, and can contribute significantly to the formation of aimed policies.

In general terms, it is possible to conclude that in a real estate market characterized by high density of values and by a high tension in prices, prices scarcely reflect the basic characteristics of the properties because of the high level of abstraction connected with a generalized   speculative perspective.

Similarly – but in the opposite sense – in a low tension market context, prices do not represent the characteristics of the properties because of an "excess of concreteness" due to the prevailing of the speculative demand of money that drains much of the liquidity of the assets and – in a generalized situation of credit crunch – of the real estate asset as well. The real estate market of Troina reflects the second situation, so that in several cases prices seem much lower than expected.

Acknowledgements. Alberto Valenti edited paragraphs 3; Salvatore Giuffrida edited paragraphs 1; Fabio Linguanti Edited paragraphs 2, carried out the market survey and edited Fig. 12-15. Paragraphs 4-6 should be attributed to both Alberto Valenti and Salvatore Giuffrida. Special thanks to "Amata Real Estate" in Troina for collaborating in providing the real estate market data.

References

1. Acciani, C., Gramazio, G.: L'Albero di Decisione quale nuovo possibile percorso valutativo. Aestimum **48**, 19–38 (2006)
2. Anderberg, M.R.: Cluster analysis for application. Academic Press, New York (1973)
3. Bertrand, P.: Structural properties of pyramidal clustering. Partitioning data sets. In: Cox, I., Hansen, P., Julesz, B. (eds.) DIMACS Series in discrete mathematics and theoretical computer science, vol. 19, pp. 35–53. American Mathematical Society, Providence (1995)
4. Bonner, R.: On some clustering techinques. International Business Machine Journal of Research and Development. **8**, 22–32 (1964)
5. Breiman, L., Friedman, J.H., Olsen, R.A., Stone, C.J.: Classification and regression trees, pp. 203–215. Wadsworth International Group, Belmont (1984)
6. Cunningham, K.M., Ogilvie, L.C.: Evaluation of hierarchical grouping techniques: a preliminary study. Computer Journal **15**, 209–213 (1972)
7. Drucker, H., Cortes, C.: Boosting decision trees. Adv Neural Inf Process Syst **8**, 479–485 (1996)
8. Forte, C.: Elementi di estimo urbano. Etas Kompass, Milano (1968)
9. Everitt, B., Landau, S., Morvene, L., Stahl, D.: Cluster Analysis, 5th edn. John Wiley and sons Ltd. (2011)
10. Fraley, C., Raftery, A.E.: Model-Based Clustering, Discriminant Analysis, and Density Estimation. Journal of the American Statistical Association **97**(458) (2002)
11. Freund, Y., Schapire, R.: A decision-theoretical generalization of on-line learning and an application to boosting. J Computer Syst Sci **55**, 119–139 (1997)
12. Gastaldi, T.: Data Mining: Alberi Decisionali e Regole Induttive per la profilazione del cliente. Science & Business **4**(5–6), 16–25 (2002)
13. Giuffrida, S., Ferluga, G., Valenti, A.: Capitalization rates and "real estate semantic chains". An application of clustering analysis. International Lournal of Business Intelligence and Data Mining (2015)
14. Hastie, T., Tibshirani, R., Friedman, J.: The elements of statistical learning. Springer-Verlag, New York (2001)
15. Hepşen, A., Vatansever, M.: Using Hierarchical Clustering Algorithms for Turkish Residential Market. International Journal of Economics and Finance **4**, 138–150 (2012). doi:10.5539/ijef.v4n1p138
16. Maitra, R., Volodymyr, M., Soumendra, L.N.: Bootstrapping for Significance of Compact Clusters in Multi-dimensional Datasets, Jasa (2012)
17. Rizzo, F.: Nuova economia. Aracne, Roma (2013)
18. Simonotti, M.: La segmentazione del mercato immobiliare urbano per la stima degli immobili urbani. Atti del XXVIII Incontro di studio Ce.S.E.T, Roma (1998)

Planning Seismic Damage Prevention in the Old Tows Value and Evaluation Matters

Salvatore Giuffrida$^{(\boxtimes)}$, Giovanna Ferluga, and Vittoria Ventura

Department of Civil Engineering and Architecture, University of Catania, Catania, Italy
sgiuffrida@dica.unict.it, gio.ferluga@virgilio.it,
vittoriaventura01@gmail.com

Abstract. Only recently seismic prevention has come up in urban planning and considered as a significant part of the social capital of the urban heritage. Among the planning tools for reducing the vulnerability of urban fabrics, the Emergency Limit Condition (ELC) defines the fields which require widespread strengthening interventions in the buildings involved in resilient town units. The extent and intensity of such interventions can be graded so as to find an optimal ELC measure, taking the needs of the urban and human capital into account. In case of the old town centre of Ragusa, a model for analysing the seismic vulnerability of the built-up areas has been integrated with a model for a cost-revenue estimation in order to identify the conditions of the ELC dimensioning and to develop the relevant incentive system.

Keywords: Old town valorisation · Seismic safety · Equalization · Emergency Limit Condition · Cost-revenue analysis

1   Introduction

The social capital of a town manifests itself by means of an urban dimension and a human dimension [17, 19]. Generally speaking, the urban capital is the form (information, organization, integration, interdependence, autopoietic capacity) of a town and, as such, the context which accounts for the human capital formation; on its part, the human capital is the "reason" or the value term of the urban form, that is the referent of its sign function. The function of the human capital value has a plethora of subjects, among which the safety in a seismic event is one of those mainly involved in the conservation strategies of the historical buildings.

Among the main aspects related to upgrading of old town centers, safety is getting more and more crucial, especially looking at the dramatic facts and in the light of two key trends, one widespread and from below, the other detailed and from above.

The former concerns the rigidity of the historical building heritage and its structural unsuitability to being transformed, which drastically selects the people applying for buildings for living, commercial, recreational and also public uses (administration, education, museums etc.).

© Springer International Publishing Switzerland 2015
O. Gervasi et al. (Eds.): ICCSA 2015, Part III, LNCS 9157, pp. 253–268, 2015.
DOI: 10.1007/978-3-319-21470-2_18

The latter regards the recent reform measure of the Civil Protection introduced by the emergency decree of 15 May 2012, converted into Act of Parliament No. 100 of 12 July 2012, which moves calamity damages to the market by means of the insurance sector and on personal initiative. This gives rise to great perplexities for a variety of reasons [21]. First and foremost, there is the difficulty of insuring buildings whose static layout could have been altered with time; the consequent long legal cases which could make the reconstruction process more complex; the ethical profile of the ensuing damage compensation scheme, which is affected by the crushing asymmetry between the insurer and the insured party in terms of probative capacity; finally, the centrality of the insurance sector which is not extraneous to the financial matter and therefore not quite replaceable by the State. The latter group of reasons concerns the difficulties in managing the reconstruction which is clearly thought as a whole and collective process; this process can hardly be managed in the presence of different conditions of owners' solvency, especially between those who have taken out an insurance policy and those who haven't; moreover, in case of complex historical fabrics, the architectural units are linked to one another in clusters which demand unified interventions and which cannot be subjected to diverse or poor insurance regimes.

Also the localization model in old town centers is affected, in that it involves economic subjects with a profile characterized by deep awareness about the heritage features of the urban real estate and a marked sensitivity to precise and contextual values expressed by historical building fabrics.

Thus, the planning has to mediate between the exigencies of the real estate and those of the social capital [17] by assuming the critical points of the former as opportunities for the promotion of the latter, but not vice versa; otherwise the renovation program would reduce or definitely exclude the increase of social housing quota [1] [16]. An excellent model for the internalization of externalities at the programming scale [18] following the seismic improvement of the buildings which are crucial for a reduction policy of global vulnerability, requires the support of a relevant evaluation model, subject of this study. It is an interscalar model which projects the evaluations carried out in the building scale on the urban scale.

Actually, since a few years the urban vulnerability has begun to be an experimented subject of study. Its importance can be promoted by developing the culture of protection and prevention. The devastating effects of the 2009 earthquake in Abruzzo, the unexpectedness occurred in Emilia Romagna in 2012, the costs and the administrative difficulties originated by them, and the perspective of structural modifications of the territories have increased the sensitivity about the seismic prevention, as established by the national legislation about seismic prevention (Article 11 of the Law 77/2009). Such investments can play a significant role in enhancing old town centers especially if some equalizing practices are adopted to combine individual and collective interests involved in renovation investments.

The study focuses on an area of the old town center of Ragusa by means of a survey and a set of evaluations made to identify the conditions of economic advantage to forming the Emergency Limit Condition (ELC), by combining the technical aspects of the vulnerability assessment [9, 10] and the economic aspects related to cost-revenue estimation.

2   Materials

2.1   The Seismic Prevention Plan of Ragusa

Ragusa is the chief town of the southernmost province of Italy and one of the eight towns of the Val di Noto in the south-eastern Sicily which are inscribed on the UNESCO World Heritage List for the innovation of their urban planning facilities, the architectural value of their late baroque monuments and the contextual value of their building fabrics. After being destroyed by the seism in 1693, its old town center re-shaped also as a consequence of some social reconfiguration of the town divided into two parts: Ragusa Ibla rebuilt on the ruins of the old town with a medieval flavour, and Ragusa Superiore where the emerging middle class settled, thus giving life to an urban fabric with a modern orthogonal system.

The urban events which in time have characterized Ragusa Superiore as it is today have been cases of property pressure which have caused violation of the eighteenth-century fabric and of the original urban landscape, especially in the second half of the '70s and, more recently, cases of large abandonment and transfer of residential and commercial activities to more suburban and newly-allotted areas.

Since the Resolution of the Regional Council No. 408 of 19 December 2003 and the Decree of the General Director No. 3 of 15 January 2004 which classify the regional territory into 4 classes according to the stratigraphic and topographic dange-rousness from the most dangerous (Zone 1) to the least dangerous (Zone 4), the territory of Ragusa is included in Zone 2 with the indication of Special Area, which requires the same in-depth analysis of the technical safety checks as in Zone 1.

The in-depth analyses of the local seismic risk are part of the analyses of Seismic Micro-zonation [20] (Fig. 1) which offer suggestions on: 1. how to direct the localiza-tion in the territory; 2. what interventions can be allowed in a certain area; 3. research planning for the next deeper levels of analysis; 4. how to direct and intervene in urba-nized areas; 5. intervention priorities.

Fig. 1. Seismic Micro-zoning of the old town center of Ragusa

The seismic vulnerability of buildings denotes their propensity to be damaged by a seismic event of a given intensity. It depends on the geometry of a building, its constitutive materials and on the system of internal restrictions. It is measured along a qualitative four-level scale, from "Moderate" to "Very high". The vulnerability of a historical town center depends on how large is the presence of vulnerable buildings (Fig. 2).

Fig. 2. Semeiotic valuation of the vulnerability of the urban fabric of the old town of Ragusa

Finally, the Exposure, which represents the value of the elements at risk (people, buildings, streets and infrastructure) in terms of economic damage, loss of cultural heritage and human lives [22] (Fig. 3).

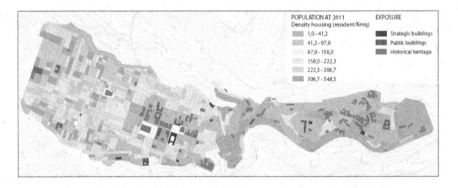

Fig. 3. Exposure of population and urban architectural heritage in the old town of Ragusa

According to the standard approach and definitions, risk can be defined as the probability of potential impacts affecting people, assets or, more generally, the habitat. Risk is a function of hazard (causes potentially triggering disasters), vulnerability (the inability to face or cope the effects of a disaster), exposure (volume and value of the infrastructural, building and human capital potentially involved by damaging effects) [14], [21]. Seismic risk is calculated combining: Hazard, Vulnerability and Exposure: $R = H \cdot V \cdot E$. Seismic risk among census sections is illustrated in Fig. 4.

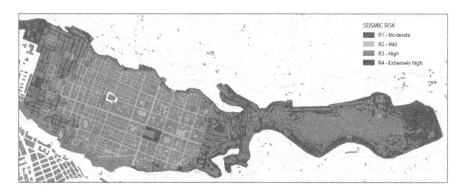

Fig. 4. Map of the seismic risk of the old town center of Ragusa

The intervention strategies outlined by the Plan need to be integrated into the ordinary planning [9] [11] – in particular the Executive detailed plan of the old town center with reference to the Technical Implementation Regulations – with measures designed to reduce the vulnerability of the building heritage and dangerousness, thus influencing the permit granting process; moreover, emergency areas and escape routes should be mapped out and specific projects should be conceived for the security implementation of critical points. The interventions should be compatible with typological and formal characteristics of the fabric and should be reversible [6].

3   Methods and Procedures

3.1   Emergency Condition Limit and Vulnerability of the Built-Up Area

The O.P.C.M. 4007-29/02/2012) rules the funding devoted by the art. 11 of law n. 77-26/06/2009 (National Fund for Seismic Prevention) available in 2011, and supposes the implementation of the Analysis of the Emergency Limit Condition (ELC).

The ELC of the urban settlement is defined "the condition overcame which the urban settlement preserves anyway, as a whole, the operation of most strategic functions for the emergency, their accessibility and connection with the territorial context, even if after the earthquake the physical and functional damage has caused the interruption of almost all of urban functions, including the residence.

The Analysis of ELC of the urban settlement is carried out, in conjunction with the seismic micro zoning studies (MS), by using standards for storaging and cartographic representation of the data that are collected through forms, specifically constituted by five tabs: SB (Strategic Building), EA (Emergency Area), AC (Accessibility/ Connection Infrastructure), SB (Structural Block) and SU (Structural Unit), provided by the technical Commission supposed by the art. 5/7-8 of the O.P.C.M. n. 3907/2010 and issued by specific decree of the Head of the Department of Civil Protection. This analysis consists of:

1. identification of buildings and areas providing strategic chances for emergencies;
2. identification of the infrastructures of accessibility and connection with the territorial context, buildings and areas referred to point 1) and any critical elements;
3. identification of structural blocks and single structural units that can interfere with
   the infrastructure for accessibility and connection with the territorial context.

The interventions offered by this tool aim at making the most effective routes safe by
linking buildings and strategic areas; ELC formation requires to identify the critical
points of buildings which can interfere with such functions [7].

ELC can have variable dimensions and schedule interventions of higher or lower
intensity in order to guarantee a more or less elevated safety level. Its dimension depends on the availability of the population to extend safety to more or less wide areas
of the urban fabric. Consequently, safety assumes the connotation of social urban
capital with an elevated intangible and perspective value. The prediction of the seismic damage was made by implementing a descriptive model of the mechanical behaviour of the façades [23] of the buildings facing the traffic streets, identified for the
ELC formation [8]. The model is able to measure the probability of the facade to
overturn under the dynamic action of the soil, with an acceleration coefficient α_{0b} and
α_{0v} depending on the basic (less favourable) or varied (more favourable) configuration
respectively taken into consideration. The coefficient measures the soil acceleration in
correspondence to which the facade overturns and therefore its dimension is inversely
proportional to the resistance to the structure overturning. The parameters considered in
the model regard: the thickness of the wall at the ground floor; the total height of the
wall; the distance between bracing walls; the total number of floors; the number of
floors devoid of chains (counted from above); the direction of the structure of the attic
(if parallel or perpendicular to the facade); the toothing with the bracing walls [23].

3.2   Evaluation Aspects and Tool

The ethics profile of the ELC planning process requires the balance between the public expense to reduce the seismic risk and the effects produced by interventions. From
the viewpoint of the values involved, this means the correspondence between the
value of the investments and the value of the produced safety, the latter being hardly
assessable because of the strong randomness of seismic events. Thus, an evaluation
model rather assumes a meta-planning function in terms of capacity to produce as
many different versions of the supposed configuration as the combinations of the
applied conditions. Such conditions are directly associated to the axiological paradigm which motivates and legitimates the preventive action.

In the renewed planning logic which characterizes the present-day urban planning,
the possibility of financing the planning process from the inside, as a matter of fact,
implies the activation of new resources and the arrangement of equalizing tools suitable to direct the surplus of the created value towards safety. A monetary measure of
the safety value allows understanding to what extent the equalizing mechanisms are
appropriate. Whenever these latter imply the activation of real estate resources as in
case of transfer of property rights, the value of these rights has to be compared to the
value of preserved urban and human heritages, net probability of the calamitous

event. The model supporting the ELC formation is divided into: description of the different architectural units giving information about their structural, material, geometric, technological and typological characteristics; allocation of the intervention type according to the level of vulnerability of each unity [3, 4]; calculation of intervention costs on the basis of an analytic model for typical packages of works; evaluation of the revenues in terms of increase in the real estate value; maximization of the revenue-cost function (Fig. 5).

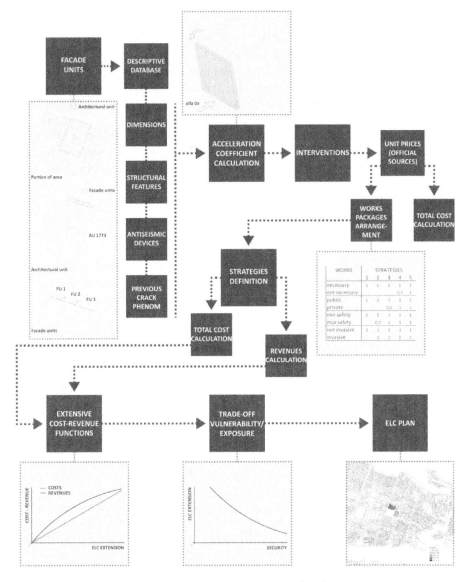

Fig. 5. General scheme of the valuation-planning pattern

The costs are referred to interventions for securing the façades of the buildings facing escape routes. In each building the façade units u_i were identified as delimited by two orthogonal walls, which are presumed to have a dynamic behaviour independent of the others belonging to the same unit or adjacent.

The ELC cost is calculated by combining an intensive with an extensive function.

An intensive function of the cost $C(b_{jk})$ is defined as the relation between the cost and intensity of intervention (higher the intensity, lower the vulnerability), which depends on the package of works b_{jk} associated to the single u_i. Each package includes works corresponding to the items of the Price List of the Public Works of the Sicilian Region – $b_{jk} \in B$, where B is the set of the whole works which can be referred to the activities for securing the registered buildings.

The works concern the insertion of tie-rods, the filling of superficial cracks, the reclamation of masonry with through-cracks, the introduction of strengthening masonry, all the external and internal finish works, related to both walls and ceilings, for a total of 36 items of the price list.

The package b_{jk} is not univocal, but it can contain a more or less elevated number of works according to the inclusion of: those strictly necessary j_n; those of primary public interest j_p; those less invasive j_v; those which provide a more or less elevated safety level j_s. Each of these can be scaled, thus obtaining a different safety level, $k_{60\%}, k_{70\%}, ..., k_{100\%}$ in terms of the acceleration coefficient calculated according to the new layout as outlined consequently to b_{jk} intervention package: by combining the five classes j with the five safety levels k, 25 different hypothetical strategies with increasing costs were formulated (Fig. 6).

The extensive function of the cost is the relation between the total ELC cost and the ELC extent: $C(l_{CLE})$. This is a discontinuous function, whose fragmented performance depends on the dimensional heterogeneity and vulnerability of the sample.

Also the revenue admits an intensive and an extensive function.

The intensive function $R = f(V)$ regards the advantages linked to the expected interventions, expressed in monetary terms with primary reference to the real estate value of the secured building, especially the market value increase due to the improvement of its static performances. The market research on a sample of the old town center encompassing the area under study has allowed to estimate the coefficient a_3 of the regressor k_t of the function $V = a_1 k_e + a_2 k_i + a_3 k_t + a_4 k_a + b$. In order to reduce the distortion effect on the calculation of the value, due to the heterogeneity of the relation between the length of the front, which makes the cost variable, and the commercial surface of the building which, on the other hand, makes the value variable, the increase in value has been index-linked to the mean incidence of the cost for façades.

The revenue extensive function $R = f(l_{ELC}, S)$ depends on the ELC dimension and on the decrease in real estate value following the progressive separation from the central area, dependent on l_{ELC}, along with a lower and lower density in population and poor architectural quality of buildings, measured by S. The coefficients of regressors k_e and k_a provide a measure of the decrement rate of this value when the l_{ELC} dimension increases. Thus, the revenue is obtained by multiplying the value increase by the dimension of the secured building fronts, thus taking the progressive decrease in real estate unitary value into account.

4   Applications and Results

4.1   The Studied Area: Data and Information

The area under study is formed by a building fabric of 246 architectural units subdivided into a total of 938 façade units listed and described in a database (Table 1). The linear development of the façades on the street is 3671 m, while the global development of the fronts which are considered for intervention is 6693 m. The characterization of each façade unit concerns the geometric and building features which are useful in calculating the two acceleration coefficients as well as characterizing the buildings and the whole context from the viewpoint of its vulnerability (Fig. 6).

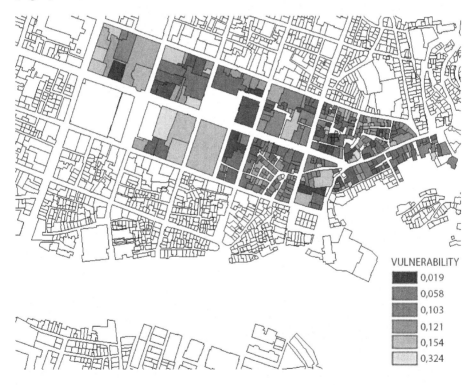

VULNERABILITY

- 0,019
- 0,058
- 0,103
- 0,121
- 0,154
- 0,324

Fig. 6. The sample: localization and seismic characterization by vulnerability degree

Table 1 shows the distribution of the sample (first column AU id. number, second column façade unit id. number) among the different classes of vulnerability expressed by the acceleration coefficients (Fig. 7).

Table 1. Calculation of the vulnerability of the façade unities of the sample

id au	id fu	wall thickness	tot h	perp wall dist	floors	fl no chains	floor direction	r perp wall	n	base/non base	index alfa0 b	alfa0 v
1773	1	1,4	7	5,3	2	2	1	0,333	72	0	0,267	0,267
1773	2	0,7	7	3,6	2	2	3	0,162	72	1	0,116	0,183
1773	3	0,7	7	4,5	2	2	3	0,135	72	1	0,114	0,179
1774	1	0,8	8,9	4,6	3	3	3	0,235	72	1	0,111	0,189
1774	2	0,8	8,9	5,6	3	3	3	0,181	72	1	0,106	0,181
1774	3	0,8	8,9	5,9	3	3	1	0,496	72	0	0,134	0,134
1774	4	0,6	8,9	5	3	3	1	0,640	72	0	0,111	0,111
1774	5	0,7	8,9	4	3	3	3	0,267	72	1	0,100	0,186
1775	1	0,8	10	4,6	3	3	3	0,235	72	1	0,099	0,183
1775	2	1	10	4,6	3	3	3	0,235	72	1	0,123	0,194
1775	3	1	10	6,5	3	3	1	0,400	72	0	0,140	0,140
1775	4	0,7	10	4	3	3	3	0,267	72	1	0,089	0,180
1776	1	0,8	7	3,1	2	2	3	0,177	72	1	0,135	0,192
1776	2	0,8	10	4,3	3	3	3	0,251	72	1	0,100	0,185
1777	1	0,9	10	7,8	3	3	1	0,192	72	0	0,107	0,107
1778	1	0,6	13	5	4	4	1	1,000	72	0	0,092	0,092
1779	1	0,7	10	3,8	3	3	3	0,277	72	1	0,089	0,182
1779	2	0,6	13	4,4	4	4	1	1,150	72	0	0,099	0,099
1780	1	0,5	7	8,5	2	2	1	0,045	72	0	0,075	0,075
1781	1	0,5	7	5,6	2	2	3	0,102	72	1	0,079	0,158
1782	1	0,5	7	3,1	2	2	3	0,177	72	1	0,084	0,169
1783	1	0,5	7	5,7	2	2	1	0,297	72	0	0,093	0,093
1800	1	0,8	13	4	4	4	3	0,417	72	1	0,087	0,195
1800	2	0,5	13	5	4	4	1	1,000	72	0	0,077	0,077
1801	1	0,7	10	3,7	3	3	3	0,283	72	1	0,090	0,183
1801	2	0,6	10	3,8	3	3	1	0,832	72	0	0,110	0,110
1801	3	0,6	10	2,6	3	3	3	0,341	72	1	0,080	0,183
1801	4	0,5	10	2,7	3	3	3	0,336	72	1	0,067	0,173
1801	5	0,6	10	1,8	3	3	3	0,384	72	1	0,083	0,189
1802	1	0,7	10	4,9	3	3	3	0,219	72	1	0,085	0,174
1802	2	0,7	10	4	3	1	3	0,067	88	1	0,075	0,233
1802	3	0,7	10	5,5	3	3	1	0,560	72	0	0,109	0,109

Fig. 7 matrix shows the selection of the intervention strategy. The less intensive strategy selects just necessary, public, minimum safety and not invasive works, the third one selects necessary, public and half of private, both minimum and maximum safety, both not invasive and invasive; the fifth strategy, the most intensive, selects the whole set of the envisaged works.

The great quantity of available data and a set of formulations carried out for other aims, combined with a detailed market research concerning a sample of the old town center encompassing 75 buildings for sale – whose accurate description is postponed to a successive specific study – have provided the elements for determining the added value due to interventions in security implementation.

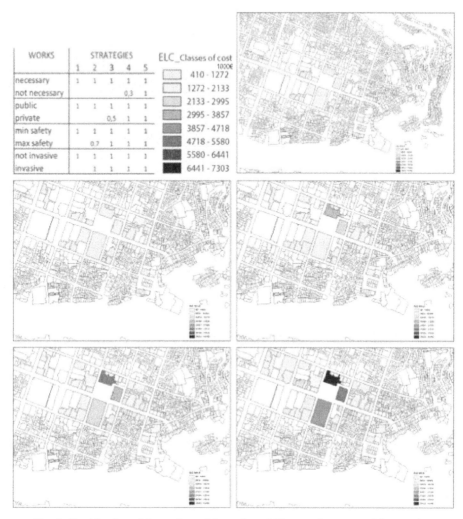

Fig. 7. Combination of the works and formation of 25 strategies: maps of 5 strategies

4.2   Results as a Support for ELC Dimensioning

In order to select the optimum ELC dimension, the 25 strategies designed by varying the type of works and the safety level were evaluated in terms of costs and revenues.

The intensive function of costs – defined in the hypothesis of the greatest ELC extent – is represented in Fig. 8.

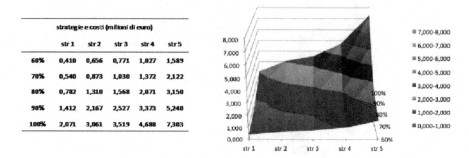

strategie e costi (milioni di euro)					
	str 1	str 2	str 3	str 4	str 5
60%	0,410	0,656	0,771	1,027	1,589
70%	0,540	0,873	1,030	1,372	2,122
80%	0,782	1,310	1,568	2,071	3,150
90%	1,412	2,167	2,527	3,373	5,240
100%	2,071	3,061	3,519	4,688	7,303

Fig. 8. Costs of the 25 strategies and a spatial representation

The intensive function of the revenue was defined by varying the safety level progressively and by estimating, on the basis of the variation of the market value, the ensuing differential in the market value, here assumed as indicator of the obtained safety value (Fig. 9).

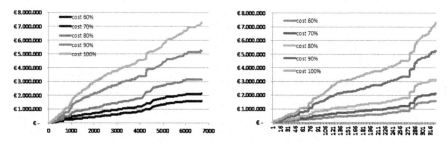

Fig. 9. Extensive cost function

The extensive function of costs and revenues, on the other hand, describes the respective variations for each safety level on varying l_{ELC}.

By maximizing the revenue-cost function (Fig. 10) in correspondence to each supposed strategy, the relevant optimal ELC extent is defined (Fig. 10). The variation of this dimension, which depends on the distance between revenues and costs, decreases when the safety level increases (a greater number of buildings will be involved in security implementation), and increases when the cost reduction at the public body's expense decreases (incentive reduction).

The graphs in Fig. 10 show the performance of the cost-revenue functions related to the ELC extent measured with regard to the development of the secured fronts, and the difference between the two functions, which allows identifying the optimal extent at the highest point.

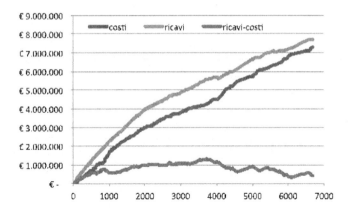

Fig. 10. Extensive revenues-cost and margin functions

By modifying the conditions and by establishing different budget limits, it is possible to build the different Pareto frontiers which describe the trade-off relation between ELC dimension and safety level.

On this basis, given the two ELC performances extent/safety, also the map of the isocost functions can be described. Fig. 11 shows the isocost functions corresponding to eight budgets (0.5 to 4 mln €).

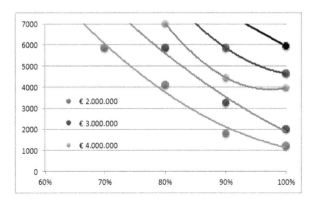

Fig. 11. Trade-off between extent and security for five different given budgets

5   Discussions

In accordance with the main objective of this study – i.e. the development of a plan to reduce the seismic risk on the basis of an evaluation model – some observations can be put forward.

The first regards the ELC configuration: it has to link strategic points of the urban fabric by means of defined routes so as to maximize the cost control, emergency safety, urban structure resilience; consequently a dimension model cannot disregard the topological pattern of the network of routes, which need to be defined before carrying out the suggested evaluations for each network hypothesis.

The second concerns the ELC dimensioning: it implies interventions for the structural improvement of buildings with very different consistency, typology and state of repair; consequently a dimensioning model of a continuous type does not work at its best. In addition, the least ELC dimension needs to be established to pursue its main aim; Fig. 10, in fact, shows indifference maps which are sometimes very fragmented.

The third refers to the need for a damage prevention strategy to be completed in the short term; ELC is a tool which works only if possible interferences between buildings and routes are reduced to the minimum, or completely eliminated. From this point of view, prevention policies can also be based on equalizing mechanisms to stimulate the owners to adhere but, on the other hand, it is quite risky to link the destiny of old towns and the safety of settled communities to incentive mechanisms based on the transfer of real property and whose success is due to the dynamics and moods of the property market.

In the light of these considerations, the results here obtained can be interpreted in a constructive way, especially with regard to the ratio between expense and expected increment of the market value, which increases when the intensity of the intervention increases and decreases when its extent increases.

Point sets of trade-off functions denote the progressive shift of the iso-expense function upwards and to the right, and a higher continuity and density in the budget ranging between 0.5 and 7.3 mnl €, which suggests a higher certainty of the solution cloud in this area of the indifference map.

The last consideration regards the nature of this function which can represent two different and, to some extent, opposite areas of action and collocation of the subject mediating between the different positions. More precisely:

— from the viewpoint of the urban system performances, it can be interpreted as an iso-utility function between the safety level of interventions and ELC extent;
— from the viewpoint of the tendency to form the fixed social capital, this functions represents the quantity and the quality of allocations in terms of urban heritage value, and measures the present vs future dialectics which defines the tendency to "real hoarding" from a settled community.

6   Conclusions and Perspectives

The old town center of Ragusa is composed of a vulnerable and a densely populated building fabric claiming a significant effort for reducing vulnerability in the perspective of a sustainable development; ELC is the tool to be adopted in that perspective of the internalization of the risk externalities [2].

From such a holistic perspective [5], the evaluation of a vulnerability reduction plan should assume as the revenue in the economic balance the avoided losses [15] [24]. Since them are very difficult to forecast, it could result an abstraction, scarcely able to involve and motivate households and municipalities. In this case, a strict individual cost revenues analysis has been carried out as the basis for further evaluations.

The strategies for the seismic damage prevention are not univocal and can be differently dimensioned in intensity and extent, according to the "future propensity" expressed by the social system; it manifests itself in a capacity of intra-seismic communication, which measures the level of unity of the social macro system, and extra-seismic communication, which measures its resilience to environmental fluctuations and therefore its capacity to differentiate from the environment, thus encompassing wider and wider areas in the internal communication [12, 13].

The case of the ELC in the old town center of Ragusa is an example of the capacity to combine public and private resources of different nature to achieve urban facilities which work only if they are fully completed.

The model for analysing, evaluating and planning the ELC of the old town center of Ragusa, referred to the sample under study, has allowed to verify the advantages of a systematic and coordinated intervention campaign on the built-up areas and the different opportunity conditions in a quantitative logic of cost-revenue type.

Further research and modelling perspectives, currently under way, concern the maximization of the safety value by means of a spatial function developed within a network model in which the distribution of the secured buildings determines the best configuration among the ELC routes. Consequently, the configuration of the route is influenced by the selected strategy which, on its turn, depends on the highest or lowest propensity of the settled community to the valorisation of the safety capital. Such a detailed analysis will offer a model subdivision so as to make it more suitable to the discontinuity and heterogeneity of the urban fabric and to systematic properties required by the ELC to fulfil its aims properly.

Acknowledgements. Salvatore Giuffrida edited paragraphs 1, 3.2, 4.2, 5, 6, drew up the valuation model, performed the calculations; Giovanna Ferluga edited paragraph 3.1 and fig. 5 and 6; Vittoria Ventura edited paragraphs 2, 4.1. *Special thanks to the Municipality of Ragusa for providing us the database of the old town of Ragusa.*

References

1. Bonafede, G., Napoli, G.: Multicultural Palermo between gentrification and real estate market crisis in: the historical center. In: From control to co-evolution. AESOP Annual congress 9-12 July 2014 Procedeengs, Utrecht/Delft: AESOP, pp.1-12 (2014)
2. Boulle, P.: Vulnerability Reduction for Sustainable Urban Development. Journal of Contingencies and Crisis Management, 179–188 (1997)
3. Carocci, C.F.: La lettura critica del costruito dei centri storici. In: Rischio sismico, territorio e centri storici, Atti del Convegno Nazionale, Sanremo (IM) 2-3 luglio 2004, pp. 257-262. FrancoAngeli, Milano (2005)
4. Carocci, C., Copani, P., Marchetti, L., Tocci C.: Vulnerability reduction procedures in ordinary urban management. The Urban Building Code of Faenza, Smart Built (2014)
5. Carreno, M.L., Cardona, O., Barbat, A.H.: Urban Seismic Risk Evaluation: A Holistic Approach. Natural Hazard **40**, 137–172 (2007)
6. Comune di Ragusa, Piano di Protezione Civile, Ragusa (2012)

7. Dolce, M., Di Pasquale, G., Speranza, E., et al.: A multipurpose method for seismic vulnerability assessment of urban areas. In: Proceedings of the 15th World Conference on Eathquake Engineering (WCEE), Lisboa (2012)

8. Dolce, M., Speranza, E., Negra, R.D.: Constructive features and seismic vulnerability of historic centers through the rapid assessment of historic building stocks. The experience of Ferrara, Italy. In: Toniolo, L., et al. (eds.) Built Heritage: Monitoring Conservation Management, Research for Development, pp. 165–175. Fondazione Politecnico di Milano, Milano (2015). doi:10.1007/978-3-319-08533-3_14

9. Hizbaron, D.R., Baiquni, M., Sartohadi, J., Rijanta, R.: Integration method of disaster risk reduction into spatial planning: case study yogyakarta special province and bantul regency. In: Proceedings of the 1st International Conference on Sustainable Built Environment, Yogyakarta, Indonesia, 4 May 2010; Faculty of Civil Engineering and Planning, pp. 311–321. Islamic University of Indonesia, Yogyakarta (2010)

10. Hizbaron, D.R., Baiquni, M., Sartohadi, J., Rijanta, R.: Urban Vulnerability in Bantul District, Indonesia - Towards Safer and Sustainable Development. Sustainability **4**, 2022–2037 (2012)

11. Hung, H.C., Ho, M.C., Chien, C.Y.: Integrating long-term seismic risk changes into improving emergency response and land use planning: a case study for the Hsinchu City Taiwan. Natural Hazard **69**(1), 491–508 (2013)

12. Luhmann, N.: Soziologie des Risikos. Walter de Gruyter Co., Berlin (1991)

13. Luhmann, N.: Social Systems. Stanford University Press, Stanford (1996)

14. Merchler, R.: Cost-benefit Analysis of Natural Disaster Risk Management in Developing Countries. GTZ (German Society for Technical Cooperation, Eschborn (2005)

15. Mileti, D.S., Gailus, J.L.: Sustainable development and hazards mitigation in the United States: Disaster by design revisited. Mitig. Adapt. Strateg. Global Chang. **10**, 491–504 (2005)

16. Napoli, G., Schilleci, F.: An Application of Analytic Network Process in the Planning Process: The Case of an Urban Transformation in Palermo (Italy). In: Murgante, B., et al. (eds.) ICCSA 2014. LNCS, vol. 8581, pp. 300–314. Springer, Heidelberg (2014)

17. Pelling, M.: The Vulnerability of Cities: Natural Disasters and Social Resilience. Earthscan Publishing, London (2003)

18. Rizzo, F.: Conservazione e Valutazione del patrimonio architettonico-ambientale nelle aree urbane a rischio sismico. Genio Rurale, 9, Bologna (2000)

19. Rizzo, F.: Il capitale sociale della città. FrancoAngeli, Milano (2003)

20. Sarris, A., Loupasakis, C., Soupios, P., Trigkas, V., Vallianatos, F.: Earthquake vulnerability and seismic risk assessment of urban areas in high seismic regions: Application to chania city, Crete Island Greece. Nat. Hazards **54**, 395–412 (2010)

21. Sawada, Y.: The Economic Impact of Eartkquake on Households. In: Guha-Sapir, D., Santos, I. (eds.) The Economic Impact of Natural Disasters. Oxford University Press, New York (2013)

22. Tiedemann, H.: Earthquake and Volcanic Eruptions – A Handbook on Risk Assessment. Swiss Reinsurance Company, Zurich (1992)

23. Tocci, C.: Vulnerabilità sismica e scenari di danno: analisi speditiva delle catene di danno. In: Blasi, C. (ed.) Architettura storica e terremoti. Protocolli operativi per la conoscenza e la tutela, pp. 113–119. Wolkers Kluwer, Italia (2014)

24. Yong, C., Ling, C., Guendel, F., Kulhànek, O.: Seismic hazard and loss estimation for Central America. Nat Hazards **25**(2), 161–175 (2002)

A Preliminary Estimate of the Rebuilding Costs for the Towns of the Abruzzo Region Affected by the April 2009 Earthquake: An Alternate Approach to Current Legislative Procedures

Sebastiano Carbonara$^{(\boxtimes)}$, Daniele Cerasa, Tonino Sclocco, and Enrico Spacone

Università G. d'Annunzio of Chieti-Pescara, Viale Pindaro, 42, 65127 Pescara, Italy
s.carbonara@unich.it

Abstract. This paper examines the preliminary cost estimate procedure followed in planning the reconstruction of the city of L'Aquila and of 56 other towns in the Abruzzo region damaged by the 2009 earthquake. As with past catastrophic events, the Italian Government has assumed full responsibility for funding repair/reconstruction of both private and public properties. A highly articulated legislative cost estimation system - developed on behalf of the national authorities in the wake of the earthquake that caused over three hundred victims - was implemented to coordinate the distribution of funding among the different municipalities and private subjects affected by the earthquake. The paper shows how the automatism of this procedure may have produced a distortion in cost estimates when compared to the costs actually needed for reconstruction. An alternate cost estimation model is proposed based on multiple linear regression analysis that uses bills of quantities from reconstruction projects funded immediately following the quake (and is based on actual structural designs rather than emergency damage assessment data). The objective of the proposed model is to achieve a more realistic reconstruction cost estimate framework, while respecting the need for a quick and rational procedure that requires no additional information beyond the post-earthquake expert survey reports available only weeks after the earthquake.

Keywords: Cost estimate · Post-earthquake reconstruction · Rebuilding costs · Multi-regression analysis

1   Introduction

The earthquake that on April 6, 2009 struck the Abruzzo region, and more specifically the city of L'Aquila and 56 smaller towns (Fig. 1), caused 308 casualties and serious damage to many buildings and infrastructures. The area highlighted in Figure 1 was later labelled the "crater". While not an extremely intense earthquake (moment magnitude of 6.3), the relatively low depth of the focus, the peculiar typology of the local historical constructions (for the most part made of load bearing, rough and irregular stone walls, in some cases constructed according to the "a sacco"

© Springer International Publishing Switzerland 2015
O. Gervasi et al. (Eds.): ICCSA 2015, Part III, LNCS 9157, pp. 269–283, 2015.
DOI: 10.1007/978-3-319-21470-2_19

method - rubble masonry construction made of double skin walls with regular stones on the outside skins, loosely filled in the middle with poorer material), coupled with pre-existing conditions of deterioration and lack of maintenance, served as multiplying factors of the event's destructive effects. More than 67,000 people were left without a home and, more than five years after the event, little over half of them have returned to their homes.

Fig. 1. Municipalities severely damaged by the April 2009 earthquake: the so called "crater" area

The earthquake attracted worldwide attention, because of the rarity of the event that hit such a large historical center, actually closing down one of the most attractive and important cultural heritage sites in central Italy. Following the event, the Italian Government announced its plan to implement a process for fixing all damaged buildings and infrastructures, using the so-called *PdRs* (*Piani di Ricostruzione* or Reconstruction Plans), regulated by Law n. 39 (April 28, 2009). These *ad hoc* urban planning instruments are designed to pursue not only the physical, but also the socio-economic recovery of damaged territories and to re-qualify inhabited areas, in order to guarantee the "rapid" return of the population to their repaired or rebuilt homes.

As a first step, the *PdRs* must identify the areas with major documented damage (inside the so-called town center *perimeter*), notably consisting of the severely damaged historical centers of L'Aquila and of the above-mentioned 56 smaller towns. In subsequent stages, the *PdRs* define reconstruction policies (both technical and administrative) and reconstruction stages. Most importantly, the *PdRs* provide preliminary cost estimates for rehabilitating private constructions and the so-called fixed social capital (roads, pipelines, parks, town halls, schools, public libraries, churches, etc). The cost estimates are supposed to support efficient programming of the reconstruction process. Given the size of the reconstruction costs (estimated for the entire region at around 15 billion Euros), and given that the reconstruction costs in Italy are by "tradition" the responsibility of the Italian Central Government [1], the Italian Government emanated a series of guidelines and decrees for the definition of the cost estimation process.

2   The Reconstruction Plans and Their Preliminary Cost Estimates

Following the identification of the perimeter that defines the damaged historical center of a given municipality, the *PdR* should subdivide the historical centres into *ambiti* (areas), intended in general terms as portions characterised by a specific configuration of the urban structure and of the buildings. Inside the *ambiti*, the focus is on integrating interventions dealing with one or more *aggregate building units*.

The *aggregate building unit* may represent a single building, or multiple buildings that can be recognised as structurally connected. In one of the official definitions by the *DPC* (*Italian Civil Protection Department*), the aggregate building unit is intended as a "... *non-homogenous group of buildings, interconnected by a more or less efficient structural connection determined by the history of their evolution, which may interact under seismic or dynamic actions* ..." [2].

The *PdRs* must include a preliminary estimate of the building rehabilitation and/or substitution costs. The criteria for developing these estimates are contained in a number of legislative instruments specifically prepared for the post-earthquake Abruzzo reconstruction. Among several normative references, the *OPCMs* (*Ordinanze del Presidente del Consiglio dei Ministri* – Prime Minister's Ordinances) are particularly relevant. Cost estimate criteria are defined based on the nature and level of damage, certified for each building by the so-called *AEDES* forms (or *Damage Survey Forms*).

The damage assessment check was based on the post-earthquake reconnaissance efforts coordinated by the *DPC*. In the weeks following the earthquake, all buildings in the "crater" area (and all buildings where owners or local authorities reported damage) were inspected by teams of experts (typically, licenced engineers and architects) that filled standard forms containing information on the structural health of the building gathered through visual inspection (often done without entering the buildings – when structural damage deemed accessing the building unsafe).

The above assessments resulted in filling a *Scheda di Rilevamento, Agibilità e Danno nell'Emergenza Sismica* or *AeDES* form (Accessibility and Damage Survey Form for Seismic Emergency) for each structural unit (or *aggregate building units*, as previously defined). Following the above checks, all buildings and structures were classified in one of the following categories:

A – Undamaged structure, usable in all parts without any danger for the occupants' lives;

B – Lightly damaged structure, temporarily inaccessible, though accessible with some emergency measures;

C – Partially inaccessible;

E – Severely damaged or collapsed structure, inaccessible.

   Two additional categories were also considered:

D – Temporarily inaccessible, to be further investigated;

F – Inaccessible due to serious external danger.

Additional factors affecting the reconstruction costs are related to the property use within the aggregate (*primary dwelling*, i.e. the owner's habitual place of residence, and *non-primary dwelling*, or building *for other uses*).

The guiding logic of the preliminary cost estimate procedure refers to maximum expenditures and maximum parametric values, some of which are prescribed by law, while others were specified by the so-called *STM* (*Struttura Tecnica di Missione* or Dedicated Technical Structure), an emergency technical and administrative office created *ad hoc* by the Italian Central Government to oversee and coordinate the kick-off of the entire reconstruction process. According to the *STM* guidelines, retrofitting and reconstruction unit costs are computed based on the outcomes of the *AeDES* forms (see above categories).

In extreme synthesis, the resulting framework is summarized hereafter.

Primary Dwellings

For *category A* buildings that have not suffered structural damage and require minor repair work, the maximum total contribution was fixed at €10,000.00; under certain conditions this amount can be increased to €12,500.00.

For *category E* buildings, with serious structural damage, the maximum contribution was fixed at € 1,276.00/sq. m. This figure was established based on the costs forecast by the Abruzzo Regional Administration for new public residential construction (when the project cost exceeds the max. cost per square meter, owners can choose between reconstruction or substitution with a newly constructed building). The above unit cost may be increased by up to 60% in the case of particularly valuable cultural heritage structures and by 100% in the case of *registered* buildings (in Italy, the National Register of Historic Places is overseen by peripheral offices of the Cultural Heritage Ministry).

As for *category B* and *C* buildings. for which local structural reinforcements are typically foreseen, cost estimate quantifications were determined by the technicians preparing preliminary designs, in collaboration with the *STM*, based on standard costs.

Non-Primary Dwellings and Buildings Used for Other Purposes (commercial spaces, warehouses, stables, etc)

For *Category A* buildings, there is no funding.

For the other categories (*B, C* and *E*) a maximum of €80,000.00 was established, with a funding cap at 80% of documented costs. Additional contributions may be allowed in special cases (when, for example, category B and E properties are present in the same building aggregate).

Public and Religious Buildings, Public and Private Utilities
The framework of reconstruction costs finally comprises public buildings, religious buildings and underground utility networks (water, sewers, electricity and telephone lines). A unit cost was established within a range of values (for each damage category) by the technicians preparing the *PdRs*, in collaboration with the *STM*.

The above principles are used to prepare, for each Reconstruction Plan, *Quadri Tecnici Economici* (Technical Economic Reports) or *QTE*. For each historical center, the *QTE* contains the costs forseen for repairing all buildings, structures, infrastructures and utility lines damaged by the earthquake. Once approved, the total cost of reconstruction for the given historical center – together with a scheduling plan - is sent to the Italian Central Authorities, which are supposed to provide the necessary funding over the following years.

The structure of this cost estimating process, described here in its main principles, may in some cases have generated estimates that differ sensibly from real market values. As previously recalled, the preliminary cost estimate procedure was configured as a calculation of the "maximum cost admissible by law". According to Carbonara [3] "*... it is important to point out the peculiar nature of this cost evaluation procedure, relative to buildings with damage evaluations defined by the AeDES forms, thus with unknown quantities and types of interventions. In other words, no design projects and/or documents of any sort were available to ensure that the cost estimation procedure is based on analytical data ...*".

Given the delay in the reconstruction process, it is still impossible to compare cost estimates with funds actually dispensed to owners: " *... Meanwhile, reconstruction has proceeded slowly.... within L'Aquila's historic city centre, reconstruction is governed by different ordinances, and, as of April 2011, the city was still cordoned off, debris had not been removed, and reconstruction had not begun, although a small number of building owners had completed self-financed repairs...*" [4]. The question remains, and merits investigation and verification.

3   Comments on the Preliminary Cost Estimate Procedure

These brief notes offer a clear understanding of the peculiarity of the cost estimates set up by the Italian Government in the wake of the Abruzzo earthquake that can be summarised as follows. The cost estimate scheme is based on the *AeDES* forms. The governing cost estimate parameters are observed damage and government-set unit repair costs. Furthermore, primary dwellings are given priority and higher reconstruction reimbursements. This is quite a peculiar framework, a sort of *unicum* that can be explained starting from two main elements.

First of all, in more mature economies, insurance companies are generally responsible for estimating the damage costs for private and public buildings. However, we often know little "about the location and character of insured value" [5]. In Italy, the situation is different. Homeowners are not currently required to insure their properties against damage by natural disasters (not even in areas where natural hazards are high). To date, the Italian State has always assumed full responsibility for the costs of reconstruction following a natural disaster, such as a major earthquake.

Secondly, there is another aspect that is no less conditioning. In the case of the Abruzzo earthquake, Central Italian Authorities estimated a maximum cost per building based on the maximum cost per square meter of public residential construction. An extremely laborious process was set in motion to determine costs based on a multiplicity of characteristics that were not only technical, but also legal and social. Hence, the result was not a *tout court* estimate of the effective costs of recovering damaged buildings, but an assessment conditioned by a number of different restrictions.

Recent literature on the economic issues related to reconstruction following major natural disasters tends to concentrate on macroeconomic aspects, focusing little attention on the issue of the reconstruction direct cost estimations and related procedures, more specifically those addressing damage and reconstruction of private properties.

For example, in the case of the Kobe (1995) and Tohoku (2011) earthquakes, both in Japan, funding for the reconstruction costs were based on the creation of a specific Reconstruction Fund. While estimated values were announced, no reference was provided to the criteria used to develop budget figures [6]. In the case of Hurricane Katrina in Louisiana, USA, Hallegatte [7] uses an input-output model to verify the indirect effects of the catastrophe on the regional economy, revealing how 107 million USD of direct losses determined a drop in pre-Katrina added values in the order of 23 billion USD. In this study, there is no indication of how the rebuilding costs were computed. Analyses of the natural devastations that have struck Sri Lanka over the past fifteen years have focused on their effects on the real estate market, highlighting a positive and very strong relationship with the increase in all housing cost indexes [8]. Greenberg et al. [9] focus on the American experience, and concentrate their attention on the need to systematically test the economic impact of catastrophic natural events at the local, national and in some cases international level, at least a decade after an event, using evaluation techniques such as the Input-Output and the Econometric time-series models.

Naturally, this does not mean that research in this field has ignored the study of methods useful to estimating the damages provoked by earthquakes [10] [11] [12] [13] [14] [15]. However, it is equally true there exists a tendency to focus on discovering those causes with the greatest impact, as opposed to operative practices and procedures for assessing damages to individual buildings: *traditionally, the assessment of damage for loss estimation studies has been based on macroseismic intensity or peak ground acceleration* [16].

In light of the above, while the nature and scope of the estimate present specific characteristics, the same cannot be said of the statistical tool selected for the development of the estimation model. In other words, the analysis of multiple linear regression, which has proven over the years to be an effective technique in numerous cases for the study, management and elaboration of prices in a wide range of markets: *a conceptually sound and most powerful analytical device that combines probability theory with calculus, thereby allowing sorting out crossed influences that affect property values* [17]. While countless studies have been conducted in the real estate market, the construction market has also revealed the need for more accurate analytical models. Kim et al. [18] have verified three diverse models for estimating construction costs, including the analysis of multiple regression, highlighting how the application of this model in studies pursuing this specific objective became more common after 1970 [19] [20] [21] [22] [23] [24]. The resulting framework of proposals may provide suggestions beneficial to estimations of building reconstruction costs in the wake of natural calamities and ensuring more realistic estimates of the costs of restoration and seismic improvement works.

4 Proposed Alternative Cost Estimation Procedure

The proposed procedure is based on the only information available for all buildings, i.e. the *AeDES* forms, which provide useful information regarding the levels and percentage of damage and the construction type for each building. Figure 2 shows two of the most important sections of the *AeDES* form: the first one (Section 3) deals with the structure type, the second one (Section 4) with the observed damage.

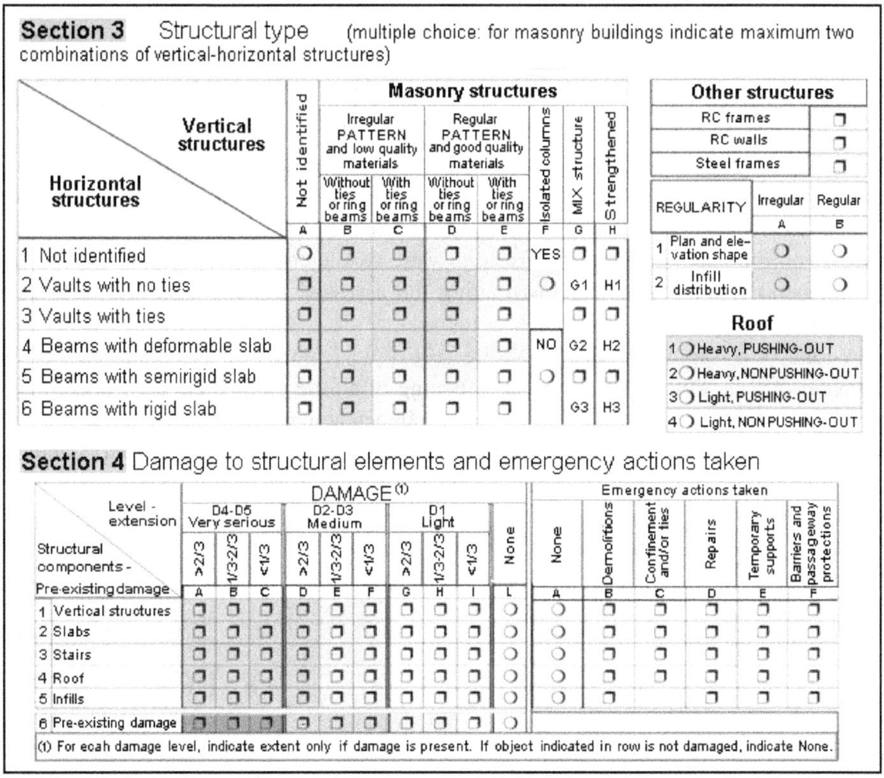

Section 3 Structural type (multiple choice: for masonry buildings indicate maximum two combinations of vertical-horizontal structures)

Horizontal structures \ Vertical structures	Not identified (A)	Masonry structures: Irregular PATTERN and low quality materials – Without ties or ring beams (B)	With ties or ring beams (C)	Regular PATTERN and good quality materials – Without ties or ring beams (D)	With ties or ring beams (E)	Isolated columns (F)	MIX structure (G)	Strengthened (H)
1 Not identified						YES		
2 Vaults with no ties							G1	H1
3 Vaults with ties								
4 Beams with deformable slab						NO	G2	H2
5 Beams with semirigid slab								
6 Beams with rigid slab							G3	H3

Other structures: RC frames; RC walls; Steel frames

REGULARITY	Irregular A	Regular B
1 Plan and elevation shape		
2 Infill distribution		

Roof
1 Heavy, PUSHING-OUT
2 Heavy, NONPUSHING-OUT
3 Light, PUSHING-OUT
4 Light, NON PUSHING-OUT

Section 4 Damage to structural elements and emergency actions taken

Structural components - Pre-existing damage \ Level - extension	DAMAGE (1) D4-D5 Very serious >2/3 (A)	1/3-2/3 (B)	<1/3 (C)	D2-D3 Medium >2/3 (D)	1/3-2/3 (E)	<1/3 (F)	D1 Light >2/3 (G)	1/3-2/3 (H)	<1/3 (I)	None (L)	Emergency actions taken: None (A)	Demolitions (B)	Confinement and/or ties (C)	Repairs (D)	Temporary supports (E)	Barriers and passageway protections (F)
1 Vertical structures																
2 Slabs																
3 Stairs																
4 Roof																
5 Infills																
6 Pre-existing damage																

(1) For ecah damage level, indicate extent only if damage is present. If object indicated in row is not damaged, indicate None.

Fig. 2. Sections 3 and 4 of the *AeDES* form. They contain the main information used to develop the proposed multiple linear regression model.

The study considers projects approved in the towns of Bussi sul Tirino, Brittoli, Civitella, Casanova, Montebello di Bertona (in the Pescara Province), Castelvecchio Subequo, Caporicano, Goriano Sicoli and Ofena (in the L'Aquila Province). These municipalities were among the first to approve their *PdRs*, whose completion involved personnel from the University of Chieti Pescara. At this date, there are still several reconstruction plans that need to be completed, thus full information over the entire area damaged by the 2009 earthquake is still missing.

In this study, category A buildings - characterised by light, non-structural damage - are not considered, given the small or zero reconstruction costs foreseen. The focus is on category B, C and E buildings, in need of extensive rehabilitation/reconstruction that require major funding.

The regression analysis starts with a set of cost estimates that required a lengthy phase of search and selection. All estimated cost data used in this study was obtained from reconstruction projects approved in the two years following the April 2009 earthquake, prior to the approval of the *PdRs*, that started in 2011. They were computed from unitary market values, and thus differ from the cost estimate calculations later imposed by the *STM* and implemented in the *PdRs*. The number of reconstruction projects actually approved and funded in this two-year span is limited but still significant from a statistical standpoint.

The objective of the initial data recovery phase was to identify the variables with the greatest influence on the overall cost of a repair or reconstruction project for a building damaged by the earthquake. This search analysed a series of documents related to the securement of national funding and available from different sources, mainly: AeDES Forms, Bill of Quantities for repairs and seismic improvement, project costs, and the chart developed by Cineas (Consorzio Universitario per l'ingegneria nelle Assicurazioni – University Consortium for Engineering in Insurances) and Fintecna. *Cineas* and *Fintecna* were two agencies in charge of setting up the financial arrangements for distributing the assessed funding once the structural projects were approved.

A total of 87 approved projects were recovered from the above-mentioned eight townhall offices. Following an initial screening of all the parameters used to estimate the projects' costs, the initial set of fifteen independent variables considered was reduced to eight. They are defined in Table 1, while Table 2 shows mean and standard deviation values computed from the 87 cases at hand, statistically insignificant variables were eliminated.

The proposed cost estimation procedure is based on the following equation:

$$\frac{cost}{m^2} = \beta_0 - \beta_1 \, S_{Cat} + \beta_2 \, L_{Vert} + \beta_3 \, P_{Vert} + \beta_4 \, L_{Stairs} +$$

$$+\beta_5 \, P_{Inf} - \beta_6 \, T_{Const} + \beta_7 \, T_{Hor} + \beta_8 \, DC + \varepsilon$$

where $cost/m^2$ is the dependent variable, β_i ($i= 0,1,...,8$) are the parameters to be computed, S_{Cat} ..., DC are the eight independent variables specified in Table 1 and ε is the stochastic component (or error). All base assumptions of the regressive multiple linear model are satisfied.

Table 1. Definitions of Independent Variables for Multiple Regression

1	Cadastral area - S_{Cat}	mq
2	**Damage level to vertical structures – L_{Vert}**	**assigned value**
	very serious	3
	medium	2
	light	1
	none	0
3	**Percentage of damage to vertical structures – P_{Vert}**	**assigned value**
	>2/3	3
	1/3-2/3	2
	<1/3	1
	none	0
4	**Damage level to stairs – L_{Stairs}**	**assigned value**
	very serious	3
	medium	2
	light	1
	none	0
5	**Percentage of damages to infills and partitions – P_{Inf}**	**assigned value**
	>2/3	3
	1/3-2/3	2
	<1/3	1
	none	0
6	**Construction Typology – T_{Const}**	**assigned value**
	Reinforced Concrete	1
	Regular, good quality masonry	2
	Irregular, poor quality	3
	Mixed structure	4
7	**Floor Construction Typology – T_{Hor}**	**assigned Value**
	Rigid floor slab	1
	Semi-rigid floor slab	2
	Deformable floor slab	3
	Vaults	4
8	**Damage Condition – DC**	**assigned Value**
	Condition B	1
	Condition C	2
	Condition E	3

Table 2. Variable Statistics

	Mean	Std. Deviation	N
$Cost\ €/m^2$	666.8253	462.86614	87
S_{Cat}	195.8287	180.76754	87
L_{Vert}	1.9080	0.69269	87
P_{Vert}	1.5632	0.74242	87
L_{Stairs}	0.5862	0.99477	87
P_{Inf}	0.9770	0.83495	87
T_{Const}	2.8506	0.69095	87
T_{Hor}	2.7126	1.06649	87
DC	1.7241	0.87191	87

Table 3 summarizes the ANOVA results. It contains the sum of the regression squares, the sum of the residual squares and the sum of the total squares. With a p-value close to zero, statistic F becomes highly significant.

Table 4 shows the values of the regression coefficient R^2 and the Durbin-Watson statistic. Figure 3 clearly shows the linearity of the regression function. Table 4 and table 5 contain the model summary and the model statistics, respectively.

Table 3. ANOVA results

Model	Sum of Squares	df	Mean Square	F	Sig.
Regression	1.66E7	8	2.07E6	87.751	0.000[a]
Residual	1.84E6	78	2.36E4		
Total	1.84E7	86			

Table 4. Model summary

Model	R	R Square	Adjusted R Square	Estimate Std. Error
1	0.949[a]	0.900	0. 890	153.69319

Table 5. Model statics

Model	Change Statistics					Durbin - Watson
	R Square Change	F Change	df1	df2	Sig. F Change	
1	0.900	87.751	8	78	0.000	1.820

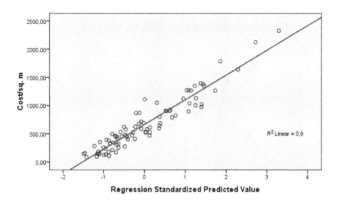

Dependent Variable: Cost/sq. m

Fig. 3. Regression Line for cost/m2 vs Standardized Predicted Value

The multiple R coefficients R^2 and adjusted R^2 assume relatively high values, indicating that the model possesses a good predictive capacity. The correlation with observed data can be deemed satisfactory. The standard error represented by an estimate of the average distance between the observed and the expected estimate is equal to 153.693 and related to the average cost/m2 value of € 667.005 by an average estimation error of 22%.

Table 6 shows the results of the least square approach, and more specifically the modelling parameters' estimates, the standard errors and the t statistics with p-values. The Sig. column indicates that the parameters are very close to zero. Table 6 also indicates the collinearity statistics, Tolerance and VIF.

Table 6. Model parameters and collinearity statistics

	Unstandardized Coefficients		Standardized Coeff.	t	Sig.	95.0% Confidence Interval for B		Correlations			Collinearity Statistics	
	B	Std. Error	Beta			Lower Bound	Upper Bound	Zero-order	Partia	Part	Tole.	VIF
β_0	-328.83	81.28		-4.04	0.000	-490.65	-167.01					
β_1	-0.17	0.09	-0.06	-1.72	0.088	-0.36	0.02	0.17	-0.19	-.06	0.85	1.16
β_2	183.78	31.66	0.27	5.80	0.000	120.74	246.82	0.56	0.54	.20	0.57	1.75
β_3	188.39	24.61	0.30	7.65	0.000	139.39	237.39	0.43	0.65	.27	0.82	1.21
β_4	82.85	21.01	0.17	3.94	0.000	41.01	124.68	0.57	0.40	.14	0.62	1.59
β_5	224.49	22.57	0.40	9.94	0.000	179.55	269.42	0.39	0.74	.35	0.77	1.29
β_6	-164.71	26.96	-0.24	-6.10	0.000	-218.40	-111.02	0.05	-0.56	-.21	0.79	1.26
β_7	59.27	17.40	0.13	3.40	0.001	24.63	93.92	0.35	0.36	.12	0.79	1.25
β_8	246.40	23.69	0.46	10.39	0.000	199.22	293.58	0.70	0.76	.37	0.64	1.55

Figures 4 and **5** illustrate the distributive form of the residuals and the non-relation between the *cost/sq. m* and the variation of the residuals.

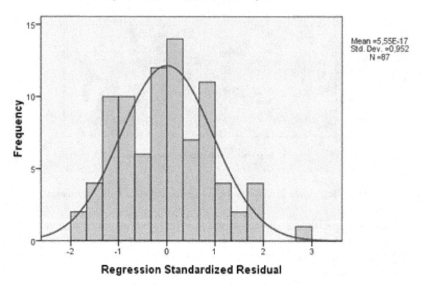

Fig. 4. Frequency Histogram

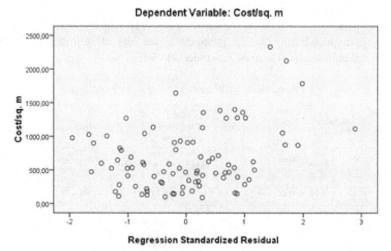

Fig. 5. Dispersion of the Residuals

The estimated β coefficients represent the increase/decrease in the dependent variable *cost/m²* when each individual independent variable increases by a unit value, while the values of the other variables remain unchanged. They are a representation of the implicit marginal prices.

The independent variables P_{Vert}, L_{Vert}, L_{Stairs}, P_{inf}, T_{Hor} and DC express coefficients entirely in line with the initial, positive assumptions. The negative sign of the Surface variable (S_{Cat}) coefficient was fully expected. The independent variable of the building typology (T_{Const}) deserves a brief discussion. The model produced a negative value of -164.71 that, according to the assigned scale, indicates a possible contradiction: an intervention of recovery and structural reinforcement involving reinforced concrete buildings is more costly than interventions involving masonry structures. It should be noted that the analysis of budgets accompanying the processed repair projects revealed a scarce quality of the concrete, more specifically for buildings erected between 1950 and 1975. The extremely high cost of repairing these structures has a significant impact on the global costs. The materials employed for repair - such as carbon fibre straps used to strengthen in bending or shear deteriorated and collapsed beams or to wrap joints and columns to increase the building overall seismic performance - belong to advanced, still costly technologies that also require skilled labour. Also, interventions on existing reinforced concrete buildings require extensive, and thus expensive, removal of non structural elements, such as infills and floor tiles. In short, the above negative parameter indicates the high cost of repairing reinforced concrete buildings.

5   Concluding Remarks

The reconstruction cost estimation procedure set up by the central Italian government in the aftermath of the L'Aquila 2009 earthquake follows an articulated framework of ad-hoc legislation emanated after the earthquake. These ad hoc rules establish unit costs and maximum allowed expenditures that depend on the reported level of damage of the building, and on the building functions (commercial, financial, public, private, apartments, etc). Often, buildings host multiple functions.

Given the conditions that characterised the entire cost estimation process, and more specifically the absence of any design project in support of the cost estimation procedure, cost estimates were viewed and assessed as maximum allowable expenditures. It follows that the actual funds required for repairing or reconstructing damaged structures may vary significantly with respect to the costs estimated in the towns' (Reconstruciton Plans).

This paper proposes a more coherent and effective procedure that could have been used for a better estimate of the reconstruction costs.

The proposed procedure is based on a multiple linear regression model that defines the cost of each reconstruction/repair project based on the costs sustained for interventions funded in the very first months following the earthquake and based on actual design projects. Even though the procedure is based on a limited set of projects (from eight of the "seismic crater" towns, for a total of fifty-seven), the approach is deemed applicable because the buildings analysed are representative of the construction typologies found in the historical centers of the Abruzzo mountain Region.

The information necessary for the regression analysis was taken from the AeDES survey tables produced during site visits to damaged buildings and was later transformed into independent variables (building typology, damage percentage level of the horizontal and vertical structural system, damage to infills and stairs, floor slab typology, etc.).

Comparisons between official cost estimate values and those obtained from the proposed model show a consistent over-estimation of the former with respect to the latter. In some cases this difference was in the order of 30%. Questions related to the proposed model's effective capacity to predict actual costs remains unanswered, though limited (and statistically still non relevant) checks on the costs of projects actually funded as part of the PdRs show very good agreement between predictive model and actual costs.

As part of ongoing studies, the model will be rerun with a more consistent number of observations, expanded to towns closer to the epicenters of the seismic shocks.

Considering that work on the Reconstrution Plans started in early 2011 and as of today not all PdRs are yet approved, there would have been ample time to set up the estimation process proposed in this paper. Following the site visits to define the structural conditions of all buildings following the 2009 earthquake, the Technical Bodies working on behalf of the Central Governemnt could have set up a cost estimation procedure based on detailed reconstruction/repair projects of a set of sample buildings defined on the basis of their representativeness in terms of accessibility, position, building, structural and functional typology and dimensional category. Such an approach would have in turn permitted a general cost estimate analysis of all damaged structures, without any additional information.

The proposed cost-estimation procedure is based on the actual characteristics of each property, its construction typology and the damage level. Furthermore, the proposed approach overcomes the rigidity of *ad hoc* rules that impose cost estimate maximum expenditures based on the building accessibility conditions only.

Finally, the proposed model would determine a more meaningful integration between two relative components of the Reconstruction Plans, mainly the interventions' programming and the relative economic framework, at this point both derived from the same analytical base, reoresented by the AeDES survey forms.

References

1. Carbonara S.: Il recupero dell'edilizia privata nell'Abruzzo post-sisma: un'analisi delle procedure di stima. Restoring private housing in post-Earthquake Abruzzo: an analysis of the estimation procedures. Territorio, (70), pp. 119–125. Franco Angeli (2014)
2. DPC - Dipartimento della Protezione Civile, ReLUIS, Linee guida per il rilievo, l'analisi ed il progetto di interventi di riparazione e rafforzamento/miglioramento di edifici in aggregato, (www.reluis.it.), in Italian (2010)
3. Carbonara S.: Il sisma abruzzese del 2009: la previsione di spesa per la ricostruzione. Valori e Valutazione, (11), pp. 67–85. DEI, Roma (2013)
4. Liel, A.B., Ross, B., Corotis, R.B., Camata, G., Sutton, J., Holtzman, R., Spacone, E.: Perceptions of Decision-Making Roles and Priorities That Affect Rebuilding after Disaster: The Example of L'Aquila, Italy. Earthquake Spectra **29**(3), 843–868 (2013)

5. Munkhammar, A.: Earthquake damage scenarios for international insurance companies. In: Tucker, B.E. (ed.) Uses of Earthquake Damage Scenarios, Proceedings of Special Theme Session Number 10 of the Tenth World Conference on Earthquake Engineering entitled Earthquake Damage Scenarios for Cities of the 21st Century Madrid, Spain, July 23, 1992
6. Hayashi, T.: Japan's Post-Disaster Economic Reconstruction: From Kobe to Tohoku. Asian Economic Journal **26**(3), 189–210 (2012)
7. Hallegatte, S.: An Adaptive Regional Input-Output Model and its Application to the Assessment of the Economic Cost of Katrina. Risk Analysis **28**(3), 779–799 (2008)
8. de Silva, L.: Forecasting of Cost Escalations in Post Disaster Construction with Special Reference to Tsunami Reconstruction in Sri Lanka. Built - Environment - Sri Lanka **9–10**(1–2), 56–63 (2011)
9. Greenberg, M.R., Lahr, M., Mantell, N.: Understanding the Economic Costs and Benefits of Catastrophes and Their Aftermath: A Review and Suggestions for the U.S. Federal Government. Risk Analysis **27**(1), 83–96 (2007)
10. Musson, R.M.W.: Intensity-based seismic risk assessment. Soil Dynamics and Earthquake Engineering **20**, 353–360 (2000)
11. Applied Technology Council, Earthquake damage evaluation data for California. Report ATC-13, Applied Technology Council, Redwood City, California (1985)
12. Powell, G.H., Allahabadi, R.: Seismic damage prediction by deterministic methods: Concepts and procedures. Earthquake Engineering & Structural Dynamics **16**(5), 719–734 (1988)
13. Calvi, G.M.: A displacement-based approach for vulnerability evaluation of classes of buildings. Journal of Earthquake Engineering **3**(3), 411–438 (1999)
14. Kircher, C.A., Nassar, A.A., Kusty, O., Holmes, W.T.: Development of building damage functions for earthquake loss estimation. Earthquake Spectra **13**(4), 663–682 (1997)
15. Ordaz, M., Miranda, E., Reinoso, E., Pérez-Rocha, L.E.: Seismic loss estimation model for Mexico city. In: Proceedings of the 12th World Conference on Earthquake Engineering, Auckland, New Zealand (2000). Paper no. 1902
16. Crowley, H., Pinho, R., Bommera, J.J.: Probabilistic Displacement-based Vulnerability Assessment Procedure, for Earthquake Loss Estimation. Bulletin of Earthquake Engineering, 2, pp. 173–219. Kluwer Academic Publishers, Netherlands (2004)
17. Des Rosiers, F., Thériault, M.: Discussion Paper presented at the Advances in Mass Appraisal Methods Seminar. Delft University of Technology, pp. 1–40, October 30–31, 2006
18. Kim, G.H., An, S.H., Kang, K.I.: Comparison of construction cost estimating models based on regression analysis, neural networks, and case-based reasoning. Building and Environment **39**(10), 1235–1242 (2004)
19. Kouskoulas, V., Koehn, E.: Predesign cost estimation function for building. Journal of the Construction Division, 589–604, December 1974
20. McCaffer, R.: Some analysis of the use of regression as an estimating tool. Quantity Surveyor **32**, 81–86 (1975)
21. Bowen, P.A., Edwards, P.J.: Cost modeling and price forecasting; practice and theory in perspective. Construction Management and Economics **3**, 199–215 (1985)
22. Khosrowshahi, F., Kaka, A.P.: Estimation of project total cost and duration for housing projects in the UK. Building and Environment **31**(4), 375–383 (1996)
23. Skitmore, R.M., Thomas Ng, S.: Forecast models for actual construction time and cost. Building and Environment **38**(8), 1075–1083 (2003)
24. Trost, S.M., Oberlender, G.D.: Predicting accuracy of early cost estimates using factor analysis and multivariate regression. Journal of Construction Engineering and Management **129**(2), 198–204 (2003)

From Surface to Core: A Multi-layer Approach for the Real Estate Market Analysis of a Central Area in Catania

Laura Gabrielli[1(✉)], Salvatore Giuffrida[2], and Maria Rosa Trovato[2]

[1] Department of Architecture, University of Ferrara, Ferrara, Italy
gbrlra@unife.it
[2] Department of Civil Engineering and Architecture,
University of Catania, Catania, Italy
{sgiuffrida,mrtrovato}@dica.unict.it

Abstract. The proposed study deals with the analysis of the real estate market in the quarter of San Cristoforo in Catania, trying to integrate different approaches to define its possible articulation in submarkets. The first one is a phenomenal type of approach that intends to represent some of the most manifest characteristics, and provides an initial hypothesis of classification of the cases (a census has been taken of) and delimitation of the segments, taking into account the ranges of prices registered inside the different classes of the characteristics. The second consists of an in-depth clustering analysis basing on three different hypotheses of three, four and five clusters respectively. The third one is a DRSA application, which is meant to extract from the studied sample a set of rules for the possible definition of a segment representing the general market rules. Given the complexity of the studied context, the results allow different interpretations and considerations of method.

Keywords: Real estate market · Complex urban context · Cluster analysis · Dominance-based rough set approach

1 Introduction

The combination of the typical characteristics of the real estate markets and their fluctuations during periods of economic and financial crisis, makes the valuations and the consequent transactions particularly difficult, affecting the perspectives of the families and the volume of private and public investments in this sector. The current paralysis of the transactions, affected by the credit crisis as well, has encouraged forms of hording affecting the implementation of recovery policies at local scale and the perspectives of fiscal equalization at national scale. The real estate market plays a role of primary importance in both the above-mentioned contexts, due to two reasons. The first is that the urban landscape is the most densely lived space, and the house is the place where the primary personal and social individual subjectivity is realized.

© Springer International Publishing Switzerland 2015
O. Gervasi et al. (Eds.): ICCSA 2015, Part III, LNCS 9157, pp. 284–300, 2015.
DOI: 10.1007/978-3-319-21470-2_20

Then the current real estate volume plays a significant part in the national economies, not only in direct and concrete terms, that means regarding the productive investments and the tax return, but in indirect terms as well, that means regarding the speculative investments and their capability to generate reserves of capital value.

The transparency of the real estate market and the possibility to divide it into specific segments becomes an indispensable requisite for every form of equalization, especially in areas like the one used as case study, severely lacking the actualization of planning policies, and demanding the intensification of local tax system actions, etc.

The sphere of public interest inside which this study, and more generally the generalist approach of the science of valuations can be ascribed, address most part of the efforts of the researchers in valuation towards the themes of the mass appraisals, requiring sophisticated approaches, and tools to scan extensive databases and analyse the data.

The consequence of that is a certain antinomy between the *normative intention* of the science of valuations, *inspired* from the top by *value*s, and the tensions characterizing the capital markets, *oriented* from the bottom, that means by the combination between the creative component of the specific investor and the panorama, often discontinuous, of the assortment of *prices*. In the real estate market, diverse forms and perspectives of investment that are not fully expressed in urban realms confront each other:

— the ones only partially structured regarding the elasticity of price vs concrete values (characteristics);
— the ones scarcely coherent on the point of view of the relation between centrality and intensity of the real estate capital.

The consolidation of these relations could attenuate the riskiness of the investments and the role played by the asymmetry between the "personal equations" of the investors (risk attitude, financial situation, and willingness to wait), more often led to act in countertendency in the attempt to anticipate the changes in the speculative trends, improving the chances of success of the town planning programming.

Therefore, a global analysis of the urban real estate market is not possible unless the segments expressing significantly the characteristics of the properties in their specific contexts are indicated and individuated. The proposed study deals with the analysis of the real estate market in the quarter of San Cristoforo in Catania, trying to integrate different approaches to define a possible articulation in submarkets.

The first approach is the phenomenal, based on a surface observation aimed at representing some of the first approximation characteristics; it provides an initial hypothesis of classification of the cases and delimitation of the segments, taking into account the ranges of prices registered in the different classes of the characteristics.

The second approach consists of an in-depth analysis basing on three different clustering hypotheses, from three to four or five clusters.

The third approach is constituted by an application of the DRSA, under diverse hypotheses as well, especially regarding the way the prices have been considered.

The analysis of this market is part of a more extensive research study concerning the hypotheses of redevelopment generated in the context of the negotiation planning, and is therefore aimed at describing its articulation in sub-markets and the relation between prices and characteristics connecting the different goods to the overall urban sub-system.

2   San Cristoforo Neighbourhood in Catania and the Real Estate Market Survey

The quarter of San Cristoforo is part of the "Centro" Municipality (the first of the ten municipalities the territory of the "Comune" of Catania consists of), comprising the quarters of San Berillo, Civita, Antico Corso and Fortino as well. It constitutes an urban sub-system characterised by a significant functional, typological and social articulation that permeates its real estate assets. The quarter is delimited by Plebiscito arch on the North, SS. Maria Assunta Street - Concordia Street axis on the South, Acquicella Street on the West and the harbour area on the East. Its northern and southern boundary areas are the most interesting concerning urban quality and vitality as well. In particular, Plebiscito Street still preserves most part of its original urban character, as the quarter was built after the 1693 earthquake, on an area outside the ancient city walls, specifically assigned for the reconstruction and a new expansion of the urban centre [4]. The quarter has a surface of ca. 0,87 sq. km, and a very high building density.

The analysed real estate sample is formed by 58 properties comprised in the residential segment. The analysis has been carried out basing on 28 characters aggregated in 6 groups: the extrinsic characteristics regarding the context k_{e1}, the ones pertaining the micro-environmental k_{e2}, the intrinsic characteristics k_i, the architectural characteristics concerning the building k_{a1}, and the architectural characteristics concerning the property unit k_{a2} [5]. The attributes are expressed in a standard scale from 1 to 5 (from the lowest to the highest quality condition).

Table 1 shows the sample and the values of the aggregated characteristics. The complexity of the sample can be considered well-represented (Fig. 1) by the different distributions of prices and characteristics corresponding to the different number of classes: the greater the number of the classes, the highest the number of waves of the frequency function, for both prices and characteristics.

Table 1. Synthesis of the market survey

id	address	floor	rooms	surface	ask price	ke1	ke2	ki	kt	ka1	ka2	k*
1	plaia	169	0-1	6,6	250	€ 240.000	3,6	3,0	3,3	2,5	3,2	3,4	2,8
2	plaia	167	0-1	11,2	300	€ 250.000	3,3	3,0	2,8	2,0	2,4	2,7	2,3
3	plaia	134	0	2,6	50	€ 28.000	3,3	2,6	1,6	2,8	2,8	2,3	2,3
4	plaia	63	1	2,6	70	€ 80.000	2,8	2,6	2,7	4,0	3,8	3,7	3,2
5	ortolani	5	0-1	6,3	130	€ 240.000	2,8	3,0	3,1	4,0	4,0	3,3	3,4
6	del principe	177	8	3,8	80	€ 110.000	2,5	2,4	1,7	2,0	1,9	2,0	1,9
7	del principe	177	6	3,3	90	€ 90.000	2,5	2,4	1,7	2,0	1,9	2,0	1,9
8	del principe	177	5	2,8	65	€ 90.000	2,5	2,4	1,7	2,0	1,9	2,0	1,9
9	del principe	117	3	3,8	80	€ 105.000	3,0	3,0	3,3	4,0	3,8	3,3	3,4
10	del faro	5	0	2,5	50	€ 80.000	2,5	2,4	1,4	2,0	2,3	2,3	1,9
11	ss. assunta	70	1	3,5	70	€ 120.000	2,7	3,0	2,4	2,8	2,6	2,0	2,5
12	villa sgabrosa	65	1	2,7	70	€ 85.000	2,6	2,6	2,3	2,8	2,6	2,0	2,5
13	del principe	37	3	2,8	65	€ 90.000	2,5	2,4	1,7	2,3	2,5	1,7	2,1
14	domenico tempio	24	1	5,1	140	€ 145.000	3,7	3,0	3,9	3,0	3,5	3,0	3,1
15	grimaldi	158	0	2,9	45	€ 48.000	2,6	2,6	1,2	1,8	1,8	1,7	1,7
16	plebiscito	597	4	6,0	130	€ 190.000	3,5	3,4	3,3	3,5	3,0	3,3	3,2
17	plebiscito	597	6	7,3	105	€ 199.000	3,5	3,4	3,6	3,3	3,0	3,0	3,2
18	plebiscito	529	2	4,6	85	€ 85.000	3,5	3,4	2,7	2,8	2,5	3,1	2,7
19	plebiscito	519	5	6,1	100	€ 185.000	3,3	3,4	3,6	3,5	3,4	2,9	3,3
20	plebiscito	415	5	4,1	100	€ 185.000	3,3	3,4	3,3	3,3	2,7	3,0	3,1
21	plebiscito	397	1	4,2	90	€ 115.000	3,3	3,4	3,1	3,5	3,0	3,7	3,2
22	plebiscito	393	0-1	2,8	60	€ 50.000	3,3	3,0	2,4	2,3	2,5	2,4	2,4
23	plebiscito	345	1	4,1	90	€ 190.000	3,3	3,4	3,1	3,5	3,0	3,7	3,2
24	plebiscito	309	0-1	8,8	170	€ 160.000	3,3	3,4	3,1	2,8	2,5	3,4	2,8
25	plebiscito	246	3	6,2	100	€ 130.000	3,3	3,0	3,1	3,5	3,0	3,3	3,1
26	s. m. delle salette	45	0-1	5,9	110	€ 90.000	2,8	3,0	2,0	2,8	2,8	3,0	2,5
27	s. m. delle salette	40	0	4,0	90	€ 110.000	2,9	3,0	1,4	3,8	2,7	2,9	2,7
28	s. m. delle salette	38	1 2	5,8	120	€ 150.000	2,9	3,0	1,9	3,0	2,7	2,6	2,5
29	s. di giacomo	44	1	2,6	65	€ 48.000	2,3	2,6	1,7	1,0	1,2	1,7	1,4
30	reitano	1	2	3,7	80	€ 150.000	2,8	3,0	2,7	3,0	3,2	3,4	2,8
31	plebiscito	148	1	2,6	40	€ 65.000	3,3	3,4	3,0	3,0	2,7	3,0	2,9
32	plebiscito	119	1	4,7	100	€ 160.000	3,3	3,4	2,8	2,3	2,9	2,4	2,5
33	grimaldi	14	6	6,2	140	€ 260.000	3,1	3,0	4,6	3,5	3,5	3,4	3,5
34	fornai	27	1	2,6	55	€ 70.000	2,6	2,4	1,8	2,8	2,4	2,7	2,3
35	g. zurria	37	1	8,4	200	€ 240.000	3,4	3,4	3,3	3,0	2,7	3,0	3,0
36	gentile	22	2	4,9	130	€ 140.000	3,0	3,0	2,3	2,3	2,4	2,4	2,3
37	scuto	32	1	1,5	35	€ 59.000	2,6	3,0	2,0	2,3	2,3	2,0	2,2
38	cristoforo colombo	94	2	3,7	80	€ 145.000	3,1	2,6	2,9	2,8	2,4	2,4	2,6
39	domenico tempio	30	1	2,9	60	€ 78.000	3,6	3,0	2,6	3,0	2,8	2,7	2,8
40	domenico tempio	30	1	2,6	65	€ 85.000	3,6	3,0	2,7	3,0	2,8	3,0	2,8
41	della concordia	68	2 3	5,5	120	€ 145.000	3,5	3,0	3,4	3,8	3,5	3,3	3,3
42	della concordia	70	4	4,9	110	€ 170.000	3,5	3,0	3,6	3,0	3,0	2,7	3,0
43	de lorenzo	200	1	2,8	45	€ 55.000	2,4	2,0	1,5	2,8	2,2	2,3	2,1
44	mulino a vento	210	1	3,6	80	€ 115.000	2,2	2,0	2,6	2,8	2,7	3,0	2,4
45	belfiore	210	0 1	4,1	80	€ 75.000	2,3	2,0	2,1	3,0	2,9	3,0	2,4
46	belfiore	218	0	1,5	30	€ 25.000	2,3	2,0	2,1	2,0	2,2	2,4	2,0
47	della concordia	126 A	1	3,6	70	€ 90.000	2,8	2,6	2,1	2,0	2,0	2,0	2,1
48	tripoli	47	0-1	7,8	240	€ 230.000	1,6	2,0	2,0	2,3	2,2	2,1	2,0
49	velis	28	1	4,1	60	€ 80.000	3,6	3,0	2,6	4,0	4,0	3,3	3,3
50	piombai	11	0 1 2	4,1	80	€ 75.000	3,6	3,0	1,7	2,8	2,2	2,3	2,4
51	zuccarelli	15	0	1,5	40	€ 40.000	3,3	3,0	1,4	2,3	2,5	2,0	2,1
52	cordai	97	0	3,0	55	€ 40.000	2,4	2,6	1,8	2,3	2,4	2,3	2,1
53	cordai	131	1	3,5	90	€ 70.000	2,4	2,6	2,4	3,0	2,5	2,3	2,5
54	delle margherite	30	2	3,5	50	€ 55.000	2,4	2,6	2,7	3,0	3,5	3,0	2,7
55	mulino a vento	116	0	4,5	100	€ 67.000	2,6	2,6	1,7	1,0	1,0	1,0	1,4
56	del principe	142	2	4,7	100	€ 50.000	2,8	3,0	3,0	3,0	3,1	3,3	2,8
57	alogna	26	1	2,6	50	€ 60.000	2,5	2,6	2,1	2,8	2,6	3,0	2,4
58	ortolani	35	2	5,9	120	€ 150.000	2,8	3,0	3,9	3,3	3,3	2,7	3,2

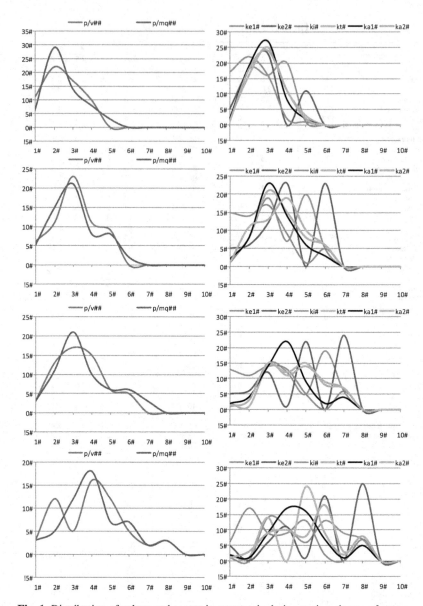

Fig. 1. Distribution of values and scores in progressively increasing classes of ranges

3   Methods and Procedures

3.1   Cluster Analysis

The term "cluster analysis" was first used in 1939, and includes a collection of algo-rithms and methods for grouping a set of observations into groups in a way that the degree of association between two objects is maximal, if they belong to the same group, and minimal otherwise. The aim of the cluster analysis is to organised observed data into meaningful structures and develop taxonomies.

Clustering algorithms can be divided into two types: hierarchical and partitional algorithms [12]. Hierarchical algorithms, e.g. hierarchical clustering, initiate with matching each object with similar ones, which are placed in a separate cluster and then merged into larger clusters. Partitional algorithms, such as K-means clustering, classify the whole dataset into smaller clusters. In research and literature, many other clustering techniques have been proposed during the years, especially with the large use of statistic software packages.

As the clustering variables of the dataset were already selected, in order to identify different groups in real estate market we proceeded determining how cluster are to be formed. In this paper, we used a partitioning method, the k-means clustering procedure.

The k-mean method groups observations into K clusters based on how close an observation is to the mean of the observations in each cluster. In such way, the procedure aims at segmenting the data that the within-cluster variation is minimized. The steps in the process are:

1. randomly assign observations to K cluster;
2. reassign the observations to other clusters to minimize the within-cluster variation, which is the squared distance of each observation from the mean of each cluster;
3. repeat the process until no observation needs to be reassigned.

The cluster affiliations of each observation can change during the process, so k-means does not build a hierarchy: the approach is also frequently labelled as non-hierarchical.

To assign an observation to the closest centroid, we need to define a proximity measure that quantifies the closest for the specific observation under consideration. In the application we used the Euclidean distance for our data points, even if there are different proximity measures could be appropriate for our type of data.

As the following step was the definition of the appropriate number of clusters, and so the measure the quality of a clustering, we use the sum of the squared error (SSE), which is the sum of the squared errors between every observations and the centroid of the cluster it belongs to. It can be used as a measure of variation within a cluster. It is possible then to compute the total sum of the squared errors.

It is preferred the cluster with the smallest SSE (the centroids of this clustering are a better presentation of the point in their clusters). The SSE will be used in the validity indexes.

The problem with the k-means method is the choice of the number of the clusters into which the data will be divided. The initial choice of k is largely subjective, and so the results can be bias by the opinion of the researcher. Successive runs of k-means can optimize the clustering of the observation for a different number of clusters. Cluster validation is a process, which tries to evaluate a particular cluster solution. Generally, the cluster validation procedures are divided into three different groups: internal, external, and relative [10].

The external validity measures (or indices) measure and analyse how close is a clustering a reference, a predetermined structure of the data. The internal validity measures analyse the goodness of a clustering structure, usually using the SSE. The relative indices are usually use to compare different attempts at clustering observations against each other and so they help to choose a proper number of clusters. The validation is crucial to a successful division of observations when there is no idea about the number of clusters. In this paper, we used the sum-of-squares based indices, which are founded on sum-of-squares within cluster (SSW) and/or sum-of-squares between clusters (SSB) values. WB-index [18] is an index where a minimum value can be attained as the number of clusters. Other indices have been proposed by Ball and Hall [2], Hartigan [11], Calinski and Harabasz [3], and Xu [17]. $X = \{x_1, ..., x_N\}$ represents the data set with N observation and D dimensions. The centroid of the clusters are $C = \{c_1, ..., c_M\}$, where c_i is the ith cluster and M is the number of clusters.

$$SSW = \sum_{i=1}^{N} \|x_i - C_{pi}\|^2 \qquad SSB = \sum_{i=1}^{m} ni\|c_i - \bar{X}\|^2$$

$$Wb\ Index = M\frac{SSW}{SSB} \qquad RSQ = \frac{SSW}{SSB + SSW}$$

$$\begin{aligned} &Calinski - Harabasz\\ &(CH) = \frac{SSB/(M-1)}{SSW/(N-M)} \end{aligned} \qquad Hartigan\ (H) = log\frac{SSB}{SSW}$$

$$Ball - Hall\ (BH) = \frac{SSW}{M} \qquad Xu = Dlog\left(\sqrt{\frac{SSW}{DN^2}}\right) + logM$$

A procedure for determining the optimal number of clusters is:

— given the dataset X, a specific clustering algorithm (K-means in our case study) and a fixed range of number of clusters $[M_{min}, M_{max}]$, are determined;
— repeat a clustering algorithm from predefined values of M_{min} to M_{max};
— obtain the clustering results (partitions P and centroids C) and calculate the index value for each of them;
— select the cluster M for which the partition offers the best result according to some criteria (minimum, maximum or knee point).

3.2   The Real Estate Market Analysis by Using a Data Mining-DRSA Approach

Given a real estate market, the analysis on a sample of properties that aimed at identifying the laws governing the dynamics of market prices can be conducted using a specific model of analysis. An analysis model that allows extracting from raw data the market rules in terms of correlation between the real estate characteristics and real estate prices on square meter can be constructed using a specific decision model and an algorithm Multiple Criteria Decision Aid (MCDA). The decision model aims at identifying the correlation between the decision to assign properties to a specific class of price on square meter and the real estate characteristics of the analysed sample.

The decision model so structured can be implemented using a specific algorithm among those proposed in the literature in the MCDA scope. This study suggests the Dominance-based Rough Set Approach (DRSA) algorithm i.e. the Dominance-based Rough Set approach [6, 7, 8, 9], in to support the decision problem for the real estate market. This algorithm has spread in the MCDA scope and suggests some features, which allow extracting from a sample of raw data a data sample that is representative for the analysed real estate market. The same algorithm permits identifying what are the most important real estate characteristics and what are their values. The implementation of this algorithm relatively to the real estate decision problem therefore allows defining a data mining using the DRSA approach. By using to the Data mining DRSA approach is possible to extract some information and then knowledge from a raw data sample. With the date mining DRSA approach, the data sample for the real estate market should not be subjected to preliminary treatments and for them, it is not necessary to define a system of weights. Furthermore, the analysis has conducted in a neutral way and ensures the robustness of the results [14, 15, and 16]. The robustness of the results is guaranteed by the algorithm DRSA that allows achieving reliable results based on the level of the accuracy and of the quality of the approximation. In particular, given the little space available, for what concerns the information about the DRSA algorithm, we refer to literature widely spread on it [6, 7, 8, 9]. However, it is important to emphasize that in this study the DRSA approach offers the following functions:

1. allows identifying the general rules for the real estate market;
2. allows training the sample on the basis of generated real estate market rules, fa-vouring the learning process for it ;
3. allows improving for subsequent steps the classification of the elements of the sample;
4. allows identifying at the end of the learning and reclassifying process the more representative data sample for the analysed market.

The convergence towards the more representative data sample is obtained based on the value of the accuracy and the value of the quality of the approximation for the classification. Once it is identified the most significant data sample for the real estate market, it is possible to get the following information:

1. the smallest subset of rules on the most representative set of market rules;
2. the most important real estate characteristics for the market;
3. the values of the real estate characteristics that are more important for the market.

4   Applications

4.1   Cluster Analysis

In our case study we fixed the number of clusters $M_{min} = 2$ and $M_{max} = \sqrt{N}$. We use the previously defined 6 variables $k_{e1}, k_{e2}, k_i, k_t, k_{a1}, k_{a2}$.

The results are presented in Table 2. In order to find the number of clusters, we used the data for knee point detection. The knee point in the graphs indicates the optimal number of cluster, even if the detection of the knee points is not very easy. The maximum value and the minimum values are the most straightforward points to detect. Some other indices are monotonous so it is not clear what the optimum value for the number of clusters.

Results presented in the tables show different optimal clusters. Some indices identify the optimal number of clusters equals to 3 (SSB/SSW). For the WB Index, RSQ, Hartigan and Ball-Hall indexes, the optimal number is 4, while for Xu index that number is 5. Calinski-Harabasz has not shown any knee, so the optimal number of cluster is not defined. The options between 3 clusters and 5 clusters are considered in the following analysis.

Table 2. Internal Cluster Validation Measures

	N. Cluster	N. Cluster						
		2	3	4	5	6	7	8
SSB/SSW	3	0,91	0,58	0,48	0,39	0,31	0,25	0,22
WB_Index	4	1,81	1,73	0,12	0,08	0,09	1,77	1,77
RS Q	4	0,52	0,63	0,67	0,72	0,76	0,90	82,00
Calinski-Harabasz	ni	61,73	47,82	37,31	33,94	33,68	33,62	32,32
Hartigan	4	0,04	0,24	0,32	0,41	0,51	0,60	0,66
Ball-Hall	4	29,14	14,91	9,97	0,69	4,82	3,53	2,77
Xu	5	-7,32	-7,49	-7,61	-7,75	-7,76	7,82	-7,88

4.2 Data Mining-DRSA Approach

A first step for the definition of a Data mining-DRSA approach to support the analysis of the real estate market is the definition of a specific decision problem.

The decision problem to support the analysis of the real estate market can be defined on the basis of a set of 28 condition criteria, namely the real estate characteristics and 4 decision classes, i.e. 4 ranges of the unit price variation for the detected properties [6]. The set of condition criteria is the following:

	1	$k_{e1}1$	centrality and settlement quality
	2	$k_{e1}2$	functional mix
	3	$k_{e1}3$	socio-economic mix
	4	$k_{e1}4$	urban maintenance
A. Location 1	5	$k_{e1}5$	equipment
	6	$k_{e1}6$	facilities
	7	$k_{e1}7$	accessibility by private transportation
	8	$k_{e1}8$	accessibility by public transportation
	9	$k_{e1}9$	internal access
B. Location 2	10	$k_{e2}1$	micro-environmental functional features
	11	$k_{e2}2$	micro-environmental symbolic features

	12	$k_i 1$	panoramic quality
	13	$k_i 2$	view
C. Intrinsic features	14	$k_i 3$	brightness
	15	$k_i 4$	exposure
	16	$k_i 5$	security
	17	$k_t 1$	plants
D. Technology	18	$k_t 2$	finishes
	19	$k_t 3$	maintenance status
	20	$k_{a1} 1$	usability
	21	$k_{a1} 2$	structural and plant quality
E. Building Architec-	22	$k_{a1} 3$	finishes and building technologies
tural quality	23	$k_{a1} 4$	stylistic coherence
	24	$k_{a1} 5$	decorum
	25	$k_{a1} 6$	internal coherence
F. Property Architec-	26	$k_{a2} 1$	size, distribution and usability
tural quality	27	$k_{a2} 2$	accessories and restrooms
	28	$k_{a2} 3$	finishes

The set of the proposed decision classes is the following:

1. $P_{min} < P_x < P_{qtl1}$;
2. $P_{qtl1} < P_x < P_a$;
3. $P_a < P_x < P_{qtl3}$;
4. $P_{qtl3} < P_x < P_{max}$;

where: P_x is the price of the generic property, P_{min} is the minimum price, P_{qtl1} and P_{qtl3} are the prices of the first and third quartile, P_a is the average price and P_{max} is the maximum price.

5   Results and Discussions

5.1   The Results of the Application of the DRSA Approach

In the first step of the analysis of the real estate market of S. Cristoforo, was defined the information table for the real estate data sample, which is constituted from 58 properties. We do not report the information table for its large size.

In the second step have been generated with the aid of the DRSA algorithm the market rules (Table 3), the "reducts" which identify a partition of the real estate characteristics in 14 fields with cardinality that goes from 12 to 14, the core that allows to find the set of criteria with the highest level of importance ($k_{e1} 8$, $k_{e1} 9$, $k_i 1$, $k_i 2$, $k_i 3$, $k_i 5$, $k_{a1} 6$, $k_{a2} 2$), those with a low level of importance ($k_{e1} 1$, $k_{e1} 3$, $k_{e1} 6$, $k_{e1} 7$, $k_{e2} 1$, $k_i 4$, $k_{a1} 1$, $k_{a1} 3$, $k_{a2} 3$) and those which haven't importance for the characterization of the real estate data sample ($k_{e1} 4$, $k_i 3$, $k_{a1} 2$).

Table 3. Internal Cluster Validation Measures

$k_i = 3$				
$k_{a1}2 = 4$				
$k_i2 = 3$	$k_{a1}5 = 4$			$P_{qtl1} < P_x < P_a$
$k_{e1}4 = 4$	$k_t1 = 3$	$k_{a1}1 = 3$		
$k_{e1}8 = 3$	$k_t3 = 2$	$k_{a1}6 = 3$		
$k_{a1}5 = 3$	$k_i3 = 2$	$k_i5 = 3$	$k_{a1}6 = 3$	
$k_{a1}3 = 4$				
$k_i3 = 5$				
$k_{e1}4 = 4$	$k_t3 = 4$			$P_a < P_x < P_{qtl3}$
$k_{e2}2 = 4$	$k_i2 = 4$			
$k_{e1}2 = 4$	$k_{e1}7 = 4$	$k_i1 = 3$		
$k_{e1}8 = 3$	$k_t1 = 3$	$k_{a1}6 = 3$	$k_{a2}1$	
$k_{a2}3 = 4$				
$k_{e1}4 = 4$	$k_i2 = 4$			$P_{qtl3} < P_x < P_{max}$
$k_{e1}7 = 4$	$k_i5 = 4$			
$k_{e1}7 = 4$	$k_i2 = 4$	$k_i1 = 3$		
$k_{e1}8 = 3$	$k_t1 = 3$	$k_{a1}4 = 4$		

In the third step, the data sample was reclassified on the basis of the general rules, allowing detecting some inconsistencies in the data.

In the fourth step, the data of the date sample, which presented in the previous step of the inconsistencies or a low degree of accuracy of the classification, were eliminated.

Then, this data sample has been reclassified in order to detect the residual inconsistencies and to converge towards the classification with a higher degree of the accuracy of the classification. The removal and reclassification process of the data for the residual data sample was conducted until it has reached a level of the accuracy and the quality of approximation of 100%. The sample thus identified is the representative sample for the studied real estate market. The representative sample of the real estate market that can be classified correctly by using the generated rules is constituted from 13 properties (id. 2, 12, 16, 17, 19, 20, 22, 24, 26, 30, 45, 56).

Once that the minimal date sample has been found, then it was possible to dissect from the set of general rules the minimal set of rules which are representative for the real estate market. The decision rules induced by the data sample show how the first decision class, namely that relating to the range of the variation price from P_{min} to P_{qtl1} in the 1st quartile $P_{min} < P_x < P_{qtl1}$, can't be represented correctly, i.e. for it, it is not possible to define on the basis of a data sample a correlation between the real estate characteristics and the decision class 1 which can be considered a robust result.

At the end, the analysis converges to the minimal set of four rules (Table 4).

Table 4. Importance level of the characteristics. Minimal set of rules for the analysed market

$k_i = 3$	$k_{a1}5 = 4$		$P_{qtl1} < P_x < P_a$
$k_{e1}8 = 3$	$k_i3 = 4$	$k_t1 = 3$	
$k_t3 = 4$			$P_{qtl3} < P_x < P_{max}$
$k_i2 = 3$	$k_{a1}5 = 4$		

Such a minimal set of rules detects that also the class decision 3, namely that relating to the range of variation price from P_a to P_{qtl3}, $P_a < P_x < P_{qtl3}$ does not appear to be representative in the stated sense previously. Finally, the representative date sample for the real estate market is characterized by a single core for the approximation, which has cardinality 1 and whose only criterion is the $k_{e1}8$ (accessibility by public transport), which in itself can be identified as the most important criterion for the analysed market.

5.2 The Results of the Application of the Cluster Analysis

Given a three options (3, 4 and 5 clusters) are commented hereafter. The solution with 3 cluster has 23, 20 e 15 observation in cluster no. 1, 2 and 3, respectively. Cluster n. 1 is quite different from cluster n. 2, and very different from cluster n. 3, while cluster 2 and 3 are not that different. The variables, which have the great impact on clustering, are k_i and k_{a2}, while both k_e variables have a small impact on clusters. In cluster n. 1 all properties with high quality and good characteristics are clustered together. In cluster n. 3, the observations show poor quality, especially for k_a1 and k_{a2}. The second hypothesis has 4 clusters with 23, 20, 12 and 3 cases each. In this case, a small cluster of 3 cases indicates properties with low value of their variables (meaning, poor quality of the characteristics). In this scenario, the variable, which has the greater impact on clustering, is k_{a2}, while the k_{e1} and k_{e2} show very small value, and so impact, measured by the value of the F-ratio. Again, the group n. 1 have all properties with high values of the variables (Mean >3), while the last one, group n. 4 have very low value of almost all variable (around 1). The level of characteristics of the properties is declining from cluster n. 1 to cluster n. 4. In the fourth scenario, with 5 cluster (of 22, 10, 12, 3 and 11 cases each) shows less difference. In fact, cluster 1 and 2 are very similar, the first group with high value of the characteristics, the second with a small reduction of values of k_i, k_t and both k_a (only k_es are > 3). The most similar clusters are the n. 3 and n. 5, which show little distance in their centroid and the means of the variables used for clustering. This is clearly the splitting of cluster n. 3 of the previous scenario. The cluster n. 3 shows smaller values of the k_i characteristic in comparison to the cases included in the 5 cluster ($k_i < 2$). The last group, already defined in the previous partitioning, has only three cases, with poor qualities (as maintenance, location, view, etc.) and so less attractiveness for the market. Again, in this last scenario, the better discriminators between observations are k_i and k_a2, which seems to be the most significant variables to cluster in all the hypothesis analysed. Market and its demand seems to appreciate particularly the characteristics intrinsic and the ones linked to the property asset unit.

Fig. 2. Display of 3 clusters

Fig. 3. Display of 5 clusters (4 clusters omitted)

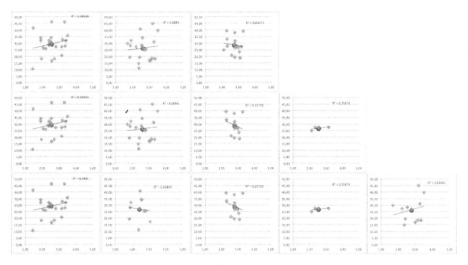

Fig. 4. Price vs overall quality for each clustering hypothesis

Fig. 5. Allocation of the different classes of prices (dark =low price, light = high price)

Fig. 6. Allocation of the different classes of overall quality scores (dark =low score, light = high score). The "reduct" sample is enlighten.

6   Conclusions

The comparison between the statistic coherence of the 3 segmentation hypotheses (3 - 5 clusters) and their spatial representation encourages some conclusive reflections.

A group dislocated along the main axes and the other two internal (more characterized by technologic and architectural homogeneity) is outlined in the 3-cluster segmentation. The detachment of the fourth cluster does not add any significant information to the subdivision, while the passage to five clusters reveals a subdivision of the second cluster basing on the architectural characteristic. This seems coherent with the check of the survey sheets, where the data are more detailed than the aggregated scores by means of which the characteristics have been represented (Fig. 2, 3).

A further verification concerns the prices. The relation between the unit price (y-axis) and the aggregated characteristic (x-axis) has been represented for each cluster, obtaining very fragmentary allocations with, in certain cases, a negative tendency line at the raising of the total quality (Fig. 4, price/sq.m only).

The spatial verification does not provide further significant indications, because a general appreciation of the centrality is observed, as the low prices are registered in the internal areas (Fig. 5, 6).

The "reduct" sample reveals a certain variability of the ask prices, per room and per sq. m as well, characterizing all bands and the whole quality range as well, while it contains the highest values (medium-high) of the aggregated characteristics.

By concluding, the real estate market analysis, carried out by using advanced statistic tools and the decision making modelling, provides helpful information for urban enhancement programs, more specifically if an equalization program is envisaged. In the proposed study case, the "reduct" sample, extracted by using the DRSA pattern, can be assumed as the most representative relationship "price vs quality", a sort of "fair sample" to be taken as a reference in the equalizing policies. Therefore, according to the cluster analysis, the low price/quality can be assumed as the cluster to be boosted by actualizing a local property taxation system able to drag the surplus coming from the renovations of the properties belonging to the best clusters.

Acknowledgements. Laura Gabrielli edited par. 3.1, 4.1 and 5.2; Salvatore Giuffrida edited par. 1, 2, 6; Maria Rosa Trovato edited par. 3.2, 4.2 and 5.1

References

1. Baarsch, J., Celebi, M.E.: Investigation of internal validity measures for K-means clustering. In: Proceedings of the International MultiConference of Engineers and Computer Scientists, IMECS 2012, Hong Kong, Vol. I, March 14 – 16, 2012
2. Ball, G., Hall, D.: ISODATA, a novel method of data analysis and pattern classification. Stanford Research Institute, Menlo Park (1965)
3. Calinski, T., Harabasz, J.: A dendrite method for cluster analysis. Commun. Stat. **3**, 1–27 (1974)
4. Dato, G.: La città di Catania. Forma e struttura, pp. 1693–1833. Officina Edizioni, Roma
5. Forte, C.: Elementi di estimo urbano. Etas Kompass, Milano (1968)
6. Greco, S., Matarazzo, B., Słowiński, R.: Rough approximation of a preference relation by dominance relations. European J. Operational Research **117**, 63–83 (1999)
7. Greco, S., Matarazzo, B., Słowiński, R.: Rough sets theory for multicriteria decision analysis. European J. of Operational Research **129**, 1–47 (2001)
8. Greco, S., Matarazzo, B., Słowiński, R.: Dominance-based rough set approach to knowledge discovery – (I) general perspective, (II) extensions and applications. In: Zhong, N., Liu, J. (eds.) Intelligent Technologies for Information Analysis, pp. 513–612. Springer, Berlin (2004)
9. Greco, S., Matarazzo, B., Słowiński, R.: Decision rule approach. In: Figueira, J., Greco, S., Ehrgott, M. (eds.) Multiple Criteria Decision Analysis: State of the Art Surveys, pp. 507–563. Springer, Berlin (2005)
10. Halkidi, M., Batistakis, Y., Vazirgiannis, M.: On Clustering Validation Techniques. Journal of Intelligent Information Systems **17**(2/3), 107–145 (2001)
11. Hartigan, J.: Clustering algorithms. John Wiley & Sons, Inc., New York (1975)
12. Jardine, N., Sibson, R.: The construction of hierarchic and non – hierarchic classifications. The Computer Journal **1**, 177–184 (1968)
13. Krzanowski, W., Lai, Y.: A criterion for determining the number of groups in a data set using sum-of-squares clustering. Biometrics **44**(1), 23–34 (1988)
14. Trovato, M.R., Giuffrida, S.: A DSS to assess and manage the urban performances in the regeneration plan: the case study of pachino. In: Murgante, B., et al. (eds.) ICCSA 2014, Part III. LNCS, vol. 8581, pp. 224–239. Springer, Heidelberg (2014)

15. Trovato, M.: A fuzzy measure of the ability of a real estate capital to increase in value. The real estate decision problem for ortigia, in appraisals. In: Evolving Proceedings in Global Change, vol. 2, pp. 697–720. Firenze University Press, Firenze (2012)
16. Trovato, M.R.: The real estate decision problem. a model to support the real estate market analysis. In: Rèsumés/Abstracts, 78 Meeting of the European Working Group "Multiple Criteria Decision Aiding" Catania 2013, pp. 65–66. University of Catania (2013)
17. Xu, L.: Bayesian ying-yang machine, clustering and number of clusters. Pattern Recogn. Lett. **18**, 1167–1178 (1997)
18. Zhao, Q., Xu, M., Fränti, P.: Sum-of-square based cluster validity index and significance analysis. In: Proc. of the 17th Int. Conf. on Adaptive and Natural Computing Algorithms, pp. 313–322 (2009)
19. Zhao, Q., Fränti, P.: WB-index: A sum-of-squares based index for cluster validity. Data & Knowledge Engineering **92**, 77–89 (2014)

The Multidimensional Assessment of Land Take and Soil Sealing

Raffaele Attardi[1], Maria Cerreta[1], Valentina Sannicandro[1],
and Carmelo M. Torre[2(✉)]

[1] Department of Architecture, University of Naples Federico II,
via Forno Vecchio 36, Naples, Italy
{raffaele.attardi,cerreta,valentina.sannicandro}@unina.it
[2] MITO Lab, Department DICAR, Technical University of Bari,
Via Orabona 4, Bari, Italy
cartorre@yahoo.com

Abstract. In 2006 European Commission stated that soil, fairly recognisable as an ecosystem structure, can be considered essentially as a non-renewable resource, thus triggering both studies for the assessment of land take phenomenon and actions for its mitigation and reduction. In last two decades, a deeper and ecosystem approach to land-use policies targeted to the sustainable development enabled a closer understanding of the complexity of urban dynamics leading to the necessity of multidimensional and integrated approaches for the assessment of the use of resources. The paper presents a multi-dimensional approach to evaluate the phenomenon of land take and soil sealing implemented on a sample of municipalities in Apulia Region, in Southern Italy. The construction of a composite indicator for comparative qualitative and quantitative measurement of land take and soil sealing among the municipalities is aimed at a better evaluation of future urbanisation scenarios and at a monitoring process of urban growth.

Keywords: Land use · Urban sprawl · Urban fringe · Impact evaluation · Composite indicator

1 Introduction: Land Take and Soil Sealing and Their Evaluation

In last two decades, land take became a relevant issue related to urban expansion models. The dichotomy between compact and sprawling city led to controversies: on the one hand the theoretical model of compact city moves towards less energy and resource consumption, even if real-world cases studies show that this model results in high urban density, congested cities and air pollution *inter alia* [1, 2]. On the other hand, the sprawling city can balance the congestion-related issues of compact city, but it has not actually been able to limit its own physical expansion, thus increasing land taking and soil sealing and smoothing the boundaries between urban and rural landscape.

© Springer International Publishing Switzerland 2015
O. Gervasi et al. (Eds.): ICCSA 2015, Part III, LNCS 9157, pp. 301–316, 2015.
DOI: 10.1007/978-3-319-21470-2_21

A number of factors influences the urban expansion, namely land morphology, urban planning approaches, environmental constraints (for risks or preservation), social and economic dynamics. In last twenty-five years, these factors led to the generation of semi-urban landscapes in regions where the sprawling city model prevailed. The "semi-urban landscape" refers to the dualism between rural and urban landscape, permeable and impermeable soils [3] thus defining urban fringes and urban sprawl.

The concept of urban fringe suggests the image of indented urban areas located at the border zone of the city centre. According to Hite [4], the urban fringe is a dynamic boundary space, where the economic returns from the new urban land-uses are comparable to the yields obtained by the traditional natural or semi-natural land-use.

The concept of urban sprawl indicates a kind of urban expansion in which the rate of land take is higher than the increase of population density [5, 6]. According to Allen [7], urban sprawl is a land-use transition outside urbanised areas with an ecological footprint higher than the minimum required for human activities. The European Environmental Agency defines urban sprawl as a physical model of low-density urban expansion, mainly located near agricultural areas [8].

While the urban fringe is still a specific place well physically linked to the urban core, urban sprawl, far from being a randomly distributed effect, is closely linked to the presence of infrastructure networks, but is less dependent on distance constraints from the urban core.

It should be noted that the semi-urban landscape does not coincide with the suburbs, rather it is an evolving area in its both spatial and structural features, and it is characterised by clusters of agricultural and non-agricultural activities [9, 10, 11].

Antrop and Van Eetvelde [12] define urban fringes as a specific landscape characterized by functional heterogeneity and are much more complex than the traditional urban landscape. Consequently, semi-urban landscapes play as transfer areas between urban and rural ecosystems, thus they are also defined as hybrid or marginal landscapes [13, 14, 15].

Semi-urban landscape is exactly where land take and soil sealing occur, primarily caused by urban sprawl in quantitative term. Soil sealing, in turn, causes the loss of natural soil in place of buildings and infrastructures [16]. Soil sealing is the most intense form of land take for urbanization purposes, and it consists in the covering of natural soil surfaces with impervious materials, such as asphalt and concrete. Urban sprawl also causes the fragmentation and of rural landscape, resulting in the loss of ecological, productive and cultural values [17] attached to natural and rural landscapes [18]. Moreover, the abandonment of agricultural activities accompanies urban sprawl, thus contributing to the loss of ecological functions, to the reduction of provisioning services and to the significant increase of hydrogeological instability [19, 20].

However, nowadays urban expansion processes cannot be considered as the only major cause for land take, but new forms land take and soil sealing arise, based on the change of local economic development models. Consequently, it would be more advisable to use the concept of land-use transition than land take. Indeed, the analysis of land-use transition trends in Apulia region in Southern Italy shows that there are three main land-use transitions emerging as relevant for land-use policies:

- Land used for energy production sites (this is usually a reversible use of soil);
- Morphological land transformation for mining activities;
- Human settlements for tourism and vacation mostly located in coastal areas.

Therefore, one can conclude that nowadays the dynamic relationship between local economic development models and urban expansion concepts highlights the issue of reducing land take as one of the most relevant aspects of the "compact city" model promoted in the Eighties by "Friends of the Earth" and the "World Environmental Forum".

Land take assessment basically depends both on the way it is defined and on the purpose of the assessment: if land take is assessed in order to evaluate urban sprawl, then at first, a basic definition of sprawl is required and at second a related land take measurement procedure can be identified (e.g. fragmentation index); on the other hand, if soil sealing is considered as the only major phenomenon for land take, then the imperviousness index is an adequate measurement. However, in complex and multidimensional approaches several issues are simultaneously taken into account and their own clear definition, measuring and modelling is required. Therefore, structuring the assessment framework is threefold:

- What is the object to be assessed;
- How it can be measured;
- How it can be evaluated.

In last two decades, a deeper and ecosystem approach to land-use policies targeted to the sustainable development enabled the closer understanding of the complexity of urban dynamics leading to the necessity of multidimensional and integrated approaches for the evaluation of planning and management policies [21, 22, 23, 24, 25]. Moreover, in 2006 European Commission stated that soil, fairly recognisable as an ecosystem structure, can be considered essentially as a non-renewable resource, thus triggering both studies for the assessment of land take phenomenon and actions for its mitigation and reduction [8].

Therefore, in environmental modelling and evaluation literature the combination of multiple measures (indices) for the assessment and monitoring of complex spatial phenomena emerges as a relevant theoretical and operative issue. Based on literature review, we observed the transition from single indicators on land-use or land-cover to the combination of several measures, such as the Landscape Metrics or the fragmentation, porosity, size, imperviousness, impermeability indices that makes the concept of land take closer to issues of ecosystem services provision and preservation [26, 27, 28, 29, 30, 31, 32].

In order to develop a multidimensional approach for land take and soil sealing, the evaluation framework has to encompass three main stages (fig.1):

- Qualification of the phenomenon: the object of the assessment (urban fringes, fragmentation of urban settlements, disconnection of the ecological network);

- Assessment of the phenomenon through the selection of appropriate indicators: measuring tools and indicators (calculation of density, fragmentation, porosity indices through spatial analysis);
- Evaluation: aggregation of single indices in order to obtain a complex map of the analysed phenomena (calculating the impacts of land-use change in terms of ecological, social and economic costs and benefits).

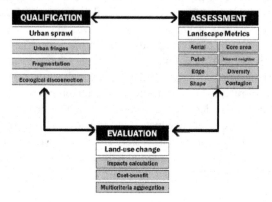

Fig. 1. Diagram of evaluation framework

The above-mentioned stages explain the steps of a cyclical/interlaced evaluation process that when applied to real-world planning and decision-making can be much more complex and fluid than a theoretical conceptualisation. Starting from this assumption, we developed a framework for the analysis of a spatial data-set referring to land take and soil sealing quantitative and qualitative indicators.

Based on those assumptions, in the following we propose a methodological approach for a multidimensional analysis of land take a soil sealing through a spatial multicriteria methodology [33]. The proposed approach enables the identification of the most influencing factors and the spatial patterns for land take and soil sealing trends. In the following, starting from the definition of research objectives (section 2), we describe the methodological approach and its application to the case study of the Italian province of Lecce (southern Italy) in section 3 and then we conclude with some final remarks based on the analysis of the output of the application and outline a few research perspectives.

2   Research objectives: The Construction of Land Take Composite Indicators

The interpretation of landscape we consider is in the sense of «a certain part of the territory, as perceived by people, whose character derives from the action of nature and/or humans and their interrelations» (European Landscape Convention, Florence, 2000, p.2, art.1.a).

Socco and Cavaliere [34] assert that the overspill beyond the threshold of environmental compatibility is a structural dynamics in the growth of semi-urban cities, which takes place not only in presence of natural and fundamental ecological components, but also in the presence of large infrastructure and settlements, thus causing land take and soil sealing. These two phenomena are defined as the definitive loss of one or more functions in the ecosystem, including both natural processes and human activities.

The main effects of the described transformation process and of land take and soiol sealing are:

- the fragmentation of rural and natural,
- the divide between urban and agricultural landscape,
- the "disorder" produced by the randomness of the locations of typically urban land uses,
- the impoverishment of ecosystems, both rural and river,
- the lack of attention to the conservation of water resources and biodiversity, in terms of quality and quantity.

These effects led the scientific community and the local administrations to reflect on the need for a methodology of analysis and evaluation of the phenomenon of land take and soil sealing. In particular, the need for methodologies integrating multiple dimensions can be envisaged, in order to consider all the multiple effects above described.

The aim of this research paper is the construction of a composite indicator for comparative evaluation of land take trends among municipalities, moving from the study of territory defined by fringe and sprawl. The composite indicator combines the measurement of both the quantity and the quality issues.

In addition, the proposed assessment methodology can be considered a valid support for the assessment and monitoring of land use evolution in local and regional land-use policies.

3   Methodological Proposal and Its Application

The methodological framework (fig.2) for the construction of a composite indicator for land take and soil sealing at municipality level involves three consecutive phases.

The first phase is related to the state of the art, as well as the definition of the peculiar features of the phenomenon to be considered (fragmentation, sprawl, imperviousness, and contribution to ecosystem services) and the real or potential impacts related to land take and soil sealing; then, an appropriate spatial dataset is gathered. The second phase is divided into two parts: the data population and the proposal of both the methodology used to assess the impacts of land-use changes and the methodology for the aggregation of single indicators into the composite one. The third and final step is the application of the model to a case study, tested on the Province of Lecce (fig.3) in the Apulia region (Southern Italy), followed by presentation of the results.

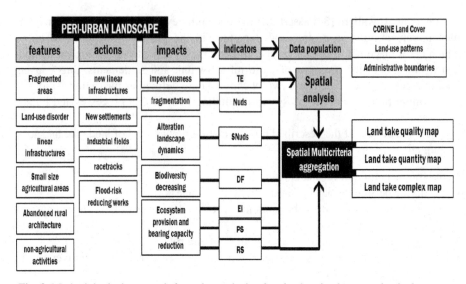

Fig. 2. Methodological proposal: from the analysis of peri-urban landscape to land take maps

3.1   First Phase. State of the Art

Which are the characteristics of the peri-urban areas?

The character of irrationality of urban design is a constant of cities whose boundaries consist of urban fringes. The disorganised aggregation of multiple urban and natural land uses manifests, in most cases, the absolute lack of a coherent planning framework, accentuated by the disorder with which the road network is planned and realized [34].

Despite peri-urban areas are constantly changing their shape, they show some easily recognisable features, described in a Territorial Plan of Milan province, the *Piano Territoriale di Coordinamento Provinciale* (PTCP) (Adaptation to Regional Law for the territorial government n.12/05), including:

- presence of small-size agricultural fields;
- fragmented areas, sometimes enclosed in built areas;
- presence of non-agricultural activities with common elements that are scarcely compatible with agriculture assets;
- accelerated dynamics of transformation and "land-use disorder";
- little interest in agriculture production;
- conspicuous presence of linear infrastructures;
- abandoned rural architecture;
- presence of disordered power lines and other elements interfering with housing;
- poor quality of residential green and open spaces;
- mix of building types and scarcity of public spaces and services;

- improper use of residual areas, with the presence of marginal activities, often abusive;
- presence of small-size agricultural connective areas, whose dominant role is to connect agricultural fields or agricultural areas with natural areas.

The main reference for the identification of the transformation actions generating impacts on the land-use patterns is the Inventory of mitigation and compensation actions for landscape and environment for the compliance of the Territorial Plan of the Province of Milan to the Lombardy Regional Law for the territorial government n.12/05.
It identifies five main categories of transformation actions:

1. linear infrastructure;
2. settlements;
3. industrial fields;
4. racetracks;
5. river flood-risk reducing works.

The present research focus is on the category of settlements, especially on development and urban expansion projects, therefore basically the impacts generated by urban expansion plans are considered.
Hereinafter we report the direct and indirect impacts related to land use, produced by low-density settlements and by compact settlements.
First, soil imperviousness influences the groundwater recharge processes, the flow rates and water quality of the final receptors, the loss of biodiversity. The fragmentation and disruption of ecological connections between distant spots affect the productive capacity of the area and the bearing capacity of the connected spots; finally, landscape alterations favour the realisation of new settlements, thus rising disturbances on ecosystems due to new settlements.
Furthermore, on a local scale, the aesthetic-perceptive impacts linked to the perceptual recognition of places have to be considered. These processes trigger the loss of landscape value, caused by: the interruption of the morphological continuity of places, the increase in the number of cars in transit on the roads and the increase of the temperature of the water drain and of the pollutant load, the loss of ecological integrity and other ecosystem functions.

3.2   Second Phase. Data population and Methodological Proposal for the Impact Assessment and Multidimensional Evaluation

In this phase, accessible and available data are collected for later analysis. In particular:

- Land Use Maps of Apulia region related to the 2006 and 2011 (www.sit.puglia.it);
- Provincial and municipal administrative boundaries provincial and municipal (www.istat.it).

Starting from the data infrastructure acquired, Table 1 illustrates the correspondence between the impacts associated to land take and soil sealing, six in total, and the indicators chosen to measure their intensity.

Table 1. Identification of impacts related to land take and soil sealing and indicators for the evaluation and the analysis of semi-urban landscape

Impacts	Indicators adopted for impacts analysis	Typology
Imperviousness and concreting of the areas surrounding new buildings	Ratio of the scattered built surface over the municipal area (TE)	Quantitative
Fragmentation and disruption of ecological connections between distant spots	Ratio of the number of non-natural land uses outside the consolidated urban fabric and the highest value of fragmentation (Nuds)	Quantitative
Alteration of landscape dynamics and increase of non-natural landscape	Ratio of the area occupied by the elements with non-natural land use external to the consolidated urban fabric over the municipal area (SNuds)	Quantitative
Biodiversity decreasing	Ratio of the extent of the reduction of natural land uses (in relation to the time series from 2006 to 2011) over the municipal area (DF)	Quantitative
	Total value of *ecological integrity* calculated on the basis of the individual values of land uses inherent municipal administrative boundaries (EI)	Qualitative
Reduction of provisioning capacity and of the bearing capacity of the spots	Total value of *provisioning services* calculated on the basis of the individual values of land uses inherent municipal administrative boundaries (PS)	Qualitative
	Total value of *regulating services* calculated on the basis of the individual values of land uses inherent municipal administrative boundaries (RS)	Qualitative

The indicators for the quality of land take, that is the ecological integrity, the provisioning services and the regulating services are defined in the "matrix evaluation of capacity expressed by land cover vs. ecosystem services" [35]. The matrix lists in rows the 44 land use determined by the Corine Land Cover classification; the columns of the matrix list the 29 ecosystem services, evaluated on the following scale: 0 = no significant capacity, 1 = low-relevant capacity, 2 = relevant capacity, 3 = medium-relevant capacity, 4 = high-relevant capacity and 5 = very high-relevant capacity. The values of these indicators are normalised with respect to the highest value. Finally, the indicators are weighted with respect to the municipal area.

It should also be noted that the value of the Ecological integrity is the result of the following components: abiotic heterogeneity, biodiversity, biotic waterflows, metabolic efficiency, exergy capture (radiation), reduction of nutrient loss, storage capacity (SOM). The value of the Provisioning services is a function of the value related to crops, livestock, fodder, fisheries, aquaculture, wild foods, timber, combustible wood, energy (biomass), biochemistry/medicine, fresh water. The value of Regulating services is a linear combination of the following parameters: local climate regulation, global climate regulation, flood protection, groundwater recharge, air quality regulation, erosion regulation, nutrient regulation, pollination.

The last step in the methodological proposal is the definition of a composite indicator, that is not an absolute index, but relative to the comparability between the municipalities of the province of Lecce. In order to calculate the composite indicator, which represents "the loss of functions of the soil", the TOPSIS (Tecnique for Order of Preference by Similarity to Ideal Solution) multicriteria methodology has been chosen, developed by Hwang and Yoon [36], which generates an "ideal positive solution" and an "ideal negative solution".

TOPSIS allows the construction of a hierarchy of preferences with respect to closeness or distance from the geometric ideal solution; it is part of the compensatory aggregation algorithms that compare a set of alternatives, assigning weights to each criterion and normalising the scores. Therefore, one needs to define in advance the decision matrix and the weight vector. Consequently, the Value of the loss of soil functions (PS) will be equal to:

$$PSi = di/(di+di),\ 0 \le PSi \le 1,\ i = 1,2,...,n \text{ (n alternatives).}$$

- PSi = 1 the solution is the worst condition;
- PSi = 0 the solution is the best condition.

3.3   Third Phase. Test on a Case Study

The selection of the province for testing the application of the methodology is evaluated on the basis of the relationship between the spread built fabric over the total built fabric on a provincial basis. The worst condition is the province of Lecce (2.50%), in the south of Apulia region (fig.4), while the best condition is the province of Foggia ties in the province of Lecce (0,23%), in the north. Therefore, for the testing phase, we selected the 94 municipalities in the province of Lecce for our exploratory application of the proposed methodology. Through spatial analysis the values of the indicators for the quality and the quantity of soil loss are calculated. For a simple and clear exposition of the results, Table 2 presents the values of the indicators related to the 94 municipalities, arranged in alphabetical order.

Figures 5 and 6 represent the output of the multicriteria analysis for the calculation of the composite indicator. Figure 5 on the left shows the distribution of the value of land take calculated in quantitative terms, thus considering only the indicators TE, Nuds, SNuds, DF, while on the right it shows the distribution of the value of land take calculated through quality indicators (EI, PS, RS). Figure 6 shows the distribution of the value of the composite indicator of land take, that includes both the quality and quantity indicators.

Fig. 3. Geographical context setting: on the left, Apulia Region, on the right, Lecce province

Fig. 4. Analysis of built areas for the municipalities of the province of Lecce

Both in terms of quantity and quality of land take, the inner central area of the province is the one where the best trends are clustered, while the coastal municipalities seem the more critical ones; this result reflects the main urban expansion trend in the province, related to new touristic and vacation properties on the coastline. When comparing the maps in figure 5 with the complex land-take map in figure 6, one can notice that both the qualitative and the quantitative assessment of land-take can be relevant, thus influencing the total ranking. The worst performance municipalities, indeed, are not due to quantitative or quantitative indicators separately, but to their combination and amplification.

Table 2. Indicators of land take and soil sealing, calculated on a municipal basis for the 94 municipalities in Lecce province

NAME	Area (mq)	TE	Nuds	S-Nuds	DF	EI	PS	RS
Acquarica del Capo	18449207,6718	0,17291	0,60870	0,17291	0,002	0,01746	0,00228	-0,00011
Alessano	28301760,9054	1,20344	0,78261	1,20344	0,015	0,02498	0,01062	0,00249
Alezio	16572254,7978	3,88621	0,65217	3,88621	0,001	0,01603	-0,00272	-0,00022
Alliste	23211930,9841	1,36362	0,69565	1,36362	0,008	0,01707	-0,00528	0,00056
Andrano	15495399,1462	1,89677	0,65217	1,89677	0,002	0,01760	0,00401	-0,00080
Aradeo	8468333,5519	12,85257	0,60870	12,85257	0,009	0,01579	-0,00153	0,00083
Arnesano	13386723,4324	7,03125	0,73913	7,03125	0,003	0,01705	0,00239	0,00048
Bagnolo del Salento	6653405,9234	0,93393	0,47826	0,93393	0,003	0,01844	0,00265	0,00045
Botrugno	9618624,2141	2,45189	0,52174	2,45189	0,009	0,02003	0,00568	0,00063
Calimera	11031459,5535	1,05801	0,56522	1,05801	0,004	0,01668	0,00237	0,00047
Campi Salentina	45276609,8076	1,04495	0,73913	1,04495	0,004	0,01912	0,00296	0,00040
Cannole	20072114,0995	0,57568	0,60870	0,57568	0,004	0,02068	0,00222	0,00073
Caprarica di Lecce	10565602,4192	1,08313	0,52174	1,08313	0,005	0,01766	0,00113	0,00062
Carmiano	23931009,5048	2,67924	0,73913	2,67924	0,008	0,01688	-0,00053	-0,00019
Carpignano Salentino	48326328,1388	1,01346	0,65217	1,01346	0,006	0,02348	0,00425	0,00310
Casarano	38216760,1990	4,76851	0,86957	4,76851	0,013	0,01546	-0,00537	0,00064
Castri di Lecce	12773205,8175	1,76170	0,56522	1,76170	0,003	0,01830	0,00316	0,00044
Castrignano de' Greci	9495537,2028	3,59025	0,56522	3,59025	0,013	0,02385	0,01006	0,00204
Castrignano del Capo	20456858,3299	1,59158	0,78261	1,59158	0,008	0,02018	0,00659	0,00045
Castro	4494352,2066	2,83677	0,56522	0,11508	0,001	0,01760	0,01899	0,00534
Cavallino	22356930,6851	1,96172	0,86957	2,83677	0,026	0,03217	0,01475	0,00172
Collepasso	12620834,6728	1,84256	0,65217	1,96172	0,019	0,02475	-0,00020	0,00237
Copertino	57780794,6480	5,05641	0,82609	1,84256	0,008	0,01991	0,01214	0,00110
Corigliano d'Otranto	28035640,4848	1,31092	0,78261	5,05641	0,014	0,02483	0,00034	0,00005
Corsano	8965028,2543	5,23093	0,52174	1,31092	0,001	0,01670	0,00761	0,00226
Cursi	8244774,2831	7,86118	0,60870	5,23093	0,011	0,02114	0,00130	0,00025
Cutrofiano	56057491,9147	1,57117	0,73913	7,86118	0,005	0,01937	-0,00006	0,00159
Diso	11266761,8408	2,14488	0,60870	1,57117	0,003	0,01952	0,00174	-0,00026
Gagliano del Capo	16365664,8354	4,25864	0,73913	2,14488	0,001	0,01700	0,00824	0,00276
Galatina	81569012,4479	8,56285	0,82609	4,25864	0,018	0,02539	-0,00290	0,00078
Galatone	46457225,9517	3,34356	0,78261	8,56285	0,011	0,01855	0,00001	0,00122
Gallipoli	40491508,7269	0,33407	0,95652	3,34356	0,003	0,01882	0,00541	0,00059
Giuggianello	10126922,8934	0,82816	0,47826	0,33407	0,004	0,02117	-0,00020	0,00029
Giurdignano	13851037,4583	0,37446	0,56522	0,82816	0,002	0,01782	0,01617	0,00223
Guagnano	37493155,8908	2,28945	0,69565	0,37446	0,019	0,02724	0,00419	0,00148
Lecce	237588845,0730	2,08983	1,00000	2,28945	0,007	0,02205	0,01234	0,00462
Lequile	36325676,3837	1,12920	0,78261	2,08983	0,024	0,02917	0,00354	0,00055
Leverano	48876744,9363	2,70834	0,73913	1,12920	0,004	0,01843	0,00481	0,00210
Lizzanello	25080794,8196	9,96533	0,73913	2,70834	0,008	0,02204	0,00731	0,00125
Maglie	22360157,6628	2,64961	0,82609	9,96533	0,010	0,02142	0,01398	0,00286
Martano	21950569,0807	1,43534	0,73913	2,64961	0,017	0,02814	0,00516	0,00142
Martignano	6405923,8247	6,40670	0,60870	1,43534	0,007	0,02171	0,00773	0,00663
Matino	26280523,9774	1,26722	0,73913	6,40670	0,021	0,02680	0,00062	0,00150
Melendugno	91015806,6998	2,30809	0,91304	1,26722	0,005	0,01935	-0,07016	-0,00274
Melissano	12384453,0857	4,16019	0,60870	2,30809	0,017	-0,00191	0,02043	0,00613
Melpignano	10954353,6902	1,33162	0,73913	4,16019	0,032	0,03086	0,00251	0,00014
Miggiano	7693757,8198	0,49708	0,52174	1,33162	0,002	0,01694	0,00109	0,00118
Minervino di Lecce	17880683,6673	5,08488	0,56522	0,49708	0,007	0,02106	-0,00156	0,00031
Monteroni di Lecce	16526933,3747	3,05133	0,65217	5,08488	0,010	0,01494	0,00278	0,00150
Montesano Salentino	8413180,0142	0,45207	0,56522	3,05133	0,005	0,02019	0,00360	0,00216
Morciano di Leuca	13377816,1187	4,69173	0,69565	0,45207	0,008	0,02094	0,00524	0,00168
Muro Leccese	16540732,4167	1,64408	0,69565	4,69173	0,010	0,02394	0,00394	0,00162
Nardò	190705871,8090	3,75955	1,00000	1,64408	0,011	0,02333	0,00710	0,00164
Neviano	16089090,0154	1,25300	0,69565	3,75955	0,009	0,02160	0,01038	0,00292
Nociglia	10976534,1306	4,72208	0,65217	1,25300	0,012	0,02662	0,01400	0,00163
Novoli	17815221,3467	1,53527	0,69565	4,72208	0,020	0,02400	0,00625	0,00120
Ortelle	10085431,0087	1,09173	0,65217	1,53527	0,008	0,02262	0,00249	0,00046
Otranto	76044991,2629	0,58425	0,78261	1,09173	0,005	0,02188	0,00087	0,00001

Table 2. (*Continued*)

NAME	Area (mq)	TE	Nuds	S-Nuds	DF	EI	PS	RS
Palmariggi	8850502,2211	12,58651	0,43478	0,58425	0,000	0,01856	0,00639	0,00287
Parabita	20816850,2108	0,68788	0,82609	12,58651	0,021	0,02250	-0,00021	0,00056
Patù	8560944,8069	1,52111	0,56522	0,68788	0,001	0,01786	0,00831	0,00214
Poggiardo	19687594,9894	0,03220	0,82609	1,52111	0,014	0,02563	0,00329	0,00013
Porto Cesareo	34223918,2472	2,53112	0,73913	1,40126	0,003	0,01942	-0,00317	0,00134
Presicce	24033936,3092	2,51970	0,65217	0,03220	0,006	0,01959	0,00432	0,00079
Racale	23959262,0639	0,63662	0,82609	2,53935	0,016	0,01802	0,01235	0,00175
Ruffano	39196047,0516	0,94145	0,78261	2,51970	0,008	0,01704	0,00288	0,00112
Salice Salentino	59032399,4029	1,47887	0,65217	0,63662	0,014	0,02519	0,01484	0,00154
Salve	32589321,7102	8,85045	0,91304	0,94145	0,007	0,02065	0,00358	0,00186
San Cassiano	8656326,1468	2,43240	0,56522	2,21760	0,024	0,03086	0,00718	0,00322
San Cesario di Lecce	7983349,1013	4,49870	0,82609	8,85045	0,010	0,01847	-0,00023	-0,00044
San Donato di Lecce	21293996,3198	2,02178	0,69565	2,43240	0,012	0,02526	0,00331	0,00056
San Pietro in Lama	8091639,0014	0,42783	0,56522	2,02178	0,005	0,01899	0,00214	0,00330
Sanarica	12846086,4453	2,51067	0,65217	1,47887	0,016	0,02716	0,00767	0,00144
Sannicola	27282660,1841	5,43806	0,73913	4,49870	0,004	0,01666	0,00145	0,00102
Santa Cesarea Terme	26448639,1935	9,27950	0,82609	0,42783	0,007	0,02651	0,00587	0,00212
Scorrano	34857513,2395	5,32954	0,69565	2,51067	0,015	0,02441	0,02635	0,00644
Seclì	8661850,4612	1,19213	0,65217	5,43806	0,014	0,01963	0,00189	0,00025
Sogliano Cavour	5260261,1543	2,31056	0,69565	9,27950	0,011	0,02100	0,00768	0,00335
Soleto	30054202,5049	0,59484	0,78261	5,32954	0,033	0,03666	0,00139	0,00000
Specchia	24757549,9055	1,78388	0,73913	1,19213	0,002	0,01772	0,00614	0,00011
Spongano	12255930,3630	1,52881	0,73913	2,31056	0,013	0,02540	0,00107	-0,00403
Squinzano	29236755,8115	1,26149	0,78261	0,59484	0,003	0,01603	-0,00036	0,00266
Sternatia	16543120,6372	0,96883	0,65217	1,78388	0,006	0,02178	0,00530	0,00135
Supersano	35920087,3677	1,40524	0,69565	1,52881	0,005	0,01449	-0,00344	0,00079
Surano	8866732,1470	1,52700	0,65217	1,26149	0,013	0,02160	0,00048	0,00071
Surbo	20513850,0384	2,09402	0,69565	0,96883	0,007	0,01991	0,00610	0,00074
Taurisano	23364642,6114	2,79297	0,69565	1,40524	0,007	0,01589	0,00125	0,00021
Taviano	21840098,1666	3,06377	0,69565	1,52700	0,012	0,01708	0,00418	0,00064
Tiggiano	7605350,6872	4,31910	0,69565	2,09402	0,011	0,02012	0,00140	0,00186
Trepuzzi	23132829,7360	0,38096	0,78261	2,79297	0,002	0,01602	-0,00367	-0,00154
Tricase	42729396,2031	0,71348	0,86957	3,06377	0,005	0,02023	0,00186	0,00157
Tuglie	8390897,1163	1,06750	0,69565	4,31910	0,008	0,01809	0,00669	0,00166
Ugento	98999657,0221	1,40189	0,86957	0,38096	0,011	0,01576	0,00118	0,00026
Uggiano la Chiesa	14259299,7173	4,07779	0,56522	0,71348	0,007	0,02072	0,01516	0,00512
Veglie	61527355,9970	2,21760	0,69565	1,06750	0,010	0,02196	0,01732	0,00439
Vernole	60372218,8856	1,15060	0,78261	1,40189	0,003	0,01939	0,00014	0,00014
Zollino	9822349,9758	1,40126	0,69565	4,07779	0,021	0,03226	0,00149	0,00110

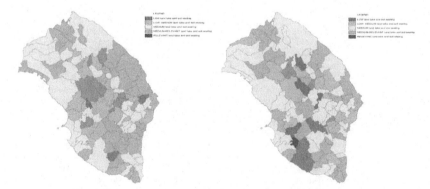

Fig. 5. Representation of the distribution of the quantitative (left) and qualitative (right) values of land take composite indicators, relative to municipalities in the province of Lecce

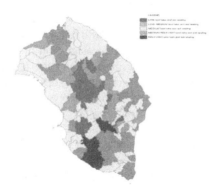

Fig. 6. Representation of the distribution of the composite multidimensional land take, relative to municipalities in the province of Lecce

4   Final Remarks and Perspectives of Research

The sensitivity analysis, carried out by assigning different weight vectors, demonstrates the validity and reliability of the methodology, because of stable results when changing weights values. In all the cases analysed, while changing the weight vector, the output overall distribution remains unchanged.

However, as shown in the maps, the indicator TE does not particularly affect the calculation. Indeed, the location of the denser built fabric surface corresponds to the municipalities of Cutrofiano, Galatone, Parabita, Galatina and Soreto. From the final analysis, however, these areas are not the "ideal negative solution", but they show an intermediate value of land take, compared all other municipalities in the province. It is necessary to remark that the composite indicator is relative and not absolute, since it is obtained from the comparisons among the municipalities, as a function of "distance" from the "ideal solution".

A careful and deep analysis for the validation of quantitative and qualitative land take assessment methodologies through their application to real case studies is the mandatory starting point for an adequate knowledge base in supporting regional and urban planning and decision-making.

The effectiveness of any policy of containment of the changes of use that determine degradation and/or loss of soils, indeed, can be evaluated and monitored only if based on the availability of data use and land cover that are up to date, comparable and scalable to different levels within which they operate choices on territorial government.

The problem on the quantification and monitoring of numerical data is given by the lack of updated and unique data on the dynamics and phenomena of the growth of urban areas; the lack of the dissemination of a shared methodology and objective analysis; issues of land government from different levels and skills; different scales of measurement adopted for the experiments carried out at regional and national level.

In Italy, there are not accessible databases sufficiently accurate informing about current and retrospective land-use trends. In particular, the studies on the "bad" use of land conducted by a number of government and no-government institutions (Legambiente, INU-Italian Institute for Urbanism, ISPRA- Urban Institute for the Protection and Environmental Research) focus on the measurement of a single typology of data, such as the land-cover, without taking into account the multiple factors that result from the land conversion from natural/agricultural to artificial/human.

The proposed methodology can be used, in this regard, for the construction of new scenarios for new policies of urban expansion; therefore, the composite indicator can support the planning process and the evaluation and monitoring of land-use trends.

The metrics we proposed for the assessment of land take and soil sealing are based on an institutional reference in Italy; for future research works, we intend to tackle with the use if Landscape Metrics for the assessment of land take, based on common methodological frameworks in the landscape ecology literature, in order to compare the new results with the previous one, thus refining the proposed model if possible.

The goal for future research perspectives is to structure new knowledge on the direct and indirect impacts of land-use change which are reflected in:

• Economy, in terms of transport and energy costs and reduction of agricultural production;

• Hydro-geology, in terms of land destabilization, irreversibility of land-use and hydraulic structural alteration;

• Climate change effects, in terms of carbon storage and spatial propagation of physic-chemical alterations;

• Ecology and biology, in terms of physical erosion and habitat destruction, fragmentation of ecosystems, loss of ecosystem services and, in general, reduction of the overall regional ecological resilience.

Acknowledgements. The paper is part of the experimental activities of MITO Lab, during the MITO-Multimedia Information for Territorial Objects Projectwork, funded by the Cohesion Action Plan 2007-2013 of European Union, at the Department of Civil Engineering and Architecture of Polytechnic of Bari.

References

1. Breheny, M., Rookwood, R.: Planning the sustainable city region. In: Blowers, A. (ed.) Planning for a sustainable environment, pp. 150–189. Earthscan, New York (1993)
2. Williams, K., Burton, E., Jenks, M.: Achieving the compact city through intensification: an acceptable solution?. In: Jenks, M., Burton, E., Williams, K. (eds.) The Compact City. A sustainable urban form? pp. 71–83, E&FN Spon (1996)
3. Gulinck, H.: Neo-rurality and multifunctional landscapes. In: Multifunctional Landscapes, Brandt, J., Vejre, H. (eds.) vol. 1: Theory, Values and History, vol. 14 of Advances in Ecological Sciences, pp. 63–74. WIT Press, Southampton (2004)
4. Hite, J.: Land Use Conflicts on the Urban Fringe: Causes and Potential Resolution, Clemson, SC (Strom Thurmond Institute, Clemson University) (1998). URL: http://www.strom.clemson.edu/publications/hite/landuse-hite.pdf
5. Fulton, W., Pendall, R., Nguyen, M., Harrison, A.: Who Sprawls Most? How Growth Patterns Differ Across the U.S. In: Survey Series, Washington, DC (The Brookings Institution) (2001). Related online version: http://www.brookings.edu/es/urban/publications/Fulton.pdf

6. Wolman, H., Galster, G., Hanson, R., Ratcliffe, M., Furdell, K., Sarzynski, A.: The fundamental challenge. In: Measuring sprawl: which land should be considered? (2005)
7. Allen, C.R.: Sprawl and the Resilience of Humans and Nature: an Introduction to the Special Feature. Ecology and Society 11(1), 36 (2006)
8. European Environmental Agency - EEA: Urban sprawl in Europe: The ignored challenge. EEA Report, 10/2006, Copenhagen (European Environmental Agency) (2006). related online version: http://reports.eea.europa.eu/eea_report_2006_10
9. Adell, G.: Theories and models of the peri-urban interface: a changing conceptual landscape, London (Development Planning Unit, University College London) (1999). Related online version: http://www.ucl.ac.uk/dpu/pui/research/previous/epm/g_adell.htm
10. Allen, A., D'avila, J.: Mind the gap! Bridging the rural-urban divide. In: id21 insights, 41 (2002). URL: http://www.id21.org/insights/insights41/insights-issu01-art00.html
11. Allen, A.: Environmental Planning and Management of the Peri-Urban Interface: Perspectives on an Emerging Field. Environment & Urbanization 15(1), 135–148 (2003). doi:10.1177/095624780301500103
12. Antrop, M., Van Eetvelde, V.: Holistic Aspects of Suburban Landscapes: Visual Image Interpretation and Landscape Metrics. Landscape and Urban Planning 50(1–3), 43–58 (2000). doi:10.1016/S0169-2046(00)00079-7
13. Meeus, S.J., Gulinck, H.: Semi-Urban Areas in Landscape Research: A Review. In: Living Reviews in Landscape Research (2008). ISSN 1863-7329: http://www.livingreviews.org/lrlr-2008-3
14. Cerreta, M., Malangone, V.: Valutazioni multi-metodologiche per il Paesaggio Storico Urbano: la Valle dei Mulini di Amalfi. BDC, Bollettino del centro Calza Bini 1, 39–60 (2014)
15. Cerreta, M., Poli, G.: A Complex Values Map of Marginal Urban Landscapes: An Experiment in Naples (Italy). International Journal of Agricultural and Environmental Information Systems 4(3), 41–62 (2013). doi:10.4018/ijaeis.2013070103
16. Balena, P., Sannicandro, V., Torre, C.M.: Spatial Analysis of Soil Consumption and as Support to Transfer Development Rights Mechanisms. In: Murgante, B., Misra, S., Carlini, M., Torre, C.M., Nguyen, H.-Q., Taniar, D., Apduhan, B.O., Gervasi, O. (eds.) ICCSA 2013, Part IV. LNCS, vol. 7974, pp. 587–599. Springer, Heidelberg (2013)
17. Attardi, R., Cerreta, M., Franciosa, A., Gravagnuolo, A.: Valuing cultural landscape services: a multidimensional and multi-group SDSS for scenario simulations. In: Murgante, B., Misra, S., Rocha, A.M.A., Torre, C., Rocha, J.G., Falcão, M.I., Taniar, D., Apduhan, B.O., Gervasi, O. (eds.) ICCSA 2014, Part III. LNCS, vol. 8581, pp. 398–413. Springer, Heidelberg (2014)
18. Perchinunno, P., Rotondo, F., Torre, C.M.: The evidence of links between landscape and economy in rural park. International Journal of Agricultural and Environmental Information Systems 3(2), 72–85 (2012)
19. Barberis, R.: Consumo di suolo e qualità dei suoli urbani. In: Qualità dell'ambiente urbano – II Rapporto Apat, APAT 2005 – pp. 703–729 (2005)
20. ARPA Emilia-Romagna: Consumo di Suolo, in Gestione delle risorse naturali e dei rifiuti; Relazione sullo stato dell'ambiente della Regione Emilia-Romagna, pp. 618–644. (2009)
21. Munda, G., Nijkamp, P., Rietveld, P.: Qualitative multicriteria evaluation for environmental management. Ecological Economics 10(2), 97–112 (1994)
22. Munda, G.: Cost-benefit analysis in integrated environmental assessment: some methodological issues. Ecological Economics 19(2), 157–168 (1996)
23. Gough, C., Castells, N., Funtowicz, S.: Integrated Assessment: an emerging methodology for complex issues. Environmental Modelling and Assessment 3, 19–29 (1998)

24. Ravetz, J.: Integrated assessment for sustainability appraisal in cities and regions. Environmental Impact Assessment Review **20**(1), 31–64 (2000)
25. Morano, P., Locurcio, M., Tajani, F., Guarini, M.R.: Urban redevelopment: a multi-criteria valuation model optimized through fuzzy logic. In: Murgante, B., Misra, S., Rocha, A.M.A.C., Torre, C., Rocha, J.G., Falcao, M.I., Taniar, D., Apduhan, B.O., Gervasi, O. (eds.) Computational Science and Its Applications - ICCSA 2014, Guimaraes 2013. LNCS, vol. 8581, pp. 161–175. Springer, Switzerland (2014)
26. O'Neill, R.V., Krummel, J.R., Gardner, R.H., Sugihara, G., Jackson, B., DeAngelis, D.L., Graham, R.L.: Indices of landscape pattern. Landscape Ecology **1**(3), 153–162 (1988). doi:10.1007/BF00162741
27. Rosini, M.D.: Applications. In: Rosini, M.D. (ed.) Macroscopic Models for Vehicular Flows and Crowd Dynamics: Theory and Applications. UCS, vol. 12, pp. 217–226. Springer, Heidelberg (2013)
28. Gomez-Sal, A., Belmontes, J.A., Nicolau, J.M.: Assessing landscape values: a proposal for a multidimensional conceptual model. Ecological Modelling **168**, 319–341 (2003)
29. Feld, C.K., et al.: Assessing and monitoring ecosystems—indicators, concepts and their linkage to biodiversity and ecosystem services. In: RUBICODE (Ed.), Rationalising Biodiversity Conservation in Dynamic Ecosystems., p. 108 (2007). http://www.rubicode.net/rubicode/RUBICODE Review on Indicators.pdf
30. Schwarz, N.: Urban form revisited. Selecting indicators for characterizing European cities. Landscape and Urban Planning **96**, 29–47 (2010)
31. Frank, S., Fürst, C., Koschke, L., Makeschin, F.: A contribution towards a transfer of the ecosystem service concept to landscape planning using landscape metrics. Ecological Indicators **21**, 30–38 (2012)
32. Hermann, A., Kuttner, M., Hainz-Renetzeder, C., Konkoly-Gyurò, E., Tiraszi, A., Brandenburg, C., Allex, B., Ziener, K., Wrbka, T.: Assessment framework for landscape services in European cultural landscapes: An Austrian Hungarian case study. Ecological Indicators **37**, 229–240 (2014)
33. Balena, P., Sannicandro, V., Torre, C.M.: Spatial multicriterial evaluationof soil consumption as a tool for SEA. In: Murgante, B., Misra, S., Rocha, A.M.A.C., Torre, C., Rocha, J.G., Falcao, M.I., Taniar, D., Apduhan, B.O., Gervasi, O. (eds.) Computational Science and Its Applications - ICCSA 2014, Springer International Publishing, pp. 446–458 (2014)
34. Socco, C., Cavaliere, A.: Il bordo della città, Osservatorio Città Sostenibili - Dipartimento Interateneo Territorio - Politecnico e Università di Torino, Working Paper P09/07 (2007)
35. Burkhard, B., Kroll, F., Müller, F., Windhorst, W.: Landscapes Capacities to Provide Ecosystem Services. A Concept for Land-Cover Based Assessments. In: Landscape Online 15, 1–22, p. 6. (2009). doi:10.3097/LO.200915
36. Hwang, C.L., Yoon, K.: Multiple Attribute Decision Making: Methods and Applications, New York. Springer-Verlag, USA (1981)

Social Balance and Economic Effectiveness in Historic Centers Rehabilitation

Carmelo Maria Torre[1,2] (✉), Pierluigi Morano[1,2], and Francesco Taiani[1,2]

[1] MITO Labs, DICAR, Polytechnic of Bari, Bari, Italy
carmelomaria.torre@poliba.it
[2] Dipartimento Di Architettura, University Federico II Naples, Naples, Italy

Abstract. The growing need to support financially the processes of urban regeneration of city centers clashes with the limited availability of public resources. Administrations are therefore forced to evaluate the priority areas of intervention, on the one hand trying to pursue goals of social equity, other actions to promote efficient financial plan. Consequently the reference institutional policy of intervention is based on regulatory frameworks that require a closer integration of programming needs of the allocation of resources, and social needs. The chapter shows an example of conciliation among the seek for efficiency and for social equality in choosing priority of intervention in the urban make up of historic centers.

Keywords: Multicriteria evaluation · Cost effectiveness · Equity · Gis

1 Accountability, Social Equity, Effectiveness in Rehabilitation plans

Development plans require a serious monitoring of policies through a consistent evaluations of results. Local governments, in order to increase governance, should set some decision points on the ability to preview the search for coherence by the new program tools. An useful example of preview method is the definition of appropriate indicators, describing the condition of social disease and environmental degradation, for a valuation method that support both land use or property plans and territorial programming, al a sort of "planology process" [1] [2], finding opportunity of land-use but looking at efficiency.

The evaluation of alternatives and choices should be made on a concept of accountability and transparency, based on principles of efficiency and equity, that can accompany not only the ex ante phases, but also the on-going and ex post evaluation of the implementation of policies [3]

In brief, we require tools to monitor and forecast results, in the light of the resources allocation; therefore, in order to check the consistency of the results a set objectives of transparency, fairness and economic sustainability should be set [4]

This search implies, in the case of the redevelopment of historic fabrics degraded, to look for methods based on cost-effectiveness, which makes visible spent efforts,

O. Gervasi et al. (Eds.): ICCSA 2015, Part III, LNCS 9157, pp. 317–329, 2015.
DOI: 10.1007/978-3-319-21470-2_22

achieved effects and measures in the light of the reduction of frequent imbalances in the nucleus of our ancient cities [5][6].

2   The Case of Study

The case study regards the assessment of intervention in the case of the urban renewal in a middle size historic center in Apulia.

The center of Monopoli is a medium-size city (about 45000 inhabitants), on the eastern coast of Bari, Apulia, and has an old town affected by various forms of degradation. Significant parts of the public and private architectural historic heritage need works, unbalanced with a significant scarcity of resources that the City Council may make available to any incentive.

If we look at the old town centre of Monopoli we can recognize the condition defined by the notation of the Italian Ministry of Public Works (dated 22/10/1997), that foster the concepts and establishes the issues of some development plans defined "Contracts for Neighbourhoods": " Neighbourhoods are often marked by widespread degradation of the built and the urban environment and by lack of services in a context of limited social cohesion and of clear criticality of housing ".

The evaluation of projects to submit for funding is based on criteria defined by several national calls, to which usually are integrated by criteria provided by Regions.

The comparison among cost and effectiveness supported the decision making and provided indications about which actions, in front of equal financial resources deployed, can pursue better targets for the reduction of the physical and social degradation in urban contexts. This information is useful not only when the public bodies have to operate with their own funds, but also when they have the need to motivate to higher-level institutions (and generally speaking all superior sources of funding) the effectiveness of the measures for which the financial support is requested.

An assessment of sustainability must somehow make consistent objectives of efficiency and social equity. In the case of the requalification on urban architectural heritage, you get trying to bring together the results in reducing the substantial deterioration in the physical environment and in the social context [7] [8].

A preliminary analysis becomes a crucial point, when it is aiming at identifying
-    the dimension of the resources to be used,
-    the conditions of the relevant social questions,
-    the conditions of the relevant physical degradation,

The relevant aspects for the physical and social degradation generate a set of evaluation criteria for a multi-dimensional assessment giving as result an index of physical degradation and an index of social deprivation.

The priority is to act where the indexes assume their relevant values.

For the purposes of the construction of the evaluation framework (Figure 1), the phases of were the following:
-    developing a survey methodology to identify and quantify the changes and situations of degradation of the historical heritage;
-    determining the costs of recovery, as measure of the cost effectiveness;

- developing a method to identify priorities for interventions supporting decisions in the light of social equity and transparency.

Create a balance between cost, and effectiveness firstly in terms in term of physical rehabilitation against decay and secondly in favour of social revitalisation.

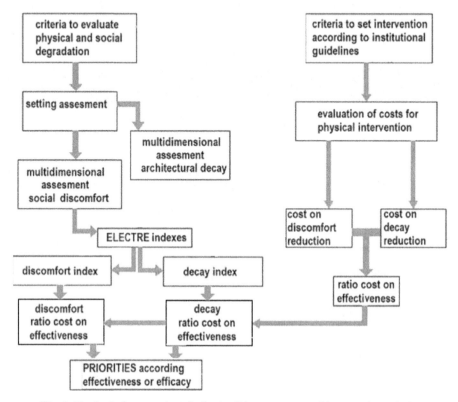

Fig. 1. The logic framework at the basis of the assessment of intervention priority

2.1   Geography of Physical and Social Criticalities

In order to organize data, a Geographic Information System has been set, containing a database of the historical Center [9].

In such database, photographic references, attributes collected inside schedules of "architectural components", and inside schedules of the "improper changes", are merged, as well as links with the stock image.

As regards details of the degradation, two sets were built.

The first set, that is the more detailed, collect data at two levels: the first refers to the small unit of cadastral parcels, the second, broader, refers to the minimum urban block, coincident with the geographic section of the Italian Population Census.

The second set, that is the broader, collect data at the level of the geographic section of the Italian Population Census.

The physical degradation and the social discomfort conditions at the end were mapped by the use of ArcMap. The geographical unit was the population census areas. Inside the urban fringe, the area was covered by census section numbered from 89 to 107.

As regards the physical cost of degradation each one of the section the absolute cost per category of refurbishment was considered as attribute, As regards the social discomfort the attribute was defined by the set of indexes of social discomfort.

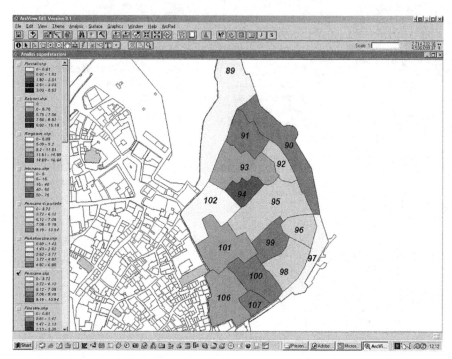

Fig. 2. The geographic context of the historic center of the town subdivided for census sections (from 89 to 107)

2.2   Cost of Urban Refurbishment and Social Disconfort

In this regard, reference was represented by the "morphological Quality" that Guidelines of District Contracts trial in the case of the "preservation and enhancement of historic fabrics."

According to such objective the preservation of the significant architectural organism, in its aspects of historical development, "guaranteeing the permanence of the figurative and material consistency is the main task. The interventions in the historic centre must comply with the following design criteria:

-    elimination of superfluous degradation; modification of the adjunctions, conservation of the original plasters, the paintworks and existing decoration (e.g. frescoes);

- preservation of the original external openings, according to their shape and position;
- construction of new windows made on the basis of schedules, featuring scores and configurations consistent with the original pre-existing typologies.

In order to identify the average cost of intervention an investigation about the typology of alteration of the built heritage has been provided. Abacuses of architectural components and their types of alterations in the built environment have been the basis of classificatory attributes of cost in each geographical units.

The obtained data were structured in a base of knowledge collecting 196 elements of cost, corresponding to the different type of intervention, and 1120 fabrics laying on parcels inside the old town.

The old town has no relevant problems of static consolidation. The urban landscape instead highlights several alterations in architectural facades, due to the insertion of improper handling, or inadequate maintenance of the buildings.

With regard especially to the architectural heritage of the private property, the restoration urban, represents a significant part of the intervention overall rehabilitation of the historic center.

The survey is divided into the following phases:
- photographic survey of the prospects of the buildings;
- construction of an "Abacus of architectural components";
- construction of an "Abacus of alterations";
- calculation of the area of land parcels and elevation of buildings affected by degradation;
- creating a database about the degradation of the external surfaces.

Thanks to the versatility offered by the used support, it was possible to recombine and aggregate data, so that they were suitable for use in relation to the goals. By queries we could examine any context through spatial relationships between intervention for each fabric.

Inside the schedule architectural elements are collected: some types of doors, windows, railings and architectural elements, to be used as models for public and private interventions to follow.

In addition to census of the valuable elements, you must also identify the factors that lead to the disqualification of the historic centre. This classification is helpful to assess the costs in the subsequent removal.

Similarly to what was done for the architectural components, were prepared then an archive called "Abacus of alterations": used non-traditional materials, colors unsuitable environment, elements of modern fashion and all those transformations that have altered the architecture of the buildings.

3   Indexing Physical Impacts and Social Impacts on Expenditure

The matrices of concordance show the results of the pairwise comparisons. They are two (reported below).

The concordance is obtained according to what Roy suggests [10], by the use of the concordance index for each alternative (ie each section enumeration).

The matrix in formula (1) is a performance table according m criteria (from A1 to Am) referring to n options. In our case criteria are related with value of the intervention, and values of the intensity of social discomfort in the area.

$$
\begin{array}{c}
& A1 & & Am \\
X_1 \\
X_2 \\
\vdots \\
X_n
\end{array}
\begin{bmatrix}
U_{X_1}(A1) & & U_{X_1}(Am) \\
U_{X_2}(A1) & & U_{X_2}(Am) \\
& & \\
& & \\
& & \\
U_{X_n}(A1) & & U_{X_n}(Am)
\end{bmatrix}
\tag{1}
$$

The Concordance index is the average of the pairwise comparison between the n options for each one of the m criteria (divided for n-1 pairwise comparison), where we assign the weight W of the k criterion to the winning pairwise comparison for the options Xi (i=1 to n).

$$
C_+ = \frac{\displaystyle\sum_{k \in I_{ij}^+ \cup I_{ij}^=} W_k \quad k=1\dots m}{n-1}
\tag{2}
$$

In our case the Intersections represent the prevalence of generic section of the census on generic section of census j, by weighed relating according to a chosen set of different criteria. The final index Ii is the concordance index obtained as the average of the heads of each section, or as the average of partial intersections reported in each row (corresponding to the value determined in the formula given above).

We have calculated two different concordance index:

- the first shows the priority according the seek for a best result in terms of physical refurbishment;
- the second shows the priority according of support looking to the most difficult social conditions, that could be in contrast with the need of refurbish of the weakest households living in the historic centre of the town.

If we privilege the first index, we can obtain a better result in terms of amount of decay that is retired (table 3).

Table 1. The table of effects in terms of incidence of costs for each intervention (economic criterion) for each sections (alternative)

criteria	% cost winters	% wooden winters	% internal doors	% external doors	% structured facade	% structured doors	% steel works	% paint works	% plaster works	% balconies	% roofs	% pvc/plastic works
weights	0,087	0,087	0,087	0,087	0,043	0,043	0,130	0,087	0,130	0,087	0,043	0,087
Sections												
89	0,00	3,26	1,43	4,89	4,89	0,61	5,09	0,00	381,13	5,30	0,82	0,61
90	3,38	8,11	3,04	7,10	13,86	3,72	13,18	5,07	402,30	8,11	4,39	0,00
91	1,47	9,19	4,78	8,46	6,25	1,84	12,87	1,47	469,89	6,62	1,84	0,00
92	2,09	5,23	2,62	6,80	3,14	2,09	11,51	2,09	521,47	5,75	6,80	2,62
93	1,11	7,08	4,87	10,40	0,66	1,55	10,40	0,22	493,37	4,65	2,66	1,77
94	0,61	10,94	1,82	12,76	0,61	3,04	16,41	0,61	726,64	7,29	6,08	1,82
95	2,84	5,52	3,01	11,04	10,21	1,51	9,20	3,18	414,17	8,20	1,00	2,51
96	4,71	6,12	3,77	16,49	3,30	0,00	10,36	1,41	630,33	6,59	16,96	1,41
97	0,00	0,00	2,29	0,00	10,30	3,43	4,58	0,00	600,69	0,00	0,00	0,00
98	5,03	5,82	4,24	7,41	3,44	0,26	7,94	2,12	502,84	7,41	5,03	2,91
99	1,18	7,87	3,54	8,65	0,39	3,93	14,16	3,54	716,53	7,08	1,97	2,75
100	2,02	8,41	3,03	8,41	2,35	2,69	9,08	4,37	687,73	4,37	3,70	2,35
101	1,26	6,55	6,05	11,60	4,54	4,03	16,64	2,02	572,90	8,82	1,51	3,03
102	4,29	3,72	2,86	4,86	2,29	1,43	10,59	0,29	642,49	3,72	1,14	2,29
106	2,13	6,62	6,85	10,16	4,96	2,60	14,89	2,13	630,37	7,56	4,02	2,13

It was analysed the uncomfortable condition of the inhabitants to privilege, in identifying priorities for action, socially weaker families.

The disease conditions were investigated by evaluating the way some situations deemed unfavourable are distributed in the area. Then they were taken into accounting the following structural characteristics of the population:

- people over 65 years and under 14;
- people looking for their first job and unemployed;
- number of entrepreneurs and professionals;
- number of employees, employers and self-employed workers;
- the number of families in relation to the size of the household;
- housing in relation to the property deed;
- territorial density (overcrowding).

Table 2. The table of effects in terms of incidence of costs for each social condition of households (social criterion) for each sections (alternative)

criteria	weak people (young elderly)	unemployed	middleclass	low income	renters households	overcrowd households
weights	0,27	0,09	0,09	0,09	0,27	0,19
Sections						
89	0,258	0,146	0,112	0,258	0,429	0,103
90	0,375	0,125	0,067	0,250	0,438	0,370
91	0,385	0,154	0,044	0,187	0,300	0,222
92	0,292	0,111	0,125	0,194	0,179	0,333
93	0,298	0,094	0,070	0,275	0,215	0,164
94	0,406	0,125	0,125	0,234	0,438	0,083
95	0,437	0,111	0,044	0,252	0,362	0,119
96	0,275	0,157	0,059	0,196	0,321	0,118
97	0,429	0,143	0,000	0,286	0,083	0,000
98	0,404	0,101	0,061	0,162	0,192	0,081
99	0,349	0,120	0,084	0,229	0,385	0,280
100	0,336	0,175	0,073	0,299	0,400	0,282
101	0,372	0,115	0,071	0,218	0,362	0,289
102	0,265	0,163	0,082	0,245	0,588	0,286
106	0,347	0,150	0,048	0,177	0,235	0,245
107	0,355	0,097	0,065	0,258	0,200	0,273

Table 3. The Concordance index on term of costs effectiveness according the ration between expenditure and decay restored for each section (alternative)

	89	90	91	92	93	94	95	96	97	98	99	100	101	102	106	107	Ii
89	0,000	0,725	0,725	0,362	1,087	0,362	0,000	0,725	4,348	0,725	0,362	1,087	0,362	1,812	0,000	1,087	0,14
90	7,609	0,000	4,348	6,159	5,072	3,623	5,072	3,986	6,522	3,986	3,623	5,072	2,899	5,797	3,986	3,261	0,71
91	7,609	3,261	0,000	4,348	4,710	2,536	4,348	4,710	5,797	3,986	2,536	4,348	2,174	5,797	1,087	2,899	0,60
92	7,971	2,174	3,986	0,000	6,159	3,623	3,623	2,899	6,522	2,899	1,449	3,986	1,812	5,797	1,087	2,174	0,56
93	7,246	3,261	3,623	2,174	0,000	1,812	4,348	3,623	6,522	3,623	2,174	3,261	1,087	3,623	1,449	2,536	0,50
94	7,971	4,710	5,797	4,710	6,522	0,000	4,348	4,710	6,884	4,348	5,072	5,072	2,899	5,797	4,348	4,710	0,78
95	8,333	3,261	3,986	4,710	3,986	3,986	0,000	2,899	6,522	3,986	2,536	4,348	1,812	5,072	3,986	2,536	0,62
96	7,609	4,348	3,623	5,435	4,710	3,623	5,435	0,000	7,609	3,986	2,899	4,710	2,899	5,072	1,812	3,986	0,68
97	2,536	1,087	1,812	1,812	1,812	1,449	1,812	0,725	0,000	1,812	0,362	0,725	1,449	0,725	0,725	1,449	0,20
98	7,609	4,348	4,348	5,435	4,710	3,986	4,348	4,348	6,522	0,000	3,623	3,623	1,812	5,797	1,812	1,812	0,64
99	7,971	4,710	5,797	6,884	6,159	3,261	5,797	5,435	7,971	4,710	0,000	5,435	2,899	7,246	3,623	3,261	0,81
100	7,246	3,261	3,986	4,348	5,072	3,261	3,986	3,623	7,609	4,710	2,899	0,000	3,623	6,522	3,623	1,812	0,66
101	7,971	5,435	6,159	6,522	7,246	5,435	6,522	5,435	6,884	6,522	5,435	4,710	0,000	6,522	3,623	3,261	0,88
102	6,522	2,536	2,536	2,536	4,710	2,536	3,261	3,261	7,609	2,536	1,087	1,812	1,812	0,000	2,536	2,899	0,48
106	8,333	4,348	7,246	7,246	6,884	3,986	4,348	6,522	7,609	6,522	4,710	4,710	4,710	5,797	0,000	2,174	0,85
107	7,246	5,072	5,435	6,159	5,797	3,623	5,797	4,348	6,884	6,522	5,072	6,522	5,072	5,435	6,159	0,000	0,85

Table 4. The Concordance index on term of costs effectiveness according the ration between expenditure and weak social conditions supported for each section (alternative)

	89	90	91	92	93	94	95	96	97	98	99	100	101	102	106	107	Ii
89	0,00	1,07	2,14	3,21	2,14	2,67	2,67	2,14	3,21	3,74	2,67	1,60	2,67	0,53	2,14	2,67	0,35
90	4,81	0,00	3,21	5,88	5,35	2,14	3,21	4,81	2,67	3,74	5,88	4,81	5,88	3,74	4,81	4,81	0,66
91	3,74	2,67	0,00	4,28	5,35	2,14	2,14	3,21	3,21	4,28	2,67	2,14	2,67	2,14	4,81	4,28	0,50
92	2,67	0,00	1,60	0,00	1,60	1,07	1,07	2,67	2,67	2,14	1,07	1,07	1,07	2,67	1,60	1,60	0,25
93	3,74	0,53	0,53	4,28	0,00	2,14	1,60	3,21	2,67	3,21	1,07	0,53	1,07	2,67	0,53	2,14	0,30
94	3,21	1,60	3,74	4,28	3,74	0,00	2,14	3,74	2,67	5,35	4,28	3,21	4,28	1,60	3,74	3,74	0,51
95	3,21	2,67	3,74	4,28	4,28	3,74	0,00	5,35	4,28	5,88	2,67	2,14	2,67	2,67	4,28	4,28	0,56
96	3,74	1,07	2,67	3,21	2,67	2,14	0,53	0,00	3,21	4,28	1,07	0,53	1,07	2,14	2,67	2,67	0,34
97	2,67	3,21	2,67	3,21	3,21	3,21	1,60	2,67	0,00	3,21	3,21	2,14	3,21	2,67	2,67	3,21	0,43
98	3,21	0,00	3,21	4,81	4,81	1,60	3,21	4,81	2,67	3,74	0,00	1,60	2,67	1,60	4,81	3,21	0,46
99	2,14	2,14	1,60	3,74	2,67	0,53	0,00	1,60	2,67	0,00	2,14	2,14	2,14	2,14	1,60	2,67	0,30
100	4,28	1,07	3,74	4,81	5,35	2,67	3,74	5,35	3,74	3,74	4,28	0,00	2,67	3,21	3,74	3,74	0,56
101	3,21	0,00	3,21	4,81	4,81	1,60	1,60	4,81	2,67	3,74	3,21	3,21	0,00	3,21	4,81	4,81	0,50
102	5,35	2,14	3,74	3,21	3,21	4,28	3,21	3,74	3,21	3,74	4,28	2,67	2,67	0,00	3,74	3,21	0,52
106	3,74	1,07	1,07	4,28	5,35	2,14	1,60	3,21	3,21	4,28	1,07	2,14	1,07	2,14	0,00	2,67	0,39
107	3,21	1,07	1,60	4,28	3,74	2,14	1,60	3,21	2,67	3,21	2,67	2,14	1,07	2,67	3,21	0,00	0,39

If we privilege the second, (table 4) we support with public funding firstly the social class that are less able to act by themselves in refurbish their own heritage.

4 Efficacy Effectiveness and Social Justice

The key to choose at this time it will be represented by equity, defined as the rule that enlarge spatially and geographically the effect in terms of social justice.

If we observe the table 5, we can see how the indexes and the costs re defined for each section. We consider most effective the solution that shares the available public fund (about 550.000 euros) in the widest part of the centre.

The first graph (figure 3) represents a classification of areas as a function of the indices of physical degradation and social distress.

A priority for action based on that classification is broadly consistent with a choice of efficacy.

This approach to the choice is desirable in the absence of constraints imposed by the need to allocate scarce resources to priority.

In fact in this case it must be remembered that each field is characterized, even in situations of not high value of the indices of degradation and discomfort from situations however deserving of intervention.

The assessment, in fact, the need to determine the list of priorities in the light of emergencies resulting from the analysis of the context.

The problem of priorities in scarcity of resources is then addressed from the evaluation of efficiency and equity of interventions.

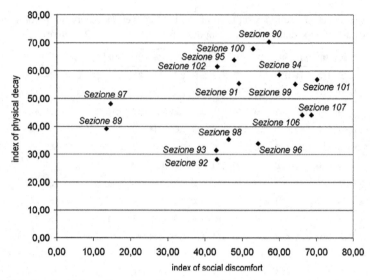

Fig. 3. Priority: the highest discomfort (horizontal axis) on the highest decay (vertical axis)

If we look at the following table, we can see the relationship between cost on physical decay and cost on social discomfort.

The first table lists the sections that have both indices of degradation and discomfort levels (visible at the farthest from the origin of the axes in the diagram degradation physical and social distress), for which the costs of intervention summed amount to approximately EUR 550,000.

The efficiency ratio measures the reduction of the rate of decay per unit monetary spending, while the "index of justice" measures the relationship between social disadvantage and financial implementation features such discomfort.

Table 5. Expenditure of more-less 550.000 euros according to the best couple (highest decay-highest social weakness) in figure 3

Sections	Intervention Costs	Rehabilitation Decay Index	Social weakness Index	Cost Effectiveness	justice index
90	€ 79.112	57,2	75,0	0,72	0,95
94	€ 58.702	60,0	58,0	1,02	0,99
99	€ 80.419	64,3	52,0	0,80	0,65
101	€ 134.001	70,2	56,0	0,52	0,42
106	€ 138.931	66,2	44,0	0,48	0,32
107	€ 51.056	68,7	44,0	1,34	0,86
Total expenditure	€ 542.221,00				
Average		*average* 64,43	*average* 54,83	*Average* 0,81	*average* 0,7

5   Sharing Effects in the Light of Equalisation

The construction of the indices of efficiency and fairness is therefore equivalent to a standard correlation between costs and results of multicriteria analysis in the light of some indication from literature, acrossing in new ways the break even point analysis [8].

The bands in figure 4 identified in the diagrams take into account the trade-off, and then indifference to choose between situations of greater physical degradation and situations of the greatest social problems in equal financial resources used, in which the indifference is given by the ratio of value of each one the other with the purpose of the expenditure of financial resources.

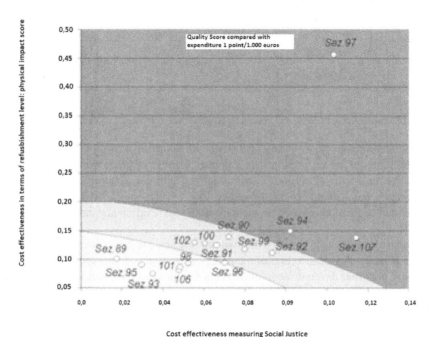

Fig. 4. Values expressing the ratio between score in figure 3, and quantity of expenditure

The second table shows instead the sections that have simultaneously high values of the efficiency index and the index of fairness (equity-efficiency in the diagram are identified by the points belonging to the first and second band of the graph), for which the amount interventions is still estimated at around 550,000 euro.

Table 6. Expenditure of more-less 550.000 euros according to the best couples (lowest ratio of cost on decay- lowest ratio of cost on social weakness) in figure 4

Sections	Intervention Costs	Rehabilitation Decay Index	Social weakness Index	Cost Effectiveness	justice index
90	€ 79.112	57,2	75,0	0,72	0,95
91	€ 46.212	49,1	56,0	1,06	1,21
92	€ 74.194	43,2	28,0	0,58	0,38
94	€ 58.702	60,0	58,0	1,02	0,99
97	€ 11.858	14,7	48,0	1,24	4,05
99	€ 80.419	64,3	52,0	0,80	0,65
100	€ 87.094	52,9	64,0	0,61	0,73
102	€ 77.776	43,3	59,0	0,56	0,76
107	€ 51.056	68,7	44,0	1,34	0,86
Total expenditure	€ 566.423,00				
Average		*average* 50,38	*average* 53,78	*average* 0,88	*average* 1,18

The choice made on the basis of the second priority list allows you to intervene with the same resources employed in a greater number of areas than those related to the first priority list, getting best distribute effects and consequently not only looking at efficiency, but also at fairness and justice.

The final evaluation is based on the definition of an "efficiency index" and a index of justice.

A further comparison between the two approaches can be inferred from the comparison between the mean values of the index of two types of intervention.

Again, note that the first mode of decision favours mainly effectiveness, being the index of physical degradation the only one to take in the first hypothesis value higher than that assumed in the second (64/100 versus 50/100) .

The changes in the index of social disadvantage not seem relevant in the transition from one approach to another (reaching its 54/100).

In the second approach ultimately mean values of the indexes of efficiency and equity related to the areas affected by the intervention are higher than those assumed in the first approach (respectively: 0.88 versus 0.81 and 1.18 against 0.7).

6   Final Remarks

The paper starts with a methodological approach

In conclusion, the study shows how the opportunity to explain the rationale based on the principles of effectiveness, efficiency and equity helpful for improving the character of the decision context. The context reminds to two different economic dilemmas:

Even without ambitious impossible hope of compensatory approaches [11], it is clear the positive contribution of integrated assessment approaches to decision making in providing an advantage in terms of accountability for choice [12][13].

They are in fact comparing two priorities that respond both to paintings of legitimacy, but prefer different principles. In this situation the transparency is fundamental to support the comparison between political views.

It promises a dimension "ethics" in institutional assessment, as well consider in a general perspective of sustainability of choice.

References

1. Archibugi, F.: La città ecologica. Urbanistica e sostenibilità. Bollati Boringhieri, Turin (2002)
2. Khakee, A.: Evaluation and planning: inseparable concepts. Town Planning Review **64**, 359–374 (1998)
3. Hausman, D.M., McPherson, M.S.: Taking Ethics Seriously: Economics and Contemporary Moral Philosophy. Journal of Economic Literature **31**(2), 671–731 (1993)
4. Torre, C., Perchinunno, P., Rotondo, F.: Estimates of housing costs and housing difficulties: an application on italian metropolitan areas. In: Szilrd, K., Balogh, I. (eds.) Housing, Housing Costs and Mortgages: Trends, Impact and Prediction, pp. 93–108. Nova Science, New York (2013)
5. Tronconi, O., Ciaramella, A., Pisani, B., eds.: La gestione di edifici e patrimoni immobiliari. Edizioni Il Sole 24 Ore, Milano, pp. 303–316 (2002)
6. Attardi, R., Pastore, E., Torre, C.M.: "Scrapping" of quarters and urban renewal: a geostatistic-based evaluation. In: Murgante, B., Misra, S., Rocha, A.M.A., Torre, C., Rocha, J.G., Falcão, M.I., Taniar, D., Apduhan, B.O., Gervasi, O. (eds.) ICCSA 2014, Part III. LNCS, vol. 8581, pp. 430–445. Springer, Heidelberg (2014)
7. Attardi, R., Cerreta, M., Franciosa, A., Gravagnuolo, A.: Valuing cultural landscape services: a multidimensional and multi-group SDSS for scenario simulations. In: Murgante, B., Misra, S., Rocha, A.M.A., Torre, C., Rocha, J.G., Falcão, M.I., Taniar, D., Apduhan, B.O., Gervasi, O. (eds.) ICCSA 2014, Part III. LNCS, vol. 8581, pp. 398–413. Springer, Heidelberg (2014)
8. Morano, P., Tajani, F.: Break Even Analysis for the Financial Verification of Urban Regeneration Projects. Applied Mechanics and Materials **438–439**, 1830–1835
9. Malczewski, J.: GIS and Multicriteria Decision Analysis. John Wiley, New York (1999)
10. High-detail damage pattern in towns hit by earthquakes of the past: an approach to evaluate the reliability of the historical sources
11. Gizzi, F.T., Tilio, L., Masini, N., Murgante, B., Potenza, M.R., Zotta, C.: High-detail damage pattern in towns hit by earthquakes of the past: an approach to evaluate the reliability of the historical sources. In: Dan, M.B., Armas, I., Goretti, A. (eds.) Earthquake Hazard Impact and Urban Planning, pp. 105–126. Springer, Berlin (2014)
12. Roy, B.: Metodologie d'Aide a la Decision. Economica, Parigi (1985)
13. Arrow, K.J.: Social Choice and Individual Values. Wiley, New York (1973)
14. Wai-Chung, L.: Neo-Institutional Economics and Planning Theory. Planning Theory **4**, 7–19 (2005)
15. Fusco Girard, L.: Sustainability, Creativity, Resilience: Toward New Development Strategies of Port Areas through Evaluation Processes. International Journal of Sustainable Development **13**, 161–184 (2010)

Geographic Data Infrastructure and Support System to the Evaluation of Urban Densification

Loreto Colombo[1], Immacolata Geltrude Palomba[1], Valentina Sannicandro[1], and Carmelo M. Torre[2(✉)]

[1] Department of Architecture, University of Naples Federico II,
Via Forno Vecchio 36, Naples, Italy
{colombo,valentina.sannicandro}@unina.it,
immapalomba@gmail.com
[2] Department of Civil Engineering and Architecture, Polytechnic of Bari,
Via Orabona 4, Bari, Italy
cartorre@yahoo.com

Abstract. In urban contexts characterized by the presence of settlements where services are lacking, densification with the transfer of development rights brings benefits, through intensification of urban areas, reduction of the vehicular mobility and containment of soil consumption.

Indeed, under certain conditions, the transfer of development rights allows to identify the volumes to be deployed by evaluating the receptive capacity of the soil.

Geographic Data Infrastructure could be a valid tool to make an Evaluation Support System to compare alternatives with different compensation criteria. These different options arising from the request for satisfaction of collective needs, due to the growing demand for supply of services to upgrade the technology and energy building.

Keywords: Land use · Urban sprawl · Urban fringe · Densification · Equalization · Integrated evaluation

1   Soil Consumption and Urban Sprawl

The territorial transformations lead to an uncontrolled consumption of soil, often diverted from the productive uses and waterproofed. Houses, sheds and buildings go to make an abacus of elements, to which we should add shopping centers that compete for visibility, size of the parking, cubic content of the building-container [1].

In these contexts, the expansion of the buildings, which only partially responds to a real housing needs, has found and still finds the support of repeated tax amnesty laws.

There are several types of urban sprawl that are based on their density, infrastructures, the old rural layout conditioning the present one, the type of buildings and the degree of authority control. The type of urban development described, is typically present in the central-southern plains of Italy, where the presence of unauthorized buildings is at its peak.

© Springer International Publishing Switzerland 2015
O. Gervasi et al. (Eds.): ICCSA 2015, Part III, LNCS 9157, pp. 330–341, 2015.
DOI: 10.1007/978-3-319-21470-2_23

Rural areas are considered as a key resource for the common good, which requires interventions to protect and enhance the importance of the ecosystem, biodiversity, landscape, for the production of primary goods, for recreational opportunities, outdoor play and leisure.

The soil consumption, generated by urban sprawl, is considered the most serious impact, so it is important to find a strategy that reduces and mitigates the effects of the phenomenon for the areas in the consolidated city and in the developing city.

The proposed actions (in the central paragraphs of the paper) to achieve this issue, for example, must recognize the priority of urban-architectural renewal and the application of densification through architectural restructuring or replacement, with increases of up to 30% of the present volumes as incentives.

Offering builders the possibility to increase volumes in exchange for the creation of public infrastructures, complying with eco-compatible building standards and aiming for architectural quality, it is possible to achieve the densification [2].

However, in recent years, it was realized that only the urban planning is not enough to stop the continuous increase of the soil consumption, but it is necessary a careful analysis of the characters of the places and especially a detailed evaluation of the cost/benefit (direct, indirect, induced and intangible) as a result of territorial transformations [3], as considered at the end of the paper.

2   Peculiarity of Urban Sprawl in the South of Italy

In the Nineties, urban sprawl, characterized by an increase of low-density buildings, becomes a new kind of metropolization [4].

This causes the disappearance of the border between city and countryside and the removal of the elements that have identified the city for many centuries: community, identity and tradition.

Urban sprawl implies the expansion of the buildings and extension of the transport networks. In this way, the landscape undergoes irreversible changes and the unsustainable phenomenon implies soil consumption, excessive use of cars and energy dissipation [5].

The expansion of the buildings, characterized by a medium-low density, becomes less dense as one moves towards the countryside. In this way, it is difficult to distinguish spaces and urban elements from rural ones.

The phenomenon involves mainly the north-east of Italy, where the zoning regulations have supported the escape by the congested city and the establishment of a new type of settlement that is an integration between home and laboratory.

Instead, in the south of Italy, the urban sprawl is closely related to illegal construction, built simply in the lack of regulatory control.

In the southern regions, urban sprawl is composed of a multitude of fragmentary architectures and held together by the infrastructure. Due to the expansion without brakes, the distances from each town center are shorter and cities can touch themselves.

Starting from the beginning to the Nineties the concept of sustainability connected the analysis of urban sprawl with the idea of soil consumption, considered an unsustainable approach in the design of urban form counterposed to the model of "Compact city", labeled as sustainable spatial organization of the urban fringe [6].

3 A Consensus-Based Plan as a Strategy for Limiting Soil Consumption via Densification

Urban sprawl in the south of Italy is the image of a individualistic society based on a dissipative model of edification, which does not perceive the damage resulting from this devastation, with a considerable erosion of agricultural areas highly productive and the consequent waterproofing of soils.

It must be identified new methods and approaches to improve the area, moving from the urban scale to that of individual buildings. Notwithstanding the enormous number of illegal buildings, we can assume that the normal measures prescribed by the law (confiscation and demolition) will not be applied.

It is necessary to ensure that, along with a rigid curtailing of any further illegal construction and a return to the rule of law, all actors that have a legal or practical role have to be involved in regulating future land use.

The objective is to prevent further illegal construction while increasing the building concentration in urban-rural areas based on precise indications. Along with the general renewal, the plan establishes clear boundaries between city and countryside and urban sprawl is to be provided with services and infrastructures that are currently absent or insufficient.

In Italy, the problem of limited public financial resources and the high value of urban property market can be overcome through the use of procedures consensual, the acquisition –not onerous- of areas for the public works, the direct execution of the works of general interest by private individuals [7].

An innovative plan is based on consensus [8]: a necessary premise of the plan program is a series of public-private agreements to be reached through debate among all parties concerned. A significant percentage of public works are financed by the private sector which, in exchange for giving land to be used for public purposes, receives more extensive building permits and tax breaks. The plan therefore regulates the development of the area leveraging on the personal interests of individuals to promote the interests of society.

Therefore, we need an alternative to the traditional urban planning. A consensual plan allows the arrest of the continuous expansion and the densification of urban-rural areas, by promoting the realization of public works, the technological upgrading and the energy efficiency improvement of the building.

Ultimate objectives of the plan are to:

- establish the boundaries between the urban sprawl and the country;
- preserve environmental, cultural, architectural quality through the preservation and promotion of natural, cultural and environmental values;

- preserve the quality of urban and extra-urban landscape as an aspect of historical and cultural identity;
- restore buildings and renovate settlements and non-urbanized areas, especially the more degraded ones, while furthering economic and social development.

4   Equalization as a Tool for the Densification

The method to compensate via equalization is now an obvious choice for the acquisition of certain areas to target public [9]. Moreover, it has been widely recognized as an essential criterion to ensure the principles of fairness and transparency of plan choices.

In the market rights for buildings and pursues the conservation objectives of the Natural Capital, it is realized the **model of the sending site/receiving site** (urban areas from which "take off" the development rights/urban areas on which "land" development rights). The transfer volumes promotes the realization of a settlement with limited environmental and landscaping impact and conservation of acquired rights, which occurs accompanying the transfer of volumes with the equalization of property values.

This system makes indifferent, to the property owners, the spatial location of the public areas (that are not building areas) inside the compartment. In fact, with acts of private nature, is aggregated the maximum volume of all lots (including those sold for public use) within areas of concentration and all owners benefit from their share of total potential volume allocated to the sector in the form of spatial index.

The owners of an equalization compartment participate, therefore, without distinction, in equal measure, to the process of the distribution of values and losses, coming from urban plans.

In summary, the equalization ensures the administration the opportunity to acquire, without cost, some soils necessary to the community with the agreement of the property at which is still recognized a share of land rent.

To think about the transfer of rights from the areas of building rights sparse to very dense area, we must first identify the volumes to be transferred and evaluate not only the accommodation capacity of the soil, but also the presence of areas that may have been partially invalidates for the construction of new buildings and for ancillary services.

In this regard, the integration between the GIS and the different methods of evaluation [10] becomes an important resource in the construction of a "Spatial Decision Support System", in which the variety of spatial information, determined by social elements, economic and environmental, can be easily combined with the different alternatives of land use.

It has demonstrated the usefulness of some models and techniques in spatial planning, where has been combined the application of GIS with the tools of evaluation and the SSP (Support System to Planning) [11].

Therefore, we can consider a "multi-methodological Spatial Decision Support System" as an integration of urban planning tools with the:

- *dynamic system*: system that considers the temporal evolution,
- *deliberative system*: system based on the will of the stakeholders,
- *comprehensive system*: system capable of considering the quantitative and qualitative aspects of the different components,
- *spatial system*: system able to identify the effects on land through their representation.

The spatial analysis, then, are excellent tools for discovery; however, it should be noted that the analysis are not predictions, but may proceed to simulations. In addition, simulation is a form of aid to the decision itself very useful, but it cannot go further. It is possible to spot trends and appreciate the likelihood that these trends have to last and to infer that some situations are "good" and "bad", but it can be done only if there are alternative hypotheses on the behavior of the actors.

Some addresses of the research investigates about how enlarge the understanding of dynamics of urban development and its impacts, by the use of open access systems [12][13]. Over all, the Geographic Information System gives the possibility to search on how to change the spatial characteristics of the status quo and those relating to possible future transformations. In this perspective, it is important the development of simulation models that allow for the construction of several possible scenarios and are able to predict the strategic implications of each of them [14].

5   The Case Study

The following case study concerns the south-western area of the County of Caserta, in the region of Campania, occupied by a sprawling and continuous, mainly unregulated urban area, which extends west from Aversa (the main center in the area) all the way to Villa Literno. It is a tentacular conurbation, which includes seven municipalities (Casal di Principe, Casapesenna, Frignano, San Cipriano d'Aversa, San Marcellino, Villa di Briano, Villa Literno) whose urban centers have progressively extended over the countryside (Fig. 1).

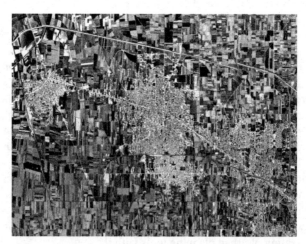

Fig. 1. Identification of the study area: a great conurbation having Aversa in the middle

Each town includes a historical center, surrounded first by a consolidated urban area and then by a sprawl that becomes progressively less dense as one moves towards the countryside. Unauthorized and unplanned construction is a typical trait of the area, especially in the peripheral urban areas. According to the PTCP (General County Plan) of the city of Caserta, since the end of the Second World War to nowadays, over two-thirds of urban areas have been built up made in the absence of the PRG (City Master Plan) and almost 90% of these fabrics have been made before 1984: about 20% of buildings made in the seven municipalities of the study area, since 1984, are illegal.

The analyses produced on the territory, focusing on the type of soil consumption, outline the geometric matrices (Fig. 2) that indicate the manner of land use, distinguishing:

- Saturated spontaneous urban area;
- Spontaneous urban area in its final phase;
- Urban periphery in the initial urbanization phase.

This research shows that urban planning promotes similar types of expansion and agreements between sellers and buyers. This problem causes expansion free and chaotic.

The tendency that emerges from recent demographic analysis shows a population increase in the seven municipalities. Based on more than one type of projection, we can foresee a population increase of 18% over the next 20 years, corresponding to 14,500 people.

This implies the need for housing and standard public infrastructures, which based on the criterion of 22 square meters per inhabitant as necessary for an acceptable quality of living, results in more than 1,825,000 square meters of surface.

Considering a unitary volume per room of 80 cubic meters, a ratio of real estate coverage of 0.3 square meters of paved surface on square meters of soil and an index of real estate building allowance of 1.3 square meters of paved surface on square meters, the total gross paved surface comes to 391,500 square meters.

At this point, the considerations marked suggest that in the absence of a planning tool suitable to meet the new needs of the population, urban sprawl would be considered directly proportional to population growth. Soil consumption would continue to increase, together with the loss of soil and the few urban fringes still free in the conurbation in "Galaxy" which revolves around the municipality of Aversa.

Having to implement an attempt to densification, transferring rights to land on hold (in the suburbs and peripheral area, and in the interstitial areas between a municipality and another, elements niche for what remains of the ecological network) to the inner areas, was carried a scenario analysis aimed at possible equalization of values crossed with the transfer of volumes. The equalization of values considers the variability of real estate values as a constraint to the sizing of the rights to be transferred [15].

The saved soil is inversely proportional to the ratio of the indices of departure of "sending sites" and the arrival of the "receiving sites". However, sending and receiving sites are located in areas where the real estate value is variable.

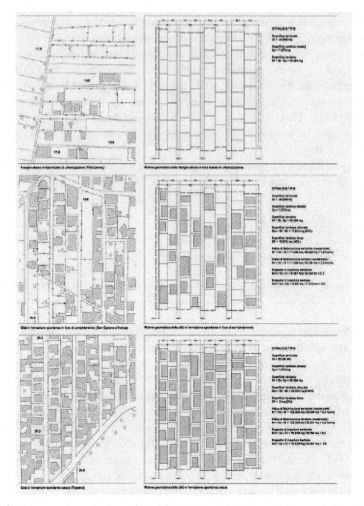

Fig. 2. The procedures for loss of land for building growth: spatial geometric matrices

In particular, in the suburbs of municipality of Aversa, that have higher real estate value and are located in the north of the present center, consist of single-family houses in tissues grain sparse.

In this research, Zone OMI identifies Receiving Sites and Sending Sites. OMI, Observatory of the real estate market of Revenue Agency (that was Territory Agency), divides the urban land in peripheral bands (identified by the letter D) semi peripheral (identified by the letter C) and central (identified by the letter B).

In each wing the possible n existing zones are identified by a number that varies from 1 to n. In the municipality of Aversa there are two central areas (B8 and B9), four semi peripherals (C8, C9, C10 and C11), and four peripherals (D5, D6, D7 and D8).

In Aversa, some outlying areas are in fact "hinge areas between Aversa and the neighboring municipalities. For them, the concept of the periphery is only partially appropriate, and should be rim-placed by that of "interstitial area" (Figure 3).

Fig. 3. Background of the study area and areas OMI, Aversa is at the center of a mosaic of values and different types of real estate assets

The transfers analyzed are the follows:

- From areas D (peripheral) to C (semi peripheral) and B (central)
- From areas C (semi peripheral) to B (central)

Transfers of rights, which should enable to get real estate values in the landing areas similar to those of the starting areas of rights, leading to a transfer with increase in volume, where the soil base has higher values of soil landing. The opposite occurs when the starting values are lower.

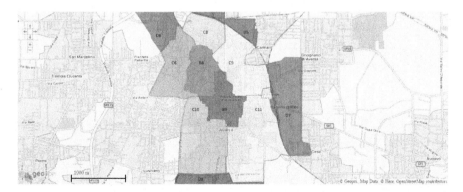

Fig. 4. Framing of the study area and OMI areas, peripheral and inner suburban stations

They are configured four scenarios, and analyzed transfers from peripheral areas, (where there is greater availability of soil and less pressure settlement) to interme-diates (expansion zones in the PRG - General Plan -, with index elevating probably until doubled), and the intermediate internal ones (where the free areas are few and therefore the possibility of transfer is limited).

It assumed a transfer with doubling index and a transfer with maintenance of the same index.

Usually the transfer development right has a main point the seek for convenience in moving rights to built volumes taking in account that the threshold for convenience should be considered. There is a break-even point to make helpful for entrepreneurs and owners the shift of the right to develop areas [16] [17].

It is assumed that the saving of soil is proportional to the increase in real estate density from outer zone to inner ones (Table 1 and 2):

- scenario 1: minimum differential estate and transfer volumes at the same density;
- scenario 2: maximum differential estate and transfer volumes at the same density;
- scenario 3: minimum differential estate and transfer volumes doubling density;
- scenario 1: maximum differential estate and transfer volumes doubling density.

The values selected are based on those of the main types for different zones, which in this case are always civilian homes.

A main constrain of the model, in this case is due to the keeping of the same use of volumes (from residential use to residential use).

In other experiences, it a more general translation of volumetric rights from an old use to a new one, such as for brown fields reclaiming, has been analyzed [18].

Table 1. saving ground on four scenarios, each receiving site

Rec. Site	C7	C8	C9	C10	C11	B8	B9	Rec. Site	C7	C8	C9	C10	C11	B8	B9
1 – pessimistic - equal density- % of saved soil								2 – optimistic - equal density- % of saved soil							
D5	17	17	31	31	26	26	44	D5	33	33	45	45	41	41	56
D6	4	4	21	21	15	15	36	D6	33	33	45	45	41	41	56
D7	-8	-8	10	10	4	4	28	D7	21	21	34	34	30	30	47
D8	17	17	31	31	26	26	44	D8	33	33	45	45	41	41	56
C7	0	0	17	17	11	11	33	C7	25	25	38	38	33	33	50
C8	0	0	17	17	11	11	33	C8	29	29	41	41	37	37	53
C9	-21	-21	0	0	-7	-7	19	C9	8	8	24	24	19	19	39
C10	-21	-21	0	0	-7	-7	19	C10	8	8	24	24	19	19	39
C11	-13	-13	7	7	0	0	25	C11	13	13	28	28	22	22	42

For example, with values ranging from 800 euro to 1000 euro, per square meter, in an OMI peripheral area, and maximum values equal to 1200 euro per square meter in the OMI interior zone, saving soil varies:

1. between 200/1200 and 400/1200 (16.66% and 33%) in the event of transfers of indexes without an increase in density, that is a ratio of 1 to 1 by sending sites to the receiving sites;
2. between 2x200/1200 and 2x400/1200 (33.33% and 66.66%) in the event of transfers of indexes with an increase in density, that is a ratio of 2 to 1 by sending sites to the receiving sites.

Tab. 2: saving ground on four scenarios, each receiving site

Rec. Site	C7	C8	C9	C10	C11	B8	B9	Rec. Site	C7	C8	C9	C10	C11	B8	B9
3 – pessimistic - Doubling Density - % of saved soil								4 – optimistic - Doubling Density - % of saved soil							
D5	33	33	62	62	52	52	89	D5	67	67	90	90	81	81	111
D6	8	8	41	41	30	30	72	D6	67	67	90	90	81	81	111
D7	-17	-17	21	21	7	7	56	D7	42	42	69	69	59	59	94
D8	33	33	62	62	52	52	89	D8	67	67	90	90	81	81	111
C7	0	0	34	34	22	22	67	C7	50	50	76	76	67	67	100
C8	0	0	34	34	22	22	67	C8	58	58	83	83	74	74	106
C9	-42	-42	0	0	0	-15	39	C9	17	17	48	48	37	37	78
C10	-42	-42	0	0	0	-15	39	C10	17	17	48	48	37	37	78
C11	-25	-25	14	14	0	0	50	C11	25	25	55	55	44	44	83

6   Final Thoughts and Perspectives of Research

The study shows that scenarios equalization of volumetric densification and of transfer of building rights are subsequent and consequent to plan choices diversified.

This requires an analysis of the distribution of volumes and types outset, that is to benefit, especially in conurbations such as Aversa municipality here examined, which has little residual areas and aspires to urban expansions supported by forecasts of growth that make the soil resource not only rare, but very fragile.

Some limitation have to be explored in the perspective.

In theory we found a way to forecast how and where is possible to transfer development rights for saving soil.

The potentials of saved soil anyway should be selected from a measure of available real residual spaces, by crossing free buildable area inside the urban fringe, considering the set of real estate values for each area [14].

It is considered that not all of the transfers are allowed, according to some constrains: the first is the real quantity of free areas, the second is the opportunity if the change of land use (from residential to commercial, or productive activities, and so on), that is allowed in a number of Countries [19] [20], but it is very difficult in Italy, due to the necessity to rebalance the quantities (that is very difficult according to norms that rebalance offer with demand, as a guarantee of soil keeping): for example, the change from brownfields to residential, increase the offer of housing stock, and reduce the offer (respect to the hypnotized demand in the City Master Plan) of productive areas.

The last consideration is the necessity of foster the economic analyses (often GIS supported) to re-interpret dynamically the forecasts of urban land development [21]and of the urban real estate market, looking in the same at impacts ad advantage, in terms of urban form, flexibility of land-uses and environmental impacts.

The mechanism of soil saving by densification, finally, is connected with Orgware (norms and procedure of plans), Civicware (the system of social involvment to make understandable the TDR process), and Ecoware (the soil as a ecological/economic resource). This aspect nowadays remind to theme of resilience [22]. A resilient city car redesign the scheme of the urban expansion with complex procedure with permuting vertical growth with horizontal, but also a single property use with multiple property uses, and so on.

Acknowledgements. The paper is the result of experimental activities of MITO Lab, during the MITO-(Multimedia Information for Territorial Objects) Project Work, funded by the Cohesion Action Plan 2007-2013 of European Union, at the Department of Civil Engineering and Architecture of Polytechnic of Bari.

The authors whish also to acknowledge Marcello Ferrara, that supported the creation of some analyses, contents and images of this paper, during his final research thesis at the "Federico II" University of Naples

References

1. Ricci, M. (ed.): Rischio paesaggio. Meltemi, Rome (2003)
2. Attardi, R., Pastore, E., Torre, C.M.: "Scrapping" of quarters and urban renewal: a geostatistic-based evaluation. In: Murgante, B., et al. (eds.) ICCSA 2014, Part III. LNCS, vol. 8581, pp. 430–445. Springer, Heidelberg (2014)
3. Burton, E., Jenks, M., Williams, K. (eds.): The Compact City: A Sustainable Urban Form? Spon Press by Taylor and Francis, Abingdon (1996)
4. Colombo, L.: Ricompattazione, marginatura, densificazione, ridisegno, consenso. Dalla diffusione edilizia alla nuova città di pianura. Planum.The journal of Urbanism **27**(2)
5. Colombo, L., De Toro, P.: Ecologically oriented urban architectural renewal: three case studies. CSE journal **1** (2014)
6. Neuman, M.: The Compact City Fallacy. Journal of Planning Education and Research **25**(1), 11–26 (2005)
7. Calavita, N., Krumholz, N.: Capturing the Public Interest Using Newspaper Op- to Promote Planning in Conservative Times. Journal of Planning Education and Research **22**(4), 400–406 (2003)
8. Urbani, P.: Urbanistica Consensuale. La disciplina degli usi del territorio tra liberalizzazione, programmazione negoziata e tutele differenziate. Bollati Boringhieri, Turin (2014)
9. Carbonara, S., Torre, C.M., eds: Urbanistica e Perequazione. Regime dei suoli, Land Value Recapture e compensazione nei piani. Franco Angeli, Milano (2012)
10. Malczewski, J.: GIS and Multicriteria Decision Analysis. John Wiley, New York (1999). ISBN 9780471329442

11. Cerreta, M., De Toro, P.: Integrated Spatial Assessment (ISA): A Multi-Methodological Approach for Planning Choices. In: Burian, J. (ed.) Advances in Spatial Planning. In Tech Publisher, Rijeka (2012)
12. Nolè, G., Murgante, B., Calamita, G., Lanorte, A., Lasaponara, R.: Evaluation of Urban Sprawl from space using open source technologies. Ecological Informatics (2014). Doi: http://dx.doi.org/10.1016/j.ecoinf.2014.05.005
13. Attardi, R., Cerreta, M., Franciosa, A., Gravagnuolo, A.: Valuing cultural landscape services: a multidimensional and multi-group SDSS for scenario simulations. In: Murgante, B., et al. (eds.) ICCSA 2014, Part III. LNCS, vol. 8581, pp. 398–413. Springer, Heidelberg (2014)
14. Amato, F., Pontrandolfi, P., Murgante, B.: Using Spatiotemporal Analysis in Urban Sprawl Assessment and Prediction. In: Murgante, B., et al. (eds.) ICCSA 2014, Part II. LNCS, vol. 8580, pp. 758–773. Springer, Heidelberg (2014). doi:10.1007/978-3-319-09129-7_55
15. Torre, C.M., Balena, P., Zito, R.: An Automatic Procedure to Select Areas for Transfer Development Rights in the Urban Market. In: Murgante, B., Gervasi, O., Misra, S., Nedjah, N., Rocha, A.M.A., Taniar, D., Apduhan, B.O. (eds.) ICCSA 2012, Part I. LNCS, vol. 7333, pp. 583–598. Springer, Heidelberg (2012)
16. Micelli, E.: Development Rights Markets to Manage Urban Plans in Italy. Urban Studies **39**(1), 141–154 (2002). doi:10.1080/00420980220099122
17. Morano, P., Tajani, F.: Break Even Analysis for the Financial Verification of Urban Regeneration Projects Applied Mechanics and Materials **438-439**, 1830–1835
18. Morano, P., Tajani, F.: The Transfer of Development Rights for the Regeneration of Brownfield Sites. Applied Mechanics and Materials **409–410**, 971–978 (2014)
19. Altermann, R.: Takings international: A comparative Perspective of Land Use Regulations and Compensation Rights. ABA Publishing, Chicago (2010)
20. Wai-chung, L.: Neo-Institutional Economics and Planning Theory. Planning Theory **4**, 7–19 (2005)
21. Mi, S., Hsiutzu, B.C.: Transfer of development rights and public facility planning in Taiwan: An examination of local adaptation and spatial impact. Urban Studies **4** (2015)
22. Girard, F.: L.: Sustainability, Creativity, Resilience: Toward New Development Strategies of Port Areas through Evaluation Processes. International Journal of Sustainable Development **13**, 161–184 (2010)

Workshop on Geographical Analysis, Urban Modeling, Spatial statistics (GEOG-and-MOD 2015)

Generalized Dasymetric Mapping Algorithm for Accessing Land-Use Change

Antonio Manuel Rodrigues[(✉)] and Jose Antonio Tenedorio

CICS.NOVA Interdisciplinary Centre of Social Sciences Faculdade de Cincias Sociais e Humanas, Universidade Nova de Lisboa, Lisbon, Portugal
amrodrigues@fcsh.unl.pt
http://www.cics.nova.fcsh.unl.pt/

Abstract. The use of multivariate micro-data, data aggregated for small-areas, allows detailed analysis of the physical and social structures of regional landscapes. Such exercises are in many cases univariate and static in nature; this happens when geometries are not coincident between datasets. Common solutions to such inconsistencies involve the use of areal interpolation techniques to build coherent information sets; when ancillary information is available, dasymetric mapping using control units may then be used. Techniques vary on the type and quality of the ancillary (or control) information. The purpose of the present article is to present a generalized tool to tackle common practical analytical problems and which produces geometrically coherent datasets. It is generalised because: (1) it is flexible, allowing distinct parametrization depending on the data; (2) it is based on Open Source tools anchored on robust database management systems (DBMS) technologies. Its aim is to provide the regular GIS user with a tool to tackle a common problem of geometric mismatch.

Keywords: Dasymetric mapping · Computational statistics · FOSS · Database Management Systems

1 Introduction

Multivariate or temporal analysis of geographical data aggregated for specific areas, polygons, is highly dependent on the geometrical coherence between these spatial units. When the shape of these units from distinct datasets does not coincide, areal interpolation and Dasymetric mapping techniques are important [2,3,7]. An areal interpolation exercise may be described as the re-allocation of numeric data according to some geometrical schema; original geometries may be called *source* spatial data and the end-geometries as *target* spatial data; the final datasets are spatially coherent because data describing all phenomena studied are aggregated according to the same set of areas (spatial units).

The characteristics of spatial data aggregated for a finite set of spatial units (regions) imply a number of problems which should be tackled, otherwise statistical results may be flawed. When using univariate statistical techniques for

© Springer International Publishing Switzerland 2015
O. Gervasi et al. (Eds.): ICCSA 2015, Part III, LNCS 9157, pp. 345–355, 2015.
DOI: 10.1007/978-3-319-21470-2_24

single variable exploratory work, the simple fact that events are aggregated into areas originate lost of information usually related to what is commonly known as *ecological fallacy* [1]. In multivariate or time-series analysis, when geometrical schema used to collect data for distinct phenomena are not the same, there is the need to use areal interpolation techniques, to which Dasymetric mapping belong to, in order to guarantee geometrical correspondence.

Although dasymetric mapping techniques are not new, there is no tool which is both general in terms of scope, free in terms of access and robust in terms of technology. The present article represents the end point in a series of research efforts where Dasymetric techniques were used with distinct types of control datasets [4,5,7,9]. The aim was to produce a tool, based only on Open Source technologies which can be used in a large variety of settings. The algorithm used is implemented in SQL using functions from PostGIS, a library which introduces spatial and geographic objects for PostgreSQL (http://postgis.net/andhttp://www.postgresql.org/).

Three distinct analytical scenarios were considered in the final design, with an increasing degree of complexity: first, when the goal is to re-allocate count data from source to target geometries and no ancillary information is available - *plain areal interpolation*. Second, when ancillary data is binary in nature, making the distinction between zones where events can and cannot occur (ex. buildings footprint which identify where individuals reside as opposed to all other areas) - *areal interpolation with binary ancillary data*. Third, when ancillary data is stratified; in these cases, a weighting scheme is designed to account for distinct probabilities that events occur in specific areas (ex. slope may influence the probability of observing certain species of plants) - *areal interpolation with stratified ancillary data*.

Practical applications are crucial to test implementation. In the final section of the article, three distinct examples are considered for all three scenarios and three distinct study areas. All cases chosen reflect one common problem, which is the non-coincidence of geometries between census micro-datasets from different years. These datasets are rich in detail and provide the best sources of demographic data, with impressive detail (small areas - census tracts) covering large areas (nation-states).

It is important to be aware of the degree of uncertainty of any Dasymetric mapping exercise. This is a function of degree of dis(agreement) in shape (*conformity*) and size (*equivalence*) between source and target geometrical datasets. In the former case, *Modifiable Areal Unit Problems* (MAUP) increase the greater the differences in shape between geometrical schemas [7]; in other words, the lesser is the degree of conformity. In cases when target geometries are smaller than source (ie. when data is pulverized), the degree of equivalence is smaller.

2   Methodology

Most phenomena, physical or human, have a spatial attribute (ie. they can be referenced to some theoretical model of, often, the globe). Hence, it is recurrent that research related to their distribution fall within the scope of the Geographical Information Sciences (GIS). Data from such phenomena are often collected in groups, defined according to some aggregating schema. Examples range from population statistics to Livestock Census. Count data is the most common data type, aggregated according to some geometrical schema which is in most cases heterogeneous in shape and size (non-conformity and non-equivalence).

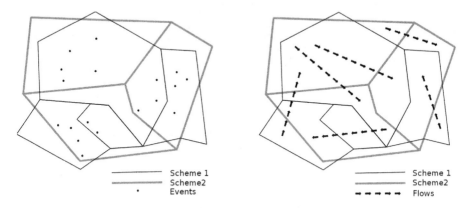

Fig. 1. Geometric aggregating schemes[1]

The adopted conceptual framework which supports areal interpolation exercises distinguishes between *form*, *structure* and *function* of spatial units [7]. Such concepts are understood as fundamental to understand the basic structure of landscapes [10]. When re-allocating counts from source to target geometries - plain areal interpolation, (primary) form is taken into account with attributes such as shape or size. When binary ancillary data is available, then the internal structure of each spatial unit is made endogenous. Finally, when multinomial or continuous data is available for relevant ancillary variables, then such attributes add functional information about the study-area.

In order to better understand the degree of uncertainty associated with areal interpolation exercises, let's assume for a specific study-area, events and flows occurring at specific locations as shown in figure 1. The spatial surface represented is partitioned according to two distinct geometrical schemas and events aggregated into counts. The resulting datasets represented by the point events are the following:

$$schema1 = \{8, 4, 2, 4\}, \ schema2 = \{5, 7, 6\} \qquad (1)$$

In terms of flows, results are significantly different when represented as movements between spatial units. Table 1 represents the flows matrices for both geometrical schemas. Regions are numbered using the following rules: for schema 1,

---

[1] Adapted from [7].

the numbering runs from left to right (matrix 1), whilst for schema 2 from top to bottom (matrix 2). Two problems emerge: first, flows whose origin or destination are outside the geometrical schema are not accounted for. Second, in most cases intra-regional flows are ignored (information is not available). Third, distinct geometries result in distinct interpretations of the same reality. The three problems just mentioned are typified respectively as *edge effects*, *ecological fallacy* and *Modifiable Areal Unit Problem* (MAUP). In the latter case, differences are extremely relevant as shown in table 1, which show the flows matrix, computed according to schema 1 (left) and schema 2 (right).

Table 1. Flows matrices

	1	2	3	4
1	0	1	0	0
2	0	2	0	0
3	0	0	0	0
4	0	0	0	1

	1	2	3
1	0	2	1
2	0	0	0
3	0	0	1

2.1 Areal Interpolation Scenarios - Algorithm Design

The algorithm developed was structured according to three distinct types of settings defined according to the type of ancillary data available (or not). These were typified into three groups: form, structure and function [7,10]. Following what was mentioned above, an exercise which relies solely upon the *form* of geometrical (aka. spatial) units is that which assumes distribution of count data is uniform within each area; this implies intra-regional isotropy: diffusion of the studied phenomena in constant whichever direction. When ancillary (or control) data is used which correspond to geometrical features in the landscape that are

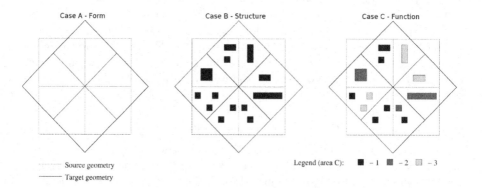

Fig. 2. Geometric aggregating schemes

themselves uniform, the *structure* of source geometries is made endogenous: the distribution of the studied phenomena becomes binomial, with parts of each spatial units where counts occur and others where values are NULL. Finally, when it is possible to discriminate between observations in the ancillary dataset, then *function* is taken into account: the distribution is then multinomial.

The following setting exemplifies the distinction between form, structure and function; figure 2 represents (as did figure 1) two aggregating schema but also exemplifying the three hypotheses implemented in the algorithm

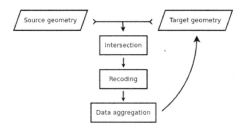

Fig. 3. Algorithm flowchart - case A (form)

Form - Case A exemplifies the case when no ancilliary/control geometric dataset is available; only areas (size) from *source* and *target* zones (polygons) are accounted for in the re-allocation exercise (figure 3).

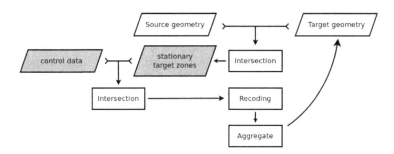

Fig. 4. Algorithm flowchart - case B (structure)

Structure - When binary ancilliary data is available - case B, the algorithm follows a two step process, taking into account the internal distribution of ancilliary polygons (represented in black). Figure 4 illustrates this process: the result of intersecting source and target geometries is named *stationary control zones* since the internal structure is not considered - zones are assumed to be homogeneous in terms of the variable of interest. This is only taken into account in the second step - when control (or ancilliary) data) is used.

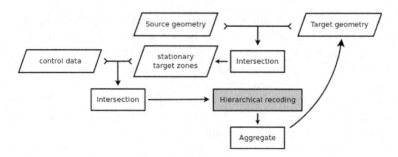

Fig. 5. Algorithm flowchart - case C (function)

Function - In the final hypotheses - case C, after intersecting the stationary target zones with the control data, the resulting weighting scheme takes into account extra information which discriminates between ancilliary polygons, for example taking into account height of the polygons if these represent the buildings' footprint. In such cases, a *hierarchical recoding* is performed taking into account an extra variable which *weights* the size of each control polygon. In the example just given, in practice, the algorithm computes the volume of each building (by multiplying area and height).

2.2   Release Strategy

Although not universally true, active research should have some practical application, or contribute to an on-going body of knowledge. Creativity builds innovation; this in turn may have a social epidemiologic effect. In Geo-science related research groups, this is particularly true, since the outcome of research potentially represent new layers of spatial information. It is the purpose of the present research to contribute with the production of a freely available tool which may, through granted access to source code, be improved. Also, it should be possible to access the tool using different techonological solutions, which should themselves also be free of access.

Following what was just said, only Free and Open Source Software (FOSS) also named as Free/Libre Open Source Software (FLOSS) technologies [1] and [2] were used. However, care has to be taken in terms of semantics. The term Free in FOSS does not mean that it is free of cost, but is referring to the software freedoms that are addressed to FOSS software [3]. These freedoms are stated in the Free Software Foundation (FSF) Website [4] and GNU Operating System Website [5] and refer to:

1. The freedom to run the program, for any purpose.
2. The freedom to study how the program works, and change it so it does your computing as you wish. Access to the source code is a precondition for this.
3. The freedom to redistribute copies so you can help your neighbour.
4. The freedom to distribute copies of your modified versions to others. By doing this you can give the whole community a chance to benefit from your changes. Access to the source code is a precondition for this.

3 Applications

All case-studies presented apply the algorithms described with the purpose of
re-allocating Census data for Continental Portugal according to some common
geometric schema (figure 6). Census exercices for distinct years reflect the problem of distinct borders (see figure 7 for a map snippet).

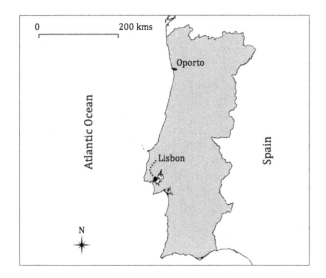

Fig. 6. Study-area

Fig. 7. Study-area

The three applications that follow represent the use of cases B and C (see
figure 2), and a hybrid application where all three methodologies are compared
for the same neighbourhood.

Structure

In application one, the area of the buildings' footprint was used as representing the internal structure of each spatial unit. As described above, a weighting scheme based on this variable was computed which served as the key to

re-allocated census data according to common end geometries (target spatial data). The study area was Lisbon [7]. Figure 8 illustrates an analysis performed with the new dataset. In this case, using data smoothing through Exploratory Spatial Data Analysis tecnhiques, it is possible to identify clusters of change over a period of 10 years.

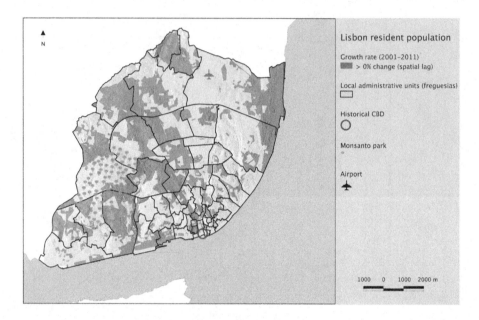

Fig. 8. Lisbon application - Structure

Function

A different application was that performed using data for the Tomar municipality, located 100 kms to the North-East of Lisbon [8,9]. In this case, information was available for each building in terms of its function (ex. residential, mix-use, commercial). This classification allowed the creation of a more complex weighting scheme, resulting in potentially greater accuracy in the re-allocation exercise. Figure 9 illustrates, for the central area of the town, the distribution of the number of residents per family in 2011. Although it is a static historical snapshot, the same data was produced (for the same geometries) for the census years 2001 and 2001.

3.1 Hybrid Solutions

The final application uses all three cases, allowing for a benchmarking exercise. The study-area was a small-area in the municipality of Amadora in the outskirts

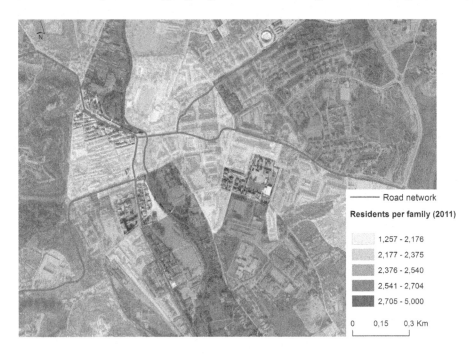

Fig. 9. Tomar application - Function

of Lisbon. Data extracted from a Unmanned aerial vehicle - UAV flight allowed the computation of volumetric information of each residential building block [11].

In figure 10, the tabular comparison of the three methods is presented [6]. The variable of interest was resident population in 2011 aggregated according to the 2001 census tracts. Each row corresponds to a target area, areal interpolation represents the case when only form is considered (intra-regional isotropy is assumed), 2D data corresponds to struture and 3D data to function. The final

zone	Aerial interpolation	Control zones		Benchmark*
		2D	3D	
TA1	4	0	0	0
TA2	14	0	0	0
TB1	374	379	380	367
TB2	327	342	356	348
TB3	312	336	321	343
TC1	486	498	498	498
TC2	350	384	384	384
Absolute deviation	17.53	3.66	6.32	

(*) Estimated from dwellings

Fig. 10. Amadora application - Hybrid

column corresponds to estimated population according to the known number of dwellings per block.

Because the area is small and there is no great differences in terms of buildings' height, coupled with the fact the benchmark dataset is itself estimated, it is difficult to say one method performs better than the other. Nonetheless, differences in computed results are important (only three target areas represent important results - TB1, TB2 and TB3), which indicate that for a larger and more heterogeneous area, significant differences can occur.

4   Concluding Remarks

Data flooding does not necessarily mean a more informed society. On one hand, data is not all structured simply because of the complex nature of the world we live in. Yet, the provision of a freely available tool which allows the non-technician to explore coherently geographical datasets aggregated according to distinct geometric schemes has important implications and potentially great value-added. It is the believe of the authors that the present article is a solid step in this direction. Also, the fact that the code produced is open-source and released under the GNU General Public License (GNU GPL) improves the odds of new developments to occur, as it increases reproducibility.

In order to test the robustness of the methods implemented, new applications should be developed. Although any areal interpolation exercise is subject to probabilistic laws, the greater the number of tests, the more robust can any inferences be.

Finally, it is worth mention that, in spite of the fact that different dasymetric mapping methods have been developed, the authors believe that they all fall under the three cases presented - form, structure and function. Hence, it is true to say that the present project represents a meta-analysis of areal interpolation and dasymetric techniques.

Acknowledgments. This article was partly funded by Portuguese national funds through FCT Portuguese Foundation for Science and Technology in the framework of projects PEsT-UID/SOC/04647/2013, SFRH/BPD/76893/2011 and SFRH/BPD/66012/2009.

References

1. Freedman, D.A.: Ecological Inference. In: Smelser, E.I.C.A.N.J., Baltes, P.B. (eds.) International Encyclopedia of the Social & Behavioral Sciences, pp. 4027–4030. Oxford, Pergamon (2001). http://www.sciencedirect.com/science/article/pii/B0080430767004101
2. Goodchild, M.F., Anselin, L., Deichmann, U.: A framework for the areal interpolation of socioeconomic data. Environment and Planning A **25**(3), 383–397 (1993)
3. Mennis, J.: Generating Surface Models of Population Using Dasymetric Mapping. The Professional Geographer **55**(1), 31–42 (2003)

4. Rodrigues, A.M., Neves, B., Rebelo, C.: Terra Communis (tComm): A free data provider for historical census micro-data. In: Proceedings of the VII Jornadas de SIG Libre. SIGTE - Servei de Sistemes d'informció Geográfica i Teledetecció, Girona (2013). http://www.sigte.udg.edu/jornadassiglibre2013/uploads/articulos_13/a9.pdf
5. Rodrigues, A.M., Rebelo, C., Tenedório, J.A.: Dasymetric mapping using volumetric information from UAV low-cost flights. In: ECTQG 2013 - European Colloquium of Theoretical and Quantitative Geography, Dourdan, pp. 197–198 (2013)
6. Rodrigues, A.M., Rebelo, C., Tenedório, J.A.: Dasymetric Mapping using 3D information: Sensitivity Analysis of UAV high resolution data in submission (2015)
7. Rodrigues, A.M., Santos, T., de Deus, R.F., Pimentel, D.: Land-use dynamics at the micro level: constructing and analyzing historical datasets for the portuguese census tracts. In: Murgante, B., Gervasi, O., Misra, S., Nedjah, N., Rocha, A.M.A.C., Taniar, D., Apduhan, B.O. (eds.) ICCSA 2012, Part II. LNCS, vol. 7334, pp. 565–577. Springer, Heidelberg (2012). http://link.springer.com/chapter/10.1007/978-3-642-31075-1_42
8. Rodrigues, A.M., Santos, T., Pimentel, D.: Asymmetrical mapping based on control zones: the construction of historical databases for the tomar local council. In: Sousa, F., Grilo, L., Grilo, H., Henriques-Rodrigues, L. (eds.) XIX Jornadas de Classificacao e Analise de Dados (JOCLAD 2012). Livro de Resumos, Tomar, pp. 159–163 (2012). http://www.fcsh.unl.pt/e-geo/sites/default/files/dl/
9. Rodrigues, A.M., Santos, T., Pimentel, D.: Socio-economic dynamics within a middle-size municipality: testing the strengths of historical micro-datasets. In: Gestão Integrada de Territórios Intermunicipais: o Papel dos Sistemas de Informaçãao Geográfica. 7as Jornadas de Gestão do Território - Livro de resumos. Tomar (2012). http://www.fcsh.unl.pt/e-geo/sites/default/files/dl/
10. Santos, M.: Espaço e Método, 4th edn. Espaços, Nobel (1997)
11. Tenedório, J.A., Rebelo, C., Estanqueiro, R., Henriques, C.D., Marques, L., Gonçalves, J.A.: New developments in geographical information technology for urban and spatial planning. In: Technologies for Urban and Spatial Planning: Virtual Cities and Territories, chap. 10, IGI Global, pp. 196–227 (2013). http://www.igi-global.com/chapter/new-developments-in-geographical-information-technology-for-urban-and-spatial-planning/104217

Comuns: An Open-Data Provider, Explorer and Analytic Toolbox Based on FOSS

Antonio Manuel Rodrigues$^{(\boxtimes)}$ and Jose Antonio Tenedorio

CICS.NOVA Interdisciplinary Centre of Social Sciences, Faculdade de Ciências Sociais e Humanas, Universidade Nova de Lisboa, Lisbon, Portugal
amrodrigues@fcsh.unl.pt,
http://www.cics.nova.fcsh.unl.pt/

Abstract. Researh efforts in the Social Sciences have a great potential impact on society in general as it promotes greater knowledge of any and every aspect of human interactions. With this in mind, a combined project - Comuns, was developed, bringing together recent technological trends in data exploration and visualization, and the production of new high-precision historical geo-demographic datasets. Using dasymetric algorithms implemented within an Free and Open Source Software (FOSS) platform, source geographical data- aggregated according to distinct geometrical schema, was re-allocated according to common areas. The innovative nature of Comuns is three-folded: (1) for the first time, a large scale comprehensive time-space database is built for Portuguese census data; (2) the fact that data is made available online contributes to the goal of making knowledge symmetrically available; (3) the fact the only FOSS is used means that, other than man-hours, the project is costs' free, accountability increases as does reproducibility. The open character of the project have potential implications in terms of the way human landscapes are perceived, given that new coherent datasets allow new explorations of the territory. Its target are primarily civil society in general; its applications range from academic to recreational, with potential uses in geomarketing projects.

Keywords: Open-data · FOSS · Database Management Systems · Dasymetric mapping · Computational statistics

1 Introduction

Research breeds innovation, which in turn spills over society and space through contagious behaviour [8]. Also, the importance of collaborative work within and between groups is justified as it promotes synergies. If, on one hand this traditionally happens within a closed system, because of agents proximity, on the other, linkages into the civil society (namely into the administrative, productive and informal sectors) are non-existent if channels are not opened between producers of new (open) knowledge and the outside.

© Springer International Publishing Switzerland 2015
O. Gervasi et al. (Eds.): ICCSA 2015, Part III, LNCS 9157, pp. 356–366, 2015.
DOI: 10.1007/978-3-319-21470-2_25

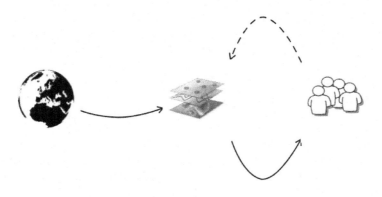

Fig. 1. Summary workflow

The locus - or focal point, of a given action, if isolated, brings no value-added into the surroundings communities. On the other hand, if channels are opened for participation of external actors, actions have a prolonged effect, which naturally depends on their significance. If the focus is on the activities of a research centre, then it is fair to say that with no channels to the outside, the end-product of research projects is of no use. Moreover, if communication between pairs is not symmetrical, research endeavours end up being redundant. It is the objective of this article to demonstrate the value-added of visibility and openness of research activities. The case-study presented has the main objective of producing a web-based digital platform, fully based on Free and Open Source Software (FOSS), which allows to visualize, explore and download in-house produced geographical layers, made available using open-data standards. The production-chain of new scientific research is discussed and means of increasing public exposure of methodologies and results presented.

Project Comuns' objectives are two-folded: first, it is intended to produce for the first time comprehensive datasets of historical information at the Portuguese census tract level, as well as other indirectly derived data. Second, by providing these information freely over the World Wide Web, a significant contribution is being made in terms of the richness of social demographic information available.

As the internet moves into a new paradigm in which agents interact easily, updating web-based information has facilitated knowledge upgrading. This happens in spite of validation issues, common in systems where unsupervised data abound. The strength and potential of Web 2.0 solutions is wide-spread as individuals are able to upload information into virtual Geography platforms [10] (Scharl and Tochtermann, 2007). In fact, with existing and emerging mobile sensor technologies, any human being is a potential contributor to new or existing layers of information (Goodchild, 2007).

A summarized workflow is shown in figure 1. One aspect which is a central motivation for the project is the purpose of closing the gap between academic research and civil society. A model was created with the objective of creating a new channel for the diffusion of new geographical knowledge. Layers of information are integrated into a geographical Database Management system (G-DBMS). Using only FOSS, layers were integrated into a common system where information is accessible through any web browser. Depending on the nature of particular information, the degree of interaction varied. Uploading capabilities for common users will be the main feature of Comuns 2.0.

The rest of the paper is organised as follows: the next two sections discuss the theoretical aspect of knowledge production and practice in Geography. Section 4 discuss implementation issues of the technological platform and presents examples of new datasets produced. Section 5 concludes.

2   On the Theory and Practice of Geography

Humans act as humans think; that is to say that human beings actions during their daily activity are a consequence of acquired knowledge, which help shape personality. To what extent interactions or genetic variables influence personality, actions/reactions, etc. is far beyond the scope of this article; the focus here is on the fact that humans build-up theories which function as pillars for representation, interpretation and interaction within a physical/social plane. Theory is in fact the root for everyday actions. It represents a metaphysical construct, which may be individual or collective. The same way humans construct bodies of rules which regulate social interactions, they root these on acquired knowledge (theories) on how society should work. Hubbard etal. (2005, p.8) distinguish between

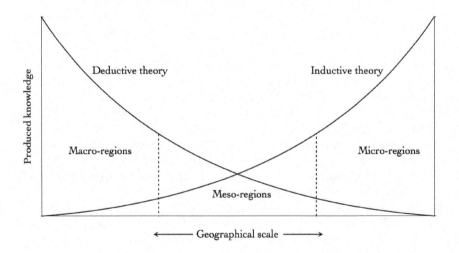

Fig. 2. Production of knowledge in Geography

a scientific and situated approach when they discuss the process of knowledge production. The former is related essentially to neutrality and objectivity in data collection, analysis and the universal validity of findings. The latter argues that knowledge production is dependent on environmental variables (where, when,) and on personal interpretations of the object of research. This discussion runs parallel to that which compares inductive and scientific methods. In Geography, such distinction is mainly related to geographical scale [6]. Figure 2 (adapted from [6, p.4]) makes the point that, at the micro-level, generalisations are difficult, as attributes are place-related. Knowledge is induced from local realities. The opposite happens as scale decreases and the units of analysis become larger.

It is important to stress that knowledge production goes well-beyond the scope of the academia. Every individual contribute in his life-span endeavour to the creation of a body of collective knowledge which shape collective values. It is through personal and collective experience that society progresses. Moreover, scientific research is of no value if its methods and results remain locked in a shelve (or stored in some piece of computer hardware). Scientific knowledge needs to be challenged, methodologies replicated. It needs to be exposed to the public.

Direct applicability varies between pure science and directly applicable science. One is not more valid than the other, and the degree of needed exposition to the civil society varies. But if society invests scarce resources into scientific research, there must be some linkage effect between the production of scientific knowledge and society in general. Yet, there is one variable which should be always accounted for: creativity. Knowledge production is an act of creativity, and when scientific research is bounded by specific goals, creativity is hampered. Hence, scientific knowledge production lives in and between the objective/subjective realms. Summing up, as with any human action, findings by researchers must be challenged by peers in order to became valid. Methods should be clear and freely available for replication. As should be results. Spillover effects into society are greater, the greater is the availability of knowledge.

3   On Theory, Geography and Perception

Harvey argues that theory is similar to maps in the sense that they both detail knowledge about the world [7]. Both use specific 'languages' to construct, store and communicate knowledge [8, p.3]. Moreover, they are both the outcome of models, which are no more than abstract representation of a given reality (again, the process of theory or map creation runs parallel to that of common daily activities).

Geographical knowledge is particular powerful as it helps shaping the way humans perceive the Earth surface. Its nature can be characterised according to two variables: scale and scope. Scale determines the size of the represented surface and to a great extent the detail of the research. Interpretations of the Earth surface depend to a great extent on acquired mental representations of such surface. For examples, Europeans grow up with images of the world such as that represented in figure 3. Europe is in the centre. This has important implications. Europeans think of Europe as the old-continent, from where the rest of the world was discovered; the focal point of the global community.

Figure 4 represents three distinct images of the globes, where different locus determine clearly distinct views of the same surface. Again, for a European, the world is centred in Europe, whilst for the Chinese in China and for the North Americans in North America.

As mentioned above, in relation to perception of the Earth, scale and scope play a decisive role. In terms of scale, as the area of analysis becomes smaller, detail generally tends to become greater. Local information/experience leads to the acquisition of knowledge through induction. At the global level, deductive knowledge is far more common.

Scope is related to aspects such as cultural variables. They may, for example, determine considerable variations between physical and cultural distances. Two simple examples illustrate this point. The physical distance between the southern tip of the Iberian Peninsula and Africa is very small. Yet, the cultural leap is

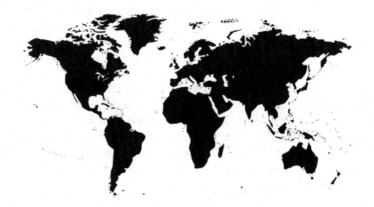

Fig. 3. Europe centered planisphere

Fig. 4. Views of the Earth - changing focal point

such, that Iberians feel much closer to other Europeans. At the urban scale, ethnic differences easily increase distances between contiguous neighbourhoods. Physical vicinity, although important to facilitate interactions and trust, is not decisive. If, on one hand, cultural borders exist in small areas, culture - or simply common interests, determine the emergence of non-spatially bounded interest groups. Examples are common within the internet realm. The internet has indeed become the primary channel for information diffusion. Hence, it is fair to say that Geography related research groups should look at virtual means of geographical representations as pivotal to their work.

4   App Commons

Using only FOSS, a framework was developed with the objective of creating a new channel for the diffusion of new geographical knowledge - Comuns. Layers of information are integrated into a Geographic Database Management system (GDBMS). PostgreSQL with spatial capabilities implemented with PostGIS (http://www.postgresql.org and http://postgis.refractions.net) proved a competent platform for both data storing and analysis. In fact, pre-processing of raw datasets and the application of dasymetric mapping algorithms would be difficult to achieve using for example desktop GIS technologies, given the size of the databank and the need to guarantee topological coherence.

Two Open Geospatial Consortium (OGC) standards allowed the publication of information for viewing and downloading purposes (respectively WMS and WFS). The R language (http://cran.r-project.org/) was the natural choice for statistical computing, whilst R-Shiny provided the tools to integrate R into a web-based application. Integration with the javascript library Leaflet (http://leafletjs.com/) allowed the production of a visual mapping interface.

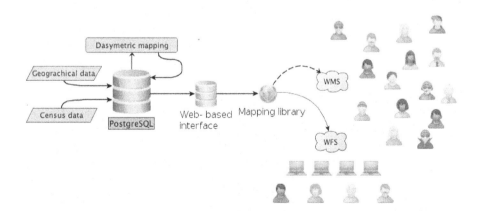

Fig. 5. Spatial data infrastructure

The tools developed are implemented within a web-interface, in a dedicated platform. R statistical computing language and libraries allow deploying over the web flexible interfaces which take advantage of all the power of R, using for example the R Shiny library. The Toolbox provides three means interacting with the DBMS: (1) search engine; (2) analytical engine; (3) Web-GIS platform using open libraries. It is expected that the the analytical engine will adapt over time, responding to feedback from users in terms of tools they would like to see implemented. The large scope of R functions facilitate the smooth integration of these new tools.

Census Historical High-Resolution Datasets. The growing availability of census micro-data - demographic data aggregated for small-areas, allows detailed analysis of the social structure of small neighbourhoods. Such exercises are in most cases static since census tracts geometries change between every census exercise. Using dasymetric mapping techniques implemented within a spatially enabled PostgreSQL database, historical datasets were built for Portuguese census information, for years 1991, 2001 and 2011 [9]. Starting form non-matching geometries (ie. distinct borders), data was re-allocated to the minimum common intersecting areas. This issue was more acute when working with the 2001 and 2011 datasets given the considerable differences between aggregating schemes (figures 6 and 8). Geometries corresponding to the three periods' census tracts were intersected to generate a weighting scheme which allowed re-distribution of data according to common geometries. The algorithm used have been developed over the past few years and applied in different geographical settings (see for exemple [9]).

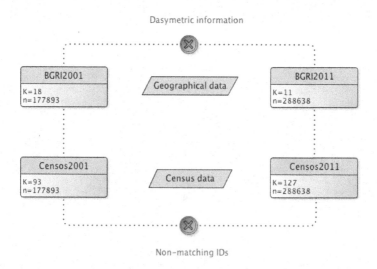

Fig. 6. Producing census historical datasets using dasymetric mapping techniques

4.1 FOSS and Open-Data

The profile of end-users (data-analysts, policy-makers, civil society) implies that the data-model will reflect open-data standards (http://open-data.europa.eu/en/data/, http://ec.europa.eu/digital-agenda/en/open-data-0) and the G8 Open Data Charter (https://www.gov.uk/government/publications/open-data-charter/g8-open-data-charter-and-technical-annex). Moreover, The SGBD is totally anchored on Free and Open Source Software (FOSS) solutions, which will guarantee that research efforts are reproducible, increasing accountability and the overall impact [2, p.174] [1]. However, care has to be taken in terms of semantics. The term Free in FOSS does not mean that it is free of cost, but is referring to the software freedoms that are addressed to FOSS software [11]. These freedoms (0 to 3) are stated in the Free Software Foundation (FSF) Website [3,4] and refer to:

0 - The freedom to run the program, for any purpose.
1 - The freedom to study how the program works, and change it so it does your computing as you wish. Access to the source code is a precondition for this.
2 - The freedom to redistribute copies so you can help your neighbour.
3 - The freedom to distribute copies of your modified versions to others. By doing this you can give the whole community a chance to benefit from your changes. Access to the source code is a precondition for this.

4.2 Implementation and Maintenance

The development of platform Comuns may be summed up in three distinct stages: (1) Implementation, (2) development and (3) projection. In stage 1, a Spatial Data and Analysis Infrastructure (SDAI) is produced, anchored on a Geographical Database Management System (G-DBMS) where all data is stored and which serves as the technological backbone of the project. The SDAI include mainly large scale high-resolution socio-economic datasets, although there is scope for future contribution of other validated spatial data. In stage 2, quantitative analytic tools are design and implemented; stage 3 is on-going as it is

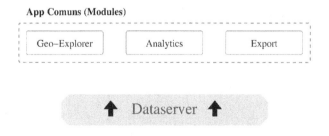

Fig. 7. Application modules

Fig. 8. Census tracts (1991, 2001, 2011)

related to the development of branch applications, which may eventually lead to parallel projects, anchored on the same datasets, which may or may not use the same SDAI.

4.3 Examples of New Available Open-Geodata

The following case-studies exemplify the use of Comuns. The added-value of both cases are the quantification of different human phenomena, which may alter perceptions at a micro and macro scales.

Place Names Dataset. One aggregating level which is useful for distinct applications, from geo-marketing to planning is local villages. The production of this layer was done again within the G-DBMS environment, using census data (see figure 9 for a dataset snippet).

Peripherality. As more research projects evolve using Comuns' datasets, the platform becomes richer. One exercise performed in-house intended to synthesize the concept of spatial peripherality to its primary element: geographical position, and how it can be measured. It is argued that by creating a singular, yet rigorous indicator, an essential tool is being provided to the geographer or to any social scientist whose experimental table is a spatial surface/phenomenon.

Fig. 9. Place names dataset with resident population (2011)

Its use together with other socio-economic indicators is of great utility when one wants to explain supply and demand along a closed surface. The indicator built is a function of geographical distance, weighted by resident population, given the fact that the place where each agent lives is, in the limit, the summary or the epitome of a large set of natural, social and economic variables.

Using GIS tools provided by Geographic Resources Analysis Support System - GRASS (http://grass.osgeo.org/), raster accessibility surfaces were built using the road network as source layer. Figure 10 shows an example of two surfaces, each with a resolution of 50 metres per cell, representing time distances to the coast, generally considered in Portugal to be a metaphoric representation of general social and economic development.

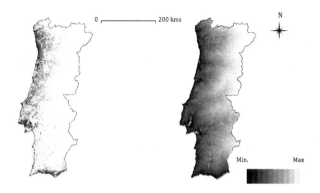

Fig. 10. Accessibility surfaces

5 Final Considerations

The preconception that an innate void exists between the Academia and civil society is rooted in the non-existence of effective bridges between the two. Web-based application-focus provides the foundations for an alternative paradigm to emerge; one dictated by accountability, no only to peers, but to any possible stakeholder.

Comuns is an initiative strongly rooted on the belief that information should be made available to the general public meeting high quality standards and through easy-to-use interfaces. The initiative became operational using only Free and open source software (FOSS). The obvious costs' reductions are perhaps the less important advantage of such strategy; FOSS guarantees accountability and promotes knowledge spillovers.

The fact that longitudinal datasets are made available for large areas, but with significant geographic detail potentially help bridging the gap between inductive and deductive produced knowledge. This is so since the analysis of general trends is now coupled with detailed information of the distribution of phenomena of interest. Mean behaviours, but also heterogeneity analysis is now possible for large areas. In other words, the estimation of deduced mathematical models is no longer restricted to mean estimated coefficients, but to the whole distribution of these coefficients.

Comuns is not an end in itself; information is only of any use if and when it is applied for the production of new knowledge and the understanding of a given reality. The scales (distinct geographical levels - from macro to micro regions) and scopes (heterogeneous rather than simply central tendencies analysis) of the databank allow its use is a multitude of applications. Also, the focus on FOSS and open-data guarantee the possibility of reproducibility and replication in other geographical settings.

References

1. Crowston, K., Li, Q., Wei, K., Eseryel, U.Y., Howison, J.: Self-organization of teams for free/libre open source software development. Information and Software Technology 49(6), 564–575 (2007)
2. De Paoli, S., Miscione, G.: Relationality in geoIT software development: How data structures and organization perform together. Computers, Environment and Urban Systems 35(2), 173–182 (2011)
3. FSF. what we do, Free Software Foundation
4. GNU. What is free software? The free software definition (2012)
5. Goodchild, M.F.: Editorial : Citizens as Voluntary Sensors : Spatial Data Infrastructure in the World of Web 2. 0. International Journal 2(2), 24–32 (2007)
6. Hagget, P.: Locational Analysis in Human Geography. Edward Arnold Ltd. (1969)
7. Harvey, D.: Explanation in geography. Edward Arnold (1969)
8. Hubbard, P., Kitchin, R., Bartley, B., Fuller, D.: Thinking Geographically: Space, Theory and Human Geography. Continuum studies in geography. Bloomsbury Academic (2002)
9. Rodrigues, A.M., Santos, T., de Deus, R.F., Pimentel, D.: Land-use dynamics at the micro level: constructing and analyzing historical datasets for the portuguese census tracts. In: Murgante, B., Gervasi, O., Misra, S., Nedjah, N., Rocha, A.M.A.C., Taniar, D., Apduhan, B.O. (eds.) ICCSA 2012, Part II. LNCS, vol. 7334, pp. 565–577. Springer, Heidelberg (2012)
10. Scharl, A., Tochtermann, K.: The geospatial web: how geobrowsers, social software and the Web 2.0 are shaping the network society. Advanced Information and Knowledge Processing. Springer, Berlin (2007)
11. Steiniger, S., Hay, G.J.: Free and Open Source Geographic Information Tools for Landscape Ecology. Ecological Informatics 4(4), 183–195 (2009)

Building 3D City Models: Testing and Comparing Laser Scanning and Low-Cost UAV Data Using FOSS Technologies

Carla Rebelo, António Manuel Rodrigues[✉],
José António Tenedório, José Alberto Goncalves, and João Marnoto

CICS.NOVA Interdisciplinary Centre of Social Sciences,
Faculdade de Cincias Sociais e Humanas, Universidade Nova de Lisboa,
Lisbon, Portugal
{crebelo,amrodrigues}@fcsh.unl.pt
http://www.cics.nova.fcsh.unl.pt/

Abstract. Presently, the use of new technologies for the acquisition of 3D geographical data on time is very important for urban planning. Applications include evaluation and monitoring of urban parameters (ie. volumetric data),indicators of an urban plan, or monitoring built-up areas and illegal buildings. This type of 3D data can be acquired through an Airborne Laser Scanning system, also known as LiDAR (Light Detection And Ranging) or by Unmanned Aerial Vehicles (UAV). The aim of this article is to use and compare these two technologies for extracting building parameters (facade height and volume). Existing literature evaluates each technology separately. This work pioneers benchmarking between LiDAR and UAV point-clouds. The basic function of LiDAR is collecting a georeferenced and dense 3D point-cloud from a laser scanner during flight. Therefore it is possible to obtain a similar 3D point-cloud using processing algorithms for stereo aerial images, obtained by large or small-format digital cameras (the small-format camera implemented in Unmanned Aerial Vehicles). The chosen study area is located in Praia de Faro, an open sandy beach in Algarve (Southern Portugal), limited west by the Ria Formosa barrier island system. The area defined has an extension of 300100m. The methodology is divided in two distinct stages: (1) building parameters extraction, (2) comparative technology analysis. Lidar point-cloud resolution is approximately 6 pts/m2 and UAV point-cloud 60 pts/m2. FOSS technologies have proven to be the most adequate adequate platform for the development and diffusion of advanced analytical tools in the Geographical Information Sciences (GISci). Data management in this paper is supported by a Geographical Database Management System (GDBMS), implemented using PostgreSQL and Post-GIS. Statistical analysis is performed using R whilst advanced spatial functions are used in GRASS.

Keywords: LiDAR · UAV · FOSS · Point-cloud · Building parameters

C. Rebelo—CICS.NOVA Interdisciplinary Centre of Social Sciences - Faculdade de Ciencias Sociais e Humanas - Universidade Nova de Lisboa.
J.A. Goncalves—Faculdade de Ciencias da Universidade do Porto.
J. Marnoto—SINFIC, S.A.

O. Gervasi et al. (Eds.): ICCSA 2015, Part III, LNCS 9157, pp. 367–379, 2015.
DOI: 10.1007/978-3-319-21470-2_26

1   Introduction

The automatic extraction of buildings parameters, such as building height and volume, can be most useful in urban planning contexts. These parameters, extracted from advanced remote-sensing technologies, allow producing 3D building models to support the monitorization of urban plans and keep track of different parameters such as illegal changes in built-up areas (new block buildings or number of floors) and a better visualization of the proposed plan in public discussion. These parameters can also be of help in gathering more precise urban indicators.

Advanced technologies such as airborne laser scanning and low-cost UAV (also called UAS - Unmanned Aerial Systems) imagery allow a higher degree of automation in acquiring 3D data, in opposition to the classical methods of digital photogrammetry. This is quite important because the classical stereo-restitution performed in a digital photogrammetric workstation (defined by a human operator) for accurate measurement in a large set of buildings is very time consuming. Both technologies produced a 3D point-cloud data which represents a set of georeferenced data points in a three-dimensional coordinate system. These dense clouds can be acquired automatically through an active aerial sensor system laser scanning or from the combination of UAV and automated dense multi-stereo image-matching processing.

The LiDAR point-cloud is acquired from a LiDAR (Light Detection and Ranging) system. The basic principle of LiDAR system is to record a set of discrete and massive elevation points above datum using a laser scanning and a direct georeferencing system (GPS/INS - Inertial Navigation System). The laser emits millions of pulses per second to the ground and part of those backscattered pulses return to the laser. At the same time each pulse can be directly georeferenced by the position and altitude of an airborne sensor to the local coordinate system (or six parameters of exterior orientation). All points of a LiDAR point-cloud are obtained from these pulses, which are classified as first and last return. The coordinates of these points are obtained from the following parameters: i) the time between the emission and reception of an energy pulse in sensor (distance value); and ii) from the six parameters of exterior orientation given by GPS/INS. The point density of LiDAR data depends of the flight height (which defines the footprint size of pulse) and the particular characteristics of Laser scanning (beam divergence and effective measurement rate of laser scanning).

The survey of urban areas for 3D modelling of buildings requires a small footprint of pulse LiDAR in tandem with high point density [7]. The UAV point-cloud requires an automated multi-stereo aerial matching processing of UAV imagery. The UAV system is a low-cost and ultra lightweight aerial photogrammetric system, which is able to collect very high-resolution imagery with a higher overlap (80%-90% in flight line). This system integrates a small-format digital camera and a miniaturized direct georeferencing system (GPS/INS). Some of the vantages of this system when compared with conventional digital airborne LiDAR and photogrammetric systems are: 1) low-cost; b) automatic pilot which allows driving the UAV automatically on the flight lines and capture the images; and c) the time between the decision to make the flight for the acquisition of aerial image and the acquiring of 3D point-cloud can be less than 24 hours. After the effective acquisition of multiple overlapping UAV imagery, a dense multi-stereo image-matching algorithm is applied to estimate the 3D point coordinates for each pixel. The point density of a UAV point-cloud depends of the resolution of aerial images and of the number of point matches found on stereo image pairs.

Over the past few years, 3D point-clouds obtained by LiDAR or automated image matching techniques have been used and tested by several authors: (i) in 3D urban models by [1,6,9]; (ii) more specifically in the extraction of building elements [4,8,10].

Regarding the accuracy of these technologies, LiDAR enables 5-10 cm of vertical accuracy [2]; the accuracy of UAV data is influenced by the resolution of the imagery and the texture and terrain through the scene [5]. The challenge of this study was to apply a (semi)-automatic extraction methodology of building parameters - building facade height and building volume - from each different 3D point-clouds data (UAV and LiDAR) using Free and Open Source Software (FOSS): GRASS GIS, PostgreSQL/PostGIS functions and R statistical language. The use of FOSS increases accountability and reproducibility, apart from reducing operational costs.

In this work we report the difficulties in acquiring these parameters from 3D point-cloud without reference data and we compare and evaluate the accuracy of building parameters extracted from different sources, UAV and LiDAR, under the same methodology.

2   Study-Area and Data Acquisition

The study area - Praia de Faro - is an island-barrier bounded north by the Ria Formosa estuary and South by the sea, located in the South Coast of Portugal - the Algarve The selected geographic area (Figure 1) has approximately 2.5 ha, with a width of 100m north to south and 250m east to west along the principal road of the island. It is a built-up area with 19 buildings. The majority are single-family dwellings with a maximum of two floors, although there is a building located north-east of the study area with four floors.

The buildings represent a diversity of architectural styles and types, with irregular shapes. The roofs are either flat, multiple-level flat, or pitched and complex (with different slopes). The degree of dissimilarity of building's shape is high.

2.1   3D Point-Clouds

The 3D point-cloud collected through the LiDAR system was performed by a TopEye MK II (Figure 2b) at a flight height of 500m above ground. The laser

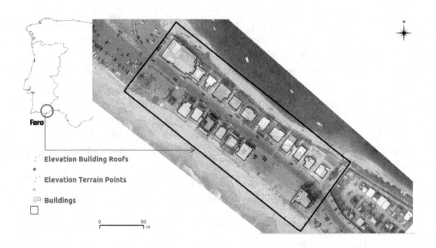

Fig. 1. Study-area: Praia de Faro

scanning used by this system has an elliptical pattern. According to the flight planning report, this point-cloud has a vertical accuracy of 10 cm. It is important to note that the point-cloud was directly acquired by the company that made the flight, without our participation.

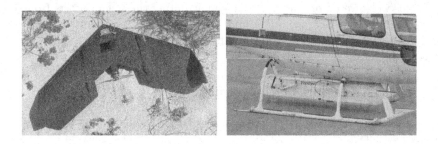

Fig. 2. Airborne systems: 2a) UAV system -Swinglet CAM; and 2b) LiDAR system TOP EYE MKII

The acquisition of UAV imagery data was performed from a Swinglet CAM produced by SenseFLY Company. This system weighted about 500 grams and has an autonomy of approximately 30 minutes of flight time. This UAV system requires at most moderate wind not above 7 m/s.

Flight Planning and Processing UAV Imagery - The flight planning lines were designed in order to acquire stereo aerial images with a 5cm resolution and a higher endlap (along flight) and sidelap (between flight lines) about 90% and 60%, respectively. This flight was performed with a wind speed bellow 10km/h.

The study area was covered by 46 aerial images (3000 by 4000 pixels) at a flight height with approximately 130m.

After the visual inspection of the quality of the images, follow-up of the multi-stereo image matching processing was performed by an automatic workflow implemented in PiX4D software, to obtain the 3D point-cloud. During this processing six Ground Control Points (GCP) were included, to generate a more accurate point-cloud. This means measuring the GCP for all images where it appears. In this particular case it was about seven images per GCP. The authors acknowledge the fact that FOSS alternatives to PiX4D do exist, and should be explored in future research. Still, existing alternatives are not as robust, which in this case justified the use of a commercial solution for pre-processing. The SFM (Structure From Motion) opensource is an alternative for getting the UAV point-cloud[?]. However, the SFM have a lower performance and still does not guarantee a higher vertical accuracy that is necessary for this type of urban applications.

The flight planning and the processing of the UAV imagery to acquire the 3D point-cloud, true orthomosaic and the digital surface model were made in a few hours.

Characterization of Point-Clouds -   The point density of the UAV point-cloud under the study area is higher than LiDAR data (Table 1).

Table 1. Characteristics of 3D point-clouds

Technology	Date acquisition	Number of points	Density points/m2	Elevation statistics
UAV	April 2013	1,142,095	61 pts/m2	Zmax=16.91; Zmin=-0.21; Zmean=5.14; Zmedian=4.08
LiDAR	November 2009	146,149	6.3pts/m2	Zmax=21.86; Zmin=-0.03 Zmean=4.92; Zmedian=3.83

However, the range of elevation values of LiDAR data is larger than the UAV data, because LiDAR recorded very tall cypress trees near the building located to the southeast. The LiDAR system is more accurate in vegetation and tree objects detection. The distribution of the UAV point-cloud is more sparse and irregular than the LiDAR point-cloud (Figure 3). On the other hand, the UAV point-cloud has gaps and low points in some building roofs (gaps). Also, vegetation or trees near the buildings is not recorded unlike in the LiDAR data, which might be an advantage in this study because this data had to be removed.

Figure 3 shows that most elevation data registered for both clouds ranged between 3 to 5 meters.

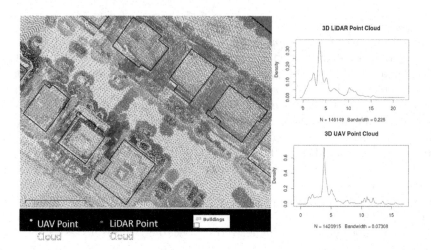

Fig. 3. Comparison of LiDAR and UAV points distribution. Density functions of the elevation values of each point-loud.

2.2   Reference Data

For this study large-scale 2D vector data (1:2000) was used as reference data to evaluate area measurements extracted for each building's. 3D vector data (points) was used to evaluate estimated data for buildings' facade height from each 3D point-cloud. These elevation points were acquired from direct field measurement with ground surveying.

The characteristics of reference data used in this study can be seen in table 2.

The 2D vector data of building outlines (Figure 1) was used to calculate the building area and the 3D vector to calculate the reference building facade height. The distribution of these 3D points can be seen in figure 1. The buildings volume reference was computed from these two reference parameters.

Furthermore, the true orthoimages produced from aerial images and Digital Surface Model (DSM), were used in this study for visual inspection, such as visualization and comparison of the building roofs extracted from 3D point-cloud data.

3   Methodology

The methodology developed for the extraction of building parameters from each point-cloud was based on the following principles: i) extraction of building parameters without vector reference data - only 3D point-cloud should be used; and ii) use Free and Open Source Software (FOSS) tools to implement a robust methodology for the acquisition of these parameters.

First, it is important to define the two building parameters that will be extracted from 3D point-cloud and which are involved in estimating a building volume. The parameters are: i) the Building facade height, which is the difference

Table 2. Description of reference data

Data	Year	Technical acquisition	Details of data
3D vector data	2012	Reflectorless Total Station (Leica TCR 705) for roof points and GPS to Ground Control Points (GCP)	Elevations points of roofs (corners and prominent points)
2D vector data	2002	Photogrammetric stereo-restitution; large scale: 1:2000	Building outlines and road network
True orthoimages	2009	Camera Rollei AIC P20 (16 MP); Data source: Aerial images from the LiDAR flight	Resolution 9 cm
True orthomosaic	2013	Data source: Aerial images from the UAV flight	Resolution 5cm; Near-infrared images (NIRGB)

between a mean elevation of the buildings' topmost limits (these are approximately the points that define the eave of the roof) and the mean elevation of the ground near the building. However, taking into consideration that a building can have different facade heights, according to its deployment on the ground, only one facade side of the building was chosen to compute this parameter - the side of building along the main public street was chosen; and ii) the area parameter is defined from the boundary of the building facade height, which is equivalent to the buildings' roof area. Thus, a building's volume is obtained from by multiplying the mean value of the facade height and the area of building roof.

The methodology developed for each point-cloud data included the following steps (Figure 4): i) selection of the set of points from the point-cloud that represents the building roofs. This filtering applied to the point-cloud was performed by a clustering CLARA (Clustering Large Applications) algorithm based on elevation values. CLARA represents a partitioning of a dataset into k- clusters around k medoids [3], which is implemented in the RCLUSTER library; ii) extraction of the building roof area was based on the generation of polygons from the points selected above, using the concave-hull algorithm implemented in GRASS 7. This algorithm creates a concave-hull from a subset of 3D points using Delaunay Triangulation. Then, the segment lines are removed if the triangular faces have a length above the threshold value defined by the user. This threshold allows to get polygons that better represent the buildings' roof area; iii) selection of the set of points that represents the edges of buildings on the top and ground using spatial analysis functions; and iv) calculate each building's volume using building facade height (mean value obtained from the points previously selected) and area.

All the steps above have been implemented in two scripts for the automation of the methodology. The scripts were developed using the R programming language (clustering CLARA step) and SQL language inside a Geographical Database Management System (GDBMS) implemented using PostgreSQL/PostGIS (iii and iv steps below). Results evaluation was performed inside the GDBMS.

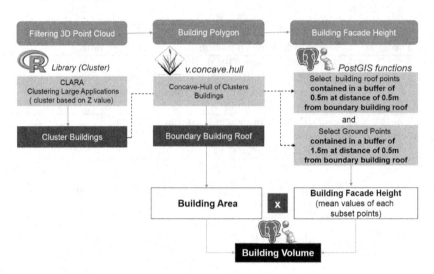

Fig. 4. Methodological approach for building volume extraction based on a 3D point-cloud

4   Results and Discussion

The accuracy of the building parameter facade height estimated is strongly dependent of the points selected during the first and third steps of the methodology. Yet, buildings' area is dependent of the success of the clustering exercises. The behaviour of LiDAR point-cloud and UAV point-cloud along the methodology is slightly different however, in general, the difficulties found in the extraction of building parameters were approximately identical. Next, the results obtained from each point-cloud will be explained in detail.

4.1   Evaluation of Building Area and Building Facade Height Parameters from LiDAR and UAV Point-Clouds

The buildings' area estimated from each point-cloud show some of the difficulties in defining building roof's boundary (Figure 5). The clustering process for LiDAR point-cloud have better results for a higher K value (number of clusters) than UAV point-cloud, respectively $K_{LiDAR} = 10$ and $K_{UAV} = 2$.

Although only some clusters were chosen (Figure 5), one and four clusters from the K_{UAV} and K_{LiDAR} values respectively.

The shapes of building roofs extracted are more regular in LiDAR point-cloud. The buildings assigned with a circle have an inaccurate area, because there are gaps where the UAV data does not have have 3D points. These gaps can be due to an inaccurate multi-stereo image matching processing of the aerial images. On the other hand, for UAV data, the threshold values chosen for concave-hull were higher (or a more concave polygon), unlike LiDAR. In this case a high-density point-cloud can influenced the behaviour of this step.

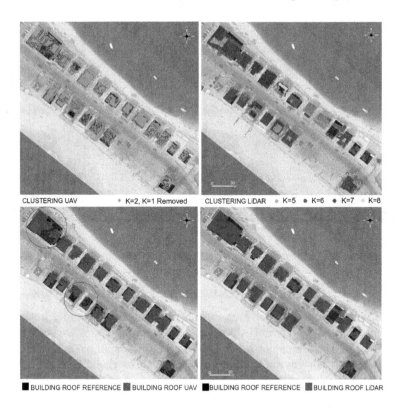

Fig. 5. Clustering results and building roofs (area) extracted from each point-cloud, LiDAR and UAV

The evaluation of the results achieved for the building facade height parameter was based on vertical error. The vertical error of the building facade height estimated corresponds to the difference between the reference value calculated from 3D vector data and the value estimated from point-cloud. The magnitude of vertical errors in the estimation of building facade height from each point-cloud can be seen in figures 6a and 6b.

Some of the best values (lower vertical errors) were obtained for buildings visible in figure 6a. About 50% of the total buildings recorded have a vertical error above 50cm. The results do not show a stronger evidence that the magnitude of the vertical error is dependent of the type of building (complex or flat). Nevertheless, the flat building roofs in figure 6a) have the same magnitude of vertical error in both point-clouds.

The worst vertical error of LiDAR (2.67m) was obtained in the buildings where balconies were considered as building roof. For UAV the worst value was obtained for the buildings that were not fully covered by UAV points.

In figure 7 is possible to visualize the behaviour of the building parameters estimated for each point-cloud when compared to the reference value parameters.

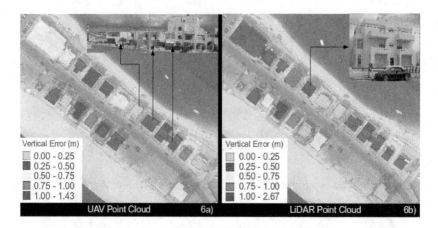

Fig. 6. Visualization of the distribution of vertical errors obtained for each building. 6a) Vertical errors from UAV point-cloud; and 6b) Vertical errors from LiDAR point-cloud.

Empirical density functions for area estimates show that generally both point-clouds approximate the reference distribution (empirical modes are similar). Yet, in regard to LiDAR, it was not possible to distinguish small irregularities, hence the larger mode. On the other hand, UAV captures the differences between buildings with too much detail (if we assume reference values as the "true" values). Yet, the estimated curve for the UAV shows a better approximation of the true values. The circle in figure 7a identifies the presence of an outlier, which represents the major error area obtained in estimation of building area from LiDAR point-cloud, i.e. the problems mentioned above for building marked in figure 5.

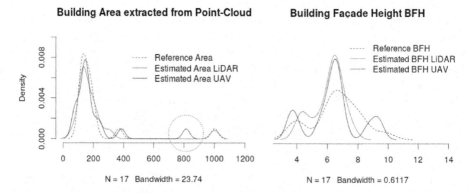

Fig. 7. Empirical density functions. a) Reference building area vs. estimated buildings area; b) The true building facade eight curve and the building facade height estimated.

The empirical density functions of each point-cloud in the estimation of the building facade height parameter is very similar (Figure 7b). Most values were overestimated by LiDAR. The estimated curve of UAV data approximates slightly better the true values.

Table 3 also shows that the errors obtained in the estimation of these parameters are very similar. The maximum error for UAV in estimation of area is an outlier (187.5m). From LiDAR there is an outlier in estimation of building facade height with a 2.67m value.

Table 3. Evaluation of building's volume from LiDAR and UAV point-clouds

Parameter	Point-cloud	Mean	Median	Minimum	Maximum
Error Area (m^2)	UAV	36.06	29.62	1.36	187.74
	LiDAR	25.24	17.79	0.40	62.45
Error Building facade Height (m)	UAV	0.61	0.51	0.01	1.43
	LiDAR	0.74	0.56	0.0	2.67

The standard error achieved for buildings' volume ranged approximately from 1% to 43% of the reference value. The magnitude of these errors is mainly due to the estimated parameter area from UAV or LiDAR data. Additionally, it is important to highlight that the error in estimating building volume with data from reference area decrease significantly, ranging approximately from 0.1% to 27%.

Comparing the vertical errors obtained in building facade height estimation with the errors obtained for the estimation of building volume computed with reference area is possible to identify three situations: a) a vertical error in estimation of building height facade up to 50cm, implies an error under 10%; b) a vertical error up to 1m, means an error in volume under 15%; and c) the vertical error between 1-2.5m, results in an error of building volume that ranges between 15% to 35%.

Table 4. Statistical measures of vertical errors for building volume estimated

Parameter	Point-cloud	Mean	Median	Minimum	Maximum
Error Volume (%) BFH*Reference-area	UAV	8.8	8.6	0.3	16.8
	LiDAR	9.8	10.1	**0.1**	27.2
Error Volume (%) BFH*Area	UAV	21.3	23.9	3.9	43.0
	LiDAR	14.4	8.9	0.7	35.3

In figure 8 we can see the behaviour of building volumes estimated from each point-cloud based on reference area. The estimated curve of UAV has a slightly better approximation to the true (reference) values.

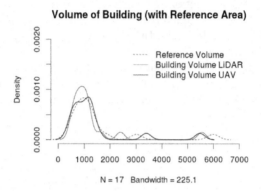

Fig. 8. Empirical density functions of building's volume estimated from each LiDAR and UAV point-clouds and building's volume relative to the reference area

The errors made by the estimation of building facade height with the UAV point-cloud have contributed to a total of volume error (computed with reference area) of all buildings equal to $1276m^3$, which corresponds to 4% of the true total volume. If we consider only the buildings with a vertical error lower then 1m in the estimation of total volume, then the percentage of error decreases to 3%. For LiDAR point-cloud the error in total volume is 9% and 4% percent more in the same situations.

5   Concluding Remarks

This work introduced a methodology for the (semi)-automatic extraction of building parameters from a 3D point-cloud, using FOSS tools. Also, it compares and analyses the accuracy and performance of different point-clouds (LiDAR and UAV imagery) in the extraction of these building parameters.

The most useful characteristics when using open source software for this study are: a) the capacity of processing dense point-clouds within a geographical spatial database; and b) the possibility of automation of some of the procedures involved in this type of studies.

The results obtained in the extraction of building parameters are very similar using both LiDAR or UAV. However, we can conclude that if the urban area has a dense vegetation and tall trees near the buildings, UAV data can be more appropriate, because it does not introduce residual information in the process. However, this is only true if the building is not surrounded completely surrounded by trees, otherwise we would have gaps in the buildings.

The major difficulty in this study is the extraction of accurate building roof (or area) data with a regular shape from point-cloud. Even facing a wide variety and complexity of building roofs (with various slopes), the results are acceptable for some stages of an urban plan.

We believe that the low-cost UAV imagery together with a robust methodology using FOSS tools can be very useful in the production of 3D buildings models for urban planning, unlike the LiDAR system. The accuracy of results shows that it can be enough for: i) a process of discussion and public participation in the planning process; ii) for the monitoring of the built-area (ex. detection of illegal changes in the height of buildings). The results clearly show that UAV technologies are a valid alternative to LiDAR. These findings have significant consequences in terms of project management in urban planning, with important methodological and financial implications.

Acknowledgments. This article was partly funded by Portuguese national funds through FCT Portuguese Foundation for Science and Technology in the framework of projects PEsT-UID/SOC/04647/2013, SFRH/BPD/76893/2011 and SFRH/BPD/66012/2009.

References

1. Hirschmüller, H., Bucher, T.: Evaluation of Digital Surface Models by Semi-Global Matching. Robotics (2010). http://www.robotic.dlr.de/Heiko.Hirschmueller
2. Hyyppä, J.: State of the art in laser scanning. In: Fritsch, D. (ed.) Photogrammetric week 2011, No. 2008, pp. 203–216. Institut für Photogrammetrie (2011)
3. Leonard, K., Rousseeuw, P.J.: Finding Groups in Data: An Introduction to Cluster Analysis. John Wiley & Sons, Inc. (2008). doi:10.1002/9780470316801.indauth
4. Khoshelham, K., Nardinocchi, C., Frontoni, E., Mancini, A., Zingaretti, P.: Performance evaluation of automated approaches to building detection in multi-source aerial data. ISPRS Journal of Photogrammetry and Remote Sensing **65**(1), 123–133 (2010)
5. Küng, O., Strecha, C., Beyeler, A., Zufferey, J.C., Floreano, D., Fua, P., Gervaix, F.: The accuracy of automatic photogrammetric techniques on ultra-light UAV imagery (2012)
6. Lafarge, F., Mallet, C.: Creating large-scale city models from 3D-point clouds: A robust approach with hybrid representation. International Journal of Computer Vision **99**(1), 69–85 (2012)
7. Lemmens, M.: Geo-information: Technologies, Applications and the Environment. Geotechnologies and the Environment. Springer (2011). http://www.google.pt/books?id=n_tUAWYg4UQC
8. Tenedório, J.A., Rebelo, C., Estanqueiro, R., Henriques, C.D., Marques, L., Gonçalves, J.A.: New developments in geographical information technology for urban and spatial planning. In: Pinto, N.N., Tenedório, J.A., Antunes, A.P., Roca, J. (eds.) Technologies for Urban and Spatial Planning: Virtual Cities and Territories, Chap. 10, pp. 196–227. IGI Global, Hershey/Pennsylvania (2013). http://www.igi-global.com/chapter/new-developments-in-geographical-information-technology-for-urban-and-spatial-planning/104217
9. Xie, F., Lin, Z., Gui, D., Lin, H.: Study on construction of 3D building based on UAV images (2012)
10. Zeng, Q.Z.Q., Lai, J.L.J., Li, X.L.X., Mao, J.M.J., Liu, X.L.X.: Simple building reconstruction from LIDAR point cloud. 2008 International Conference on Audio, Language and Image Processing (2008)

Discrete Simulation of Pedestrian Dynamics on a Triangulated Ring Structure

Minjie Chen[1], Günter Bärwolff[2], and Hartmut Schwandt[1(✉)]

[1] MA 6-4, Institut für Mathematik, Technische Universität Berlin, Berlin, Germany
{minjie.chen,schwandt}@math.tu-berlin.de
[2] MA 4-5, Institut für Mathematik, Technische Universität Berlin,
Straße des 17. Juni 136, 10623 Berlin, Germany
baerwolff@math.tu-berlin.de

Abstract. We propose a new modelling method for the simulation of pedestrian dynamics when the walking direction of the pedestrians cannot be represented by straight lines. The geometry of the simulation is approximated on a special triangular grid. We also study the pedestrians' step execution for the general case of multi-position velocities and the possible interaction among them. We discuss the model on a ring-formed environment with periodic boundary.

Keywords: Pedestrian dynamics · Path-oriented coordinate system · Triangular grid · Periodic boundary

1 Introduction

Situations and environments which are characterized by the presence of numerous people like public places, shopping malls, airports, railway stations, sport stadiums, manifestation areas etc. require a proper management and regulations ensuring trouble-free functioning, i.e. a preferably smooth movement of all persons and, in particular, precautionary measures under security aspects for special situations like jam.

In this sense, the knowledge and the understanding of pedestrian behaviour and pedestrian movements has become an important scientific issue in the past twenty years. Pedestrian behaviour and pedestrian dynamics cannot be described by a closed mathematical model or equation as they are mostly governed by typical behavioural rules, situation-dependent aspects, i.e. stochastic events, but also physical phenomena and individual human decisions and actions. Consequently, the description of pedestrian dynamics mostly relies on mathematical simulation models. In view of the great variety of influence factors, there is no unique simulation strategy. We distinguish two important classes of simulation models. *Microscopic* models focus on the description of the individual pedestrian as a particle of a crowd. They consider individual behaviour and the interaction of individual particles. Typically, microscopic approaches are based on discrete simulation models. *Macroscopic* models consider pedestrian crowds as large number

O. Gervasi et al. (Eds.): ICCSA 2015, Part III, LNCS 9157, pp. 380–389, 2015.
DOI: 10.1007/978-3-319-21470-2_27

of particles acting mostly like a mass, i.e. revealing global mass effects rather than individual behaviour. Physical principles like mass, momentum and energy balance become predominant in large crowds. It is well-known, for example, that pedestrians in very large crowds behave similarly to particles in gas flows. Therefore, macroscopic simulation models are usually derived form appropriate (partial) differential equations. At ICCSA 2013, we presented a macroscopic model in this context [9]. An overview of both vehicular and pedestrian traffic modelling and simulation approaches, in particular macroscopic models, can be found, for example, in [5].

In the category of discrete microscopic approaches, the so-called *cellular automaton (CA) models* play a prominent role. Simulation of pedestrian dynamics is mostly carried out on geometric settings in two dimensions. Therefore, the geometry is generally represented by a two-dimensional grid. Regular grids have been adopted in CA models in which pedestrians are simulated as agents subject to a set of universal rules [2], [8], [6]. However, regular grids suffer from some drawbacks. Regular grids do not allow for the modelling of individually varying space requirements or for a sufficiently precise physical modelling of the pedestrians trajectories, in particular, when the latter cannot be represented or approximated directly by straight line segments.

In [3] we discussed the possible application of triangular grids. Part of our previous work finds its origin in [4] where the notion of a fictional *path force* had been introduced indicating the influence under which the pedestrians are drawn toward the centre line of the path. [3] proposed a specific discretization scheme called *path-oriented coordinate system*. In the current text, we construct a ring-formed structure by connecting multiple path-oriented coordinate systems to study the dynamics of pedestrians in a circular environment with periodic boundaries.

2   Model

2.1   Geometry and Triangulation

We briefly sketch the construction of the path-oriented coordinate system for our modelling with correction and revision of the formulation in our previous work [3].

Assume the orientation of a path S and the maximum deviation $\delta > 0$ a pedestrian can be away from the orientation to be known. Without loss of generality, we request that the sign of the curvature of S remains unchanged, since any arbitrary path can be decomposed into path segments satisfying this condition. It is then important to choose a suitable grid cell size in the discrete modelling respecting this geometry. Starting with an adequate even number $m > 0$, we can divide the path into m stripes separated by $m + 1$ equidistant curves. These curves will be addressed by $S_0, S_1, \ldots, S_m$ starting from the inner side of the original path. In nontrivial cases (with a non-zero curvature of the path orientation), the equidistant curves in the inner side always have shorter lengths. Observing this, we divide, by arc length, the equidistant curve S_i $(i = 0, \ldots, m)$

into $n + i$ parts and further approximate them by straight line segments. In such a way, the original path can be represented by a set of triangles. In can be reorganized into the Euclidean coordinate system for a better visual control of the subsequently described process. We notice that the two triangles bounded by two neighbouring equidistant curves sharing a common edge now form a quadrilateral. These quadrilaterals can be labelled by their x- and y-positions, for $x = 0, \ldots, m - 1$ and $y = 0, \ldots, n + x$, cf. Fig. 1. In these quadrilaterals, we associate the triangles in the negative direction of x with an additional bit 0, and in the positive direction, an additional bit 1, respectively, thus all the triangles in the path-oriented coordinate system can be easily denoted by 3-tuples.

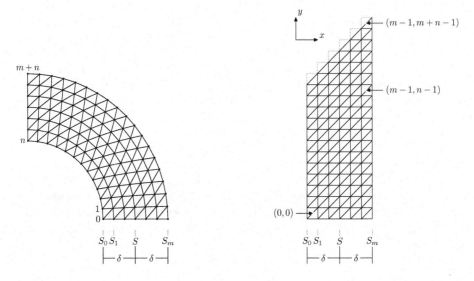

Fig. 1. Discretization of an arc-shaped area by means of a series of equidistant curves and its representation in the Euclidean coordinate system. The innermost curve S_0 of the path is divided into n parts, whereas for the outermost curve S_m, there will be $m+n$ parts. The maximum span of the path is 2δ. The representation in the Euclidean coordinate system in the right sugfigure enables the correct labelling of the triangles in the original geometry.

Although m and n are independent parameters in the discretization process, m and n should be selected in a way that the quadrilaterals composed of neighbouring triangles should have a form of close to a square whenever possible. This feature is essential in the discrete modelling of pedestrian dynamics, since generally speaking pedestrians are able to cover similar distances in various walking directions. These directions may or may not be axis-parallel in the case of a regular grid. Now as in a special case, we assume that S is an arc with a central angle of α and radius of R; in fact, an arbitrary path can always be approximated by arcs of this kind.

Since the centre of the path S (which is identical to $S_{\frac{m}{2}}$) will be divided into $n + \frac{m}{2}$ parts, we can now request

$$2R \cdot \sin \frac{\alpha}{2\left(n + \frac{m}{2}\right)} \doteq \frac{2\delta}{m}, \tag{1}$$

this gives

$$n \doteq \frac{1}{2}\left(\frac{\alpha}{\arcsin \frac{\delta}{mR}} - m\right). \tag{2}$$

The length of each of the $n+i$ approximating line segments on S_i $(i = 0, \ldots, m)$ reads

$$2 \cdot \left(R + \left(\frac{2i}{m} - 1\right) \cdot \delta\right) \cdot \sin \frac{\alpha}{2n + 2i}. \tag{3}$$

The difference of (3) to the specified length $\frac{2\delta}{m}$ in (1) can be calculated and checked to see whether it lies within a certain tolerance range. Concerning the geometric structure to model, α, R and δ are already known, whereas the length $d = \frac{2\delta}{m}$ should resemble the usual step length of the pedestrians in discrete modelling, which had been commonly suggested as 0.4 m (see [2], for example), thus m can be decided in this way. On the other hand, we shall see that in common discrete models based on regular grids, the pedestrian enjoys an minimum exclusive space of d^2; in our case, however, in each quadrilateral composed of two small triangles two pedestrians are allowed to be present. Hence, an additional factor $\sqrt{2}$ should be considered and we choose $d = 0.6$ m. Also d can be used to denote the one-dimensional grid cell size in the geometry.

It remains to check whether n derived from (2) in combination with α suffices the condition of the tolerance in length, if not, the parameter choice of α can be adjusted.

Hence, it is possible to decompose a ring-shaped area into four $\frac{\pi}{2}$-arcs and apply the above discretization scheme on each of them to acquire a triangulation describing the geometry settings of the simulation, see Fig. 2. Obviously, 2δ will be the maximum span (or width) of this area for the pedestrians in the simulation. For $d = 0.6$ m, Fig. 2 (left subfigure) corresponds to a configuration of $m = 6$, $n = 12$, the width of the path is 3.6 m and the radius at the centre of the path is 12 m.

2.2   Position Transition

Definition. Position changes in the system can be represented by two types of position transition which take place in the direction of the orientation of the local path and in the direction of its deviation.

In the primary direction, the position transition takes place in the same stripe bounded by two neighbouring curves S_i and S_{i+1} $(i = 0, \ldots, m - 1)$. For a forward position transition we have:

$$\mathrm{Fwd}(x, y, b) = \begin{cases} (x, y, & 0), & \text{if } b = 1, \\ (x, y + 1, 1), & \text{if } b = 0. \end{cases} \tag{4}$$

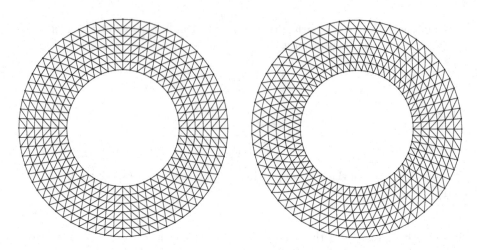

Fig. 2. Triangulation of a ring. Left: the triangulation is further composed of four pieces of triangulation of a $\frac{\pi}{2}$-arc. Right: the triangulation is carried out on the circle area defined by a 2π-arc; the length difference among the line segments is too large.

This is illustrated in the first two subfigures of Fig. 3. The backward position transition Bkwd can be formulated in analogy.

The case with deviation is different. In the third subfigure of Fig. 3 we see positive deviation from a position with additional bit 1 or negative deviation from a position with additional bit 0, this can be written as:

$$\begin{aligned}
\mathrm{Dev}^+(x,y,1) &= (x+1,y,0), \\
\mathrm{Dev}^-(x,y,0) &= (x-1,y,1).
\end{aligned} \tag{5}$$

(5) involves the position change into a neighbouring stripe. In contrast, positive deviation from a position with additional bit 0 and negative deviation from a position with additional bit 1 take place in the same stripe, as shown in the last subfigure of Fig. 3. However, the centre of the stripe must be paid attention. If $n+x$ is an even number, we will leave $\mathrm{Dev}^-(x,\frac{n+x}{2},1)$ undefined. For an odd $n+x$, $\mathrm{Dev}^+(x,\frac{n+x-1}{2},0)$ is allowed to take either $(x,\frac{n+x-1}{2},1)$ or $(x,\frac{n+x+1}{2},1)$. In Fig. 3, for example, we see that Dev^- is for $(0,6,1)$, $(2,7,1)$ and $(4,8,1)$ undefined, whereas concerning $(1,6,0)$, $(3,7,0)$ and $(5,8,0)$, there exist two possibilities for the execution of Dev^+. In summary, we define

$$\begin{aligned}
\mathrm{Dev}^+(x,y,0) &= \begin{cases} (x,y,\quad 1), & \text{if } y < \frac{n+x-1}{2}, \\ (x,y+1,1), & \text{if } y \geq \frac{n+x}{2}, \end{cases} \\
\mathrm{Dev}^-(x,y,1) &= \begin{cases} (x,y,\quad 0), & \text{if } y \leq \frac{n+x-1}{2}, \\ (x,y-1,0), & \text{if } y \geq \frac{n+x+1}{2}. \end{cases}
\end{aligned} \tag{6}$$

Connecting two neighbouring paths for the construction of complex geometric settings should be obvious.

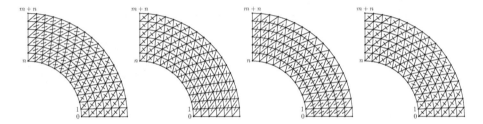

Fig. 3. First two subfigures (counting from the left): Position transition Fwd in the primary direction of the path, the start position has additional bit 1 and 0 respectively. Third subfigure: Position transition Dev$^+$ from a start position with bit 1 and Dev$^-$ starting from bit 0. Fourth subfigure: Position transition Dev$^-$ starting from a position with bit 1 and Dev$^+$ starting from bit 0.

Effective Length in the Position Transition. The operation Fwd implies a partial position change in the secondary direction (that is, deviation). However, in two consecutive Fwd-operations this deviation will be compensated; in fact, two consecutive operations have roughly an effective position change of d. Similarly, two consecutive Dev$^+$- or Dev$^-$-operations manifest an effective position change of d or $-d$ in the secondary direction, respectively. For a sequence of these elementary operations, the effective position change can be evaluated by summing up the individual position transitions.

Deviation Under the Influence of the Fictional Path Force. In a more general sense, an extra computing module should be responsible for the navigation of all participants in the simulation. In our (over-)simplified model, it is assumed that the pedestrians have pre-configured walking speeds obeying a Gaussian distribution. As an important system parameter, this speed v can be located through adequate measurements of empirical data (see [7], for example).

On the other hand, to demonstrate the impact of the fictional path force, a velocity component $\boldsymbol{v}_x$ in the secondary direction can be defined in proportion to the pedestrian's deviation from the centre of the path

$$\boldsymbol{v}_x = (\bar{x} - x) \cdot a \tag{7}$$

with a constant parameter a, $\bar{x}$ stands for the x-position of the centre line of the path. a can be chosen in a way that in the primary and secondary directions the pedestrian has the same possible maximum speeds, this means

$$|a| = \frac{v_{\max}}{\delta}. \tag{8}$$

3   Update Scheme

[1] presented interesting results about the so-called *frozen shuffle update* for *totally asymmetric exclusion process* (TASEP) applied on pedestrian dynamics.

The term "total asymmetry" refers to the requirement that backward movement is not allowed in the modelling; "exclusion" stands for the basic rule regarding conflicts among the pedestrians: at any time at any position at most one pedestrian is allowed to be present, which is a common assumption in discrete modelling of pedestrian dynamics. In [1], the pedestrians are labelled by index i, every of them is assigned with a constant phase number $\tau_i \in [0,1)$, and the update of the pedestrians will be carried out successively in the order of $\{\tau_i\}$. This procedure is in fact the update of the pedestrians in a sequence of a given permutation of the indices of the pedestrians (hence, the update scheme is coined "shuffle") and owing to the once assigned constant phase numbers, the update is "frozen" at the same time. In addition, the model is called *simple*, if position change in every simulation cycle is limited to be 1. [1] showed that at a critical density free flow state of the system with $v_{\max} = 1$ changes into the jammed state. The system of [1] is composed of a single traffic lane of an infinite size with periodic boundary conditions. The structure of this system is very close to a ring-formed traffic system.

In our model, the position transition in the simulation cycle is not limited to be $v_{\max} = 1$ and the environment is a triangulated ring structure in the geometric sense. Our system applies the same periodic condition. In our model, the step execution in the simulation cycle is composed of a series of Fwd- and Dev-operations. With multiple operations in the simulation cycle it is possible to simulate pedestrians with heterogeneous characteristics (for example, walking speed). A proper update scheme should take care of the execution of all the elementary operations of all the participants in the simulation. For each pedestrian there will be $|\boldsymbol{v}| + |\boldsymbol{v}_x|$ operations in the simulation cycle. Let $v_{\max}$ denote the maximum speed. With (7) and (8) it is implied that there will be $2 \cdot v_{\max}$ elementary operations of Fwd and Dev in the simulation cycle at most. Now we divide the simulation cycle further into u equal subintervals with $u \in \mathbb{N}$, $u \geq 2 \cdot v_{\max}$. Each pedetrian will now execute $|\boldsymbol{v}| + |\boldsymbol{v}_x|$ operations in the u subintervals, this is shown in the following code fragment.

```
procedure OPERATE:
parameter: a collection of participants

    mark all participants as "unprocessed";

    i ⇐ u;
    while i ≥ 1
    do
        process all participents marked "unprocessed" by calling
            procedure OPERATE_PARTICIPANT sequentially with parameter i;
        i ⇐ i − 1;
    enddo

    return
```

For the individual pedestrians we have the update:

procedure OPERATE_PARTICIPANT:
parameter: participant, i

    if all elemenary operations executed
      mark as "processed";
    else
      count (in the current simulation cycle) the successful operations
      in the main direction s and in the secondary direction s_x;
      if $\frac{v-s+v_x-s_x}{i} \leq 1$;
        execute an operation in the main direction with probability $\frac{v-s}{i}$;
        execute an operation in the secondary direction with probability $\frac{v_x-s_x}{i}$;
      fi
    fi

    return

In both procedures the parameter i serves as a time stamp. In the first procedure OPERATE_PARTICIPANT, unsuccessful operations in the current simulation cycle contribute to a higher probability that operations should be tried again, since v, v_x, s and s_x remain unchanged, but i is decremented. This reduces the occurrence of conflict notably and is not unlike human behaviour. In the procedure OPERATE_PARTICIPANT, with $\frac{v-s+v_x-s_x}{i} \leq 1$, that is, an operation of Fwd or Dev should be performed in a subinterval in the simulation cycle, we still need to decide which operation is to be carried out exactly. Therefore, the Fwd-operation in the main direction is assigned a probability for $\frac{v-s+v_x-s_x}{i} \cdot \frac{v-s}{v-s+v_x-s_x} = \frac{v-s}{i}$, and naturally for Dev as deviation, $\frac{v_x-s_x}{i}$.

Since the pedestrians in the simulation are not expected to have a homogeneous velocity, the execution of the operations in OPERATE_PARTICIPANT varies from pedestrian to pedestrian substantially. We consider this flexibility as an advantage of the model, since the model is no longer "simple" ($v_{\max} > 1$). In comparison to TASEP, our model implements a frozen shuffle update for symmetric exclusion process with generalized velocity. We argue our model to be "symmetric", because backward state (position) transitions—in both primary and secondary directions—have been defined in elementary directional operations of Fwd, Bkwd, Dev$^+$ and Dev$^-$.

Fig. 4 illustrates the effect of the simulation algorithm by some snapshots from first simulation runs. These screenshots are taken after the random start has evolved into a stabilized state. A higher magnitude of a causes the pedestrians inevitably to rally in the middle of the path. It can be recognized that the maximum magnitude of a is much too large for the simulation, this conforms to the fact that pedestrians have higher mobility in their main moving direction.

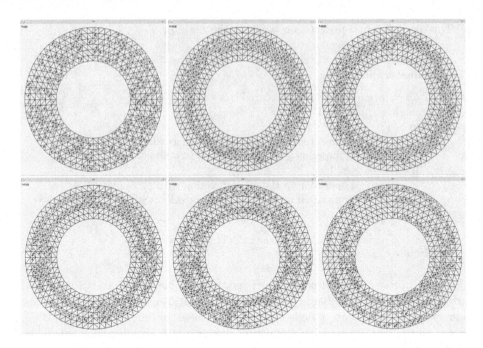

Fig. 4. First row left: Random start of a simulation. From first row middle to second row right: the snapshots show results for several values of the constant a - the maximum magnitude according to (7) and (8), one half, one fourth, one eighth of its maximum and finally zero.

Conclusion and Future Work

The present paper is intended to be a further improvement of grid-based discrete microscopic models for the simulation of pedestrian dynamics. The present contribution avoids some of the drawbacks of classical cellular automata models. In particular, more flexibility and a better approach of realistic physical contexts can be achieved introducing modelling concepts for more general geometries. The circular geometry considered in this contribution only represents a first illustrative example. In addition to the geometry the proper choice of parameters like, in particular, the speed notion from (7) and (8) significantly influence the simulation results.

References

1. Appert-Rolland, C., Cividini, J., Hilhorst, H.-J.: Frozen shuffle update for an asymmetric exclusion process on a ring. Journal of Statistical Mechanics: Theory and Experiment **07**, P07009 (2011)
2. Burstedde, C., Klauck, K., Schadschneider, A., Zittartz, J.: Simulation of pedestrian dynamics using a two-dimensional cellular automaton. Physica A **295**, 507–525 (2001)

3. Chen, M.-J., Bärwolff, G., Schwandt, H.: Modeling pedestrian dynamics on triangular grids. In: Transportation Research Procedia (The Conference on Pedestrian and Evacuation Dynamics 2014), vol. 2, pp. 327–335 (2014)
4. Gloor, C., Stucki, P., Nagel, K.: Hybrid techniques for pedestrian simulations. In: Sloot, P.M.A., Chopard, B., Hoekstra, A.G. (eds.) ACRI 2004. LNCS, vol. 3305, pp. 581–590. Springer, Heidelberg (2004)
5. Helbing, D., Farkas, I., Vicsek, T.: Traffic and related self-driven many-particle systems. Reviews of Modern Physics **73**, 1067–1141 (2001)
6. Kirchner, A., Klüpfel, H., Nishinari, K., Schadschneider, A., Schreckenberg, M.: Discretization effects and the influence of walking speed in cellular automata models for pedestrian dynamics. Journal of Statistical Mechanics: Theory and Experiment **10**, P10011 (2004)
7. Plaue, M., Chen, M.-J., Bärwolff, G., Schwandt, H.: Multi-view extraction of dynamic pedestrian density fields. Photogrammetrie, Fernerkundung, Geoinformation **5**, 547–555 (2012)
8. Schadschneider, A.: Cellular automaton approach to pedestrian dynamics - theory. In: Schreckenberg, M., Sharma, S.D. (eds.) Pedestrian and Evacuation Dynamics, pp. 75–85. Springer, Heidelberg (2002)
9. Schwandt, H., Huth, F., Bärwolff, G., Berres, S.: A multiphase convection-diffusion model for the simulation of interacting pedestrian flows. In: Murgante, B., et al. (eds.) ICCSA 2013, Part V. LNCS, vol. 7975, pp. 17–32. Springer, Heidelberg (2013)

A Decision Support System for the Analysis of Mobility

Mauro Mazzei[(✉)] and Armando Luigi Palma

National Research Council, Istituto Di Analisi Dei Sistemi Ed Informatica Antonio Ruberti,
Via Dei Taurini 19 00185, Rome, Italy
mauro.mazzei@iasi.cnr.it, palma@arpal.it

Abstract. We present in this paper the analysis of mobility in the provinces of Salento, in the modal split and their reasons for moving. We used data on commuting (OD matrix municipal disaggregated) recognized by 'ISTAT in the 2011 census and published in December 2014. The results of the analysis are represented by thematic maps showing estimated levels of air and noise pollution on each municipal areas. The analysis software is designed for automatic search as a decision support system for the intervention policies for the smooth flow of traffic and noise and air pollution.

Keywords: GIS · Spatial interaction models · Spatial data analysis · Spatial statistical model · Urban models

1 Introduction

During the 2011 census, were detected by Istat data on the mobility of respondents, according to the modal split of journeys, time slots, the duration of the shift, the reasons for the move, and other parameters that are described in detail in website Istat. These data were published in December 2014 and the file has records related to the Italian municipalities ordered by code of the province and the municipality, for a total of 110 provinces and 8100 municipalities. The Origin-Destination matrix of travel for work or study refers to the population living in family or cohabitation detected at the 15th General Census of Population (reference date: October 9, 2011). The file contains data on the number of people moving between areas or within the same municipality - classified, as well as the reasons for traveling, for sex, the means of transport used, the time of departure and the journey time. The basis of calculation are the 28.871.447 people who said they go daily to the usual place of study or work, starting from the house of residence. Of these, 28.852.721 are residing in the family and 18.726 are living in cohabitation (convent school in prison, institutional care etc.). The OD matrix is stuctured in a methodological document that, in addition to describing the structure of data, provides guidance for the use of the matrix, with particular reference to the variables obtained with sampling method (method of transport, time slot of departure and travel time). The variables considered in this study are structured according to Table 1.

© Springer International Publishing Switzerland 2015
O. Gervasi et al. (Eds.): ICCSA 2015, Part III, LNCS 9157, pp. 390–402, 2015.
DOI: 10.1007/978-3-319-21470-2_28

Table 1. MATRIX STRUCTURE OF THE ORIGIN AND DESTINATION – ISTAT 2014

DESCRIPTION OF VARIABLE	VALUE
Province of residence	See: List of Italian municipalities January 1, 2011 - Istat.
Municipal of residence	See: List of Italian municipalities January 1, 2011 - Istat.
Reasons for traveling	1 – Study 2 – Work
Place of work or study	1 - In the same municipality of residence 2 - In other municipality Italian 3 - Foreign
Province usual work or study	See: List of Italian municipalities January 1, 2011 - Istat.
Municipal habitual study and work	See: List of Italian municipalities January 1, 2011 - Istat.
Transport	01 train 02 tram 03 underground 04 city bus, trolley bus 05 bus, extra urban bus 06 bus company or school 07 private car (as driver) 08 private car (as a passenger) 09 motorcycle, moped, scooter 10 bicycle 11 other means 12 walk
Check out	1 before 7:15 2 from 7:15 to 8:14 3 from 8:15 to 9:14 4 after 9:14
Number of individuals	Count variable

In particular, we have considered the following values of the variables under consideration, see Table 2, which represent the highest rate of movement.

Table 2. VARIABLES EXAMINED BY O-D MATRIX – ISTAT 2014

DESCRIPTION OF VARIABLE	VALUE
Province of residence	Cod. 073 – 074 - 075 See: List of Italian municipalities January 1, 2011 - Istat
Municipal of residence	See: List of Italian municipalities January 1, 2011 - Istat
Reasons for traveling	1 – Study - 2 – Work
Place of work or study	1 - In the same municipality of residence - 2 - In other municipality Italian
Province usual work or study	See: List of Italian municipalities January 1, 2011 - Istat
Municipal habitual study and work	See: List of Italian municipalities January 1, 2011 - Istat
Transport	07 private car (as driver)
Check out	2 from 7:15 to 8:14
Number of individuals	Count variable

2   Case Study and Data

The case study is located in the province of Brindisi, Lecce, Taranto in southern Italy. For this work have been extracted from general file published by ISTAT in 2014, all the data collected in 2011 on movements during rush hour in the morning, by private car (as driver) and in the time slot from 7:15 at 8:14, relative to the municipalities of the three provinces of Salento: Taranto, Brindisi and Lecce.

For each joint of each province were calculated displacements of private cars during rush hour in the morning, both within the town itself, which outbound to other municipalities. Were also calculated for each municipality the flows from an to other towns in the same province or other neighboring provinces. The tables numbered from 3 to 5 are shown below, in the last column of the same table, there are the percentage values of the movements of morning rush hour calculated in the resident population.

Table 3. DATA 2014 ISTAT – PROVINCE BRINDISI

ZONES	MUNICIPAL	ORIGIN	DESTINATION	TOTAL DISPLACEMENT	POPULATION ISTAT 2009	PROPENSITY DISPLACEMENT
1	Brindisi	6244	352	6596	89735	7,4%
2	Carovigno	805	38	843	16138	5,2%
3	Ceglie Messapica	1008	60	1068	20671	5,2%
4	Cellino San Marco	323	38	361	6753	5,3%
5	Cisternino	891	42	933	11894	7,8%
6	Erchie	313	31	344	9011	3,8%
7	Fasano	2722	98	2820	38493	7,3%
8	Francavilla Fontana	1770	145	1915	36619	5,2%
9	Latiano	763	54	817	15052	5,4%
10	Mesagne	1655	140	1795	27827	6,5%
11	Oria	723	49	772	15385	5,0%
12	Ostuni	2453	90	2543	32453	7,8%
13	San Donaci	457	35	492	7074	7,0%
14	San Michele Salentin	259	28	287	6390	4,5%
15	San Pancrazio Salent	538	52	590	10367	5,7%
16	San Pietro Vernotico	777	95	872	14430	6,0%
17	San Vito dei Normann	1015	47	1062	19884	5,3%
18	Torchiarolo	307	28	335	5156	6,5%
19	Torre Santa Susanna	425	38	463	10584	4,4%
20	Villa Castelli	383	30	413	9180	4,5%

Table 4. DATA 2014 ISTAT – PROVINCE LECCE

ZONES	MUNICIPAL	ORIGIN	DESTINATION	TOTAL DISPLACEMENT	POPULATION ISTAT 2009	PROPENSITY DISPLACEMENT
1	Acquarica del Capo	251	56	307	4966	6,2%
2	Alessano	414	89	503	6558	7,7%
3	Alezio	373	48	421	5540	7,6%
4	Alliste	311	37	348	6704	5,2%
5	Andrano	341	39	380	5049	7,5%
6	Aradeo	556	53	609	9802	6,2%
7	Arnesano	288	52	340	3929	8,7%
8	Bagnolo del Salento	112	24	136	1879	7,2%

Table 4. (*Continued*)

ZONES	MUNICIPAL	ORIGIN	DESTINATION	TOTAL DISPLACEMENT	POPULATION ISTAT 2009	PROPENSITY DISPLACEMENT
9	Botrugno	154	40	194	2916	6,7%
10	Calimera	659	67	726	7287	10,0%
11	Campi Salentina	736	87	823	10857	7,6%
12	Cannole	107	21	128	1768	7,2%
13	Caprarica di Lecce	215	31	246	2575	9,6%
14	Carmiano	817	71	888	12308	7,2%
15	Carpi. Salentino	260	30	290	3852	7,5%
16	Casarano	1606	278	1884	20593	9,1%
17	Castri di Lecce	195	24	219	3065	7,1%
18	Castrignano G	244	40	284	4125	6,9%
19	Castrignano Capo	285	55	340	5424	6,3%
20	Cavallino	1156	152	1308	12149	10,8%
21	Collepasso	372	57	429	6478	6,6%
22	Copertino	1432	114	1546	24452	6,3%
23	Corigl. d'Otranto	419	75	494	5858	8,4%
24	Corsano	321	31	352	5740	6,1%
25	Cursi	255	47	302	4290	7,0%
26	Cutrofiano	661	68	729	9262	7,9%
27	Diso	204	33	237	3175	7,5%
28	Gagliano del Capo	272	57	329	5502	6,0%
29	Galatina	2214	286	2500	27317	9,2%
30	Galatone	1028	89	1117	15850	7,0%
31	Gallipoli	1151	171	1322	21038	6,3%
32	Giuggianello	82	21	103	1248	8,3%
33	Giurdignano	128	19	147	1897	7,7%
34	Guagnano	355	37	392	5980	6,6%
35	Lecce	7885	924	8809	94949	9,3%
36	Lequile	701	64	765	8550	8,9%
37	Leverano	815	70	885	14173	6,2%
38	Lizzanello	967	55	1022	11647	8,8%
39	Maglie	1352	312	1664	14982	11,1%
40	Martano	601	126	727	9484	7,7%
41	Martignano	140	23	163	1775	9,2%
42	Matino	742	126	868	11821	7,3%
43	Melendugno	700	60	760	9894	7,7%
44	Melissano	389	65	454	7374	6,2%
45	Melpignano	162	78	240	2241	10,7%
46	Miggiano	188	51	239	3702	6,5%
47	Minervino di Lecce	241	49	290	3830	7,6%
48	Monteroni di Lecce	905	203	1108	13947	7,9%
49	Montesano Salen- tino	194	27	221	2731	8,1%
50	Morciano di Leuca	178	27	205	3484	5,9%
51	Muro Leccese	398	58	456	5138	8,9%
52	Nard=	2167	157	2324	31195	7,4%
53	Neviano	289	28	317	5568	5,7%
54	Nociglia	143	43	186	2502	7,4%
55	Novoli	523	47	570	8227	6,9%
56	Ortelle	161	17	178	2428	7,3%
57	Otranto	364	148	512	5531	9,3%
58	Palmariggi	114	27	141	1579	8,9%
59	Parabita	705	89	794	9414	8,4%
60	Pat¿	77	23	100	1737	5,8%
61	Poggiardo	443	92	535	6137	8,7%
62	Presicce	281	59	340	5627	6,0%
63	Racale	719	71	790	10839	7,3%
64	Ruffano	557	87	644	9732	6,6%
65	Salice Salentino	528	47	575	8772	6,6%
66	Salve	247	43	290	4699	6,2%
67	Sanarica	122	15	137	1484	9,2%
68	San Cesario di Lecce	749	110	859	8254	10,4%
69	San Donato di Lecce	421	53	474	5869	8,1%
70	Sannicola	341	46	387	5959	6,5%
71	San Pietro in Lama	238	38	276	3655	7,6%
72	Santa Cesarea Terme	242	77	319	3070	10,4%
73	Scorrano	450	100	550	6989	7,9%
74	Secl²	121	27	148	1953	7,6%
75	Sogliano Cavour	292	30	322	4143	7,8%
76	Soleto	448	52	500	5630	8,9%

Table 4. (*Continued*)

ZONES	MUNICIPAL	ORIGIN	DESTINATION	TOTAL DISPLACEMENT	POPULATION ISTAT 2009	PROPENSITY DISPLACEMENT
77	Specchia	318	46	364	4942	7,4%
78	Spongano	304	38	342	3799	9,0%
79	Squinzano	950	62	1012	14631	6,9%
80	Sternatia	190	23	213	2478	8,6%
81	Supersano	221	56	277	4521	6,1%
82	Surano	129	74	203	1723	11,8%
83	Surbo	1206	132	1338	14621	9,2%
84	Taurisano	737	92	829	12674	6,5%
85	Taviano	899	78	977	12642	7,7%
86	Tiggiano	185	21	206	2927	7,0%
87	Trepuzzi	931	57	988	14702	6,7%
88	Tricase	1323	212	1535	17803	8,6%
89	Tuglie	367	40	407	5281	7,7%
90	Ugento	608	109	717	12195	5,9%
91	Uggiano la Chiesa	313	33	346	4414	7,8%
92	Veglie	760	79	839	14352	5,8%
93	Vernole	497	51	548	7409	7,4%
94	Zollino	172	32	204	2077	9,8%
95	San Cassiano	130	37	167	2133	7,8%
96	Castro	153	33	186	2511	7,4%
97	Porto Cesareo	238	45	283	5573	5,1%

Table 5. DATA 2014 ISTAT – PROVINCE TARANTO

ZONES	MUNICIPAL	ORIGIN	DESTINATION	TOTAL DISPLACEMENT	POPULATION ISTAT 2009	PROPENSITY DISPLACEMENT
1	Avetrana	260	35	295	7117	4,1%
2	Carosino	476	35	511	6659	7,7%
3	Castellaneta	813	75	888	17229	5,2%
4	Crispiano	756	43	799	13621	5,9%
5	Faggiano	135	35	170	3535	4,8%
6	Fragagnano	236	36	272	5464	5,0%
7	Ginosa	1194	52	1246	22683	5,5%
8	Grottaglie	1817	129	1946	32845	5,9%
9	Laterza	778	40	818	15203	5,4%
10	Leporano	606	27	633	7674	8,2%
11	Lizzano	354	49	403	10266	3,9%
12	Manduria	1327	128	1455	31757	4,6%
13	Martina Franca	4245	117	4362	49756	8,8%
14	Maruggio	236	31	267	5539	4,8%
15	Massafra	1644	95	1739	32210	5,4%
16	Monteiasi	280	26	306	5514	5,5%
17	Montemesola	180	35	215	4162	5,2%
18	Monteparano	90	20	110	2390	4,6%
19	Mottola	765	54	819	16349	5,0%
20	Palagianello	222	24	246	7896	3,1%
21	Palagiano	587	46	633	15991	4,0%
22	Pulsano	613	39	652	10904	6,0%
23	Roccaforzata	78	6	84	1845	4,6%
24	San G. Ionico	1019	79	1098	15987	6,9%
25	San Marzano	320	51	371	9223	4,0%
26	Sava	705	70	775	16863	4,6%
27	Taranto	13560	413	13973	193136	7,2%
28	Torricella	159	40	199	4219	4,7%
29	Statte	679	47	726	14488	5,0%

To get the results of the operations we used the methods of spatial interactions. The term spatial interaction, it indicates the movement of the means of transport, of peoples, ideas and products within the geographical areas and between them. Such movements are simple demonstrations of the principles that are the basis of each spatial interaction [21]. The flow of movement, that is, the connections between places offered and places of application, are the answer to such differences.

3   Spatial Interaction Models

Spatial interactions cover a wide variety of movements such as journeys to work, migrations, tourism, the usage of public facilities, the transmission of information or capital, the market areas of retailing activities, international trade and freight distribution. Economic activities are generating (supply) and attracting (demand) flows. The simple fact that a movement occurs between an origin and a destination underlines that the costs incurred by a spatial interaction are lower than the benefits derived from such an interaction. As such, a commuter is willing to drive one hour because this interaction is linked to an income, while international trade concepts, such as comparative advantages, underline the benefits of specialization and the ensuing generation of trade flows between distant locations.

In order to study the spatial interactions of urban aggregates (flows of people, goods, money, etc.), the gravity models are the ones most commonly used as a form of mathematical model. According to these models (traffic flows, money, goods, etc.), the force of attraction that is exerted between two urban centers is directly proportional to the product of their "masses", they are measured as the number of residents, jobs, bank deposits and inversely proportional to a power of the distance between them [5]. This law can be expressed in general as:

$$T_{i,j} = k \, O_i \, D_j \, d_{ij}^{\,-\alpha}$$

where T_i, j is the intensity of interaction from urban centers i and j, O_i and D_j express the intensity of attraction exerted by the two centers as places of origin and destination α is the exponent of the distance between the centers of the i and the center j (not necessarily equal to 2) and k is a proportionality constant that ensures the internal consistency of the model [13][14].

In this general formulation of the gravity model, you can add the formulation of the gravity model in the source and destination bound:

$$T_{i,j} = A_i \, O_i \, B_j \, D_j \, d_{ij}^{\,-\alpha}$$

where A_i and B_j are two separate sets of scale factors defined as follows:

$$A_i = 1 / (\Sigma_j \, B_j \, D_j \, d_{ij}^{\,-\alpha})$$

$$B_j = 1 / (\Sigma_i \, A_i \, O_i \, d_{ij}^{\,-\alpha})$$

below is the formulation of the gravity model for specific purposes:

$$T_{i,j} = W_i \, B_j \, D_j \, d_{ij}^{\,-\alpha}$$

with

$$B_j = 1 / (\Sigma_i W_i d_{ij}^{-\alpha})$$

$$\Sigma_i T_{i,j} = D_j$$

that W_i is an estimate of the attraction exerted by the urban center as i-th origin. Finally, we report the formulation of the gravity model in origin bound:

$$T_{i,j} = A_i O_i W_j^* d_{ij}^{-\alpha}$$

with

$$A_i = 1 / (\Sigma_j W_j^* d_{ij}^{-\alpha})$$
$$\Sigma_j T_{i,j} = O_i$$

that W_j^* is an estimate of the attraction exerted by the urban center as the seat of the j-th destination.

Often, a value of 1 is given to the parameters, and then they are progressively altered until the estimated results are similar to observed results. Calibration can also be considered for different O/D matrices according to age, income, gender, type of merchandise and modal choice. A part of the scientific research in transport and regional planning aims at finding accurate parameters for spatial interaction models. This is generally a costly and time consuming process, but a very useful one. Once a spatial interaction model has been validated for a city or a region, it can then be used for simulation and prediction purposes, such as how many additional flows would be generated if the population increased or if better transport infrastructures (lower friction of distance) were provided.

4   Conclusions

For each municipality, with the calculated values of internal displacement, output and inbound, relative to private cars, it has been possible to estimate the contribution of pollutants in the atmosphere, such as CO_2, considered an average emission of 200g/km. This value, in fact, being a constant multiple of the number of vehicles circulating, during rush hour, in the home town, can be neglected, then being able to estimate the level of emissions into the atmosphere through the number of vehicles circulating in the hour of tip. The thematic maps 1, 2 and 3 show each province levels of air pollution due to the component of urban traffic.

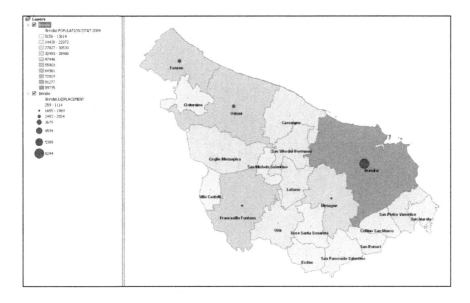

Fig. 1. Province of Brindisi

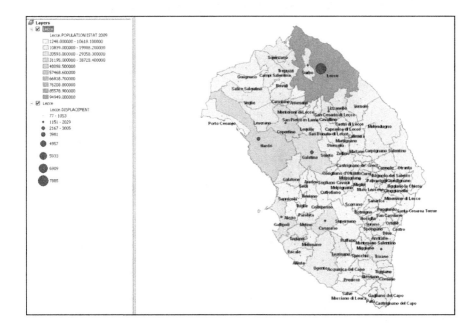

Fig. 2. Province of Lecce

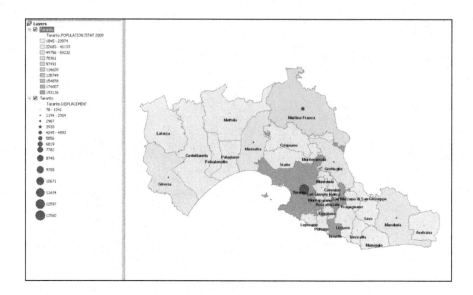

Fig. 3. Province of Taranto

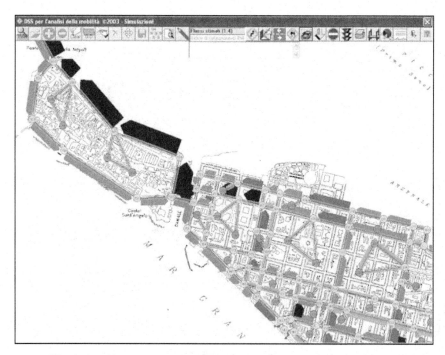

Fig. 4. Graphical representation of the flows in the municipality of Taranto

For mobility analysis was used a Decision Support System DSS owner (DSS2000) in Windows environment can estimate the matrix OD displacement based on census data or direct observations.

To estimate the demand for mobility (matrix displacement OD) by the gravity model was necessary to have the number density of the population in each area (or source) that comprise the territory. Each zone is represented by a central node (centroid) which is divided the population, while they are used as attractors (D destinations) the jobs available in the various areas (work movements residence) or the number of customers served. At this point, we have moved the allocation to the transport network through cost functions arcs, using the equation Festa and Nuzzolo (1990) to calculate the running time and Highway Capacity Manual (HCM 1994) for the calculation of time of waiting.

Determined the shortest path for every displacement OD software allows you to assign the question to the transport network in two different models: incremental assignment or Method of Successive Averages (MSA) [19] [20].

The result of assignment (Figure 4) is displayed on the map base of reference where the different color intensity of the arc represents the different degree of saturation of each arc and the different thickness represents the flow of each arc, this result allows the designer to have an immediate perception of flows and congestion on the entire network.

We have examined the conditions of mobility in some of the most crtical areas in the provinces of Salento, on the basis of data collected by Istat on commuting in the last census, focusing mainly on residential areas, since the recognition of graphs and direction for the three provinces and the simulations for assigning traffic showed that there were no obvious signs of congestion on the road network of the three southern provinces of Puglia.

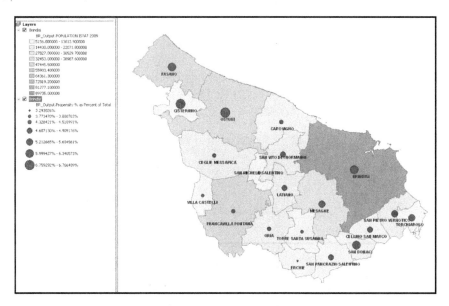

Fig. 5. Province of Brindisi

In figures numbered from 5 to 7, are the thematic maps obtained for domestic travel, outbound and inbound, each municipality of the three provinces in percentage of journeys in the morning rush hour calculated in the resident population.

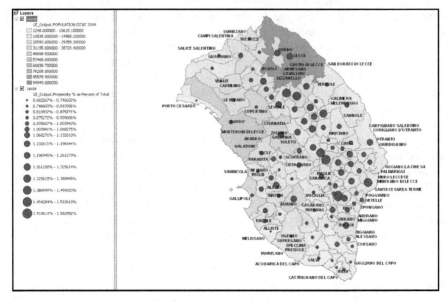

Fig. 6. Province of Lecce

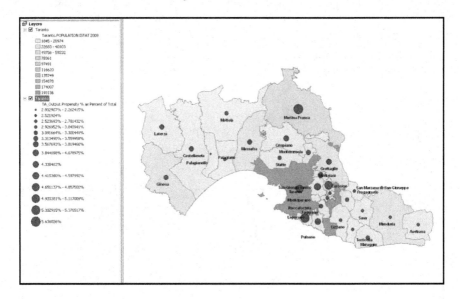

Fig. 7. Province of Taranto

Such values can be interpreted as indirect indicators of socio-economic conditions of a joint. In fact, with reference to the province of Taranto, it may be noted that the town of Martina Franca has a value of 8.5% which means that about one-tenth of the population, daily moves with own car for work or study, while the municipality of San Marzano di San Giuseppe, with a 3.5%, marking the lowest value of the Ionian province. It should be noted, moreover, that the two municipalities of Leporano and Carosino, with private mobility values higher than 7%, are placed in the top of the list, right after Martina Franca, while the capital is among the first places in this ranking of wellness . It should also be noted that the municipality of Palagianello is placed at the bottom of this list, with a value of 2.8% of journeys in the morning (for work and study) compared to the population residing in the town. That is, in Palagianello, only three out of 100 people come out between 7:15 and 8:14 with a private car to go to work or to study.

As for the province of Brindisi, the highest value of this indicator, corresponding to 7.6%, can be attributed to the municipality of Ostuni, while the lower value is up to Erchie, with 3.5%. The capital, Brindisi, marks a good 7% while the town of Cisternino with a value of 7.5% highlights an expectation of socio-economic development, albeit with some problems of environmental.

In the province of Lecce, there is the highest value of all the Great Salento with San Cesario di Lecce to which is ascribed the 9.1% followed by other municipalities as Calimera and sweaters that stood on the threshold of 9%. Lecce make a good 8.3%, with a resident population of 94.949 inhabitants at the end of 2009. The lowest value is marked by Porto Cesareo with a 4% and a resident population of just over 5.500 inhabitants in 2009.

References

1. Batty, M.: An activity allocation model for the Notts/Derby sub-region. Regional Studies **4**(3) (1970)
2. Broadbent, T.A.: Zone size and spatial interaction. Centre for Environmental Studies, Working Note **106** (1969)
3. Carroll, J.D., Bevis, H.W.: Predicting local travel in urban regions. Papers and Proceedings of Regional Science Association, vol. 3 (1957)
4. Hayes, C.: Retail Location Models. Centre for Environmental Studies, Working Paper **16** (1968)
5. Cripps, E.L.: Limitations of the Gravity Concept. In: Styles (1968)
6. Cripps, E.L., Carter, E.: The Empirical Development of a Disaggregated Residential Location Model: Some Preliminary Results. Urban Systems Research Unit, University of Reading, Working Paper **9** (1971)
7. Fox, F.: Wilbur Smith and Assoc: London Traffic Survey, vol. II. Greater London Council (1966)
8. Hansen, W.G.: How accessibility shapes land use. Journal of the American Institute of Planners, Maggio (1959)
9. Isard, W.: Methods of Regional Anaiysis. MIT Press (1960)
10. Lakshmanan, T.R., Hansen, W.G.: A retail market potential model. Journal of American Institute of Planners, Maggio (1965)

11. Lewis, J.P. The invasion of planning. Journal of the Town Planning Institute, maggio (1970)
12. McLaughlin, J.B., et al.: Regional Shopping Centres in North West Englattd, Part 11. University of Manchester (1966)
13. Shcneider, M.: Gravity models and trip distribution theory. Papers and proceeding of the Regional Science Associution, vol. 5 (1959)
14. Styles, B.J. (Ed.) Gravity Models in Town Planning. Lanchester Polytechnic (1968)
15. Tanner, J.C.: Sotne Factors afecting the Amoicnt of Travel. Road Research Laboratory Paper No. 58 (1961)
16. Wilson, A.G.: The use of entropy maximising methods in the theory of trip distribution. Journal of Transport Econoinics and Policy 3(1) (1969a)
17. Wilson, A.G.: Disaggregating Elementary Residential Models. Centre for Environmental Studies, Working Paper 37 (1969b)
18. Wilson, A.G.: Entropy in Urban and Regional Modelling. Centre for Environmental Studies, Working Paper 26 (1969c)
19. Johnson, M.P.: A spatial decision support system prototype for housing mobility program planning. Journal of Geographical Systems 3(1), 49–67 (2001)
20. Johnson, M.P.: Spatial decision support for assisted housing mobility counseling. Decision Support Systems 41(1), 296–312 (2005)
21. Mazzei, M., Palma, A.L.: Comparative analysis of models of location and spatial interaction. In: Murgante, B., Misra, S., Rocha, A.M.A., Torre, C., Rocha, J.G., Falcäo, M.I., Taniar, D., Apduhan, B.O., Gervasi, O. (eds.) ICCSA 2014, Part IV. LNCS, vol. 8582, pp. 253–267. Springer, Heidelberg (2014)

Seismic Risk Analysis at Urban Scale in Italy

Alessandro Rasulo[1(✉)], Carlo Testa[2], and Barbara Borzi[2]

[1] University of Cassino and Southern Lazio, Cassino, Italy
a.rasulo@unicas.it
[2] European Centre for Training and Research in Earthquake Engineering, Pavia, Italy
{carlo.testa,barbara.borzi}@eucentre.it

Abstract. Seismic risk maps are a useful tool researchers use for representing to stakeholder and decision makers the adverse outcomes a seismic event can have over the territory. Generally, in those studies, urban areas, where the human activities are concentrated, focuses major attention. Main concerns are about the existing building stock, mostly composed by structures not compliant with modern seismic design criteria. The production of a seismic risk map is a complex task that involves the combination of data coming from different field of expertise. The aim of the study is to show how the already available information can be combined together in a Geographical Information System (GIS) tool. The results provide a reliable representation of the seismic risk at urban scale to be used when planning the mitigation measures to be undertaken in order to improve the level of preparedness in case of an earthquake. The analysis has been applied for demonstration purposes to the town of Cassino, Central Italy.

Keywords: Risk analysis · Reliability engineering · Earthquake engineering · Socio-Economic modeling · Seismic hazard · Seismic fragility functions

1   Introduction

The public awareness about the painful consequences that a moderate to strong earthquake [6],[17] can induce to a community in terms of loss of lives and damages is generally accompanied by the expectation that the modernstandards of living would be set in such a manner to prevent that harm. It is, however, a matter of fact that the knowledge about how the structures would be constructed in order to prevent collapses and reduce damages has evolved at a highly faster speed than the renovation rate of the building stock and sometimes the interventions needed to increase the structural safety of existing buildings, infrastructures, and critical facilities located in seismically prone areas are extremely expensive so that any decision about the mitigation measures to be undertaken is necessarily a trade-off between the cost-effectiveness of preparing for risks and that of coping with their consequences.

Seismic risk maps are usually employed in order to represent the expected loss an earthquake can produce over a territory taking into account the uncertainty that are involved into the forecast. It is worth noticing that during the risk evaluation some of the uncertainties are inherent (the randomness of the seismic phenomena is such that

© Springer International Publishing Switzerland 2015
O. Gervasi et al. (Eds.): ICCSA 2015, Part III, LNCS 9157, pp. 403–414, 2015.
DOI: 10.1007/978-3-319-21470-2_29

no one can say when, where and how intense will be the next earthquake), others uncertainties are epistemic and can theoretically be reduced, but equally practically persistent (the wider is the area object of the study, the looser will necessarily be the inventory of all the goods subject at risk considered in the analysis).

Usually the developing of such a map is a complex task that involves many disciplines including geophysics and geology (in order to take in account past seismicity, seismo-tectonic framework, wave propagation as well as soil effects), survey (in order to collect data about the building stock), structural analysis (in order to assess the building response under seismic loads) and social and economic sciences (in order to evaluate socio-economic consequences of an earthquake) [8],[9],[15],[16].

The standard definition of seismic risk is the probability or likelihood of a damage, due to an earthquake, and consequent loss to a specifiedclass of elements at risk over a specified period of time. In order to keep the problem of computing the risk tractable, it is tackled initially decomposing the task in specialized (simpler) components, conditionally independent and conventionally referred as hazard (pertaining to the likelihood of the seismic shaking on ground), vulnerability (pertaining to the susceptibility to damage of the built environment) and exposition (containing the socio-economic evaluation of the loss) and then recursively applying the total probability theorem in order to aggregate together the separate components. Hence the risk can be expressed by a convolution integral [5].

2   Seismic Hazard

Seismic Hazard analysis is aimed at estimating a measure of the intensity of the ground motion at a site considering the characteristics of surrounding seismic sources. This kind of study is restricted to the shaking felt at the ground level and does not consider the action on the built environment. Therefore in hazard analysis the core aspects investigated are the source modeling (i.e. mechanism at the epicenter that produces the shaking), the wave attenuation (along the path between the source and the site of interest) and the local ground amplification (through the ground layers around the site). The probabilistic assessment of seismic hazard involves determining either the probability of exceeding a specified ground motion, or the ground motion that has a specified probability of being exceeded over a particular time period. Accordingly, output of the hazard analysis is either a curve showing the exceedance probabilities of various ground motions at a site, or a hazard map that shows the estimated magnitude distribution of ground motion that has a specific exceedance probability over a specified time period within a region.

Despite the fact that several studies on seismic hazard were undertaken in Italy before, only after 2004 this kind of analysis assumed official recognition in technical community, since the seismic classification was compulsory associated with the likelihood of reaching some levels of seismic accelerations at site.

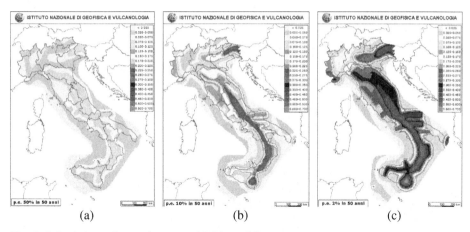

Fig. 1. Seismic hazard maps in terms of PGA at different return period,Tr: (a) 72, (b) 475, (c) 1000 years

Therefore the probabilistic hazard analysis conducted by the INGV [14], has become the Italian national reference in engineering applications. The results have been mapped on national scale over a 0.05° grid for various annual frequencies of exceedance (the reciprocal of the return period: T_r, varying from 30 to 2500 years) presenting peak ground acceleration (PGA) and spectral ordinates in acceleration for various natural periods (T_n: varying from 0.1 to 2.0 sec.); in total 90 maps have been produced, three of which are presented in Figure 1.

3   Seismic Vulnerability

Seismic Vulnerability represents the susceptibility to damage of the object at study, givena measure of the seismic input. Methods applied in representing the vulnerability analysis vary greatly depending on the complexity of the approach and the available data about exposure (see next section). Generally when in a vulnerability analysis it is considered a single item (like a specific building) the study can reach a very fine level of detail, defying the modality of damage and/or the number and type of components damaged [10,11]; on the other hand when it is under scrutiny a bulk of items, like a building stock, the vulnerability may necessarily been defined in looser terms as the damage potential of a class of similar structures, using as classification a broad identification (as for example the same structural type, number of floors, age, technique of construction …). Vulnerability of structures to ground motion effects is often expressed in terms of fragility curves (or damage functions) that take into account the uncertainties in the seismic demand and capacity.

In the present study the fragility curves have been built according to the SP-BELA approach [2,3]. According to this methodology the displacement capacity of the buildings at different damage levels (limit states) is produced, relating the displacement capacity to the material and geometrical properties. Three limit state conditions have been taken into account: slight damage (LS1), significant damage (LS2) and collapse (LS3). The slight damage limit condition refers to the situation where the building can be used after the earthquake without the need for repair and/or strengthening. If a building deforms beyond the significant damage limit state it cannot be used after the earthquake without retrofitting. Furthermore, at this level of damage it might not be economically advantageous to repair the building. If the collapse limit condition is achieved, the building becomes unsafe for its occupants as it is no longer capable of sustaining any further lateral force nor the gravity loads for which it has been designed. The aforementioned limit states can be assumed equivalent to the definitions contained in Eurocode 8, as follows: LS1: Damage Limitation (DL), LS2: Significant Damage (SD) and LS3: Near Collapse (NC).

In order to fit fragility functions to exposure data, in the case of masonry buildings, four separate building classes have been defined as a function of the number of storeys (from 1 to 4), whilst for reinforced concrete the building classes have been defined considering the number of storeys (from 1 to 4) and the period of construction. The year of seismic classification of each municipality has then been used so that the non-seismically designed and seismically designed buildings could be separated. In this way, the evolution of seismic design in Italy and the ensuing changes to the lateral resistance and the response mechanism of the building stock could be considered.

4   Exposure

Exposure is a representation on the population of items object of the study and their relevant aspects in relation to the risk analysis (this kind of information has necessarily to interact with hazard and vulnerability components of the study). Depending on the extension of the scope of the analysis, exposure may include a single building with its occupants and contents, or may include the entire constructed environment in a specified area, inclusive of buildings and lifelines (infrastructural systems forming networks and delivering services and goods to a community). In order to facilitate information collection about the existing facilities in a region, a standardization of the inventory is deemed, providing a systematic classification of the structures according to their type, occupancy and function.

In Italy the general characteristics of the building stock are provided by the Census. The data utilized in the present study are obtained from the 14th General Census of the Population and Dwellings (ISTAT 2001) [12]. The Census data are collected and aggregated at different levels: the basic unit for data collection is the single household and dwelling, but each dwelling is classified as being located within a building, of a given construction type (RC, Masonry, Other), with a given number of storeys (1, 2, 3, 4+) and age of construction ($\leq$1919, 1919/1945, 1946/1961, 1962/1971, 1972/1981, 1982/1991, $\geq$ 1991). In order to protect privacy, the collected data are disclosed only in aggregated format whose minimum territorial extension is the Census tract (a small, relatively permanent statistical subdivision of a geographical region, designed to be relatively homogeneous with respect to population characteristics, economic status and living conditions). In highly urbanized areas, like the Cassino town centre, a census tract has the dimensions of a building block. Further details about the elaboration of the exposure data are discussed in the next section.

5   Application Results

The case analyzed in this paper is represented by Cassino, a small sized town (35'000 inhabitants) located in southern Lazio, in a seismic prone area classified as at medium hazard level.

Local seismicity is characterized by active faults surrounding the town (even at a very close distance). The historic events that have hit Cassino are shown in Figure 2, where for each earthquakeyear and local seismic intensity are reported. Intensityis the classification of the strength of the earthquake shaking based on the observed effects (e.g. building damage) and in the graph it is measured according to the Mercalli-Cancani-Sieberg (MCS) scale, spanning from I=1 not felt, to XII=12 total destruction.

The main feature of the built environment of Cassino, that differentiates this town from similar Italian municipalities, is the fact that the town was almost completely destroyed at the end of World War II during the so called 'Battles of Monte Cassino' (January-May 1944) [21, 22] and then rebuilt, at the end of the war, in a relatively short time [23, 24, 25, 26].

In Figure 3 it is shown the evolution of the building activity at Cassino over time, together with the more relevant legislative measures that can be of interest in a seismic risk analysis. It is, indeed, of great interest for the aims of this study that the building stock of Cassino is relatively younger than the Italian average and that reconstruction began when the municipality was already classified in seismic zone after the Avezzano earthquake (January 13, 1915, $M_w=7.0$) [24], so that the first structures built during the reconstruction are supposed to be designed according with the seismic principles commonly applied at the time (elastic design relying over the allowable stress principle and using horizontal forces about 7% of the weight). Cassino was subsequently declassified in the 20 years span period since 1962 until 1982, when the economic boom was associated with the maximum rate of the building activity. It was, indeed, felt that the enforcement of seismic rules was an impediment to economic activities and urban development and therefore it was not so uncommon that municipalities, after some time since the last seismic event that justified their insertion in the seismic zone list, petitioned to be removed. In the case of Cassino, the cancellation was 'de facto', since it was sufficient not to be included in the new list prepared in 1962, while in the case of the nearby Pontecorvo town an 'at hoc' decree was issued in 1959 to selectively declassify the periphery (awaiting to be urbanized) whilst the already constructed urban centre was kept seismic. Cassino was then re-classified in 1983, after the Irpinia earthquake (November 23, 1980, $M_w=6.9$). Only after the Molise earthquake (October 31/November 2, 2002, $M_w=6.0$), a fundamental revision of the seismic classification as well as of the seismic design rules was undergone, redefining the seismic classification on the basis of a probabilistic hazard analysis rather than on an historical basis and incorporating in the new recommendations the limit state approach, with load and resistance safety factors and capacity design principles.

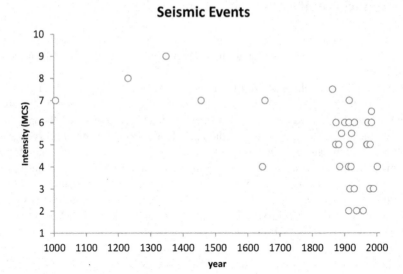

Fig. 2. Historic seismicity around Cassino

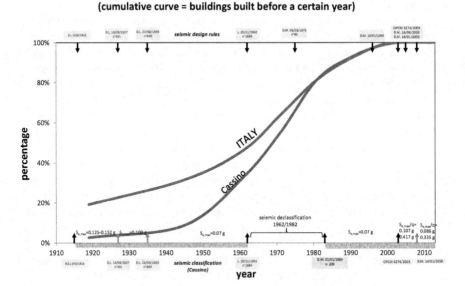

Fig. 3. Evolution of the Building Stock at Cassino

The information about the geotechnical setting has been obtained by a recent study on micro-zonation [18,19], from which emerges that the town of Cassino is settled in an alluvial plain, characterized by the presence of soft soils.

For privacy purposes the relevant data about buildings contained in census tracts (Cassino municipality is subdivided in 780 tracts) are made available through their marginal frequency, without disclosing the underlying joint distribution (this kind of data is available in aggregate format only for provinces and big cities). The problem of reconstruct the joint distribution from the marginal seven if it is known the correlation structure, it is not an easy task, since no unique solution exists. A possible approach is to consider a possible model representing the stochastic dependence structure among variables. In such a case Copula functions provide a useful tool to generate joint distributions by combining given marginal distributions.

The data made available by the Census have been surveyed in order to check the affordability and eventually introduce corrections. The following operations have been undertaken:

1. Extensive use of Google Earth all over the municipality to verify the correctness of the number of buildings and the story distribution within a single Census Tract;
2. Visual survey of the town centre in order to resolve dubious question left from point 1 above and to verify the construction technique (masonry, reinforced concrete, other).
3. Examination of some of the information available at the local office of the Civil Engineering Corps (Genio Civile). It is important to notice that since 1971 all the Reinforced Concrete structural projects (L.November 5, 1971, n. 1086) were subjected to be filed at Civil Engineering Corps before the construction started. The review of 15 complete structural projects (inclusive of technical drawings, relations and calculations) permitted to have an insight about the implementation into practice of design rules and construction standards;
4. Examination of the documentation available at the Technical Office of the Municipality of Cassino. On May 7, 1984 the Lazio-Abruzzo earthquake (Mw=5.9) hit the region, with epicenter located approximatively near San Donato-Val Comino, 27 km away from Cassino. In Cassino the damages to buildings (classifiable according to the EMS-98 scale as negligible to moderate, I=VI-VIII MCS) have to be reported to the Municipality by the property owners in order to accede to financial contributions for repairs. Each of the 86 requests examined were accompanied by a technical report signed by a local engineer, providing insight about the structures and the damages (usually the reports were accompanied by structural drawings, photographs and sketches of the crack patterns).

All the aforementioned components of the seismic risk have been handled within a Geographical Information System (GIS). A GIS represent the ideal environment for the management of spatial information, since it permits to archive, handle, compute, and display very large amount of data both in graphic or tabular format. The system, with its ability to be linked to external resources (like computational programs or high level database management systems) has also the feature to provide the required information interoperability, making possible to manage the great volume of data involved and the numerous processes needed in the calculations.

The seismic risk analysis has been carried initially performing the calculations over the 84 classes of buildings and then combining the results on tracts (in order to keep the output format consistent with the one provided by the Census) considering the effective composition of each tract through a weighted average. The results of the analysis are shown in Figures 3 and 4.

Figure 4 reports, both on the entire municipality and on a significant quadrant of the town centre, the probability of exceedance in a 50 years period of the three limit states considered (LS1: slight damage, LS2: significant damage and LS3: collapse). In order to have a term of comparison, the probability of exceedance calculated on the existing buildings, $P_{ex,50}(LS_i)$ (i=1,2,3), has been dividedby the probability of occurrence of the seismic action used in the design of the new residential buildings (and the assessment of the existing ones), $P_{new,50}(LS_i)$. According to Italian seismic rules NTC-08 [5] this probability is given for the three limit states as follows:

$$P_{new,50}(LS_1)=0.63 \qquad P_{new,50}(LS_2)=0.10 \qquad P_{new,50}(LS_3)=0.05.$$

Therefore the obtained index, $I_1=P_{ex,50}(LS_i)/P_{new,50}(LS_i)$, represents a comparative measure between the expected capacity (numerator) and the expected demand (denominator) in terms of probability of exceedance (the highest is the index, the less safe is the structure). Obviously the new structures, which have at least to comply with the indicated demand, are designedwith additional conservative measures (represented by load and resistance safety factors, capacity design rules, minimum design requirements), so that the few cases where the ratio is $I_1<1.0$, do not necessarily imply that an existing structure is safer than a new one.

As shown in figure 4, while the differential between capacity and demand is acceptable for LS1, it deepens as the level of damage increases (for LS2 or LS3). This kind of result was somehow expected, since the slight damage (LS1) is conditioned mostly by the quality of the details of non-structural components (whose design is controlled by architectural or climatic rather than seismic or structural considerations), while the occurrence of significant damage (LS2) and collapse (LS3) is conditioned by the presence in the design of seismic provisions and considerations about the expected mechanism of collapse under seismic actions.

The variability that can be observed, for each limit statethrough the entire municipality, is mostly due to the differences in structural type and age of construction of the buildings, since the soil conditions are constant for the whole town centre.

Finally figure 5 represent an index, $I_2=E(L)/R$, aimed at comparing the expected monetary lossesin 50 years due to an earthquake, E(L), and the cost for the retrofit of the structure, R.

The expected loss, E(L), has been defined considering for each limit state the associated costs for repairing or rebuilding (the more severe is the damage, the higher are the costs) and the probability of occurrence of the limit states.On the other hand, the retrofit cost, R, has been assumed as a deterministic value and independent of the existing structural conditions.

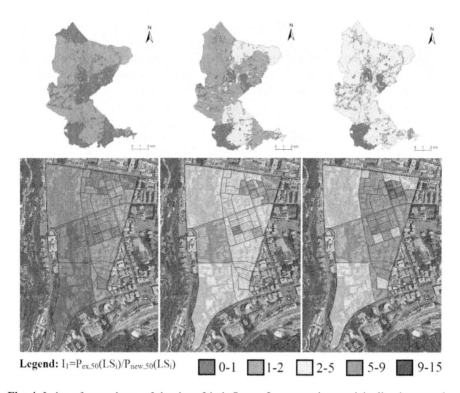

Legend: $I_1 = P_{ex,50}(LS_i)/P_{new,50}(LS_i)$ ▮ 0-1 ▮ 1-2 □ 2-5 ▮ 5-9 ▮ 9-15

Fig. 4. Index of exceedance of the three Limit States, I_1; top: entire municipality, bottom: details of the South-West quadrant of the town centre; left: slight damage (LS_1), middle: significant damage (LS_2), right: collapse (LS_3)

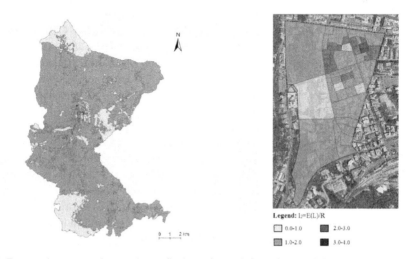

Legend: $I_2 = E(L)/R$ □ 0.0-1.0   ▮ 2.0-3.0   ▮ 1.0-2.0   ▮ 3.0-4.0

Fig. 5. Expected monetary loss and retrofitcost ratio, I_2. left: entire municipality, rigth: details of the South-West quadrant of the town centre

Obviously the reference economic values required for this kind of analysis depend highly upon many factor such as the method used for the cost estimation (this task can be performed either analytically considering typical standardized cases and analyzing the breakdown of the works deemed and then multiplying their quantity for their unit cost or historically considering how much was spent in the past in similar circumstances), the local conditions (construction costs have a significant regional variation), the quality of building finishes. In this analysis, the monetary values assumed have been selected consistently with international [13],[20] and national literature [1],[4].

Therefore the cost of construction of a new building has been assumed as 1'200 €/m^2 (1'280 €/m^{2}is the maximum contribution the State pays for reconstruction in L'Aquila after the 2009 earthquake [4]) whilst the one for retrofit using a traditional technique is around 500 €/m^2 (the State pays for the retrofit of public buildings a maximum of 150 €/m^3, that is 450 €/m^2 when considering a typical 3 m inter-story height, OPCM 3362/2004).

Obviously the index graphed in Figure 5 wants to represent the order of cost-effectiveness of undertaking measures of reduction of seismic risk even if it does not consider the possible utility associated with the market value of the real estate.

6   Conclusions

The work presented herein consisted in the assessment of theseismic risk map of the town of Cassino using the state-of-the-art evaluation procedure. The study, even if focused on a particular case for demonstration purposes, can be usefully extended to any other Italian urban agglomerate since the basic ingredients used in the analysis are already made available at national scale and the procedure can be easily standardized using modern computing tools like GIS. The study permitted to evaluate the level of affordability of the input ingredients and thus to evidence the aspects requiring a better refinement in a possible extension of the study at national scale.

It is important to point out that when tackling a small town, like Cassino, an extensive verification of the quality of the information utilized in the analysis was possible and reasonably not onerous. On the contrary, at national level, the availability of a very large amount of data, coming from different institutions and not necessarily collected for the scopes of a seismic risk analysis, poses the problem of harmonization of the pieces of information. The problem is quite arduous when considering the extreme variety of construction techniques (Italian building stock has a not indifferent percentage of vernacular and heritage architecture built following local traditions) and the different implementation of design rules and construction standards throughout the country. On the other hand the development of risk analysis through regional at hoc studies poses the problem of not-consistencies, especially when it comes to transform descriptive information about quality of construction and level of damage (usually expressed through verbal expressions) into measurable results (such as a quantification of the probability of occurrence or the monetary losses).

Coming to the specific aspects of the application, the results of the study have highlighted that, although a century has passed since the devastating 1915 Avezzano earthquake and although seismic design rules have been introduced after the event (but suspended during the construction boom), the seismic risk is still unacceptably high: a large number of buildings would suffer significant damage and collapse, causing loss of life, damages and business interruptions.

References

1. ANCE Catania: Sicurezza strutturale per gli edifici residenziali in cemento armato costruiti prima dell'entrata in vigore della normativa sismica [in Italian], p. 56, Catania (2012)
2. Borzi, B., Crowley, H., Pinho, R.: Simplifiedpushover-basedearthquakelossassessment (SP-BELA) for masonrybuildings. Int. J. Archit. Heritage **2**(4), 353–376 (2008)
3. Borzi, B., Pinho, R., Crowley, H.: Simplified pushover-based vulnerability analysis for large scale assessment of RC buildings. Eng. Struct **30**(3), 804–820 (2008)
4. Comune de L'Aquila: QR1 - Sisma Abruzzo 2009. Quaderno per la ricostruzione n. 1. Il modello parametrico e la scheda progetto esecutivo ricostruzione per l'Aquila [in Italian], L'Aquila (2013). http://www.usra.it/wp-content/uploads/studi-pubs/QR1-A-23maggio2013.pdf
5. Cornell, C.A., Krawinkler, H.: Progress and challenges in seismic performance assessment. PEER Center News **3**(2) (2000). http://peer.berkeley.edu/news/2000spring/performance.html
6. Decanini, L., De Sortis, A., Goretti, A., Langenbach, R., Mollaioli, F., Rasulo, A.: Performance of masonry buildings during the 2002 Molise, Italy, earthquake. Earthquake Spectra **20**(S1), 191–220 (2004)
7. Decree of Ministry of Infrastructure: Nuove norme tecniche per le costruzioni [inItalian]. Istituto Poligrafico dello Stato, Roma (2008)
8. Dolce, M., Masi, A., Marino, M., Vona, M.: Earthquake damage scenarios of the building stock of Potenza (Southern Italy) including site effects. Bulletin of Earthquake Engineering **1**(1), 115–140 (2003)
9. Faccioli, E., Pessina, V. (eds): The Catania Project: earthquake damage scenarios for high risk areas of the Mediterranean. CNR—Gruppo Nazionale per la Difesa dai Terremoti, Rome, p. 225 (2000). ftp://ftp.ingv.it/pro/gndt/Pubblicazioni/Faccioli_copertina.htm
10. Grande, E., Rasulo, A.: A simple approach for seismic retrofit of low-rise concentric X-braced steel frames. Journal of Constructional Steel Research **107**, 162–172 (2015)
11. Grande, E., Rasulo, A.: Seismic assessment of concentric X-braced steel frames. Engineering Structures **49**, 983–995 (2013)
12. Istituto Nazionale di Statistica (ISTAT): 14° Censimento della popolazione e delle abitazioni 2001 [in Italian], Roma (2005)
13. Liel, A.B., Deierlein, G.G.: Cost-Benefit Evaluation of Seismic Risk Mitigation Alternatives for Older Concrete Frame Buildings. Earthquake Spectra **29**(4), 1391–1411 (2013)
14. Meletti C., Montaldo V., Stime di pericolosità sismica per diverse probabilità di superamento in 50 anni: valori di ag. Progetto DPC-INGV S1, Deliverable D2 (2007). http://esse1.mi.ingv.it/d2.html
15. Nuti, C., Rasulo, A., Vanzi, I.: Seismic safety of network structures and infrastructures. Structure & Infrastructure Engineering: Maintenance, Management, Life-Cycl **6**(1–2), 95–110 (2010)
16. Nuti, C., Rasulo, A., Vanzi, I.: Seismic safety evaluation of electric power supply at urban level. Earthquake engineering & structural dynamics **36**(2), 245–263 (2007)
17. Rasulo, A., Goretti, A., Nuti, C.: Performance of lifelines during the 2002 Molise, Italy, earthquake. Earthquake Spectra **20**(S1), 301–314 (2004)

18. Regione Lazio: Studio di Livello I di Microzonazione Sismica dell'Unità Amministrativa Sismica di Cassino (FR). Determination of December 21, 2012 [in Italian] (2012)
19. Saroli, M., Lancia, M., Albano, M., Modoni, G., Moro, M., Mugnozza, G.S.: New geological data on the Cassino intermontane basin, central Apennines, Italy. Rendiconti Lincei 25(2), 189–196 (2014)
20. Smyth, A.W., Altay, G., Deodatis, G., Erdik, M., Franco, G., Gulkan, P., Kunreuther, H., Lus, H., Mete, E., Seeber, N., Yuzugullu, O.: Probabilistic benefit-cost analysis for earthquake damage mitigation: Evaluating measures for apartment houses in Turkey. Earthquake Spectra 20(1), 171–203 (2004)
21. Herbert, B.: The bombardment of Monte Cassino. (February 14-16, 1944) A New Appraisal. In: Benedictina, Grottaferrata, vol. 20, pp. 384–424 (1973)
22. Caddick-Adams, P.: Monte Cassino: Ten armies in Hell. Oxford University Press, London (2013)
23. Pelliccio, A., Cigola, M.: Cassinonei piani regolatori del Novecento, Tipogr, p. 70. Francesco Ciolfi, Cassino (2010)
24. Pelliccio, A.: Edilizia residenziale pubblica del secolo scorso aCassino: case popolari del 1911 e case asismiche del 1915. Studi Cassinati 5(4), 222–230 (2005). Tipografia Ugo Sambucci, Cassino www.studicassinati.it/db1/jupgrade/images/stories/pdf/2005-4.pdf
25. Pelliccio, A., Cigola, M., Gallozzi, A.: La ricostruzione post-bellica della città di Cassino. Rilievo e catalogazione documentale attraverso procedure GIS. In: Proceedings of Workshop "Tecnologie per Comunicare L'architettura", pp. 457-461. Università Politecnica Delle Marche, Ancona (2004)
26. Cigola, M., Pelliccio, A.: Cassino nel XX secolo, Storia e rappresentazione di un tessuto urbano. In:Proceedings of 1st Conference "Saper Valorizzare - Unicittà, l'Università incontra la città". Maniaci, M., Orofino, G. (eds) pp. 207-227. Edizioni Università di Cassino, Cassino (2006)

Big Data in Civil Security Research: Methods to Visualize Data for the Geovisual Analysis of Fire Brigade Operations

Julia Gonschorek[✉], Hartmut Asche, Harald Schernthanner, Benjamin Bernhardt, Anja Langer, Marius Humpert, and Caroline Räbiger

Department of Geography, University of Potsdam,
Karl-Liebknecht-Str. 24/25, 14476 Potsdam, Germany
{julia.gonschorek,gislab,harald.schernthanner,
bebernha,anlanger,mahumper,craebige}@uni-potsdam.de
https://geoinfo.geographie.uni-potsdam.de/

Abstract. This article gives insight in a running dissertation as well as a finished project of the Masterprogram Geoinformation and Visualization at the University in Potsdam. Point of discussion is the spatial and temporal distribution of emergencies of German fire brigades that have not sufficiently been scientifically examined. The challenge is seen in Big Data: enormous amounts of data, that exist now (or can be collected in the future) and whose variables are linked to one another. These analyses and visualizations can form a basis for strategic, operational and tactical planning, as well as prevention measures. The user-centered (geo-) visualization of fire brigade data accessible to the general public is a scientific contribution to the research topic 'Geovisual Analytics and geographical profiling'. It may supplement antiquated methods such as the so-called pinmaps as well as the areas of engagement that are freehand constructions in GIS. Considering police work, there are already numerous scientific projects, publications, and software solutions designed to meet the specific requirements of Crime Analysis and Crime Mapping. By adapting and extending these methods and techniques, civil security research can be tailored to the needs of fire departments. In this paper, a selection of appropriate visualization methods will be presented and discussed.

Keywords: Geovisual analytics · Big data · Visualization · Explorative (Data-) analysis · Civil security

1   Civil Security as a Research Topic

Civil security is a public good. In the United States of America, Canada, and the United Kingdom computer-based research on civil security has been conducted, and the result published, for more than two decades, especially in the field of criminology: "Crime analysis refers to the set of systematic, analytical process that provides timely, pertinent information about crime patterns and crime trend correlations. It is primarily a tactical tool. [...] Analyzing and comparing data on file with those on current cases can give

© Springer International Publishing Switzerland 2015
O. Gervasi et al. (Eds.): ICCSA 2015, Part III, LNCS 9157, pp. 415–425, 2015.
DOI: 10.1007/978-3-319-21470-2_30

patrol officers important information on activities in their beat areas. [...] Using this information, patrols can better deploy resources." (BOBA, 2005, 5) The focus is on solutions that lead to perpetrator identification and successful judicial prosecution. Crime scenes and hotspots of criminal activity are mapped with computers and then evaluated by analysts. In Germany, the responsible test facility is that of the federal police in Lübeck. For the fire department, a central responsible institution does not exist. In the Länder as well as the professional fire services, individual programs for demand-oriented research and development are implemented. Civil security research has been on the agenda of the federal government for several years. With the program "Research for Civil Security 2012-2017" of the Federal Ministry of Education and Research, a previous program of the same name is continued. It defines civil security as "the basis of free life and [as] an important factor of economic prosperity in Germany" (BMBF, 2012 2). The aim of the program and of the financial support of R&D projects between universities, research institutions, companies, and government institutions, is to find solutions for the protection of public goods such as infrastructure and economy, as well as for the citizens. Civil security and research in this area be understood as cross-cutting issues. One focus will be on finding solutions for the protection and rescue of people that fit the operational routine of responsible organizational units.

Data from the Fire Brigade in Cologne will be visualized in terms of their spatial, temporal and spatio-temporal distributions to support tactical, strategic and operational planning in the field of civil security. Highly specialized software products, e.g. individual solutions from Siemens, SAP and others, support the operational planning of the police. Examples include Geovista, RIGEL Analyst and Compstat. Analytics and visualization systems customized to the needs of the fire brigade are rare, particularly considering the fact that no products for decision support or mobile services are to be designed, as there are already solutions available that support the dispatchers in the control center.

Evaluating past, but also ongoing modeling and forecasting operations can help with strategic, tactical, and operational planning. This specifically concerns decision support of operational planning, the demand for equipment, risk assessment and the identification of any potential dangers. For this purpose, samples of the data available to a Ph.D. Project are analyzed (approx. 500.000 entries), using selected tools within the open-source statistics platform R and the JavaScript library D3.

Thus, fire brigade operations can be visualized in terms of their type, location, duration and frequency in the test area Cologne. Objectives of data visualization are:

1. to create a deeper understanding of the information embedded in the data,
2. to establish a basis for statistical analyses to identify hotspots (clustering method and kernel density estimation) and
3. to identify spatial and temporal movement patterns and contexts.

2   Geovisual Data Analysis

In the english language, geo-visualization, geo-visual analyses and geo-visual analytics are often all referred to as geo-visual analytics, even though their meanings differ. This is caused by the fact that there is no adequate German term.

Geo-visual analytics are interpreted as range of methods that enable the cognitive process of data analysis and decision-making by visualizing information. The goal is to confirm knowledge and experience, and to discover and communicate new understanding (compare KEIM/ANDRIENKO/FEKETE, 2008). However, Geovisual Analytics is more than "just" visualization: geo-visualization and geo-visual analysis methods are combined with data analysis and user experience to derive semi-automated processes using one or more (software) systems.

This is applied in Geographic Profiling Analysis (short: Geoprofiling) as part of an investigative analysis. Hereby, spatial and temporal profiles of emergencies can be created. Visible, solid structures in the form of clusters or hot spots of certain operation types, that ultimately contribute to profiling, are required.

This raises the fundamental question whether or not there is a connection between the spatial and temporal localization of an emergency and the next one. An emergency happens in one place, and another emergency nearby shortly after. One can easily imagine that this is of great interest for the prediction of emergencies and subsequently for a focused operational planning.

To answer this question, hotspot analyses of past firefighting operations are used. These use clustering methods and kernel density estimates with variable time windows. To apply these methods in a justifiable way, a solid preanalysis is necessary, that is to be made with the help of select, partly dynamic visualization techniques.

2.1   Data: Basics and Preprocessing

For this project, the Cologne Fire Service gave us access to two archived databases that provide information about operations and the equipment used from 1.1.2007 to 31.7.2011. The database contains around 530,000 entries that include type, time and geo-coordinates of the operation. 810,000 entries in the equipment database describe the operational materials as well as the travel time to the scene. An operational key connects both databases. These databases are set up and run according to the law for the fire protection and assistance (cf. FIG. FSHG, 1998) of the government of North Rhine-Westphalia. The control center for fire and rescue service must be manned at all times, and all operations of the voluntary, compulsory, and professional fire departments must be reported here. When an emergency call reaches the control center, it is ensured that the emergency services will be alarmed. The dispatcher takes the call and keys information on the nature of the emergency, place, persons involved, etc. into the dispatch system. The address information is geocoded in the system. The type of emergency defines, what emergency services and resources are alerted. According to the predefined operational chain, the guard on duty and a fully equipped team are alarmed. The operation team uses a reporting system in the vehicle (e.g., "assignment accepted", "arrival at site", "ambulance").

In the database, these messages, along with the time, are recorded automatically. When the dates were collected, the chain of communication between teams ran on BOS radio. This is prone to errors as well as incorrect address information provided by the person requiring assistance on the phone. Partially missing data are ignored in the analysis, and amount to less than 3% of the total data volume.

The data records must be converted into file formats that can be read by the free statistical analysis environment R and the JavaScript library D3, before it can be edited. Preprocessing is very time-consuming, because the data must be converted to different formats. The main steps are:

1. Selecting the test sample(s) from the databases and possibly merging multiple data tables. Selected data samples out of about 140 different emergency cases:
   (a) internistic emergency (INTERN1 and INTERN2)
   (b) surgical emergency (CHIRU1)
   (c) crash (UNF1)
   (d) helpless person (HILO).
2. Extracting the numerical time information and either converting it into a common date/ time format, or re-encoding it completely.
3. Cumulating results and integrating statistical values.
4. Storing the data tables as CSV or TSV files.

2.2   Methods to Visualize Data

"Visual exploration, in particular, enables the user to visually investigate the data. In short, the visual exploration enables the user to try out some parameters, instantly view the results of the parameter change, manipulate the data through selection and highlight operations and relate that information to other sources and visualizations." (ROBERTS, 2008, 26)

In the recent past, common visualization techniques such as choropleth maps, bar graphs, heat maps and statistical charts underwent development, particularily in the ever-changing environment of free and open-source software. The linking of different data views and levels of information that are either focused on the individual or on target groups, can be implemented. The main objective is the implementation of sophisticated graphical visualization projects of an interdisciplinary and diverse group of developers and users with FOSS products.

Examples of available, open source software, libraries and tools are Leaflet, Data-DrivenDocuments (D3), Processing, R and Highcharts. From the broad spectrum of (geo)visualization methods, we select and present below four visualization types appropriate for the user and the anticipated application.

1. The stacked bar chart is a special form of the bar chart that displays the frequency distribution of a specific feature. The stacked column chart shows a total value, which is composed of individual values. Changes in the total value are visible, but variations within the individual values are difficult to discern. By implementation into D3 however, one can switch between two views. This means that selected features can be seen stacked in one column or as separate individual columns. That is how the two different data views are achieved. To simplify the comparison, the same features are uniformly color-coded and divided according to a predefined time window.

2. The choropleth map is used to visualize the spatial distribution of quantitative traits. By using a monochromatic coloring, the pre-classified data are encoded in the saturation of a single color. This is how the increase or decrease of the variable's values are coded.
3. Heatmaps are convenient tools to visualize multidimensional data sets. To map the levels of each specific value, warm colors are selected and defined by using the Colorbrewer. Thus a concentration of events can be visualized and hotspots can be identified in a spaceless map.
4. Tree diagrams are suitable for hierarchically structured data. They allow a deeper understanding of coherences, affiliations of variables, and event frequencies. The chord diagram or sunburst chart is a relatively new adaptation of these, which shows a relationship between data groups with at least two different parameters. The similarities are defined by the varying thickness of lines. If a certain combination occurs several times, the thickness changes accordingly. Since the data used do not have positive or negative notation, a neutral color scheme is used (in this case blue instead of green or red).

2.3   Realization with FOSS Products

The staple diagram created with D3 shows the five main types of application over the period from March to June 2007 and is based on the example *Stacked-to-Multiples* (cf. BOSTOCK, 2013B). Using a smaller portion of the data described above, emergencies with disabled persons, accidents, as well as internal and surgical operations, that have differentiated after months, are displayed.

With D3, a dynamic form of representation is possible, whereby the operations are visible as a continuous column (cf. Fig. 1) or individually (cf. Fig. 2). By using the stack view, it becomes clear that INTERN1 operations take up the majority within the chosen operational types. The INTERN2 and CHIRU1 operations follow with fair margin and approximately equal frequency. In total, these two operational types occur about as often as INTERN1. UNF1 and HILO operations approximately take place the same number of times, however, even together they do not reach the frequency of the CHIRU1 or INTERN2 operations. By using the stack view, temporal changes become visible. In March, the frequency of operations was significantly higher than in the following month. In June, however, the March level was reached again.

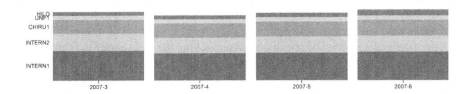

Fig. 1. Stack representation of the main operation types during the period of investigation (own source)

The single view permits a comparative month-by-month analysis of a specific operational type. Changes within the period of investigation become evident. It is striking, that the frequency of CHIRU1 and UNF1 operations show little variation, whereas the INTERN1 and INTERN2 operations in April decrease compared to the previous month. The next to doubling of HILO operation in June, which is visible with the single view, is likewise notable. (cf. Fig. 2).

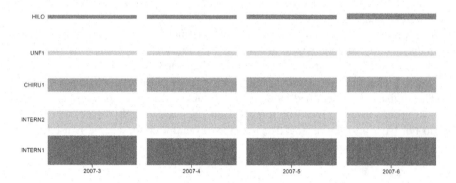

Fig. 2. Individual representation of the main application types during the period of investigation (own source)

Using Processing, a choropleth map at the district level of Cologne is produced (cf. Fig. 3). An essential step is the import of the GeoMap library and input data to be included in TSV and SHP format. Then, the emergency data are to be standardized, the color values are to be set, and the mouse-over function is to be implemented. Thereby, the surface coloring illustrates the frequency of INTERN1 operations in different city districts.

The mouse-over feature permits one to display the district name and the specific number of operations. The coloring of the districts demonstrates that the majority of operations took place in the city-center. In the five neighboring districts, only about half as many emergencies occurred.

During preprocessing for the presentation in a heatmap, the data set is pooled on the city districts of Cologne as well as on the months of January to June. In addition, five classes are formed according to Sturge's rule of thumb. Their limits are determined with the natural fraction method. The city districts are shown on the x-axis and the y-axis represents months. In addition, a mouse-over function is integrated, which indicates the exact number of operations when moving the mouse over the value field. An example of the distribution of surgical emergencies is visualized (cf. Fig. 4). Especially in the months of March, April and May, emergencies in the city-center of Cologne were very frequent. The fire department was called for service in the city center a total of 400 times. The following example code provides insight into the core of the Java script and shows the creation of the heat map in D3 (cf. FIG. Bostock, 2013A).

```
import org.gicentre.geomap
org.gicentre.geomap.Table("INTERN1.tsv", this)
geoMap.readFile("stadtbezirk")
float normINTERN1 = norm(float(intern1), dataMin,
dataMax)
fill(lerpColor(cbBluesC, cbBluesI, normINTERN1
int id1 = geoMap1.getID(mouseX, mouseY);
if (id1 != -1)
  {
String name = geoMap1.getAttributes().getString(id1, 3)
String nummer = geoMap1.getAttributes().getString(id1,
9)
```

Fig. 3. Choropleth map of the internist emergencies in the city of Cologne (own source) and code snippets in Processing

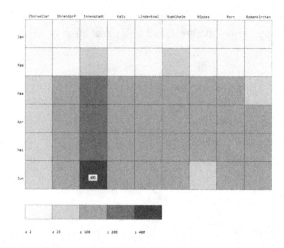

```
d3.tsv("data.tsv",
function(d) {
return {
Monate: +d.Monate,
Stadtteil: +d.Stadtteil,
Anzahl: +d.Anzahl}});
var heatMap = svg.selectAll(".Stadtteil")
.data(data)
.enter().append("rect")
.attr("x", function(d) { return (d.Stadtteil -1 ) *
gridSize; })
.attr("y", function(d) { return (d.Monate-1 ) *
gridSize; })
.attr("rx", 0)
.attr("ry", 0)
.attr("class", "Stadtteil bordered")
```

Fig. 4. Heatmap of surgical emergencies in the district of Cologne per month (own source) and the code snippets in D3

The chord chart, which shows coherences between operations and the alerted fire station for surgical emergencies, is established in D3 as well. The line thickness indicates how often the fire brigades accepted a call from the control center and moved out. In addition, the equipment in operation of each fire station is noted (vehicle numbers are used for this). Other parameters can be read using the mouse-over function. For example Wache01 was alarmed for 19 operations in the period from January to June 2007. Their activities make up 48.1 percent of the total chart (cf. Fig. 5).

Fig. 5. Chord Chart showing the alerted stations and the operational equipment for surgical emergencies (own source)

3   Discussion of the Results and Outlook

These data visualizations demonstrate the high flexibility that large data sets can have. The stacked bar chart shows a chronological sequence of the emergency numbers for the five main operational types over a course of four months. This is variable and can easily be transferred to a more detailed level, such as weekly or hourly.

A mouse-over function is conceivable for a more intuitive handling with larger data sets. The choropleth map of Cologne shows internist emergencies. By implementing additional choices and adding the locations of the fire station, the user can gain in-depth information.

One may choose from all the 140 types of operations in different time windows, such as time of day, day of week, month or year. For detection and subsequent statistical analyses, e.g. kernel density estimates, the combination of spatial and temporal information in a cartographic representation is suitable. Additional benefit lies in adding the fire stations in the area that were under investigation, as well as by matching the finished operations. In this way, a cartographic counterpart to the non-cartographic chord diagram might be realized.

The dynamically designed heatmap as type of visualization is well adapt to the given data. The color coding illustrates, in which city districts and in what months a high number of operations took place (hotspots). Retrieving information, e.g. the frequency, with the mouse-over function increases the content visualized. Again, the implementation of a user selected function of the spatial segment, time window and the application mode is possible.

The heatmap matrix would be enlarged, but the level of detail would increase significantly. Chord diagrams enable the decision maker for the first time to oversee coherences between operations and operational equipment in a non-numerical way. Again, many of the approaches for optimization that were discussed above are conceivable.

Each of the data visualizations discussed in this paper gives deep insight into the confusing masses of data of the operations of professional fire departments in Cologne. Taken together, they constitute a scientific basis for statistical analyses and for the creation of geo profiles as well as visualization of information.

The information that can be drawn from the data visualizations leads to an improved selection of analysis methods and helps focus on spatial, temporal and spatio-temporal events of interest in the data. The expertise of operational forces and decision-makers can be validated with the approach presented here. In addition, these visualization methods facilitate the exchange of information and knowledge and their use for strategic, operational and tactical planning in the field of civil security.

References

1. Boba, R.: Crime analysis and crime mapping. Sage Publications Inc, Thousand Oaks/California (2005)
2. BMBF, Bundesministerium für Bildung und Forschung: Forschung für die zivile Sicherheit 2012-2017: Rahmenprogramm der Bundesregierung (2012). http://www.bmbf.de/pub/Rahmenprogramm_Sicherheitsforschung_2012.pdf (accessed 2013-02-13)
3. Bostock, M.: Day/Hour-Heatmap (2013a). http://bl.ocks.org/tjdecke/5558084 (accessed 2015-01-20)
4. Bostock, M.: Stacked-to-Multiples (2013b). http://bl.ocks.org/mbostock/4679202 (accessed 2015-01-22)
5. D3 Data Driven Documents (2013). http://d3js.org/ (accessed 2015-01-31)
6. FSHG, Gesetz über den Feuerschutz und die Hilfeleistung (2015). https://recht.nrw.de/lmi/owa/br_bes_text?anw_nr=2&ugl_nr=213&bes_id=4725&menu=1&sg=0&aufgehoben=N&keyword=FSHG#det0 (accessed 2015-01-31)

7. giCentre, geoMap: Easy drawing of geographic maps in Processing (2015). http://www.gicentre.net/geomap/ (accessed 2015-01-22)
8. Hey, B.: Präsentieren in Wissenschaft und Forschung. Springer, Heidelberg/Dodrecht (2011)
9. Highcharts: Make your data come alive, http://www.highcharts.com/ (accessed 2015-01-31)
10. Keim, D.A., Andrienko, G., Fekete, J.-D., Görg, C., Kohlhammer, J., Melançon, G.: Visual analytics: definition, process, and challenges. In: Kerren, A., Stasko, J.T., Fekete, J.-D., North, C. (eds.) Information Visualization. LNCS, vol. 4950, pp. 154–175. Springer, Heidelberg (2008)
11. Leaflet: An Open-Source JavaScript Library for Mobile-Friendly Interactive Maps (2014), http://leafletjs.com/ (accessed 2015-01-31)
12. Processing (2015). https://processing.org/ (accessed 2015-01-31)
13. The R Project for Statistical Computing (2015). http://www.r-project.org/ (accessed 2015-01-31)
14. Roberts, J.C.: Coordinated Multiple Views for Exploratory GeoVisualization. In: Dodge, M., McDerby, M., Turner, M. (eds.) Geographic Visualization: Concepts. Tools and Applications. John Wiley & Sons Ltd., Chichester (2008)

Application of SLEUTH Model to Predict Urbanization Along the Emilia-Romagna Coast (Italy): Considerations and Lessons Learned

Ivan Sekovski[1,3] (✉), Francesco Mancini[2], and Francesco Stecchi[3]

[1] Department of Earth Sciences, CASEM, University of Cadiz, Puerto Real, Spain
ivansekovski@gmail.com
[2] DIEF, University of Modena and Reggio Emilia, Modena, Italy
francesco.mancini@unimore.it
[3] Department of Biology, Geology and Environmental Science,
University of Bologna, Ravenna, Italy
francesco.stecchi2@unibo.it

Abstract. Coastal zone of Emilia-Romagna region, Italy, has been significantly urbanized during the last decades, as a result of a tourism development. This was the main motivation to estimate future trajectories of urban growth in the area. Cellular automata (CA)-based SLEUTH model was applied for this purpose, by using quality geographical dataset combined with relevant information on environmental management policy. Three different scenarios of urban growth were employed: sprawled growth scenario, compact growth scenario and a scenario with business-as-usual pattern of development. The results showed the maximum increase in urbanization in the area would occur if urban areas continue to grow according to compact growth scenario, while minimum was observed in case of more sprawled-like type of growth. This research goes beyond the domain of the study site, providing future users of SLEUTH detailed discussion on considerations that need to be taken into account in its application.

Keywords: SLEUTH · Urban growth · Land use planning · Cellular automata · Scenarios

1   Introduction

Uncontrolled urbanization can lead to series of environmental issues, such as encroachment of natural habitat and agricultural land, high energy or water consumption and waste generation, among others [1]. Worldwide, urban growth is particularly taking part in coastal areas [2]. For instance, between 1990 and 2000 the urbanization rates of European coastal regions were approximately 30% higher than in inland areas [3]. One of the hotspots of urbanization in Europe is the coastal zone of Mediterranean region, where urban growth is driven to large extent by tourism development [4]. A relevant example is the coastal zone included within the administrative boundaries of Emilia-Romagna region, Italy. It was heavily urbanized in decades following the Second World War, mainly due to the development of beach related tourism [5].

© Springer International Publishing Switzerland 2015
O. Gervasi et al. (Eds.): ICCSA 2015, Part III, LNCS 9157, pp. 426–439, 2015.
DOI: 10.1007/978-3-319-21470-2_31

Since this area is characterized by low lying setting and sandy beaches, it is suscepti- ble to inundation and erosion, caused mainly by marine flooding [6].

In order to study Emilia Romagna's vulnerability to coastal flooding in dynamic manner, a previous study [7] compared flooding scenarios with outputs of different scenarios of urban growth. For marine flooding scenarios this was done by applying functions implemented in the Cost-Distance tool of ArcGIS® to a high resolution Digital Terrain Model [8]. The urban growth scenarios were estimated by applying the SLEUTH model [9]. SLEUTH belongs to the group of cellular automata (CA) models, known for the ability to capture complex non-linear behaviour in growth patterns and self-organization emerging from the local interaction between cells and their neighbours.

This study faces issues related to the use of the SLEUTH model to project urban growth on regional level by using geographical dataset at high spatial resolution and all available information related to the environmental management policy. This in- cludes a detailed step-by step discussion on SLEUTH application throughout all of its phases, highlighting some potential considerations and summarizing the lessons learned. The output of this paper could serve all researches that consider the applica- tion of SLEUTH to introduce the best practice in the field of present and future envi- ronmental management and planning.

2   SLEUTH Model

SLEUTH is a self-modifying probabilistic cellular automata model. It is a public do- main C-language source code that runs under UNIX or UNIX-based operating sys- tems, structured into two modules that can be activated independently. One module is the Urban Growth Model (UGM) that simulates the urban growth, and the other is Land Cover Deltraton Model (LCD) that simulates the changes in land use. The code is publicly available on Project Gigalopolis website [10], a project born from collabo- ration between the University of California of Santa Barbara (UCSB) and United States Geological Survey (USGS).

SLEUTH's acronym is derived from its input requirements: Slope, Land use, Ex- clusion, Urban, Transportation and Hillshade. In brief, it can be described as a scale- independent CA model with Boolean logic, since each cell can be categorized only as urbanized or non-urbanized. Whether or not a cell becomes urbanized is defined by four transition rules of urban growth: spontaneous, diffusive (new spreading centre), edge growth and road-influenced growth. These rules are controlled by five coeffi- cients with values ranging from 0 to 100: dispersion (DI), breed (BR), spread (SP), road gravity (RG) and slope resistance (SR) coefficient [11]. Relationship between the growth types and growth coefficients is schematized in Figure 1.

Growth coefficients do not necessarily remain static throughout the model applica- tion. If growth rate exceeds or falls short of limit values, a self-modification process is applied. Without this feature, the growth could appear as linear or exponential, which is unrealistic [12, 13].

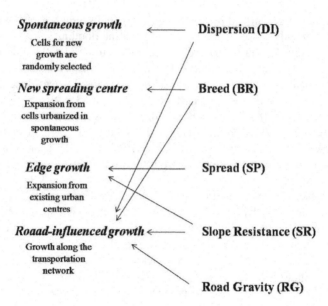

Fig. 1. Relationship between growth types and growth coefficients in SLEUTH (adapted from [14] and [15]

The model is implemented in two general phases: calibration phase, which simulates historic growth; and prediction phase; which uses patterns of historic growth to derive scenarios for future growth. The output of the model is a series of GIF images showing the predicted urban growth scenarios for each year.

All details on SLEUTH phases and other considerations (e.g. input data), are explained in the following section, through an example of our application of the model on the coastal area of Emilia-Romagna region.

3   Application of SLEUTH on the Coastal Area of Emilia-Romagna (IT)

This section describes the overall methodology related to: (i) selection, quality and overall characteristics of input data; (ii) calibration process; and (iii) design of three scenarios used for prediction.

3.1   Input Data

SLEUTH requires five types (or six, if land use is included) of input data: historic urban cover of at least four time periods, historic transportation network of at least two periods, slope, hillshade, and exclusion layers. All raster or vector-based layers in the end need to be attributed to a reference grid and successively converted into GIF images of same number of rows and columns.

In our case all input layers were prepared in SAGA (System for Automated Geoscientific Analyses) and ArcGIS 10.1 software. They were clipped to the same extent (rectangle of approximately 76 km of length and 26 km), representing a portion of the coastal zone of Emilia Romagna.

Urban layers were digitized for the years 1978, 1990, 2000 and 2011, from different sources. The 1978 urban layer was derived from 1:5000 scale topographic map provided by the Regione Emilia Romagna. The 1990 urban layer resulted from the digitizing of a LANDSAT image at 30mx30m spatial resolution, obtained from the United States Geological Survey (USGS) Global Visualization Viewer (GloVis) web service (http://glovis.usgs.gov). The 2000 urban layer was based on the aerial photogrammetric surveys after the 1999-2000 Istituto Geografico Militare Italiano (IGMI) flight (1:29000 scale, 0.65 m spatial resolution). Finally, for the year 2011 urban areas were digitized by using the World Imagery Basemap feature in ArcGIS 10.1, with high resolution (0.3 m) imagery of Western Europe provided by Digital Globe®.

Two transportation layers were prepared for years 1978 (the same topographic map as for 1978 urban layer), and for 2011 layer (from Web Mapping Service of Italian National Geo-portal: www.pcn.minambiente.it). Both layers took into consideration provincial and national level of roads, as well as highways.

Regarding Slope and Hillshade layers, they were both created from a 10mx10m Digital Terrain Model (DTM) provided by Regione Emilia Romagna after being resampled to a 20 m resolution by using the nearest neighbour method. As required by the model, the slope was extracted in percentage values.

Special attention was given to the Exclusion layer. Two different exclusion layers were considered: the first one referring to historical settings, used for calibration purposes; and the second one, used for the prediction stage. The need for such an approach deserves further explaining. Calibrating the model with current exclusion layer could be erroneous since many areas have received their protection status in period between the first historical year of calibration and the most recent one [16]. Early periods of calibration would therefore get informed by the actual distribution of currently excluded areas, leading to better fit in calibration in sort of "manipulative" manner [15].

Both of our exclusion layers had joint exclusion areas which remained unchanged, such as the sea and inland water bodies. The present exclusion layer contains additional zones where construction is prohibited, such as different zones of protected natural areas on a regional level; national reserves; sites of community importance related to the Natura 2000 network of the EU Habitats Directive (92/43/EEC); zones of special protection related to the EU Birds Directive (79/403/EEC), different protection levels of archaeological sites; and 150 m buffer zones around river banks and 300 m buffer around shorelines (see [7] for more details). Historic exclusion layer included only areas which were known as areas with prohibited construction since the very beginning of the time period used for calibration.

All input layers were converted into 20 m resolution raster grids of 1323 columns by 3816 rows using SAGA software and saved as greyscale GIF images, as required for the calibration stage.

Input layers are shown in Figure 2.

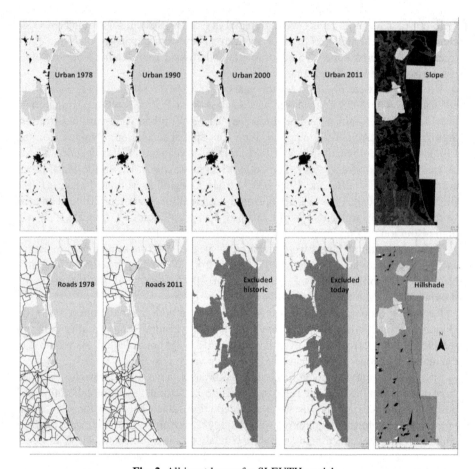

Fig. 2. All input layers for SLEUTH model

3.2   Model Calibration

The main goal of the calibration phase is to determine the values of growth coefficients that simulate urban growth for certain historic time periods. SLEUTH calibration is carried out through a "brute force" method which consists of three phases: coarse, fine and final [17]. Growth is simulated multiple times by using the Monte Carlo method, an iterative procedure used for computation of different spatial statistics [18-19].

In the coarse calibration phase, the widest range (1-100) of coefficient values is used, incremented of 25 at a time. The range of coefficients values used in calibration's subsequent phases (fine and final) is narrowed based on coefficient values that best replicate the historical growth in each preceding phase.

Self-modification constraints are causing coefficient values to be constantly altered from the first date to the last date of the run. For that reason, the best coefficient set for forecasting is actually derived by averaging the resulting values coming out from the final phase. In our case, this was done through 100 Monte Carlo simulations (with 1 step increment) so that an average value for each coefficient could be derived (read more in [14]). We named this additional phase "derive", according to [15]. This procedure is also recommended in the official website of the Gigalopolis Project.

In order to obtain the coefficient range of each successive step of calibration, we used the goodness-of-fit metric called Optimal SLEUTH Metric (OSM). The OSM is a product of the most relevant metrics offered by the code - compare, population, edges, clusters, slope, X-mean, an Y-mean metrics [20][16].

The resulting values of the calibration parameters for our study site, concerning all calibration phases, are visualized in Table 1. The coefficient range for successive steps of calibration was selected by examining the top three rankings of the OSM values, as indicated on the official Project Gigalopolis website. The highest OSM value increased with each calibration step (from 0.38 in coarse to 0.397 in final phase), meaning that the resemblance between modelled and observed data improved as calibration progressed.

Table 1. SLEUTH calibration parameters for 1978-2011 historic urban growth of Emilia-Romagna coastal area (from [7])

Growth coefficients	COARSE		FINE		FINAL			DERIVE		
	Monte Carlo iterations = 4		Monte Carlo iterations = 7		Monte Carlo iterations = 9			Monte Carlo iterations = 100		
	Range	Step	Range	Step	Range	Step	Final	Range	Step	Final
DI	1-100	25	0-20	5	0-5	1	1	1-1	1	1
BR	1-100	25	0-20	5	0-5	1	1	1-1	1	1
SP	1-100	25	15-35	5	20-30	2	24	24-24	1	30
SR	1-100	25	0-75	10	0-10	2	10	10-10	1	1
RG	1-100	25	0-50	10	10-50	5	50	50-50	1	52

The low final values of the DI (1) and the BR (1) coefficients imply that there was very little sprawled growth in the coastal area of Emilia Romagna for the 1978-2011 period. The value of the SP coefficient (30) indicates that growth occurred in a more compact manner, around existing urban areas. High value of the RG coefficient (52) implies that the road network played an important role in attracting urban development. Low value of the SR coefficient (1) was somewhat expected, since the study area is characterized by an extremely low slope variations and, therefore, slope does not represent a limiting factor for growth. It appears that self-modification parameters did influence the coefficient values that came out of the final phase, since they changed towards derive phase for SP, RG and especially for the SR coefficient.

3.3   Model Prediction – Development of Urban Growth Scenarios to the Year 2050

There are three different approaches to develop scenarios of urban growth within SLEUTH: (i) changing the values of growth parameters obtained through the calibration phase (e.g. [21][14]), (ii) assigning different protection levels to the exclusion layer (e.g. [22, 23]), and (iii) manipulating the self-modification constraints (e.g. [24]).

In our study, we used the combination of the first two approaches: both growth coefficients and exclusion levels were modified to establish different scenarios of urban growth up to 2050. Three growth scenarios were designed - the Business As Usual (BAU) scenario, which assumes that future urban development will follow the same pattern as in history, and two "alternative" scenarios: the Sprawled Growth Scenario (SGS) and the Compact Growth Scenario (CGS). The characteristics of all three scenarios are summarized in Table 2.

Prior to decision making which exact value to assign to different growth coefficients to fit different scenarios, a sensitivity analysis was performed. This was done in order to examine how each single coefficient affects urban growth in our case. The prediction (100 runs) was executed by assigning high value (80) to each coefficient while keeping the values of other coefficients as low as possible (1) (similar to [13]). The results revealed that the SP coefficient had by far the highest impact on urban growth (increase of urban cover by 11.25%), while DI, BR and RG coefficients proved to be less significant in influencing the increase of urban cover (0.35%, 0.14% and 0.11% respectively). Keeping in mind the results from this sensitivity analysis, the coefficient values for two alternative scenarios were established in following manner:

— In SGS scenario the values of DI, BR and RG were increased by 25, while SP coefficient was decreased by 10 ("only" by 10, because of high affinity of urban increase to changes in SP values shown in sensitivity analysis). Exclusion levels were arbitrarily weighted with a value of 80, meaning that there is an 80% probability that the exclusion level will remain as such in areas where urban development can be permitted under certain conditions.
— In CGS scenario the DI and BR remained at minimal values while RG was decreased by 25. The SP was decreased by 10. Maximum exclusion levels were assigned to all polygons within the Exclusion layer (100).
— The SR coefficient was not modified in any of the alternative scenarios since it was shown not to be a limiting factor for the urbanization in the area
— The BAU scenario remained with the same values of growth coefficients that resulted from the calibration. Exclusion levels were weighted with a value of 80 for the same reason as in SGS

Table 2. Scenarios used for SLEUTH prediction

Scenario	Main characteristics	Impacts on values
Business As Usual (BAU)	The parameter values are the same as ones resulted from calibration were used	
Sprawled Growth Scenario (SGS)	Dispersive growth: new sub-urban and peri-urban centres likely to emerge (mainly in existing agricultural and forested areas).	Higher values for DI and BR coefficients
	The "infilling" growth is expected to be minimal	Reduced value of the SP coefficient
	Higher probability for growth along the road network since sprawled growth could result in greater travel distances	Higher value for RG
	Spatial planning is more aimed at satisfying high demand for urban areas	Flexible exclusion levels
Compact Growth Scenario (CGS)	Compact-like growth	The DI and BR values were lowered while SP was increased
	Reduced travel distances	Lower RG value
	Less demand for urbanization outside already urbanized areas and hence, less rationale to allow construction in areas that are currently protected	Maximum exclusion levels assigned

Finally the values for the "alternative" scenarios were: for SGS: DI (25), BR (25), SP (10), SR (1), and RG (77); and for CGS: DI (1), BR (1), SP (40), SR (1), and RG (27). With established coefficient values for all scenarios, the prediction was executed by running 100 Monte Carlo iterations. The results for urbanization up to 2050 were the following: the SGS predicted minimum increase in urbanized areas: 0.76 %. On the contrary, the maximum increase in urbanization was predicted by the CGS: 7.26%. The BAU predicted an increase of urbanized areas by 3.7%. More details on results are shown in Table 3. It seems that a transition from compact to sprawled type of urban growth is not likely to occur in the study area according to SLEUTH predictions for the future. In other words, if urbanization continues to take place in the area, it will most probably happen around the existing urban areas, in compact manner. Prediction for 2050 is illustrated in Figure 3 for all three scenarios. The figure depicts only a portion of the study area for visualization reasons.

Table 3. Share of the number of urban pixels in the total number of pixels (%urban) and the percent of the new urban pixels in one year divided by the total number of urban pixels (grw_rate) for all three scenarios up to 2050

Year	BAU %Urban	BAU Grw_rate	SGS %Urban	SGS Grw_rate	CGS %Urban	CGS Grw_rate
2012	10.29	1.06	10.27	0.88	12.07	1.36
2020	11.05	1.17	10.67	0.42	13.31	1.63
2030	12.11	1.11	10.89	0.19	15.14	1.53
2040	13.19	0.99	10.99	0.09	17.13	1.44
2050	13.99	0.44	11.03	0.05	19.33	1.39

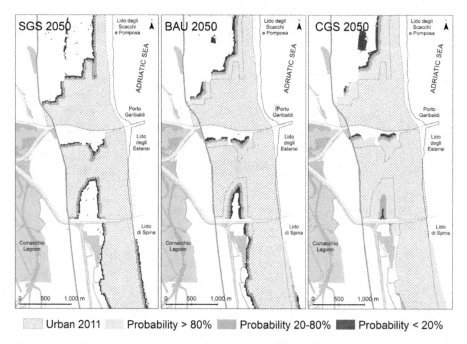

Urban 2011 ▨   Probability > 80%   Probability 20-80%   Probability < 20%

Fig. 3. Probability of urban growth for 2050 for part of the Emilia-Romagna coastal region according to CGS, BAU and SGS scenario

It seems that more compact urban development in the past made a mark on the prediction for the study area. This is particularly evident in the example of SGS scenario. Although some sparse urbanized areas appear in this scenario, their probability of occurrence is less than 20%. The fact that the SGS showed the lowest to-urban conversion even though the DI and BR values were increased by 25, implies how dominant was the SP coefficient in the control of urbanization, even though it was changed by a lower value. Moreover, it could be that the sprawl was hindered by great share of Exclusion in the area (see Figure 2). All in all, the results indicate that urbanization levels until 2050 will be relatively low in the area. This could be related to the fact mention above – considerable share of area that is either urbanized or excluded from development. In addition, the scenarios were driven by the information coming from calibration, which in our case, initiated with 1978, i.e. after the period in which the biggest boom of urbanization took place (1950s and 1960s [5]). The quality data for the pre-1978 period were not available to "capture" this "boom" in the calibration.

4   Discussion and Conclusions

Like any other model, SLEUTH has its own limitations, as well as uncertainties that can emerge during its application. Like many urban models, it cannot capture the driving forces behind urban growth [12]. Which factors will drive urbanization in the future, and to what extent, is dependent on complex interrelationship between uncertain future demographics and socio-economic aspects [25]. However, it seems that SLEUTH is deliberately focused more on form and dynamics, i.e. "where" could the development take place, not "why" [26]. In addition, SLEUTH is sensitive to time resolution and spacing, geographical resolution and scale, and the classification scheme applied to land use. It is neither capable to capture the interior structure of cities, nor to create destiny estimates within them. Finally, there is no explicit model of uncertainty in SLEUTH, although it is accounted for [26].

Some uncertainties were also encountered in our study, during different steps of SLEUTH application. Initial concern was the appropriateness of input data. All of the available maps for historic urban extent were from different sources, with different spatial resolution. This can be of particular significance since SLEUTH performance can be sensitive to different sources of input data, even when they cover the same geographical area [27]. On the other hand, the input datasets used were relatively recent with no big gaps between time series. This is highly important since SLEUTH seems to show temporal sensitivity regarding historic datasets. Using more dense historic observation points, i.e. shorter time series, may produce better agreement between the simulated and observed urban growth [18].

One of main uncertainties regarding calibration was the choice of goodness-of-fit metric that implies the most relevant coefficient values for each successive phase. Although many earlier applications of SLEUTH used some other metrics (Lee-Sallee metric in particular), we employed the OSM since it is believed to be the most robust measurement of model accuracy with the past data [20]. However, since OSM is a product of several values ranging from 0 to 1, even small differences from 1.0 can quickly compound downward. It is important to remember that the values of component metrics of OSM are not in correlation in rankings, i.e. the highest OSM value does not correspond with the highest of any of its constituent metrics [16].

In this study we did not alter the resolution of the input images throughout the different phases of the calibration process. Lowering the resolution of input images is a common practice to reduce computation intensiveness [28]. However, changing the resolution of input layers may influence the growth rules, impact the overall performance of the model and finally, lead to inaccurate representation of growth [29]. By taking these observations into consideration, the resolution of input images was kept the same (20m cell size using the nearest neighbor resampling algorithm) throughout the whole calibration process.

Other concerns were mainly related to the design of the scenarios. There are numerous issues that need to be taken into consideration when developing scenarios of land use change [30]. We opposed the two alternative scenarios sprawled vs. compact type of growth, a concept that was also used, in similar manner, in some earlier applications (e.g. [21] [31, 32]).

One uncertainty related to the scenarios was how much to increase/decrease the coefficient values in order to represent scenarios. We have performed a sensitivity analysis, based to some extent on [13], to help us in making such a decision. However, it needs to be said that this kind of sensitivity analysis perhaps oversimplifies the connection between single coefficient and certain type of growth. In other words, all growth coefficients are highly correlated to each other and certain type of growth is a product of their interaction. Different coefficients do not necessarily need to be inversely related within a scenario. For example, even if a scenario aims to reflect a sprawled growth (i.e. DI and BR coefficients increase) this does not mean that the compact growth should automatically decrease, and vice versa. Some studies used different methods to estimate whether an urban area is passing through a sprawl phase or a coalescence phase during certain time period. These include calculating a ratio between the number of clusters and average cluster size [33] or computing metrics such as number of patches, patch density, Euclidian nearest-neighbor distance, mean patch size etc [34].

The other uncertainty related to scenario-design lies in the designation of exclusion levels. In our study the excluded value for certain areas in BAU and SGS scenarios was set to 80%. The question arises which number is the appropriate one since, if the demand for urban land increases and protection gets treated "more loosely", the excluded areas could be assigned with even lower values. Finally, all these values had to be set arbitrarily. After all, scenarios are "possibilities, not predictions" [35].

Despite these uncertainties, SLEUTH has proven to be successful in providing us with insights regarding historic urban growth in the coastal area of Emilia-Romagna, as well as being practical in deriving scenarios of future urban growth. We found the code relatively easy to operate, and the fact that it is freely available online should not be disregarded. Although there is always a certain level of uncertainty in trajectories of future urban development, we believe that the outputs from SLEUTH scenarios can assist coastal planners in taking into account where urban development could take place in future. SLETH outputs could have an important role on the decision-making in coastal planning: GIF maps are highly compatible with Geographic Information Systems (GIS) and hence, suitable for further quantitative analysis. This compatibility can be taken as an advantage and utilized as a visualization tool with a potentially high l impact on different types of end-users.

Acknowledgments. Authors would like to thank Prof. Claudia Ceppi from the Technical University of Bari for all advice regarding the SLEUTH model. Ivan Sekovski would like to thank his supervisors, Prof. Giovanni Gabbianelli from the Environmental Sciences Department of the University of Bologna, and Prof. Laura Del Rio from the Department of Earth Sciences, University of Cadiz, for their help and guidance. Ivan Sekovski was financially supported by the Erasmus Mundus foundation [specific grant agreement number 2011-1614/001-001 EMJD].

References

1. United Nations Populations Fund (UNFPA): State of the World Population; Unleashing the Potential of Urban Growth. UNFPA, New York (2007)
2. Wong, P.P., Losada, I.J., Gattuso, J.-P., Hinkel, J., Khattabi, A., McInnes, K.L., Saito, Y., Sallenger, A.: Coastal systems and low-lying areas. In: In: Field, C.B., et al (eds.) Climate Change 2014: Impacts, Adaptation, and Vulnerability. Part A: Global and Sectoral Aspects. Contribution of Working Group II to the Fifth Assessment Report of the Intergovernmental Panel on Climate Change, pp. 361–409. Cambridge University Press, Cambridge (2014)
3. European Environment Agency (EEA): Urban Sprawl in Europe: The ignored challenge. European Environment Agency report 10. Office for Official Publications of the European Communities. Luxembourg (2006)
4. UNEP/MAP: State of the Mediterranean Marine and Coastal Environment, UNEP/MAP – Barcelona Convention, Athens (2012)
5. Cencini, C.: Physical processes and human activities in the evolution of the Po Delta, Italy. J. Coastal Res. **14**(3), 774–793 (1998)
6. Armaroli, C., Ciavola, P., Perini, L., Calabrese, L., Lorito, S., Valentini, A., Masina, M.: Critical storm thresholds for significant morphological changes and damage along the Emilia-Romagna coastline. Italy. Geomorphology. **143–144**, 34–51 (2012)
7. Sekovski, I., Armaroli, C., Calabrese, L., Mancini, F., Stecchi, F., Perini, L.: Coupling scenarios of urban growth and flood hazard along the Emilia-Romagna coast (Italy). Nat. Hazard. Earth Sys. Sci. Discuss. **3**, 2149–2189 (2015). doi:10.5194/nhessd-3-2149-2015
8. Armaroli, C., Perini, L., Calabrese, L., Ciavola, P., Salerno, G.: Evaluation of coastal vulnerability: comparison of two different methodologies adopted by the Emilia-Romagna Region (Italy). Geophysical Research Abstracts **16**. EGU General Assembly EGU2014-11299 (2014)
9. Silva, E.A., Clarke, KC.: Calibration of the SLEUTH urban growth model for Lisbon and Porto, Portugal. Comput. Environ. Urban Syst. **26**, 525–552 (2002)
10. Project Gigalopolis website http://www.ncgia.ucsb.edu/projects/gig/
11. Clarke, K., Gaydos, L.: Loose-coupling a Cellular Automaton Model and GIS: Long Term Urban Growth Prediction for San Francisco and Washington/ Baltimore. Int. J. Geo. Inform. Sci. **12**(7), 699–714 (1998)
12. Jantz, C.A., Goetz, S.J., Shelley, M.K.: Using the SLEUTH urban growth model to simulate the impacts of future policy scenarios on urban land use in the Baltimore-Washington metropolitan area. Environ. Plan. B. **30**, 251–271 (2003)

13. Caglioni, M., Pelizzoni, M., Rabino, G.A.: Urban Sprawl: A Case Study for Project Giga-lopolis Using SLEUTH Model. In: El Yacoubi, S., Chopard, B., Bandini, S. (eds.) ACRI 2006. LNCS, vol. 4173, pp. 436–445. Springer, Heidelberg (2006)

14. Rafiee, R., Mahiny, A.S., Khorasani, N., Darvishsefat, A.A., Danekar, A.: Simulating urban growth in Mashad City, Iran through the SLEUTH model (UGM). Cities 26(1), 19–26 (2009)

15. Akın, A., Clarke, K.C., Berberoglu, S.: The impact of historical exclusion on the calibration of the SLEUTH urban growth model. Int. J. Appl. Earth Obs. Geoinf. 27, 156–168 (2014)

16. Onsted, J., Clarke, K.C.: The inclusion of differentially assessed lands in urban growth model calibration: a comparison of two approaches using SLEUTH. Int. J. Geogr. Inf. Sci. 26(5), 881–898 (2012)

17. Goldstein, N.C.: Brains vs. brawn – comparative strategies for the calibration of a cellula automata–based urban growth model. In: Atkinson, P., Foody, G., Darby, S., Wu, F. (eds.) Geo Dynamics. CRC Press, Boca Raton (2004)

18. Candau, J.T.: Temporal calibration sensitivity of the SLEUTH Urban Growth Model. M.A. thesis. Department of Geography, University of California, Santa Barbara (2002)

19. Syphard, A.D., Clarke, K.C., Franklin, J.: Using a cellular automaton model to forecast the effects of urban growth on habitat pattern in southern California. Ecol. Complex 2, 185–203 (2005)

20. Dietzel, C., Clarke, K.C.: Toward Optimal Calibration of the SLEUTH Land Use Change Model. Transactions in GIS 11(1), 29–45 (2007)

21. Leao, S., Bishop, I., Evans, D.: Simulating Urban Growth in a Developing Nation's Region Using a Cellular Automata-Based Model. J. Urban Plan. D-ASCE 130, 145–158 (2004)

22. Oguz, H., Klein, A.G., Srinivasan, R.: Using the Sleuth Urban Growth Model to Simulate the Impacts of Future Policy Scenarios on Urban Land Use in the Houston-Galveston-Brazoria, CMSA. Research Journal of Social Sciences 2, 72–82 (2007)

23. Jantz, C.A., Goetz, S.J., Donato, D., Claggett, P.: Designing and implementing a regional urban modeling system using the SLEUTH cellular urban model. Comput. Environ. Urban Syst. 34(1), 1–16 (2010)

24. Yang, X., Lo, C.P.: Modelling urban growth and landscape changes in the Atlanta metropolitan area. Int. J. Geogr. Inf. Sci. 17(5), 463–488 (2003)

25. Herold, M., Goldstein, N.C., Clarke, K.C.: The spatiotemporal form of urban growth: measurement, analysis and modelling. Remote Sens. Environ. 86, 286–302 (2003)

26. Clarke, K.C.: Why simulate cities? GeoJournal 79, 129–136 (2014)

27. Syphard, A.D., Clarke, K.C., Franklin, J., Regan, H.M., Mcginnis, M.: Forecasts of habitat loss and fragmentation due to urban growth are sensitive to source of input data. J. Environ. Manage. 92(7), 1882–1893 (2011)

28. Dietzel, C., Clarke, K.C.: Spatial differences in multi-resolution urban automata modelling. Transactions in GIS 8(4), 479–492 (2004)

29. Jantz, C.A., Goetz, S.J.: Analysis of scale dependencies in an urban land-use-change model. Int. J. Geogr. Inf. Sci. 19(2), 217–241 (2005)

30. Xiang, W.N., Clarke, K.C.: The use of scenario in land-use planning. Environ. Plan. B: Planning and Design 30, 885–909 (2003)

31. Solecki, W.D., Oliveri, C.: Downscaling climate change scenarios in an urban land use change model. J. Environ. Manage. 72, 105–115 (2004)

32. Mahiny, A.S., Gholamalifard, M.: Linking SLEUTH Urban Growth Modeling to Multi Criteria Evaluation for a Dynamic Allocation of Sites to Landfill. In: Murgante, B., Gervasi, O., Iglesias, A., Taniar, D., Apduhan, B.O. (eds.) ICCSA 2011, Part I. LNCS, vol. 6782, pp. 32–43. Springer, Heidelberg (2011)
33. Martellozzo, F., Clarke, K.C.: Urban Sprawl and the Quantification of Spatial Dispersion. In: Borruso, G, Bertazzon, S. Favretto, A., Murgante B. Torre, C.M. (eds.) Geographic Information Analysis for Sustainable Development and Economic Planning, ch 9, pp. 129–142. New Technologies. IGI Global (2013)
34. Dietzel, C., Oguz, H., Hemphill, J.J., Clarke, K.C., Gazulis, N.: Diffusion and coalescence of the Houston Metropolitan Area: evidence supporting a new urban theory. Environ. Plan. B. **32**(2), 231–246 (2005)
35. Schwartz, P.: The Art of the Long View: Planning for the Future in the Uncertain World. Doubleday, New York (1996)

The Integration Between Cultural Value, Historical Understanding and Urban Energies, to Assess Existing Real Estate Suitability for Intervention: The Case of Cremona

Pier Luigi Paolillo[1] and Umberto Baresi[2(✉)]

[1] Politecnico di Milano, Milano, Italy
pierluigi.paolillo@polimi.it
[2] University of Queensland, Brisbane, Australia
u.baresi@uq.edu.au

Abstract. The building stock of Cremona is considered here as good opportunity to provide potential socioeconomic and physical/architectural analyses to assess the suitability for intervention in existing structures. The treatment of alpha-numeric, vector databases is integrated with the physical dimension of buildings that, in this article, refers to living characteristics useful to assess the suitability for change of each statistical unit analysed. However, single variables can only outline the geographies of specific phenomena. The use of multi-dimensional appraisals allows understanding how these phenomena interact, showing complex latent potentialities. This enables the identification of units requiring strategic interventions for urban regeneration, using innovative analyses based on thematic paths geographies within the built space. These techniques are useful to enhance the preservation and overall quality of city cultural heritage and the urban building stock. Finally, a 'suitability for intervention' map which considers the relationships between economic activities, cultural/monumental heritage, vegetation structures is presented.

Keywords: Historic centres · Regeneration intervention · Suitability for intervention · Multi-dimensional analysis

1 The Bases of Suitability for Intervention Within the Existing Building Stock

1.1 Some Unsolved Issues About Building Management Within Historic Centres

The safeguard and enhancement of historic buildings is an unsolved issue within the Italian context, a problem ambiguously debated for decades without the support of an adequate legislative system. At national and regional level, sufficient detail is lacking to adequately safeguard historic centres, whereas planning processes have struggled to tackle this issue by producing suffocating and unmotivated constraints within urban

© Springer International Publishing Switzerland 2015
O. Gervasi et al. (Eds.): ICCSA 2015, Part III, LNCS 9157, pp. 440–456, 2015.
DOI: 10.1007/978-3-319-21470-2_32

planning tools. At the same time, many attempts to address this problem have failed to define an effective way to govern historic buildings management [18], producing uncertain or truculent regulations that increased the distance with a hypothetical solution. To be clear, what problem (or which one, among the problems) needs to be faced and solved? Within such uncertainty, there are in fact a variety of factors of social, cultural, monumental, artistic, and historic nature [8,9], [10], [17], in addition to those related to environment and hygienic issues, social security, order and residential needs. As all these aspects seemingly require dedicated treatments, it appears unthinkable to organise a tailor-made process to intervene on existing built fabrics. This is especially true if we consider the double perspective of i) the government of the conservation, considering characters and distinctive memories [30] and ii) the regeneration of the existing building stock, to be pursued through the identification of different intervention degrees according to nowadays real-estate market needs. Furthermore, the abstract and unproductive scholarly efforts – together with the inner complexity of this topic – has tilted the balance toward the economic side. Such approach was introduced in the 1968 Ancsa conference, detailed within the Paper of Gubbio two years later, and considered again in 1971 at the VI Bergamo Convention. Despite the evolution of this debate, ongoing difficulties are still preventing the possibility to reach a satisfactory solution as many cultural models have shown their inner fragility, bringing reduced support to the urban planning experience. The question of what should be done with the historical building stock still remains: to confirm it? Sure, but how much can be changed? If this change has to derive from the interpretation of current needs, not only in residential terms but also in quality-living ones, how can urban planning proceed in such historically problematic built areas? As the examination of some relevant planning experiences in historic centres did not help enough in finding an innovative analytical and evaluation methodology, it seems necessary to develop a new polyhedric vision of this issue [42,43]. This evolution should include economic, demographic, constructive, historic, artistic, stylistic and architectural issues in a progressive way, well knowing how difficult it is to identify a unique characterization of historic centres. Nonetheless, this approach does not seem that appropriate as characterising models within historic centres have shown their limits being too abstract and not connected to operational planning processes, thus leaving unresolved the key issue about what to do in nowadays historic centres [35], as shown in previous relevant experiences [2], [11,12]. Nevertheless, it seems that the extraordinary computing capacity of Geographical Information System (GIS), to deal with complex databases, could finally allow the construction of evaluation procedures useful to interpret the complex reality stratified not only within historic centres but also in other areas of the existing building stock. In this way, the analytical process could be supported by the use of statistical matrixes and elaborations, effectively backing up the decision-making process towards better scenarios.

1.2   The Intervention Appraisal Within the Built Fabrics

Recent planning experiences [33], [38] have shown several analytical procedures, outlining useful approaches to evaluate the possibility of intervention within the built space. In detail, the methodology structured for the General Variation of the Rules Plan for the city of Cremona (Fig. 1) was particularly useful. It involved the following steps:

1. identification of relevant themes, coming from legislative factors within the Unique Text of Building, the planning legislation, the reports, maps and rules of previous urban plans, and within some useful planning experiences in historic centres since the 1950's;
2. transformation of the previous themes in detailed variables, according to the information available from the dataset archives of the specific context analysed;
3. construction of consequent synthetic indeces, useful to describe the phenomena characterising the build landscape both from in socioeconomic and historic/architectural terms;
4. complex elaborations of the aforementioned indeces, using multi-dimensional analyses [1], [5,6], [26,27], [16], [29], to identify suitable intervention categories [34] for specific parts of the built space.

Within the following diagram it is possible to recognise the basic steps structuring the procedure used to intervene within the historic centre in the city of Cremona. However, before establishing how to develop this analysis and how to classify the analytical units as well as define the possible levels of intervention, it is necessary to identify the 'units of intervention'. These consist of: i) the constructive element (the simple object with its own structural or service functions), ii) the organism (the object composed by several constructive elements and characterised by its own functions, related to specific uses), iii) the real estate unit (group of all the constructive organisms linked to the specific use), iv) the built fabric (group of constructive organisms and/or functional units being part of a unique organism morphologically defined), v) the compartment (group of built fabrics and public/private areas required for a unique urban re-structuring intervention). Besides, these are temporary categories that can be changed, implemented and grouped into simple geometric units on where to locate information about the existing building stock. Because such units could be used to locate real estate and building rights, it is still preferable to use more extended units like blocks and census sections. This should allow avoiding several issues experienced in cases where smaller units (e.g. single buildings) were identified in numerous ways according to a variety of analyses (involving historic centre settlements, services supply, and non-residential activities geographies). At the same time, particular attention is paid to establish the best way to proceed with the analyses of built fabrics and social phenomena by discovering the relationships between space and population (configuration of services and community spaces, ways of living residential areas, non-residential activities dynamics). Thus, the following diagrams show the single steps involving the identification of urban phenomena, the definition of degrees to modify built fabrics, the distribution of specific value, non-value and risk factors within the relational space.

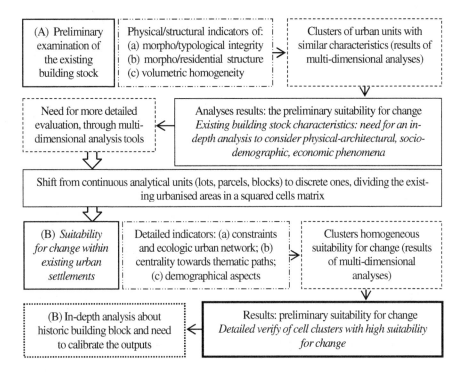

Fig. 1. Overview on the analytical and classifying procedure, applied to the historic building stock in Cremona

2   The Preliminary Studies on the Existing Building Stock

This analysis aims to identify the degrees of preliminar suitability for intervention, assessed considering the characteristics of built fabrics mainly involving: i) the urban structure (the building morphology, the building types, the volumetric organisation, the centrality inclination, the location of green areas), ii) the socioeconomic conditions (the population distribution, the vitality of non-residential activities, the demographic stability, and so on). The aforementioned themes are considered as follows.

2.1   The Indicators Built to Assess the Preliminar Suitability for Intervention

In this case, seven blocks are identified to explain meaningful phenomena within the urban system of Cremona, useful to identify a same number of synthetic indicators (later used for the multi-dimensional analysis [7], [19]) coming from thematic packages such as Multiple Centrality Assessment Tool [13], [41] and some existing tools within GIS, used to structure urban geographies while estimating the complex relationships among variables.

Table 1. The logic blocks and the structure of single indicators used within this survey

Logic blocks	Indicators	Specific indeces and variables used	
1. Morpho/ typological integrity	a1 – Historic and functional stability of the existing building stock	i)	building historic persistence index
		ii)	historic level of existing building
	a2 – Typological and functional homo-geneity analysis	i)	typological homogeneity index
		ii)	functional homogeneity index
2. Morhpo/ residential structure	b1 – Perimeter de-structuring analysis	i)	perimeter de-structuring index
	b2 – Settlement compactness analysis	i)	settlement compactness index
3. Volumetric homogeneity	c1 – Height homogeneity analysis	i)	height homogeneity index
	c2 – Building distribution homogeneity analysis	i)	building distribution homogeneity
	c3 – Volumetric homogeneity analysis	i)	volumetric homogeneity index
4. Potential ecological connectivity	d1 – Current connectivity analysis	i)	presence of green areas
	d2 – Potential connectivity analysis	i)	potential connectivity between greeen areas
5. Non-residential vitality	e1 – Functional heterogeneity analysis	i)	Shannon index applied to non-residential activities
	e2 – Non-residential activity density analysis	i)	non-residential activity index
6. Centrality inclination	f1 – Global centrality analysis	i)	Global Closeness index
	f2 – Global structure analysis	i)	Betweenness index
7. Demographic stability	g1 – Residential density analysis	i)	residents index
	g2 – Old population rate analysis	i)	old population rate index
	g3 – Foreign population rate analysis	i)	foreigner rate index

In order to illustrate the general structure of the seven indices developed in the orginal database, it is here explained the structure of the non-residential vitality index. Shown as example, this indicator was useful to evaluate both the preliminary suitability for intervention and the detailed one, the latter focused on Cremona historic centre. This index is the result of the mean value between the $\sum$ of the partial indeces of functional heterogeneity (e1) and density (e2), respectively:

$$e_1 = E_{st} = \frac{H_j}{H_{max}} \tag{1}$$

$$e_2 = \sum_{j=1}^{n} x_j * w_j \tag{2}$$

where:
Est = index of structural Evennes [36];
H = structural heterogeneity of urban system = Hst;
Hmax = value obtained with the same probability of all the categories to be present;
n = categories considered to calculate the diversity xj;
xj = areal extension (sqm) of the i cell with 25 m side, included within the surface of the spatial unit considered (lots, parcels, blocks);

wj = density of non-residential activities in the single cell, obtained with Kernel Density [39].

These indices allow the analysis of several configurations within the urban settlements, both in terms of the internal structure and functions/volumes, and the relationships with the built environment, internal and external to the building stock.

2.2   The Identification of Preliminary Suitability for Intervention, Through Multi-dimensional Analysis

At this point, it was necessary to build data matrices for multi-dimensional statistical analysis developed with AddaWin software [20].

The first step focuses on the existing correlations among the seven synthetic indicators considered (the morpho/typological integrity, the morhpo/residential structure, the volumetrical homogeneity, the potential ecological connectivity, the non-residential vitality, the centrality inclination, the demographic stability), within a statistic sample of approximately 4,000 units corresponding to the urban area of Cremona. For such, it is useful to consider how the distinction between active and supplementary variables in the Addawin software.

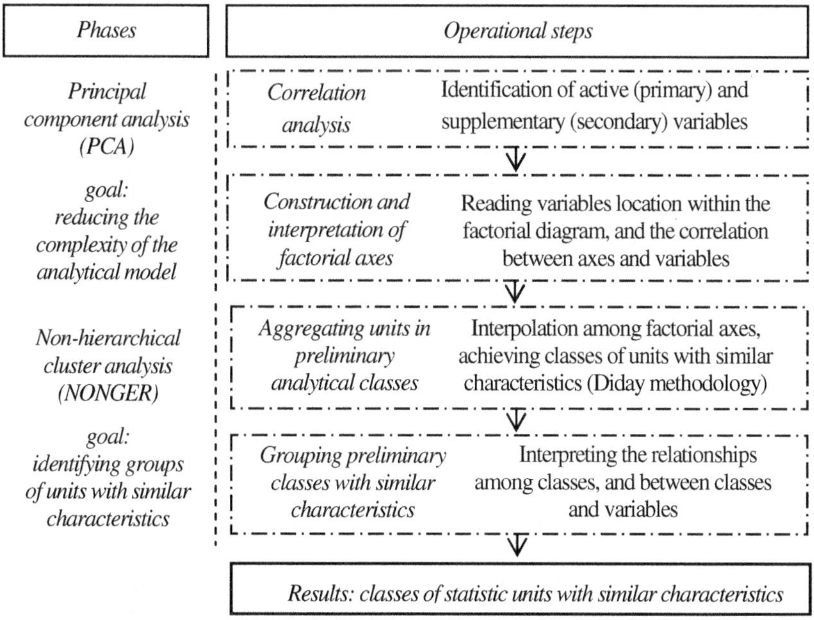

Phases	Operational steps	
Principal component analysis (PCA)	*Correlation analysis*	Identification of active (primary) and supplementary (secondary) variables
goal: reducing the complexity of the analytical model	*Construction and interpretation of factorial axes*	Reading variables location within the factorial diagram, and the correlation between axes and variables
Non-hierarchical cluster analysis (NONGER)	*Aggregating units in preliminary analytical classes*	Interpolation among factorial axes, achieving classes of units with similar characteristics (Diday methodology)
goal: identifying groups of units with similar characteristics	*Grouping preliminary classes with similar characteristics*	Interpreting the relationships among classes, and between classes and variables
	Results: classes of statistic units with similar characteristics	

Fig. 2. Multi-dimensional analysis procedure, applied to territorial data within the software AddaWin

This step allows recognizing the most important ones to describe the statistic sample, placing the supplementary ones in the condition of back-up information. The results showed relevant correlation between physical/structural and architectural characteristics, while the variables related to relational phenomena (dealing with the way people live urban areas) were placed in a supplementary role.

The following steps were then applied to reduce the complexity of the whole phenomenon, classifying the 4,000 units in 11 classes whose profiles allowed identifying 5 spatial clusters used for the preliminary suitability for intervention map. In detail, the level of approximation used in this multi-dimensional analysis was reduced to 7%, referring to the 93% of total inertia to explain the characteristics belonging to single lots and blocks. At the same time, the lack of variables strongly correlated among the four considered (the three with strong correlation have been treated in a preliminary way) allowed refining the results without major information duplication.

The result is the construction of preliminary suitability for intervention geographies, coherent with the existing building stock analyses developed before the multi-dimensional analysis [40]. For instance, it is useful to outline how the more relevant suitability to transform nowadays settlements lies within industrial and peri-urban areas, while intermediate values are located in the areas around the historic centre, because of local functional and typological heterogeneity.

3   The In-Depth Analysis Within the Existing Urban Building Stock

The following part deals with the process of structuring urban thematic paths, using advanced socio-economic indicators.

Some of them, like the socio-demographic stability and the non-residential vitality, were previously assessed in the matrix of value, non-value, risk factors [32], whereas some others were instead calculated 'ad hoc' for this second analytical step, like the constraints intensity and the vacant flat incidence. All these indices were considered for the second multi-dimensional analysis, aimed at detailing the real opportunity to intervene in the real estate market, especially within the historic centre. The main contribution of this step is the identification of values (cultural heritage), non-values (physical degradation, marginal location) or risks (vacant flat incidence), supporting the contribution of the suitability for intervention map to Cremona urban plan strategies [37].

The information dataset related to existing constraints within built fabrics shows that the historic centre contains most of the buildings to be protected in terms of relevant importance for public interest, or cultural and monumental values. This step is useful as it confirms that the historic centre itself deserves to be classified with low suitability for intervention, within the aforementoned preliminary analysis. Nevertheless, this output outlines the need for a more sophisticated analysis in order to recognise and separate the buildings with real constraints and values from the remaining structures.

This means, for instance, to identify in detail other indicators of value, non-value and risk factors, like the vacant flat incidence. Precisely, this indicator was developed referring to water and energy consumption communal databases, that showed how the 40% of vacant flats were located within the historic part of the city. Thus, the previously developed indicators of sociodemographic stability and non-residential vitality have been considered as main indicator as well, after being only identified as supplementary ones in the preliminary evaluation. In the end, in order to facilitate mathematical operations of sum among indicators, Natural Breaks (based on Jenks algorithm in GIS, that identify the natural interruption of a series recognizing the groups to be used) classifications were used to obtain homogeneous number of classes with values ranging between 0 and 1.

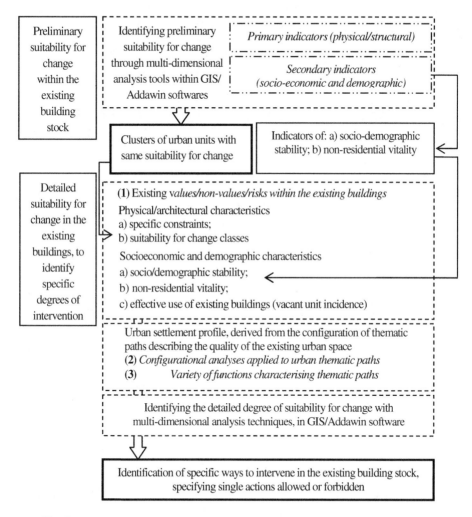

Fig. 3. The connections between preliminary and detailed suitability for intervention

4   The Contribution of Configurational Analyses to Identify Urban Thematic Paths

4.1   The Selection of Useful Indicators Within Configurational Analyses

The distribution of values, non-values and risks within the statistic analytical units has outlined the potentialities linked with the lack in use of a consistent range of the existing building stock, with the consequent need to recognise the most needing areas in terms of urban regeneration. In this sense, the reading offered by the use of thematic paths within the urban area could help understanding the geographies linked with physical, historical, cultural values for Cremona urban community.

In order to approach the thematic paths topic, it is useful to consider the configurational models [4], [22], [29], [47,48], as the Multiple Centrality Assessment – MCA, Visual Graph Analysis – VGA [49], Axial Analysis – AxA [25]. In the Cremona case, the Mca was applied in a more consistent way to what concern the urban scale analysis, whereas Vga and AxA are preferrable while considering cycling and walking movements [14,15]. Unlike MCA, based on the network of connected edges and nodes to represent the whole urban system, VGA and AxA approach treat public space (streets, places) as a complex structure of convex polygons. In this sense, the more relevant is the relationship between open spaces, buildings and perspectives, the more probable is the chance that a person moving in that space will keep walking in an established direction [15], [23,24]. The following comparison of the indicators characterising the Visual Graph Analysis and the Axial Analysis, outlined two main indicators that are useful to structure thematic paths within Cremona historic centre. The first one is the Connectivity (number of lines and vertices directly connected to another line/vertex) and the second one is the Integration (medium accessiblity of a line/vertex referring to the whole system), both calculated with the UCL Depthmap software [46].

4.2   The Construction of Thematic Paths and the Related Centrality Degrees

In the Cremona experience, the development of thematic paths could allow a wider knowledge of the cultural heritage within the historic centre, useful especially for people (tourists and residents) intensely using urban spaces. This process could involve a broader use of built and open spaces connected to specific themes: i) archeology; ii) religion; iii) museums; iv) violins; v) beauty; vi) memory; vii) historic/architectural values; viii) production; ix) nature.

The procedure involved at first the geo-referencing of aerial and punctual elements structuring the specific paths (e.g. museums, churches, historic palaces, gardens), and then the identification of specific connections between the elements constituting the infrastructure network of the public space, using GIS (Fig. 4a).

A number of different analyses were then used to treat paths located only within the historic centre, and the paths located both inside and outside of the historic city. The latter were in fact structured using applications within the GIS software (Kernel Density tools, [3], [44]), based on the density of specific phenomena in the urban space. A different approach was instead developed for historic city paths, were the close nature of Cremona urban structure allowed the implementation of configurational analyses, associating the perception and permeability of public spaces with the geographies of cultural thematic structures and the related paths. The following step involved the medium values between Connectivity and Integration indices (standardized considering the maximum possible value), using the Axial Map lines as analytical units (Fig. 4b). These values were then expanded to the historic centre using the Kernel Density tool (de-structuring urban space in a 5m side cells matrix), in order to evaluate the intensity of each path in a wider area (Fig. 4c), involving not only single streets but also the adjacent buildings.

This procedure allowed calculating the value of every urban unit (lots, parcels, blocks), coming from the medium values of all the cells included, thus reducing the incidence of buildings/structures characterising thematic paths. In this way, the characteristics of public spaces were considered to outline the potential penetration of each path, not only referring to the primary street network, but also to the secondary one (Fig. 4d). Hence, once spatialized the influence of every path within the historic centre, the sum of each value was calculated to identify the thematic centrality degree of the existing building stock. This indicator was later used within the multi-dimensional analysis, in order to assess the suitable degree of intervention within each building located in the historic centre.

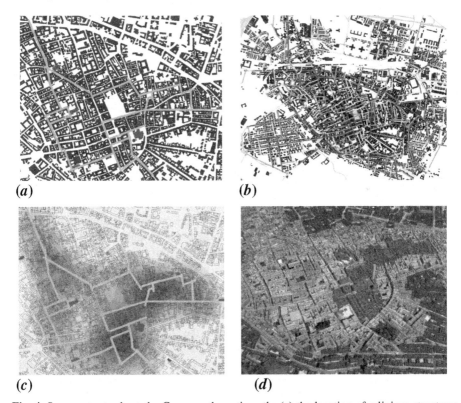

(a)   (b)

(c)   (d)

Fig. 4. Some outputs about the Cremona thematic paths:(a) the location of religious structures and the primary ways of the "faith" path;(b) the representation of axial Connectivity index within the historic centre;(c) the Kernel Density of the "faith" path (radius 250 m, cells side 5 m);(d) the classification of the buildings close to the path, with color scale from red (major) to blue (minimum intensity)

5   The Detailed Suitability for Intervention of the Existing Building Stock

The final step is to link the results with the strategies to be adopted to manage the historic buildings, so to enrich the preliminary suitability for intervention map. Hence, the goal is to identify the main characteristics within several parts of the existing building stock, as the suitability for intervention is an important factor to support urban plan decisions, identifying homogeneity, peculiarities, and specific features.

As previously outlined, the early step involved de-structuring urban analytical units in a 5m-side cells matrix, linking in it every building's information. Considering how data collected were extremely detailed and complete in each cell identified (x, y), instead of a generic Overlay procedure involving every Layer, the multi-dimensional analysis was chosen as a better approach. In detail, the principal components analysis and the non-hierarchical one were used (within the software AddaWin) to identify clusters of cells with similar characteristics, after reducing the total complexity of the whole model.

5.1   The Principal Components Analysis, to Reduce the Phenomenon Complexity

The correlation analysis, preliminar to the application of the multi-dimensional analysis, didn't show any particular link among the 22 variables considered, corresponding to the 6 final indicators structured: i) preliminary suitability for intervention; ii) thematic paths centrality; iii) value density; iv) non-value density; v) risks density; vi) ecologic significance.

The last step involved the identification of principal components, with the consequent definition of 13 main factorial axes, impacting for approximately the 91.4% of the total model inertia: this relevant threshold was identified to assure the continuity with the previously developed analysis, focused on the preliminar suitability for intervention. The following analysis, involving the factorial weights (Factor Pattern Matrix) of the principal components obtained, allowed identifying which variables were more explicatory in the relationship between initial indicators and statistical units considered (the 5m side cells).

The Plot of every variable was then built within AddaWin software, considering each pair of principal axes, as a supporting tool to interpretate the principal components and the factorial weights matrix. In detail, the Plots were basically considered on the main plan generated, whose coordinates represent the correlation coefficients showed in the aforementioned factorial weights matrix. Among the results, it was evident that low preliminary suitability for intervention (basically in historic structures), was located in areas where quite intense are both non-values phenomena and the incidence of thematic paths, the latter slowly declining outside the inner city.

5.2   The Non-hierarchical Analysis, the Variables Considered and the Clusters of Cells with Similar Characteristics

The second step was developed within the software AddaWin, focusing on the identification of similar unit clusters through a non-hierarchical analysis. First, the objective function was considered in order to identify the amount of inertia to use, and the related number of classes necessary to re-aggregate all the statistical units. The 72% of total inertia was assumed as an acceptable threshold, corresponding to 11 classes of cells with similar characteristics in terms of variables profile. The following analysis of these classes profiles (this means the relationships between classes and variables) confirmed cells characterization through the R coefficient (that concerns the relationship between the frequency of each variable in a class and its global frequency, as in Tab. 2). The related descriptive table was therefore analysed to identify the variables at maximum (++++) and high (++) level of qualification per each cluster; these outputs were considered to interpret every cluster profile, eventually proceeding with the further aggregation of specific, similar groups.

Table 2. Relationships between classes and variables per frequency ratio

Class	Units	Weight (%)	R > 2.00 (++++)	1.2 < R < 2.00 (++)
1	2.859	28.3	IMA	VN, DA, RN, PN
2	2.043	20.2	IM	VN, DA, RN, PN
3	1.816	18.0	IA	VN, DA, RN, PN
4	1.711	16.9	IMB, DM, REP	IB, RN, PN
5	391	3.9	RB	IM, IMB, VN, DM
6	231	2.3	IB, VA, VB, DB, DN, PA	RA
7	398	3.9	DB, PB	
8	156	1.5	RM	IMA, IM, VM
9	167	1.7	VM	IMA, DM
10	140	1.4	VA, PB	IMB, IB, DM, RN, REP
11	200	2.0	IB, VA, DB, DN, PM	VB, RB

5.3   The Draft of the Final Suitability for Intervention Map

To finalise, the results were processed with specific spatial queries developed in GIS software, aiming at correcting (basing on the direct knowledge of Cremona urban system) eventual aberrations within specific lots or buildings inside the historic centre. Hence, the eventual isolation of wrongly classified buildings was considered in order to identify five classes of suitability for intervention (Fig. 5), later used to detail specific actions (Tab. 3) about the need or opportunity to requalify the existing building stock.

Table 3. Strategies and actions to apply within specific suitability for intervention classes

Suitability for intervention	Actions
High	Recovery and change of destination for the 'Tamoil' oil refinery area; relocation of industries from mixed-residential areas to marginal ones; re-organization of mono-functional settlements; transformation of buildings incoherent with the surroundings, especially in the historic centre.
Medium/ high	Re-organisation of built spaces, improving relationships between built and green areas in peripheral areas; inside the historic centre, reduce nowadays building underuse or misuse.
Medium	Interventions on built fabrics aimed at reaching a better balance with the open space.
Medium/ low	Building of a suburban green belt in natural areas of relevant ecological connectivity, to relate built and open space. Enhancement of living standards within the existing fabrics, considering thematic paths influence, so to subsidize a full development of residential and tertiary geographies.
Low	Limited interventions of maintenance on buildings with specific constraints.

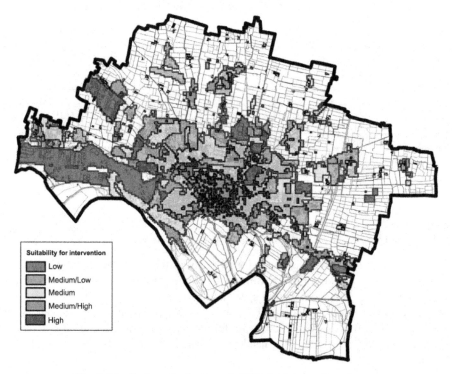

Fig. 5. The final synthesis: the suitability for intervention classes

6   The Importance of the Thematic Paths, and the Future Perspectives for the Sustainable Reshaping of the Existing City

Recognizing the characteristics of the existing settlements allowed planning the most suitable actions to adopt in order to value the underused building stock, while recovering the vitality of the historic centre. Operationally, the primary goal was to identify less constraining degrees of intervention, for buildings without relevant valuable features. This choice will allow the buildings' owners to intervene not only to redefine the external facades, but also to re-organize the internal spaces according to current real-estate market expectations. In this regard, the owners' will to invest in the existing building stock is a crucial factor [28], in a national scenario where the real-estate market recently dropped in a dramatic way. These things considered, the thematic path strategy can be decisive.

The related use of Space Syntax strategies, applied to the Italian historic centers' renewal, is a major contribution to the planning practices and the related literature recently grown in the Lombardia regional context [33], [35], [37]. On one side, there are previous experiences primarily based on MCA techniques to identify the proper features of the relational space [21]. On the other one, the case illustrated outlines the connections between the existing building stock and the open space. This is pursued considering this connection a primary indicator to define the intensity of the transformations allowed in the historic buildings.

The strategy to introduce thematic paths of historic and artistic value relates to the multifaceted population that daily lives Cremona, namely residents, commuters, and tourists. Because of this, the joined introduction of thematic paths and updated degrees of intervention within the building stock aims both at re-shaping the whole historic city, and at limiting potentially negative outcomes deriving from a not adequately planned, rapid increase of tourism phenomena [45].

The contrast of citizens' disaffection for the historic city, pursued through the development of latent potentialities, is not the only goal addressed. The use of multidimensional and GIS procedures as "room for thoughts" [31] is actually evolved in Lombardia after the legislative introduction of GIS. In terms of techniques, if the space Syntax is a solid basis to analyse the configurational space, further developments moving from this case study might focus on the physical/architectural dimension of the historic centre.

This trend should be supported by the multiplication of updated database, which should become available once the digitalization process within Italian public administration is complete. At the same time, the evolved necessities at global level are requiring deeper connections between requalification and sustainability of build spaces. From this perspective, the renewal of historic centres is linked to the reduction of free land consumption and to sustainable building techniques. The management of these factors with GIS tools might involve the development of new methodologies aiming at redefining the nature and the features of the existing city.

7   Conclusions

The experience of Cremona testify how the Space Syntax techniques and the use of GIS tools can contribute in a relevant way to the development of the information usable within planning processes. This information, properly processed through multidimensional analysis tools, allowed identifying detailed criteria and ways to intervene in the historic part of the city. In a scenario characterized by the continuous diminishing of the resources available, this contribution might be decisive to requalify the existing building stock. For these reasons, this study can be identified at one time as a relevant reference and a starting point for further explorations. Moving from the framework identified, new inter-disciplinary techniques might be considered to develop more detailed analyses dealing with the recovery and the valorization of existing urban settlements.

References

1. Anderson, T.W.: An Introduction to Multivariate Statistical Analysis. Wiley, New York (1958)
2. Astengo, G.: Assisi: salvaguardia e rinascita. Urbanistica **24–25**, 1–124 (1958)
3. Bailey, T.C., Gatrell, A.C.: Interactive Spatial Data Analysis. Longman Higher Education, Harlow (1995)
4. Batty, M.: Exploring Isovists Fields: Space and Shape in Architectural and Urban Morphology. Environment and Planning B: Planning and Design **28**, 123–150 (2001)
5. Benzécri, J.P.: L'analyse des données. Leçons sur l'analyse factorielle et la reconnaissance des formes et travaux. Dunod, Paris (1982)
6. Benzécri, J.P.: Histoire et préhistoire de l'analyse des donneés. Dunod, Paris (1982)
7. Bolasco, S.: Analisi multidimensionale dei dati. Metodi, strategie e criteri d'interpretazione. Carocci, Roma (1999)
8. Bonfanti, E.: Tempo di demolire. Edilizia Popolare **110**, 29–32 (1973)
9. Bonfanti, E.: Architettura per i centri storico. Edilizia Popolare **110**, 33–54 (1973)
10. Carozzi, C., Rozzi, R.: Centri storici questione aperta. De Donato, Bari (1972)
11. Cervellati, P.L., Scannavini, R.: Interventi nei centri storici: Bologna, politica e metodologia del restauro. Il mulino, Bologna (1973)
12. Coppa, M.: Vicenza nella storia della struttura urbana: piano del centro storico. Cluva, Venezia (1969)
13. Crucitti, P., Latora, V., Porta, S.: The Network Analysis of Urban Streets: A dual Approach. Physica A, Statistical Mechanics and its Applications **369**(2), 853–866 (2006)
14. Cutini, V.: Spazio urbano e movimento pedonale Uno studio sull'ipotesi configurazionale. In: Cybergeo: European Journal of Geography (1999). http://cybergeo.revues.org/2236
15. Desyllas, J., Duxbury, E.: Axial Maps and Visibility Graph Analysis. In: Proceedings of the 3rd International Space Syntax Symposium. University of Michigan, Ann Arbor (2001)
16. Everitt, B.S., Dunn, G.: Applied Multivariate Data Analysis. John Wiley & Sons, New York (1991)
17. Giovannoni, G.: Vecchie città edilizia nuova. Utet, Torino (1931)
18. Gorio, F.: Critica dell'idea di "centro storico", Rassegna di Architettura e Urbanistica, XVI, vol. 46 (1980)

19. Griguolo, S., Palermo, P.C., Vettoretto, L.: Le analisi multidimensionali. In Aa. Vv. (eds.) Analisi. Parte II, collana: Enciclopedia di urbanistica e pianificazione territoriale. FrancoAngeli, Milano (1988)
20. Griguolo, S.: AddaWin. Un pacchetto per l'analisi esplorativa dei dati – Guida all'uso. Istituto Universitario di Architettura di Venezia, Venezia (2008)
21. Harvey, D.: Spaces of global capitalism. Verso, London (2006)
22. Hillier, B., Hanson, J.: The social logic of space. Cambridge University Press, Cambridge (1984)
23. Hillier, B., Penn, A., Hanson, J., Grajewski, T., Xu, J.: Natural Movement: or, Configuration and Attraction in Urban Pedestrian Movement. Environment and Planning B: Planning and Design **20**, 29–66 (1993)
24. Hillier, B., Iida, S.: Network and Psychological Effects in Urban Movement. In: Cohn, A.G., Mark, D.M. (eds.) Spatial Information Theory, pp. 475–490. Springer, Berlin. (2005)
25. Intelligent Space: Pedestrian movement modelling and consulting. Intelligent Space, London (2010). http://www.intelligentspace.com/
26. Johnson, R.A., Wichern, D.W.: Applied Multivariate Statistical Analysis. Prentice Hall, New York (2007)
27. Jolliffe, I.T.: Principal Component Analysis. Springer, Berlin (2002)
28. Jones, G.A., Bromley, R.D.F.: The relationship between urban conservation programmes and property renovation: evidence from Quito Ecuador. Cities **13**(6), 373–385 (1996)
29. O' Sullivan, D., Turner, A.: Visibility Graphs and Landscape Visibility Analysis. International Journal of Geographical Information Science **15**, 221–237 (2001)
30. Pane, R.: Attualità dell'ambiente antico. La Nuova Italia, Firenze (1967)
31. Paolillo, P.L.: Informazione disinformazione piano: percorsi possibili di teoria dei SIT. Territorio **16**, 123–132 (1993)
32. Paolillo, P.L.: Una modalità descrittivo – classificatoria di individuazione dei "bacini d'intensità problematica ambientale" alla scala regionale. In: Paolillo, P.L. (ed.) Terre lombarde. Studi per un Ecoprogramma in Aree Bergamasche e Bresciane, pp. 103–153. Giuffrè, Milano (2000)
33. Paolillo, P.L. (ed.): Il piano di governo del territorio di Como. Maggioli, Rimini (2011)
34. Paolillo, P.L., Benedetti, A., Baresi, U., Terlizzi, L., Graj, G.: An Assessment – Based Process for Modifying the Built Fabric of Historic Centres: The Case of Como in Lombardy. In: Murgante, B., Gervasi, O., Iglesias, A., Taniar, D., Apduhan, B.O. (eds.) Computational Science and Its Applications – ICCSA 2011, pp. 162–176. Springer, Berlin (2011)
35. Paolillo, P.L.: L'urbanistica tecnica. Costruire il piano comunale. Maggioli, Rimini (2012)
36. Paolillo, P.L., Baresi, U., Bisceglie, R.: The Construction of Landscape Mapping Using Gis Applications: the Case of Cremona. In Niglio, O. (ed.): Paisaje cultural urbano e identitad territorial. pp. 892–905. Aracne, Roma (2012)
37. Paolillo, P.L. (ed.): Il nuovo piano di governo del territorio di Cremona. Maggioli, Rimini (2013)
38. Paolillo, P.L.: La tecnica paesaggistica. Stimare il valore dei paesaggi nel piano. Maggioli, Rimini (2013)
39. Paolillo, P.L.: Limbiate, dalla condizione indifferenziata di 'corea' alla scoperta delle nuove centralità. Territorio **66**, 81–91 (2013)
40. Paolillo, P.L., Baresi, U., Bisceglie, R.: Classification of Landscape Sensitivity in the Territory of Cremona: Finalization of Indicators and Thematic Maps in Gis Environment. International Journal of Agricultural and Environmental Information Systems (IJAEIS) **4**(3), 892–905 (2013)

41. Porta, S., Latora, V., Strano, E.: Networks in Urban Design. Six Years of Research in Multiple Centrality Assessment. In: Estrada, E., Fox, M., Higham, D.J., Oppo, G.L. (eds.) Network Science, pp. 107–129. Springer, Berlin (2010)
42. Ricci, M.: Piccoli centri storici Recupero espansione ambiente. Edilizia Popolare **52**, 66–77 (2010)
43. Ricci, M.: I migranti nei centri storici minori: criticità e risorsa. Urbanistica **142**, 24–29 (2010)
44. Silverman, B.W.: Density Estimation for Statistics and Data Analysis Monographs on Statistics and Applied Probability. Chapman and Hall, London (1986)
45. Simpson, F.: Tourist Impact in the Historic Centre of Prague: Resident and Visitor Perceptions of the Historic Built Environment. The Geographical Journal **165**(2), 173–183 (1999)
46. Space Syntax Network, Ucl Depthmap (Original). London: Ucl, University College London (2011). http://www.spacesyntax.net/software/ucl–depthmap/
47. Turner, A., Doxa, M., O'Sullivan, D., Penn, A.: From Isovists to Visibility Graphs: a Methodology for the Analysis of Architectural Space. Environment and Planning B: Planning and Design **28**, 103–121 (2001)
48. Turner, A., Penn, A., Hillier, B.: An Algorithmic Definition of the Axial Map. Environment and Planning B: Planning and Design **32**, 425–444 (2005)
49. VR Center for Built Environment: Depthmap: a Program to Perform Visibility Graph Analysis. University College London, London (2010) http://www.vr.ucl.ac.uk/research/vga/.

Sensitivity of Nutrient Estimations to Sediment Wash-off Using a Hydrological Model of Cherry Creek Watershed, Kansas, USA

Vladimir J. Alarcon[1(✉)] and Gretchen F. Sassenrath[2]

[1] Civil Engineering School, Universidad Diego Portales, 441 Ejercito Ave., Santiago, Chile
vladimir.alarcon@udp.cl
[2] Southeast Agricultural Research Center, Kansas State University, Parsons, KS, USA
gsassenrath@ksu.edu

Abstract. This paper presents a hydrological and water quality model for Cherry Creek watershed, located in southeastern Kansas, USA. The Cherry Creek catchment drains approximately 88220 ha and it is a main contributor of water to the Neosho River. Hydrological modeling of the Cherry Creek watershed is performed using the Hydrological Simulation Program Fortran (HSPF). Simulated results for total ammonia (TAM) concentrations occurring at the Cherry Creek watershed outlet for four land use scenarios and a 2-year simulation period, are presented. Sensitivity analysis of total ammonia estimations ($TAM=NH_3+NH_4$) to the unbounded HSPF parameter POTFW is subsequently presented. POTFW represents the ratio of a water quality constituent yield to sediment wash-off outflow. Results showed that small perturbations to a 50 mg/Kg POTFW base value produce the largest normalized sensitivities. Peak sensitivities reached up to 248%, with the -1% perturbation producing the most dramatic sensitivity response in TAM estimations. Results showed a strong relationship between normalized sensitivities to river flow regime. Non-linearities in sensitivity for small ±1% perturbations were detected. These non-linearities are more evident for high stream flow values (strong flood events). For larger perturbations (±60%, ±40%, ±20%, and ±5%) the sensitivity response was shown to be linear.

Keywords: Hydrological modeling · HSPF · Sensitivity analysis · Total ammonia · Sediment wash-off

1 Introduction

Agriculture and mining in the State of Kansas (USA) is extensive. Environmental and human impacts of those activities in the region are well known. For example, the Tri-State Mining District (TSMD) in southeastern Kansas, southwestern Missouri, and northeastern Oklahoma, covering approximately 6,475 km^2, has records of mining-related hazardous substances since the late 1800s which caused elevated concentrations of metals in the environment of mined areas, streams, and downstream rivers [1]. A watershed management plan for the region states that it is essential to identify

© Springer International Publishing Switzerland 2015
O. Gervasi et al. (Eds.): ICCSA 2015, Part III, LNCS 9157, pp. 457–467, 2015.
DOI: 10.1007/978-3-319-21470-2_33

and quantify all pollutant sources in the basin, including both nonpoint and point sources through the use of computer models that would incorporate data soil type, topography, land use practices, climate, and so on, for estimation of pollutant loadings and transport throughout the watersheds of the region [2]. The Middle Neosho watershed is located in the state of Kansas (USA) in the middle of TSMD. It begins in Neosho County in southeast Kansas and ends at the point at which the Neosho River crosses the Kansas-Oklahoma state line, being the primary drainage area for the Neosho River and its numerous tributaries [3]. The Neosho River drains into Grand Lake, an important water body located in northeastern Oklahoma. The Middle Neosho Watershed does not meet state water quality Standards, i.e., it fails to achieve aquatic system goals related to habitat and ecosystem health, and it has been assigned a priority category for restoration [3].

Cherry Creek is one of the main tributaries to Neosho River (Figure1). The middle Neosho Watershed Restoration and Protection Strategy study [3] lists Cherry Creek watershed as a targeted area for agricultural Best Management Practices (BMP) to meet sediment and nutrient load reductions. Also, the Grand Lake Watershed Plan [2] lists Cherry Creek as a high-priority stream for a Total Maximum Load study oriented towards reducing nutrient and sediment loads that cause the stream to be listed for a Dissolved Oxygen TMDL.

Hydrological models usually pass a calibration process in which flow and water quality estimations are compared to measured data. The model is considered calibrated when model estimations fit measured data trends and values. During calibration, the modeler adjusts model parameters until a statistical indicator of fitness (such as R^2, Nash-Sutcliff, or other) indicates that the model output resembles measured data. The Hydrological Simulation Program Fortran (HSPF) is one example of models that rely heavily on a calibration process [4]. In HSPF most of the parameters have ranges of values in which their values may fall. Those ranges are specified in the HSPF user's manual. However, there are a number of parameters that are unbounded. The "monthly wash-off potency factor" (POTFW), that is the ratio of a water quality constituent yield to sediment wash-off outflow, is one of those parameters. This parameter has a minimum value of 0.0 with no specified maximum value.

This paper presents a hydrological and water quality model for Cherry Creek watershed. Since some water-quality constituent estimations are highly dependent on sediment wash-off parameters, a sensitivity analysis of total ammonia nitrogen estimations (TAM=NH_3+NH_4) to POTFW is also presented in this paper. This research seeks to identify how sensitive TAM estimations are to POTFW.

2   Methods

2.1   Study Area

The Cherry Creek watershed (that drains to the stream of the same name) located in southeastern Kansas was the focus of this study (Figure 1). The Cherry Creek catchment drains approximately 88220 ha and is a main contributor of water to the Neosho River, with an average stream flow of 24.5 m^3/s.

2.2   Hydrological Code

Hydrological modeling of the Cherry Creek watershed is undertaken using the Hydro-logical Simulation Program Fortran (HSPF). The HSPF code is a computer model designed for simulation of non-point source watershed hydrology and water quality. Time-series of meteorological/water-quality data, land use and topographical data are used to estimate stream flow hydrographs and polluto-graphs. The model simulates interception, soil moisture, surface runoff, interflow, base flow, snowpack depth and water content, snowmelt, evapo-transpiration, and ground-water recharge. Simulation results are provided as time-series of runoff, sediment load, and nutrient and pesticide concentrations, along with time-series of water quantity and quality, at any point in a watershed. Additional software (WDMUtil and GenScn) is used for data pre-processing and post-processing, and for statistical and graphical analysis of input and output data [4]. The BASINS 4.1 GIS system [5] was used for downloading basic data and delineating the watershed included in this study. The creation of the initial HSPF models was done using the WinHSPF interface included in BASINS 4.1.

2.3   Topographical, Land Use and Meteorological Datasets

The National Elevation Data (NED) dataset was used for characterizing the topography of the Cherry Creek watershed. The NED dataset provides a seamless mosaic elevation data derived from the 7.5-minute elevation data for the conterminous United States [7]. NED has a consistent projection (geographic), resolution (1 arc second, approximately 30 m), and metric elevation units [8].

The National Land Cover Data 2001 (NLCD) derived from the early to mid-1990s Landsat Thematic Mapper satellite data is a 21-class land cover classification scheme applied consistently across the United States [Alarcon and O'Hara]. The spatial resolution of the data is 30 meters and mapped in the Albers Conic Equal Area projection, NAD 83.

2.4   Land Use Scenarios for Water Quality Simulations

An example of the output from the watershed model was performed in the Upper Cherry Creek watershed. Four scenarios of land use were simulated and the total ammonia (TAM) concentrations at the outlet of the Cherry Creek watershed were calculated (using HSPF) for those scenarios:

- Base (BASE): land use areas in the watershed were characterized using the NLCD 2001 land use map.
- Historical (HIST): agricultural and pasture areas were decreased and forest, grasslands, and wetlands areas were increased proportionally in sub-basins 1, 4, and 16 (Figure 2 and Table 1).
- Agricultural (FUT): crop and pasture land use areas were increased in sub-basins 1, 4, and 16 (Figure 2 and Table 1), and forest, grasslands, and wet-lands were decreased proportionally.

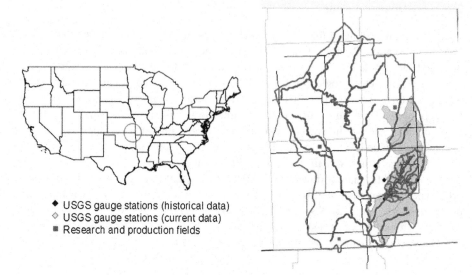

Fig. 1. Cherry Creek watershed. The watershed (outlined in red) is located in southeast Kansas, near the border to Oklahoma, Missouri and Arkansas. The current and historical USGS stream flow gauge stations used for setting up the hydrological model are shown.

- Base management practice application (BMP2): a Filter Strip BMP was applied to 50% of crop and pasture land use areas, in sub-basins 1, 4, and 16 (Figure 2 and Table 1).

2.5   Sensitivity Analysis

The hydrological model developed for Cherry Creek watershed was used to study the sensitivity of total ammonia (TAM) estimations to numerical perturbations applied to the HSPF parameter "monthly wash-off potency factor" (POTFW), that represents the ratio of a water quality constituent yield to sediment wash-off outflow [6]. POTFW is a parameter with a minimum value of 0.0 and an unbounded maximum value. A review of the literature on HSPF-based water quality modeling reveals that values as low as $4.4*10^{-6}$ [7], 0.4 [8], and as large as 10.0 [9], 340.0 [10], are used by the modeling water quality community. However, comparison of values and establishment of ranges for the POTFW parameter is challenging, because the value is dependent on the water quality constituent being modeled, the units used in the actual modeling, and the reporting of the modeling results. POTFW values in HSPF are introduced in qty/ton or qty/tonne, where "qty" could be either in US pounds (lb) or kilograms (kg), and "ton" is the US short ton (2000 pounds, or 907.185 kilograms), and "tonne" is the metric tonne (1000 kg). This becomes even more confusing because POTFW are converted to mg/kg when modeling methodologies and modeling results are reported (water quality constituents content in soil are usually measured in mg/kg). In this paper, a sensitivity analysis of total ammonia estimations ($TAM=NH_3+NH_4$) to POTFW was performed as follows.

Table 1. Scenarios of land use areas in sub-basins 1, 4 and 16

AREA IN ACRES

Land use	Sub-basin	Base	Historic	Future	BMP2
Agriculture – Cropland	1	2640	2640	2940	1320
Agriculture – Pasture	1	1630	1630	1930	815
Filterstrip BMP	1	0	0	0	2135
Forest	1	279	279	279	279
Grass Land	1	28	28	28	28
Urban	1	131	131	131	131
Water/Wetlands	1	603	603	3	603
Agriculture – Cropland	4	4224	1224	4624	2112
Agriculture – Pasture	4	2467	467	2867	1233
Filterstrip BMP	4	0	0	0	3345
Forest	4	905	3905	105	905
Grass Land	4	131	2131	131	131
Urban	4	356	356	356	356
Water/Wetlands	4	98	98	98	98
Agriculture – Cropland	16	1960	1960	2160	980
Agriculture – Pasture	16	973	973	1173	486
Filterstrip BMP	16	0	0	0	1466
Forest	16	79	79	79	79
Grass Land	16	6	6	6	6
Urban	16	116	116	116	116
Water/Wetlands	16	404	404	4	404

The HSPF model of the Cherry Creek watershed was used to estimate TAM concentrations (in mg/L) at the stream outlet of the watershed (see Figure 2). These estimations were performed for POTFW values ranging from 0.01 to 10000. Maximum normalized sensitivity coefficients were calculated for each of those estimations. A range of POTFW values for which the highest maximum normalized sensitivities were calculated was defined. For that selected range, an additional sensitivity analysis was performed using perturbations of ±60%, ±40%, ±20%, ±5%, ±1% to the POTFW parameter.

The combination of small and big perturbations was done to identify if non-linear sensitivities occurred. Normalized sensitivity coefficients represent the percentage change in the output variable resulting from a 1 percent change in each input variable Brown and Barnwell (1987). They were calculated with:

$$S_{ij} = \frac{\dfrac{\Delta y_j}{y_j}}{\dfrac{\Delta x_i}{x_i}}$$

Where: S_{ij} is the normalized sensitivity coefficient for output y_j to inputs x_i, x_i = base value of input variable (POTFW), Δx_i = magnitude of input perturbation to the base value ($\Delta POTFW$), y_j = base value of output variable (TAM concentration at the outlet), and Δy_j = sensitivity of output variable (ΔTAM).

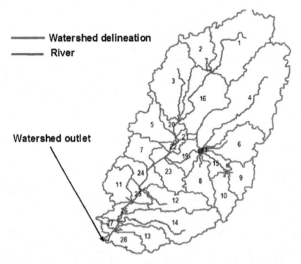

Fig. 2. Streams, outlet, and watershed delineation of Cherry Creek watershed

The USGS stream flow gage stations that were used for calibration and validation are shown in Figure 1 . The stations were chosen based on the extensive period of measured data that they have. The periods of calibration and validation are in most cases from 1970 to 1985, and 1986 to 1995, respectively. However, the meteorological stations near gage stations 02443500 and 02425000 do not provide as extensive a rainfall record as is available for the rest of the area. For that reason, the stream flow calibration period in those cases is smaller.

3   Results

3.1   Hydrological Modeling and Land Use Scenarios

Watershed delineation was performed using the National Elevation Dataset (NED) described in section 2.2. The DEM pre-processing (fill, flow direction, etc.) was set up to be burned-in to the stream network. The threshold area for delineation was specified as 300 ha. Several outlets were established manually to avoid the creation of "large" sub-basins and to provide outlets near USGS stream gages. Figure 2 shows the delineated watershed. The HSPF model corresponding to the Cherry Creek watershed (Figure 3) was set up such that the "Individual" option was specified in the Model Surface Segmentation section. This option assigns to each sub-basin individualized land use categories. In this way, each sub-basin has its own table for land use. Having a HSPF model with these characteristics also allows the user to specify other model parameters for each sub-basin. Figure 3 shows the resulting HSPF application and illustrates the correspondence between physical sub-basins and HSPF hydrological units (sub-basins generated by the delineation process).

Figure 3 shows simulated results for total ammonia (TAM) concentrations (calculated using the HSPF model of Cherry Creek) occurring at the watershed outlet for four scenarios for a 2-year simulation period (1970/09/09-1972/08/11). The change in TAM (total ammonia) concentrations demonstrate how the HSPF model can be used to explore the environmental impact of management practices on water quality. A shift towards more agricultural acres and loss of forests, grasslands and wetlands (FUT scenario) results in higher TAM concentrations in the stream in comparison to the BASE scenario, especially for high rainfall events during the crop growing season (March – July). Conversely, implementation of a filter strip (agricultural management practice) on half of the crop and pasture acreage in the FUT scenario (BMP2) would decrease TAM concentrations at the watershed outlet to the level of the concentrations occurring in the BASE scenario. This decrease in TAM concentrations in the stream persists throughout the water-year. However, the implementation of the agricultural management practices do not achieve reductions in the TAM concentrations to those estimated for the HIST scenario.

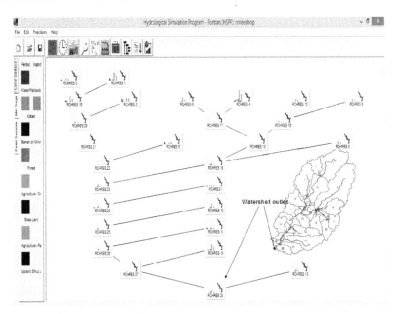

Fig. 3. The HSPF model for the Cherry Creek watershed and delineated catchment (embedded). Each sub-basin in the delineated catchment is represented by a hydrological unit in the HSPF model. Numbers in the boxes (HSPF reach/reservoir representation of sub-basins and stream segments) correspond to numbers in each of the sub-basins.

464   V.J. Alarcon and G.F. Sassenrath

3.2   Sensitivity Analysis

The HSPF model of Cherry Creek watershed was used to estimate TAM concentrations (in mg/L) for POTFW values ranging from 0.01 to 10000. Maximum normalized sensitivity coefficients were calculated taking the following POTFW values sequentially: (0.01, 0.1, 0.5, 1, 5, 10, 50, and 100. Figure 5 shows results of this exploration.

The results presented in Figure 5 illustrate that the highest maximum normalized sensitivity coefficients are those corresponding to 5, 10, 50, and 100 base input POTFW values. Figure 5 clearly shows that a POTFW value of 50 mg/Kg requires further sensitivity analysis of TAM estimations. For this POTFW value, maximum normalized sensitivities are above 100% and higher POTFW values do not produce

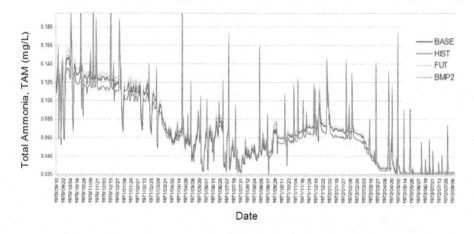

Fig. 4. Scenarios of land use and resulting total ammonia (TAM) estimations for a 2-year simulation period (1970/09/09-1972/08/11)

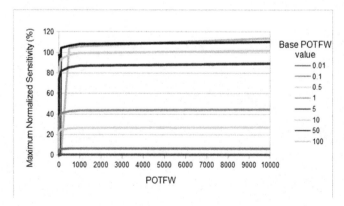

Fig. 5. Maximum normalized sensitivity of total ammonia, TAM, in mg/L (output variable) for POTFW values (in mg/Kg) ranging from 0.01 to 10000, calculated with the following POTFW values: 0.01, 0.1, 0.5, 1, 5, 10, 50, and 100, as base input values for the normalized sensitivity calculations

higher sensitivity in TAM estimations. Also the Kansas Risk-Based Standards RSK Manual [13] requires that nitrate plus ammonia (NO_3+NH_4) as N be below 40 mg/kg for remediation purposes. Choosing POTFW=50 mg/Kg for TAM (NH_3+NH_4) seems to be reasonable since TAM content in soil after fertilization will be in the range of Once a reference POTFW value was determined for which the highest maximum normalized sensitivities were estimated and which was also consistent with realistic TAM contents in soil (50 mg/Kg), perturbations of ±60% , ±40%, ±20%, ±5%, ±1% to POTFW were performed, as described in section 2.5.

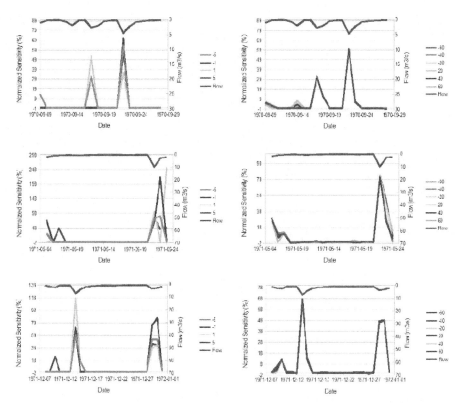

Fig. 6. Sensitivity analysis of TAM estimations for a base input POTFW value of 50 mg/Kg for several different flood events. Normalized sensitivities are shown for ±60%, ±40%, ±20%, ±5%, ±1% perturbations to the base input POTFW value. The secondary vertical axis (right-hand axis) shows stream flow values (m^3/s) in reverse scale.

Figure 6 shows the percent normalized sensitivities of TAM estimations from the Cherry Creek HSPF model for a base input POTFW value of 50 mg/Kg to which ±60% , ±40%, ±20%, ±5%, ±1% perturbations were applied. All charts include a secondary vertical axis (right-hand axis) with stream flow values in m^3/s (in reverse scale), showing the relationship of normalized sensitivities to the river flow regime.

Results show that small perturbations to the POTFW base value produce the largest normalized sensitivities. Peak sensitivities for ±5% and ±1% perturbations range from 69% to 248%, with the -1% perturbation producing the most dramatic sensitivity response in TAM estimations. Peak normalized sensitivity of TAM concentrations to ±60%, ±40%, and ±20% perturbations to POTFW range from 58% to 79%. In general, normalized sensitivities follow similar trends, with the trends more alike for bigger ±60%, ±40%, and ±20% perturbations. Peak flows, however, are associated with the highest differences in normalized sensitivity values. Most of the perturbations produce positive sensitivities for all flood events included in the analysis.

4 Conclusions

Simulated results for total ammonia (TAM) concentrations occurring at the Cherry Creek watershed outlet calculated using the HSPF model for four scenarios in a 2-year simulation period, demonstrate how the HSPF model for Cherry Creek can be used to explore the environmental impact of management practices on water quality.

The HSPF model of the Cherry Creek watershed was used to estimate the sensitivity of TAM estimations to POTFW values ranging from 0.01 to 10000 mg/Kg. Maximum normalized sensitivity coefficients (calculated sequentially) showed that the highest sensitivity occur for POTFW values within 5 to 100 mg/Kg. Additional sensitivity analysis of TAM estimations taking 50 mg/Kg as base input POTFW value was performed for several different flood events. This value was chosen because for POTFW values higher than 50, maximum normalized sensitivities asymptotically approach 110%, and previous research results found a similar value as reference for remediation purposes [14]. Results showed that small perturbations to the POTFW base value produce the largest normalized sensitivities. Peak sensitivities for ±5%, ±1% perturbations reached up to 248%, with the -1% perturbation producing the most dramatic sensitivity response in TAM estimations. All results showed a strong relationship between normalized sensitivities to river flow regime, with the highest sensitivities associated with peaks in the river hydrographs. Most of the perturbations produced positive sensitivities. Non-linearities in sensitivity are evident for small ±1% perturbations since differences in normalized sensitivities range between 30% to 100% were calculated. These non-linearities are more evident for high stream flow values (strong flood events). For ±60%, ±40%, and ±20% perturbations to POTFW, normalized sensitivity is almost constant, (for all practical purposes, indicating that the sensitivity response is linear. This means that when assigning values to the POTFW parameter (e.g., during calibration or validation of a hydrological model) careful choosing of values is required.

Acknowledgements. This is contribution number 15-328-J from the Kansas Agricultural Experiment Station. This research was funded by a grant from Vicerrectoria Academica, Universidad Diego Portales (travel fund).

References

1. Manders, G.C., Aber, J.S.: Tri-State Mining District legacy in northeastern Oklahoma. Emporia State Research Studies **49**(2), 29–51 (2014). http://academic.emporia.edu/esrs/vol49/manders_aber.pdf
2. Grand Lake O' the Cherokees Watershed Alliance Foundation, 2008. Grand Lake Watershed Plan: For improving water quality throughout the Grand Lake Watershed. Draft. http://www.ok.gov/conservation/documents/Grand_Lake_%20WBP_DRAFT.pdf
3. Kansas Center for Agricultural Resources and the Environment. Middle Neosho Watershed Restoration and Protection Strategy, final draft plan (2011). http://www.kswraps.org/files/attachments/middleneosho_plansummary.pdf
4. Alarcon, V.J., Hara, C.G.: Scale-Dependency and Sensitivity of Hydrological Estimations to Land Use and Topography for a Coastal Watershed in Mississippi. In: Taniar, D., Gervasi, O., Murgante, B., Pardede, E., Apduhan, B.O. (eds.) ICCSA 2010, Part I. LNCS, vol. 6016, pp. 491–500. Springer, Heidelberg (2010)
5. EPA. Better Assessment Science Integrating Point and Nonpoint Sources (BASINS) (2013). http://water.epa.gov/scitech/datait/models/basins/index.cfm
6. Bicknell, B.R., Imhoff, J.C., Kittle Jr., J.L., Donigian Jr., A.S., Johanson, R.C.: Hydrological Simulation Program – FORTRAN, User's Manual for Version 11 (1997)
7. Lent, M., McKee, L.: Guadalupe River Watershed Loading HSPF Model: Year 3 final progress report. San Francisco Estuary Institute. Richmond, Califormia (2011). http://www.sfei.org/sites/default/files/Guad_HSPF_Model__forSPLRev_17Feb2012.pdf
8. Deliman, P.N., Pack, W., Nelson, E.J.: Integration of the Hydrology Simulation Program—FORTRAN (HSPF) WatershedWater Quality Model into the Watershed Modeling System (WMS). Technical Report W-99-2, September 1999, US Army Corps of Engineers (1999)
9. Tetra Tech, Inc. Appendix VII: LSPC Watershed Model Development for Simulation of Loadings to the Los Angeles/Long Beach Harbors. Prepared for: USEPA Region 9 Los Angeles Regional Water Quality Control Board (October 2010). http://www.waterboards.ca.gov/losangeles/board_decisions/basin_plan_amendments/technical_documents/66_New/10_1217/
10. Jobes, T.H., Kittle, Jr. J.L., Bicknell, B.R.: A Guide to Using Special Actions in the Hydrological Simulation Program—FORTRAN (HSPF). AQUA TERRA Consultants (1999) ftp://hspf.com/hspf/Special_actions_v11.doc
11. Brown, L.C., Barnwell, T.O.: The enhanced stream water quality models QUAL2E and QUAL2E-UNCAS: documentation and user manual. Env. Res. Laboratory. US EPA, EPA /600/3-87/007, Athens, GA. p. 189 (1987)
12. North Carolina Department of Environment and Natural Resources, NCDECNR. Falls Lake Nutrient Response Model: Final Report. Modeling & TMDL Unit, Division of Water Quality, North Carolina Department of Environment and Natural Resources (2009). http://portal.ncdenr.org/c/document_library/get_file?uuid=33debbba-5160-4928-9570-55496539f667&groupId=38364
13. KDHE. Guidelines for investigating and remediating nitrate/ammonia contamination from agricultural chemical releases. Kansas Bureau of Environmental Remediation/Remedial Section Policy. BER policy # BER-RS-050 (January 2007) http://www.kdheks.gov/ber/policies/BER_RS_50.pdf

Assimilation of TRMM Precipitation into a Hydrological Model of a Southern Andes Watershed

Vladimir J. Alarcon$^{(\boxtimes)}$, Hernan Alcayaga, and Enrique Alvarez

Civil Engineering School, Universidad Diego Portales, 441 Ejercito Ave., Santiago, Chile
{vladimir.alarcon,hernan.alcayaga,enrique.alvarez}@udp.cl

Abstract. This paper presents results of assimilation of TRMM precipitation in-to a hydrological model of Los Almendros River watershed, assessed by the comparison of simulated stream flow values to observed stream flow data. Los Almendros River is a tributary to the Clarillo River, located in Reserva Nacio-nal Rio Clarillo (National Reserve Clarillo River), Chile, South America. Los Almendros basin, covering approximately 4.9 km^2, is a typical Andean water-shed with an average slope of 46.3. It drains an average of 18.5 L/s from a catchment area covered predominantly by grasslands (91%), while forests and savanna land cover types are less predominant (3% and 6% respectively). The hydrological model of Los Almendros watershed was developed using the Hy-drological Simulation Program Fortran (HSPF). Results showed that raw TRMM precipitation time-series overestimated disaggregated precipitation val-ues for the whole period of analysis. TRMM data required time-averaging for the monthly and annual values to be in the same range as those of disaggregated data. Time-averaging produced daily precipitation time-series consistent to dis-aggregated data (R^2=0.85). The use of TRMM-enriched precipitation time-series for hydrological modeling of stream flow at Los Almendros watershed outlet, slightly improved a previous simulation in which only disaggregated precipitation dataset was used. When comparing simulated and observed data, the statistical fit coefficient improved from R^2=0.64 (corresponding to only dis-aggregated precipitation data introduced into the hydrological model) to R^2=0.68 (corresponding to TRMM-enriched precipitation data).

Keywords: Hydrological modeling · HSPF · TRMM precipitation · Stream flow simulation

1 Introduction

The Reserva Nacional Rio Clarillo (National Reserve Clarillo River), located in cen-tral Chile, was created to protect biological species particular to the zone [1] and it is also used for recreation and research [2]. Clarillo River is the main river in the Clarillo reserve and it's a tributary to the Maipo River. The basin is a typical Andean valley in which the main river is fed by several perennial and intermitent streams. Thirteen main sub-basins (and their corresponding main and secondary streams) con-form the whole Clarillo River basin. Within those sub-basins, Los Almendros River

© Springer International Publishing Switzerland 2015
O. Gervasi et al. (Eds.): ICCSA 2015, Part III, LNCS 9157, pp. 468–476, 2015.
DOI: 10.1007/978-3-319-21470-2_34

catchment (covering approximately 4.9 km^2) is one of the few catchments for which some precipitation and stream flow published data exist, albeit outdated.

Los Almendros River is not currently gauged for stream flow neither well documented for land use, topography, and meteorology. There are some technical reports on Los Almendros watershed (mostly outdated). For example, a report published decades ago [3] reports monthly total precipitation and air temperatures (maximum, minimum, mean) for years 1998 to 2002. Discontinue instantaneous stream flow data for the Almendros River are also reported in [3]. A recent study [4] has used the existing sparse meteorological and stream flow data reported in [3] for developing a hydrological model for Clarillo and Los Almendros rivers, but the authors report inconsistencies in the calibration and validation of the model, mostly due to the inexistence of high-resolution temporal precipitation time-series.

The purpose of this research is to assimilate Tropical Rainfall Measurement Mission daily precipitation data (TRMM) into an existing Los Almendros watershed model [4] and assess its validity for improving the calibration and validation of the model. The Hydrological Simulation Program Fortran (HSPF) is the hydrological code used for modeling the watershed.

2   Methods

2.1   Study Area

Los Almendros River watershed (denominated Los Almendros watershed in the rest of this document for brevity) is a sub-watershed of the Clarillo River watershed located in central Chile. The watershed is within a natural reserve (Reserva Nacional Rio Clarillo), fifty kilometers away from Santiago City (Figure 1).

Los Almendros is an ungauged watershed that covers approximately 490 hectares with an average slope of 46.3 %. Los Almendros River extends for approximately 4.46 km, having nearly 16.9% of longitudinal slope. It drains an average of 18.5

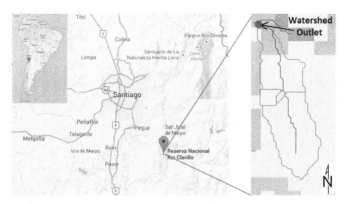

Fig. 1. Los Almendros watershed. The watershed (outlined in red) is located in central Chile, within the Clarillo River National Reserve and watershed. The watershed outlet is shown.

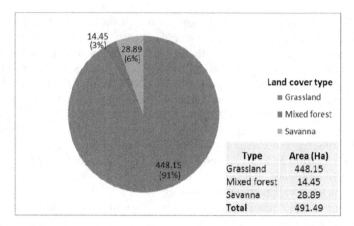

Fig. 2. Land cover types in Los Almendros watershed

L/s from a catchment area where the predominant land cover category is grasslands/shrub-lands (91%), while forests and savanna land cover types are less predominant (3% and 6% respectively). Figure 2 and Section 2.3 give more details on land-use/land-cover characterization.

2.2   Hydrological Model

A hydrological model of Los Almendros watershed was developed using the Hydrological Simulation Program Fortran, HSPF [5]. The HSPF code is a computer model designed for simulation of non-point source watershed hydrology and water quality.

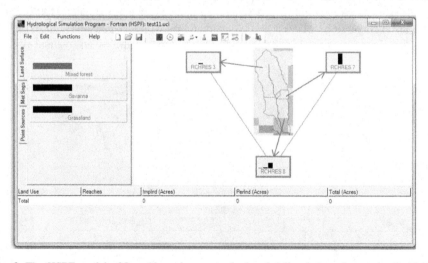

Fig. 3. The HSPF model of Los Almendros watershed and delineated catchment (embedded). Each sub-basin in the delineated catchment is represented by a hydrological unit in the HSPF model

Time-series of meteorological, land use and topographical data were used to develop a first basic watershed model (Figure 3). Stream flow simulation results in the form of time-series are generated for the watershed outlet location (Figure 1). Software (WDMUtil and GenScn) were used for data pre-processing and post-processing, and for statistical and graphical analysis of input and output data [6] during the hydrological calibration process. The BASINS 4.1 GIS system [7] was used for all geoprocessing and delineation of the watershed. The creation of the initial HSPF models was done using the WinHSPF interface included in BASINS 4.1. Figure 3 shows the hydrological model for Los Almendros watershed.

2.3   Physiographic Datasets

The NASA MODIS MCD12Q1 Land Cover Product (MODIS Aqua/Terra Land Cover, 500 m spatial resolution) for year 2002 provided by NASA through [8] was used to characterize land-use/land-cover in Los Almendros watershed. The dataset was reprojected into UTM coordinates (19 South) and then reclassified into aggregated land cover categories for introducing the parameterized land use data into the HSPF model. It was assumed that since the watershed is within a natural reserve where human-induced change is prohibited, the 2002 MODIS land cover is representative of the period for which hydrological simulation was performed (1998-2002).

In this research, topographical data sets from the Shuttle Radar Topography Mission (SRTM), publicly available at the International Centre for Tropical Agriculture website [9] were used. Approximate spatial resolutions for the study area varied between 70 m to 80 m. The raster datasets were geo-processed to parameterize topographical information as required by the hydrological code used for modeling (HSPF).

2.4   Meteorological Data

As detailed in [4], monthly accumulated precipitation data reported in [3] were disaggregated for building an initial hydrological model of Los Almendros watershed. Figure 4 shows the disaggregated monthly accumulated precipitation data.

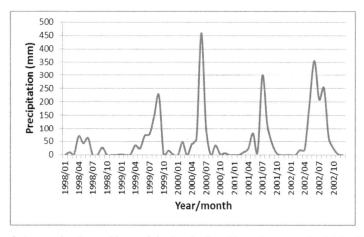

Fig. 4. Accumulated monthly precipitation for Los Almendros watershed (after [3])

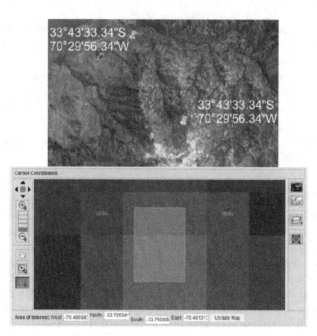

Fig. 5. Geographical container for Los Almendros watershed introduced in the NASA GES DISC [10] Giovanni online system [11] and corresponding TRMM grid

Since the statistical fit indicators (for measured stream flow vs simulated stream flow) reported in [4] were not optimal, Tropical Rainfall Measurement Mission (TRMM) daily precipitation data (Figure 5) were assimilated into the hydrological model by combining TRMM data and [thesis] data at critical dates where measured stream flow and estimated values were the most dissimilar. The following paragraphs illustrate the process.

The acquisition of daily TRMM rainfall data for the period of analysis and study area was performed through the NASA Goddard Earth Sciences (GES) Data and Information Services Center (DISC) [10] and its Giovanni online data system [11] (developed and maintained by the NASA GES DISC). The geographical container for which TRMM daily precipitation time-series are desired has to be specified in geographical coordinates and introduced in the interactive online system for visualization and analysis. The system fetches the data files using the spatial and temporal constraints, retrieves daily precipitation from TRMM_3B42_daily.007 preprocessor, re-format the data cubes to HDF-4 format, applying scaling factors and preset filtering. Precipitation values are computed as the differences between anomalies and a selected climatology over the grids of the selected region, then the values are area-averaged over the selected spatial area (with area weighting = 1), and finally rendered as a time-series [11]. The time-series generated by the Giovanni data system for Los Almendros watershed is shown in Figure 6.

Area-Averaged Time Series (TRMM_3B42_daily.007)
(Region: 70W-70W, 33S-33S)

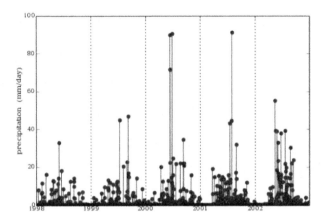

Fig. 6. Precipitation time-series generated by the NASA GES DISC Giovanni online data system [11] for Los Almendros watershed

3   Results

3.1   Comparison of Precipitation Time-Series

As shown in Figures 5 and 4, the disaggregated precipitation dataset for hydrological modeling in [4] was somewhat different to the time-series produced through the NASA SEAC DESC TRMM data portal. Figure 6 further illustrates the differences by comparing monthly and annual precipitations for Los Almendros watershed Area.

Clearly, TRMM time-series overestimate precipitation values for the whole period of analysis. Hence, TRMM data was time-averaged such that monthly and annual values were in the same range as those of the disaggregated data. Figure 7 shows results of this process.

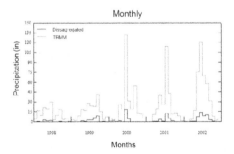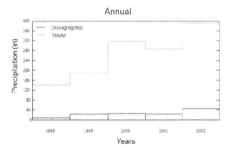

Fig. 7. Comparison of disaggregated and TRMM monthly and annual precipitation time-series for Los Almendros watershed

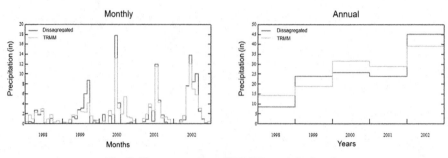

Fig. 8. Time-averaging of TRMM precipitation data

An additional check on the actual daily data used for hydrological modeling was performed (Figure 8) to assess the consistency of the TRMM data with respect to the disaggregated data used in Alarcon et al. 2015.

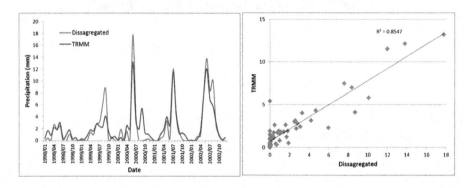

Fig. 9. TRMM and disaggregated precipitation data used for hydrological modeling

3.2   Hydrological Simulation

The HSPF model of Los Almendros watershed, fed with disaggregated precipitation data enriched with TRMM precipitation, was used to estimate stream flow values at the outlet of the watershed. Figure 9 shows results of this exploration.

Figure 9 shows that using TRMM precipitation data slightly improves stream flow estimated values, as reflected by a better agreement of simulated stream flow values with measured values. The statistical fit coefficient improved from $R^2=0.64$ to $R^2=0.68$. This improvement is due mainly to the decrease in simulated stream flow peaks after January 10, 2000 up to the end of the simulation period. Also, simulated stream flow data from September 13, 2000, through December 27, 2000, increase to reach similar to observed stream flow values.

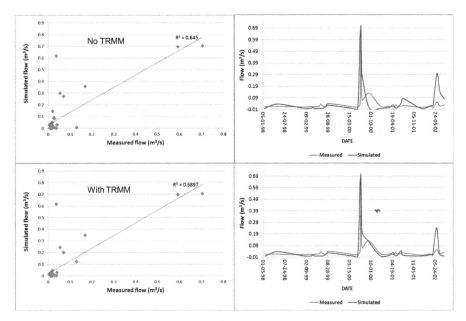

Fig. 10. Comparison of measured and simulated stream flow at Los Almendros watershed outlet

4   Conclusions

The TRMM precipitation time-series overestimates disaggregated precipitation values for the whole period of analysis. TRMM data required time-averaging for the monthly and annual values to be in the same range as those of the disaggregated data. After this artifice, a comparison of daily data showed statistical consistency as revealed by a calculated R^2 value of 0.85. When using the TRMM-enriched precipitation time-series for hydrological modeling of stream flow at Los Almendros watershed outlet, results showed that using TRMM precipitation data slightly improves stream flow estimated values, as reflected by a better agreement of simulated stream flow values with measured values. When comparing simulated and observed data, the statistical fit coefficient improved from $R^2=0.64$ to $R^2=0.68$. This improvement was due mainly to decreases in simulated stream flow peaks after January 10, 2001 up to the end of the simulation period, and the increase of simulated stream flow data from September 13, 2000, through December 27, 2000.

Acknowledgment. This research was funded in part by seed grant 1040308002: Fondo de Produccion del Conocimiento de la Vicerrectoria Academica, Universidad Diego Portales, and by a travel grant from Vicerrectoria Academica, Universidad Diego Portales.

References

1. CONAF, Comision Nacional Forestal, Reserva Nacional Río Clarillo (2014). http://www.conaf.cl/parques/reserva-nacional-rio-clarillo/
2. Niemeyer, H., Bustamante, R.O., Simonetti, J.A., Teillier, S., Fuentes-Contreras, E., Mejia, J.: Historia natural de la reserva nacional rio clarillo: un espacio para aprender ecología. Impresos Socias, Santiago (2002)
3. Villarroel, A.: Aplicación del Sistema Hidrológico Europeo (SHETRAN) en una Microcuenca cordillerana a problemas de Inundación, Thesis. Universidad de Chile, Santiago (2003)
4. Alarcon, V.J., Alcayaga, H., Alvarez, E.: Hydrological Modeling of an Ungauged Watershed in Southern Andes. In: AIP Conference Proceedings of ICCMSE 2015 (in print, 2015)
5. Barnwell, T.O., Johanson, R.: HSPF: A Comprehensive Package for Simulation of Watershed Hydrology and Water Quality. In: Nonpoint Pollution Control: Tools and Techniques for the Future. Interstate Commission on the Potomac River Basin, Rockville, MD (1981)
6. Hummel, P., Kittle, Jr. J., Gray, M.: WDMUtil, Version 2.0, A Tool for Managing Watershed Modeling Time Series Data. User's Manual. AQUA TERRA Consultants, Decatur, Georgia (2001). http://hspf.com/ftp/basins3/doc/WDMUtil.pdf
7. EPA, Better Assessment Science Integrating Point and Nonpoint Sources (BASINS) (2013). http://water.epa.gov/scitech/datait/models/basins/index.cfm
8. Broxton, P.D., Zeng, X., Sulla-Menashe, D., Troch, P.A.: A Global Land Cover Climatology Using MODIS Data. J. Appl. Meteor. Climatol. **53**, 1593–1605 (2014). doi:10.1175/JAMC-D-13-0270.1
9. Jarvis A., Reuter, H.I., Nelson, A., Guevara, E.: Hole-filled seamless SRTM data V4, International Centre for Tropical Agriculture (CIAT) (2008). http://srtm.csi.cgiar.org
10. NASA GES DISC, Goddard Earth Sciences Data and Information Services Center (2015). http://disc.sci.gsfc.nasa.gov/
11. NASA GES DISC, Giovanni - Interactive Visualization and Analysis (2014). http://disc.sci.gsfc.nasa.gov/giovanni

Conceptualizing, Modeling and Simulating Sustainability as Tools to Implement Urban Smartness

Maria-Lluïsa Marsal-Llacuna[✉]

University of Girona, Girona Smart City Chair Campus Montilivi.
Politècnic Building III. Room 121, 17003 Girona, Spain
luisa.marsal@udg.edu

Abstract. The smart cities initiative does not consist of filling our cities with tech gadgets, cameras and sensors for the purpose of monitoring and controlling both citizens and urban environment. The smart cities movement is not a marketplace for companies and businesses to sell their ICT products and services to governments. The smart cities objective is not to target the use of urban technology as an end in itself but as a tool towards more citizen-centric cities. Therefore, smart cities are not a big brother watching citizens but an inclusive and participatory arena for local governments and all interested stakeholders (including citizens) to co-create communities that can offer development opportunities to everyone and quality of live, no matter social group, age, or gender. This is the true objective of smart cities, which sometimes gets lost, forgotten or becomes shadowed by an excess of technology. But, how to achieve such a challenging objective? Fortunately, just recently, a definition on smart cities has been agreed by international standardization bodies which, so far, are the ones paving the way regarding concepts, terminology and general understanding of the smart cities movement. This novel definition builds on the sustainability concept, promotes urban development in good harmony with resilience, and uses technology only as a way for cities to measure their smartness and improve their performance. The author, aware of the big step forward that this internationally acknowledged definition is, takes the opportunity that this new piece of formally structured knowledge offers to elaborate a first-of-a-kind framework to conceptualize, model and simulate sustainability, and its practical implementation while promoting urban development and resilience to achieve higher performance levels of urban smartness.

Keywords: Smart cities · Sustainable city indicators · Urban modeling and simulation · Standardization technologies

1 Introduction. State-of-the-Art on Definitions and Understanding of the Smart Cities Initiative

This is the first time ever that an urban initiative is raising interest and stimulating the participation of all urban stakeholders. Industries are developing IT solutions and services to help cities optimize natural and fiscal resources. Local governments buy

© Springer International Publishing Switzerland 2015
O. Gervasi et al. (Eds.): ICCSA 2015, Part III, LNCS 9157, pp. 477–494, 2015.
DOI: 10.1007/978-3-319-21470-2_35

these solutions to increase the economic efficiency of their investments. Activities of citizens are being tracked, and monitored through their smartphones and other devices, turning them into veritable sensors which provide IT companies with a wealth of valuable information. Scientists are doing research, which is strongly oriented towards industrial interests, and making inventions (and filing new patents), for more economically efficient and environmentally friendly solutions. Universities are already offering official master's degree programs in smart cities[1] that also lead to doctoral degrees in this topic. Professional bodies of architects and engineers offer specific practical training to their members to help them incorporate technical solutions in their projects. Trans-national institutions are launching specific programs with significant funding to support innovative multi-stakeholder research[2], mainly in the areas of clean energy, efficient mobility, and future IT. Even standardization bodies have joined this group of active stakeholders in the smart cities domain. Never before has an urban initiative been normalized, and it is taking place now with smart cities. Trans-national and national standardization bodies are all already carrying out standardization works, and some standards are already in use.

The smart cities initiative doesn't start from scratch. It builds on previous initiatives of the sustainable city, a major movement which mainly lasted between 1995 and 2005 [1], along with other minor movements within the same period, such as the creative city [2], the digital city [3], and the knowledge city [4]. However, none of these initiatives generated technological solutions, competitive research programs, industrial interest, the involvement of citizens, scientific researchers, educational institutions creating master´s degree programs, and professional training bodies, or standardization works.

The International Standards Organization (ISO) was the first trans-national body to initiate standardization works in the domain of smart cities, in 2011. It is now the most advanced institution in this field, setting the stage for national normalization. Also at the trans-national level, the International Telecommunications Union (ITU) started standardization works in 2013, followed by the International Electro-technical Committee (IEC), which also created a committee for smart cities standardization in 2013. At the European level, CEN-Cenelec, the European trans-national standardization body, also initiated its own works at the end of 2013. National standardization bodies in different parts of the world are also very active. Some of them began working more or less in parallel with ISO, such as AENOR in Spain and AFNOR in

---

[1] Taking an official master's degree program in smart cities, like the one offered at the University of Girona, http://www.udg.edu/mastersmartcities, makes it possible to subsequently obtain a doctorate in the same field: http://www.udg.edu/tabid/17111/Default.aspx?ID=350130813& language=en-US&any=2014

[2] European Commission H2020 Call Smart Cities and Communities (H2020-SCC) , 92M€: http://ec.europa.eu/research/participants/portal/desktop/en/opportunities/h2020/calls/h2020-scc-2014.html
JPI Urban Europe's Call ERA-NET Smart Cities and Communities (ENSCC), 20M€: http://jpi-urbaneurope.eu/new-joint-call-for-proposals-era-net-cofund-smart-cities-and-communities/

France. The first standard has already been completed and launched, *ISO 37120 Sustainable Development and Resilience. Indicators for City Services and Quality of Li*fe [5]. International standardization bodies define smart cities as communities designed to promote urban development and resilience, while safeguarding environmental, social and economic sustainability, making use of intelligent technologies [5], [6]. Accordingly, ISO entitles its committee for the standardization on smart cities ISO/TC 268 "Sustainable Development in Communities" and uses "smartness" when referring to metrics. This is because IT-based measurements make a process or an action smart[3].

The ISO/TC 268 has three lines of work: on Management of Sustainable Development (Working Group 1 within ISO/TC 268), on Indicators for Sustainable Development (Working Group 2 within ISO/TC 268), and a specific Subcommittee on "smartness" in Infrastructures which focuses on developing criteria for smart urban infrastructure metrics. ISO/TC 268 WG1 is working on the elaboration of the standard 37101 on *Management Systems Standards for Sustainable development in communities.* ISO/TC 268 WG2 has already published the 37120 standard on *Sustainable Development and Resilience. Indicators for City Services and Quality of Li*fe [5]. The ISO/TC 268/SC1 is about to publish, as soon as the public presentation of the standards ends, the 37150 on *Smart Community Infrastructures. Review of existing activities relevant to metrics.*

At European level, CEN-Cenelec created the "Smart and Sustainable Cities and Communities Coordination Group (SSCC-CG)". In comparison with ISO, we note the addition of *smart* to the *sustainable* terminology and the addition of *cities* to *communities.* Accordingly, the more general definition on smart cities given by CEN-Cenelec is *"A Smart City is a city seeking to address sustainability issues via ICT-based solutions on the basis of a multi-stakeholder, municipally based partnership"*[4],[5]. National standardization institutions follow the steps of international standardization bodies and, therefore, for the smart cities standardization, they also understand that sustainable urban development supported by IT-based solutions is the purpose of the smart cities initiative. According to the Brundtland report [7], sustainable development is the *"development that meets the needs of the present without compromising the ability of future generations to meet their own needs"*. Hence, the smart cities initiative, thanks to its IT-based approach, is in a unique position to achieve not only sustainable but also objectively measurable urban development.

We can state, however, that the target for standardization bodies will be to elaborate standards measuring the performance of cities in the three pillars of sustainability; economic, environmental and social. As already mentioned, ISO was the first

---

[3] ISO/TC 268/SC 01 "smart urban infrastructure metrics". Project PWI37151. General principles and requirements.

[4] Directorate General for Internal Policies in the EU. Mapping smart cities in the EU., IP/A/ITRE/ST/2013-02. Jan. 2014.

[5] European Innovation Partnership on Smart Cities and Communities Operational Implementation Plan: First Public Draft.

body to begin standardization works, mostly covering the environmental[6] and economic[7] aspects of sustainability. Although some aspects of social sustainability are also covered in the recently approved standard ISO 37120[8], in the opinion of the author, social sustainability requires wider standardization coverage. It follows a list of the 100 indicators included in the 37120 standard. The author classified them according their social, environmental or economical predominant nature. This classification exercise shows the bias of the standard towards measuring non-social aspects of sustainability:

INDICATORS FOR THE MEASUREMENT OF SOCIAL SUSTAINABILITY:
6. Education
[c] 6.1 Percentage of female school-aged population enrolled in schools
[c] 6.2 Percentage of students completing primary education: survival rate
[c] 6.3 Percentage of students completing secondary education: survival rate
[c] 6.4 Primary education Student/teacher ratio
[s] 6.5 Percentage of male school-aged population enrolled in schools
[s] 6.6 Percentage of school-aged population enrolled in schools
[s] 6.7 Number of higher education degrees per 100 000 population

12. Health
[c] 12.1 Average life expectancy
[c] 12.2 Number of in-patient hospital beds per 100 000 population
[c] 12.3 Number of physicians per 100 000 population
[c] 12.4 Under age five mortality per 1 000 live births
[s] 12.5 Number of nursing and midwifery personnel per 100 000 population
[s] 12.6 Number of mental health practitioners per 100 000 population
[s] 12.7 Suicide rate per 100 000 population
13. Recreation
[s] 13.1 Square meters of public indoor recreation space per capita
[s] 13.2 Square meters of public outdoor recreation space per capita

14. Safety
[c] 14.1 Number of police officers per 100 000 population
[c] 14.2 Number of homicides per 100 000 population
[s] 14.3 Crimes against property per 100 000
[s] 14.4 Response time for police department from initial call
[s] 14.5 Violent crime rate per 100 000 population

---

[6] ISO/TC 268/SC 01 "smart urban infrastructure metrics". Project PWI37151. General principles and requirements.
[7] ISO/TC 268/WG 01 "Management Systems standards for sustainable development in communities" Project ISO 37101 Internal Design Specifications.
[8] ISO/TC 268/WG 02 "Urban indicators" ISO 37120 Sustainable Development and Resilience. Indicators for City Services and Quality of Life.

15. Shelter

[c] 15.1 Percentage of city population living in slums

[s] 15.2 Number of homeless per 100 000 population

[s] 15.3 Percentage of households that exist without registered legal titles

INDICATORS FOR THE MEASUREMENT OF ENVIRONMENTAL SUSTAINABILITY:

7. Energy

[c] 7.1 Total residential electrical use per capita (kWh/year)

[c] 7.2 Percentage of city population with authorized electrical service

[c] 7.3 Energy consumption of public buildings per year (KWh/m2)

[c] 7.4 The percentage of total energy derived from renewable sources, as a share of the city's total energy consumption

[s] 7.5 Total electrical use per capita (kWh/year)

[s] 7.6 Average number of electrical interruptions per customer per year

[s] 7.7 Average length of electrical interruptions (in hours)

8. Environment

[c] 8.1 Fine Particulate Matter (PM 2.5) Concentration

[c] 8.2 Particulate Matter (PM10) Concentration

[c] 8.3 Greenhouse gas emissions measured in tons per capita

[s] 8.4 NO2 (nitrogen dioxide) concentration

[s] 8.5 SO2 (sulphur dioxide) concentration

[s] 8.6 O3 (Ozone) concentration

[s] 8.7 Noise Pollution

[s] 8.8 Percentage change in number of native species

10. Fire and emergency response

[c] 10.1 Number of firefighters per 100 000 population

[c] 10.2 Number of fire related deaths per 100 000 population

[c] 10.3 Number of natural disaster –related deaths per 100 000 population

[s] 10.4 Number of volunteer and part-time firefighters per 100 000 population

[s] 10.5 Response time for emergency response services from initial call

[s] 10.6 Response time for fire department from initial call

16. Solid waste

[c] 16.1 Percentage of city population with regular solid waste collection (residential)

[c] 16.2 Total collected municipal solid waste per capita

[c] 16.3 Percentage of city's solid waste that is recycled

[s] 16.4 Percentage of the city's solid waste that is disposed of in a sanitary landfill

[s] 16.5 Percentage of the city's solid waste that is disposed of in an incinerator

[s] 16.6 Percentage of the city's solid waste that is burned openly

[s] 16.7 Percentage of the city's solid waste that is disposed of in an open dump

[s] 16.8 Percentage of the city's solid waste that is disposed of by other means

[s] 16.9 Hazardous Waste Generation per capita (tons)

[s] 16.10 Percentage of the city's hazardous waste that is recycled

19. Urban planning
[c] 19.1 Green area (hectares) per 100 000 population
[s] 19.2 Annual number of trees planted per 100 000 population
[s] 19.3 Areal size of informal settlements as a percentage of city area
[s] 19.4 Jobs/housing ratio

20. Wastewater
[c] 20.1 Percentage of city population served by wastewater collection
[c] 20.2 Percentage of the city's wastewater that has received no treatment
[c] 20.3 Percentage of the city's wastewater receiving primary treatment
[c] 20.4 Percentage of the city's wastewater receiving secondary treatment
[c] 20.5 Percentage of the city's wastewater receiving tertiary treatment

21. Water and Sanitation
[c] 21.1 Percentage of city population with potable water supply service
[c] 21.2 Percentage of city population with sustainable access to an improved water source
[c] 21.3 Percentage of population with access to improved sanitation
[c] 21.4 Total domestic water consumption per capita (liters/day)
[s] 21.5 Total water consumption per capita (liters/day)
[s] 21.6 Average annual hours of water service interruption per household
[s] 21.7 Percentage of water loss (unaccounted for water)

INDICATORS FOR THE MEASUREMENT OF ECONOMIC SUSTAINABILITY:
5. Economy
[c] 5.1 City's unemployment rate
[c] 5.2 Assessed value of commercial and industrial properties as a percentage of total assessed value of all properties
[c] 5.3 Percentage of city population living in poverty
[s] 5.4 Percentage of persons in full-time employment
[s] 5.5 Youth unemployment rate
[s] 5.6 Number of businesses per 100 000 population
[s] 5.7 Number of new patents per 100 000 population per year.

9. Finance
[c] 9.1 Debt service ratio (debt service expenditure as a per cent of a municipality's own-source revenue)
[s] 9.2 Capital spending as a percentage of total expenditures
[s] 9.3 Own-source revenue as a percentage of total revenues
[s] 9.4 Tax collected as a percentage of tax billed

11. Governance
[c] 11.1 Voter participation in last municipal election (as a percentage of eligible voters)
[c] 11.2 Women as a percentage of total elected to city-level office
[s] 11.3 Percentage of women employed in the city government workforce
[s] 11.4 Number of convictions for corruption/bribery by city officials per 100 000 population
[s] 11.5 Citizens' representation: number of local officials elected to office per 100 000 pop.
[s] 11.6 Number of registered voters as a percentage of the voting age population

17. Telecommunication and innovation
[c] 17.1 Number of internet connections per 100 000 population
[c] 17.2 Number of cell phone connections per 100 000 population
[s] 17.3 Number of landline phone connections per 100 000 population

18. Transportation
[c] 18.1 Kilometers of high capacity public transport system per 100 000 population
[c] 18.2 Kilometers of light passenger transport system per 100 000 population
[c] 18.3 Annual number of public transport trips per capita
[c] 18.4 Number of personal automobiles per capita
[s] 18.5 Modal split (% of commuters using a travel mode to work other than a personal Vehicle)
[s] 18.6 Number of two-wheel motorized vehicles per capita
[s] 18.7 Kilometers of bicycle paths and lanes per 100 000 population
[s] 18.8 Transportation fatalities per 100 000 population
[s] 18.9 Commercial air connectivity (number of non-stop commercial air destinations)

The distribution in percentages of ISO 37120 indicators within the 3 pillars of urban sustainability is the following:

Indicators for the measurement of social sustainability, 24%
Indicators for the measurement of environmental sustainability, 47%
Indicators for the measurement of economic sustainability, 29%

2   Discussion. On the Need for a Theoretical Basis to Ground the Smart Cities Initiative

Since the smart cities initiative builds on the safeguarding of urban sustainability while promoting the development and resilience of cities making use of urban technology, it will be very important for the consolidation and expansion of the smart cities movement to make sure that all aspects of sustainability are well covered so that cities joining the smart cities movement can develop in greater harmony with resilience. The author researched the State-of-the-Art on instruments acknowledged and endorsed by the international community to find established and widely accepted references on urban sustainability theories and concepts. Researched instruments have been international treaties, charters, and relevant trans-national declarations. We found two Charters fulfilling these requirements, the European Charter for the Safeguarding of the Human Rights in the City [8] and the Global Charter-Agenda for Human Rights in the City [9]. Both documents cover all aspects of urban sustainability and place the citizen at the heart of the city by presenting sustainability as a citizens' right, but also as a duty since safeguarding of sustainability involves us all.

Both documents were acknowledged by United Nations at the 50th and 60th anniversaries of the Universal Declaration of Human Rights of 1948 [10]. Both Charters have the purpose of ensuring sustainability in the broad and comprehensive "right to the city" of citizens proclaimed by Lefebvre in the late sixties [11, 12, 13], and followed by other urban thinkers such as Marcuse [14,15]. The protection of human rights on an international scale was advanced greatly by the Universal Declaration of Human Rights in 1948. Many other conventions and treaties followed this seminal document, endorsing, expanding, or simply acknowledging its fundamental principles. Both Charters contribute to this positive trend by adding specific urban rights, and linking human rights and citizens' rights. Therefore, cites which sign The Global Charter-Agenda also endorse instruments which follow the 1948 Declaration, such as the International Covenant on Civil and Political Rights of 1966 [16], the International Covenant on Economic, Social and Cultural Rights of 1966 [17], the Millennium Declaration of 2000 [18], the Declaration for the 60th Anniversary of the United Nations of 2005 [19], and the Vienna Declaration of 1993 [20]. For the European Charter the following documents are also endorsed by signatory cities when signing the Charter: the European Convention for the Protection of Human Rights and Fundamental Freedoms of 1950 and 2010 [21,22], and the European Social Charter of 1961 and 1996 [23,24].

The holistic spatio-temporal understanding of the right to the city of Lefebvre and Marcuse which consists of a practical and comprehensive view of the right of citizens to be urban dwellers is safeguarded by both International Charters (but mainly by the European text). The Charters add to the general right of being a citizen (Article I. Right to the city, Chapter I) a set of articles in the areas of General Provisions (Chapter I), Civil and Political Rights in the City (Chapter II), Economic, Social and Cultural Rights in the City (Chapter III), Rights relative to Democratic Local Administration (Chapter IV), and Mechanisms for the Implementation of Human Rights in the City (Chapter V).

Taking as a basis the European Charter for the Safeguarding of the Human Rights in the City the author elaborated indicators for the 28 rights included in the Charter. We choose the European Charter instead of the Global Charter-Agenda because of its greater coverage of urban sustainability since this Charter is aimed to be endorsed by European cities, where urban complexity is more developed and therefore it covers a more sophisticated reality of urban sustainability. The indicators we elaborated are intended to be KPI (Key Performance Indicators) of urban sustainability since they measure all rights of the European Charter. Therefore, since smart cities are to safeguard urban sustainability while promoting the development and resilience of communities we can affirm that the proposed indicators together with an agenda for the implementation of sustainability benchmarks in combination with policies for the promotion of development and resilience would be the perfect instrument to make cities smarter. In the last section of this paper, we will discuss further about mechanisms to promote urban development and resilience since this will be next steps and future work of this research.

3   Results: First-of-a-kind Conceptual Framework for the Modelling and Simulation of Urban Sustainability

The European Charter for the Safeguarding of the Human Rights in the City was promoted by a group of cities during preparatory work initiated in Barcelona in 1998, in the framework of the conference "Cities for Human Rights", which commemorated the 50th Anniversary of the Universal Declaration of Human Rights. Hundreds of mayors and political representatives participated in the event and united their voices to call for stronger political acknowledgment of their role as key actors in safeguarding human rights in a highly urbanized world [8]. The final document was approved in Saint Denis in 2000. During the timeframe between the "Barcelona Engagement" (1998) and approval of the European Charter (2000), a wide-ranging and fluent dialogue was established between European cities, civil society representatives, and human rights experts to develop the draft of the European Charter. Since then, a European conference has been held every two years to share the progress made by signatory cities, now numbering more than 400, in the implementation of the European Charter. To date, the following conferences have been organized: 1998, Barcelona (Spain); 2000, Saint Denis (France); 2002, Venice (Italy); 2004, Nuremberg (Germany); 2006, Lyon (France); 2008, Geneva (Switzerland); and, 2010, Tuzla (Bosnia-Herzegovina). It was only after the conference in Geneva (2008) that the most active cities of the network (Barcelona, Saint Denis, Lyon, Geneva, and Nantes) offered the world organization of cities, United Cities and Local Governments (UCLG), through its Committee on Social Inclusion, Participatory Democracy and Human Rights, to entrust the European Charter for a larger promotion [8]. Since 2006 UCLG started elaborating a municipal charter, but with a worldwide scope, the Global Charter-Agenda for Human Rights in the City. UCLG formally presented the Global Charter-Agenda at its World Council in Florence, Italy, which was attended by over 400 mayors from all over the world [9].

As mentioned in the previous section, the European Charter presents a more sophisticated view on the right of citizens as urban dwellers, in comparison with the Global Charter-Agenda, since it needs to address the more developed conditions in Europe. The European Charter includes 28 rights, each corresponding to one article, while the Global Charter-Agenda covers 12 more basic rights presented in 12 articles. The 28 articles of the European Charter are grouped in five Parts or Chapters. For comparability purposes, we did the same exercise as for the standard 37120 and classified the rights of the European Charter into rights safeguarding social, environmental and economical sustainability, as follows:

RIGHTS SAFEGUARDING SOCIAL SUSTAINABILITY:
Part I: General Provisions
– Article I. Right to the City
– Article II. Principle of Equality of Rights and Non-Discrimination
– Article III. Right to Cultural, Linguistic and Religious Freedom
– Article IV. Protection of the most Vulnerable Groups and Citizens
Part II: Civil and Political Rights in the City
– Article VIII. Right of Political Participation
– Article IX. Right of Association, Assembly and Demonstration
– Article X. Protection of Private and Family Life
– Article XI. Right to Information
Part III: Economic, Social and Cultural Rights in the City
– Article XII. General Right to Public Services of Social Protection
– Article XIII. Right to Education
– Article XIV. The Right to Work
– Article XV. Right to Culture
– Article XVI. Right to a Housing
– Article XVII. Right to Health

RIGHTS SAFEGUARDING ENVIRONMENTAL SUSTAINABILITY: (Part III cont')
– Article XVIII. Right to the Environment
– Article XIX. Right to Harmonious City Development
– Article XX. Right to Movement and Tranquility in the City
– Article XXI. Right to Leisure
RIGHTS SAFEGUARDING ECONOMIC SUSTAINABILITY: (Part I and part III cont'):
– Article V. Duty of Solidarity
– Article VI. International Municipal Cooperation
– Article VII. Principle of Subsidiarity
– Article XXII. Consumers' Rights

Part IV: Rights relative to Democratic Local Administration
– Article XXIII. Efficiency of Public Services
– Article XXIV. Principle of Transparency
Part V: Mechanisms for the Implementation of Human Rights in the City
– Article XXV. Local Administration of Justice
– Article XXVI. Police in the City
– Article XXVII. Preventive Measures
– Article XXVIII. Taxation and Budgetary Mechanisms

The percentages of the rights of the European Charter classified in the 3 pillars of urban sustainability are:

Rights safeguarding social sustainability, 50%
Rights safeguarding environmental sustainability, 14%
Rights safeguarding economic sustainability, 36%

From the comparison between the distribution of indicators of the standard 37120 and the distribution of rights of the European Charter we can observe opposite weights regarding social and environmental sustainability and similar importance given to economic sustainability. While environmental sustainability has the higher number of indicators in the standard, it has the lowest number of rights in the European Charter. On the other hand, social sustainability has the lowest number of indicators in the standard but the highest number of rights in the European Charter. Because of the complementarity between the standard 37120 and the Charter, it seems that a combination of both would provide a complete spectrum of indicators covering urban sustainability. The author did this combinatory exercise and it is presented in Figure 1:

Article VI: International Municipal Cooperation

Article VII: Principle of Subsidiarity

Part II : Civil and Political Rights in the City

Article VIII: Right of Political Participation

Article IX: Right of Association, Assembly and Demonstration

Article X: Protection of Private and Family Life

Article XI: Right to Information

	h) 2 The **municipal authorities offer free open and easy access to information.** With this in mind the **learning, facilitation of access to and regular updating of Information Technology skills is to be encouraged.**	No of different municipal communication channels (such as local radio, local newspaper, local TV, etc.) (*)
		No2 of public free wifi coverage
		N° of access points to public Internet
		N° of places for the learning of new technologies per 1000 inhabitants
		(*) If yes, their type and if possible, the number of TV and radio programs, and the number of newspaper supplies printed

Part III : Economic, Social and Cultural Rights in the City

Article XII. General Right to Public Services of Social Protection

	XII.1 The signatory cities consider **social policies** a decisive part of their policies for the **protection of human rights** and they undertake to **guarantee these rights within the limits of their authority.**	m2 of surface of all public facilities belonging to Nat and Loc Gov used for social care purposes per 1000 inhabitants
		% of municipal budget devoted to social policies
		same indicators as art. X.3
	XII.2 Citizens are entitled to **enjoy free and smooth access to general municipal services.** For this reason the signatory cities **oppose the privatisation of personal support services** and undertake to ensure that **good quality basic services are made available in other areas of public life the lowest possible and stable prices.**	N° of financial aids (adapted to N° of financially-enabled families) provided by Nat, Loc Gov and NGOs to pay utilities (**)
		N° of social care personal services (*) provided by Nat, Loc Gov and NGOs over the total of personal services provided
		% of unmet (*) (internalised /outsourced) over the total of city services
	DEFINITIONS: (*) we understand as social care personal services the following ones: home assistance, accompaniment of elderly, help and support to handicapped, service of adapted transport, etc. (**) we understand as utilities the following urban services: water provision, sewage, waste collection, public facilities electricity, gas provision, etc.	% of budget of Loc. Gov and NGOs devoted to the most disadvantaged for social social services including stigmatization and equality policies
	XII.3 The cities commit themselves to **develop social policies, aimed expressly at the most disadvantaged,** which would **reject any form of exclusion** to **champion human dignity and equal rights.**	% of population below poverty threshold

Article XIII. Right to Education

13.2 % of students completing primary education: survival rate 13.3 % of students completing secondary education: survival rate 13.4 Student/teacher ratio 13.6 % of school-aged population enrolled in schools 13.7 Number of higher education degrees per 100 000 population	XIII.1 Citizens have the right to an education. The municipal authorities **provide access to basic education for all children** of school age. They encourage a considered education which corresponds with the values of democracy	m2 surface of public educational facilities per 1000 (children or profiles at educational level)
		N° of municipal educational officers over the total of municipal officers
		N° of school -age population enrolled in public educational facilities over the total of school-age population (*)
		N° of non school-age population enrolled in public educational facilities over the total of non school-age population (*)
	TO CONSIDER: (*) mandatory school-age may vary between countries. In Spain is up to 16.	N° of languages in which a traditional and cultural offer is admired and reproduced
		see related art. X.4
	XIII.2 The cities **contribute to social integration and be multi-culture** by **making available to everyone** without distinction public spaces, as well as spaces in further education establishments: schools and cultural centres.	N° of kids of foreign families attending well-consolidating classrooms over the total of kids of foreign families
		see related art. X.5
		N° of kids of foreign families learning in public centres over the total of school-age population (*)
		N° of adults and youth of foreign families learning in public centres over the total of non-school-age population (*)
	XIII.3 The municipal authorities **raise public consciousness through education,** particularly with regard to the **struggle against sexism, racism, xenophobia and discrimination.**	% of budget of Loc Gov and NGOs devoted to the struggle against sexism, racism, xenophobia and discrimination

Article XIV. The Right to Work

14.4 Jobs/housing ratio 6.1 City's unemployment rate 6.3 % of city population living in poverty 6.4 % of persons in full-time employment 6.5 Youth unemployment rate 6.6 Number of businesses per 100 000 population 6.7 Number of new patents per 100 000 population per year 6.2 Commercial/industrial Assessment as a percentage of total assessment	XIV.1 Citizens have a right to **enjoy sufficient financial recompense for a worthwhile occupation** which guarantees a satisfactory quality of life	% unemployment over the total of working age population (*)
		% inactivity (*) over the total of working-age population (*)
		% part-time contracts over the total of contracts
		% temporary contracts over the total of contracts
		Average salary
		% self employment over the total of working age population (*)
		% public employees over the total of employees
		% of employees earning minimum wage over the total of employees
	XIV.2 The municipal authorities strive towards the **creation of full employment** within their possibilities. To make the right towards an achievable goal, the signatory cities encourage the **matching of supply and demand in the job market,** and resolve to **promote further education and the updating of skills in this work force.** They **develop measures with the unemployed in mind.**	% of budget of Loc Gov and NGOs devoted to help unemployed (education, orientation, advice, insertion, support to entrepreneurs, etc.)
		% of budget of Loc Gov and NGOs devoted to help employed (promotion, training, recertification, self-promotion, specialisation, etc.)
		N° of enriched in Loc Gov assoc.NGOs attending labor causes over the total of personnel in Loc Gov and NGOs
	XIV.3 The signatory cities undertake **not to sign any municipal contract without including clauses - rejecting child labour** and - **illegal employment,** whether involving nationals or foreign workers.	% of existing contracts signed after (Jiang the Charter, specifically including these clauses)
		existence of mechanisms to ensure compliance and institutions follow hardship and labour integration laws (*)
	XIV.4 The municipal authorities **develop mechanisms,** in collaboration with other labor institutions and companies, **- to ensure equality for everyone at work,** and **- to prevent any discrimination** on the grounds of nationality, gender, sexual orientation, age or disability in matters of salary, working conditions, right of participation, professional promotion or wrongful dismissal. They **promote equal opportunities in the workplace for women** by studying various structures available to them such as fixturies, **and for the handicapped** by the installation of appropriate amenities and equipment.	N° of places in leader-garden owned by Loc Gov and by NGOs over the total of infants-aged population
		N° of complaints received for discrimination in the access to work and services (*) over 1000 inhabitants and per year
		N° of handicapped individuals working in the City Council over the total of City Council employees
		N° of work placements for disables in Special Work Centres owned by Loc Gov and by NGOs over the total of working-age population
		N° of complaints received by the Protector of Citizens (Ombudsman) for labor discriminative reasons over the total of complaints received
	DEFINITIONS: in this category we understand mechanisms for fair wages, fair work conditions, non-discrimination, right to participation and opinion, protection against dismissal, etc. (*) e.g. in Spain, companies over 50 employees, for each %B employees have to hire 1 handicapped employee	N° of consultants received by the Protector of Citizens (Ombudsman) for labor discriminative reasons over the total of population

Article XV. Right to Culture

	XV.1 The citizens have a **right to culture in all its expressions, forms and manifestations.**	% of municipal budget devoted to actors and cultural expressions
		% of municipal budget devoted to subsidies for cultural NGOs
	XV.2 The municipal authorities, in co-operation with cultural associations and the private sector, **promote the development of urban cultural life** with a respect for diversity. **Suitable public spaces are at the disposal of citizens to use for cultural and social activities** as they see fit, with a sole condition applying to all	m2 of surface of museums, exhibition centers and cultural spaces, in general owned by Nat Gov and by Loc Gov per 1000 inhabitants (*)
		m2 of surface of libraries and reading centers owned by Nat Gov and by Loc Gov per 1000 inhabitants (*)
		m2 of surface of concert halls and auditoriums owned by Nat Gov and by Loc Gov per 1000 inhabitants (*)
		m2 of surface in municipal buildings at disposal of cultural associations (*)
		IN ADDITION (*) if available, include the proportion of those listed surfaces which is adapted to personal and physical disability

Article XVI. Right to a Housing

15.1 % of city population living in slums
10.3 Areal size of informal settlements as a percentage of city area
7.1 Total residential electrical use per capita (kWh/year)
7.3 Percentage of city population with authorised electrical service
15.2 Number of homeless per 100 000 population
14.3 Crimes against property per 100 000 population
15.3 % of households that exist without registered legal titles

XVI.1 All citizens have the right to **proper, safe and healthy housing.**

XVI.2 The municipal authorities endeavour **to ensure appropriate supply of housing and infrastructure for all their inhabitants,** without exception, within the limits of their financial resources. They must include measures encompassing the minorities which will guarantee their safety and dignity, as well as safeguard structures for women who are victims of violence or who are attempting to escape from prostitution.

XVI.3 The municipal authorities guarantee the **right of nomads to stay in the city** under conditions which are **compatible with human dignity.**

Article XVII. Right to Health

12.1 Average life expectancy
12.2 Number of in-patient hospital beds per 100 000 population
12.3 Number of physicians per 100 000 population
12.4 Under age five mortality per 1 000 live births
12.5 Number of nursing and midwifery personnel per 100 000 population
12.6 Number of mental health practitioners per 100 000 population
12.7 Suicide rate per 100 000 population

XVII.1 The municipal authorities encourage **equal access for all citizens to medical and preventive health care.**

XVII.2 The authorities view, through their actions in the spheres of **economy, culture and social town planning,** contribute in a general way to the **promotion of health for all their inhabitants,** with their active participation.

Article XVIII. Right to the Environment

16.4 % of the city's solid waste that is disposed of in a sanitary landfill
16.5 % of the city's solid waste that is disposed of in an incinerator
16.6 % of the city's solid waste that is burned openly
16.7 % of the city's solid waste that is disposed of in an open dump
16.8 % of the city's solid waste that is disposed of by other means
16.9 Hazardous Waste Generation per capita (tonnes)
16.10 % of the city's hazardous waste that is recycled
9.1 Fine Particulate Matter (PM 2.5) Concentration
9.2 Particulate Matter (PM10) Concentration
9.3 Greenhouse gas emissions measured in tonnes per capita
9.4 NO2 (nitrogen dioxide) concentration
9.5 SO2 (sulphur dioxide) concentration
9.6 O3 (Ozone) concentration
9.8 Percentage change in number of native species

XVIII.1 The citizens have a **right to a healthy environment which seeks a sustainable balance between industrial development and the environment.**

XVIII.2 The cities and the municipal authorities take precautionary measures by **creating policies of pollution prevention** (including that of noise pollution). They promote energy saving, recycling of re-usage and guarantee refuse disposals. They extend and protect the green zones of the cities.

XVIII.3 The municipal authorities put into practice all the actions necessary for citizens to appreciate and **care for the countryside surrounding their city.** At the same time they **enable the citizens to be consulted over any changes affecting this landscape.**

XVIII.4 The municipal authorities develop specific **teaching strategies for presenting the theme of environmental protection, particularly aimed at children.**

Article XIX. Right to Harmonious City Development

18.1 Green area (hectares) per 100 000 population
18.3 Annual number of trees planted per 100 000 population

XIX.1 The citizens have a right to an **ordered town planning development** which guarantees a **harmonious relationship between residential areas, public services and amenities, and green areas.**

XIX.2 The municipal authorities, **with citizen participation, deliver a system of town planning** and administration which sustains a **balance between urban development and the environment.**

		% of budget devoted to the rehabilitation and recovery of architectural and historical heritage
	XIX.3. In this context they pledge to **respect the natural, historic, architectural, cultural and artistic heritage of the cities** by actively seeking the **restoration and reuse of existing buildings.**	% of budget invested in the rehabilitation and recovery of natural heritage
		% of budget devoted to the rehabilitation and recovery of cultural and artistic heritage

	Article XX. Right to Movement and Tranquillity in the City	
10.1 Km of high capacity public transport system per 100 000 population		% of architecture barrier-free of pedestrian sidewalks
10.2 Km of light passenger transport system per 100 000 population		km² of streets served by public transport (*)
10.3 Annual number of public transport trips per capita	XX.1. The local authorities recognise the **right of the citizens to have a transport system in keeping with the desired tranquillity of the city.** To this end they develop a **public transport system accessible to all and incorporating a system of city and inter city links. They manage motor traffic and see to it that it runs smoothly and in harmony with the environment.**	% of public transports stops per km² of street served by public transport (*)
10.4 Number of personal automobiles per capita		% of interested exchangers (urban and inter-urban) per km² of street or motorway served by public transport
10.5 Modal split (% of commuters using a travel mode to work other than a personal vehicle)		Time average frequency of service per each type of public transport
10.6 Number of two-wheel motorized vehicles per capita		Time average of travel per km and type of public transport
10.7 Km of bicycle paths and lanes per 100 000 population		Time average of journey per type of public transport
10.8 Transportation fatalities per 100 000 population		(*) provide distribution per day and night services
10.9 Commercial air connectivity (number of non-stop commercial air destinations)	XX.2. The municipalities approve, control the **emission of all types of noise and vibrations. They define these permanent pedestrian areas and those restricted to certain times of the day and encourage the use of environmentally friendly vehicles.**	% of electric vehicles in public fleet over the total of vehicles in public fleet
6.7 Noise Pollution		% of non-electric vehicles of non-carbon fuel over the total of vehicles in public fleet
		% of electric recharging points per km²
		% of public bikes per 1000 inhabitants
	XX.3. The legislator does undertake to **set aside resources to help fulfil these rights,** calling upon where appropriate, financial cooperation between local authorities, private business and society in general	existence of PPP to support the implantation of electric vehicle

	Article XXI. Right to Leisure	
13.1 Square meters of public indoor recreation space per capita	XXI.1. They also recognise the **right of citizens to leisure time**	% average of working hours per week in full time contracts (*)
13.2 Square meters of public outdoor recreation space per capita		m2 of play centres per 1000 inhab.
		m2 of youth centres and/or jointly per 1000 inhab
	XXI.2. The local authorities guarantee **quality leisure spaces for all children without discrimination.**	% of municipal budget devoted to after school activities in public schools over the total of budget devoted to public schools
		% of students participating in after school activities over the total of students in public schools
		rate shaded art X.5.
	XXI.3. The **municipal authorities facilitate participation in sports activities, and provide accessible facilities for all.**	m2 of public sports facilities per 1000 inhab
		maximum distance to a sports facility
		daily average of users of municipal sports facilities over 1000 inhab.
		daily average of opening hours in municipal sports facilities
	XXI.4. The municipal authorities encourage **sustainable tourism** and promote the **balance between city tourism** on the one hand **and the social and ecological wellbeing** of the citizens on the other	Annual taxh collected through the tourists tax
		% of tourists per year and 1000 inhabitants
		% of overnight stays per m² of tourists air d'year

	Article XXII. Consumers' Rights	
	XXII. Within the limits of their authority, the **cities uphold consumers' rights.** To this end, and with relevance to foodstuffs, they **guarantee the supervision of weights and measures,** the **quality and composition of products** and the **accuracy of information with regard to sell-by dates.**	% of complaints filed in consumer care offices and police
		% of complaints filed in consumer care offices per 1000 inhabitants
		% of consumer complaints filed in citizen care offices per 1000 inhabitants
		average time to resolve complaints for both consumer care and citizen care offices

	Part IV : Rights relative to Democratic Local Administration	
	Article XXIII. Efficiency of Public Services	
7.5 Electrical use per cap. (kWh/Y)	XXIII.1. The local authorities ensure the **efficiency of public services,** and that these to the requirement of users. Every measure will be taken to **avoid any form of discrimination or abuse.**	existence of quality plans/checks in the delivery of services in facilities (*) (focusing in guaranteeing universal access and in fair conditions)
7.3 Energy cons. of public buildings as a % of city cons.		existence of a quality plans/checks in the delivery of utilities (**), (focusing in guaranteeing universal access and in fair conditions)
7.4 % of total energy derived from renewable sources, as a share of the city's total energy cons.		
7.6 Average number of electrical interruptions per customer/year		
7.7 Average length of electrical interruptions (in hours)	DEFINITIONS: (*) we understand as services in facilities the following ones: education, sports, health, social care, culture, administration, etc. (**) we understand as urban services the following: water provision, sewage, waste collection, public facilities, electricity, gas provision, etc	
21.6 Average annual hours of water service interruption per household		
21.7 % of water loss		
21.1 % of city population with potable water supply service		
21.2 % of city pop. with sustainable access to an improved water source		
21.3 % of pop. with access to improved sanitation		
16.2 Total collected municipal solid waste per cap.		
16.3 % of city's solid waste that is recycled		existence of consolidation mechanisms to assess the performance of service delivery in facilities (*)
20.1 % of city pop. served by wastewater collection		execution of consolidation mechanisms to assess the performance of urban service (**) delivery
20.3 % of the city's wastewater that has received no treatment	XXIII.2. The municipal authorities will have a their disposal various **means of evaluating their activities** and will take on board the results of this evaluation	% of complaints filed regarding services offered in municipal facilities (*) per 1000 inhab. and year
20.2 % of the city's wastewater receiving primary treatment		% of complaints filed regarding urban services (**) per 1000 inhab. and year
20.4 % of the city's wastewater receiving secondary treatment		% of quality control officers to oversee the quality of the delivery of services in public facilities (*) and of urban services (**)
20.5 % of the city's wastewater receiving tertiary treatment		
21.4 Total domestic water consumption per cap. (L/day)		
21.5 Total water consumption per capita (L/day)		
10.1.4 Firefighters/100 000 pop.		
10.2 # fire related deaths per 100 000 pop.		
10.3 # natural disaster –related deaths per 100 000 pop.	DEFINITIONS: (*) we understand as services in facilities the following ones: education, sports, health, social care, culture, administration, etc. (**) we understand as urban services the following: water provision, sewage, waste collection, public facilities, electricity, gas provision, etc	
10.4 # volunteer and part-time firefighters per 100 000 pop.		
10.6 Response time for emergency response services from initial call		
10.6 Response time for fire department from initial call		
14.4 Response time for police department from initial call		

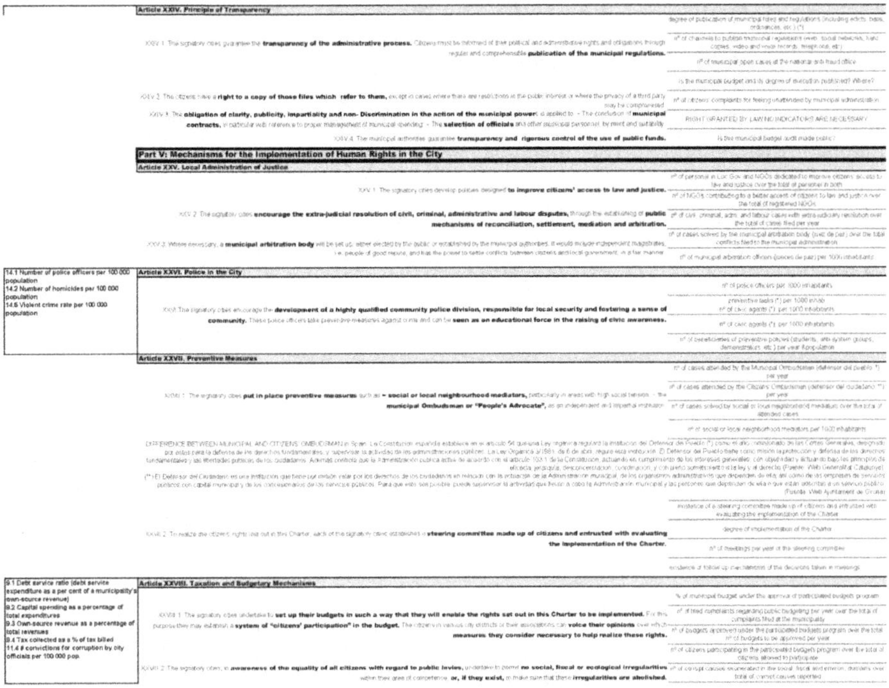

Fig. 1. Set of indicators measuring urban sustainability, which results from the combination and integration in one single framework the indicators measuring urban sustainability in the standard 37120 and the indicators measuring the rights safeguarding urban sustainability in the European Charter for the Safeguarding of the Human Rights in the City

ISO 37120 indicators are shown in the left side column while indicators elaborated for the EU Charter on rights are presented in the right side column

To the best of our knowledge, this set of indicators is the first and a novel attempt to build a comprehensive tool to measure urban sustainability and constitutes the basis of a conceptual framework to model and simulate urban sustainability. By testing the framework in different cities, a repository of sustainability benchmarks could be created and different sustainability situations of cities, modelled. Cities being measured under this framework would be first classified according to the basic traits and features highly affecting sustainability performance (e.g. a low-density city cannot perform at the same level of environmental sustainability as a high-density city). Next, for cities willing to improve their sustainability, higher sustainability performance situations could be simulated using the repository of sustainability benchmarks.

4   Conclusions and Next Steps: Towards a Framework for Making Cities Smarter

In this paper we presented a first-of-a-kind framework to conceptualize, model and simulate sustainability. The practical implementation of simulated sustainability benchmarks in combination with policies promoting development and resilience of cities are the perfect instrument to make cities smarter. Therefore, the framework presented in this paper is midway through this final instrument to help cities become smarter.

VI. Right to Accessible Public Services

1. All city inhabitants have the right to a socially and economically inclusive city and, to this end, to have access to nearby basic social services in acceptable technical and financial conditions.

2. The city creates, or promotes the creation of, quality and non-discriminatory public services that guarantee the following minimum to all its inhabitants: training, access to health, housing, energy, water, sanitation and sufficient food, under the terms outlined in this Charter-Agenda.

   Particularly, in countries with rapid urban growth, the cities take urgent measures to improve the quality of life and opportunities of its inhabitants, especially those of lesser means as well as people with disabilities.

   The city is concerned with the protection of the rights of the elderly and encourages solidarity among generations.

   The city takes the necessary measures to ensure a fair distribution of public services over its entire territory, in a decentralized manner.

3. City inhabitants use social services responsibly.

Suggested Action Plan

Short-term

a) Establishment of a social participation system in the design and monitoring of services delivery, especially with respect to quality, fee setting and front office operations. The participation system should focus in particular on the poorest neighborhoods and the most vulnerable groups in the city.

b) Immediate abolition, where existing, of the legal, administrative and procedural requirements that bind the provision of basic public services to the legal status of city inhabitants.

c) Review of local procedures and regulatory provisions, to provide low income people with greater access to basic services.

d) Establish an equitable system of local taxes and fees that takes into account the people's incomes and their use of public services. Provide information to users on the cost of public services and the source of the funds paying for them.

e) Careful monitoring of the needs of transients and other nomadic populations in the area of basic public services.

Mid-term

a) Establish efficient measures to make sure that private sector players who manage social or public interest services respect the rights guaranteed under this Charter- Agenda fully, and without discrimination. Contracts and concessions of the city shall clearly set forth its commitment to human rights.

b) Adoption of measures to ensure that public services report to the level of government closest to the population, with the participation of city inhabitants in their management and supervision.

c) Encourage access of the elderly to all public services and to life of the city.

Fig. 2. Example of Article VI, namely Right to Accessible Public Services, and its implementation agenda or action plan

For that instrument to be completed, a policy agenda promoting resilience and development of cities, in good respect with urban sustainability, should be added. The next steps of our research are targeted at completing this instrument which has already been initiated with the framework presented in this paper. To that end we will use the Global Charter-Agenda for Human Rights in the City as a reference. In the Global Charter-Agenda each right has an agenda or action plan to help its implementation. In figure 2 we show an example of a given right and its agenda.

The challenge for an instrument aimed at helping cities to become smarter is to include an agenda or action plan to promote the development and resilience of cities on the top of the sustainability pillars presented in this paper. Resilience and development need to be experimented in parallel and in good harmony with sustainability, otherwise they can become antagonists. And, considering that the promotion of urban development and resilience has an strategic (and political) component, which varies from city to city since that promotion materializes through individual city policies, we can affirm that the achievement of urban smartness in the terms of the international definition presented in this paper poses a great challenge to the research community: the three engines of urban smartness namely, sustainability, development and resilience to walk together and at the same pace.

Acknowledgments. This research is consuming a huge amount of work and dedication. The elaboration of indicators would not have been possible without the valuable contributions and help of this team, who disinterestedly met with me once a week during months, always active, helpful, and bringing lots of great ideas: Dr. Pere Soler (University of Girona, director the Director of the Joint Master's Program in Youth and Society (MIJS); Dr. Imma Boada (University of Girona, director of the Institute of Informatics and its Applications); Dr. Joaquim Meléndez (University of Girona, director of the Doctoral Program in Technology); Ms. Anna Serra (Lawyer at Red Cross Girona); Mr. Fran Quirós (Responsible of Cooperation Programs at Charity Girona); Mr. Lluís Puigdemont (Responsible of the Rights Department at Charity Girona); Dr. Montse Aulinas (Project Manager at Grup Fundació Ramon Noguera); Ms. Yolanda García (Responsible of social programs at Grup Fundació Ramon Noguera).

Special thanks to Mr. Mark Segal, international consultant on democratization issues, for his valuable comments and general editing support during the elaboration of this research.

References

1. Jenks, M., Jones, C. (eds.): Dimensions of the Sustainable City. Springer Science+ Business Media, Future City, 2 (2010)
2. Florida, R.: The rise of the creative class. Basic Books. Perseus Books Group, New York (2002)
3. Ishida, T., Isbister, K. (Eds.): Digital Cities: Technologies, Experiences, and Future Perspectives. Lecture Notes in Computer Science, vol. 1765. Springer Science+Business Media (2000)
4. Carrillo, F.J.: Knowledge Cities: Approaches, Experiences and Perspectives. Elsevier: Butterworth Heinemann (2005)

5. International Standards Organization 37120 Sustainable Development and Resilience. Indicators for City Services and Quality of Life. Standard briefing note and outline (available on-line and standard on purchase) [WWW document] (2014). URL http://www.iso.org/iso/catalogue_detail?csnumber=62436 (last accessed: 18/12/2014)
6. International Electrotechnical Commission. White paper. Orchestrating Infrastructure Smart Cities [WWW document] (2014). URL http://www.iec.ch/whitepaper/smartcities/?ref=extfooter (accessed: 18 December 2014)
7. United Nations, World Commission on Environment and Development: Our Common Future. Oxford University Press, Oxford (1987)
8. United Cities and Local Governments, Committee on Social Inclusion, Participatory Democracy and Human Rights UCLG- CISDP. The European Charter for the Safeguarding of the Human Rights in the City. [WWW document] (1998). URL http://www.uclg-cisdp.org/en/right-to-the-city/european-charter (accessed 18 December 2014)
9. United Cities and Local Governments, Committee on Social Inclusion, Participatory Democracy and Human Rights. Global Charter-Agenda for human Rights in the City. [WWW document] (2006). URL http://www.uclg-cisdp.org/en/right-to-the-city/world-charter-agenda (accessed 18 December 2014)
10. United Nations, General Assembly. Universal Declaration of Human Rights. [WWW document] (1948). URL http://www.un.org/en/documents/udhr/ (accessed 18 December 2014)
11. Lefebvre, H.: Writings on Cities. Blackwell, Oxford (1996). edited and translated by Kofman, E. and Lebas, E. (1968)
12. Lefebre, H.: Introduction to Modernity. Verso, London (1995). Translated by J. Moore
13. Marcuse, P.: The Politics of Public Space/The Right to the City: Social Justice and the Fight for Public Space. Journal of the American Planning Association 73(1) (2007)
14. Marcuse, P., Mayer, M., Brenner, N. (eds.): Cities for People, Not for Profit: Critical Urban Theory and the Right to the City. Routledge, Abingdon/New York (2011)
15. United Nations, Office of the High Commissioner for Human Rights. International Covenant on Civil and Political Rights [WWW document] (1966). URL http://www.ohchr.org/en/professionalinterest/pages/ccpr.aspx (accessed 18 December 2014)
16. United Nations, Office of the High Commissioner for Human Rights. International Covenant on Economic, Social and Cultural Rights [WWW document] (1966). URL http://www.ohchr.org/EN/ProfessionalInterest/Pages/CESCR.aspx (accessed 18 December 2014)
17. United Nations, General Assembly (2000) United Nations Millennium Declaration [WWW document]. URL http://www.un.org/millennium/declaration/ares552e.htm (accessed 18 December 2014)
18. United Nations, Office of the High Commissioner for Human Rights. Declaration for the 60th Anniversary of the United Nations [WWW document] (2005). URL http://www.ohchr.org/en/udhr/pages/60udhr.aspx (accessed 18 December 2014)
19. United Nations, Office of the High Commissioner for Human Rights. Vienna Declaration and Programme of Action [WWW document] (1993). URL http://www.ohchr.org/EN/ProfessionalInterest/Pages/Vienna.aspx (accessed 18 December 2014)
20. Council of Europe. Convention for the Protection of Human Rights and Fundamental Freedoms. [WWW document] (1950 and 2010). URL http://conventions.coe.int/treaty/en/treaties/html/005.htm (accessed 18 December 2014)
21. Council of Europe. European Social Charter. [WWW document] (1961 and 1996). URL http://www.coe.int/T/DGHL/Monitoring/SocialCharter/ (accessed 18 December 2014)

Wave Energy Potential in the Mediterranean Sea: Design and Development of DSS-WebGIS "Waves Energy"

Maurizio Pollino[✉], Emanuela Caiaffa, Adriana Carillo,
Luigi La Porta, and Gianmaria Sannino

ENEA National Agency for New Technologies,
Energy and Sustainable Economic Development,
UTMEA Energy and Environmental Modelling Technical Unit,
C.R. Casaccia, Via Anguillarese, 301, 00123 Rome, Italy
{maurizio.pollino,emanuela.caiaffa,
adriana.carillo,luigi.laporta,gianmaria.sannino}@enea.it

Abstract. GIS technologies are able to provide useful tools for estimating the energy resource from the sea waves, assessing whether this energy is exploitable and evaluating possible environmental impacts. The idea to convert the energy associated with the marine wave motion (both off-shore and coastal) into exploitable electrical energy is not new and over time several projects have been developed, aiming at the implementation of devices for electrical energy generation from the sea. However, compared to other well-established renewable sources (such as wind, solar or biomass), the exploitation of the tidal power, is currently only in prototype form. Nevertheless, it has shown very promising potentiality, as also emerges from the activities currently carried-out by public institutions and private stakeholders.

This paper describes an approach, based on Geomatics, developed to assess the marine energy resource, evaluating if it is possible to exploit such resource and estimating also environmental and socio-economic impacts in blue water and/or in the seaboard. In this framework, a DSS-WebGIS application has been developed (called "Waves Energy"), as tool for displaying and sharing geospatial data and maps related to the potential use of energy from the sea, as well as a valuable support for new installations planning, forecasting systems and existing infrastructure management.

Keywords: GIS · DSS · Renewable Energy Sources · WebGIS · Wave Energy

1 Introduction and Objectives of the Work

In recent years, among the different disciplines that regulate the alternative energy sector, a significant growth in the development of new technologies for exploiting potential energy from sea waves has been recorded. As other typologies of Renewable Energy Sources (RES), wave energy has the potential to generate considerable amounts of clean and renewable electricity and predictable power, as an additional source to be exploited in order to cope with the growing energy demand [1].

© Springer International Publishing Switzerland 2015
O. Gervasi et al. (Eds.): ICCSA 2015, Part III, LNCS 9157, pp. 495–510, 2015.
DOI: 10.1007/978-3-319-21470-2_36

Numerous research projects have been carried-out in recent years, especially focused on the energy production from waves and tidal currents at various scales and by means the design and implementation of specific devices, which in most cases are currently at the prototype stage. Just to mention the main ones, various studies at global and regional scales have been conducted in the west coasts of North America [2] and in Europe [3, 4] (for example, in Scotland, Ireland, Portugal and Spain): those experimental activities are meaningful in terms of the large amount of energy to be potentially generated. Although many of the conversion devices have shown their potential exploitability, they haven't yet reached an operational stage, due to problematic operating conditions. However, considering the growing costs of energy produced from conventional fuels, wave energy can be considered economically achievable in the future [5].

To extract energy from waves, there are many typologies of conversion devices, based on different technologies depending on the local marine conditions. For example, Pelamis (developed by Pelamis Wave Power[1]) is an attenuator-type device used where wave heights are high, such as the region offshore of Ireland and Scotland [6]. Moreover, Pelamis devices are currently operative in the Rance Tidal Power Station (North-West of France, installed capacity: 240 MW) and in the Shihwa Tidal Power Station (South Korea, installed capacity: 254 MW). Alternatively, in areas where waves with longer periods are prevalent (typically, for swell) [7], terminator-type devices are used, such as the oscillating water column device produced by Energetech[2].

To this end, it is fundamental to support public planners and private stakeholders in their decision processes, by providing methodologies and tools able to evaluate the capability of specific sites to produce energy and, contextually, to analyse the potential impacts on the surrounding ecosystems and activities [8]. In fact, despite the intrinsic characteristics of cleanness of such typology of RES, wave energy conversion facilities could conflict with existing activities (e.g., navigation, fishery, tourism, etc.) or ecosystems (e.g., natural habitats, marine protected areas, etc.) [9].

In this paper we present and describe the research activities carried-out by ENEA (the Italian National Agency for New Technologies, Energy and Sustainable Economic Development, "Energy and Environmental Modelling" Technical Unit) in order to evaluate the potential energy produced by waves in the Mediterranean Sea and, in particular, along the Italian coast, through the use of oceanographic numerical models and Geomatics techniques (GIS spatial analyses and Web-Mapping) [10, 11]. The main goal was to estimate the energy resources available in Mediterranean Sea and, in the near future, to support the identification of the most suitable sites to host the installation of power generators.

In particular Geomatics methodologies and tools are very useful to assess how and where the energy resource from the sea is exploitable, supporting the spatial planning related to the installation of conversion devices. Such approach requires the development of ad hoc applications (e.g., Decision Support Systems, DSS) allowing a wide-ranging analysis of environmental, technological and socio-economic features that are situated in the areas where production facilities could potentially be placed. To this

---

[1] http://www.pelamiswave.com/

[2] http://www.energetech.com/

end, there are different parameters and information to be considered, such as the wave energy flux exploitable, the characteristics of conversion devices, the cost-effectiveness of the facility installation, the interaction with the surrounding environment, possible limitations and specific compatibility issues, etc. [5].

Thus, a comprehensive approach is required to assess the impacts of planned installations, considering both the blue water conditions and the coastal environment characteristics.

2   Materials and Methods

2.1   Marine Waves Energetic Potential Evaluation

For of the purposes of the present study, the Mediterranean Sea has been considered, taking into account the followings three types of natural phenomena:

- tidal currents;
- surface waves, generated by wind;
- underwater waves, produced by the interaction between current and seabed.

Their possible exploitation, from the energetic point of view, is closely connected to detailed knowledge of the physical quantities related, such as: the speed of the current, the height of the waves, the intensity of the tides and, subsequently, the technological development of suitable devices for the conversion of this kind of energy [12, 13, 14, 15, 16].

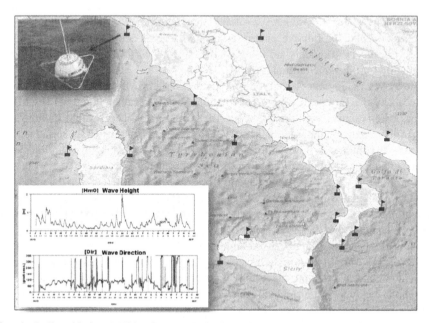

Fig. 1. Italian National Tide Gauge Network (RON stations) by ISPRA (http://www. idromare.it)

Almost all information, related to the waves energetic potential along the Italian coastline, is currently evaluated on the basis of data recorded by 15 mareographs of the "Italian National Tide Gauge Network" ("Rete Ondametrica Nazionale" - RON and "Rete Mareografica Nazionale" - RMN), active since 1989 and managed by the Italian Institute for Environmental Protection and Research (ISPRA). In particular, RON stations are distributed along the Italian coast as shown in Figure 1.

In each station (constituted by an instrumented buoy) is installed a localization system that uses the Advanced Research and Global Observation Satellite (ARGOS) for the continuous control of the position. Each buoy, anchored at a depth of about 100 meters, by following the movement of the water surface allows determining wave height and direction.

It is possible to calculate the available energy flux per unit crest [17, 18, 19] by using the following equation:

$$P = \frac{\rho g^2}{64\pi} \ T_e H_s^2 \ \ [kW/m] \tag{1}$$

where:

ρ represents the sea water density (assumed to be 1025 kg/m^3)
g is gravity (9.8 m·s^{-2})
T_e is the wave significant period (s)
H_s is the significant wave height (m)

The data obtained from the RON sensors, although they constitute a valuable source of information, do not provide adequate spatial coverage for the identification of suitable sites for wave energy exploitation. Therefore, for the description of the wave motion and the relative energy associated with it and eventually exploitable, high-level numerical models are used, such as the WAM (WAve prediction Model), specifically for the wave motion [20].

The WAM model was forced with the dataset obtained from the analysis of wind produced by the European Centre for Medium-Range Weather Forecasts (ECMWF) having a horizontal resolution of about 40 km [21, 22]. The results of the WAM model (numerical simulations), unlike the data recorded from the RON sensors, allow to obtain a continuous information layer, describing the spatial distribution of the physical quantity simulated throughout the Mediterranean Sea.

First of all, a series of seasonal-average maps of the significant wave height were produced, calculated on the basis of data obtained from the above described WAM simulations: such climatology information was extracted for the 1980-2004 time span (25 years) and for the 2001-2010 time span (10 years).

Using these data, it has been possible to assess [23] the potential energy of the waves (energy flux) in the Mediterranean Sea (Figure 2), by the equation shown in the formula (1) and to calculate the potential energy producibility index from waves (an example is depicted in the right side of Figure 3). The same information has been also produced by subdividing it into quarterly intervals (Jan-Mar, Apr-Jun, Jul-Sept and Oct-Dec) (Figure4).

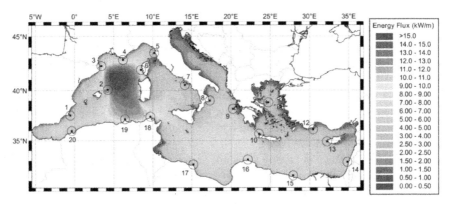

Fig. 2. Energy Flux distribution: distribution of average power per unit crest in the Mediterranean between 2001 and 2010, calculated with WAM model at 1/16° resolution

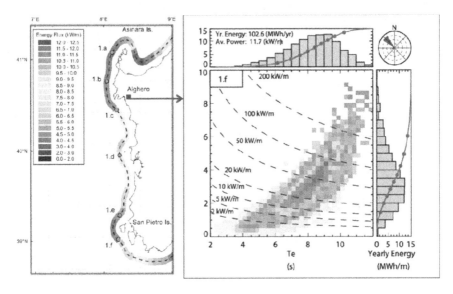

Fig. 3. On the left: distribution of average wave power flux per unit crest (2001-2010) on western Sardinia coastline (Italy). Values are calculated on a line located 12 km off the coast. On the right: spatial variability of the distribution of wave power among heights, periods and directions (energy producibility index) for the Alghero site.

For instance, considering the map reported in Figure 2 and focusing on the Italian coast, the highest values of energy potential are located along the west coast of Sardinia and, to a small degree, the southwest coast of Sicily.

In parallel, the operative model is also daily exploited to produce five days of hourly forecasts [24, 25, 26] of the Mediterranean Sea waves (120 forecasts in total). In this case, the WAM model processes the whole Mediterranean Sea at 1/32° resolution, which is about 3 km (made of 1333x501 surface points). Since the model represents each geographic point with a 32x36 matrix of spectral values, we have WAM makes its computations on about 769 million points, executing 120 times over each geographic point to cover the period of 5 days with a time step of one hour. To achieve these goals, an UV2000 computer (Silicon Graphics, 128 cores and 512 GB RAM) is used: 96 cores are shielded from the system and dedicated to the execution of the model. WAM is started soon after the preliminary steps, getting input data from the SKIRON system (developed by the Atmospheric Modelling and Weather Forecasting Group, University of Athens, Greece) [27] and pre-processing them. WAM, using 96 cores at 100%, takes about one hour to produce its 120 forecasts. Data daily produced is about 25 GB by WAM.

Then, the subsequent forecasting steps are performed by means the SWAN third-generation wave model (developed at Delft University of Technology) [28, 29]. The SWAN zooms into ten sub-areas (at 1/200° resolution, which is about 800 m) of particular interest for their characteristics. For example, for the Sardinia island zoom, SWAN makes its computations on about 93 million points. Also this model executes 120 times over each geographic point to cover the 5-days period with hourly steps. At the end of WAM runs, ten executions of SWAN are launched, each on a zoomed area, in such a number of cores and order to maximize the usage of the 96 cores while minimizing the wall clock time. SWAN, using 48 cores at 100%, takes about 40 minutes to produce its 120 forecasts for each sub-area considered.

After WAM and SWAN runs are finished, is then executed a plotting procedure that produces, for each physical quantity considered (energy flux in kW/m, wave height, wave period and wave direction) [30], all the files used in the following GIS analyses and available by the WebGIS application.

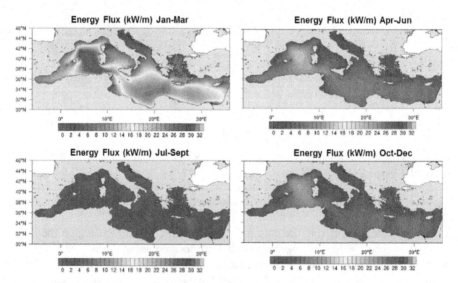

Fig. 4. Seasonal Energy Flux distribution (2001-2010 period), calculated with WAM model at 1/16° resolution

2.2   GIS Analysis

The results obtained from the WAM Oceanographic model are stored as NetCDF (Network Common Data Form) files. Through appropriate procedures, supported by Geomatics tools, the NetCDF files are converted in a GIS compatible format and organised into a geospatial database for the further processing steps. Such spatial processing allows to combine the layers containing marine energetic potential data with other types of geospatial information, which are required to assess if, how and where the exploitation of marine energy is possible. In fact, the choice of the most suitable zones for wave energy production must be made by considering, in addition to the available energy, also a set of specific environmental and socio-economic parameters, which may constitute an obstacle to the installation of the devices.

Then, from the gridded map of energy for the whole Mediterranean Sea was extracted an additional information layer, describing the energy values along the 8,000 km of Italian coastline and calculated at 12 km from the shoreline (Figure 5).

Nevertheless, detecting an area along the coast characterised by high energy potential, is not a sufficient criteria to plan the exploitation of such resource. In fact, as stated before, it is also necessary to combine energy potential values by overlaying with other types of data, such as, for example: the position of main commercial ports and hubs, marine traffic routes (large vessels and main ports), maritime parks and protected areas (Figures 6). In such a way, it is possible to perform a more accurate and comprehensive assessment of the most suitable areas for future installations.

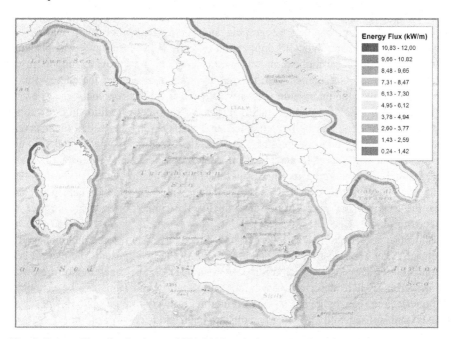

Fig. 5. Energy Flux distribution on 2001-2010 period, calculated at 12 km from the shoreline

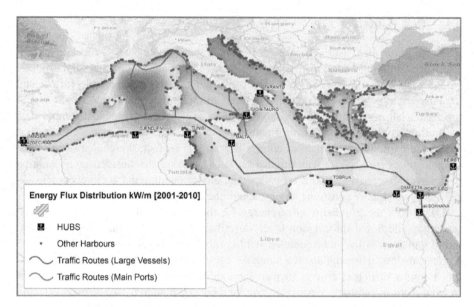

Fig. 6. Energy Flux distribution (2001-2010 period) on Mediterranean Sea and thematic layers describing some anthropogenic activity

2.3   DSS-WebGIS "Waves Energy": Design and Implementation

In order to store and manage geographic and spatial data concerning marine and coastal areas, a specific WebGIS application (called "Waves Energy") was designed and implemented. This application was mainly conceived to provide support in the estimation of energy resources from the sea, by assessing whether the energy flux is exploitable as well the interaction with marine and coastal environments.

Thanks to the use of a WebGIS applications, GIS projects, traditionally developed for stand-alone users, can be implemented on a web-server (also called map-server), in order to allow, through the internet network, the access to the geospatial data and the interaction with thematic maps and information associated. WebGIS applications are exploitable by common internet browsers and among the benefits of using such technology there are: the global sharing of geographic information and geospatial data, the usability and accessibility and, finally, the capability to reach a wider audience of users [31].

The basic geospatial data and the maps produced have been stored and managed in a repository ad hoc structured. In such a way, the WebGIS represents the natural geographical interface of the Decision Support System (DSS) envisaged in the framework of the activities here described: basic local information and maps can be visualized and queried via web, by means a standard internet browser or by mobile devices (e.g., tablets) and, consequently, the main results are open and accessible online.

To this end, the specific objectives of the DSS-WebGIS "Waves Energy" are to:

- define and characterise the marine and coastal areas investigated;
- support the integrated analysis of the area of interest, with the identification of specific environmental indicators (e.g., during the phases related to the design of new facilities);
- provide support in monitoring and forecasting activities;
- share data, maps and information via Web.

The implementation of the above listed features has required an advanced and integrated management of:

- basic geo-spatial data, necessary for marine and coastal zones characterisation (e.g., natural features, infrastructures, etc.);
- new geospatial data, produced to support current analyses and planning activities (e.g., weather and marine forecasts, etc.).

Among the advantages of using the WebGIS approach and technology, it is possible to include:

- the global sharing of geographic information and geospatial data;
- the large usability (the WebGIS application is exploitable by any common internet browser);
- the widespread availability and the capability to reach a broader audience of users.

Concerning the implementation of the WebGIS application, a client-server architecture was adopted, using Free/Open Source Software (FOSS) packages. Such architecture has been properly pictured to allow the interchange of geospatial data over the Web and to provide to the users a user-friendly application, characterised by accessibility and versatility.

The WebGIS architecture is shown in Figure 7 and can be outlined by the following logical chain:

- *Data Repository* -> *Web Server* (GeoServer suite) -> *Libraries* (OpenLayers environment) -> *Map Viewer* (WebGIS)

The Data Repository identifies the storage area containing the data-set used (in GIS format or in interoperable one). The access is allowed only to devices physically defined at the level of the Storage Area Network (SAN), in order to ensure the absolute integrity and consistency of the data themselves.

The Web Server represents the hardware/software environment that allows organizing information and making it accessible from the network. In the present case, it was decided to adopt the GeoServer suite [32]: it is a largely used open source application server, which plays a key role within the Spatial Data Infrastructure (SDI). It allows sharing and managing (by means of different access privileges) the information layers stored in the repository; it also supports interoperability (e.g., reads and manages several formats of raster and vector data).

In the specific framework of the "Waves Energy" application, thanks to the above mentioned characteristics, GeoServer has represented an effective tool to manage the layers (thematic maps, basic information and data, etc.) stored in the geospatial data-base and to accomplish their publication via web within the WebGIS front-end, according to the standards defined by the Open Geospatial Consortium (OGC) [33], such as, for example, the Web Map Service (WMS).

OpenLayers [34] is an open source JavaScript library, used to visualize interactive maps in web browsers. OpenLayers provides a so-called Application Programming Interface (API) allowing the access to various sources of cartographic information on the Web, such as WMS and WFS (Web Feature Service) protocols, commercial maps (Google Maps, Bing, etc.), different GIS formats, maps from the OpenStreetMap project, etc.

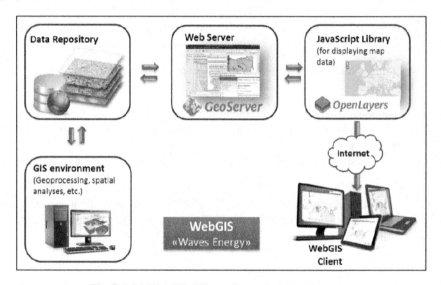

Fig. 7. DSS-WebGIS "Waves Energy": SDI architecture

By accessing the WebGIS interface through his own web browser, the user (not necessarily with specific GIS skills) can gather and view thematic maps and charts representing the results produced. In particular, to visualize the data available, the WMS standard is exploited, by means a map-server approach that allows producing thematic maps of geo-referenced data and responding to queries about the content of the maps themselves.

3   Results: "Waves Energy" Web-Application

"Waves Energy" is a web-based geographical application, which makes available various basic features typical of a WebGIS, such as zoom, pan, transparency, linear and areal measurements, etc. In addition, by clicking anywhere on a selected layer, the relative information is displayed (qualitative or quantitative attributes, according the so-called "inquiry" function). The "Waves Energy" application is fully operational and can be accessed at the following address: http://utmea.enea.it/energiadalmare.

The data provided by WebGIS have been grouped, on the basis of their characteristics and specifications, in the following three typologies:

a) "Forecasting" (outcomes from simulation/prediction models, for the whole Mediterranean Sea and for specific sub zones)
b) "Climatology" (energy flux time series)
c) "Other Layers" (basic geographical information and thematic layers).

The first set of data, produced for the entire Mediterranean Sea (from WAM model, with spatial resolution of about 3 km, 1/32°), provide the thematic maps of the 5-days forecasts, at hourly intervals, related to the following physical quantities:

1. wave energy flux;
2. wave height;
3. wave direction;
4. wave period.

By clicking a point of interest on the sea map, it is possible to obtain and display a specific graph (runtime produced), showing the temporal trend of the selected physical variable (height, direction, etc.) for the following five days, at hourly intervals (Figure 8).

Furthermore, using the SWAM outputs, the same typology of data is available and queryable at a greater detail for the ten specific sub-areas selected (Figure 9), which are considered of particular interest for their characteristics and for the energy potential exploitability.

As described in Paragraph 2.1, the "Climatology" data (listed at previous point b) are derived from recorded time series and are related to the potential energy flux from waves; they contain the average values of energy flux in kW/m for the time span 2001-2010, subdivided into quarterly periods (Figure 10).

In particular, in the DSS-WebGIS are included historical climatology data for the Mediterranean Sea, as well as additional layers representing a specific focus along the Italian coast in a range of 12 km from the shoreline (Figure 11), for each of which it is also available and displayable the respective chart containing the local trends of energy producibility index calculated for the 2001-2010 interval.

The third category ("Other Layers", point c of the list) includes a series of basic geospatial data and environmental information. Such ancillary layers, in addition to those previously described, allow providing a better geographical and thematic characterisation of the marine areas considered for the application purposes (Figure 12). The most significant are: the bathymetry of the Mediterranean Sea (source: GEBCO, General Bathymetric Chart of the Oceans) [35]; a bathymetry subset, with depth values ranging from 0 to 200 m (layer specifically derived from GEBCO data); the distance from the coast, articulated in two zones: 0-25 km and 25-50 km; harbours and hubs; Italian Marine Protected Areas (source: SINANET, Italian Ministry of the Environment and Protection of Land and Sea of Italy).

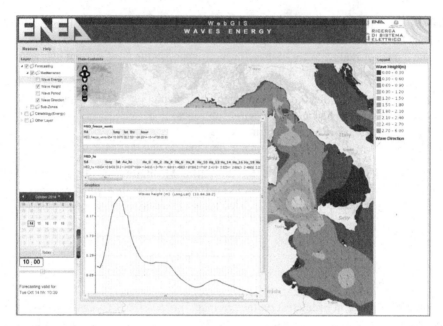

Fig. 8. Forecasting: thematic maps related to height and direction of waves (GIS overlay), with the relative daily trend chart (Source: "Waves Energy" application, http://utmea.enea.it/energiadalmare)

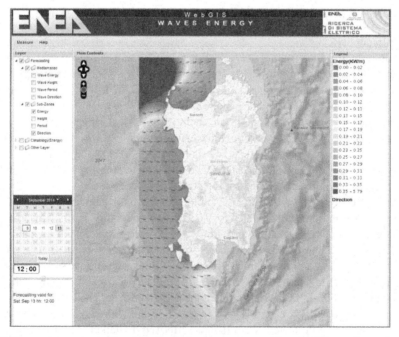

Fig. 9. Forecasting: subset map of the Sardinian western coast (Source: "Waves Energy" application, http://utmea.enea.it/energiadalmare)

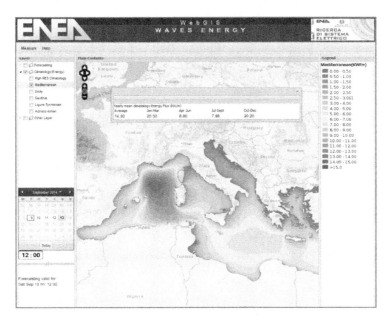

Fig. 10. Thematic map of potential energy waves, derived from climatological data (2001-2010) (Source: "Waves Energy" application, http://utmea.enea.it/energiadalmare)

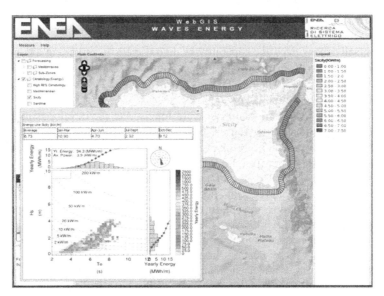

Fig. 11. Average energy values (kW/m) observed during the period considered (2001-2010) and related graphs for the Sicilian coast (Source: "Waves Energy" application, http://utmea. enea.it/energiadalmare)

Fig. 12. Italian Marine Protected Areas and major Italian harbors. GIS overlay with distance from coast and bathymetry (Source: "Waves Energy" app., http://utmea.enea.it/energiadalmare)

4   Conclusions

It is universally recognized the capability of digital maps in providing a comprehensive overview of environmental phenomena. Through appropriate descriptions and thematic maps, it is easier to understand environmental features and characteristics, as well to point out patterns and interactions.

The activities described in this paper were finalised to the development of a specific GIS-based application, having the target to publish forecasting and climatology layers, related to the potential use of energy from the sea, and to share, through the network, a specific set of geospatial information.

The produced geospatial data and the thematic maps were properly structured within the DSS-WebGIS application, not only to show a range of information about the areas of interest, but also to support management and monitoring policies related to the exploitation of energy flux from the sea.

To consult and interrogate adequately these maps, it has been designed and developed the specific web-application interface, called "Waves Energy" accessible directly to the URL: http://utmea.enea.it/energiadalmare.

From the practical point of view, the WebGIS has been realized using Free/Open Source Software environments, which encompass a set of application solutions suitable for the purposes and implementable in a well-integrated and easy to use platform. This solution has allowed publishing on the Web geospatial information following the standard required by the Open Geospatial Consortium (OGC), through a series of specific features for viewing and consulting of thematic maps in an advanced framework, specifically tailored for the "Waves Energy" application.

Acknowledgments. The development of the "Waves Energy" WebGIS application is part of the activities related to the Italian National Programme *"Ricerca di Sistema Elettrico"* ("Research on Electric System", Project B.1.5), carried-out in the framework of an agreement between the Italian Ministry of Economic Development and ENEA.

References

1. Barstow, S., Mørk, G., Mollison, D., Cruz, J.: The wave energy resource. In: Cruz, J. (ed.) Ocean Wave Energy: Current Status and Future Perspectives, pp 93–132. Springer, Heidelberg (2008)
2. Dunnett, D., Wallace, J.S.: Electricity generation from wave power in Canada. Renewable Energy **34**, 179–195 (2009)
3. Dalton, G., Gallachòir, B.P.O.: Building a wave energy policy focusing on innovation, manufacturing and deployment. Renewable and Sustainable Energy Reviews **14**, 2339–2358 (2010)
4. Iglesias, G., Carballo, R.: Wave energy and nearshore hot spots: the case of the SE Bay of Biscay. Renewable Energy **35**, 2490–2500 (2010)
5. Kim, C.-K., Toft, J.E., Papenfus, M., Verutes, G., Guerry, A.D., et al.: Catching the Right Wave: Evaluating Wave Energy Resources and Potential Compatibility with Existing Marine and Coastal Uses. PLoS ONE **7**(11), 1–14 (2012)
6. Dalton, G.J., Alcorn, R., Lewis, T.: Case study feasibility analysis of the Pelamis wave energy convertor in Ireland. Portugal and North America. Renewable Energy **35**, 443–455 (2010)
7. Previsic, M., Bedard, R., Hagerman, G., Sìddiqui, O.: System level design, performance and costs - San Francisco California Energetech offshore wave power plant, p. 69. The Electric Power Research Institute, Palo Alto. E2I EPRI - 006B – SF (2004)
8. White, C., Halpern, B.S., Kappel, C.V.: Ecosystem service tradeoff analysis reveals the value of marine spatial planning for multiple ocean uses. Proceedings of the National Academy of Sciences **109**, 4696–4701 (2012)
9. Langhamer, O., Haikonen, K., Sundberg, J.: Wave power–Sustainable energy or environmentally costly? A review with special emphasis on linear wave energy converters. Renewable and Sustainable Energy Reviews **14**, 1329–1335 (2010)
10. Carillo, A., Bargagli, A., Caiaffa, E., Iacono, R., Sannino, G.: Stima del potenziale energetico associato al moto ondoso in regioni campione della costa italiana. Proj. Report "Ricerca di Sistema Elettrico" ("Research on Electric System"), RdS/2012/170 (2012) (In Italian)
11. La Porta, L., Lombardi, E., Pollino, M., Carillo, A., Caiaffa, E., Sannino, G.: Sviluppo di modelli meteo-marini per la previsione del moto ondoso in aree portuali italiane: produzione di mappe energetiche e visualizzazione web-gis. Project Report "Ricerca di Sistema Elettrico" ("Research on Electric System"), Rds/2013/229 (2013) (In Italian)
12. Cavaleri, L., Malanotte-Rizzoli, P.: Wind wave prediction in shallow water: Theory and applications. J. Geophys. Res. **86**(C11), 10961–10973 (1981)
13. Komen, G.J., Hasselmann, S., Hasselmann, K.: On the existence of a fully developed wind - sea spectrum. J. Phys. Oceanogr. **14**, 1271–1285 (1984)
14. Janssen, P.A.E.M.: Wave induced stress and the drag of air flow over sea waves. J. Phys. Oceanogr. **19**, 745–754 (1989)
15. Janssen, P.A.E.M.: Quasi-linear theory of wind-wave generation applied to wave forecasting. J. Phys. Oceanogr. **21**, 1631–1642 (1991)

16. Mastenbroek, C., Burgers, G., Janssen, P.A.E.M.: The dynamical coupling of a wave model in a storm surge model through the atmospheric boundary layer. J. Phys. Oceanogr. **23**, 1856–1866 (1993)
17. Cornett, A.M.: A global wave energy resource assessment. In: Proceedings of International Offshore and Polar Engineering Conference, pp. 318–326 (2008)
18. Waters, R., Engström, J., Isberg, J., Leijon, M.: Wave climate off the Swedish west coast. Renewable Energy **34**(6), 1600–1606 (2009)
19. Vicinanza, D., Cappietti, L., Ferrante, V., Contestabile, P.: Estimation of the wave energy along the Italian offshore. Journal of Coastal Research **64**, 613–617 (2011)
20. The WAMDI-group: Hasselmann, S., Hasselmann, K., Bauer, E., Janssen, P.A.E.M., Komen, G.J., Bertotti, L., Lionello, P., Guillaume, A., Cardone, V.C., Greenwood, J.A., Reistad, M., Zambresky, L., Ewing, J.A.: The WAM model - A Third Generation Ocean Wave Prediction Model. Journal of Physical Oceanography **18**, 1775–1810 (1988)
21. Janssen, P., Bidlot, J.R.: ECMWF Wave Model Operational: implementation April 9, 2002 – IFS Documentation Cy25R1 (2002)
22. Günther, H., Behrens, A.: The WAM model validation document version 4.5.3. Tech. Rep. Institute of Coastal Research Helmholtz-Zentrum Geesthacht (HZG) (2011)
23. Liberti, L., Carillo, A., Sannino, G.: Wave energy resource assessment in the Mediterranean, the Italian perspective. Renewable Energy **50**, 938–949 (2013)
24. Pierson, W.J., Moskowitz, L.: A proposed spectral form for fully developed wind seas based on the similarity theory of S.A. Kitaigorodskii. J. Geophys. Res. **69**(24), 5181–5190 (1964)
25. Hasselmann, K., Barnett, T.P., Bouws, E., Carlson, H., Cartwright, D.E., Enke, K., Ewing, J.A., Gienapp, H., Hasselmann, D.E., Kruseman, P., Meerburg, A., Müller, P., Olbers, D.J., Richter, K., Sell, W., Walden, H.: Measurements of wind-wave growth and swell decay during the Joint North Sea Wave Project (JONSWAP). Dtsch. Hydrogr. Z. Suppl. **12**, A8 (1973)
26. Battjes, J.A., Janssen, J.P.F.M.: Energy loss and set-up due to breaking of random waves. In: Proc. 16th Int. Conf. Coastal Engineering, ASCE, pp. 569–587 (1978)
27. Kallos, G.: The regional weather forecasting system SKIRON. In: Proceedings of Symposium on Regional Weather Prediction on Parallel Computer Environments, October 15-17, Athens, p. 9.7 (1997)
28. SWAN - Scientific and Technical documentation. Delft University of Technology, Environmental Fluid Mechanics Section. http://www.swan.tudelft.nl
29. Van der Westhuysen, A.J., Zijlema, M., Battjes, J.A.: Nonlinear saturation based whitecapping dissipation in SWAN for deep and shallow water. Coast. Engng. **54**, 151–170 (2007)
30. Thornton, E.B., Guza, R.T.: Transformation of wave height distribution. J. Geophys. Res. **88**(C10), 5925–5938 (1983)
31. Pollino, M., Modica, G.: Free web mapping tools to characterise landscape dynamics and to favour e-participation. In: Murgante, B., Misra, S., Carlini, M., Torre, C.M., Nguyen, H.-Q., Taniar, D., Apduhan, B.O., Gervasi, O. (eds.) ICCSA 2013, Part III. LNCS, vol. 7973, pp. 566–581. Springer, Heidelberg (2013)
32. Geoserver. http://geoserver.org/display/GEOS/Welcome
33. Open Geospatial Consortium (OGC). http://www.opengeospatial.org/
34. OpenLayers. http://www.openlayers.org/
35. GEBCO (2010). http://www.gebco.net/data_and_products/gridded_bathymetry:data/

Spatial Cluster Analysis Informing Policy Making in Ireland

Harutyun Shahumyan[1(✉)], Brendan Williams[2], and Walter Foley[3]

[1] University of Maryland, College Park, MD 20742, USA
harut@umd.edu
[2] University College Dublin, Belfield, Dublin 4, Ireland
brendan.williams@ucd.ie
[3] Eastern and Midland Regional Assembly, Dublin 9, Ireland
wfoley@dra.ie

Abstract. This research aims to provide a robust evidence base contributing to improving the quality of policy formation from local to national level in Ireland. The distributions of businesses within key economic sectors in Ireland are explored aiming to find clustering effects occurring across the country. The density mapping and hot spot analysis approaches were applied to find statistically significant clusters of companies for specific business sectors. The research was implemented in collaboration with Dublin Regional Authority to inform key policy makers in Ireland. It assists in assessments of the nationwide spatial distribution of economic activities adding to the overall body of evidence on business intensive regions.

Keywords: Cluster analysis · Spatial clusters · Hot spots · Business clusters · Economic activities · Decision making

1   Introduction

The National Spatial Strategy (NSS) of Ireland was launched in 2002 as a framework for spatial planning until and beyond 2020 [1]. Its review in 2010 described spatial planning as a "critical instrument to inform and assist prioritisation and coordination of scarce resources long-term resilience into the economy" [2]. The NSS in particular sought to achieve balanced regional development by redistributing economic activity from "core" areas experiencing diseconomies of scale to areas experiencing underdevelopment but with capacity for growth. A critical mass of strategically located settlements for economic development and investment selected as growth gateways and hubs were identified (Fig. 1). In turn the NSS recognised that there was a need to develop critical infrastructure in key larger urban centres.

The ongoing rapid urbanization within an open world economy implies an increasing importance for large cities as units for national and international trade [3]. The 'cluster model' has been advanced to promote competitiveness with cities seen as the places of economic development [4]. Modern clustering has been defined as value-creating geographical concentrations of companies, suppliers, associated institutions

© Springer International Publishing Switzerland 2015
O. Gervasi et al. (Eds.): ICCSA 2015, Part III, LNCS 9157, pp. 511–524, 2015.
DOI: 10.1007/978-3-319-21470-2_37

and specific skills in relative proximity [5]. Regardless of the complexity of analysing the advantages of cluster models, the benefits of higher levels of innovation and productivity are implicit. Modern clusters offer new ways for economic development and are prominent features in advanced economies globally [6, 7].

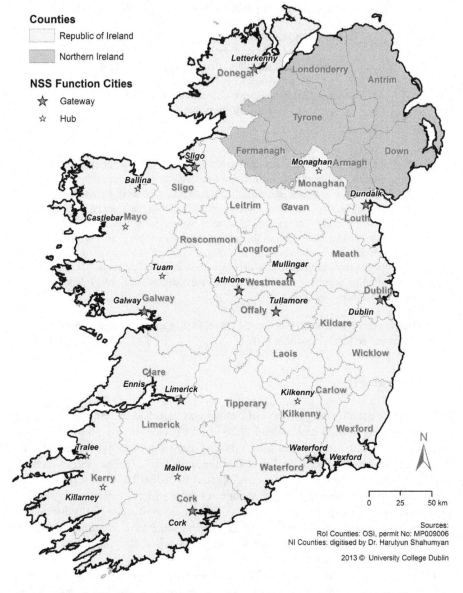

Fig. 1. Map of Ireland with National Spatial Strategy gateways and hubs

There are several studies exploring economic geography in Ireland [8-11], but none have provided a detailed analysis of the spatial clustering patterns of business sectors. In this research, the distributions of businesses within key economic sectors in Ireland are explored aiming to find significant clustering effects occurring across the state. The research aims to provide a robust evidence base contributing to improving the quality of policy formation from local to national level. It was implemented in collaboration with Dublin Regional Authority to inform key policy makers in Ireland. This type of analysis can be used in future iterations of National Spatial Policy.

2   Methodology

2.1   Data

The main source of the data used was the Irish address database GeoDirectory[1], which is developed by An Post and Ordnance Survey Ireland combining accurate postal and geographic addresses for whole country. There were about 300,000 organisations registered in the Republic of Ireland (RoI) in 2012 according to GeoDirectory. And these are classified based on NACE Rev2 classes [12] including 19 broad sections (listed later in Fig. 3) and more than 1300 individual classes.

In addition to the organisation addresses provided by GeoDirectory, the Industrial Development Agency's (IDA) database on the companies with Foreign Direct Investment (FDI) was also analysed. All of the FDI site locations were extracted from the IDA address database. In total there were 771 records of which 756 were mapped to a specific location. The address information was matched against the GeoDirectory. Almost 70% of the records were matched using a geocoding software provided by GIS and Forward Planning unit of Department of Education. The remaining 30% of the records were matched manually. Geocoding the records allowed for a county, enumeration districts and coordinates to be added to the dataset and mapped.

2.2   Density Analysis

Density mapping is an effective approach for revealing where the highest concentration of features is. It is particularly useful for looking at patterns rather than locations of individual features. In this case, GeoDirectory provides point coordinates of all buildings in the RoI, including the addresses of businesses. This allowed us to explore the density of spread of different business sectors over the country.

There are different methods of density analysis. For example, the widely used Kernel Density Estimation (KDE) is a non-parametric method of estimating the probability density function of a continuous random variable. Its output is a raster with pixels values matching to point intensity; and particularly useful for visualisation since areas of high intensity can be identified [13]. However, the KDE is based on the concept that the pattern has a density at any location in the study region and not just at locations where there is an event [14]. This is not applicable to our case. Therefore, for

---

[1] http://www.geodirectory.ie

general visualisation we have applied simpler but more relevant approach – a dot density mapping, which uses a pattern of dots to indicate the areal density of a phenomenon [14]. It is particularly useful for understanding global distribution of a discrete phenomenon and comparing relative densities of different regions on the map. We have applied the boundaries of local level administrative areas known as electoral divisions (ED) as bounding layer to reflect the drawback of locating dots arbitrary in dot density maps.

2.3   Hot Spot Analysis

Hot spot analysis compares the distribution of values associated with the geographic features to a hypothetical random distribution to discover statistically significant clustering. Here, it is used to find statistically significant high and low concentration of businesses in Ireland. The numbers of companies have been aggregated to EDs. The null hypothesis was that companies are randomly distributed among the EDs. *z-scores* (G_i^* statistic) for every ED were calculated based on equation 1:

$$G_i^* = \frac{\sum_{j=1}^{n} w_{i,j} x_j - \overline{X} \sum_{j=1}^{n} w_{i,j}}{S \sqrt{\frac{n \sum_{j=1}^{n} w_{i,j}^2 - \left(\sum_{j=1}^{n} w_{i,j}\right)^2}{n-1}}}, \text{ where } \overline{X} = \frac{1}{n} \sum_{j=1}^{n} x_j, \ S = \sqrt{\frac{1}{n} \sum_{j=1}^{n} x_j^2 - \overline{X}^2}, \tag{1}$$

where x_j is the attribute value (number of companies) for feature (ED) j, $w_{i,j}$ is the spatial weight between features i and j, n is equal to the total number of features [15]. Very high/low *z-scores* are found in the tails of the normal distribution; and indicate significant deviation from a random pattern. For such cases the null hypothesis can be rejected.

The resultant *z-scores* and *p-values* pointed to where EDs with high/low values cluster spatially. More specifically, Hot Spot Analysis examines each ED within the context of neighbouring EDs. An ED with a high number of companies can be a statistically significant hot spot if it is surrounded by other EDs with high numbers. When the difference of the local sum from the expected local sum is too large to be the result of random chance, a statistically significant z-score results. The larger (lower) the *z-score* is, the more intense the clustering of high (low) values. A *z-score* near zero indicates no apparent concentration.

A model was build (Fig. 2) to implement calculations through the following steps:

1. Selection of the companies based on their class.
2. Spatial join of the selected companies with the ED layer to get their count in each ED.
3. Running Spatial Statistics Hot Spot Analysis.

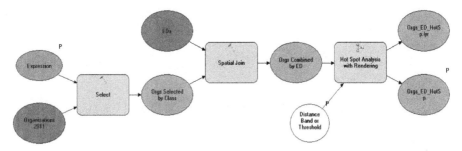

Fig. 2. A model of Hot Spot analysis developed in ArcGIS Model Builder

The parameters of the *Hot Spot Analysis* tool were defined based on the characteristics of the region. Particularly, the ED areas vary essentially between different parts of Ireland. They are much wider in rural areas and much smaller in urban areas, with very small EDs in Dublin city centre. Therefore, the "Fixed distance" method was used, as it works well where there is a large variation in polygon sizes. Clustering is an evidence of an underlying spatial process and the distance band that exhibits maximum clustering is the distance where that spatial process is most "active". Therefore, *Global Moran's I Spatial Autocorrelation* tool was applied for each business sector to find the fixed distance when the *z-score* reaches its peak.

3   Results

Spatial analyses have been implemented for the economic sectors in Ireland which have more than 4% share in the overall number of organisations in the Republic (Fig. 3). In spite of their smaller share "Information & communication" and "Finance & insurance" sectors were also included in the analysis due to their importance to Irish economy. Collectively, these sectors cover about 85% of all companies in Ireland (organisations with missing sector values in GeoDirectory were not counted). Though the "Information & communication" sector covers only about 1% and the "Financial & insurance" sectors covers 1.35% of all organisations in the State, they are of particular interest as they respectively include multinational companies such as IBM, Microsoft, Google, Ebay, Yahoo, and others in the former sector and the activities of the Ireland's International Financial Services Centre (IFSC) in the case of the latter sector. In addition, the IDA reports that 9 of the top 10 Information and Communications Technology (ICT) companies globally have a presence in Ireland and contributes over 20% to total turnover in industry and services in Ireland[2] and 16% of total GDA. Furthermore, the employment statistics from the Central Statistics Office (CSO) of Ireland indicates that the ICT sector directly employs over 5% of total services employment nationally with the majority of this employment concentrated in Dublin and a few other key locations [16].

---

[2] http://www.ictireland.ie

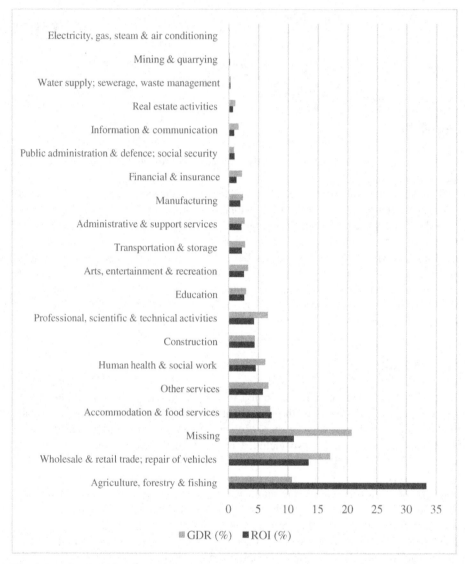

Fig. 3. Distribution of organisations by their broad NACE sections in Republic of Ireland (RoI) and the Greater Dublin Region (GDR)

Dot density map of company locations from the covered business sections shows density differences in geographic distributions across the country and indicates some concentration in the main cities (Fig. 4). However, even if showing agglomeration patterns, such density maps do not reveal the significance of clustering.

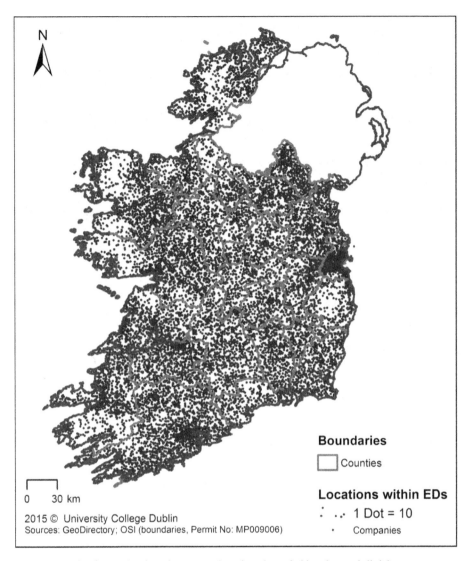

Fig. 4. Dot density of company locations bounded by electoral divisions

The results of hot spot analysis reveal with statistically proven significance the scale with which the main gateway settlements and in particular Dublin, are centres for the key business sectors in Ireland (Fig. 5). The exception is the agriculture sector, for which there are significant hot spots along peripheral areas of the country but not near major cities, as might be expected. Dublin's functional business area was found to extend beyond the metropolitan area for Construction, Wholesale & retail, ICT, Finance & insurance, Professional, scientific & technical and Health sectors. The importance of the economic corridor model was confirmed by the results. There is a significant clustering along the Dublin-Belfast (Dundalk) corridor across Professional, scientific & technical; Wholesale & retail and Accommodation & food sectors.

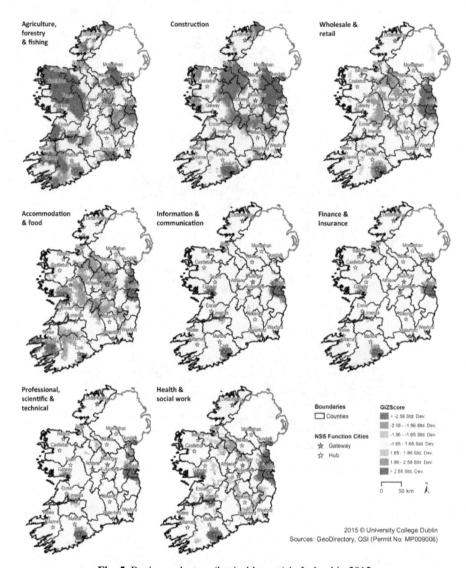

Fig. 5. Business clusters (hot/cold spots) in Ireland in 2012

In general the cities with most extensive clustering patterns outside of Dublin are Cork, Galway and Limerick. While in some cases it can be explained with the population concentration in those cities, in a few cases the patterns are raising questions and require further analysis. For example, Cork exhibits a statistically significant hotpot for the Accommodation & food services which is however limited in geographic scope compared to areas concentrated around Kerry, Galway and Dublin. For a city of its size and reputation this may suggest that it is underutilising its broader geographic potential in this sector especially as this sector is likely to have a potentially high proportion of small and medium enterprises (SME). Further investigation would nonetheless be required to confirm this assumption. Comparison with other datasets (i.e. Failte Ireland[3] data) can be used to uncover if there is a relationship between business location and visitor travel and accommodation patterns.

Ireland is responsible for the attraction and development of foreign investment in Ireland. Its strategy has focused on increasing the local potential for winning FDI through a national program of strategic investment in critical infrastructure, properties and large sites. The locations benefiting from overseas investment are spread over the country as shown in Fig. 6.

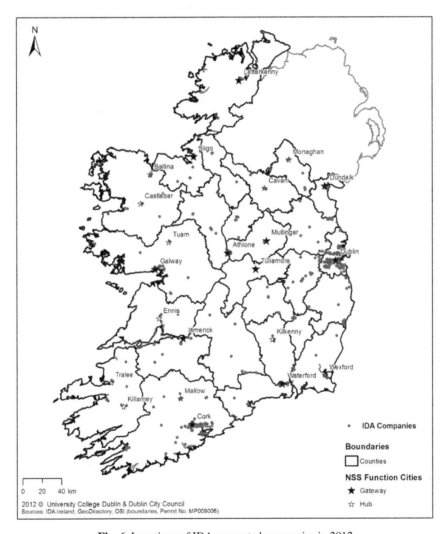

Fig. 6. Locations of IDA supported companies in 2012

---

[3] National Tourism Development Authority (http://www.failteireland.ie)

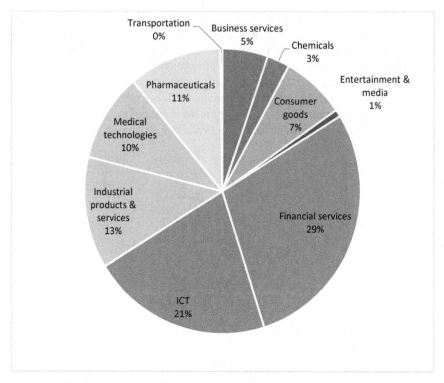

Fig. 7. Sectoral spread of IDA assisted companies in Ireland in 2012

Financial Services (29%) and ICT (21%) in particular account for half of all IDA supported sites in Ireland (Fig. 7).[4] However hot spot analysis revealed that statistically significant clustering is evidenced only in Dublin, Cork and Galway cities (Fig. 8). Regional spreads of specific sectors were also explored. Dublin and Cork dominate in the area of Business services (with 48% and 25% of the national totals), Chemicals (with 30% and 25%) and Pharmaceuticals (with 29% and 27%). Unsurprisingly Dublin dominates in the Financial services and Entertainment & media sectors with an over 80% national share. Over half of all the ICT sector locations were in the Dublin region. Cork leads in location of Industrial products & services while Galway leads in the area of Medical technologies accounting for 17% of the national total.

While national policy [17] officially favours locating 50% of new industry outside both Dublin and Cork many investors in the knowledge based sectors look to Dublin first [18]. This is evidenced by the fact that over 50% of national FDI investment is located in the Dublin city region and a statistic from the Department of Jobs, Enterprise and Innovation which outlined that 44% of 341 approaches/itineries received in 2012 from potential foreign direct investors were for Dublin [18]. In this regard spillover along and between economic corridors gains added significance in terms of distributing economic growth nationally.

---

[4] Employment numbers by site was not accessible for this research. Such information would enrich the quality of the analysis.

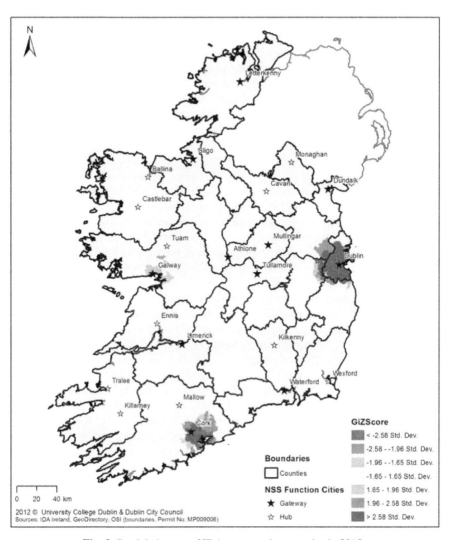

Fig. 8. Spatial clusters of IDA supported companies in 2012

4   Discussion and Summary

This study is aimed to provide new approaches for assessments of the nationwide spatial distribution of business formation and investment in services, that is, to determine best locations for return on investment, business formation and where demand for social infrastructures are highest now and may be highest in the future. The results show the statistically proven significance of the main gateways and in particular Dublin as key centres for the main business sectors in Ireland. A clear pattern emerged whereby Dublin and the other gateways across the State were the centres representing those broad geographic areas of greatest opportunity in the country.

Dublin is evidently central to the health of the entire Irish economy. Business Demography headline figures released by the CSO in 2012 revealed that there were over 195,000 enterprises active in the private business economy in Ireland in 2010, with nearly 1.24 million persons engaged (employees and family members)[5] [19]. The data show that Dublin has maintained levels of resilience in terms of balance between enterprise creation and destruction. Dublin's share of active enterprises as a proportion of the national total increased in the five years to 2010. This trend is evident across all enterprises of all sizes. For example, Dublin increased its share of total business of under 10 persons (micro-enterprises) from 26.6% to 29% and its share of large enterprises of over 250 persons from 50.5% to 60.1% (which equates to 70% of total employment).

Though the hot-spot analysis described here does not have an employment quantum built in, it does point up the economic importance of the Dublin region. Coupled with other indicators of economic viability such as human capital and infrastructure Dublin is evidently a gateway region with mutually reinforcing factors which have allowed to grow to a critical mass across a number of key economic sectors.

While Dublin is both the global economic focal point for Ireland and the key component of a broader spatial distribution of gateways and hubs nationally the results can also be viewed in the context of Dublin and other key gateways having a combined critical mass of business occurrence across key economic sectors which may be attractive to inward investment. The significant business clustering within key cities Dublin, Cork, Galway and Limerick point to a need to build on the comparative advantages of these areas through investment in infrastructure.

The research also indicates a demand to prioritize a selected number of locations regionally. It highlights regional strengths across business sectors and the potential to develop economic corridors and investment. Thus, the Dublin-Belfast corridor is important to the overall health of the wider national economy and offers a lot of opportunity for enterprise and employment creation with critical mass clustering along it already within the aforementioned sectors. This pre-eminent corridor should be a forerunner to the development of similar regional networks across the State.

Though the spatial outputs from the density mapping and hot spot analysis are described in relative isolation here, they have been examined in conjunction within qualitative and quantitative review of other relevant reports and datasets [20-23]. These outputs show the value of a quantitative evidence base which is based on distribution of economic activity. Future iterations or a complete revision of spatial policy in Ireland could in theory integrate these data derived from analysis of economic activity with analysis and projections of population growth. This would provide a more rounded view of current levels of activity on the spatial scale and enable more targeted planning.

One of the outcomes of this work was to show consensus of opinion in terms of the usefulness, need for both provision and use of more in-depth evidence in the formation of future spatial planning. The National Spatial Policy as originally conceived aimed to provide balanced regional development. Some evidence suggests that it has failed to achieve anything close to it primary objective. It can be argued that the

---

[5] The release covers the whole business economy, i.e. NACE Rev 2 sectors B - N, plus sector P (Education). It does not include: Agriculture; Public Administration; Health; Arts, Entertainment, Recreation; Other Service Activities, Activities of Households; and Extraterritorial Bodies sectors.

Planning and Development Act (PDA) 2000 [24] did not facilitate the implementation of the NSS through its main strategic arm the Regional Planning Guidelines (RPGs) until its amendment in 2010. Prior to this, according to PDA 2000, a planning authority was obliged to 'have regard to' any RPGs in force for its area when making and adopting a development plan.[6] The amendment in 2010 means that planning authorities have to, when making a development plan, ensure that the plan is consistent with any RPGs in force for the area in question and that each development plan includes a core strategy outlining this consistency.

Evolving economic geography has since seen key locations such as Dublin and the other main gateways to a lesser extent consolidate their economic base and centrality within the spatial hierarchy and do so at the expense of smaller urban centres. While emerging trends such as technological advancement and broadband infrastructure provision may theoretically result in more people working from home/remote locations this evidence points to a continued trend of clustering and the emergence of large urban centres as focal points of national and global trade.

The spatial cluster analysis of economic activity in Ireland shows the importance of key urban centres, the potential to develop economic corridors similar to the M1 Dublin-Belfast motorway, specialisms across the broad economic geography of Ireland by location and potentially the need for targeted investment. While settlement size and population density affect the range of services and jobs that can be supported, these factors are linked to and reinforced by current levels and clustering of businesses. Future revision of spatial policy in Ireland could integrate these results with population and job projections by area enabling more targeted planning and investment.

This research assists in assessments of the nationwide spatial distribution of economic activities adding to the overall body of evidence on business intensive regions. In spite of using standard/traditional GIS methodology approaches it provides a new evidence base heretofore not available to key policy-makers in Ireland and raises questions concerning the most appropriate locations for future investment.

Acknowledgement. This research was implemented in scope of the project "Dublin's Role in the Irish and Global Economy" funded by Dublin Regional Authority and supported by a Marie Curie International Outgoing Fellowship within the 7th European Community Framework Program.

References

1. NSS, *National spatial strategy for Ireland 2002-2020: People, Places and Potential*, D.o.t.E.a.L. Government, Editor. Stationery Office, Dublin, p. 160 (2002)
2. DoEHLG, Implementing the National Spatial Strategy: 2010 Update and Outlook. Department of the Environment, Heritage and Local Government: Dublin (2010)
3. Fujita, M., Krugman, P., Mori, T.: On the evolution of hierarchical urban systems. European Economic Review **43**, 209–251 (1999)

---

[6] The objective of RPGs is to provide a long-term strategic planning framework for the regions of Ireland following NSS population targets and the guidelines should not be less than 12 years or more than 20 years under s.23(1)(a) of the PDA 2000.

4. Asheim, B., Cooke, P., Martin, D. (ed.): Clusters and regional development critical reflections and explorations, NY, Routledge (2006)
5. Porter, M.: On Competition. Harvard Business School Press, Boston (1998)
6. Ciccone, A.: Agglomeration effects in Europe. European Economic Review, 213–227 (2002)
7. Krugman, P.: Innovation and agglomeration: Two parables suggested by city-size distributions. Japan and the World Economy **7**(4), 371–390 (1995)
8. Morgenroth, E.: Exploring the Economic Geography of Ireland. Journal of the Statistical and Social Inquiry Society of Ireland **38** (2008)
9. Curran, D., van Egeraat, C.: Defining and Valuing Dublin's Creative Industries. Dublin City Council (2010)
10. CSO, Census 2006 - A Profile of the Working Population of Large Towns, Central Statistics Office & Economic and Social Research Institute (2009)
11. Tol, R., et al.: Towards Regional Environmental Accounts for Ireland. Journal of the Statistical and So-cial Inquiry Society of Ireland **38** (2009)
12. EUROSTAT, NACE Rev. 2 Statistical classification of economic activities in the European Community. In Methodologies and Working papers. EUROSTAT (2008)
13. Plug, C., Xia, J.H., Caulfield, C.: Spatial and temporal visualisation techniques for crash analysis. Accident Analysis and Prevention **43**(6), 1937–1946 (2011)
14. O'Sullivan, D., Unwin, D.J.: Geographic Information Analysis. John Wiley & Sons, Inc. (2010)
15. Getis, A., Ord, J.: Local spatial autocorrelation statistics: Distributional issues and an application. Geographical Analysis, 287–306 (1995)
16. CSO, Quarterly National Household Survey. Central Statistics Office, Dublin (2013)
17. DoJE&I, Action Plan for Jobs, E.a.I. Department of Jobs, Editor. Dublin (2012)
18. Deegan, G.: IDA admits Dublin receives lion's share of FDI itineraries. In Irish Examiner (2012)
19. CSO, Business Demography 2010. Central Statistics Office, Dublin (2012)
20. Foley, W., et al.: Opinion and Evidance. In: Dublin's Role in the Irish & Global Economy. University College Dublin, Dublin (2012)
21. Shahumyan, H., Williams, B., Foley, W.: Spatial Analytic Approaches Assessing Socio-Economic Development of the Dublin Region Compared with Other Regions in Ireland. In: Dublin's Role in the Irish & Global Economy. University College Dublin, Dublin (2013)
22. Shahumyan, H., Williams, B., Petrov, L., Foley, W.: Collating and assessing the availability and applicability of socio-economic data and information relating to development of the dublin region. In: Dublin's Role in the Irish & Global Economy 2012. University College Dublin, Dublin (2012)
23. Williams, B., et al.: Synthesis Report: Dublin Ireland's Flagship. In: Dublin's Role in the Irish & Global Economy. University College Dublin, Dublin (2012)
24. OAG, Planning and Development Act, O.o.t.A. General, Editor. Dublin (2000)

The Use of Territorial Information Systems to Evaluate Urban Planning Decisions in Transformation Areas: The Case for Parco della Valle del Lambro in Lombardy, Italy

Pier Luigi Paolillo[✉], Massimo Rossati, Luca Festa, and Giuseppe Quattrini

Dipartimento di Architettura e studi urbani, Politecnico di Milano,
Via Bonardi 3, 20133 Milano, Italy
pierluigi.paolillo@polimi.it

Abstract. The construction of the territorial information systems, provided by the recent law for territorial governance of Lombardy (the most industrious Northern Italian region), has become an essential procedure to evaluate of environmental sustainability of urban planning. This procedure is necessary for the handing of descriptive and relevant data that is needed to represent the environmental complexity of a region. In addition, territorial information systems allow for the composition of useful indicators to identify the intrinsic peculiarity of a territory (as in our area application, Parco della Valle del Lambro, 35 municipalities on a North-South axis of 25 km, extending across 4,081 hectares). The proposed case for the Lambro Valley and the related Strategic Environmental Assessment represents a positive example of how territorial governance and land use is often conflict.

Keywords: Geostatistical multivariate analysis · Geographic information systems · Environmental planning · Govern the territorial transformation · Indicators · Strategic environmental assessment

1 Environmental Emergencies, Good Governance of the Territory and the Digital Cartographic Instrument to Support Decisions

The complex municipalities of Regional Park of the Lambro Valley lies in an intensely settled area. New residents are encouraged. To a large extent families who moved in the last 30 years from the metropolitan area were searching for wide open green spaces looking for a better quality of life within the park. This alternative lifestyle compared favorable to the progressive deterioration of living standards in metropolitan Milan. The very existence of the Regional Park offered a wide range of incentives for families to move.

The northern Milan basin of Brianza, where the municipalities of the Regional Park of the Lambro Valley are located, is the busiest, most active area in Lombardy with small to medium sized successful companies. High housing demands make

O. Gervasi et al. (Eds.): ICCSA 2015, Part III, LNCS 9157, pp. 525–539, 2015.
DOI: 10.1007/978-3-319-21470-2_38

Brianza attractive for investors. Because of the attraction of the pre-alpine lakes natural beauty and proximity to Milan, second homes and tourism predominate. Therefore a better infrastructure is required to compensate for the urban sprawl and commuting problems.

Our environmental Assessment document for the Spatial Plan of the Park must (**a**) take into account the settlement needs without losing the physical-environmental, geo-pedological landscaping character; (**b**) consider macroeconomic factors including town planning; (**c**) compare data to identify: *i*) the municipal settlement needs; *ii*) the amount of land which can be developed; *iii*) the classification of the physical resources and the space for the effective use of the soil types; *iv*) and classify and resolve the optimum choices between multiple uses of soil and physical resources; *v*) the range of possible options to diminish conflicts while causing the least damage to the physical-environmental heritage including eco-choices and taking into account the already existing structures.

It goes without saying that, Gis are irreplaceable to create the data bank, to display the results and transform the descriptive cartographic base in n quantitative matrices. The statistics apply and calculate the interdependencies. In this way, qualitative factors could be replaced by quantitative elements for statistics, whose symbolic space is physical space representing m cells for n matrices and calculates numerical mass.

Thus, the first step represents the construction of a cognitive frame, constituted by a core-set of indicators inferable and/or calculable starting from the data base. In different phases of the environmental assessment process, it is necessary to include basic information of the territory and population, essential for the specific objectives and the cognitive frame of the environmental assessment. The Strategic Environmental Assessment (Vas) and the functional analysis to the Plan are to be considered essentially a single product.

2   The Strategic Environmental Assessment (Vas) Structure of the Lambro Valley Regional Park

The Regional Park controls the Lambro River Valley, which intersects Milan, Lecco and Como Provinces, involving 35 municipalities. The diverse character, high density town planning and low density in the foothills has spoiled the overall plan and balance for green spaces.

The Vas analyzes the sustainability of the choices introduced by the new Plan of coordination of the Park as a cyclical open process, able to identify the alternatives corresponding to the changing of the evolutionary trends and the distinction of the invariances (of environmental, historical and cultural nature) which form a reference point for comparison to evaluate the eligibility of the projects and the intervention of transformation.

2.1   The Setting of the Environmental Report and the Methodology

Scientific research identifies the environment amongst the complex systems: a system structured by various physical and man-made changing components and continuously interacting, with endo-exogenous interference. We are concentrating: *i*) on pre-existing links between the environmental parts of the examined space, *ii*) with a key to read the degrees of complexity, *iii*) checking the outcome via planned measures.

The methodology starts with the following premises: **x**) the environmental complexity depends on interactions of local phenomena; **y**) the necessity to discover specific physiognomies and substantial differences of this complex rural environment; **z**) to research the problems of the local sub-areas and to expose the environmental damage to protect the future.

We still have time to control the many degenerative phenomena and the resulting irreparability of the physical-environmental resources and, perhaps, the possibility of considering the territory in the complexity of its elements and the dynamic relationships dealing with is fully manifesting, without longer be content of mechanistic simplifications but considering the whole environment.

In this work, the adopted methodological approach refuses the assumption of rigid explanatory models due to the complexity and the fluctuation of its evolutionary process and favors the "data analysis" [1-14] application, a complex of descriptive-explorative methods which have shown for some time some relevant advantages: *a*) the capacity of producing a large amount of data, often heterogeneous, at the cost of a loss of information; *b*) the further capacity of synthesizing masses of data in a limited number of variables and indicators, through factor analysis or (cluster analysis), together with the comparability of the new variables to the pre-existing spatial-environmental objects; *c*) discriminating analysis in the form of matrices or Cartesian diagrams; *d*) the opportunity, contrary to the premises for the use of the classical models (which favor the aprioristic explicit force of the assumed pre-judgment), of formulating weak hypothesis of the phenomena under consideration.

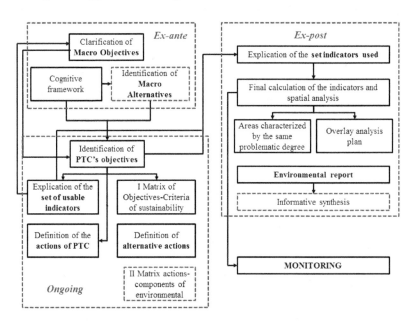

Fig. 1. Path of preparation of the Environmental Report for the Regional Park of Lambro Valley

In this context the environmental assessment consists of many methodological steps and the path of the Vas environmental report can be classified in the diagram below. Particularly, the following characteristics of the assessment made by the Lambro valley Environmental Report are to be stressed: *i*) the interpretative schema assumed, of the Pressure–State–Answers [15, 16] type; *ii*) the calculation, of the informative division in analytical matrices to defined cells; *iii*) the classification of the environmental effects identified on the bases of the Values/Detriments/Risks model; *iv*) this methodology estimates the degree of interdependence of data from different origins.

The analytical path provides: *i*) the application of geostatistics techniques for effective informative treatment of the available databases; *ii*) the discovery of relationships between data through multivariate analysis; *iii*) identification of homogeneous groups of local phenomena; *iv*) the final identification of homogeneous basins, with respect to the sustainability of the rules of the municipal consortium partners clarify the Plan of Lambro Valley Park.

2.2   The Database Being Used and the Various Components in the Construction of Knowledge Matrix

This is key information for the articulation of general and specific objectives, to analysis the cognitive frame, to construct the related scenario and for the assessment – through a complex indicators' set core – of the environmental effects of the plan the knowledge derive from the studies already put in place for the Lambro Valley Territorial Coordination Plan (Ptc), together with the Lombardy Region and affected Provinces database, constitute a fundamental reference.

The physical status of each cartographic work was stored in the Park's archives and, in addition to that, the relevant degree of mapped information: since there are analytical topics which need current data and, since the maps go back approximately to the first half of the Eighties (even if the Ptc was approved in 2000), the information is often due to irregular updating.

But, apart from these limits (which were overcome in the best way), the cartographic set of the Park's Plan represented a good starting base for the initiation of the territorial analyses.

We considered the following components taken from the European directives on the plans and projects' environmental assessment, whose presence appeared justified by the complexity of Lambro Valley Regional Park territory.

The investigative systemic approach favored the organizing of the information through viable and transparent analysis.

The method adopted to work on data, identifying the phenomena's intensity in the social-economic and physical sub-components' classification (macro-indicators), is based on a linear process of aggregation which contemplates: *a*) for each sub-component, the identification and the calculation of the significant indicators of the researched phenomenon, selected from the check list of the indicators available in the literature; *b*) the normalization and standardization of the set of the indicators adopted, in order to make the indicators mutually comparable; *c*) a following horizontal aggregation of the indicators standardized with the method of the geometric mean and the production of a vector – column index synthetic of values expressing the intensity of the macro-indicator; *d)* the estimate of the phenomenal intensity classes (High/Middle/Low) through the application the column aggregation vector; *e)* the estimate of the synthetic indicator of each declaratory objective through the intensity column vectors re-encoding, and the following aggregation through the geostatistics analysis.

The area is considered as a systemic group of homogeneous units (cells) spatially defined, equipped with inherent properties, carrying information and able to exchange it with adjacent units, able to assume different states of biological information and able to interact in the "transactional space".

2.3 From the Knowledge Matrix to the Synthetic Moment

The environmental Assessment process ends with the synthetic moment consisting in the judgment of the acceptability of the stimulus received by the municipal consortium members and the subsequent assessment of the specific scenario through the indicators resulting from the analysis made: *i)* on the social-economic form, *ii)* on the atmospheric environment, *iii)* on the aquatic environment, *iv)* on the soil environment, *v)* on the physical environment, *vi)* on natural factors, *vii)* on the landscape, *viii)* on the stringency level of the restriction system.

The results of the analysis are explained in the *Map of the multidimensional environmental value.* But this map is not the only column vector for the assessment of the demands, as there is the morph-settlement factor which forms the second column vector. We used information to identify if the typologies were aimed at the consumption of new soil or in the protection of areas currently not included in Park which would play a relevant role.

We needed to estimate the acceptability of the demands which would not be complete if it didn't include the assessment of the current provincial planning. It was necessary to overlap in Overlay the provincial planning to assess and verify the legitimacy.

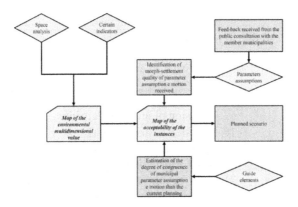

Fig. 2. The orientation logical block

We need to highlight an effective *Planned scenario.*

3   The Acceptable Environmental Level to Sustain the Lambro Valley Regional Park

3.1   The Morph-Settlement Classification of the Transformation Motions Received from the Park Administration

The urban transformation requests received from the Park Municipalities involve 157 areas for a total of 8.34 kmq. Requests received unfortunately, in most cases, were focused on destinations which would increase environmental pressures (58,05% requested not to be under the jurisdiction of the Park, and more than the 20% asked for a change of destination to insert areas in urban aggregates).

Changes were subjected to the assessment of the morph-settlement quality to judge them, while limiting the soil abuse and the pressure on the agro-forestry natural systems. Morphological characterization was identified by indicators and reclassified and interpreted.

A set of indicators was identified and selected aimed at describing and measuring the problematic situation of each single request, composed by: *i*) form coefficient; *ii*) impact of the area of the request; *iii*) localization of the request compared to the

Table 1. Subdivision of the categories and request of motions received

Category	Required destination	Area (ha)	Weight (%)
Motions aimed to obtain a classification not contained in the Rules of the Park's Plan	1. Obtaining settlements/productions areas	16.68	2.00
	2. Obtaining functions accommodation	11.54	1.38
	3. Obtaining variation of viability	1.28	0.15
	4. Obtaining parking	1.49	0.18
	5. Obtaining public services	3.11	0.37
	6. Removal from the Park	484.48	58.05
	7. Adding territories in the Park	26.42	3.17
	8. Obtaining environmental recovery	40.06	4.80
	9. Adjustment of areas identified by Park's Plan	14.51	1.74
Motions related to areas specifically classified in the Rules of the Park's Plan	11. Inclusion of agricultural areas	7.18	0.86
	12. Inclusion of urban settlements	38.31	4.59
	18. Inclusion of historic Park	2.25	0.27
	20. Elimination of elements of industrial archeology	9.04	1.08
	21. Entering into settlement areas	145.10	17.39
	22. Inclusion in areas of settlement redevelopment	1.39	0.17
	23. Entering into areas for sports and recreational facilities	31.69	3.80
Total areas (ha)		*834.58*	*100.00*

borders of the Regional Park; *iv*) localization of the request compared to the borders of the natural Park; *v*) localization of the request compared to the closest urban centre.

A) *Form coefficient* – The first indicator used in the morphological analyses determines their degree of compactness or intended perimeter of the urban centers. The perimeter morphology of the transformation areas was considered to assess the impact that the transformation requests could have on the current perimeter forms to minimize compromises of the settlement perimeters. Those transformation areas presenting a compact geometry were assessed with a low problematic nature because they didn't distort the sensitivity of the urban perimeter already compromised by an exaggerated urban growth;

B) *Impact on the requested area* – Indicator which calculated the impact of single requests on the area;

C) *Localization of individual requests compared to the borders of the Regional Park;*

D) *Localization of the requests compared to the borders of the Park;*

E) *Localization of the requests compared to·the closest urban centre* – To estimate the local requests in relation to the closest urban centre.

With the six column vectors the evaluation of the degree of morph-settlement quality of the single requests was made, based on the following scheme.

Table 2. Classification of the type of criticality

Type of critical issues	Morphological quality	Class
No critical nor high prevalence of low criticality		1
Presence of high criticality (maximum 1) and balanced presence of critical medium and low		2
Presence of high criticality (maximum 2) and the presence of medium criticality (maximum 2)		3
Prevalence of criticality high (over 3)		4

To evaluate the sustainability of the urban planning transformation's requests received, to bring the results of the investigation to a single synthetic vector in order to have a uniform multidimensional classification of the environmental matrix of the Park.

Below is the methodological model, built to obtain firstly the definition of the susceptibility to the transformation of the considered areas and, secondly, to address the sustainable planning of the soil source in the territory of the Park.

3.2 The Model Used to Identify the Extent of the Physical-Territorial Sensitivity

The classification of the multidimensional environmental value selected evidence gathered synthetic index and the further research detailed, for each investigation unit (a 25 x 25 m cell), of the physical-environment: (i) the high natural values, to conserve flora, fauna and traditional landscapes; (ii) the degree of alteration and abuse of lands and natural forms, with a view to better conserve the original morphological features; (iii) the important of the virginal territory; (iv) a condition of sustainability and suitability of current use; (v) maintaining ecological continuity; (vi) to protect the environmental, landscape and historical-cultural elements.

Before identifying the gathering of the analytical space and the data georeferencing, we applied the Value/Detriment/Risks model: x) the values, adequacies and positive prerogatives; y) the detriments negative aspects; z) the risks, uncertainties restricted use limitations of the resources, identifying a graduation of the sensitivity of the Park basins regarding the recurrence of the natural and man-made transformations, so that the protection and valid policies could take advantage of a spatial framework aimed at the orientation of the choices and at the definition of the priorities; only then we began a true and proper "eco-friendly project", whose conceptual terms are located in a cyclical dimension, justified by the permanent control of the transformations.

In the assessment of the physical sensitivity level in any investigation cell belonging to the Park, great importance was given to: (a) the ecological-environmental dimension; (b) the landscape, morph-cultural matrix historically layered in places; (c) on the sustainability of the soil, to reach situations, necessary to influence the local choices of transformation.

After the assessing of the links analyzed to the best combination. The correlation analysis shows a direct significant interdependency between nature, alteration and risk of compromising the soil resource index.

Now the following can be assumed as evidence for the multi-dimensional characterization of the environmental structure of the Parco della Valle del Lambro.

    a)   Natural state of the ecosystem index (Nat);
    b)   Alteration of the territorial forms index (Alt);
    c)   Landscape sensitivity index (Paes);
    d)   Risk further deterioration of the soil resource index (CompD);
    e)   Ecological potential of the natural factors index (CompF);
    f)   Restrictive system index (Vinc).

The non-urbanized environments are of two types: the intensive farming areas; the natural areas of forest, shrubs and trees functioning as connecting corridors between urbanized and agricultural areas (expression of the portions of territory not yet corroded by man-made or urbanizing activities).

In the Park, the areas characterized by a medium-high landscape mostly of agricultural use enhance and strengthen, the importance together with the ecotonal branches of transition and more important natural basins.

The urban dimension is high and the urbanized areas are located mostly located outside the Park perimeter.

The low presence (only 6%) of natural areas at high physical and environmental value are symptomatic and worrying, which shows high levels of man-made erosion of the natural habitats of the Park caused by urban development.

Almost 5% of the territorial surface of the Park suffered as a result of human action and a recovery program was implemented.

Finally, the aquatic areas are of substantial importance located near the Alserio and Pusiano lakes. In other parts of the Park historical villas with large gardens and vast green spaces exist.

The fifteen classes were classified, according to the present environmental values and peculiarities, according to six synthetic classes of physical sensitivity.

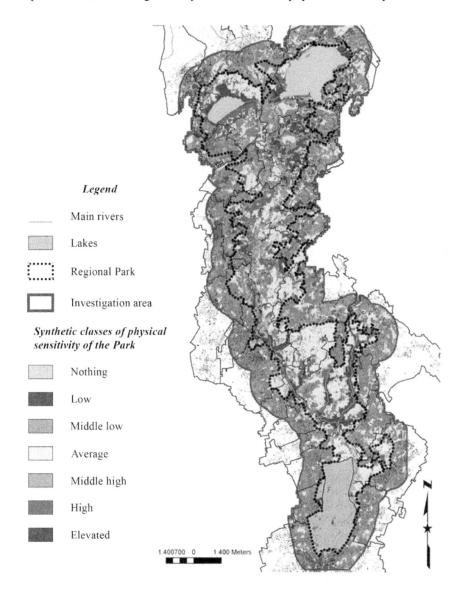

Legend

Main rivers

Lakes

Regional Park

Investigation area

Synthetic classes of physical sensitivity of the Park

Nothing

Low

Middle low

Average

Middle high

High

Elevated

Fig. 3. Map of the synthetic classes of physical sensitivity of the Park

Table 3. The characterization of the sensitivity classes for the physical territorial hired

Territori al areas	Units	Classes of physical sensitivity of the park						
		Elevated	High	Middl e high	Aver age	Middle low	Lo w	Urba n soils
Park	Ha	708	1090	2087	2494	710	454	1133
	%	8.2%	12.6%	24.1%	28.8%	8.2%	5.2%	13.1%
Outer band (within 1 km)	Ha	0	545	160	1069	1200	276	4542
	%	0.0%	7.0%	2.1%	13.7%	15.4%	3.5%	58.3%
Study area	Ha	708	1635	2247	3563	1910	730	5675
	%	4.3%	9.9%	13.6%	21.6%	11.6%	4.4%	34.5%
% of the Park outer band	%	100%	67%	93%	70%	37%	62%	20%

We found a clear dominance of the clusters characterized by a middle natural state, instead of those higher, because of the widespread facilities in the Park, also the farmers to the detriment of the original green spaces.

3.3   The Transformation Susceptibility as Compared to the Socio-Economic Factors

The investigation found the municipalities in the Lambro Valley Park needed to: (a) to evalue the service sector; (b) difficulties for business opportunities, (c) multifunctional development of the primary sector; (d) innovation capability and businnes development; (e) quality of life of the citizens dictated by the best employment and financial situation as well as by the growth of the demographic factors (average life, births etc.); (f) inter-communal capacity for interaction.

The identified territorial competition was divided into five classes based on the interpretation of the socio-economics where local structures would welcome new growth and provide incentives to strengthen existing ones and create new entrepreneurial opportunities.

The relationships between the socio-economical tendencies and the physical-environmental methods emerge in the Overlay results among maps: i) of the physical sensibility of the territory and ii) of the territorial competitiveness, where it emerges the Park transformation susceptibility level, intended as combined evaluation of suitability and identified physical-environmental sensitivities, as well as the predisposition to adopt change in relation to the structure of the local socio-economical system: through the transformation susceptibility Map it is thus possible for conflicts for the use of resources not yet mobilized to emerge, where the attractiveness and competitiveness of the local economic system determine suitable situations for the insurrection of transformations able to induce the establishment of antagonist uses and of change of use (it should be noted how the physical sensibility and competitiveness of the socio-economic structure are not necessarily antagonists, because a good socio-economical system can positively have an impact on the compensation, mitigation and environmental change of modus operandi.

3.4 The Overlay Between the Physical Sensibility and the Territorial Viability, in Order to Estimate the Transformation Susceptibility Degree

The *overlay* operation resulted in thirty different types, each expressing one of the eight possible different transformation susceptibility, arising from the crossing between environmental suitability and predisposition to the change of the socio-economic structure; then, the resulting types were reduced in terms of complexity reaching the five synthetic classes of High, Medium-High, Medium, Medium-Low, and Low transformation susceptibility.

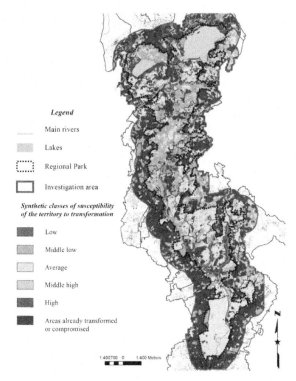

Fig. 4. Map of the synthetic classes susceptibility of the territory transformation

Around the 50% of the space of the Park is characterized by a medium-low transformation susceptibility, and the 40% is medium; it is important note – on the bases of the allocation of a value to this susceptibility – an explicit judgment on the admissibility of building and/or transforming the spaces isn't expressed, but rather to address the condition of balance liable for identifying an suitability to transform and begin legitimizing economic factors, present in the territory.

Table 4. The summary table

Territorial areas	Units	Classes of susceptibility to the territorial transformation				
		High	Middle high	Average	Middle low	Low
Park	Ha	56	695	3147	2412	1232
	%	0.7%	9.2%	41.7%	32.0%	16.3%
Outer band (within 1 km)	Ha	87	1001	1194	716	252
	%	2.7%	30.8%	36.7%	22.0%	7.7%
Study area	Ha	143	1696	4341	3127	1484
	%	87	1001	1194	716	252
% of the Park outer band	%	39%	41%	72%	77%	83%

4   The Evaluation of the Acceptability of the Demands and Amending of the Regional Lambro Valley Park Borders

The path of the Evaluation Report on the status and on the existing environmental pressures on the Lambro Valley Park resources found a relevant phase to judge the acceptable demands received by the municipalities. All municipalities and private motions have been included in the drafting of this Environmental Report.

First of all it was necessary to judge the location suitability of the proposals put forward with regard to the basins of transformation susceptibility involved. the municipal proposals were in this sense assessed in order to identify the types in which they fell (namely if they appeared to be finalized to the new soil consumption or, on the other hand, to the protection of areas currently not included in the Park, if the perimeter of the spaces involved would generate continuity or fragmentation, if their dimension and their proximity to the Park borders could create criticism, and so on).

To evaluate their acceptability, the transformation demands were then compared with the provincial planning in order to identify any conflict and, thus, recall in synthetic sheets which would indicate any need to change the Park borders; indeed, from their assessment it emerged a significant stimulus share received by the municipal Administrations in order to correct the Regional Parco della Valle del Lambro borders making emerge in this way how, as compared with the requests complex, less than the 20% of them found a justification and, thus, for more than the 80% of the demands of perimeter reduction the negative opinion of the strategic environmental Assessment was reiterated by the judgment of the Regional Park Administration, by rejecting almost every one of those demands which would have created a reduction of the Park area; as a consequence, the areas with changed destination and their corresponding variations would be as follows.

Table 5. The changed destination area summary table

Changed areas	Recorded surface previously in force in the Park Plan	Surface on which was assumed the change of destination	Magnitude of change hypothesized
	ha	ha	ha
Areas of historic Park	369,25	+ 2,49	+ 0,67%
System of river and lake areas	1.451,60	− 5,68	− 0,39%
System of predominantly agricultural areas	2.011,96	+ 14,50	+ 0,72%
Areas for sports and recreational facilities	77,08	− 17,66	− 22,90%
Deteriorating areas	110,33	+ 2.33	+ 2,11%
Areas of redevelopment settlement	55,45	+ 1,69	+ 3,04%
Settlement areas	488,45	+ 3,26	+ 0,67%
System of urban centers	632,92	− 0.12	− 0,02%
Wooded areas	1.143,45	− 2,44	− 0,21%

From this first hypothesis emerges: (a) an adjustment of the perimeter involving a substantial increase in the system of predominantly agricultural areas and areas of the historical Park, and requirements for sport and leisure facilities (in some spaces of the Park indeed oversized) to mostly agricultural areas, leading in this case to a reduction of the infrastructures of about the 23% of the area concerned; (b) this observation is applicable also to the urban aggregates system, interested by 9 variation requests among which 8 are conversion in urban aggregates and 1 new agricultural destination (the most vigorous, causing the reduction of the 0,02% in favor of the mostly agricultural areas of the Park); (c) particularly significant is also the 4% increase of the areas that need intervention of both environmental recovery and settlement redevelopment, that the Park should look after in order to pursue the set objectives of environmental and landscape quality; (d) all of this, taking into account a little increase of the surfaces of the settlement areas (only 0,67% of the adopted ones) showing a strong environmental footprint impressed at the revision of the Park territorial Plan.

Further perimeter variations of the areas were hypothesized after: (x) the inconsistencies identified among the destinations of the current territorial Plan and its new laid down rules, showing the existence of polygons/cartographic areas without the corresponding discipline; (y) the transposition of new statistical studies on the Lambro Valley (among them the Plan related to forestry that the Park would adopt) which, inter alia, generated a new delimitation of the existing forests, therefore, comparing the forest of the current Plan and that of the 2010 study, the positive result of the forest variation is +1.046 ha (+91%), about twice the forestry of the current Plan.

5   The Relevance of the Methodology for the Sustainable Land Government

Thanks to the application of multivariate techniques the peculiarities of the territory of the Lambro Valley Regional Park have become recognizable, thus providing directions and requirements for the selection of actions designed to promote and protect the environment and the exiting landscape. This approach allowed us to objectively classify the individual proposals requested by the municipalities belonging to the consortium, giving them a specific class of susceptibility to transformation and thus ensuring the best choices of land government. The ability of the Public body to guide the future changes is undoubtedly a crucial factor, especially in densely urbanized urban areas such as the central Brianza. The Park therefore should no more be seen as an element of static constraint, unable to enhance the existing heritage but, through a clever stitching of the lacerations that for too many years characterized the territory – not only of the Brianza – must be able to drive scrupulously local bodies and private operators in an articulated and complex enhance process of environmental, economic and social resources. The use of GIS tools together with the application of multivariate techniques, facilitated the emergence of *values* or propensities and positive prerogatives, *disvalues* intended as unsuitability and negative specificity and *risks* or uncertainties and limitations on the use of resources, thus identifying a graduation in the sensitivity of the Park's basins towards the support of the natural and anthropic transformations, so that the policies of protection and enhancement can take advantage of a spatial framework aimed at the orientation of the choices and the definition of priorities. These priorities are the guide for the local communities interested at enhancing the ecosystem services that their territory offers, trying to develop the positive peculiarities and contain the negative ones. The GIS emerges in this context as a central tool through which it is possible to create a dialogue at different levels, not only environmental issues, but also social and economic, favoring the continuous exchange of data and information that are essential for knowledge and action.

6   Conclusions

The work of the Lambro Valley Regional Park showed how the techniques of multivariate geostatistical analysis represent an important tool to bring out the many features of an area so complex and articulated as the central Brianza. The method used and the results obtained have allowed, in a difficult economic period like the current one object of multiple economic pressures, to support the choices of land government on both municipal and regional scale by identifying more or less convertible areas and ensuring thus greater attention to the environmental system, its conservation and enhancement.

References

1. Anzaldi, C., Mirri, L.: Databases and mathematical models: a support for decisional problems. Urban Management, Padova (1984)
2. Arbia, G., Espa, G.: Statistica economica territoriale. Cedam, Padova (1996)
3. Bellacicco, A., Labella, A.: Le strutture matematiche dei dati. Feltrinelli, Milano (1979)
4. Bellacicco, A.: Metodologie statistico-matematiche per l'identificazione di aree subregionali per le politiche dei servizi sociali. In: Palermo, P.C. (ed.) Modelli di Analisi Territoriale, pp. 349–362 (2001)
5. Benzécri, J.P.: L'analyse des donnèes. La taxonomie. Dunod, Paris (1973)
6. Benzécri, J.P.: L'analyse des donnèes. Analyse des correspondances. Dunod, Paris (1973)
7. Borachia, V., Boscacci, F., Paolillo, P.L.: Analisi per il governo del territorio extraurbano. Angeli, Milano (1990)
8. Chadule, G.: Metodi statistici nell'analisi territoriale. Clup, Milano (1983)
9. Lee, C.: I modelli nella pianificazione. Marsilio, Venezia (1974)
10. Matthews, J.A.: Metodologia statistica per la ricerca geografica. Angeli, Milano (1981)
11. Palermo, P.C., Griguolo, S.: Nuoi problemi e nuovi metodi di analisi territoriale. Angeli, Milano (1984)
12. Paolillo, P.L.: Mesoscala e analisi del processo insediativo nell'evoluzione della rete locale. In: Borachia, V., Boscacci, F., Paolillo, P.L. (eds.) Analisi Per il Governo del Territorio Extraurbano. Angeli, Milano (1990)
13. Racine, J.B., Reymond, H.: L'analisi quantitativa in geografia. Marsilio, Venezia (1983)
14. Turer, M.M., Gardner, R.: Quantitative methods in landscape ecology. Springer, New York (1991)
15. Paolillo, P.L.: Una modalità descrittivo-classificatoria di individuazione dei "bacini d'intensità problematica alla scala regionale". In: Paolillo, P.L. (eds.) Terre Lombarde. Studi per un Ecoprogramma in Aree Bergamasche e Bresciane, pp. 103–153 (2000)
16. Paolillo, P.L.: L'estrazione dei bacini di intensità problematica ambientale in bergamasca e bresciana. In: Paolillo, P.L. (eds.) Terre Lombarde. Studi per un Ecoprogramma in Aree Bergamasche e Bresciane, pp. 287–247 (2000)

Geomatics in Climate Services and Local Information: A Case Study for Mediterranean Area

Emanuela Caiaffa[✉], Luigi La Porta, and Maurizio Pollino

ENEA, Italian National Agency for New Technologies, Energy and Sustainable Economic Development, UTMEA Energy and Environmental Modelling Technical Unit, C.R. Casaccia, via Anguillarese, 301, 00123 Rome, Italy
{emanuela.caiaffa,luigi.laporta,maurizio.pollino}@enea.it

Abstract. This paper describes a new idea and methodology developed for overcoming problems linked to the climate changes science and how to make them effectively accessible by end users. In today's scientific approach, communication, on climatic information and on climatic changes condition, has to be clear, accessible and perceptive to everyone, to definitely address the right interest on such problem. Results coming from climatic changes Research and Development projects, in addition to being accepted by policymakers, that have to take future decisions and practical actions to mitigate their possible negative effects, must also be recognized and believed by ordinary citizens that have to understand and accept restrictions, for example in energy consumptions, potentially affecting their habit of life. Accordingly, it will be necessary to regenerate policy-making so as to make it more direct and potentially open to networking interactive communication tools.

The present work substantially deals with the problem of the outcomes communication both in the field of climatic studies both of climatic trend conditions, also at a long-term scale.

Furthermore, it has been observed that one of the most widely discussed issues, in addition to that of the uncertainty of the climatic condition predictions, is the data/results communication to policy makers, stakeholders, environmental agencies, citizens, etc.

Keywords: Geomatics · Climate local information · Climate services · Communication and dissemination

1   Introduction and Objectives of the Work

In the framework of CLIMRUN project (Climate Local Information in Mediterranean region Responding to User Needs) (2011-2014, http://www.climrun.eu/) funded under the European Commission's Seventh Framework Programme (FP7), it was set up a case study aiming to develop:

- a methodology able to provide climate information, at regional and local scale, more adequate and inherent to reality and functional for the needs of different segments of society, ranging from simple citizens to stakeholders, policymakers, industry, services, etc. [1, 2, 3];

© Springer International Publishing Switzerland 2015
O. Gervasi et al. (Eds.): ICCSA 2015, Part III, LNCS 9157, pp. 540–555, 2015.
DOI: 10.1007/978-3-319-21470-2_39

- a bottom-up tool, in order to involve stakeholders and policymakers from the early phases of the research process, with the aim to have a good fit with their specific needs at regional and local scales [4,5,6];
- a web-based geographical application (WebGIS) through case studies applications in different sectors of society such as tourism, energy, natural hazards (e. g. wild fires) for descriptive target areas that in this case is the Mediterranean area [7].

The work described in this paper is part of the more general objective in creating a "Mediterranean area of a Climate Services Network" laying the basis for the development of a Mediterranean-wide network of climate services that would eventually confluence into a pan-European network.

In particular, during the "First CLIMRUN Workshop on Climate Services" (Trieste, Italy, October 2012) [8], it was also focused the CLIMRUN role on the development of, and training for, a new research expertise lying at the interface between climate science and stakeholder application within the Climate Services framework.

Since the running of the "World Climate Conference-3" (Geneva, 31 August- 4 September 2009) it was set up the following Declaration:

> *"We, Heads Of State end Government, Ministers and Heads Of Delegation Present at the High level Segment of the World Climate Conference-3 (WCC-3) in Geneva, noting the findings of the Expert Segment of the Conference:*
> *DECIDE*
> *to establish a Global Framework for Climate Services to strengthen the production, availability, delivery and application of science-based climate prediction and services"*

1.1   Climate Services: What They Are?

During the WMO congress (Geneva, Switzerland, May 2011) it was carried out a definition of a climate services. The climate services are:

- climate information prepared and delivered to meet user's needs,
- climate information provided in order to assist decision making by individuals and organizations. A service requires appropriate engagement along with an effective access mechanism and must respond to user needs.
- a tool to fill the Last Mile gap between large amount of climate data and user needs.

The High-level Taskforce full report [8] uses the following definition:

"Climate Services encompass a range of activities that deal with generating and providing information based on past, present and future climates and on its impacts on natural and human systems. Climate services include the use of simple information like historical climate data sets as well as more complex products such as predictions of weather elements on monthly, seasonal or decadal timescales, also making use of climate projections according to different greenhouse gas emission scenarios. Included as well are information and support that help the user choose the right product for the decision they need to make and that explain the uncertainty associated

with the information offered while advising on how to best use it in the decision-making process".

1.2   Uncertainties

Recently one of the most debated issue, involving stakeholders using climate services, is the uncertainty concept intrinsic in future climate change scenarios. Among the causes there are climate natural variability, climate model uncertainty as well as socioeconomic scenario uncertainty. Some of the uncertainty causes are nowadays identified by increasingly sophisticated climate models that, for example, stated that uncertainties in climate change projections strongly depends on the temporal horizon considered. The latter consideration is of primary importance, since the decisions of stakeholders and policy makers can strongly depend on the time range proposed by the forecasts.

1.3   Climate Services: Communication and Visualization

Among the well-known problems linked to scientific climatology research, we have to take in to account the problem of data communication and visualization, because the Climate Service goal is to deliver tailored climate information, consultable and customizable, in order to provide decision tools to end users, aiming at reducing the vulnerability of their activities and optimize investments in view of climate variability and change.

Climatology research is based on past, present and future climates and produce weather predictions on monthly, seasonal or decadal timescales, also making use of climate projections according to different greenhouse gas emission scenarios. These predictions are perceived by policymakers, citizens, etc., with a certain degree of uncertainty that becomes crucial when stakeholders/policymakers have to formulate mitigation measures that must be understood and adopted.

As Doubuais pointed out during the "First CLIMRUN Workshop on Climate", a serious problem to challenge is the uncertainty: what types they are, how to study them, how to mitigate them, etc. and finally, how to communicate them? [9].

The way how to communicate climate data/results to non-climatologists is a long term debated theme that, in the last time, has assumed a very relevant weight. The solution to the problem requires arguments and communication method able to capture people interest and to address their sensitivity on climate changes problems and on what it is necessary to know and to do at policy makers' level as well at citizen common life level, in order to mitigate them.

The communication of one result, one fact, one reality, is closely associated to how it is showed, in order to make it usable by non-specialists as stakeholders, policymakers, citizens, etc.

The present work, besides having stimulated the application of new methodologies in utilizing climatological models results, for the provision of adequate climate information at regional to local scale, proves to be a good starting test case, to exploit geographic and informatics (in a single word: Geomatics) tools, in order to territorially characterize the phenomena involved.

In this perspective, considering the territorial nature of the information to process and manage, it seemed useful to create a web-based geographical application (WebGIS) to communicate data and share results.

2   Materials and Methods

Climate models represent all physical processes of a region and are useful to explain the variations of temperature and precipitation, changes of solar radiation and wind speed. The Mediterranean area is particularly interested in climate change and for the future an increase in temperature with drier summers is expected. It is possible to imagine how this fact is of great importance because the Mediterranean area is densely populated.

Furthermore, a crucial issue is represented by the creation a genuine and effective link between scientists and administrators also including entities such as the Civil Protection and all those institutions involved in preparing prevention plans and classifying the impacts of climate change. For this reason, it is also very important to identify which are the targets most affected by climate change.

2.1   Starting Data

The coupled Regional Model PROTHEUS, developed at ENEA (the Italian National Agency for New Technologies, Energy and Sustainable Economic Development, "Energy and Environmental Modelling" Technical Unit), is composed by regional atmospheric model RegCM3 and MITgcm ocean model, coupled by the coupler OASIS3 [10]. The temporal horizon of interest for the study is the 2010-2050 interval, in order to encompass the contributions of both inter-decadal variability and greenhouse-forced climate change.

To characterise data coming from simulations conducted with MED44i_ERAINT and PROTHEUS climatological models Data the following climate indexes have been taken into account and used: Tourism Climatic Index (TCI) [11, 12, 13], Wind Power (WP) [14, 15] and Fire Weather Index (FWI) [16, 17, 18]. The results of the modelling simulations performed for the Mediterranean area are stored as files in format NetCDF (Network Common Data Form). These numerical files, due to their nature, are static and do not allow directly the superposition of other types of information, to combine them with different type of data in order to perform additional data combination/elaboration (e.g., spatial analysis).

2.2   GIS and WebGIS Methodologies

Through appropriate processing carried out within GIS software environment, the NetCDF files were processed in a GIS compatible format and loaded into a GIS project for spatial processing. In this way, it was possible to exploit all the advantages of being able to display and edit data in a spatial way, adding greater new significance from the point of view of climate indexes, jointly with the possibility to combine this information with other types of data characterizing the area from the socio-economic point of view. Using the mapping power, a GIS (Geographic Information System) environment allows visualizing, querying and analysing data to understand relationships, patterns and trends in order to improve communication and decision making processes.

A common GIS project is aimed at spatial data management, spatial analysis development (*geoprocessing*) and thematic maps representation. Geospatial data base and maps are processed, stored and managed in an appropriate repository: they can be viewed and interrogated by means of a desktop GIS suite or via web (through a WebGIS application). The latter is the interface of some of the major geographical results produced in various research projects, making them usable in an open way and accessible online. The term WebGIS refers to a collection of technologies allowing to take advantage of GIS capabilities via the Web (Internet/Intranet).

Thanks to the use of a web application, GIS projects, traditionally developed for stand-alone users, can be implemented on a web server (also called map-server), in order to allow, through the Internet network, the interaction with thematic maps and data associated with it. WebGIS applications are exploitable by common internet browsers (Mozilla Firefox, Google Chrome, Internet Explorer, etc.)

Among the benefits of using WebGIS technology there are: the global sharing of geographic information and geospatial data, ease to use by the Client, data network dissemination and the ability to reach a wider audience of users [7]. The WebGIS tool can be used, therefore, as an information consultation tool enriched by the geospatial component, query and analysis of geographic data and thematic maps. The WebGIS is not a simple extension of a GIS a Desktop suite, but it is part of the larger category of Web Oriented Software. The network is the way for the interchange of data through the Web browser and the communication is based on client-server architecture in which two independent modules interact to perform a task.

The development of a WebGIS as well as the implementation of an ICT architecture (hardware and software), should also refer to a set of standards, rules and procedures designed to facilitate the availability, consistency and access to geospatial data. This is the case of a Spatial Data Infrastructure (SDI), which is a common platform for research, publication and use of geographic data. An SDI is something more than a set of data: it handles data and its attributes, metadata, tools for searching and viewing data. An SDI provides an ideal environment to connect applications to data, at the same time influencing the creation of data and the development of applications based on standards and appropriate procedures.

3   CLIMRUN WebGIS

In order to make accessible and available on-line a series of results obtained in the framework of CLIMRUN project, a WebGIS application has been designed and developed. Another significance of the present work is that, just because the results should be manageable to all, much of the information available on a territorial area, have been organized and stored together in a single geodatabase. This means that the available data are consistent with each other and can easily be accessible by all users: in other words, a *Data Repository* for the end users was built. The geodatabase can be used at various levels by exploiting the potential offered by GIS tools, starting with the effective support for the operational management in various sectors of interest.

The main purpose of CLIMRUN WebGIS is to make territorially compatible and easy to handle the climatic models outputs, which provide data on the current and future scenarios concerning some themes of interest that have been individuate by the climate indexes considered. For the development of this application the basic concept was to take advantage of Free Software/Open Source (FOSS) packages[1]. The logical architecture of the WebGIS is shown in Figure 1 and consists of the operational chain:

— Data Repository -> Web Server (e.g. GeoServer[2]) -> Library (e.g. OpenLayers[3]) -> Map Viewer (the proper WebGIS application interface).

The Data Repository identifies the storage area that contains the set of data used (in general, in GIS format) and that allows accessing only to the devices physically defined at the level of the Storage Area Network (SAN), in order to ensure the absolute integrity and consistency of the same.

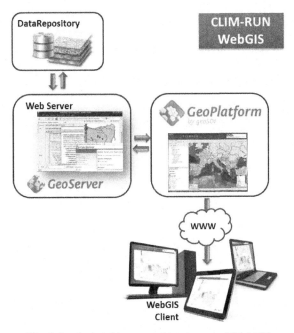

Fig. 1. Logical architecture of implemented WebGIS

---

[1] http://freeopensourcesoftware.org/

[2] http://geoserver.org/

[3] http://www.openlayers.org/

The Web Server is a set of equipment and related software that allow the system to organize information and make it accessible to the network. The web server can provide maps and data, from a variety of formats, to standard clients, such as web browsers and desktop-like GIS software. This makes it possible to store spatial data in almost any format users prefer. From the technical point of view, GeoServer is the reference implementation of the standards defined by the Open Geospatial Consortium[4] (OGC): Web Feature Service (WFS), Web Coverage Service (WCS) and Web Map Service (WMS). The Open Geospatial Consortium (formerly the OpenGIS Consortium, http://www.opengeospatial.org/) is an international no-profit organization based on voluntary consent, which is responsible for defining the technical specifications for geospatial and location services (*location based*). OGC is made up of over 280 members (governments, private industry, universities) with the objective to develop and implement standards for the content, services and the exchange of geographic data that are "open and extensible" and which promotes the interoperability. The specifications defined by OGC are public (PAS) and available for free.

In order to build the WebGIS for CLIMRUN project, it was then decided to use open source software, whose use is governed only by the license GNU General Public License. In particular, as web server and web application, two software environments used are: GeoServer and Geo-Platform[5] by geoSDI[6].

3.1   WebGIS Application Interface

The CLIMRUN WebGIS application (Figure 2) is fully operational and can be accessed at the following address: http://www.climrun.eu/webgisclimrun/index.html (using the credentials: username: climrun; password: climrun).

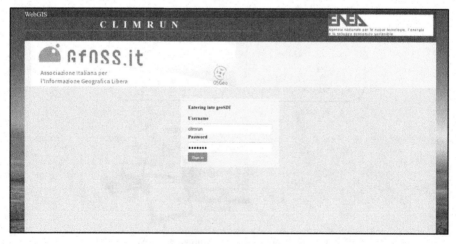

Fig. 2. CLIMRUN WebGIS access page

---

[4]   http://www.opengeospatial.org/standards/
[5]   https://github.com/geosdi/geo-platform
[6]   http://www.geosdi.org/

The first step of the carried out workflow was the elaboration of numeric files (one example for TCI is reported in Figure 3), coming from mathematical model providing present (1961-1990) and future (2021-2050) scenarios for the topics of interest: TCI, FWI and WP. These numeric files, once inlayed inside GIS environment, were transforming in geographic layers in order to become manageable through geo-processing functions.

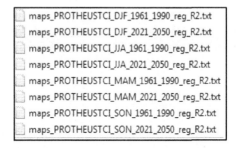

Fig. 3. Starting data example (TCI, Protheus elaboration)

Having available these scenarios in a GIS format, the subsequent purpose of the research was to map the temporal evolution and then, through the geo-processing functions, to calculate changes and differences and mapping it on the Mediterranean area. In other words, it was possible to map the future predictable changes of important index as TCI and FWI, as well as in WP trend.

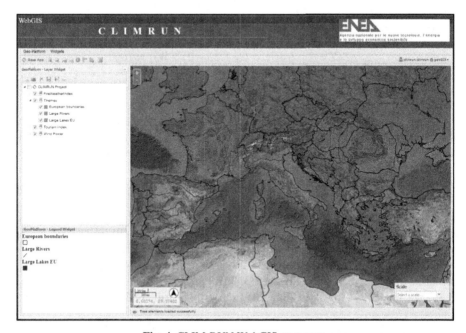

Fig. 4. CLIM-RUN WebGIS start page

For example, in order to be processed in the GIS environment, it was produced the corresponding geographical layer for each file representing the following time periods: December-January-February (DJF), June-July-August (JJA), March-April-May (MAM), September-October-November (SON).

The purpose was to observe and highlight the differences on territory between 2021-2050 and 1961-1990 scenarios, succeeding in seeing where and how large were the areas showing improvements respect to those affecting by worsening. TCI and WP were evaluated for all the four seasonal periods; FWI only for June-July-August (JJA). In order to make the result more readable, some reference layers were also loaded, like the European Administrative Boundaries, major rivers, large lakes, cities, etc. (Figure 4).

3.2   TCI: Tourism Climatic Index

The Tourism Climatic Index assessment is based on a methodology developed by Mieczkowski [19] to quantify and classify "favourable climate", for typical sightseeing tourism. The aim of the TCI evaluation is to determine how touristic resources might change in Europe due to climate changes.

The TCI evaluation is realized through the application of climate model projections that include high spatial resolution, high temporal resolution and for the major part of climate model, a reliability analysis. In order to quantify the TCI index, several variables are weighted in accordance with the relative importance that they are supposed to bear in tourist well-being. The problem of weighting climatic variables in the TCI formula is still under discussion.

The Mieczkowski formula is:

$$TCI = 2 \cdot [(4C_{dt} + C_{dl} + 2R + 2S + W)] = 100 \tag{1}$$

in which C_{dt} is the daytime comfort, C_{dl} is daily comfort, R is precipitation, S is the sunshine and W is the wind.

In the TCI are combined five climatic characteristics important for tourism. As shown in the formula (1), all variables are calculated in their own specific units and then, in a very empirical way, are rated on a scale from -3 to 5 (or 0 to 5 for sunshine and wind). As specified, all sub-indices have a maximum score of 5, that leads to an overall maximum score of 100, with suitable score lying above 40, good score above 60 and excellent score above 80 (see Table 1) [19].

To get the changes that will affect the TCI, data processed by the climatological models have been taken into account, for the observed period 1961-1990 and for the predicted scenario 2021-2050.

In Figure 5 it is possible to see the TCI values for June-July-August in 1961-1990 scenario, distributed in different classes. In Figure 6 it is possible to see the TCI values for June-July-August in 2021-2050 scenario, distributed in different classes.

Table 1. Rating categories of the Tourism Climatic Index (Mieczkowski 1985)

TCI score	Category	Mapping category
90–100	Ideal	Excellent
80–89	Excellent	
70–79	Very good	Very good and good
60–69	Good	
50–59	Acceptable	Acceptable
40–49	Marginal	
30–39	Unfavourable	
20–29	Very unfavourable	Unfavourable
10–19	Extremely unfavourable	
< 10	Impossible	

To assess the variation of the TCI between the past scenario (1961-1990) and the predicted scenario (2021-2050), by means the tools provided by the Esri ArcGIS suite[7], a simple map algebra calculation was performed (pixel by pixel subtraction between the two raster files relative to the scenarios). In Figure 7 the layer indicated as "JJA DIFF 2150 6190" describes the difference between TCI values, for the scenario 2021-2050 and the scenario 1960-1990 (JJA seasonal period).

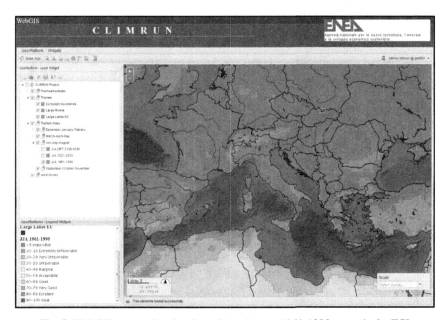

Fig. 5. WebGIS output showing June, July, August 1961-1990 scenario for TCI

---

[7] http://www.esri.com/software/arcgis

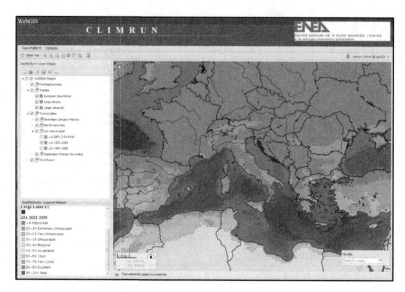

Fig. 6. WebGIS output showing June, July, August 2021-2050 scenario for TCI

The map provides a visual result from which is possible to individuate, in a territorial way, where the changes are located and how much these areas are changed. In this way, the amount of quantitative and territorial change is perceived (Figure 7).

As shown also by the legend of the Figure 7, highlighted separately in upper-right corner of the same Figure, the areas in which a TCI class increase is expected (green colour) are clearly and geographically defined.

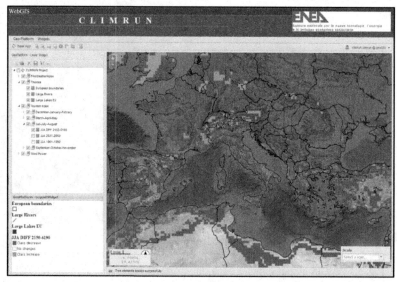

Fig. 7. Example of output combination showing changes in TCI (in red: class decrease; in green: Class increase; transparent colour: No changes)

3.3   WP: Wind Power

Wind fields can be considered as climate variables of interest, both for study on energy for renewable production [14] and in forecasting natural hazards for wild fires [20]. Besides the now-casting and the seasonal forecasts, in the energetic sector is necessary to understand wind modelling capabilities at a longer time scale in order to have a good siting evaluation and to assess risks that may affect the return on investments on time.

Figure 8 shows the trend of the scenarios, obtained by analysing WP data set over the entire European and Mediterranean area. The question to answer is: *what changes we expect in wind power? It will decrease, or will increase in the next forty years around?* By mapping the WP scenarios for the two time interval considered (thematic maps named: "WP_1961_1990" and "WP_2021_2050") and performing their difference, it was possible to calculate and then visualize areas in which WP decreases and areas in which it increases.

In particular in Figure 8 is reported the WP thematic map (layer indicated as: "DIFF 2021-2050 1961-1990") showing the difference between the two scenarios of reference. The chromatic representation scale provides the entity and the position on the map of the decrease/increase of wind power in GWh/year. It is possible to clearly observe some zones, in the central Europe and in some Mediterranean areas (in red colour), in which the difference is in positive: this indicate an expected increase of WP of about 3 GWh/year.

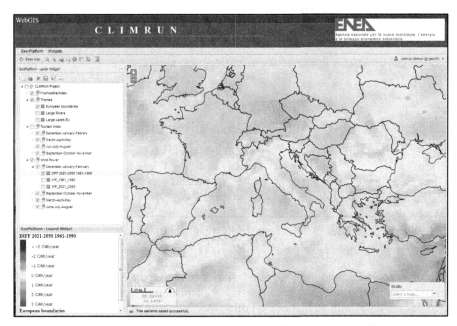

Fig. 8. The layer indicated as "DIFF 2021-2050 1961-1990" is showing the difference between the two scenarios

3.4   FWI: Fire Weather Index

Another fundamental task is to understand the issues related to forest fires occurrence and intensity and their implication with the climate changes. Taking into account the observed and predicted trends of the Mediterranean climate, it is possible to point out that more summer periods characterised by droughts are expected, in some case even longer respect the typical seasonal climate.

In order to investigate the relationship between wildfire risk and meteorological conditions, one of the most accredited methods is the Canadian Fire Weather Index (FWI) [21]: a meteorologically based index designed in Canada and used worldwide, including the European and Mediterranean areas. To calculate daily values of FWI, was used a generalized approach based on the following meteorological observations: mean daily values of air maximum temperature, relative humidity, wind speed and rainfall data (24 hours accumulated).

In the present study the KNMI-RACMO2 regional climate model, provided by the Royal Netherlands Meteorological Institute [22, 23] (KNMI), was forced with output from a transient run conducted with the ECHAM5 Global Climate Model.

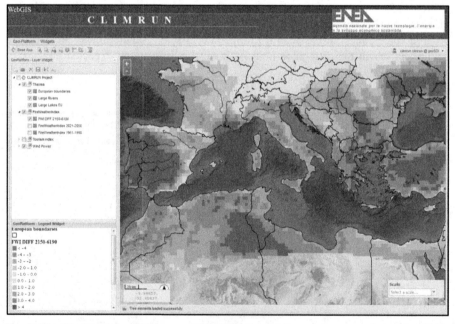

Fig. 9. Fire Weather Index differences: 2021-2050 June, July and August period minus 1961-1990 June, July and August period

In order to understand the FWI trend, shown in the Figure 9, as the FWI strongly depends from meteorological parameters (e.g., temperature), it is reasonable to deduce that the areas with an higher FWI difference are those more affected by drought, air temperature increase, rain scarcity. Thus, a TCI class change, in a positive or negative way, indicates those territorial zones where a future transformation (in the time interval 2021-2050) is potentially expected.

The FWI can be exploited in activities related to fires risk assessment and – in general – to Civil Protection services, allowing more effective strategies for wildfires prevention and forest protection management, all the year around, but in particular, during the most risky periods (like summer).

However, it is important to underline that the FWI was calculated taking into account only the meteorological parameters of the Mediterranean area, not even considering other important variables able to trigger wildfires, such as land cover, vegetation categories and altitude.

4   Conclusions

To narrow the distance between citizens [24], policy makers and scientists, in a new type of knowledge-based society, it is time to think out a new way to approach information. The geographic component embedded in a large part of commonly used data can turn these data into "geo-knowledge", a new type of territorially-linked information. Continuous transformations of economic and social institutions have changed policy-makers approaches to socio-economic problems. The need has now emerged to develop future-oriented tools to support scientific works and technological activities, and to conduct integrated research aimed at identifying the tools most helpful to building a European knowledge-based society.

The CLIMRUN WebGIS has revealed its effectiveness in publishing and sharing results coming from climatic mathematical models with the possibility to be coherently linked to the territory. Climatic models data, once uploaded and organized into a GIS project, become suitable geographic information and from this point onwards can be treated as GIS layers to all-round. By organizing a wealth of data from various agencies and sources, in a unique/coherent format and in the same geographic reference system, it has been possible to overlap each specific layer performing typical GIS functions in order to find the seeking data and analyse them.

The usefulness of the work is to understand where, territorially, insist the differences between diverse scenarios and then where are expected variations, changes in temperature, in wind power availability, etc., which may affect, for example, human activities such as tourism, etc.

The exploitation of open-source software, finally, was a key factor able to simplify the way for the dissemination and sharing of results produced within the research activities.

Acknowledgements. This work was supported by CLIMRUN project (www.climrun.eu, grant agreement no. 265192), funded under the European Commission's Seventh Framework Programme (FP7). Regional climate model data were provided by the EU FP6 project ENSEMBLES (www.ensembles-eu.org).

References

1. Smith Jr, W.J., Liu, Z., Safi, A.S., Chief, K.: Climate change perception, observation and policy support in rural Nevada: A comparative analysis of Native Americans, non-native ranchers and farmers and mainstream. America Environmental Science & Policy **42**, 101–122 (2014)

2. Spruijt, P., Knol, A.B., Vasileiadou, E., Devilee, J., Lebret, E., Petersen, A.C.: Roles of scientists as policy advisers on complex issues: a literature review. Environmental Science & Policy **40**, 16–25 (2014)

3. Hamilton, L.: Education, politics, and opinions about climate change: evidence for interaction effects. Clim. Change (2010). http://dx.doi.org/10.1007/s10584-010-9957-8

4. Caiaffa, E.: Geographic Information Science in Planning and in Forecasting. Institute for Prospective Technological Studies (eds.) in cooperation with the European S&T Observatory Network. The IPTS Report, vol. 76, pp.36–41. European Commission JRC, Seville, (2003)

5. Caiaffa, E.: Geographic Information Science for geo-knowledge-based governance. In: Proc. AGILE 8th Conference on Geographic Information Science, Estoril Portugal, 26-28-May (2005)

6. Caiaffa, E., Cardinali, S., Screpanti, A., Valpreda, E.: Geographic Information Science: a Step Toward Geo-governance Solutions. In: Proc. of International Conference on Information & Communication Technologies: From Theory to Applications, Damascus Syria, 7-11 April (2008)

7. Pollino, M., Modica, G.: Free web mapping tools to characterise landscape dynamics and to favour e-participation. In: Murgante, B., Misra, S., Carlini, M., Torre, C.M., Nguyen, H.-Q., Taniar, D., Apduhan, B.O., Gervasi, O. (eds.) ICCSA 2013, Part III. LNCS, vol. 7973, pp. 566–581. Springer, Heidelberg (2013). doi:10.1007/978-3-642-39646-5_41. ISBN 978-3-642-39645-8

8. Godess, C.: Climate Services. Climatic Research Unit University of East Anglia UK c.goodess@uea.ac.uk, First CLIM-RUN workshop on climate services (2012)

9. Doubuais, G.: Using ENSEMBLES to communicate uncertainty CLIMRUN School on climate services. ICTP, Trieste (2012)

10. Artale, V., Calmanti, S., Carillo, A.: Dell'Aquila A., Herrmann M., Pisacane G., Ruti P.M., Sannino G., Struglia M.V., Giorgi F., Bi X., Pal J. S., Rauscher S. The Protheus Group: An atmosphere–ocean regional climate model for the Mediterranean area: assessment of a present climate simulation". Clim. Dyn **35**, 721–740 (2009)

11. Perch-Nielsen, S.L., Amelung, B., Knutti, R.: Future climate resources for tourism in Europe based on the daily Tourism Climatic Index. Climatic Change **103**, 363–381 (2010). doi:10.1007/s10584-009-9772-2

12. Amelung, B., Viner, D.: Mediterranean tourism exploring the future with the Tourism Climatic Index. J Sustain Tour **14**, 349–366 (2006)

13. Hamilton, J.M., Maddison, D.: RSJ T.: Climate change and international tourism: a simulation study. Glob. Environ. Change **15**, 253–266 (2005)

14. Caiaffa, E., Pollino, M., Marucci, A.: A GIS Based Methodology in Renewable Energy Sources Sustainability. International Journal of Agricultural and Environmental Information Systems **5**(3) (July-September, 2014)

15. Baban, S.M.J., Parry, T.: Developing and applying a GIS-assisted approach to locating wind farms in the UK. Renewable Energy **24**(1), 59–71 (2001)

16. DaCamara, C.C., Calado, T.J., Ermida, S.L., Trigo, I.F., Amraoui, M., Turkman, K.F.: Calibration of the Fire Weather Index over Mediterranean Europe based on fire activity retrieved from MSG satellite imagery. International Journal of Wildland Fire (2014). doi:10.1071/WF13157

17. Amraoui, M., Liberato, M.L.R., Calado, T.J., DaCamara, C.C., Coelho, L.P., Trigo, R.M., Gouveia, C.M.: Fire activity over Mediterranean Europe based on information from Meteosat-8. Forest Ecology and Management **294**, 62–75 (2013)

18. Brotons, L., Aquilué, N., de Cáceres, M., Fortin, M. j., Fall, A.: How fire history, fire suppression practices and climate change affect wildfire regimes in Mediterranean landscapes. Wildfire Regime Change in Mediterranean Landscapes PLOS ONE **8**, e62392 (2013)
19. Mieczkowski, Z.: The Tourism Climatic Index: a method of evaluating world climate for tourism. The Canadian Geographer / Le Géographe canadien **29**(3), 220–233 (1985)
20. Cane, D., Ciccarelli, N., Gottero, F., Francesetti, A., Pelfini, F., Pelosini, R.: Fire Weather Index application in north-western Italy. Advances in Science and Research, Copernicus Publications **2**, 77–80 (2008)
21. Turner, J.A., Lawson, B.D.: Weather in the Canadian Forest Fire Danger Rating System. A user guide to national standards and practices. Environment Canada, Pacific Forest Research Centre, Victoria, BC. BC-X-177 (1978)
22. Van den Hurk, B., Van den Hurk, B.J.J.M., Co-Authors: KNMI Climate Change Scenarios 2006 for the Netherlands. KNMI-publication: WR-2006-01, p. 82 (2006)
23. Lenderink, G., van den Hurk B., van Meijgaard, E., van Ulden, A.P., Cuijpers, H.J.: Simulation of present-day climate in RACMO2: first results and model developments. KNMI Technical Report **252** (2003)
24. Modica, G., Zoccali, P., Di Fazio, S.: The e-participation in tranquillity areas identification as a key factor for sustainable landscape planning. In: Murgante, B., Misra, S., Carlini, M., Torre, C.M., Nguyen, H.-Q., Taniar, D., Apduhan, B.O., Gervasi, O. (eds.) ICCSA 2013, Part III. LNCS, vol. 7973, pp. 550–565. Springer, Heidelberg (2013)

Sharing Environmental Geospatial Data
Through an Open Source WebGIS

Grazia Caradonna[1], Benedetto Figorito[2], and Eufemia Tarantino[1(✉)]

[1] Politecnico di Bari, via Orabona 4 70125, Bari, Italy
eufemia.tarantino@poliba.it
[2] ARPA Puglia, Via Trieste 27 70126, Bari, Italy

Abstract. In recent years, people's need to participate to decision making, especially when it concerns inalienable human rights such as health and living in a healthy environment, has become increasingly manifest. In order to meet the request for environmental information sharing on the web and to make citizens feel "partakers" in the development of environmental policies, the Physical Agents Simple Operative Unit of ARPA Puglia, developed an open source WebGIS as a communication, participation and working tool for both Citizens and Technicians.

This paper proposes an efficient approach to customize and integrate an open source WebGIS system based on MapServer and Pmapper. The layout of the WebGIS was customized by filling pages in Cascading Style Sheets (CSS) to make it intuitive and easy to use. The features offered are those commonly provided by a WebGIS system, in particular: geographical navigation (pan, zoom, zoom to selection), query time and multiple layers, transparency level options, printing and exporting of current image views or pdf files. Environmental data results from a query can be downloaded in pdf, kml and shp formats. The possibility to download files is a key component of the system as it allows the average expert user to find data in an easily and processable format.

Keywords: WebGIS · MapServer · Pmapper · Open data · Shared information

1 Introduction

Access to environmental information and public participation in decision-making are today two basic principles, which should inspire the work of public administrations [4]. With such activity, considerable amounts of territorial data are collected, including them into more or less digitally formatted archives showing a huge potential they remain an unknown and unexploited resource. In less than a decade, we have moved from a situation of seeking to understand the world with scarce and costly data to an overabundance of open data [10]. Many national and international initiatives have been devised for the purpose of adding value to public sector information. These include the European INSPIRE directive considered to lead to the creation of an Infrastructure for Spatial Information in the European Community and numerous directives in the field

© Springer International Publishing Switzerland 2015
O. Gervasi et al. (Eds.): ICCSA 2015, Part III, LNCS 9157, pp. 556–565, 2015.
DOI: 10.1007/978-3-319-21470-2_40

of e-government and re-use of public documents in the public sector, some of which have been incorporated into Italian regulations [6], [8], [1].

Interactive mapping or Internet GIS were launched about twenty years ago [14] and they are now effectively applied in many fields [7], [11], [13], [16], [19]. These considerable WebGIS applications for acquiring heterogeneous datasets, performing spatial analysis and providing flexible data visualization capabilities all prove that WebGIS is a valuable technology in disseminating geospatial data through open tools. On the other hand, there are some primary difficulties in working with open source when compared to commercial products. Open source applications often require greater amounts of time and computational skill and the professional support available often depends on the maturity of software and the size of the user community [18]. These complexities aside, however, it is the low cost and customizability of open source that still makes it an appealing alternative for many users. At first, the web was only used as a one-way channel for publishing passive geographic information and processed data were not shared in editable formats. Only in recent years, we have been witnessing the development of the Web as a channel for two-way transmission offering the opportunity to make geographic information available to a wide audience [20]. Because of an extensive use of graphics, sometimes it loses functionality in speed, i. e. long waits for consultation in cases of heavy data and not optimized interfaces [5], and limitations in updating of data. Performance analysis becomes an essential issue when developing architectures of Web-based GIS platform to overcome these drawbacks [12].

This paper proposes a customized architecture that integrates two open source platforms: the web GIS server MapServer, for publishing spatial data and interactive mapping applications on the web, and the Pmapper automatic mapping technology, a GUI front-end written in PHP and Javascript allowing for a dynamic control of MapServer. The result is exhaustive, proving the system to be very competitive compared to proprietary software [14]. The features offered are those commonly provided by a WebGIS system, in particular: geographical navigation (pan, zoom, zoom to selection), query time and multiple layers, transparency level options, printing and exporting of current image views or pdf files. Environmental data result from a query that can be downloaded in pdf, kml and shp, allowing the average expert user to find data in an easily processable format.

2   Methods

2.1   The Study Case and Data

Physical agents are sources of energy that may cause injury or disease, e. g. noise, vibration, radiation (ionizing and non-ionizing), etc. Their presence in living environments and work determines the level of input energy prejudicial to human health. In order to meet the growing demand for shared knowledge about polluted areas and localization of monitoring instrumentations by external users, the Physical Agents Unit of ARPA Puglia implemented a geographic web platform (http://www.webgis.ARPA.puglia.it/) to deliver data and inform citizens on the state of the environment and monitor whether stressors respect the limits established by the law. The geographic hosting system is able to work both with data stored directly on the internal server and with information accessible via web services. Data are saved in the geodatabase by default, but the system is able to also work with data in individual files of different formats. Data repositories are represented by the File Repository and the GeoDatabase. All data were pre-processed with Quantum GIS (QGIS) open desktop software and stored in the geographic coordinate reference system EPSG 4326 (WGS84). MapServer supports numerous OGC standards, allowing users to publish and use/access data through neutral service interfaces. WebGIS data components are divided into two big categories: Environmental and Cartographic data (Table 1). Most of data were produced by the ARPA Agency, while the Geologic and Administrative Maps were open data provided by the Joint Research Centre ISPRA and the National Institute of Statistics (ISTAT) respectively. The WebGIS also uses external Web Map Services (WMS) i.e. information sources such as OpenStreetMaps.

Table 1. Data used as WebGIS layers

Category	Thematism	Source	Type
Environmental	Strategic noise maps for Bari and Taranto agglomerations	ARPA Puglia	Vector polygon
Environmental	Weather stations	ARPA Puglia	Vector polygon
Environmental	Radon Gas Monitoring	ARPA Puglia	Vector polygon
Environmental	Electromagnetic field Monitoring	ARPA Puglia	Vector polygon
Environmental	Post-activation controls of radio telecommunications systems.	ARPA Puglia	Vector polygon
Cartographic	Base map	OpenStreetMap	WMS Connection
Cartographic	Geologic map	ISPRA	Vector polygon
Cartographic	Administrative boundaries (2011)	ISTAT	Vector polygon
Cartographic	Geographic grid	ARPA Puglia	Vector line

2.2   Application Architecture

The Web site runs on an Apache HTTP Server version 2.2.22 [15] by a Quad-core Intel® XEON processor E 7330 (3.2 GHz), running Windows server 2007 with one SCSI virtual disk of 80 GB capacity and 4 GB of RAM. WebGIS is based on the MapServer platform (version 6.4.1) and modules of Pmapper for publishing spatial data on the web. The user, on the client's side, sends a query, commonly trough a browser (Firefox, Chrome, etc.), to the web server. The Apache Web Server receives and processes HTTP requests, and invokes MapServer throw a Pmapper infrastructure (Fig.1.). On the server side, there is a common gateway interface (CGI) software (e.g., PHP script) using a map file and GIS input data (e.g., shapefiles, databases) to produce an image-map and send it back to the client [9].

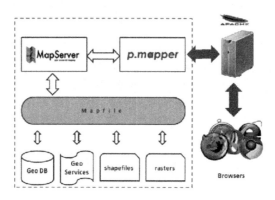

Fig. 1. WebGIS Architecture

2.2.1   MapServer

UNM MapServer is a development environment which allows the creation of Web applications for uploading, viewing and accessing data, cartographic and of other types, from GIS systems. It is the GIS engine of the WebGIS project. MapServer consists of three different components: map file, template files and CGI program. The CGI is an executable program; the map file is to configure data retrieval and representation modes and the template file is the web page that serves as an interface between the users and the application.

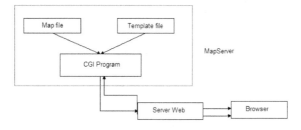

Fig. 2. MapServer elements

The ability of MapServer to serve geographical information through the Web is based on a single ASCII file: the "mapfile". In a "mapfile", all available geographical information is indexed, and the relationships between the geographical information and the display objects are defined. The established relationships uniquely specify the layout, organization, style and scale of publication of geographical information. In a "mapfile", geographical information is organized in separate layers. For each layer, classes are defined and styles of visualization are associated to each class. Each information layer may have different geographic coordinate systems. To follow, the map server reprojects the maps in the reference system indicated in the first part of the mapfile. To this purpose, MapServer uses the library Proj4 "Geographic Projection Library". Its structure is flexible, e.g. the publication of maps uses different styles (colors, patterns, symbols) depending on the scale of visualization of geographical information. The template file is a graphical interface that connects MapServer and users and it is constituted by an HTML page equipped with Javascript. It contains the CGI parameters and the appropriate template variables as reference. The reference template is enclosed in square brackets in which the Web server replaces the user-defined settings for every work session and the incoming results of CGI output procession.

MapServer has three operation modes:

- Map;
- Browse;
- Query.

In Map mode, it produces an image that is displayed in the browser. In browser mode, it produces a temporary image that is then included in the HTML template. In query mode, it runs a query on attributes.

2.2.2   The Web Page

The web mapping client was based on Pmapper 4.3.2 framework. It consists of PHP scripts that provide access to mapping-related "widgets" and functions. The Web page was customized based on the GUI, integrating and setting Pmapper plugins and tools of interest (Fig. 3.). Each layout file in the root directory of a page corresponds to a page that is shown in ARPA Puglia WebGIS. The layout of the main entry page contains the map elements to be shown. These panels are contained in *<div> tags* that generally define left, center or right columns of the page. The layout displays:

- a map in the center;
- a reference map at the bottom right;
- a table of content for each layer   on the right;
- a vertical toolbar containing navigation tools (zoom in, zoom out, pan, zoom on coordinates, etc..) and advanced query (Identify, Select, point marker, etc.) on the right of the map area;
- a dropdown menu with help, print, download map (georeferenced) and home functions.

Template-based customizations were entirely performed using the CSS. The web pages comply with the standards defined by the W3C Organization having two types of authorizations for the eXtensible HyperText Markup Language (XHTML) and the CSS for visual layout. Two icon checks are placed in the lower left. The contents of the home page are mainly divided into Javascript files (Fig.4.). The WebGIS uses client-side scripts, i.e. it uses a js. file to execute instructions on a Web page.

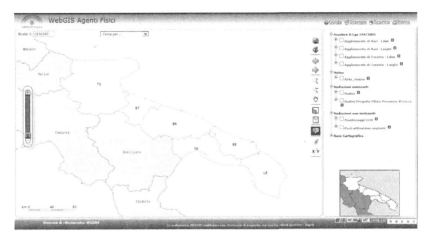

Fig. 3. WebGIS Homepage

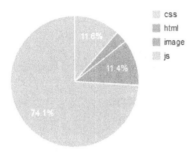

Fig. 4. WebGIS homepage computer typesetting

3   WebGIS Functions and Testing

The major components of the ARPA Puglia WebGIS are categorized into navigation tools, search tools, layer visualization and export results. The vertical toolbar, to the left of the map, shows various zoom tools. The toolbar also contains select/deselect functions to query multiple features of interest and identify widgets for investigating a single entity. The marker tool and the navigation help the user mark and localize points of interest on the map.

The information is organized by themes. For each theme, a legend is provided. The order of the themes specifies the sequence used to prepare the map through overlays of selected themes. (Fig 5.). Every layer can be selected through a click on the map: the result is a table with layer's attribute that can be downloaded (Fig 6 -7), exported in three different formats: xls, pdf and shp or printed (Fig 8.).

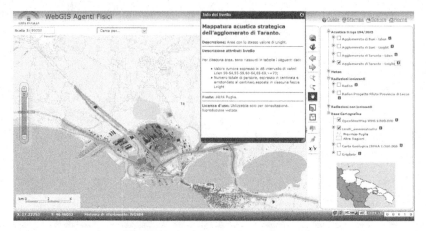

Fig. 5. Interactive layer description

Fig. 6. Selection by attribute

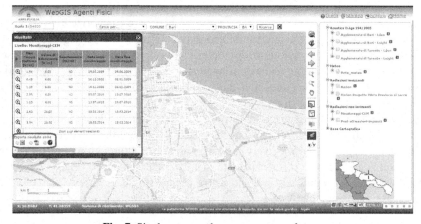

Fig. 7. Single query and export query results

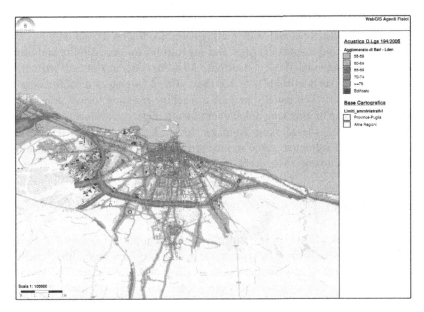

Fig. 8. Printed page of the selection results

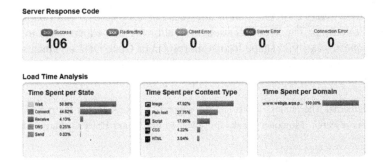

Fig. 9. Test results obtained from the pingdom site

The WebGIS efficiency wastested using the pingdom site (http://tools.pingdom.com) which examines all parts of a web page (Fig 9): file sizes, load times and other details about every single element of a web page (HTML, JavaScript and CSS files, images, etc.). The tests showed a performance grade of 73/100 which seems to be satisfactory. The main disadvantage of the implemented project is the dependency on internet connection [14].

The WebGIS was tested on a Notebook (Win8.1, 1.7 GHz, 4 GB of RAM) with a domestic Wifi ensuring peak rate of 150 Mbps and on a last generation smartphone with 3G and 4G mobile broadband connection. The difference between 3G and 4G mobile broadband connections is the speed: the 3G connection ensures peak rates up to 42 Mbps, while the 4G reaches peaks of 100-150 Mbps. The map generation and display of layers very much depends on the connection speed and not on the device being used. While with 3G network the user must wait about 3 seconds before starting viewing, with a 4G network and Wifi waiting is limited to one second only.

Any website should respect the rules and suggestions for accessibility and usability issued by the international W3C consortium. Most Web documents are written using markup languages, such as HTML or XHTML. These languages aredefined by technical specifications, which usually include a machine-readable formal grammar (and vocabulary). The act of checking a document facing these constraints is called validation, and this is what the Markup Validator does. The realized WebGIS was tested through the markup validator, a free service provided by W3C that helps checking the validity of Web documents. The WebGIS passed the W3C test.

4   Conclusion

This paper, presents a WebGIS intentionally simple and attractive aimed at the diffusion of information regarding environmental monitoring and modelling results produced by the Physical Agents Simple Operative Unit of ARPA Puglia. The web system realized consists of various user-friendly open source GIS tools for spatial data visualization, analysis, querying and lastly production of maps in the form of map prints and shape format [3]. This system combines two approaches. It offers the possibility to publish proprietary data and use metadata catalogues to share information with other portals. It also offers easy integration and reuse of user derived information sources like OpenStreetMap. The WebGIS strength is allowing non-expert users to download structured queries easily in shape format, in respect of Open Data arguments by which data must be open, reusable and in a processable format. Future developments are foreseen in implementing open tools for acquiring data from citizens.

Acknowledgments. The authors appreciate the support of ARPA Puglia for providing material related to this research. Moreover thanks are given to the Agenti Fisici Operative Unit of Arpa Puglia.

References

1. Borga, G., Castelli, S., Dalla Costa, S., Di Prinzio, L., Picchio, S., Sau, A.: A prototype system for monitoring information demand and data. In: Urban Data Management: Urban Data Management Society Symposium 2007, Stuttgart, Germany, 10-12 October 2007, p. 333. Routledge, October, 2007
2. Borruso, G.: Cartografia e Informazione Geografica "2.0 e oltre", Webmapping, WebGIS. Un'introduzione Cartography and Geographic Information "2.0 and Beyond", Webmapping, WebGIS. An introduction. Bollettino dell'Associazione Italiana di Cartografia **147**, 7–15 (2013)
3. Boulos, M.N., Honda, K.: Web GIS in practice IV: publishing your health maps and connecting to remote WMS sources using the Open Source UMN MapServer and DM Solutions MapLab. International Journal of Health Geographics **5**(1), 6 (2006)
4. Carver, S., Evans, A., Kingston, R., Turton, I.: Accessing Geographical Information Systems over the World Wide Web: Improving public participation in environmental decision-making. Information Infrastructure and Policy **6**(3), 157–170 (2000)

5. Chandramali, E.A.G., Wijesekera, N.T.S.: Identification of WebGIS Development Potential and Issues–A land and Water case Study Application for Moratuwa, Sri lanka (2014)
6. Directive, I.N.S.P.I.R.E.: Directive 2007/2/EC of the European Parliament and of the Council of 14 March 2007 establishing an Infrastructure for Spatial Information in the European Community (INSPIRE). Published in the official Journal on the 25th April (2007)
7. Doyle, S., Dodge, M., Smith, A.: The potential of web-based mapping and virtual reality technologies for modelling urban environments. Computers, Environment and Urban Systems **22**, 137–155 (1998)
8. Giannola, E.: Mappe online e processi partecipativi innovativi per la costruzione di una nuova immagine del territorio. In: 17 Conferenza nazionale ASITA, pp. 769–775 (2013)
9. Gkatzoflias, D., Mellios, G., Samaras, Z.: Development of a web GIS application for emissions inventory spatial allocation based on open source software tools. Computers & Geosciences **52**, 21–33 (2013)
10. Kitchin, R.: Big data and human geography Opportunities, challenges and risks. Dialogues in Human Geography **3**(3), 262–267 (2003)
11. Lazzari, M., Danese, M., Masini, N.: A new GIS-based integrated approach to analyse the anthropic-geomorphological risk and recover the vernacular architecture. Journal of Cultural Heritage **10**, e104–e111 (2009)
12. Li, Z., Yang, C., Sun, M., Li, J., Xu, C., Huang, Q., Liu, K.: A high performance web-based system for analyzing and visualizing spatiotemporal data for climate studies. In: Liang, S.H., Wang, X., Claramunt, C. (eds.) W2GIS 2013. LNCS, vol. 7820, pp. 190–198. Springer, Heidelberg (2013)
13. Murgante, B., Tilio, L., Lanza, V., Scorza, F.: Using participative GIS and e-tools for involving citizens of Marmo Platano-Melandro area in European programming activities. Journal of Balkan and Near Eastern Studies **13**(01), 97–115 (2011)
14. Netek, R., Balun, M.: WebGIS solution for crisis management support – case study of olomouc municipality. In: Murgante, B., et al. (eds.) ICCSA 2014, Part II. LNCS, vol. 8580, pp. 394–403. Springer, Heidelberg (2014)
15. PHP: Hypertext Preprocessor. http://www.php.net/
16. Pirotti, F., Guarnieri, A., Vettore, A.: Collaborative Web-GIS Design: A Case Study for Road Risk Analysis and Monitoring. Transactions in GIS **15**, 213–226 (2011)
17. Plewe, B.: GIS Online: information retrieval, mapping, and the Internet. OnWord Press (1997)
18. Steiniger, S., Bocher, E.: An overview on current free and open source desktop GIS developments. International Journal of Geographical Information Science **23**(10), 1345–1370 (2009)
19. Wang, N.-H., Li, D., Pan, H.: Information service platform of forest pest forecast based on WebGIS. Journal of Forestry Research **20**, 275–278 (2009)
20. Zheng, K.G., Rahim, S.T., Pan, Y.H.: Web GIS: Implementation issues. Chinese Geographical Science **10**(1), 74–79 (2000)

Geodata Discovery Assistant: A Software Module for Rule-Based Cartographic Visualisation and Analysis of Statistical Mass Data

Hartmut Asche[✉], Carolin Kucharczyk[✉], and Marion Simon[✉]

Geoinformation Research Group, Department of Geography,
University of Potsdam, Karl-Liebknecht-Str. 24/25 14476, Potsdam, Germany
{hartmut.asche,carolin.kucharczyk,marion.simon}@uni-potsdam.de

Abstract. This paper presents a novel software approach termed "Geodata Discovery Assistant" (GDA) to visualise and visually analyse heterogenous statistical mass data with a spatial component. Only recently the potential of digital geospatial information has been recognised in large to small scale enterprises. This is partly caused by the fact that software systems used in the industry to model and process economic or business information lack functions to handle the geospatial attributes properly. The GDA is currently developed in an ongoing R&D project between commercial and academic partners. When fully implemented, the GDA will contribute to the full utilisation of statistical mass data with a spatial component thus facilitating meaningful and informed decision making for users of classical data discovery or business intelligence and analysis systems. In this contribution, we focus on its core component, the rule-based map construction process.

Keywords: Geovisualisation · Quality maps · Rule-based map construction · Map construction assistant · Business intelligence systems · Data discovery systems

1 Introduction

It has been estimated that 95 per cent of all digital data include a spatial reference (Hamilton, in Perkins, 2010), thereby turning them into geospatial data. Positional information in 3D space, such as geospatial data, has been found to provide additional value for a broad range of applications in environmental, societal, and economic studies. Recently, the potential of geospatial information for businesses of all size has been recognised. To date, this potential has not been fully exploited for data research and analysis, particularly considering statistical mass data. This is partly caused by the fact that software systems used in the industry to model and process economic or business information lack functions to handle the geospatial attributes properly. Since the 1990s, the software systems commonly used to store, manage, and process data are of the business intelligence (BIS) or data discovery (DDS) type. The majority of presently available BIS or DDS solutions provides either the most basic functions or none at all to represent geospatial data in map form. Application of such rudimentary BIS or DDS functions result in simple map illustrations lacking the cartographic quality that is required to efficiently display and extract the mapped information.

© Springer International Publishing Switzerland 2015
O. Gervasi et al. (Eds.): ICCSA 2015, Part III, LNCS 9157, pp. 566–575, 2015.
DOI: 10.1007/978-3-319-21470-2_41

2   Combined Data and Map Presentation

These shortcomings have not gone unnoticed in the industry. That is why additional software systems for geospatial data processing, namely geo information systems (GIS), are used in parallel to process and visualise geospatial attributes after the data have been processed in BIS or DDS. To do so, additional investment (e.g. software systems, interface programming, expert operators etc.) is necessary. GIS or other graphics systems are mostly used either loosely coupled to the existing BIS or DDS, or completely separate. Software links, if any exist, are unidirectional from BIS/DDS to GIS. As things stand at present, the spatial dimension of economic and statistical mass data is not effectively used in the industry. The following deficiencies can be noted:

(1) Enterprises lack the ability to produce quality maps that conform to the principles of thematic cartography or that offer spatial analysis services for analytical BIS or DDS, and wider geospatial data processing;
(2) BIS or DDS do not support functions for professional geovisualisation of heterogeneous, statistical data in cartographic modeling quality, that facilitates visual data research and analysis;
(3) Bidirectional, dynamic software links between BIS/DDS and GIS allowing for parallel, interactive use of non-graphic statistical data and graphic statistical maps via an overlapping graphical user interface (cockpit), are not available in the (geo) information market;
(4) Integrated software systems combining core functions of BIS or DDS with professional map construction capabilities to analyse and professionally visualise heterogeneous statistical mass data do not exist to the authors' knowledge, although few attempts are known anecdotally.

These deficiencies become all the more evident considering the growing need for analytical BIS/DDS with professional cartographic functionalities in the industry. In this paper, we will present a novel software approach to combine numerical and geovisual data analysis in a BIS/DDS map construction environment called "Geodata Discovery Assistant" (GDA). We will concentrate on its core component, a rule-based map construction process.

The BIS/DDS map construction assistant above is currently being developed in an R&D cooperation project between commercial and academic partners. Its global objective is the design and prototypical implementation of a professional map visualisation and analysis component (GDA) coupled to open-source BIS/DDS and, at a later stage, to commercial systems. For this project, the industry partners WhereGroup (Bonn, Germany) and GraS Graphics Systems (Berlin, Germany) cooperated with the Geoinformation Research Group at the University of Potsdam (Potsdam, Germany). The project was funded by the German Federal Ministry for Economic Affairs and Energy (BMWi) within the framework of its ZIM programme format (Central SME Innovation Program).

3   Principles of Geospatial Data Visualisation

A map is a classical medium to store, represent and communicate spatial information in visual form in science, politics, the economy, and society. Maps are not just images or illustrations, but graphical models of geospatial reality. Like any other model, maps simplify complex geographical phenomena by reduction and abstraction. This is achieved by transforming non-graphic statistical data into graphic map representations. Any professional transformation is based on a set of cartographic principles at the core of which lie icon reference to objects and generalisation. The objective of map modeling is to facilitate optimal perception and mental construction of individual spatial models of real-world information. Effectiveness of this process is a prerequisite for understanding the complex geographical and thematic structures of reality. Data tables, however, do not facilitate the perception of spatial structures, as they focus on the presentation and statistical analysis of single values.

Rule-based map modeling creates cartographic maps that display meaningful and thus usable graphic models of geospatial reality. In fact, this rule-based transformation of non-graphic geospatial data into graphic maps accounts for the map's professional cartographic quality and fitness-for-use (cf. Fig. 1-1). These professionally modeled maps are given the term "quality maps", highlighting their distinctive cartographic quality that facilitates meaningful and effective map use.

Map presentations produced with statistics software lack essential elements that professional map models have. They are simple, map-like visualisations, but not models of geodata. These inadequate and naive map graphics reflect the ignorance of cartographic map modeling (cf. Fig. 1-2). Their map-like appearance prevents visual communication and hinders effective, task-oriented map use, making the presentations inapt for use (e.g. Monmonier/Johnson, 1991).

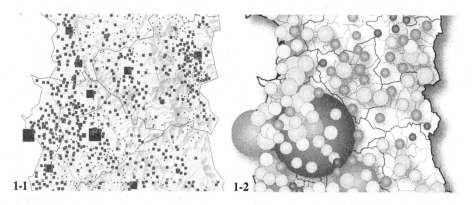

Fig. 1. Population distribution of Albania in 2001; 1-1 Professional cartographic map model (Bërxholi et al 2003); 1-2 Inadequate map presentation (INSTAT 2004)

The GDA provides rule-based quality map construction functionalities for BIS and DDS. Developed and programmed as a software assistant, it is linked with the BIS/DDS, which is operated by a user. Its integration into the processing chain of statistical (mass) data facilitates professional map construction and thus ensures that the full potential of these data is exploited for economic purposes. By developing a common user interface in the form of a cockpit, the two software types are integrated. This provides parallel access to both non-graphic and map data in separate, but linked windows (data view vs. map view). This allows simultaneous interactive manipulation and in-depth analysis of both data sets.

4   Rule-Based Map Construction with the GDA

As with any software system, the functionality and field of application of BIS/DDS is determined by the methods that are implemented for the analysis and visualisation of mass data. We have shown that professional map visualisation is only supported to some extent, if any. As a consequence, quality map construction will have to be provided by the GDA introduced above.

Against this background, the GDA aims to provide rule-based map construction to process and analyse (geo)statistical mass data. To do so, the development and implementation of a rule set based on algorithms is required. To our knowledge, such algorithms are not available on a non-proprietary base. Neither are they implemented in GIS which lack extensive graphical modeling functions required for this cartographic application. In fact, GIS do not support independent graphical modeling of non-graphic data at all.

For a start, a rule base has been defined on a conceptual level. These rules facilitate the professional construction of choropleth and line maps, which are among the most important map types. On a logical level, the respective rules are then converted into processes that are driven by algorithms using open-source software components.

Fig. 2 depicts the design of the GDA and its BIS/DDS environment. The GDA is completely based on and composed of open-source software components. The geo database component, e.g., is based on the well-known PostgreSQL database management system with its spatial extension PostGIS, while the rendering component is based on the R software.

The GDA is conceived as a plug-in software module that is connected to a range of open-source and proprietary BIS/DDS, such as Spotfire, Yellowfin, Tableau, Jaspersoft or R, by API. The numerical and/or semantic data is processed by the BIS/DDS available. This statistical data processing does not exploit the full potential of the source data, particularly not their spatial component. The processed non-graphic data are transferred to the GDA plug-in for cartographic visualisation. First, the GDA filters the imported data according to the requirements defined. A second step applies the rule set selected for the map type to the filtered data. The result is a preliminary geospatial visualisation in simple vector graphics. Then, this graphic representation is transformed into a professional map model consistent with the relevant principles of thematic cartography. The end result is an application-specific graphic map visualisation complementing the non-graphic data analysis of the BIS/DDS in use.

The processes above are driven by the following GDA functionalities:

- Data input of BIS/DDS data, additional geodata and metadata are required for the construction of a defined map visualisation;
- Rule-based data filtering of the imported data (e.g. selection, supplement, abstraction, classification) to produce a preprocessed dataset for subsequent mapping according to the selected cartographic representation method;
- Rule-based graphic representation of the filtered data, which can be viewed with a preview function. This allows the user to check and manipulate the proposed map visualisation. Possible data modification will be channeled back into the filtering component before being forwarded to the mapping component.
- Rule-based rendering of the previewed map presentation, i.e. cartographic generalisation and map composition of defined map type with preselected symbolisation and lettering.

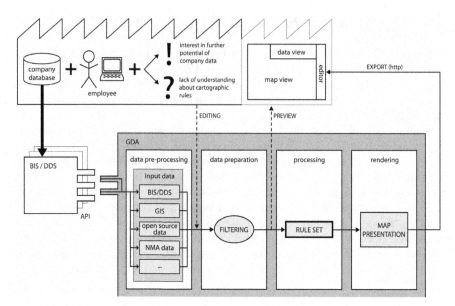

Fig. 2. Architectural design of the GDA and its BIS/DDS environment (own source)

The map dataset is then exported to BIS/DDS. For that purpose, BIS/DDS is enhanced by an interactive map/data editing and visualisation module, the GDA. Its user interface can be designed as a cockpit which enables simultaneous interaction, as well as manipulation with BIS/DDS data and map data. When fully implemented, the combined BIS/DSS and GDA provide the user with a synchronised data and map view in interconnected interactive cockpit windows.

5   GDA Application: Three Exemplary Use Cases

5.1   Selecting the Appropriate Map Projection

Map projections depict the mathematical basis of any map. They represent the curved 3D surface of the earth in a flat two-dimensional map plane. Thus, any map construction starts with the selection of a map projection that meets the relevant input data features, as well as the defined map purpose. This is an important step, considering the fact that more than 400 map projections exist for different areas, scales and geometrical properties. Selecting the optimal map projection for a particular map purpose requires specific cartographic expertise. When it comes to map construction from BIS/DDS data, this expertise is given by the respective GDA rule base. That is why the GDA contains a range of map projections relevant for recurring applications. To visualise statistical data in area maps, such as choropleth maps or cartograms, both equivalence[1] and shape fidelity[2] will have to met. Based on the requirements of map type,

---

[1]   An equal-area projection preserves area relationships of any parts of the globe.
[2]   Shape fidelity means a realistic graphical representation of the shape and contour of a spatial object in the map. In contrast to equivalence this is a non-geometrical property.

scale and audience, the GDA automatically selects and applies the map projection most appropriate for the map purpose (Fig. 3). This automated selection process is based on Young's rule (Maling 1992), a formula for choosing the appropriate class of map projections. Originally developed to select azimuthal projections vs. conic or cylindrical projections, the selection process now includes conic and cylindrical projections. This process is implemented in the GDA. Depending on the presentation map scale, the appropriate geodetic (large scale) or cartographic projection (small scale) is chosen.

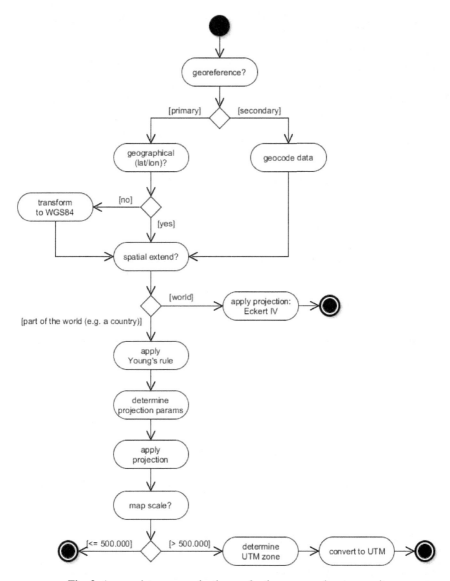

Fig. 3. Appropriate map projection - selection process (own source)

5.2   Selecting the Appropriate Cartographic Representation Method

Following the choice of the appropriate map projection, the cartographic representation method most suitable to the input data features and map requirements is chosen. Cartographic representation methods, about a dozen of which exist, are essential for the transformation of non-graphic spatial data into effective graphic map representations. A map representation method defines the basic structure of the map graphic. This, in turn, predetermines the effectiveness of visual perception and analysis of the spatial structures of the data communicated (Dent 2008, Slocum et al. 2009, Robinson et al. 1995). Based on the respective data properties, the GDA selects the appropriate representation method (cf. Fig. 4). Relevant data features include geospatial properties (discrete or continuous phenomena), geometric properties (point, line, area or surface data), semantic characteristics (qualitative, i.e. non-numeric data or quantitative data) and attribute value (absolute or relative). Note that there is no one-to-one relation between data attributes and representation methods. More than one cartographic representation method might be suitable for the construction of the respective statistical map.

It has been mentioned that the data in question are statistical mass data with a geospatial component. The majority of these data are quantitative data with a positional or areal geospatial reference. In a first implementation stage, the range of representation methods available can be limited to those that are applicable to the data. A large amount of statistical data typically processed in BIS/DDS systems are area-related. These quantitative data sets at a ratio level of measurement include mostly density values or ratios (e.g. inhabitants per square kilometer) and only seldom absolute quantitative values. In the majority of applications, the areas of reference are real-world administrative areas. These can be of different types- political divisions, e.g. districts or provinces, judicial districts, economic districts etc. Although areal reference to a regular grid is logical and also applicable, it is rarely found.

Automated processing starts by choosing the representation method that matches the characteristics of the input data and the map target best. This process is termed icon reference. If more than one cartographic representation method is selected (usually two are chosen), the data can be visualised in both ways. The preview function is then used to determine the appropriate representation method for rendering the final map graphic into an interactive dialogue with the user. When it comes to the cartographic representation of quantitative areal data, the diagrammatic map method is found to be the appropriate cartographic mode of representation. With the help of a detailed feature analysis of the respective area-related quantitative data, one of the two variants of this method is selected. Map construction of area-related, relative values requires the use of the choropleth mapping method. Absolute values are visualised by the diagram map method.

5.3   Constructing a Choropleth Map

Given a quantitative dataset with an areal spatial reference, the GDA analyses the respective value properties in the preprocessing component. When the value features are identified as relative, the GDA determines the choropleth mapping method as the

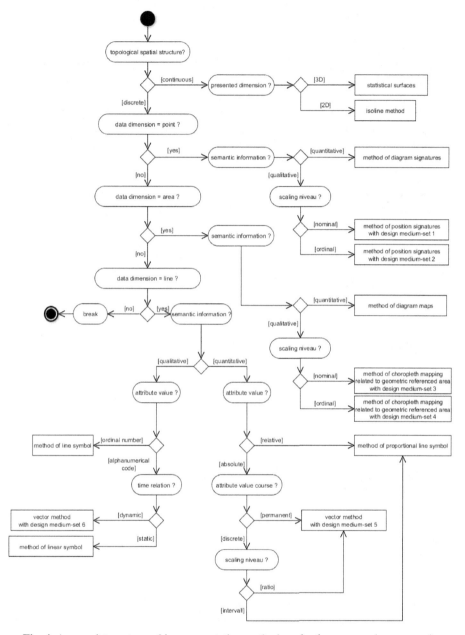

Fig. 4. Appropriate cartographic representation method – selection process (own source)

appropriate mode of map visualisation. For this representation method, the numeric data must be classified. The resulting map is thus termed a classified map. Depending on the amount of data in the data set, the GDA first calculates the number of classes based on Sturges' Rule[3]. Secondly, depending on the data distribution, data values are allocated to the classes by employing one of three classification methods implemented in the GDA: standard deviation (in case of normal distribution), quantiles (equal distribution), and Jenks' natural breaks (unequal distribution). As cartographic data representations show the geographical distribution of the data values, the GDA makes sure that more than one data value is assigned to each class. Then, the GDA analyses the value range, i.e. whether the data set contains only positive (or negative) values, both positive and negative values, or values above or below an average.

Once data classification has been determined, the GDA selects a colour scheme matching the classification executed. Assuming that data values are unipolar, i.e. that they progress from the lowest to highest value (or vice versa), are above zero, or are average, the appropriate fit is a sequential scheme. In this scheme hue and saturation are kept constant. The sequential steps in the data (classes) become evident with the variation of brightness, e.g. a progression from light to dark blue. Low data values are assigned light colours, high values are assigned dark colours. For the map-specific selection of colours, the GDA accesses the relevant colour scheme. Finally, the GDA chooses the best-fitting area colouring from a range of implemented colour schemes. For specific applications, the well-known Colorbrewer colour scheme selection tool (http://colorbrewer2.org) can also be accessed from the GDA.

6　Conclusion

This article presents a technological concept of a software assistant with professional map construction capabilities to visualise and visually analyse heterogenous statistical (mass) data with a spatial component. It has been argued that visualisation of geospatial data significantly enhances the utilisation of statistical data by providing users a complementary function, i.e. a spatial view on the characteristics and distribution of existing data. This additional feature is appreciated by users primarily employing BIS/DDS to process and analyse data. At this stage, development and implementation work of the GDA has proved that the implementation of the technology concept on an open-source base is viable. The GDA has been constructed as a plug-in for mainstream BIS/DDS. It collaborates primarily with open-source systems but also interfaces with proprietary systems, such as TIBCO's Spotfire.

The R&D effort discussed here has also shown that an extended preparatory period is required prior to implementation. This phase includes the development and programming of a rule set for map construction in accordance with cartographic representation principles. To date, the algorithm-based map construction rules have been implemented in an automated map visualisation process to generate professional choropleth, isopleth and diagram maps. As a prototypical version, the GDA is operational and stable. Next, a trial operation with selected BIS/DDS under practical conditions is required. This step will include an evaluation in terms of practicality, performance and business efficiency. The final phase will integrate selected BIS/DDS

---

[3]　$k = 1 + 3,32 * \log n$, with k = number of classes, n = number of observations.

under a uniform cockpit user interface, allowing for synchronised data and map view interaction and analysis. When available, the GDA will contribute to the full utilisation of statistical mass data with a spatial component, thus facilitating meaningful decision making of BIS/DDS users.

Acknowledgement. This R&D cooperation project is funded by a grant from the German Federal Ministry for Economic Affairs and Energy (BMWi) within the Central Innovation Program SME (ZIM) 2012-2015. This support is gratefully acknowledged. Thanks are also due to the ZIM project teams at WhereGroup (Bonn) and the Geoinformation Research Group of Potsdam University (Potsdam) for assisting in the preparation of this article.

References

1. Asche, H., Stankute, S., Mueller, M., Pietruska, F.: Towards developing an integrated quality map production environment in commercial cartography. In: Murgante, B., Misra, S., Carlini, M., Torre, C.M., Nguyen, H.-Q., Taniar, D., Apduhan, B.O., Gervasi, O. (eds.) ICCSA 2013, Part IV. LNCS, vol. 7974, pp. 221–237. Springer, Heidelberg (2013)
2. Bërxholi, A., Doka, D. Asche, H. (Eds.): Atlas of Albania. Demographic Atlas of Albania. Akademia e Shkencave, Tiranë. Distribution of Settlements 2001, p. 33 (2003)
3. Dent, B.D.: Cartography. Thematic Map Design, 6th edn. McGraw-Hill, Boston (2008)
4. INSTAT (Ed.): Albania Census Atlas 2001, Population and Housing Census, Research Publication. ISBN 99927-840-1-6, Population distribution, p. 11 (2004)
5. Maling, D.H.: Coordinate Systems and Map Projections. Pergamon Press, Oxford (1992)
6. Monmonier, M., Johnson, B.B.: Using qualitative data gathering techniques to improve the design of environmental maps. In: Rybaczuk, K., Blakemore, M. (eds.) Mapping the Nations, London, vol. 1, pp. 364–373 (1991)
7. Perkins, B.: Have you mapped your data today? Computer World (2010). http://computerworld.com/s/article/350588/Have_You_Mapped_Your_Data_Today (access March 2014)
8. Robinson, A., Morrison, J.L., Muehrcke, P.C., Kimerling, A.J., Guptill, S.C.: Elements of cartography. Wiley, New York (1995)
9. Slocum, T.A., McMaster, R.B., Kessler, F.C., Howard, H.H.: Thematic Cartography and Geovisualization. Pearson Prentice Hall, Upper Saddle River (2009)

Spatio-Temporal Modeling as a Tool of the Decision-Making System Supporting the Policy of Effective Usage of EU Funds in Poland

Robert Olszewski[1], Jedrzej Gasiorowski[2(✉)], and Magdalena Hajkowska[3]

[1] Faculty of Geodesy and Cartography, Department of Cartography,
Warsaw University of Technology, Warsaw, Poland
r.olszewski@gik.pw.edu.pl
[2] Institute of Geodesy and Cartography, Warsaw, Poland
jedrzej.gasiorowski@igik.edu.pl
[3] Ministry of Infrastructure and Development, Warsaw, Poland
magdalena.hajkowska@mir.gov.pl

Abstract. Spatial data mining, space-temporal modelling and visual exploratory data analysis are tools that are useful not only for the analysis of multi-characteristics spatial data, but can also be used for the development of Spatial Decision Support Systems. Such system enables the optimisation of decision-making based on a thorough Spatial Multicriteria Decision Analysis. The authors of the present study have developed a set of multicriteria analyses with use of spatial data mining (SDM) techniques for the analysis of the spatial distribution of the allocation and spending of EU funds in Poland. The ten-year period of Poland's membership in the EU enables not only the analysis of spatial differentiation of EU subsidies in different regions of the country, but also the dynamics of changes in this differentiation in time.

The proposed analytical system based on information technologies combines the possibilities offered by GIS packages and advanced statistical software, thus enabling to conduct highly complex analyses. One of the methods to carry out such analysis is the application of so-called data mining and data enrichment to detect patterns, rules and structures "hidden" in the database.

Keywords: Spatial data mining · Space-temporal modelling · Spatial data analysis · Spatial concentration · Visual exploratory analysis · EU funds · Time series · Kriging

1 Introduction

The main aspect of spatial data mining is the development of methods that enable "indirect" analysis of the surrounding space. This process is conducted by means of analysing the model, which is a correctly designed spatial database [1]. Such an analysis methodology directly contributes to knowledge discovery [2]. Expanding the statistical methods by operators and spatial relations enables the development of innovative algorithms and rules. Analyses of large sets of data realised in such a way,

© Springer International Publishing Switzerland 2015
O. Gervasi et al. (Eds.): ICCSA 2015, Part III, LNCS 9157, pp. 576–590, 2015.
DOI: 10.1007/978-3-319-21470-2_42

so-called data mining, allow us to "discover" spatial knowledge and to obtain "added value" in form of knowledge (data mining). This issue is particularly important in the case of the analysis of spatial and space-temporal data, as it enables to "discover" patterns and geometrical structures basing on the analysis of spatial distribution and to model the dynamics of changes of such distribution, which is the basis for the development of decision making systems [3].

Searching for the relations between variables is the fundamental aim of numerous scientific studies [4,5,6], as it allows to analyse the dependencies and to create statistical models that explain complex phenomena. An interesting subject of research is such analysis of multi-characteristic data that takes into account not only classical descriptive attributes, but also information of a spatial nature [7,8].

Digital spatial and statistical data collected in Poland in the course of realisation of various geoinformational projects are typically only visualised and/or processed in an elementary way. In order to use them more widely, a deeper statistical analysis is required, taking into account the spatial and temporal aspect, to transform the "raw" data included implicitly in the database to the explicit form, to visualize spatial patterns, discover trends and to demonstrate large-scale spatial and temporal relations between specific objects and phenomena [9,10]. Detecting the rules "hidden" in a geographical database enables to create a decision making system that will allow, for example, to forecast the development of agriculturally used areas or demographical and social changes in various regions of the country, to evaluate the correctness of income redistribution policies or to assess whether the application of intervention policy is justified. This type of analyses thus enable both government institutions and academic entities to obtain direct benefits.

Spatial data mining techniques (in particular spatial statistical techniques) may be applied e.g. for the creation of digital tools enabling the analysis of optimal space use for such purposes as infrastructure investments (motorways, railways), precision farming, location of various types of investments (industrial areas, wind farms, leisure areas etc.). However, the combined application of strictly statistical methods and spatial analyses requires both a large degree of expertise and possessing information tools of highly varied functionality [11]. The authors of this study have attempted to develop the concept of a geoinformational tool that would integrate the possibilities of a GIS platform with spatial data mining techniques. The concept also involves the development of a creator of analyses that would facilitate the application of data mining techniques for users who are not specialised in the field of spatial information and statistics. Existing IT solutions force potential users on the one hand to develop their own methods of including the spatial data in statistical analyses and on the other hand to use several IT environments (GIS and statistical ones) more or less effectively in order to achieve the desired outcome [12].

2   Study Area and Source Data

Due to the political and socio-economic transformations that took place in Poland after 1989 it became quite interesting to investigate the causes of the uneven spatial distribution of numerous characteristics, including the degree of accumulation and distribution of EU funds, or the degree of support for the idea of European integration itself in various regions of the country. The aim of the conducted studies is to develop research regimes based on spatial statistics that would enable to analyse the realisation of the strategic objectives of Poland and the EU, defined in such documents as: Europa 2020, Cohesion Policy, Development Policy, Common Agricultural Policy, National Strategic Reference Framework, connected with the distribution and spending of EU funding.

Within the realisation of the European Cohesion Policy, whose aim is to ensure economic, social and territorial cohesion of Europe and which is an element of the EU development strategy: Europa 2020, in the years 2007-2013 Poland received subsidies from structural funds in the amount of EUR 67.5 billion.   This amount accounts for 19% of the EU budget for the realisation of cohesion policy in the 2007-2013 financial perspective. The received funding originate from three European funds: European Regional Development Fund (ERDF), European Social Fund (ESF) and the Cohesion Fund (CF). Pursuant to the assumptions provided in the strategic document entitled National Strategic Reference Framework, these funds have been allocated to operational programmes. ¼ of the funds allocated to Poland have been allocated for the realisation of 16 regional programmes implemented by voivodeship local territorial government units. The remaining funds were allocated to national programmes. Human Capital, Infrastructure and Environment, Innovative Economy, Development of Eastern Poland etc. Apart from the structural funds, Poland has received EUR 13.4 billion from the European Agricultural Fund for Rural Development and EUR 0.7 billion from the European Fisheries Fund.

Following wide community consultations, the distribution of subsidies from ERDF and ESF to individual regions was based on various algorythms. For the ERDF resources foreseen for 16 regional programmes, the so-called 80/10/10 algorithm was applied. This algorithm is based on the assumption that it will enhance the competitiveness of all regions, at the same time providing the highest support for the poorest areas. Pursuant to this principle, all regions participated in the distribution of 80% of the funds from the total pool of resources, basing on the population. The remaining 20% of funds were distributed among these regions where GDP per capita was lower than 80% of the national average and where the rate of unemployment recorded on poviat level exceeded 150% of the national average. The ESF funds in the regional component Operational Programme Human Capital were distributed basing on several criteria of varied weight, including, among others: population, number of registered unemployed, number of SME, number of individuals obtaining their revenues from farming.

Pursuant to EU regulations, Member States are obliged to ensure management and control systems of operational programmes. With the aim to fulfil this obligation, the Ministry of Infrastructure and Development created and continues to maintain the KSI SIMIK 07-13 database. Information included in this database covers the financial data of all operational programmes. A similar IT system is also operated by the Ministry of Agriculture and Rural Development.

Data collected in the databases of both government bodies specified hereinabove (after appropriate processing and aggregation) constituted the basis for the conducted analysis. Additionally, data gathered in the official databases of the Central Statistical Office and the Head Office of Land Surveying and Cartography were used. Other data used for the realisation of the analysis included also historical data characterising the division of Poland as a result of the annexations in the 19th and 20th centuries as well as the data of the State Election Commission documenting the degree of support of the inhabitants of Poland for Polish accession to the EU in 2004.

3   Research

Due to the vast amount of the collected source data, characterized by nearly 600 features (attributes) on three levels of generalization corresponding to the administrative division of the country into 16 voivodeships, 379 poviats and 2479 communes, this study discusses only a few selected spatial data mining, spatial statistics and space temporal modelling algorithms. Data analysis was conducted with use of spatial concentration algorithms, dynamics of spatial and temporal changes, correlations determined in a moving window and correspondence analysis. Due to the enormous amount of characteristics describing individual territorial units, too high degree of simplification resulting from the aggregation of data to the level of 16 voivodeships and excessive specificity connected with the atomization of the administrative division of Poland into nearly 2500 communes, 27 features characterizing 379 poviats in Poland were selected for the experiment.

3.1   Correspondence Analysis

The statistical technique which can be useful either for evaluating the influence strength of explanatory (exogenous) variables on a response variable or for looking for additional explanatory variables is multivariate correspondence analysis (MCA) [13]. This method concerns qualitative variables expressed in a nominal scale. It allows for an easy and intuitive illustration of relationships between some sets of objects (observations) against selected characteristics (attributes). The result of multivariate correspondence analysis can be presented in form of a perception map (which is not a map in the cartographical sense), which shows the location of groups of objects in a two- or three-dimensional coordinate system on the background of defined characteristics, which are defined by the values of attributes [14]. The relative position of groups of objects and specific characteristics in the coordinate system allows for the inference of conclusions concerning the relationships between individual groups of objects and characteristics.

In the present study multivariate correspondence analysis was used to illustrate the influence of specific factors on the level of support for European Union accession in Polish districts in 2003 and, moreover, to look for some other variables which might potentially better explain the variety of the support level in the European Union accession referendum among the districts. As the basis of the analysis the following variables for districts were used:

- Degree of support of inhabitants of different regions of Poland to Poland's accession to the EU in 2004 (high, medium, low);
- Historical division of Poland resulting from the annexations (Austrian, Prussian and Russian annexation zones);
- Total amount of EU funding (under all programmes) per capita as of 2010 (high, medium, low);
- Degree of accumulation of the EU funds under "agricultural" programmes per capita, accumulated in the years 2007-14 (high, medium, low).

The first step of the analysis was data preparation. All of the variables, except for annexation membership which was expressed in nominal scale, were originally expressed in the ratio scale, so – in order to unify the scale among all variables – they had to be converted into the nominal scale. For each of the converted variables three final values (classes) were generated. This was done basing on histograms of the values of the variables, where some natural breaks in the distribution could be observed. After such a creation of qualitative data, the contingency table was created by means of multivariate correspondence analysis, and finally the perception map was generated (Fig. 1).

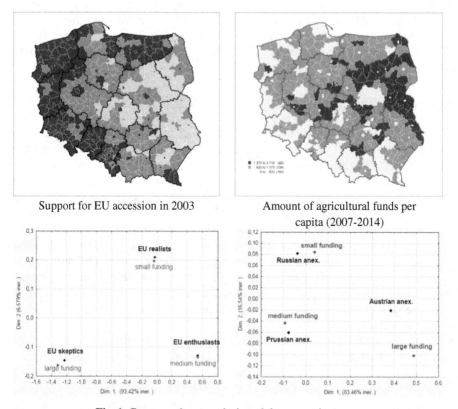

Support for EU accession in 2003   Amount of agricultural funds per capita (2007-2014)

Fig. 1. Correspondence analysis and the perception maps

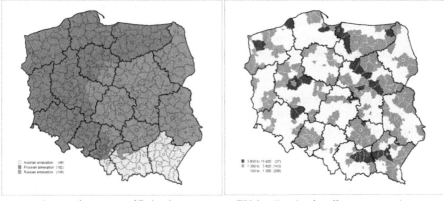

| Annexation zones of Poland | EU funding (under all programmes) per capita in 2010 |

Fig. 1. (*Continued*)

The intuitive analysis of the perception map allows us to draw the following conclusions:

• EU funds under agricultural programmes (European Agricultural Fund for Rural Development, Rural Development Programme) were used to a highest extent in those regions, where the degree of support for EU accession was the lowest. This may prove either purposeful support for areas with a large percentage of population with a sceptical attitude towards European integration under the cohesion policy or a gradual change in the attitudes of the rural population in these regions, resulting from the noticeable benefits from accession to the EU.

• In the initial period of the financial perspective 2007-14 the degree of usage of EU funds per capita was strongly correlated with the division of Poland resulting from the annexations. However, the correlation noted in 2010 (explained by the continuing differences in thinking between inhabitants of various regions of Poland) may be of a purely apparent nature or it may result from other factors. One of such factors may be e.g. the costly construction of A4 motorway in the South of Poland (former Austrian annexation zone), as correspondence analysis conducted in the subsequent years (2011-14) does not show such strong correlations any more. However, the conducted tests and developed analysis regimes as well as the cartographic visualisations thereof in form of thematic maps may be an important source for both econometric and sociological research.

3.2   Spatial Statistics – Local Correlation in Moving Window

The Pearson correlation coefficient is used to determine the level of linear correlation between variables expressed on quantitative scales when their distribution is similar to normal. Depending on the strength of correlation between such variables, the value of this coefficient may fall between -1 and 1, where the extreme values mean a functional correlation between variables, "zero" refers to lack of linear correlation and intermediate values reflect statistical correlations of a stronger or weaker nature. However, the determination of a unified value of the correlation coefficient for all analysed data (e.g. for all 379 poviats in Poland) enables only to obtain a single numerical value. This number reflects a specific type of average quantitative correlation between two analysed variables, although it does not differentiate the strength of such correlation with respect to spatial location in any way. The essential aspect of the approach proposed by the authors is the application of the moving window methodology, which allows for a spatial differentiation of the correlation strength [15]. The resulting cartographic study that presents the spatial distribution of the determined statistical measure in form of a topical map allows to transform raw data into useful information and knowledge (data enrichment).

The conducted tests involved the analysis of the spatial differentiation of the correlation (measured by the correlation coefficient determined locally in the moving window) between the following variables expressed in ratio scale (Fig. 2):

- the population of the given region and the global amount of EU funding used in the period 2007-14.
- the value of expenditure per capita under the Human Capital programme and the percentage of the unemployed and changes in this correlation in the years 2009 and 2014.
- tangible fixed assets of enterprises and the amount of subsidies under the Innovative Economy programme and changes in this correlation in the years 2009 and 2014.

Due to the fact that statistical calculations require the analysis of at least several dozen objects to be credible and basing on the average surface area of voivodeships in Poland, the authors adopted a moving window of the size of a circle of 100 km diameter. A window defined in such a way enables to analyse the local correlation between phenomena and at the same time to aggregate (generalise) the obtained results to a level corresponding to the scale of a voivodeship. Conducting the analysis also required the verification whether the distributions of specific variables are normal and the development of geoinformational tools that enable the automation of the process of local determination of the correlation coefficient as well as the interpolation of the obtained results and the development of the resulting topical map.

The obtained results prove that:

- in spite of significant expenditure of EU funding (Human Capital) designed for "soft" activities such as trainings, changing qualifications etc., the level of unemployment in the North-western regions of Poland remains high. This may be interpreted as a type of permanent structural unemployment that deepens with social benefits, although it may be caused by a series of other factors, such as the transformation of the labour market in this region. Thus, the presented method of quantitative data analysis and visualisation of results should be treated as only an introduction to a specialist qualitative analysis, both in the economic and sociological and social aspects.
- strong positive correlation between tangible fixed assets and the amount of subsidies under the Innovative Economy programme in South-western Poland proves the development of companies that have a strong equipment base. On the other hand, negative correlation in the North and East of Poland may point to the development of innovations connected, for example, with the promotion of tourism.

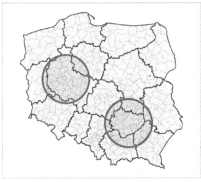

The size of the moving window.

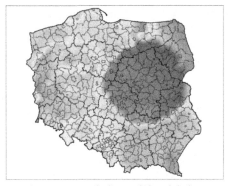

corr. between population and the global amount of EU funding

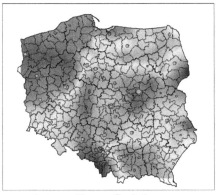

corr. between Human Capital programme and the percentage of the unemployed in 2009

corr. between Human Capital programme and the percentage of the unemployed in 2014

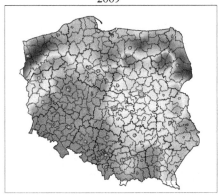

corr. between assets of enterprises and the amount of subsidies under the Innovative Economy programme in 2009

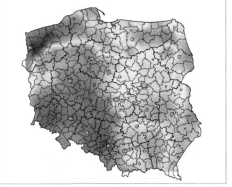

corr. between assets of enterprises and the amount of subsidies under the Innovative Economy programme in 2014

Fig. 2. Local correlation in moving window

- extremely strong correlation (close to functional correlation) between the accumulated amount of EU subsidies and population within the central Masovian voivodeship reflects the specificity of the province (capital, central offices, the concentration of high-tech industries).

It should be emphasised that the main objective of the authors of the present study was to develop algorythms and analysis regimes. as well as the methods of visualisation of the obtained results and of geoinformational tools that enable the realisation of these concepts, but not an expert economic opinion on the underlying causes of existing discrepancies and differences. The developed methodological concepts may however be a useful analytical tool for specialists in the areas of economics or culture anthropology as well as for decision makers on ministry level who are in charge of allocating and distributing EU funding.

3.3   Spatial Concentration

This method, which is based on the notion of spatial concentration, enables the analysis of the uneven spatial distribution of phenomena or objects within an adopted system of spatial reference units [16]. Such system may consist of such administrative division units as voivodeships or poviats. Basing on the cumulative series of the given phenomenon and the Lorenz concentration curve a map of spatial concentration is prepared, showing the degree of inequality of the spatial distribution of a given phenomenon. This disproportion is also measured by the so-called Gini index [17,18].

The conducted tests involved the analysis of temporal changes in the spatial concentration of EU funding under two extremely different operational programmes: Innovative Economy and Human Capital. The objective of the first programme is to support innovative business activities, while the other one is aimed at supporting social development. The research focused on defining the degree of spatial concentration (or scattering) of the IE or HC funds throughout the country and on verifying whether the degree of spatial concentration varies in time (Fig. 3).

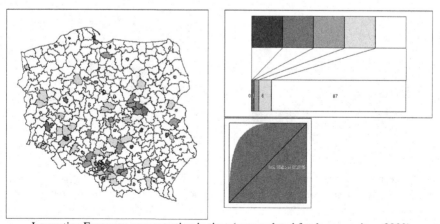

Innovative Economy concentration in time (accumulated funds per capita – 2009)

Fig. 3. Spatial concentration

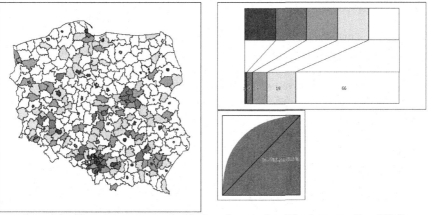

Innovative Economy concentration in time (accumulated funds per capita – 2014)

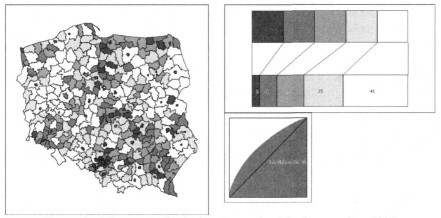

Human Capital concentration in time (accumulated funds per capita – 2009)

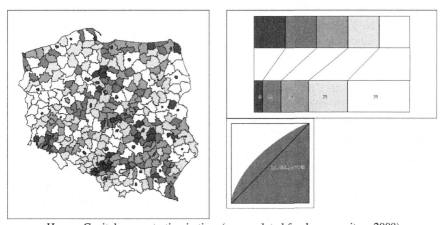

Human Capital concentration in time (accumulated funds per capita – 2009)

Fig. 3. (*Continued*)

The obtained results prove that:

- EU subsidies under the Human Capital programme were distributed proportionally to various regions of the country throughout the 2007-2014 period, which fosters the enhancement of professional competences and building social capital.
- EU funds under the Innovative Economy operational programme were distributed basing on the level of innovativeness in specific regions, which means that in 2009 20% of the funds were allocated to only 14 poviats, while 287 poviats (87% of the country area) also received only 20% of the funding. In 2014, the distribution of accumulated investments financed from the Innovative Economy programme was slightly more even, although 190 poviats (66% of the country area) still received only 20% of the EU funds. Subsidies under the Innovative Economy programme were addressed mainly to entrepreneurs, universities and academic research centres. The observed distribution of funding, concentrated in large cities and industrial areas is a natural consequence of this fact.
- It is worth noting that in the case of both the aforementioned programmes the presented disproportions in obtaining EU funding may be a good starting point for a deeper analysis of the potential and activity of selected groups of entities applying for subsidies in various regions. Some valuable information might be provided as a result of an analysis with respect to type of beneficiaries, e.g. large companies and SME, universities, research and academic institutions, NGOs etc.

3.4 Spatial Generalization of Time Series Models

Spatio-temporal modeling includes also time series analyses across the space. Individual time series models irregularly distributed on a given area can be generalised to one, global, continuous time series model for such area. The framework of such an approach was outlined by Kyriakidis et al. [19,20]. The idea of spatial generalisation of time series models is based on the independent determination of parameters of time series components (such like trend, seasonal or cyclic parameters) [21,22] in each location. The next step is to interpolate the value surface for every time series parameter using kriging interpolation method [23]. Such surface model for each time series parameters corresponds to a global, generalised time series model of the given area and it enables the determination of values of time series parameters in any point of this area.

The implementation of this approach will be shown by the use of data obtained from the Central Statistical Office of Poland, particularly the data of percentage of people using sewage system in Polish districts over the years 2002–2013. From the overall number of 381 districts in Poland, we have chosen 66 districts characterised by similar area, population, socio-economic features and distributed throughout the country. The districts are represented by point objects, determined as centroids of districts areas.

The scatterplot of the phenomenon for an example district is presented in Fig. 4. The nature of the time series is similar in every district, but there is a differentiation in the dynamics of changes. So the time series model includes only trend component (there are neither seasonal or cyclic components, nor interventions), and the trend is of a linear nature. Therefore only one parameter has to be determined for each trend

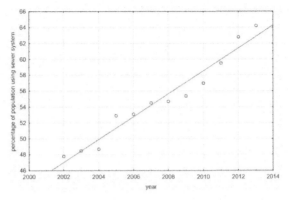

Fig. 4. The scatterplot of percentage of population using sewer system over the years 2002–2013 for an example district

component. To compute the trend parameter, the simple regression model for each localisation was determined which has the form of following equation:

$$y = \alpha + \beta t \tag{1}$$

where y is the dependent variable of the percentage of population using sewage system, α is an absolute term denoting an initial value of the phenomenon, β is the slope of the regression straight — the trend parameter to determine, and t is the independent variable of time. 66 regression models for each analysed district were calculated and the trend parameter — β coefficient of the regression equation (1) — was assigned for each district point.

The next step was to perform the kriging interpolation method to develop a global, continuous model of the time series trend component. The total set of 66 districts was divided into two subsets: training (56 districts) and test (10 districts). The training subset was used to generate the surface model, while the test subset — to validate the model. In this case the ordinary kriging method was used and as a result, trend parameter surface was generated (Fig. 5a).

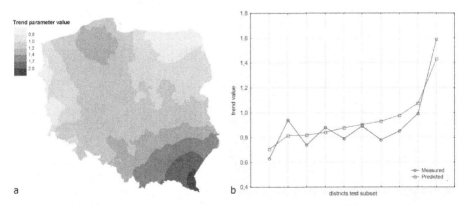

Fig. 5. a – Trend parameter surface for the global time series model, b – Kriging model validation result — plot of measured and predicted trend values

The model validation result as plot is shown on Fig. 5b. To assess the model quality, the mean absolute percentage error (MAPE) was calculated and it amounted to 10.5%. The authors believe that, although the estimation is fairly accurate, it can be even enhanced by using co-kriging model with some additional independent variables. In addition to potential predictive application of the described method, an analysis of the trend surface obtained by kriging interpolation may provide some knowledge about the spatial pattern of dynamics of the given phenomenon as an "added value". Fig. 5a shows, that south-eastern part of Poland is characterised by the highest growth of the sewer system (highest trend parameter), while western and north-eastern — the lowest. There is also some "island" of higher growth in north-central part of the country, which can be the subject of further analysis with the use of various variables.

3.5   Analysis of the Dynamics of Space-Temporal Changes

The conducted analyses were based on the assumption that a linear regression model would be determined independently for each of 379 poviats, together with the standard error as a measure of adjustment of this model. The determined direction coefficient of simple regression reflects the pace of accumulation of EU funds per one inhabitant in each subsequent year. For example, the value "1000" means an additional PLN 1000 of EU funds per one inhabitant each year, while "200" means that each inhabitant 'generates' only PLN 200 of additional income. All administrative units and all operational programmes were subject to iteration analysis in accordance with the adopted scheme. The obtained results (Fig. 6) show that the region of eastern Poland is characterised by a strong dynamics of accumulation of agricultural subsidies, while in the northern region very strong increase in funds under "soft" operational programmes, such as Human Capital is observed. In some poviats, this increase is close to PLN 250 per capita annually, while in the southern regions this value is five times lower (sometimes even lower than PLN 30 per capita annually). Similar disproportions may also be observed in the agricultural programmes: from over 600 PLN per capita per year to as little as less than 10 PLN per capita per year).

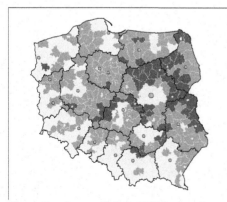   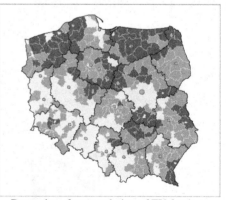

Dynamics of accumulation of EU funds under
agricultural programmes in time (2009-14)

Dynamics of accumulation of EU funds
under Human Capital in time (2009-14)

Fig. 6. Dynamics of accumulation of EU funds under different programmes

4   Conclusions

General access to data (including spatial data) does not necessarily mean obtaining useful knowledge. As Manuel Castels pointed out in "The Information Age" [24], "the paradox of the great civilization change consists in the fact that we have practically unlimited access to information and data and yet we are nearly unable to use it in any way". Thus, knowledge acquisition based on available information (including spatial information) is essential in the era when economic value is not generated in factories any more but is produced by media and IT and telecommunication networks instead.

The conducted research demonstrates that appropriately processed data contained in generally available databases enable to obtain useful information about the spatial structure and dynamics of space-temporal changes. Such "enriched" information carries useful value that may be applied in practice, e.g. to improve the use of the enormous EU subsidies that were transferred to Poland in the last 10 years. The concept of a geoinformational system that would unify the capacities of the spatial information systems and statistical tools proposed by the authors will enable to deepen the spatial analyses conducted on varied sets of data, due to the applied methodology of spatial data mining. Spatial data mining methods, in particular spatial statistics, will enable a wide group of users to analyse the correlations between phenomena, to detect spatial and temporal trends and patterns and to forecast changes basing on the obtained spatial regression models.

References

1. Gotlib, D., Iwaniak, A., Olszewski, R.: GIS. Obszary zastosowań [GIS. Application fields]. Wydawnictwo Naukowe PWN, Warszawa (2007)
2. Olszewski, R.: Kartograficzne modelowanie rzeźby terenu metodami inteligencji obliczeniowej [Cartographic modelling of terrain relief using computational intelligence methods], Prace Naukowe - Geodezja, z. 46, Oficyna Wydawnicza Politechniki Warszawskiej (2009)
3. Fiedukowicz, A., Gąsiorowski, J., Kowalski, P., Olszewski, R., Pillich-Kolipińska, A.: The statistical geoportal and the cartographic "added value"– creation of the spatial knowledge infrastructure. Geodesy & Cartography 61(1), 47–70 (2012)
4. Miller, H.J., Han, J.: Geographic Data Mining and Knowledge Discovery. Taylor & Francis, London (2001)
5. Cabena, P., Hadjinian, P., Dtadler, R., Verhees, J., Zanasi, A.: Discovering Data Mining: From Concept to Implementation. Prentice Hall, Upper Saddle River; Journal of Laws of 1995, No. 88, item 43, New York (1998)
6. Hand, D., Mannila, H., Smyth, P.: Principles of Data Mining. MIT Press, Cambridge (2001)
7. Gatnar, E.: Symboliczne metody klasyfikacji danych [Symbolic methods of data classification]. Wydawnictwo Naukowe PWN, Warsaw (1998)
8. Sokolowski, A.: Metody stosowane w data mining [Methods used in data mining]. In: Data mining – metody i przykłady [Data mining – methods and examples]. StatSoft, Cracow (2002)

9. Franzese, R.J., Hays, J.C.: Spatial econometric models of cross-sectional interdependence in political science panel and time-series-cross-section data. Political Analysis **15**(2), 140–164 (2007)
10. Getis, A., Mur, J., Zoller, H. (eds.): Spatial Econometrics and Spatial Statistics. Palgrave Macmillan, New York (2004)
11. Kuijpers, B., Paredaens, J., Van den Bussche, J.: Lossless representation of topological spatial data. In: Egenhofer, M., Herring, J.R. (eds.) SSD 1995. LNCS, vol. 951, pp. 1–13. Springer, Heidelberg (1995)
12. Bao, S., Anselin, L., Martin, D., Stralberg, D.: Seamless integration of spatial statistics and GIS: The S-PLUS for ArcView and the S+Grassland Links. Journal of Geographical Systems **2**(3), 287–306 (2000)
13. Greenacre, M.: Correspondence Analysis in Practice, 2nd edn. Chapman & Hall/CRC, London (2007)
14. Everitt, B.S.: The Cambridge Dictionary of Statistics. Cambridge University Press, Cambridge (1998)
15. Shiklomanov, N.I., Nelson, F.E.: Active-Layer Mapping at Regional Scales: A 13-Year Spatial Time Series for the Kuparuk Region. North-Central Alaska. Permafrost Periglac. Process. **13**, 219–230 (2002)
16. Sadahiro, Y., Kobayashi, T.: Exploratory analysis of time series data: Detection of partial similarities, clustering, and visualization. Computers, Environment and Urban Systems **45**, 24–33 (2014)
17. Holmes, M.J., Otero, J., Panagiotidis, T.: Modelling the behavior of unemployment rates in the US over time and across space. Physica A: Statistical Mechanics and its Applications **392**(22), 5711–5722 (2013)
18. Bronars, S.G., Jensen, D.W.: The geographic distribution of unemployment rates in the U.S.: A spatial-time series analysis. Journal of Econometrics **36**(3), 251–279 (1987)
19. Kyriakidis, P.C., Miller, N.L., Kim, J.: A Spatial Time Series Framework for Modeling Daily Precipitation at Regional Scales. Journal of Hydrology **297**(1–4), 236–255 (2004)
20. Kyriakidis, P.C., Journel, A.G.: Stochastic modeling of atmospheric pollution: A spatial time-series framework. part I: Methodology. Atmospheric Environment **35**(13), 2331–2337 (2001)
21. Shumway, R.H.: Applied Statistical Time Series Analysis. Prentice Hall, Englewood Cliffs (1988)
22. Papritz, A., Stein, A.: Spatial prediction by linear kriging. In: Spatial Statistics for Remote Sensing. Remote Sensing and Digital Image Processing, vol. 1, pp. 83–113 (2002)
23. Keogh, E., Kasetty, S.: On the need for time series data mining benchmarks: a survey and empirical demonstration. Data Mining and Knowledge Discovery **7**(4), 349–371 (2003)
24. Castels, M.: The Rise of the Network Society. The Information Age: Economy, Society and Culture. Wiley-Blackwell (2009)

Coastal Transport Information System (Co.Tr.I.S.): System and Subsystems Description

Vassilis Moussas, Dimos N. Pantazis[✉], Panagiotis Stratakis, Elias Lazarou,
Eleni Gkadolou, Charalambos Karathanasis, and A.C. Daverona

Research Laboratory SOCRATES (Society for Organizations, Cartography,
Remote Sensing/Road Design and Applications Using Technology/
Transport Engineering on Earth and Space),
Surveying Engineering Department, School of Technological Applications,
Technological Educational Institution (TEI), Athens, Greece
vmouss@teiath.gr, dnpantaz@otenet.gr

Abstract. Co.Tr.I.S is a multifunction information system that will be used for the effective design of coastal transportation lines. Co.Tr.I.S incorporates six subsystems (S1-S6) which include models, tools and techniques that may support the design of improved coastal networks. A major contribution expected by Co.Tr.I.S is to support the decision making process of the policy makers and the involved players (Ministry, Maritime companies, Local Authorities), towards an improved Coastal Transport System. In order to support this functionality, Co.Tr.I.S is equipped with subsystems for data retrieval (S1, S2), statistical analysis (S3), data visualization (S4), as well as subsystems with specialized modules for network generation, scenario validation (S5), solution optimization and decision support (S6). All these subsystems can have access to the vast amount of the up-to-date data classified and modeled by Co.Tr.I.S, and therefore, the system user (coastal line designer or decision maker) will benefit from the improved representation and the improved solution proposals offered by Co.Tr.I.S. The aim of this paper is to present the Co.Tr.I.S subsystems with emphasis to subsystems S5 & S6.

Keywords: Coastal transport · Modeling · Optimization

1 Introduction

Coastal transportations lines may affect not only the development of a coastal city in many ways but also the development of an entire country. Especially in a country like Greece which is considered an important maritime crossroad with more than 400 inhabited islands, the formation of urban networks and city-hierarchies under the conditions of economic globalization and European integration, is strongly affected by the coastal transportation network.

Although a lot of models for road / land transport networks exist, only very few proposals for the coastal transport (e.g. motorways of the sea, EC; Wright and Bartlett, 2000) or multi-modal transport networks are mentioned in the literature (Xie, 2009). Given the lack of literature and similar systems, the main innovation of this system is

© Springer International Publishing Switzerland 2015
O. Gervasi et al. (Eds.): ICCSA 2015, Part III, LNCS 9157, pp. 591–606, 2015.
DOI: 10.1007/978-3-319-21470-2_43

that it gives integrated solutions to the coastal design networks. Co.Tr.I.S uses original and multi-parametric tools which incorporates spatial, statistical analysis, cartographical visualization, games theory and modified the salesman problem. C.o.Tr.I.S. is developed in the frame of a research project co-funded by the European Union and the Greek Government. This system could be used in any island environment taking into account topics like geography, fleet composition, volume of traffic, network design, port infrastructure, and other system parameters (Pantazis et. al, 2013). Co.Tr.I.S. includes six major subsystems (see Fig. 1): S1: Subsystem for information retrieval and queries from DBs, S2: Subsystem for information retrieval from relative sites, S3: Subsystem for statistical treatment and forecasting scenarios, S4: Subsystem for thematic mapping and graphic representations, S5: Subsystem for new lines design, line modifications, what-if scenarios and cost/benefit calculations, S6: Subsystem for decision support using optimization & games theory techniques.

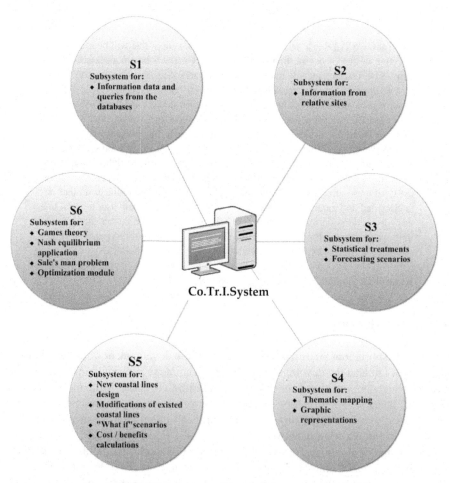

Fig. 1. Co.Tr.I.S. subsystems (*Source:* Moussas et al., 2015)

This paper presents all Co.Tr.I.S. subsystems with special focus on subsystems S5 and S6. The next section presents the overall Co.Tr.I.S. System, sections three to nine present the Co.Tr.I.S. subsystems S1 to S6 respectively, section ten analyze the Optimization module with reference to the optimization tools, and the last section presents future steps and research perspectives.

2   System Description

Co.Tr.I.S. is based on a Geographic Information System (GIS) software, adds-on traffic management and other (mapping, statistics, design) applications and it is divided in six different subsystems (Fig. 1). The system design was based on MECOSIG method (Pantazis & Donnay, 1996). The system's functionality combines network (vector) analysis, spatial analysis of raster surfaces (e.g. representing the cost), game and graph theory including Nash equilibrium and the salesman problem (Pantazis et al., 2013). Coastal transport network is stored in a geodatabase using a custom transportation data model. The system user requirements are met by the system's functions using the necessary data (Fig. 2).

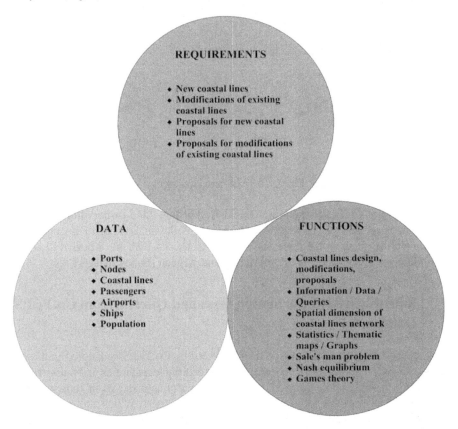

Fig. 2. Co.Tr.I.S. requirements data and functions

The system's users are divided into five different categories (groups). Each group will use a slightly different version of the application that is adjustable (data & functions) to their needs (Fig. 3).

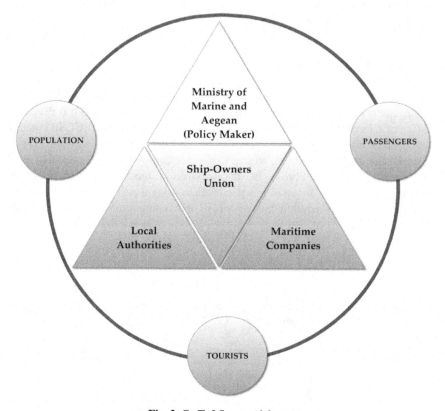

Fig. 3. Co.Tr.I.S. potential users

The potential Co.Tr.I.S users are: Local Authorities, Ship-owners union, Coastal lines companies, Ministry of Marine and Aegean (Policy makers) and Passengers. For the moment we focus in the three major potential users which are: Local Authorities, Coastal lines companies, Ministry of Marine and Aegean (Policy makers).

3   Subsystem S1: Information Data and Queries from Co.Tr.I.S. Databases

Subsystem S1 hosts the information data collected by Co.Tr.I.S and stored in the system's database. Subsystem S1 provides extended editing capabilities and a user friendly interface to facilitate the queries into the system. Some examples of supported queries (see Fig. 4) are the following (Pantazis et. al, 2013): Which are currently the coastal lines?, Which are the ports that had a specific number of passengers (set by the user) disembarked last year?, Which are the coastal lines (or itineraries) that start from e.g. Piraeus and have a ticket fare less than e.g. 40 euros?, Which are the coastal lines (or itineraries) that have Paros as an intermediate port (or destination or final destination)?, Which are the coastal lines (or itineraries) that have Chania as destination (or intermediate port or final destination)?

The system has extended capabilities concerning the automatic generation of reports and the display of summarized numerical tables with integration or not of schemas or maps. These modules are mainly designated to the local authorities, helping them to create proposals for new costal lines including a preliminary cost-benefits analysis. S1 also includes the functions of databases updating.

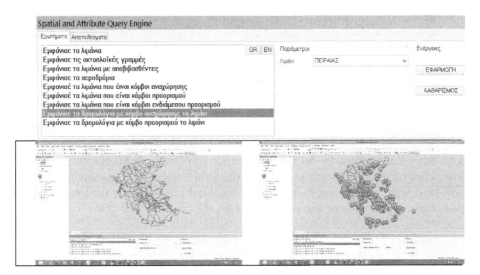

Fig. 4. Spatial and Attribute Query Engine of Co.Tr.I.S.

4   Subsystem S2: Information from Relative Sites

Information concerning the marine traffic and the Vessel positions tracking based on Automatic Identification System (AIS) data are available in the Co.Tr.I.S interface. Data concerning the available itineraries between islands or ports in general, the ships available, the coastal lines between specific ports and fare prices are also available via the system's connection with the relative sites (i.e. www.petas.gr web site). S2 functions give an additional control of the stored information facilitating the error identification and the future proposals creation (for new coastal lines).

5   Subsystem S3: Statistical Treatments, Forecasting Scenarios

Co.Tr.I.S. has capabilities for monitoring, evaluating and analyzing coastal transport lines and the relative infrastructure facilities (e.g. ports), and supports: a) strategic queries (e.g. given a number of passengers at a specific line, how many itineraries could simultaneously operate in a week base?), and, b) operational tasks (e.g. how many and what kind of ships can simultaneously arrive to a port?), based on a number of statistical results and data which are integrated into the system. The system uses statistical analysis results for an effective decision making during the process of new costal lines design. Data concerning the Greek coastal transport network have been organized, analyzed and interpreted through statistical processes.

For statistical purposes, after the collection and the initial classification of the data, the following assumption has been made: The numerical data correspond only to the disembarked passengers (residents/visitors) per each port, and not to those who have embarked on the same ship of the same line at the same time. There is no any significant quantitative difference between these two categories. The applied statistical treatments are: -Calculation of central tendency measures (arithmetic mean, median) and variation measures as well (standard deviation, minimum, maximum, range), for disembarkations per each year and season, -Calculation of percentage change between the three years per quarter, shipping line and port, in order to have comparable results between the same intervals of different years where data are available, -Creation of linear graphic charts presenting the coastal traffic of disembarked passengers per quarter, shipping line and port, -Creation of diagrams showing distributions of passengers per quarter, shipping line and port. The statistical treatments are performed with Statistical Package for Social Science (S.P.S.S.) and with Microsoft Excel. The integration of statistical process in ArcGIS platform is under development. Co.T.Ri.S. will be able to produce reports, charts and statistical treatment, e.g the total distribution of passengers per year and per port (see Fig. 5).

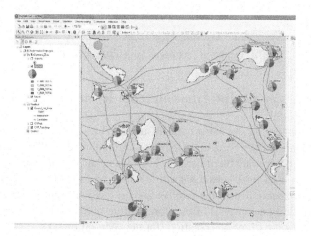

Fig. 5. Statistical treatment and graphic representation by Co.T.Ri.S

6   Subsystem S4: Thematic Mapping, Graphic Representations

Co.Tr.I.S. gives the possibility of various maps creation and visualization such as: a) digital and paper maps representing e.g. the coastal lines, the ports and the types of ships (with photos), b) thematic maps based on the islands and coastal lines characteristics such as: island populations, island tourists/passengers number, frequency of ship itineraries, etc., and, c) animated maps. The most important question to answer for the development of the subsystem S4 was: what kind of maps are necessary to improve the costal lines design? The visualization of all the spatial entities related with the system (ports, hubs/nodes, coastal lines) may be realized in classic (with topographic background) and schematic (thematic) maps helping the decision making.

This section presents the thematic maps created by Co.Tr.I.S.. The cartographic representation helps to understand the current situation in costal transport network and reviles its weaknesses. This information is used by the system in order to identify potential connections which are missing and support the process for new lines proposals in the costal network. The spatial data used for the thematic maps are the following: a) Coastline from the Organization of Cadastre of Greece (data type: line), b) Administrative boundaries of municipalities of Greece according to the last reform of the Ministry of Interior called "Kallicrates"(data type: polygon), c) Location of the ports in Greece (data type: point), d) Location of the airports in Greece (data type: point), e) Location of the hospitals and health centers (data type: point).

The creation of the coastlines were based on a map of the Hellenic Military Geographical Service (HMGS), 1:100.000 scale, which was scanned at a high resolution and georeferenced at the Greek Geodetic Reference System (EGSA 87). The coastlines were also verified with data from the Ministry of Maritime and the Aegean. Passenger's number in each line and each port, provided by the Hellenic Statistical Authority (EL.STAT.), were joined with the above mentioned spatial data in order to be used for map creation.

There are three different categories of maps created based on the main spatial data type. The first refers to the polygon data type where the attributes have a different color graduation according to the population number. The second refers to the point data type where there is a different symbol size based according to the number of the passengers at each port. The third category of the maps refers to the schematic representation of the costal lines. There are 23 thematic maps that have been created in order to describe the current situation of the coastal transport traffic network.

Co.Tr.I.S. provides a specific interface to the user, in order to permit the on-the-fly creation of those thematic maps. This means that the user will be able to create spatial and attribute queries regarding the coastal network, and the system will respond with the resulting map satisfying the query. These queries are kept in a separate table and each one of them has an SQL statement assigned to it. The parameters of the SQL statement are managed via the ArcMap interface.

7   Subsystem S5: New Coastal Lines Design, Modification of Existing Coastal Lines, What if Scenarios, Cost/Benefit Calculations

Subsystem S5 handles the main Co.Tr.I.S. functions: creation and /or modification of coastal lines. The system's user may create new costal lines or generally interact with the network in order to determine whether a new connection between ports is needed to be created or deleted or altered and modified. The system substantially supports the decision making process concerning the costal transportation issues, eg.: • Must be removed a specific costal line? Why? How? When? • Must be changed an existing coastal line? • Must be created a new costal line? • How is it possible to reduce the cost of a specific costal line? (Pantazis et al, 2013).

The new coastal lines creation or modifications may be realized manually (manual generation) or semi automatically or automatically. In the first case the user chooses the ports involved in the itinerary directly from the geographic interface in order to create the new coastal line (the system will provide all the relative information stored in the databases), in the second case the user chooses the ports from a popup menu and in the third case the system may suggest new lines, having specific characteristics and according to selected rules, that will be displayed in digital maps. Based on these rules the system could also generate all coastal lines connections that are relevant.

8   Subsystem S6: Decision Support with Games Theory, Nash Equilibrium Application, Sales Men Problem

The coastal transport problem in the Aegean Sea is very complex and multivariable. It can be generally described with the classic problem of salesman. Given a collection of (a) cities/ports/ harbors etc. and (b) the cost of travel between each pair of them, the travelling salesman problem (TSP) is to find the cheapest way of visiting all the cities and return to the starting point. In the standard version of the problem, the travel costs are symmetric in the sense that travelling from port X to port Y costs just as much as travelling from Y to X. In our case a) the latter does not apply, b) the salesman does not wish to visit all the customers, c) the salesman visits some customers only in specific periods through the year, d) the salesman needs different transportation means to reach every customer, e) the salesman agrees to visit some customers if they pay a specific price, f) exists many salesman g) some salesmen wish to collaborate with other salesmen or wish not etc. (Pantazis et al., 2013)

Elements of Game theory can be also used and integrated in our conceptual framework. Traditional applications of game theory attempt to find equilibriums in these games. In equilibrium, each player of the game has adopted a strategy that they are unlikely to change.

The Subsystem S6 is addressing to the coastal lines companies in order to help them in the strategic choices of coastal lines design and creation. It is also addressing to the ship-owners union in order to avoid unnecessary conflicts if possible. This module is not yet completely developed. For example it is still under investigation the fact that Nash equilibrium application may lead to "cartel" situations. S6 and S5 are the main subsystems for Decision Support and they are, in turn, supported by a set of tools including: simulators, route generators, cost evaluators, and optimizers as described in the next section.

9   Optimization in Co.Tr.I.S Subsystems S5 and S6

A major contribution expected by Co.Tr.I.S is to support the decision making process of the policy makers and the involved players (Ministry, Maritime companies, Local Authorities), towards an improved Coastal Transport System in the area. In order to support this functionality, Co.Tr.I.S subsystems S5 & S6 include several specialized modules for network generation, scenario validation, solution optimization and decision support. All these modules will have access to the vast amount of data classified and modeled by Co.Tr.I.S and therefore, will offer globally improved solution proposals.

The size and complexity of the entire Aegean costal transport system prohibit any manual or interactive search for better or optimal solutions. The overall Coastal Transport System complexity increases exponentially with the number of lines, destinations and realistic parameters included, and any typical enumeration or Branch and Bound method would require an unacceptable long time to reach an optimal solution. As a result, a tool using Evolutionary Methods was needed to provide optimization support on user demand and provide a ranked list according to the user's requirements and priorities of the best alternative scenarios.

During the last decades, many researchers studied the complex problem of the Aegean Sea coastal transport system. Several methods for coastal transport optimization and simulation have been proposed in the literature, but they usually tackle only a small part or a reduced version of the entire problem (Chainas 2012, Giziakis et al. 2006, Darzentas and Spyrou, 1996). Works on coastal network optimization demonstrated the promising use of evolutionary/heuristic techniques when applied to solve the container ship fleet problems such as cost, consumption, distance and deadlines (Karlafatis et al., 2009, Sun and Li, 2006), the hub positioning problem the liner shipping problem (Tsilingiris and Psaraftis, 2006, Khaled Al-Hamad et at. 2012) or, to consider environmental issues (Windek, 2013). Evolutionary techniques were also successfully combined with methods from Graph theory or Game theory especially in a search of equilibrium for network design with competing goals (Dinu and Odagescu, 2011, Fagerholt, 2004).

The Co.Tr.I.S user is usually a decision maker trying to find the solution/scenario that will best satisfy several competing goals such as: cost, consumption, user satisfaction, etc. Co.Tr.I.S database (ArcGIS) offers all kinds of current and past information about the Aegean Sea transport system, such as: connection lines, ships, schedules, geographical & thematic maps, passenger & vehicle demand, prices, etc.

These multifold and detailed data and their numerous alternative values, create a huge solution space that contains many potential coastal transport scenarios. By using Co.Tr.I.S the decision maker is supported by the decision/optimization modules to come up with a proposal containing a ranked set of the best solutions. The use of Co.Tr.I.S framework accelerates the entire procedure and the search concludes in a fraction of the time usually required.

9.1   Modeling for the Co.Tr.I.S. Optimization Tools

The detailed information on the Aegean Sea coastal transport offered by Co.Tr.I.S, the realistic and complex relationships employed, and, the many and contradicting goals create a Non-deterministic Polynomial (NP-hard) optimization problem that cannot always be solved within acceptable computer time by the exact algorithms. Therefore, in Co.Tr.I.S the optimization module is based on heuristics/evolutionary techniques i.e. the Genetic Algorithms that are more suitable, they search only a part of the vast solution space and converge faster to or near the optimal solution.

Optimization Methodology

The problem of Coastal Transport System can be seen as a multi-level or multi-stage problem consisting of several interconnected stages such as: 1) The Geometry of the network with all the nodes (ports) and their connections. At this stage we define the mainland ports, the potential hub ports, principal and secondary connections, fixed connections set by the state, forbidden connections due to safety or environmental issues, redundant connections for reliability, and alternative connection combinations. At this stage constrains and goals may include the overall distance, the total number or the reliability of the connections. 2) The Routes serviced. The routes & trips as well as the sequence of ports per trip are defined here. Different types of routes like shuttle trips, circles, or hub & spoke structures can be selected. At this stage constrains and goals may include the number of ports per trip, the allowed deviation, the overall distance, capacity of ports concerning the type of ships, etc. 3) The Schedule and Frequency of Service. The frequency of service is defined by the Routes and their corresponding demand. This stage is highly dependent on the previous stage 2 decisions and at the same time its results may lead to a route rearrangement, in order to cover demand. Other goals may include min frequency per destination, reduction of bottlenecks, min connection delays, etc. 4) The Ship Allocation. For each route/schedule defined in 2&3 the company must assign a ship with the appropriate characteristics of type, speed, capacity The overall Fleet deployment must satisfy constrains and goals such as max demand coverage, max ship utilization, min ship cost, ports posed by the maritime companies. 5) Operational issues. This stage is based on the results of 3 & 4. Ship operation issues like speed,   port   queues,   and   time   windows   for   arrival/departure or load/unload are defined here, and the results may lead to a ship reallocation at 4 or re-scheduling at stage 3.

The optimization tools in Co.Tr.I.S will be able to face either single stage or multi-stage problems. When facing a simple network problem or when the optimization focuses in one of the above stages, the use of a simple optimization technique is usually sufficient to provide the required solution. But, the typical problems that Co.Tr.I.S user (decision maker) is about to face will cover almost all five stages of the large and complex Aegean Sea coastal network, thus, resulting to a multi-level hierarchical system. In this case a bi-level optimization may be required where by using game theory techniques equilibrium is reached (e.g. Stackelberg, Nash).

By taking in to account all possible variations in each of the aforementioned five stages, the resulting alternative scenario combinations create a huge solution space that can be efficiently searched only by an evolutionary technique. The optimization tool employed is based on the Genetic Algorithm method. An Input Processing/Validation module validates all user inputs and collects any required information from the Arc-GIS database, a Scenario Generation module creates valid routes & scenario solutions, a Scenario Evaluation module evaluates the scenario performance based on selected performance indices, and the GA Optimization module implements the GA algorithm until convergence is reached (Moussas et. al, 2015).

Evolutionary methods and GAs represent each member of the population by a chromosome. In our case this chromosome is created by the variables set by the user and the ranged variables retrieved from the database. The entire set of these variables will create the search space for the optimal scenario. Any unfeasible solutions or con-strain violations are either discarded or they receive a higher penalty based on a quadratic function.

Coastal System Modeling for Optimization

In order to achieve globally optimal solutions and search flexibility, the system model contains a large number of coastal transportation parameters and variables. All together they build a set of Measured Values (MV) of our system. From the full set of MVs another set of Calculated Values (CV) is also produced. Altogether MVs and CVs create a phenotype a system solution/scenario. Phenotype contents are then combined to form the Performance Indices (PI) for each scenario. Finally, all the PIs are combined to a single KPI (Key Performance Index). The KPI corresponds to the overall score of the fitness function of the optimization algorithm. There is no unique KPI, as it is a weighted sum of all PIs and the corresponding weights are not fixed but they are defined by the user (KPI_u).

$$KPI_u = \sum_{k=1}^{n} w_{u/k} \cdot PI_k$$

Each user may have a separate set of weights W_u that best represent is user objectives and goals. Currently four different KPIs have been designed, corresponding to three different actors i.e., the ministry, the maritime companies and the local authorities. An indicative list of the model variables (MV, CV, PI & KPI), is presented in the following Table 1.

For a typical user request many of the above quantities will be fixed (constant value), others will be free to change inside an acceptable range (discrete or continuous) and the rest of them will be set/adjusted/restricted by the user. The non-fixed variables will create the GA chromosome that defines the search space that contains all the alternative solutions (scenarios).

Table 1. Indicative List of tUhe Problem (Scenario) Model Variables

Measured Values (MVs)	Calculated Values (CVs)
Island characteristics per Island (population, autonomy, hospitals and public services, airports, local population transportation demand, visitors demand, etc.)	Calculated from the available MVs
Port characteristics - per Port (capacity constrain for ships – size and number, infra-structure for refueling, waste, passenger accommodation, load/unload delays, .etc.)	Per Implemented Route:
Connections-All possible combinations (geographical distance, nautical distance, shuttle line or cyclic route, forced route, number of stops, alternative routes, etc.)	- travel time, - delays - over-length - productivity, - reliability
Demand - per Route (passengers, cars, trucks, seasonality, etc.)	- demand coverage - fleet capacity coverage,
Fleet – per Ship (type, capacity, speed, various costs, various fares, etc.)	- cost, - income,
Schedule (frequency of each route, depar-ture, travel time, delays, time windows, bad weather delays, winter/ summer adjust-ments, waiting queues, etc.)	etc.
Demand statistics and Forecasts, Weather statistics, etc.	

Scenario Performance Indices (PIs)
Total travelling distance
Total delay,
Total cost,
Total revenues,
Total demand coverage,
Total fleet coverage,
Coverage of local demand
Coverage of state demands,

Scenario Key Performance Indices (KPIs)
Per User preferences/goals: $KPI_u = \sum_{k=1}^{n} w_{u/k} \cdot PI_k$

In addition to the model variables & parameters a large number of constrains is always imposed to the designer and/or decision maker. For the coastal transport system under development, an indicative list of constrains follows: available fleet (set of available ships) and shipping companies, min/max capacity per route (required/permitted), min/max speed per route (required/permitted), min wind speed (for delays, for cancelation), max over-length (final route length over direct connection), max time per route (sum of travel+delays time allowed to complete a trip), max number of stops (number of intermediate ports), min allowed frequency of service (required number of trips per week), aver/max load/unload delay (acceptable delays per port/stop), aver/max waiting for connection line (waiting time in a hub port), max number of hubs, min demand coverage, min capacity coverage, etc. The example in the following section demonstrates the functionality of Co.Tr.I.S decision support and its optimization modules.

9.2   Sample Scenario Selection

A sample of the Aegean islands was selected to demonstrate the basic workflow during the optimization & decision making phase that is supported by the Co.Tr.I.S tools. The sample contains 6 ports: the mainland port and 5 islands in the Cyclades, Greece area: Syros, Paros, Naxos, Mykonos, Amorgos.

The data are retrieved from the database and they contain the required information such as: the distances between all ports, the passenger demands from one port to the other, etc., mainly in a form of symmetric matrices. Distance, demand and infrastructure information is used to classify and select the ports more suitable to act as hubs (the mainland port is always a hub). For each hub setup, all possible port combinations are detected and classified as: mainland lines, hub lines or local lines. For the 6 ports of the above example, and by considering the use of up to 3 hubs, a total of 523 different permutations was found, creating a large pool of routes from where, each scenario should select the ones to implement.

A first optimization step is applied to reduce the number of routes, by selecting the optimal one between permutations of the same combination. The Shortest Path optimization is used to select the shortest route servicing the same set of ports, leading to the best 106 port combinations. A classification step is then applied to further reduce the number of routes by removing the less efficient ones. The efficiency of each route is calculated as a function of its demand, distance, time number of stops and other characteristics. The less efficient routes to ports that are already serviced by more efficient ones can be discarded, thus reducing the available lines to 59 or: a) 23 using 1 hub, b) 20 using 2 hubs, or, c) 16 using 3 hubs. These numbers are still large as they can create hundreds & thousands of scenarios to implement for just 6 islands.

From the available set of routes, hundreds or thousands of valid scenarios can are created. In this optimization step, a Genetic Algorithm is implemented to cope with the huge number of alternative in a more manageable way. The GA population is consisted of different implementation scenarios. Each scenario will contain a number of routes in order to serve the 6 ports. For each scenario, the network implementation is represented by a chromosome containing 1s and 0s indicating if a route of the set is included or not in the scenario (Fig. 6). The GA starts with an initial population of randomly generated scenarios, and, after a number of generations concludes to a ranked list of the best individuals survived.

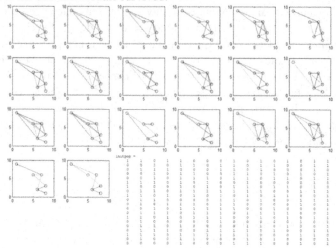

Fig. 6. The GA's random initial population of scenarios (networks) and their chromosomes

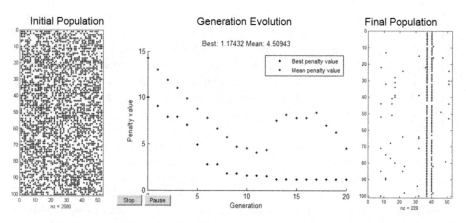

The Best Network Scenario (the final population member with the lowest penalty)

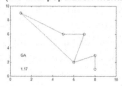

Fig. 7. The GA's result (lower), its convergence after 20 generations (up-middle), the chromosomes of the random initial (left) and the last (right) generation.

The fitness function calculates the overall efficiency of each scenario and the members of each generation are classified and reproduced according to their score. In this demo sample the overall score of a scenario is a function of the number and the efficiency of the implemented routes, their redundancy and of any penalty due to missing connections. The scenario heaving the highest score when convergence is met, it is considered as the best or optimal (Fig. 7).

The above described scheme demonstrates only the workflow towards the selection of an optimal Aegean Sea coastal Network scenario. It is not the complete solution of the problem as, another complementing part with chromosomes representing the traffic of the network (ships & schedules) is also under development. Depending on the case under consideration and the user selections, the two Genetic Algorithms will function either separately in sequence (first network then traffic) or in collaboration as the Bi-level optimization scheme described in section 9.

10   Future Steps and Research Perspectives

A presentation of the system will be realized in the next months to the three categories of the potential users. The under-development system presentation will include both "power point" presentation of the system's functions and the operational use of the system prototype. Views and critical analysis on the proposed system from the potential users are expected in order to improve the functions and complete its role and functionality. The main role of the system for each potential user, will determine the final crucial functions of the system for each user, which currently are: a) for the Ministry, the existence of coastal transport lines during the whole year for all islands, b) for the local authorities, the possibility of new proposals development concerning new coastal lines, c) for the maritime companies, the "discovery" of new profitable costal lines.

The system presentation will provide a feedback with proposed changes in order to improve the functionality of the system. Data, spatial database, system's rules, system's interface, system's functions, etc. will be modified if necessary. We also consider and analyze the system's contribution in the event of a natural disaster by developing the following functions and databases: Information about the number and type of ships which can arrive at the same time in a port, Information about the airports and choppers fields, Information about the time / distance of the nearest hospital / health center for each island. Automatic production of useful maps representing: location of schools, hospitals, area of natural disaster, etc. with possibility of crowd sourcing mapping capabilities. The module is under development.

An important number of islands lack health infrastructure like medical centers, clinics or hospitals. Many islands have not any doctor or dentist. In serious diseases or accidents it is necessary to transfer the patient by helicopter or ship or special boats rented by the state for this purpose. The Co.Tr.I.S. will provide the necessary information for the location (ports) of the special boats for patient transport, permitting a better and overall management of their fleet.

Acknowledgements. This research has been co-funded by the European Union (European Social Fund) and Greek national resources under the framework of the "Archimedes III: Funding of Research Groups in TEI of Athens" project of the "Education & Lifelong Learning" Operational Programme.

References

1. Chainas, K.: The Optimization Of The Greek Coastal Shipping Transportation Network, Tourismos: An International Multidisciplinary. Journal Of Tourism **7**(1), 351–366 (2012)
2. Darzentas, J., Spyrou, T.: Ferry Traffic in the Aegean Islands: A Simulation Study. Journal of the Operational Research Society **47**(2), 203–216 (1996)
3. Dinu, S., Odagescu, I.: Research on an Optimization Model for Transportation Network Design (February 12, 2011). SSRN: http://ssrn.com/abstract=1760663 or http://dx.doi.org/10.2139/ssrn.1760663
4. Fagerholt, K.: Designing optimal routes in a liner shipping problem. Maritime Policy & Management **31**(4), 259–268 (2004)
5. Giziakis, K., Paravantis, J.A., Michalochrista, M., Tsapara, A.: Optimal operation of passenger shipping. In: The Aegean, International Conference "Shipping in the era of Social Responsibility" Argostoli, Cephalonia, Greece, September 14–16, 2006
6. Karlaftis, M.G., Kepaptsoglou, K., Sambracos, E.: Containership routing with time deadlines and simultaneous deliveries and pick-ups. Transportation Research, Part E **45**(1), 210–221 (2009)
7. Al-Hamad, K., Al-Ibrahim, M., Eiman Al-Enezy, A.: Genetic Algorithm for Ship Routing and Scheduling Problem with Time Window. American Journal of Operations Research **2**, 417–429 (2012)
8. Moussas, V., Pantazis, D.N., Stratakis, P.: Modeling methodologies for optimization and decision support on coastal transport information system (Co.Tr.I.S.). In: Proceedings of the ICGIS 2015, 17th International Conference on Geographic Information Systems, Paris, France, May 18–19, 2015
9. Pantazis, D.N., Stratakis, P., Daverona, A.C., Gkadolou, E., Lazarou, E., Babalona, E.: Coastal transport integrated system: spatial database schema, metadata and data dictionary. In: Proceedings of 2nd International Conference on Advances in Computing, Electronics and Electrical Technology - CEET 2014, December 20–21, 2014, Malaysia, pp. 111–115 (2014)
10. Pantazis, D.N., Stratakis, P., Karathanasis, C., Gkadolou, E.: Design of a coastal transport integrated system: preliminary system specifications and data collection for the Aegean Sea islands. In: Murgante, B., Misra, S., Carlini, M., Torre, C.M., Nguyen, H.-Q., Taniar, D., Apduhan, B.O., Gervasi, O. (eds.) ICCSA 2013, Part IV. LNCS, vol. 7974, pp. 268–283. Springer, Heidelberg (2013)
11. Sun, X., Li, N.: A new immune genetic algorithm for the large scale ship routing problems. In: Presented at the International Conference on Management of Logistics and Supply Chain Chang Sha/Chine, Sydney, Australia (2006)
12. Tsilingiris, P., Psaraftis, H.: A multi-stage optimization-based approach for the liner shipping problem. In: 3rd International Congress on Transportation Research in Greece, Thessaloniki, Greece, May 2006
13. Windeck, V.: A Liner Shipping Network Design: Routing and Scheduling Considering Environmental Influences, Wiesbaden. Produktion und Logistik, Springer Gabler, Germany (2013)
14. Wright, D., Bartlett, D. (eds.): Maritime and Costal Geographical Information Systems. Taylor & Francis, London (2000)

How Regulation Affects Energy Saving: Smart Grid Innovation in Tall Buildings

Valentina Antoniucci, Chiara D'Alpaos[✉], and Giuliano Marella

Department of Civil, Architectural and Environmental Engineering,
School of Engineering, University of Padova, Padua, Italy
valentina.antoniucci@dicea.unipd.it,
{chiara.dalpaos,giuliano.marella}@unipd.it

Abstract. The economic problems involved in new high-rise buildings are mainly approached from the developer's perspective, especially in private-public partnerships, popular in Italian urban planning. Interest in energy savings and sustainability for buildings has recently increased considerably. Italian regulations have generally approached these problems from the viewpoint of materials and structures; zoning regulations do not cover energy from the viewpoint of investment projects and their externalities, either in urban development or from the economic viewpoint.

This paper examines the economic problems of energy consumption, with specific reference to tall buildings and high-density areas. Due to their typology of construction, dimensions and complexity, tall buildings may be viewed as urban developments in their own right. In more detail, this paper describes how the energy demand and consumption of single buildings can affect the energy trade-off of entire cities.

Keywords: Building height · Skyscrapers · Smart city · Smart Grid · Urban planning · Sustainability

1 Introduction

Italian town planning legislation still lacks proper methods for managing the energy costs and negative externalities of new developments, mostly in high-density urban regeneration. Urban planning in Italy does not take into account the problems of saving energy and consumption in highly advanced systems in terms of norms, on either local or regional scale.

Nation-wide and local norms merely focus on a building's scale of energy consumption, taking into account construction materials and plant performance. At urban level, town planning regulations cover only air pollution and natural environmental protection. There are no requirements covering the performance of new developments in terms of energy savings.

Energy consumption is a critical aspect of EU policy and research, especially concerning EU Smart Cities funding to European States by the year 2020. One of the most innovative areas of research is the Smart Grid system, which exploits alternative

© Springer International Publishing Switzerland 2015
O. Gervasi et al. (Eds.): ICCSA 2015, Part III, LNCS 9157, pp. 607–616, 2015.
DOI: 10.1007/978-3-319-21470-2_44

energy sources, promotes more efficient use of existing power grids, and reduces the environmental damage caused by traditional power plant. These objectives are important not only from a generic sustainability perspective; they are also related to the scarcity of natural and economic resources in Italy, which means that current practices of production and use of electric energy throughout the country must be revised.

In recent years, each EU state member has improved and upgraded its national, regional and local legislation regarding energy savings and has promoted academic studies and pilot projects. Italian norms focus solely on individual buildings and do not consider the interaction between existing buildings and new developments as a whole in terms of energy consumption.

The aim of this paper is to explain the potential of taking into account innovative distributed generation systems and new devices for saving electric energy consumption, together with the crucial role of town planning laws and norms as this process develops. Section 2 describes the Italian approach to current town planning regulations regarding the economic feasibility of real-estate investments and energy-saving regulations. This is followed by a discussion of high-rise properties in terms of energy use and consumption in other contexts. Section 4 presents the Smart Grid System and the way it has been applied to tall buildings. Our paper ends with some comments on how local urban planning legislation has the potential to create entire new developments and reduce energy consumption at the same time.

2   The Role of Town Planning

In Italy, the economics which determine how new building developments are established, are based on norms and laws which mainly focus on the developers' perspective, especially in densely populated urban areas. When considering the economic sustainability of large urban areas, the public sector only considers the gains for the developers, who must demonstrate the economic feasibility of a given project and the structure of their business model. Even when examining what benefits a new development might bring to a city or town, whether greenfield or brownfield, the local authority only checks that the required standards for public services such as parks, schools and community services, and how the development will meet the basic requirements for it to become an integral part of the city. If these requirements are not met, the public authority obliges private developers to pay the local administration an amount equal to the work which would be involved in the case of a public works construction.

In other words, economic feasibility evaluation is well-known and is determined by verifying that the revenues from the sale of a development which has transformed the local environment are large enough to pay for production costs, industrial and market risks, financial costs and developers' profits during the time-span of the construction work involved. In the case of significantly large new developments, the local administration may often have an interest in sharing the project with the developers, in order to manage the social conflicts which often arise from local stakeholders, who are aware of the possible future consequences of major developments in terms of green-field consumption and of the change in land use from the existing one to a new situation. In Italy, this approach and its debate is widely shared across the relevant literature, from urban policy-making to urban economics and, more in general, to the social sciences [1, 2, 3, 4].

However, there appear to be several weak points in the legislation of public administrations, when we examine the useful life of new developments promoted as parts of urban projects, especially as regards the social costs, both public and private, of urban development in terms of energy consumption. In addition, urban planning and building norms have no specific rules as regards the energy sustainability of large developments, but rather focus solely on the energy sustainability of single buildings in terms of a reduction in energy consumption. That is, these norms do not take into account the various types of buildings and how developments will influence the entire urban area.

To date, at an urban planning level, both normative and based on incentives, Italian norms have only dealt with containing energy consumption and working to reduce environmental pollution.

The first class of norms basically promotes two objectives: limiting land use and protecting environmental resources (e.g., rivers, forests, etc.). The former is usually dealt with by establishing the maximum amount of agricultural land which can become building land. The latter poses two problems: limitations as to how areas which are environmentally important can be developed, and how to choose the best types of facilities or buildings which will conserve both the environment and the landscape.

At national, regional and local levels[1], Italian legislation offers incentives for retrofitting buildings by allowing increases in their volumes and reducing the administrative costs of developing them. The local authority does not analysis the effects of these choices on market demand, nor does it verify whether the incentives are proportional to the number of on-going energy retrofitting operations or to the cost which such changes will have in terms of increased development in urban areas. Another aspect of the question is whether public incentives to reduce costs for developers are truly proportional to the collective benefits in terms of reduced energy consumption. The choice to use incentives has never been and still is not accompanied by any business model which enables analysis of the effects which such incentives might have on the real-estate market.

The relevance of this legislation becomes even more significant in the case of new high-density developments. If urban sprawl leads to increased atmospheric pollution due to congestion and the greater use of private means of transport as well as increased land consumption, then tall buildings and skyscrapers are by their very nature

---

[1] See the main references concerning on Italian nation-wide laws, which regional laws are based on: L. 10/91 - *Norme per l'attuazione del Piano Energetico Nazionale in materia di uso razionale dell'energia, di risparmio energetico e di sviluppo delle fonti rinnovabili dell'energia*; D.Lgs. n. 192 del 19.08.2005 *Norme per l'attuazione del Piano Energetico Nazionale in materia di uso razionale dell'energia, di risparmio energetico e di sviluppo delle fonti rinnovabili dell'energia*; D.M. 26.06.2009 *Linee Guida Nazionali per la certificazione energetica degli edifici*.

extremely complex entities requiring increased energy consumption simply for their management (water, waste management, electricity, gas, etc.).

Current Italian building and urban legislation does not take these aspects into account; on the contrary, it appears to have decided to support limited building from a "one-off" perspective and solely in the short term. To date, all legislation supports investments enabling existing buildings to reduce their energy consumption, rather than promoting norms which will help reduce costs via alternative management and a more in-depth approach to consumption, which could lead to significant reductions in both individual and collective costs, especially for tall buildings.

3   Tall Buildings as Prototype of Energy Innovation

Tall buildings are extremely costly, both to develop and to manage. The economic sustainability of building them can be analyzed by examining their construction costs [5, 6, 7, 8], both technological and economic-financial, but the problems of energy management and cost control are only viewed from the former perspective, not the latter. In the current scientific literature, progress in technical solutions for construction and energy savings do not appear to be accompanied by economic evaluation of building management costs.

At the same time, interventions to contain and/or reduce consumption are increasingly common in both new and restructuring projects [9]. Deep retrofit is not only considered necessary for existing buildings which have to meet new energy standards, but is also considered as a significant investment for improving property portfolio performance, due to the increased costs of regular maintenance, which greatly affect both rents and market values [10]. This becomes even more important in fragile markets such as those in southern Europe, where developers must maintain profitable investments in a marketplace which has suffered from slumps in profits and real-estate values from 2008 to the present, and also in extremely volatile markets such as those in the English-speaking countries and the Far East. In the former case, the significant decrease in demand leads to a shift in investments towards more sustainable buildings, i.e., those with lower management costs; in the latter, competition for innovation in the real estate market is limited by the efficiency of investments which require substantial capital expenditure. Consequently, energy requalification and keeping costs down in tall buildings become two of the ways of guaranteeing both investments and an increase in a building's efficiency.

The Empire State Building in New York is a notable example: $90 million were spent [11] on its energy requalification, leading to annual savings of management costs for both tenants and owners of $4.4 million, i.e., 38% of the building's total consumption. This example is ideal because, despite being a "trophy building", i.e., one considered to have secure investments in terms of profitability, energy costs clearly needed to be reduced although solid profitability was made with prime tenants. Based on the Life Cycle Cost Analysis (LCCA), investments aimed at reducing heat loss, improving the efficiency of air conditioning systems, and monitoring consumption for Heating, Ventilation and Air Conditioning (HVAC) in real time.

Adopting digital systems for controlling real-time HVAC is one of the basic approaches to innovation in the field of electricity production and consumption, according to the so-called Smart Grid approach. This field of research is less well developed, although it has enormous potential on both micro and macro scales, ranging from building projects to full urban developments.

There are varying definitions of Smart Grid (SG), but there are several basic aspects which it shares with the scientific community: distributed energy production based on renewable sources at local level, exploiting power stations which are smaller than those traditionally used; a bidirectional flow of energy instead of the traditional unidirectional flow; real-time production in response to energy needs; and a new role for the end-user, from consumer to prosumer [12, 13].

A new model for distributed energy production [14] offers new opportunities for both urban areas and the buildings they contain. This is especially relevant in buildings which, in terms of the numbers of people they house and their size, are comparable to neighborhood developments.

The technologies used to develop SGs are well-known and readily available as regards both production and distribution. However, whereas technical factors have been taken into account, what is missing is economic valuation, that is, a consolidated business model for diverse investments (e.g., SGs, micro grids, differences in terms of cost depending on what is needed) and estimates of the break-even point of investments with regard to savings. All these points depend, on one hand, on the fact that the approach is still experimental and therefore the number of cases to refer to is limited and, on the other, on the significant numbers of production and consumption variables which depend on the size of the network but, even more importantly, on prosumers' attitudes and actions.

The most interesting aspect in terms of innovation is that, the more widespread the prosumer network, the greater its potential to be efficient in terms of energy production and use of resources. Although there are clearly benefits to individual homeowners who develop SG homes, this solution does not fully exploit the decrease of auto-consumption of the energy produced. The network is effective only if it connects a large number of users who can coordinate their consumption in relation to their production. From an economic viewpoint, the efficiency of an SG increases with the number of buildings in it. This is why tall buildings, whether residential or not, are the best type of building for this kind of technological innovation.

Proof of the huge potential of tall buildings in improving energy-saving systems is, once again, the Empire State Building's deep retrofit. Attention was focused on the tenants' sense of involvement and their attitude. During the planning phase, the steps defined as "Identify Opportunities" were carefully examined [11]. Three key programs were prepared to involve tenants in building renewal: the tenant pre-build program, tenant design guidelines, and a tenant energy management program. The first is a green pre-built design which foresaw savings of $ 0.70-0.90/square foot in operating costs per year and respected tenant guidelines. Due to the turnover of nearly 40% of tenant space during the retrofit, the facility management needed an immediate, definitive and new approach to incoming tenants.

The most important program involving tenants in an active role was the tenant energy management program: each space was sub-metered and a feedback/reporting system was set up to inform tenants about their individual energy use. Monitoring of energy consumption is the first essential step in changing user behavior. The installation of occupancy sensors for light control educates tenants to a more aware use of electricity: the feedback system helps tenants to learn about their own use of energy. These devices reduce tenant loads and give them a benchmark for saving electricity. The total capital cost for this measure was $ 24.5 million and the energy savings are about $ 941,000/year. Other smart devices were installed, such as Direct Digital Control (DDC) and demand control ventilation, measuring CO_2 concentrations inside the building, improving air quality and reducing energy use. The installation of smart devices generated savings in energy costs of about $ 117,000/year. The Variable Air Volume (VAV) handling units, unlike Constant Air Volume (CAV) systems - which supply a constant airflow at a variable temperature - provide variable air volume control by changing the airflow at a constant temperature. This device, which allows more precise temperature control, compressor wear reduction and lower energy consumption thanks to system fans, enhances optimal HVAC operation and provides more accurate sub-metering of energy use. This measure cost $ 47.2 million and energy savings of $ 702,000/year.

All these devices, linked to smart electric energy control and consumption, guaranteed 16% of annual energy savings on the 38% total.

Although it is beyond the scope of this paper, we emphasize that a substantial financial analysis was undertaken to determine exactly which measures maximize energy savings while providing economic benefits. The interaction between the models of energy consumption and Net Present Value (NPV) package enables not only evaluation of the financial feasibility of a project, but also its profitability. The robustness of this approach is testified by the increased estimated energy-saving target for the deep retrofit, now in its second year.

The Empire State Building deep retrofit is not only a success story but also a model for advanced energy refurbishment in a high-density building context.

As already noted, tall buildings and skyscrapers are major consumers of energy and resources. The people who live in them and/or use their spaces represent a significant segment of market demand in quantitative terms. This demand is thus marked by substantial expenditure[2] which can justify investments in building innovation. There are two other important aspects, one technical, and one linked to regulations. In the former, two points stand out: energy production and energy consumption. The very nature of tall buildings means that they have large vertical surfaces which, if they face south (in the northern hemisphere) or north (in the southern hemisphere), can produce energy from photovoltaic panels. This well-known technology has been extensively experimented and implemented. In addition, if large numbers of consumers live near a power station, energy distribution costs can be reduced.

---

[2] The literature [6] states that tall buildings and skyscrapers are part of the real-estate luxury market.

The fact that prosumers are close to a grid is not solely a technical factor: more importantly, it means that the grid can be exploited to maximum effect as regards consumption. This point is clearly not a technical aspect of the system, but it does demonstrate the importance of proper regulations in the potential success of regulating SGs.

Tall buildings and skyscrapers are regulated by community contracts [15] or homeowner associations [16]: in English-speaking countries, especially the United States, very precise regulations exist regarding how people should behave inside and outside buildings, in both private and public areas. These regulations may be implemented by owners and enforced by building managers, so that tenants must follow a set of rules which are much more specific than those required by local legislation. Similarly, rules and regulations can be agreed upon and implemented by groups of private homeowners.

In such systems, the use of private spaces is regulated in agreement with the owners of neighboring buildings, making this an excellent opportunity for extending norms for the use and consumption of energy to experimenting with SG approaches. Adopting shared and distributed technology energy production and consumption can become part of existing regulations much more easily and effectively than in places where consortia must be created to bring together buildings physically distant from each other, as is the case in urban sprawl, and in areas where setting up energy production plant may be more expensive [17].

It is also worth noting that introducing specific regulations regarding electricity can be facilitated when the number of property owners is limited: in English-speaking countries, many residential buildings have a single owner, like many company buildings in Southern Europe.

The innovation in energy consumption in multi-tenant income properties (MTIPs) is not just an opportunity granted by a context of prosperity; it also has a theoretical framework. Lowi and MacCallum very clearly explain the potential of community technology [17] and how the spread of this particular kind of self-organization is restricted by laws, both planning and procurement, and not by technology know-how or physical conditions.

The authors clarify how "requirement for power [...] can be design and managed as one integrated system at enormous savings and enhancement of user's satisfaction" [17]. If an MTIP community behaves as a "community entrepreneur", there will be more efficient use of energy and economic advantages, due to the development of a specific market demand. However, the success of community entrepreneurs depends to a great extent on the ability of the community itself to build and design an efficient contractual format.

4   The Role of Norms on Smart and Micro Grids Implementation

How can norms and regulations support the technological innovation of SGs? In the field of SG research in the EU, one of the main goals to be reached by 2035 is a properly functioning energy market based on distributed energy production and regulatory innovations supporting it [18].

With specific reference to the Italian market, can regulations support the implementation of distributed energy production? It is beyond the scope of this paper to discuss the role national legislation might play in adapting existing high-voltage power-er plants, when our aim is to reflect on the implications of updating local urban planning and building norms, after having highlighted energy problems related to different classes of population density, differing needs and potential benefits of specific types of construction.

The greatest opportunities currently available in the Italian real-estate market in terms of investments lie in the revitalization and redevelopment of brownfields. These are more frequently found in large urban areas, and the type of restoration chosen may also be planned by increasing building volumes which, beyond certain thresholds, require special types such as tall buildings and skyscrapers, the costs of which are different from those of low-density contexts.

From an economic point of view, the incentives for the energy redevelopment of existing buildings or the construction of new low-consumption ones only provide for initial investments, whereas SGs require long-term vision, which includes the day-to-day costs of the building - the only way of demonstrating the actual economic benefits of adopting such grids in the first place. Current urban planning laws do not deal with energy on an urban scale, such as shared energy management in neighborhoods or in tall buildings. Examples of more advanced urban energy efficiency legislation do exist, such as the English Leed Protocol (which, however, does not focus specifically on electricity production).

In any case, in order to exploit existing legislation, local governments must allow owners greater power to manage shared facilities and resources in order to develop a network and possibly a consortia of owners, when building both new power stations and redeveloping energy needs in a group of already existing buildings. Currently, Italian legislation regarding the management of residential buildings does not allow for such private or local regulation, as the relationships between owners of units are regulated by nation-wide norms.

In addition, regarding current incentives, local governments could offer energy providers subsidies, currently only regulated nation-wide. Clearly, in order to maximize efficiency, a local system of electric energy production would have to be regulated by local norms and regulations.

Lastly, it would benefit all players if local authorities were to evaluate buildings which have already received economic incentives, in order to assess whether there have actually been savings in consumption. Such a system should determine the relationship between the cost of the public investment to promote energy consumption reduction with actual energy consumption.

5   Final Remarks

Reducing energy consumption is one of the main topics under discussion in the development of the European Urban Agenda [19], and innovation in the field of energy is one of the main areas to receive funding from the European Union until 2020. There are many ways of solving these problems at urban level: one of the most significant, for both urban areas and individual buildings, is to use Smart Grids to produce shared energy and manage production and consumption in real time.

In densely populated areas, especially those with tall buildings and skyscrapers, the potential of SGs to make a difference is greater, because of the high costs of energy management and the proximity of a large number of consumers. These types of buildings also have norms and regulations regarding the use of shared and private spaces which could serve as a basis for developing shared energy consumption. However, in Italy at present, local urban planning legislation is not sufficiently flexible to promote this kind of innovation. The only incentives currently available are investing in the energy requalification of existing buildings, and no evaluation process exists to assess whether or not such interventions actually lead to reduced consumption and costs. Consequently, there are many opportunities for research into how to improve this situation. At the present time, there are no case studies of cost analyses regarding the potential benefits of SGs on different scales, and no business models.

An economic analysis of the effectiveness of distributed production systems could aid local legislators to improve their choices in urban planning. Legislation regulating urban planning and building management must support not only technological innovation but, even more importantly, the behavior and attitudes of local inhabitants, who currently do not benefit from any norms or regulations which could promote the radical change in consumption habits granted by Smart Grids.

Acknowledgment. Funding for this study was supplied by the University of Padova, project no. CPDA133332/13.

References

1. Urbani, P.: Urbanistica consensuale. La disciplina degli usi del territorio tra liberalizzazione, programmazione negoziata e tutele differenziate, Bollati Boringhieri, Torino (2000)
2. D'Alpaos, C., Marella, G.: Urban planning and option values. Applied Mathematical Sciences **8**(157–160), 7845–7864 (2014)
3. Morano, P., Tajani, F., Manganelli, B.: An application of real option analysis for the assessment of operative flexibility in the urban redevelopment. WSEAS Transactions in Business and Economics **11**(1), 465–476 (2014)
4. De Mare, G., Nesticò, A., Tajani, F.: Building investments for the revitalization of the territory: a multisectoral model of economic analysis. In: Murgante, B., Misra, S., Carlini, M., Torre, C.M., Nguyen, H.-Q., Taniar, D., Apduhan, B.O., Gervasi, O. (eds.) ICCSA 2013, Part III. LNCS, vol. 7973, pp. 493–508. Springer, Heidelberg (2013)
5. Watts, S., Kalita, N., MacLean, M.: The economics of Super – tall towers. The Structural Design of Tall and Special Buildings **16**, 457–470 (2007)
6. Barr, J.: Skyscraper height. Journal of Real Estate Finance and Economics **45**, 723–753 (2012)
7. Lau, E., Yam, K.S.: A study of the economic value of high-rise office buildings, strategic integration of surveying services. In: FIG Working Week, Hong Kong (2007)

8. Antoniucci, V., Marella, G.: Torri Incompiute: i costi di produzione della rigenerazione urbana in contesti ad alta densità. Scienze Regionali **13**(3), 117–124 (2014)
9. Collins, A., Watts, S., McAlister, M.: The economics of sustainable tall buildings. In: Proceedings of CTBUH 8th World Congress, Dubai (2008)
10. de Jong, P., Wamelink, H.: Building cost and eco-cost of tall buildings. In: Proceedings of CTBUH 8th World Congress, Dubai (2008)
11. Rocky Mountain Institute. http://www.rmi.org/Content/Files/ESBCaseStudy.pdf
12. Fang, X., Misra, S., Xue, G., Yang, D.: Smart Grid – The New and Improved Power Grid: A Survey. IEEE of the Communications Surveys & Tutorials **14**(4), 944–980 (2011)
13. Tang, S., Huang, Q., Li, X.-L., Wu, D.: Smoothing the energy consumption: peak demand reduction in Smart Grid. In: Proceedings of INFOCOM. IEEE, Turin (2013)
14. Marsen, J.: Distributed generation systems: a new paradigm for sustainable energy. In: Green Technology Conference. IEEE (2011)
15. Brunetta, G., Moroni, S.: Contractual communities in the Self-Organising City. Freedom, Creativity, Subsidiarity. Springer, Netherlands (2012)
16. McKenzie, E.: Privatopia. Homeowner associations and the rise of residential private government. Yale University Press, New Haven (1994)
17. Lowi, A., MacCullum, S.: Community technology: liberating community development. In: Andersson, D.E., Moroni, S. (eds.) Cities and Private Planning. Property rights, Entrepreneurship and Transaction Costs, pp. 106–134. Edward Elgar, Cheltenham (2014)
18. European Technology Platform Smart Grids: Smart Grids SRA 2035. Strategic Research Agenda. Update of the Smart Grids SRA 2007 for the needs by the year 2035, EU (2012)
19. European Commission: The Urban Dimension of EU Policies – Key Features of an EU Urban Agenda. Communication to the Council, EP, CoR, EESC, Brussels (2014)

Estimating the Biotope Area Factor (BAF) by Means of Existing Digital Maps and GIS Technology

Roberto De Lotto$^{(\boxtimes)}$, Vittorio Casella, Marica Franzini, Veronica Gazzola, Cecilia Morelli di Popolo, Susanna Sturla, and Elisabetta Maria Venco

DICAr, University of Pavia, via Ferrata 3 27100, Pavia, Italy
{uplab,roberto.delotto}@unipv.it

Abstract. The problem of excessive soil sealing and the consequent macro-microclimate degradation process need specific measures.

The use of standards and indexes finalized to the regulation of the soil permeability permits flexible and resilient solutions that allow environment to find a new balance.

The application of the Biotope Area Factor (BAF), as index focused on the integration of urban planning and ecology, is a tool with a dual function: it allows to address a structural choice giving more space to planning and to get better consolidated areas.

The main scope of the paper is presenting a way to calculate BAF which is automatic and based on existing digital cartography and GIS technology.

First authors present a description of the general problem; then they focus on the BAF index as urban parameter computable from direct analysis and web images. The classical way to manually measure BAF is presented with application to the city of Pavia.

Then authors introduce the automatic GIS-based procedure starting from a description of the specification of the Lombardy Region cartography and illustrate all the steps and issues of the developed procedure.

Finally authors rigorously assess the developed procedure by means of manually surveyed test areas and discuss results and further steps of the research.

Keywords: GIS based system · Urban-ecological standards · Environmental quality · Topographic database standards

1   Introduction

In a constantly changing urban context, due to new and different needs of its inhabitants, it is trying to find a dynamic balance with the natural systems: thus, it is fundamental the quality improvements of urban development processes through the conscious use of resources (energy, land use and environmental quality) [1].

A part of the processes of macro-microclimate degradation is caused and increased by new covered areas and soil sealing, phenomena that are constantly increasing with sprawl (expansion of peripheral low-density urban areas) that characterizes many urban territories. The new surfaces increase air mass heating and the convective

© Springer International Publishing Switzerland 2015
O. Gervasi et al. (Eds.): ICCSA 2015, Part III, LNCS 9157, pp. 617–632, 2015.
DOI: 10.1007/978-3-319-21470-2_45

motions produce a recirculation of dust released into the atmosphere, one of the major cause of urban pollution. A consequence of the accumulation and irradiation of sun heat is the temperature increase because of the lack of the natural mitigation effect given by the vegetable evapotranspiration [2]. To assess the quality of urban environment there are different parameters that in most of cases try to synthesize different factors in unique values using multicriteria style calculation. Looking at the block scale, and considering the sum of effects that in a wider scale some behaviours (such as soil sealing and evapo-transpiration) may have, since the '90 some new parameters emerged. Among them, the Biotope Area Factor, that has been defined in the Landscape Plan of Berlin (see Chapter 3 for a more complete explanation).

In new settlements, "quality parameters" can be considered as main goals of urban development and they are easily manageable. In existing city they need to be calculated with automatic procedure because manual calculation is too laborious.

Authors developed a methodology capable of estimating BAF in an automatic and objective way. It is based on the application of GIS technology to municipalities' existing digital maps. This methodology is really quick, doesn't require any dedicated field work and can be applied to large areas as well as to a whole city. The authors applied the procedure, as a first case study, to Pavia's city, and also performed detailed validation of the developed approach by comparison to manually surveyed test areas.

2   Ecological Issues in Urban Planning

SEA (Strategic Environmental Assessment) [3] is an overall evaluation system of monitoring and evaluation that seeks to optimize relationship between resource, their utilization and environmental protection. In European context, since 2001, it is the procedure that forced to insert environmental and ecological issue from the first steps of planning process. In SEA and in all related examples of integration between urban planning and ecology [4, 5, 6, 7, 8], environmental indicators are the starting point for the creation of composite indices that contain urban parameters and environmental element. Although new standards do not yet codified by the law (at the European level), they are introduced and defined in the more innovative plans with particular reference to: permeability of soil, allowable (sustainable) urban load, parameters expressing the environmental carrying capacity of transformation areas, definition of standards and landscape indicators for urban and suburban territory, etc.

Instead of comprehensive and complicate indexes, an effective index should have a solid scientific base, should have tested different and complex case studies, should be able to direct a structural choice and not tamper with the design phase and should also refer to the regulatory phases of planning processes.

Among the various indices available in bibliography [9, 10, 11], the Biotope Area Factor (BAF) presents all these characteristics. BAF is useful for Urban Planning because many ecological parameters can be controlled throughout classical planning indexes (such as covered ratio); moreover the ecological effectiveness of settlements depends also on design decisions related to, in example, buildings' materials.

3   BAF Definition

In 1984, in Berlin, was introduced the Landscape Programme (Environmental Landscape Plan) with the objective of the protection of nature, natural resources, landscape and collective green areas in urban areas. The Landscape Programme operating mode tries to identifying new ways of spatial planning implementation for city functional element that increase their environmental performance, without losing the typical characteristics of existing city. In 1994 the method called BAF - Biotope Area Factor was developed [12]. All areas that potentially contribute are included in the BAF and an assessment differentiated value is assigned, referring to the quality of evapotranspiration, permeability, providing habitat for plants and animals. With reference to the general goals of sustainability, the BAF is able to provide a parametric measure about: preservation of the microclimate and the atmosphere, check of land and water use, improvement of the quality of plants' life and animals' habitat, improvement of living space for the human being. BAF is defined as follows:

$$BAF = \text{ecologically effective surface area} / \text{total land area} \qquad (1)$$

The ecologically effective surface area (EESA) is given by the sum of the surfaces multiplied by an ecological coefficient, assigned according to the specific characteristics of those surfaces and summarized in tabulated values. Criteria that guide the assignment of values are: high efficiency of evapotranspiration, ability to powder's fixation with a reduction of suspended dusts, ability to capture water from soil and its storage, conservation and long-term development of soil functions (filtering, buffering and transformation of pollutants-hazardous substances), availability of suitable habitats for plants and animals.

Table 1. BAF in Berlin. Different types of surfaces' weights

Surface type	Weight factor
Sealed surface	0
Partially sealed surfaces	0.3
Semi-open surfaces	0.5
Surfaces with vegetation unconnected to soil below (less than 80 cm of soil covering)	0.5
Surfaces with vegetation unconnected to soil below (more than 80 cm of soil covering)	0.7
Surfaces with vegetation connected to soil below	1
Rainwater infiltration per sqm of roof area	0.2
Vertical greenery up to 10 m in height	0.5
Green roofs	0.7

This index was studied for areas with different land use (commercial, residential, infrastructure, industrial). Considering green planted or lawn areas, green roofs and walls, areas with not sealing soil and shading: there, the value of BAF target is between 0.3 and 0.6 depending on the function established and the Degree of Coverage. In Berlin specific BAF targets were settled as function of Degree of Coverage and land use. Berlin's Plan provides a set of ecological values for each type of surface [Table 1]. After Berlin's Plan, other European and US cities applied similar coefficients in planning phase [13, 14, 15].

3.1   Application in Pavia

The hypothesis authors consider is to use the same Berlin's indexes in Pavia.

The climate conditions of the two cities is moderately continental and the average temperature trend does not differ significantly; the level of precipitation is the factor that most differs and it is higher for the city of Berlin. The main criteria to define the indexes, as already described, are assigned considering evapotranspiration and permeability of soil. Evapotranspiration is a value used particularly in agro-meteorology and, in the time unit, represents the amount of water lost by the soil and wasted in the atmosphere by the combined effect of plants transpiration and water evaporation from the soil. Evapotranspiration is a complex phenomenon that involves different disciplines; it depends on: nature of soil, and the force with which retains water, climate and related parameters as solar radiation, temperature, air humidity and wind speed, physiology of greenery cover, cultivation techniques adopted in a specific area. The interaction of numerous and variables factors and the possible differences in the calculation's methods make the assignment of a defined and univocal value extremely complex. In particular the higher scores depends on the presence of surfaces with vegetation, ignoring in part the permeability of the soil and the evaporation of the same. In this sense, the parameters that compose BAF are not only quantitative as the calculation of a precise index based on evapotranspiration or permeability, but also qualitative as the improvement of environmental, ecological and landscape quality of a place.

Using two calculation methods (Penman Monteith's one and Hargreaves' one) [16, 17], it has been verified by authors that, the annual average of evapotrasfiration value calculated in Pavia is between 11% and 22% more than in Berlin. It means that, using Berlin's indexes in Pavia, the targets defined in Berlin are surely reached.

3.2   Operative Issues in Manual Calculation

BAF calculation (and the creation of related thematic maps) involves a big amount of editing work including the use of orthophotos and bird view images (sources: Google Maps, Bing Maps, Google Street View) as well as direct surveys on the field.

This operating mode, suitable for specific cases' application, is more expensive, in terms of time, widening the field of investigation to the urban scale (existing city).

In this perspective, and because of the lack of uniformity and objectivity, it is fundamental the trial of automatic calculation.

The following images refer to the manual calculation of BAF in two sites of the city of Pavia. This procedure was extended to quite the entire city, while in 12 areas the automatic procedure was applied to asses its precision (Fig. 1).

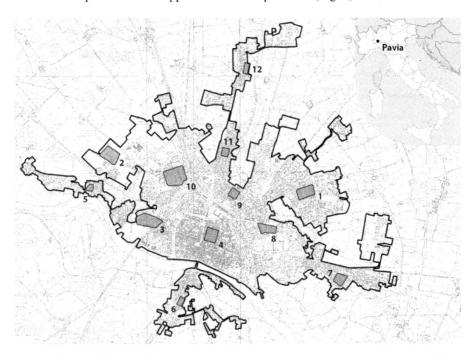

Fig. 1. The city of Pavia with twelve test areas used for the assessment and its location in Italy

Fig. 2. Area 1: bird view

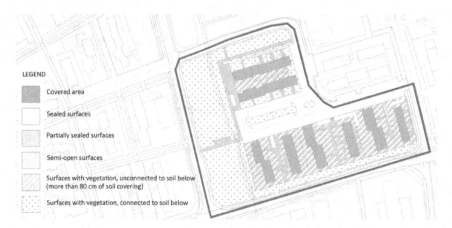

Fig. 3. Area 1: CAD elaboration indicating different surface types

Table 2. Area 1 manual BAF calculation

Description of surface types	m^2 of surface type	Weight factor	EESA (m^2)
Covered area	4,703.65	0	0
Sealed surface	10,674.55	0	0
Partially sealed surfaces	2,819.7	0.3	845.91
Semi-open surfaces	0	0.5	0
Surfaces with vegetation unconnected to soil below (less than 80 cm of soil covering)	7,124	0.5	3,562
Surfaces with vegetation unconnected to soil below (more than 80 cm of soil covering)	0	0,7	0
Greenery on rooftop	0	0,7	0
Surfaces with vegetation, connected to soil below	12,415	1	12,415
Total land area (m^2)			37,736.9
BAF			0.45

Fig. 4. Area 2: bird view

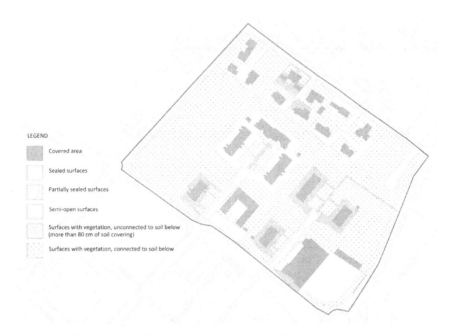

Fig. 5. Area 2: CAD elaboration indicating different surface types

Table 3. Area 2: BAF maual calculation

Description of surface types	m² of surface type	Weight factor	EESA (m²)
Covered area	26,930.46	0	0
Sealed surface	24,100	0	0
Partially sealed surfaces	5,744	0.3	1,723.2
Semi-open surfaces	436	0.5	218
Surfaces with vegetation unconnected to soil below (less than 80 cm of soil covering)	0	0.5	0
Surfaces with vegetation unconnected to soil below (more than 80 cm of soil covering)	0	0.7	0
Greenery on rooftop	0	0.7	0
Surfaces with vegetation, connected to soil below	50,067	1	50,067
Total land area (m²)			107,277.46
BAF			0.57

4   Structure of Pavia's Digital Map

Pavia is a medium sized city having 72,000 inhabitants and an area of 63 square kilometers; it is located in Northern Italy, 35 km South of Milan (Fig.1) [18]. The currently used city map is owned by the Municipality and was created in the early eighties as a paper map having the 1:2,000 ratio scale. The map was successively digitized and progressively updated and transformed according to the guidelines issued by the Lombardy regional Government. The map can be currently visualized at [19] and can also be downloaded at [20]. For the present research, we had the whole map in ESRI shape format directly from the Municipality, which is here acknowledged.

As anticipated, the Lombardy Region published guidelines (which can be downloaded at [21] and [22]) for the creation and updating of the municipalities' large scale maps. Noticeably, the Lombardy Region has co-financed the creation from scratch of new maps or the updating of existing ones, in order to stimulate the Municipalities to adopt the regional specifications. Up to now, 1,244 municipalities out of 1,544 (80%) have an updated and conformal map, according to [23] and [24].

Interestingly, the regional guidelines are inspired to a sort of national standard, which is published at [25]. The Regions had the power to integrate the national standard by adding details, but leaving the general design untouched. In summary, Pavia's map is a good representative of the large-scale maps which are owned by many Italian municipalities.

Pavia's map is stored in an ESRI geodatabase and is preferably exported in the ESRI shape format. It consists of 74 layers of the punctual, linear or areal type. The primary key is 6-digits long, is composed by three two-digit subkeys and has a tree structure. The first two digits can take the values shown in [Table 4] below together with their meaning.

Table 4. The first level of the map tree-structured classification

00	**Geodetic, cartographic and photogrammetric information**
01	Roads, mobility and transport
02	Buildings and man-made features
03	Street names and house numbers
04	Hydrography
05	Orography
06	Vegetation
07	Technological networks
08	Toponyms
09	Administrative borders

Once chosen the '02' branch, five further classes are planned, which are listed in [Table 5]. Finally, five more options are available for the '0201' branch, which are listed in [Table 6].

Table 5. The admissible values for the second level of the '02' branch

02 01	**Constructions**
02 02	Small man-made structures
02 03	Transportation infrastructures
02 04	Retaining and soil conservation structures
02 05	Hydraulic constructions

Table 6. The admissible values for the third level of the '0201' branch

02 01 01	**Volumetric unit**
02 01 02	Buildings
02 01 04	Roofs
02 01 05	Architectonic details
02 01 06	Minor buildings

For some categories, there is a secondary key for a better specification of the features' nature; the name of the secondary key and its allowable values change from one category to another. For the 020101 (*Buildings*) primary key value, the secondary key is named EDIFC_TY and can take a tenth of values.

A second interesting example is related to a feature type named *Vehicle circulation areas* having 010101 as primary key value; its secondary key is named AC_VEI_FON and allows the user to distinguish between: asphalt, cement, stones, gravel, slabs of stone, pavè, ground, grass, opus incertum, and interlocking paving bricks.

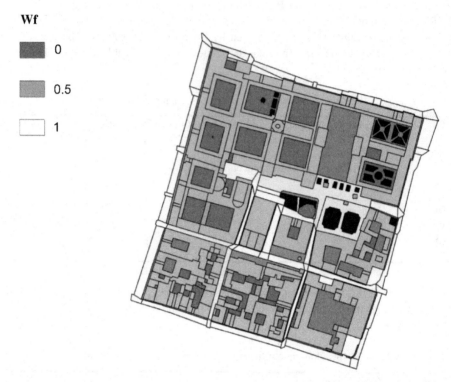

Wf

■ 0

▨ 0.5

□ 1

Fig. 6. Illustration of the classification detail for test area 4: the various grey levels shown correspond to different classification assigned to the features

In summary, the map has a very detailed classification of its content, as illustrated by [Fig. 6], which can be used to effectively associate a BAF coefficient to each map's feature.

Nevertheless, the map presents some issues which were accurately tackled in our work. First of all, there are overlapping features such as overpassed roads, courtyards and ramps, as well as courtyards and underground garages. Moreover, courtyards are classified in a less detailed way than other features and the nature of the paving (ground, gravel, cement and pavement) is not accounted for.

5   The Developed Methodology

As anticipated in the previous section, our developed methodology is based on the observation that the existing digital map of the city of Pavia is sufficiently detailed (with respect to geometry and classification) to perform a reliable BAF mapping.

The final product of our approach is a simple surface made of adjacent, non-overlapping polygons, without holes, covering all the city and giving the BAF values for every location.

First of all, a selection was performed among the existing map layers to keep only those really interesting for BAF evaluation: punctual and linear features were discarded, as well as unnecessary or redundant polygonal ones. In conclusion, a set of 29 layers, out of 74, were identified, stored in the ESRI shape format and used for subsequent analysis.

All the editing and processing work was performed with ESRI ArcMAP, ver. 10.1. Implementation details are omitted in the present publication and will be presented in a further one.

A grand-key was added to each feature, coming from the concatenation of primary and secondary keys. A look-up table was created associating each value of the grand-key with the corresponding BAF code (between 1 and 9) and weight factor (having values between 0 and 1, see [Table 1]). An excerpt of the created table is shown by [Table 7], for the sake of clarity.

Table 7. An excerpt of the prepared look-up table

Grand-key	Key1	Key2	BAF code	Weight factor	Description
010101 0221	010101	0221	1	0	asphalt
010101 0228	010101	0228	2	0.3	pavè
010101 0212	010101	0212	6	1	grass
020204 0102	020204	0102	6	1	soccer field
020204 0103	020204	0103	2	0.3	tennis court
050393 0105	050393	0105	3	0.5	gravel
050393 0116	050393	0116	1	1	sand

The look-up table was used to associate BAF code and weight factor to each feature of each considered layer. Then all the layers were merged to form a unique large topographic database containing all the classified features.

Furthermore, overlapping polygons were taken into consideration and properly processed, as they generally have different BAF coefficients: courtyards and underground garages for instance. The layers were transformed by a complex application of various Boolean polygon operators: multi-part polygons were converted to single-part ones; two partial layers were created, that of non-overlapping polygons (or part of polygons) and that of overlapping parts of polygons; the latter was further transformed by keeping, among the multiple overlapping sub-polygons, the one with the lowest BAF coefficient, in order to be conservative; the so-transformed partial layers were finally merged to form the final BAF map. [Fig. 7] shows an example of the obtained map.

Weight factor

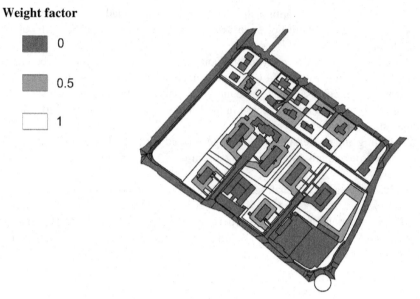

Fig. 7. Feature BAF classification for Test Area 2

Finally, a report procedure was created to calculate the synthetized BAF values for a certain area, producing a table like [Table 8].

Table 8. Synthetized BAF values for test area 2

Weight factor	# Features	Area [sqm]	Area [%]	EESA[sqm]	EESA[%]	BAF
1.0	63	48,551.36	45.26	48,551.36	86.00	
0.7	0	0.00	0.00	0.00	0.00	
0.5	21	15,808.18	14.73	7,902.59	14.00	
0.3	0	0.00	0.00	0.00	0.00	
0.0	173	42,920.92	40.01	0.00	0.00	
Area 2	257	107,277.46		56,453.95		0.53

[Table 8] refers to Test area n. 2 (see next Section) and shows: in col. 1 the BAF coefficient considered: not all the values listed in [Table 2] are visible, but only those applicable to Pavia city; in col. 2 the number of features having a defined BAF value; in col. 3 the total area of the features; in col. 4 the corresponding percentage, with respect to the total area; in col. 5 the BAF equivalent area (col. 1 times col. 3); in col. 6 the corresponding percentage value.

The last row summarizes results and shows the test area name, the total number of features, their total area, the total equivalent area and, finally, the BAF value for the whole area considered, coming from the ratio between the two previous values.

The reporting procedure can run in a three-fold way, as it can be applied: to the whole city map; to a partial file, obtained by cutting the general one, as we did for test areas; to a limited region of the global map, which can be interactively chosen by the user through a selection.

6   Methodology Assessment

The developed procedure was assessed, in order to check its capability of properly estimating the BAF value. To this goal, twelve test areas were chosen, characterized by different building typologies, whose total area roughly equals 5% of the overall built up area of the municipality of Pavia. Test areas are shown [Fig. 1] as darker polygons over the background of the city map; the thick line encloses the built-up area.

For each test area, three kind of maps were prepared, which are listed and explained below.

Level 1 - *The map as it is*. The same map (same geometries, same classification) which can be downloaded from the web site of the Municipality,   modified according to the above described methodology.

Level 2 - *The map as it should be*. Level-1 maps could contain mistakes in polygon drawing or in feature classification, which are fixed in Level 2, according to the regional Guidelines [20, 21].

Level 3 - *The true BAF map*. Level-2 maps could be different from the ideal BAF maps not for compilation mistakes, but for specific aspects of the Guidelines. For instance, courtyards (which are considered less important than buildings and roads from a pure cartographic point of view) must be represented as a unique polygon even if they contain a grass part and a paved one. Thus mixture courtyards, although correctly represented with respect to the map's specifications, could compromise a good BAF calculation. To overcome these issues, Level-3 maps were created by enriching Level-2 ones, splitting existing polygons and manually assigning BAF codes and coefficients. Level-3 maps were considered as the true BAF maps.

Level-2 and -3 maps were created by manually editing the existing cartography. To properly carry out this step the existing digital orthophoto of the Municipality (having a ground resolution of 10 cm roughly) was used as a support, as well as the various information sources which are available on the web (Google Street View, the Bing Maps oblique imagery). Finally, an on-the-field check was performed.

Table 9. Final results for four test areas and for all the three levels considered

	Area [sqm]	BAF 1	BAF 2	BAF 3	Delta 1-3 [%]	Delta 2-3 [%]
Test Area 1	81,727	0.08	0.11	0.11	-23.89	0.00
Test Area 2	102,922	0.53	0.47	0.49	7.20	-3.46
Test Area 3	132,390	0.29	0.29	0.28	3.68	2.98
Test Area 4	31,099	0.29	0.27	0.31	-7.40	-14.38
Total	**348,138**	**0.31**	**0.30**	**0.30**	**3.33**	**-1.88**

[Table 9] shows final results for four test areas and for the three considered levels. Column 2 shows the area (extent measurement) of each test area; column 3 to 5 report the average BAF value for the considered test area and for the three depicted levels; columns 6 and 7 shows the BAF percentage variations between level 1 and 3 and level 2 and 3, respectively. Last row shows results for all the considered test areas. Classification and editing of the remaining eight areas is almost concluded and only the final field check is still ongoing.

7   Discussion

Figures reported in the *BAF 3* column refers to level-3 maps and must be considered as true. *BAF 1* column reports values calculated on level-1 map (the current municipal map), and *BAF 2* is related to the level-2 map, which was amended from possibly existing mistakes.

BAF variations between levels 1 and 2 are really limited: in absolute terms, the highest difference is 0.06; in relative terms, area 1 shows a difference of 24%, corresponding to a variation of 0.03: it might seem high only because BAF values are very low for that area. We can therefore conclude that Pavia's map is well done. The comparison between BAF 1 and BAF 3 figures brings to an overall percentage difference of 3% which is very low and highlights that BAF values can be calculated by existing digital maps with a very good quality.

8   Conclusion

A novel methodology has been proposed to measure the BAF value for a municipality or a part of it. It is based on performing sophisticated analysis on the city's digital map.

The methodology is automatic, objective and proved to be very accurate. The proposed methodology is significantly replicable as it can be applied to almost all municipalities of the Lombardy region. It could also be applied to most of the Italian cities with minor adaptations.

The proposed methodology is feasible because the considered digital map is very detailed, in terms of geometry and classification. Its replicability outside Italy depends on the characteristics and the degree of detail of the maps used in the various countries.

Our described experience could also be used in a reverse way, to highlight which requirements would be necessary for a map to be a good support for BAF evaluation.

Acknowledgements. About the authors' contributions. The paper originates from a common idea of Roberto De Lotto and Vittorio Casella about the possibly fruitful contamination between urban planning and GIS science for BAF evaluation. Vittorio Casella developed the classification methodology described in Sec. 5 and conceived the assessment methodology illustrated in Sec. 6. Vittorio Casella and Marica Franzini jointly wrote Sections 4-7 of the paper. Section 8 comes from the joint contribution of all the authors. Giuseppe Girone and Paolo Marchese (technicians working at the Geomatics Laboratory at the University of Pavia) made an impressive editing work by preparing level 2 and 3 maps; they are gratefully acknowledged here. Paolo Marchese also prepared the look-up table described in Sec. 5.

The calculation of BAF of the city of Pavia started in the teaching of Urban Planning and of Strategic Environmental Assessment to whic Veronica Gazzola, Cecilia Morelli di Popolo, Susanna Sturla and Elisabetta Venco participated as assistants. They also developed the manual calculation to almost the whole city. Images and tables referred to the manual calculation derives from their contribution.

References

1. De Lotto, R.: Assessment of development and regeneration urban projects: cultural and operational implications in metropolization context. International Journal of Energy and Environment **2**(1), 24–35 (2008). NAUN
2. Akbari, H.: Shade trees reduce building energy use and CO2 emission from power plants. Environmental Pollution **116**, 119–126 (2002)
3. European Commission Strategic Environmental Assessment (SEA). http://ec.europa.eu/environment/funding/cei_en.htm
4. Bernini, M., Campeol, A., Felloni, F., Magoni, M. (eds.): Aspetti ecologici nella pianificazione territoriale. Grafo, Brescia (1993)
5. Bregha, F., et al.: The Integration of Environmental Factors in Government Policy. Canadian Environmental Assessment, Research Council (1990)
6. Camagni, R., Capello, R., Nijkamp, P.: Towards sustainable city policy: an economy-environment technology nexus. Ecological Economics **24**, 103–118 (1998)
7. Ricci, L.: Piano locale. Nuove regole, nuovi strumenti, nuovi meccanismi attuativi. ISPRA Qualità dell'Ambiente Urbano, IX rapporto, pp. 130–170 (2013)
8. Steiner, F.: The living landscape: an ecological approach to Landscape Planning. McGraw Hill Professional, Milano (2000)
9. City of Bolzano, Environment and territory protection office. http://www.comune.bolzano.it/urb_context02.jsp?ID_LINK=512&page=10&area=74&id_context=4663
10. De Lotto, R., Venco, E.M.: Efficacia e attuabilità di indici ecologico-ambientali nella pratica urbanistica. Urbanistica Informazione, pp. 75–77 (2013)
11. Sturla, S., Venco, E.M.: Qualità ambientale nel progetto urbano: strumenti applicativi. In: De Lotto, R., di Tolle, M.L. (eds.) Elementi di progettazione urbanistica. Rigenerazione urbana nella città contemporanea, pp. 131–152. Maggioli Editore, Santarcangelo di Romagna (2013)
12. Senate Department of Urban Development and the Environment (Berlin). http://www.stadtentwicklung.berlin.de/umwelt/landschaftsplanung/bff/en/l_plan.shtml
13. City of Seattle Department of Planning & Development http://www.seattle.gov/dpd/cityplanning/completeprojectslist/greenfactor/whatwhy/
14. Kruuse, A.: GRaBS expert paper 6: the green space factor and the green points system. In: GRaBS (eds.) The GRaBS project. Town and Country Planning Association & GRaBS, London (2011)
15. NB-LEED, LEED for Neighbourhood Development Rating System. http://www.usgbc.org/
16. Allen, R.G., Pereira, L.S., Raes, D., Smith, M.: Crop evapotranspiration - Guidelines for computing crop water requirements - FAO Irrigation and drainage paper 56. FAO - Food and Agriculture Organization of the United Nations, Rome (1998)
17. Hargreaves, G.H., Samani, Z.A.: Estimating potential evapotranspiration–Tech. Note. J. Irrig. and. Drain Eng. ASCE **108** (1982)
18. Pavia's Wikipedia page. http://it.wikipedia.org/wiki/Pavia (visited last on 18.03.2015)
19. WebGIS of Cartography of Pavia. http://webgis.comune.pv.it/PV_FLEXESTERNO/ (visited last on 18.03.2015)
20. Pavia's website page for Cartography Download. http://www.comune.pv.it/site/home/dai-settori-e-servizi/settore-pianificazione-e-gestione-del-territorio/s.i.t.-sistema-informativo-territoriale/download-open-geodata.html (visited last on 18.03.2015)
21. Regional Technical Standard for Topographic Database – Content Specifications. http://www.regione.lombardia.it/shared/ccurl/888/28/DBT2009SpecificheDiContenuto_1.0_20090215.pdf (visited last on 18.03.2015)

22. Regional Technical Standard for Topographic Database – Attachment A of Content Speci-
    fications.
    http://www.regione.lombardia.it/shared/ccurl/942/780/Allegati_SpecificheDiContenuto_1.
    1_4.0_20090302.pdf (visited last on 18.03.2015)
23. Regional website page on the Progress of Topographic Database Project. http://www.
    regione.lombardia.it/cs/Satellite?c=Redazionale_P&childpagename=DG_Territorio%2FDe
    tail&cid=1213616799512&packedargs=NoSlotForSitePlan%3Dtrue%26menu-to-render%
    3D1213277392613&pagename=DG_TERRWrappere%3dDG_TERRWrapper"/> (visited
    last on 18.03.2015)
24. WebGIS on the Progress of Topographic Database Project. http://is.gd/geobandi (visited
    last on 18.03.2015)
25. National Technical Standard for Topographic Database. http://www.centrointerregionale-
    gis.it/public/DB_Topografici/1n1007_1-2_vers2006_3-3.pdf (visited last on 18.03.2015)
26. Lakes, T., Kim, H.O.: The urban environmental indicator "Biotope Area Ratio"- An
    enhanced approach to assess and manage the urban ecosystem services using high resolu-
    tion remote-sensing. Ecological Indicators 13(1), 93–103 (2012)

Environmental and Urban Spatial Analysis Based on a 3D City Model

Katarzyna Leszek

Department of Cartography, Warsaw University of Technology, Warsaw, Poland
kasia.leszek@yahoo.com

Abstract. In this paper possible analyses and simulations are presented based on a city modeling process in order to manage realistic approach for city planning within the Smart City concept. The 3D city model was constructed for City of Plock in Poland on Level of Detail 1, 2 and 3. The model consisted of almost 18,000 buildings. The proposed analyses within both environmental and urban fields were directly addressed to issues that the City of Plock is currently struggling with, such as air pollution, solar panels, flood risk, landscape analysis and city management. The paper concludes with the presentation of technological trends as well as data and policies that needs improvement to provide accurate 3D city model analyses.

Keywords: 3D city model · City planning · Flood, energy · Solar panels · Air pollution · Analysis · Data mining

1   Introduction

Urban metropolises are extremely complex, multidimensional, ever changing social ecosystems that remarkably contribute to disturbances within sustainable development. Undoubtedly, municipal areas are the future of humanity, as it is shown by the progressive urbanization. Currently, more than 50% of Earth population lives in cities. It is estimated that by 2050, 66% of humanity will be settled in urban areas [14]. Thus, overall analysis on mutual impact of high density cities with its components and the environment is greatly required.

At the same time, the availability of spatially referenced data and information is rapidly growing. A wide number of users are collecting, gathering, updating and processing data sets in order to perform spatial analysis, modeling or visualization of both raw data and analytical results. Proper spatial analysis is a process that facilitates the decision process through explanation of complex issues, the relationship between its components and hidden patterns that influence our reality. Those subjects are strongly connected to the field of Smart City which systematize existing technologies, merge them with the newest, resulting with the solutions enhancement performance.

In this context, 3D city modeling that includes both semantic and geometrical data for the whole city have proved significant possibilities in Smart City context in terms of spatial planning and environmental issues, e.g. air movement between buildings,

© Springer International Publishing Switzerland 2015
O. Gervasi et al. (Eds.): ICCSA 2015, Part III, LNCS 9157, pp. 633–645, 2015.
DOI: 10.1007/978-3-319-21470-2_46

flood risk, landscape management or energy distribution. The number of cities built in a virtual reality as 3D models is increasing, as the costs and time required to create such a model is significantly decreasing.

In this paper, various methods of spatial analysis are introduced based on a created 3D city model of Plock in Mazovian Voivodeship (region), Poland. The model was built within the author's master thesis on three Levels of Details established by CityGML standards. The data needed to form the virtual city model were derived from CODKIG databases as well as architectural projects needed for the most detailed models. The propositions of analysis were conducted on the whole city of Plock (17.675 buildings).

In the last part some technological aspects as well as data collection issues connected with formal policies are described, which have major impact on models as well as spatial analysis results accuracy and realism.

2   Description of the Integrated Process

Propositions of environmental and urban analysis described in this paper are based on a virtual 3D city model of Plock. The city model was created within ESRI software both on CityEngine platform for modeling and creating a virtual city and in ArcGIS environment for data edition and analysis.

2.1   CityGML Understanding

The virtual 3D city model of Plock was created within the OGC Standard CityGML that supports a multiscale model with 5 well-defined consecutive Levels of Detail (LOD). LOD are diversified by geometric complexity, 3D objects precision and integrity. This kind of division allows creating model tailored to the actual needs of chosen analysis.

The basic LOD 0 consist of a terrain model, without any object implemented. The simplest geometric model, LOD 1, creates rectangular blocks as buildings. The second Level of Detail (LOD 2) contains also the information about the roof shape and roughly represents the façade of a building. LOD 3 adds vast amount of detail creating very realistic virtual model of a building, up to 10 centimeters correct. Level of Detail 4 incorporates the indoor space modeling.

Virtual 3D city model can be generated based on a various input data, either aerial photography, laser scanning or digital cadaster combined with information about building properties (height, construction material, roof type). Most of the environmental analysis can be performed on the Level of Detail 1, taking into account only information about building placement, form and height. Nevertheless, city planning or crisis management require more detailed model of a city. That is why it is extremely important to define the purpose and expected results in order to create suitable city model first.

LOD1 – representation of building

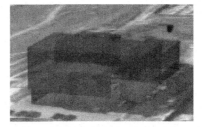

LOD2 – representation of building

LOD3 – representation of building

Fig. 1. Three Level of Detail of CityGML (Source: Leszek, K., 2015)

2.2   Building Information Sources

In order to create virtual city model the following data were used:

- CODKIG (Main Centre of Geodetic and Cartographic Documentation)
- Architectural projects of buildings built in LOD 3
- Aerial and ground photographs
- 3D model libraries

The CODKIG database collects, stores and manages both raster and vector data at the national scale. In order to create the virtual 3D city model, topographic maps (TBD) were acquired, containing the vectors of administrative units, road network, buildings, land use, land cover, the network of watercourses, protected areas and address points. CODGIK database also contains indispensable to the project ortophotographs, as well as digital elevation model and laser scanning data.

Architectural projects of buildings were acquired in order to collect detailed information on buildings:

- height of buildings components,
- roof type and shape
- position and dimensions of basic elements such as windows, door and architectural details

3D model libraries consisted of pre prepared components created in 3D modeling software. Most of chosen elements were selected from ESRI 3D Library and other open source databases, though some of them were created by the master thesis author.

Aerial and ground photographs were taken with the dual purpose. Firstly, it provided spatial recognition with object physical properties as well as position information in relation to another element. Secondly the photographs were used as a material specimen for building's façade creation.

Although the above data are sufficient to create the 3D city model, additional databases may be required to conduct more complicated analysis on environmental issues. Depending on the purpose of research, additional data can significantly increase model value – every piece of information added to the dataset can result in multiple analysis and evaluation of dependencies between variables.

2.3   3D City Model of Plock

Based on the previously described data, virtual 3D city model of Plock was created on respectively 1, 2 and 3 Level of Detail. However, before modeling and generating Plock buildings, the virtual environment of the city ought to be established.

Fig. 2. 3D city model of Plock on Level of Details 1, 2 and 3 (Source: Leszek, K., 2015)

The 3D city was bestead on a groundwork created out of set of 1-meter/pixel resolution color satellite photos (orthophotomaps) which were elevated to the height acquired from Digital Terrain Model (DEM). Implemented terrain was supplemented with vector layers containing ground floor projections of buildings, street network, water bodies and points representing locations of trees. Each of those layers was elevated to DEM. The following step of 3D city modeling required writing a unique set of rules, each one of them linked to the particular vector layer. After the implementation of both rules and layers to the modeling program, the software generated every shape as indicated.

Street network was created automatically in CityEngine environment, based on existing vector data. The street rule allowed distinguishing lanes widths determined by street categories derived from attribute table of street network shapefile.

```
StreetWidthDefault = 7
attr StreetClass = getObjectAttr("StreetClass",false)
# D G GP I L Z
attr StreetWidth =
case StreetClass == "GP" : 7
case StreetClass == "G" : 7
case StreetClass == "D" : 5
case StreetClass == "L" : 5.5
case StreetClass == "Z" : 6
else : streetWidthDefault
```

3D trees models were generated based on point layer, reflecting trees location in the model, together with rule determining trees height and type: broadleaf or deciduous. For water bodies only color and transparency was set in the rule environment.

Based on localization, set of buildings were categorized on LOD1, LOD2 and LOD3 and respectively three layers were generated and embed within the model. Each of Level of Details was based on a different layer as well as had a different spatial reach. LOD 1 was created for the whole city of Plock (88 square kilometers) whereas LOD 2 was generated for old town and adjacent areas (0.8 square kilometers) and LOD 3 was created for historical unit of Malachowianka School. Each building has been associated with the attributes of the property parcel that it falls within.

For each objects in LOD1 category the assigned rule was determining height acquired from attribute table. Additionally, depending on the usage of model, LOD1 buildings could represent zoning of Plock, building function or ownership, distinguished respectively by color.

LOD2 rule also consisted of buildings heights, though additional data were implemented as well. For every building, unique façade and roof type was determined. Facades were seized from ground photos, subsequently each one of the façades was manually edited and placed on adequate building. Roof types were generated based on information acquired from table of attribute and aerial photos.

The most detailed model was generated on LOD3, where every single building had a unique, complicated and time consuming rule, nesting architectural projects information and a set of 3D models. All the heights and dimensions were taken from architectural plans, and coded within the rule file. Firstly, rule gathers implemented to the project objects: 3D models, photos, textures, dimensions, colors, attributes. Afterwards, the user constructs a code, explaining step by step to the program how it should generate a building. A piece of code is presented beneath, next to the example of implemented 3D model of window and arch. A rule can be implemented to multiple objects, resulting in a reproducibility of a code to various buildings. In the presented model, every building had various façade, thus every building had a unique code.

```
FloorSplit -->
split(y){groundfloor_height:Floor
(split.index)
| floor_1 : Floor(split.index)
| floor_2 : Floor(split.index)}
Floor(floorindex) -->
case floorindex == 0 :
 split (y) {0.81 : Wall|0.98: Brick
 |2.13 : Tile(floorindex)|1.79 : Brick}
case floorindex == 1:
split(y){2.07:Tile(floorindex)|1.89:Brick}
else :
split(y){1.68:Tile(floorindex)|0.57:Brick}
```

Fig. 3. 3D models of window and arch implemented to the building
(Source: Leszek, K., 2015)

Physical properties of buildings were presented in the visualization generator, but were also nested in the attribute table. Thus, all the parameters could be used for further analyses and simulations.

Fig. 4. Model created within LOD3 standard (left) and the original building (right) (Source: Leszek, K., 2015)

The values assessed first through benchmarking or standard values can be updated both in ArcGIS editor or CityEngine toolbar that allows real time changes and immediate model – shape response.

3   Environmental Analysis Propositions

Based on a created 3D model, there are various analyses that have the possibility to provide information on the changing situation of environmental as well as urban issues. An environmental analysis includes air quality, movements and transmission channels within the city, flood simulation and prediction or energy-based topic. An urban analysis includes viewshed for visual impact assessment, placement of buildings shadows and any other topics connected to the urban tissue and decision – making process. All of the above mentioned analyses possible to conduct within 3D city model of Plock are directly connected with current issues that the city is facing.

3.1   Air Quality, Movement and Transmission Channels in Plock

In most urban areas there is often an agreement that the major pollution source is contributed by vehicles' emission. In Plock not only this issue is a pollution concern. One of the major investments in Plock is Petrochemia Plock (PKN Orlen), a state-run firm in charge of the refineries and at the same time the largest complex of this kind in Poland. The refinery and petrochemical complex have an extensive impact on a city taking into account large amounts of stored hazardous substances (oil, gasoline, flammable gases etc.) on the area of approximately 8.5 square kilometers. This area is exactly a 10% of a total city area; in addition the localization of a company is in a relatively small distance to the housing area, with the minimum value of approximately 1500m in a straight line. Those issues directly indicate the need of spatial simulations and analyses in order to establish prevention, preparedness and response of the city council in case of potential major industrial accident.

Plock air quality and movement analysis conditions are unique, due to the point-based pollutant represented by Petrochemia Plock. Many of researchers focus on the roadway as a major source of air pollution in cities which indicates linear approach of studies and implementation of models based on a street geometry. Thus in this particular situation it is necessary to take into consideration models that aim attention at city as a set of points.

A very promising study was established by Ioannili and Rocchi [5] who conducted research on Urban Canopy Parametrization (UCP). This approach acknowledges geometric, morphological and physical characteristics of urban agglomerations by selecting parameters derived from the analysis of high resolution databases. This issue has a great importance in the ongoing research on meteorological models used for analyzing air quality and pollution spread in urban areas.

UCP allows evaluating the effects of the city on the atmosphere using the length of the roads arcs, the area and perimeter of the buildings and the rooftop height of the different building units which compose a building block. The important objective of the proposed solution is to determine the roughness length (z_0) and the displacement height (z_d) for an urban context. There are various methods that are calculating these urban roughness parameters, being as follows: Height building methods, Lettau's model, MacDonald's model, Duijm's model, Raupach's model, Bottema's model and Bottema – Mestayer's application.

3.2   Energy Consumption Prediction and Solar Exposition in Plock

In order to maintain the municipal energy policies development, especially within defining the refurbishment priorities for indicating possible energy savings, it is important to construct models that manage realistic energy analyses of the building stock. In the recent light of the discussion on alternative sources of power, it is possible to estimate the best placement for solar panels, which would provide permanent access to energy. Solar panels are an investment that needs to be deliberately placed due to the many requirements, restrictions and financial profitability. One of the basic elements of energy calculation is energy demand. Based on this information, it is possible to establish "*hot spots*" of areas where energy is most needed and used.

Quasi-static monthly energy balance is a standardized algorithm, used worldwide by energy standard organizations to calculate the energy demand for indicated areas in a very simple but reliable way [10]. It has limited input requirements, hence the computing time of calculation is adequate to the model and is able to compare long-term urban energy scenarios for vast districts.

Based on calculated specific energy demand of the city it is possible to indicate areas where energy is most needed and the solar panels could have an actual use. To select proper areas for panel placements, several issues should be discussed:

- Aspect
- Building ownership
- Flood risk
- Sector affiliation
- Visual impact

The 3D city model of Plock is based on vector layer with attribute table, thus the many essential parameters are already implemented and ready to use. Information such as the ownership of a building, height, or building function and sector affiliation are already given.

For solar panel placement estimation it is important to properly assess the availability of buildings in terms of the ownership and its function. The city can set up this kind of installation only on buildings that officially and fully belong to the municipality. Also, the roof area should be taken into account so that the maintenance of the solar panels is affordable, and costs are lowered to the necessary minimum.

Models created on Level of Detail 2 are presenting the general shape of a building with the actual roof type. This information is crucial when calculating the aspect of building which is one of the most important parameter in a solar panel installation site selection. With GIS tools estimation of the roof exposure to the sun, meaning both length of exposure and intensity of solar radiation taking into account the time of a year, the calculations are relatively simple.

Avoiding installation of solar panels on buildings that are exposed to possible flood risk is another necessary aspect that needs to be taken into account (see chapter 3.3). The 3D city model allows selecting the buildings that even in the event of a flood still have roofs above the surface of a water body.

Having many features influencing decision making on a solar panels installation site, it is worth taking into consideration using data mining process, which is designed to explore data in search of patterns and possible systematic relationships between variables. It aims at validating the outcomes by applying the detected pattern to the new set of data. Using a data mining engine automates the process of finding suitable roofs for placing solar panels. The predictive model finds the correlation between height, flood risk estimation, aspect, area and other parameters, and when implementing new data it can more effectively indicate and recommend the best placement for solar panels.

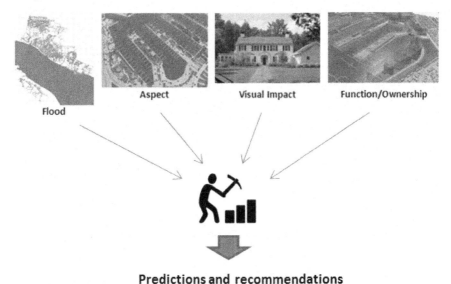

Predictions and recommendations

Fig. 5. Data Mining (Source: Leszek, K., 2015)

3.3   Flood Risk Analysis in Plock

Due to the geographical features of Plock, the city is repetitively at risk of being affected by flood. A considerable part of Plock is located on lowlands, being directly in jeopardy of water damage, whereas the rest of the city lies on a slope. Simulating and predicting flood and its effects has a great impact on decision making as well as it is a provider of a powerful visual portrayal of an citizens' lives as well as the safety of their belongings. Figure 5 presents the 500-years flood in the city of Plock, covering almost all of the southern part of the urban tissue.

High – rise in water levels could affect technical infrastructure such as electrical, water, communication facilities or even sewage systems. City Managers guided by 3D city model simulation can construct flood preparedness and a response both for the city as well as its citizens. Those actions include such operations as closing roads, government buildings and organizing pre-flood evacuation, which can help avoiding a lot of loss and expenses on rebuilding the infrastructure. Those simulations can be a clear message on where planning residential areas should be avoided or restricted in terms of specifications for buildings. That kind of approach should be used especially for locations that are not included yet in city zoning, land – use planning documents.

The 3D city model allows estimating and visualizing the water level as well as the information up to which floor the buildings will be flooded, which provide a valuable guidance for rescue teams.

Fig. 6. Flood reach simulation for 500 years flood (Source: Leszek, K., 2015)

Fig. 7. Flood simulation in 3D city model of Plock (Source: Leszek, K., 2015)

3.4   Urban Analysis in Plock

The virtual 3D city model has the possibility to simulate how new investment would impact the urban tissue in terms of landscape aesthetics, buildings shadows or powerful visualization of proposed changes and plans of city's administration for the future. Figure 7 shows the possible presentation of current zoning in Plock. Together with other useful information such as building height, ownership, permissible building height, floor area ratio etc. the model can be an extremely helpful visualization both for citizens, investors and administrative bodies.

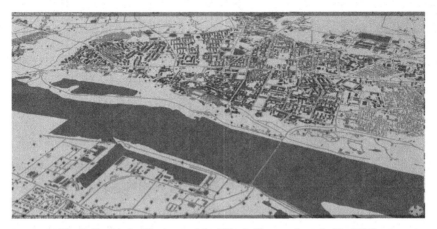

Fig. 8. Zoning in 3D city model of Plock (Source: Leszek, K., 2015)

A landscape study can be performed with a viewshed analysis which is commonly used method adopted in spatial analysis. It enables to illustrate which pixel is visible from the indicated observation point, where morphology is already established. The input data consist of sets of points, whereas the outcome is a grid layer that contains the numerical information about the visibility of a chosen object. In most cases visibility evaluation from only one point is not satisfactory, this is why in order to increase the analysis functionality multiple, viewshed was established. This approach allows producing a binary grid where:

- 0 means not visible from the union of single viewshed raster
- 1 means visible from the union of single viewshed raster.

Another method examines cumulative viewshed analysis, where union operator is used together with a counter insertion. The final result indicates the number of observation points or objects for each cell. However, this approach does not fully exploit all the possibilities. Method does not identify which target is visible from indicated cell. Thus, new viewshed analysis was established, named identifying viewshed, which shows which target is visible for each cell.

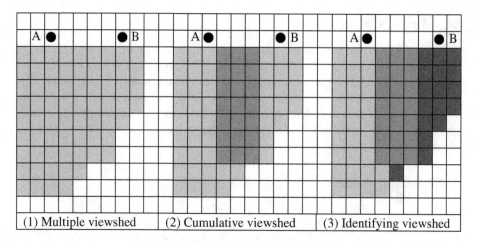

	Not visible		
	Visible(1) // Visible for A(3)		
	Visible for both A and B		
	Visible for B		

Fig. 9. Main differences among Multiple, Cumulative and Identifying Viewshed in the case of targets A and B (Source: Based on: Murgante, Nole, Danese, 2015)

Apart from a calculated analysis a simple visualization of proposed investments and spatial development are as important. In the software environment the view angle can be changed as well as the position, which allows for fully and freely controlling the camera. Based on the visual evaluation it is possible to assess the project of expected development.

Fig. 10. New buildings in urban tissue; landscape analysis (Source: Leszek, K., 2015)

4   Conclusion

In this paper the 3D city model of Plock municipality is presented with suitable propositions of environmental and urban analyses. Many of the ideas discussed in this paper need further examination. Nevertheless, presented application could be extended to other cities, on condition that they have at their disposal a virtual 3D city model with relevant database. Many uncertainties of the usability of algorithms come from the technological obstacles as well as policy decision and data access restrictions.

However, the usability of 3D city models will continue to increase exponentially. The further development of this issue will lead to new algorithms and ways to implicate them, as well as predict and simulate the complexity of municipalities.

References

1. Adda, P., Mioc, D., Anton, F., McGillivray, E., Morton, A., Fraser, D.: 3D Flood – Risk Models of Government Infrastructure. Commission VI, WG VI/4, Promotion of Regional Cooperation and Regional Capacity Development in Geoinformatics
2. Carrion, D., Lorenz, A., Kolbe, T.H.: Estimation of the Energetic Rehabilitation State of Buildings for the City of Berlin Using a 3D City Model Represented in CityGML. International Archives of the Photogrammetry, Remote Sensing and Spatial Information Sciences, Volume XXXVIII-4/W15
3. Falconer, G., Shane, M.: Smart City Framework. Cisco Internet Business Solutions Group (IBSG) (2012)
4. Gröger, G., Kolbe, T.H.: CityGML – A GML3 Application Profile for virtual 3D City Models. Institute for Cartography and Geoinformation, Univ. of Bonn, Germany (2005)
5. Heljula, A.: Oracle BI and Geo – Spatial Big Data. Oracle Spatial Summit (2015)
6. Ioannilli, M., Rocchi, E.: Urban Roughness Parameters Calculation in the City of Rome by Applying Analytical and Simplified Formulations: Comparison of Results. In: Geocomputation and Urban Planning, pp. 150–169 (2009)
7. Kolbe, T.H., Gröger, G.: Towards unified 3D city models. In: Proceedings of the ISPRS Comm. IV Joint Workshop on Challenges in Geospatial Analysis Integration and Visualization, September 8–9, 2003 in Stuttgart, p. 8
8. Leszek, K.: Utilization of 3D city models at LOD1, LOD2 and LOD3 for Smart City concept development in terms of spatial planning for Plock. Master thesis (2015)
9. Li, D.: Using GIS and Remote Sensing Techniques for Solar Panel Installation Site Selection. Ph.D Thesis, Canada (2013)
10. Murgante, B., Nole, G., Danese, M.: Visual Impact Assessment in Urban Planning. In: Geocomputation and Urban Planning, pp. 133–146 (2009)
11. Nouvel, R., Zirak, M., Dastageeri, H., Coors, V., Eicker, U.: Urban Energy Analysis Based on 3D City Model for National Scale Applications. In: Fifth German-Austrian IBPSA Conference. RWTH Aachen University (2014)
12. Szadkowski, A., Izdebski, W.: Wirtualne Miasta. Geodeta NR 2(165), 13–16 (2009)
13. Ujang, U., Anton, F., Rahman, A.A.: Unified Data Model of Urban Air Pollution Dispersion and 3D Spatial City Model: Groundwork Assessment towards Sustainable Urban Development for Malaysia. Journal of Environmental Protection 4, 701–712 (2013)
14. United Nations. World Urbanization Prospects (2014)

Modeling the Propagation of Forest Insect Infestation Using Machine Learning Techniques

Mileva Samardžić-Petrović[(✉)] and Suzana Dragićević

Spatial Analysis and Modeling Laboratory, Department of Geography,
Simon Fraser University, 8888 University Drive, Burnaby, BC V5A 1S6, Canada
`{msamardz,suzanad}@sfu.ca`

Abstract. Infestations caused by the mountain pine beetle (MPB) can be seen as complex spatio-temporal process with severe ecological impacts on the forest environment. In order to manage and prevent the insect infestation and reduce significant forest loss it is necessary to improve knowledge about the infestation process. The main objective of this research study is to design and implement a model based on decision trees (DT) mashie learning (ML) technique to forecast the spatial propagation of MPB infestation. The study is implemented in the Bulkley-Nechako region of British Columbia, Canada using data sets for the three time points 2004, 2008 and 2012. The results indicate that the derived DT can accurately characterize the relationships between the considered factors and MPB propagation. The developed DT method can be used to estimate future spread patterns of MPB infestations.

Keywords: Machine learning · Decision tree method · Geographic information system · Spatial modeling · Mountain Pine Beetle · Insect infestation

1   Introduction

Forest cover has important ecological and socio-economic influence on the environment. The total loss of global forest area was 2.3 million km^2 in the period 2000-2012 (Hansen et al. 2013) arising from changes in land use, forest exploitation, wildfires, pathogens and insects infestations. The bark beetle insects caused massive forest losses across Europe, Asia and North America and are expected to increase in the future due to factors such as climate change (Robinet et al. 2011; Latifi et al. 2014; Karvemo et al. 2014). One of the North American forests most destructive bark beetle insect is the mountain pine beetle (MPB) (Coops et al. 2006). Since 1990, MPB caused mortality of approximately 50% of the total volume of commercial lodgepole pine in British Columbia, Canada (NRC, 2015) and is also impacting new forest areas in northern Alberta (Bone and Altaweel 2014). In addition to significant economic losses, the MPB epidemic has an influence on landscape, surface fuel, wildfire hazards, carbon sources, nutrient cycling, water quality, and regional climate (Liang et al. 2014; Kurz et al. 2008; Lamers et al. 2014). In order to mitigate spreading of forest infestation and reduce future significant losses of forest it is necessary to improve knowledge of the infestations, examine the causal factors and to develop

© Springer International Publishing Switzerland 2015
O. Gervasi et al. (Eds.): ICCSA 2015, Part III, LNCS 9157, pp. 646–657, 2015.
DOI: 10.1007/978-3-319-21470-2_47

modeling approaches that are capable of forecasting possible locations of MPB infestation to assist in forest management.

In the last decade various methods have been used to analyze and model the spatial distribution of MPB infestation and other pine beetle infestations across the world. These include: regression (Robertson et al. 2008; Negron et al. 2008; Preisler et al. 2012), cellular automata (Bone et al. 2006; Perez and Dragicevic 2012; Pukkala et al. 2014) and agent-based models (Perez and Dragicevic 2010; Fahse and Heurich 2014). Further, the MPB epidemic is a complex spatial process and depends on various factors ranging from individual tree conditions to climate (Bone and Altaweel 2014; Safranyik et al. 2010). As such, models of MPB epidemics were developed to focus on different factors such as forest fragmentation (Bone et al. 2013) and climate (Coops et al. 2012).

Machine learning (ML) techniques are modeling approaches that have the capability to extract information and reveal patterns by exploring unknown relations between input and output variables. ML techniques, such as Decision Tree (DT), can generate models that are not significantly biased by human experts. The DT is a class of simple but very powerful ML techniques that produces models as a set of IF – THEN rules. These rules are generated by splitting the attribute values of each considered factor in a way to better understand the relations and influence of those factors. The DT technique was implemented for remote sensing data classifications used for monitoring forest (Pouliot et al. 2009; Liu et al. 2014) and detection of bark beetle infested areas (Fahse and Heurich 2014; Marx 2010). However, DT has not been sufficiently investigated in the context of modeling the spread of MPB infestation, particularly to forecasting the spatial propagation of MPB infestation. Consequently, the objective of this research study is to design and implement a model based on DT to forecast the spatial propagation of MPB infestation using factors such as elevation, slope, aspect, and solar radiation. The study was implemented using data sets for three time points, 2004, 2008 and 2012, and for the Bulkley-Nechako region of British Columbia, Canada.

2   Methodology

The stages of applying the DT technique for modeling the propagation of MPB infestation are: appropriate training and test datasets identification, factor selection, DT based model building, and model validation. A brief theoretical background for each of these stages is presented in the sections that follow.

2.1   Training and Test Datasets

Since that MPB infestation is a complex spatial process influenced by various spatial factors it is very convenient to manage all the data within a GIS database where each considering factor as well as infested areas can be represented as one GIS data layer. The study area is represented by a square grid of cells were each cell is presented as an instance c_i, which contains corresponding values of all considered spatial factors

a_j and MPB infestation information k. The considered factors, usually called attributes, are the real values such as distances to road, elevation and slope. While the MPB infestation information, k, is represent as a class of binary values, were 0 represents uninfested and 1 represents infested instances. The main goal of the DT is to discover pattern of spatial propagation of MPB by finding a function which maps considering factors of a grid cell c_i^t at time t to its information of MPB infestation at time $t+1$; $c_i^t \{a_1, a_2,...,a_j, k^t\} \rightarrow k_i^{t+1}$. In order to build and validate a DT based model to forecast the spatial propagation of MPB infestation it is necessary to prepare two data sets. The first is the training dataset, $[(c_i^{t-1}, k_i^t), i=1,..., N]$, used to build the model and learn the relationship between the considered factors and MPB infestation, $c_i^{t-1} \{a_1, a_2,...,a_j, k^{t-1}\} \rightarrow k_i^t$. The second is the test dataset, $[(c_i^t, k_i^{t+1}), i=1,..., N]$, used to validate the model by comparing the simulated and real datasets of MPB infestation at time $t+1$.

2.2   Decision Tree Based Model

Using the assumption that the relationships between the considered factors and MPB infestation are stationary, it is possible to create and validate the DT-based MPB infestation model providing valuable information about the examined relationships. There is a number of different DT learning algorithms used to build the tree but due to performance abilities the C4.5 decision tree algorithm (Quinlan, 1993) was used in this study. Generally, DT consists of internal nodes (test nodes) and leaf (terminal) nodes (Fig. 1). The tree construction process performs a greedy search in which decision trees are constructed in a top-down recursive, divide and conquer, manner. The tree construction is performed over the instances from the training dataset $\{a_1, a_2,...,a_j, k^{t-1}\} \rightarrow k_i^t$. The process starts by selecting an attribute which defines the first internal node in the tree as a root node and makes one branch for each possible attribute value (categorical attribute) or range of values (numerical attribute). The training dataset is further split into subsets, one for every attribute value. The process is repeated recursively for each branch, using only those instances that reach the branch. Specifically, DT classifies instances by testing the value of one particular attribute a_i per node, following a certain path in the tree structure, which depends on the tests in previous nodes, and finally reaches one of the leaf nodes (MPB infestation). A new test node is added below a branch if the instances following the branch are partitioned after the candidate attribute is tested in such way that the distinction between the classes becomes more evident. The C4.5 algorithm uses the Gain Ratio (GR) measure to choose between the available attributes and to define which attribute will be populating each particular internal node, starting from the root node and until reaching the leaf nodes. The GR effectively measures the capacity of an attribute to split the input set into sets with lower entropy concerning class labels of containing instances (Shannon 1948). The C4.5 uses a 'post-pruning' technique to reduce the size of the tree. After 'growing' the initial tree, iterations are performed to remove and/or join some nodes yielding to a tree with good classification capability based on the training dataset and for general prediction outside the training datasets.

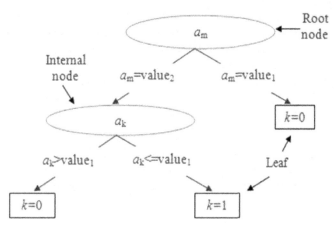

Fig. 1. Structure of the decision tree

From the generated decision tree it is possible to extract a set of rules for each leaf by making a conjunction of all the tests encountered on the path from the root to that particular leaf. Consequently, the DT is able to provide meaningful information about the influence of each attribute on the process being examined.

2.3   Selection of Factors

It is important to perform the factors (also called attributes) selection process since factors with little or irrelevant information many confuse the algorithm learning process and lead to incorrect conclusions. The main goal of attribute selection is to choose a subset of informative attributes by eliminating those with little or no information relevant for the MPB infestation process (Kim et al. 2003). The Correlation-based Feature Subset (CFS) (Hall and Smith 1998) selection method was used in this research study.

The CFS method automatically determines a subset of m relevant attributes (factors) that are highly correlated with the MPB infestation but uncorrelated with each other. The CFS ranks the subset of attributes according to a correlation based heuristic evaluation function. After sorting all attributes according to their respective correlation with the MPB infestation, attributes are added to the subset beginning with the most correlated one. The next attribute is added if it has a higher correlation with the MPB infestation than with any other attribute already in the subset. In this manner the method automatically determines a subset of relevant attributes.

2.4   Model Validation

Model validation can be performed by comparing the model outcomes with real MPB infestation data at time $t+1$. In this research study, the validation measures used are Area Under Receiver Operating Characteristic Curve (AUC) (Bradley 1997), and the well-known kappa statistics (Cohen 1960). Kappa has values between -1 (no agreement) and 1 (perfect matching of two raster maps). The kappa index having values less than 0 indicate no agreement and values in the ranges of (0.00–0.20) are categorized as slight, (0.21–0.40) as fair, (0.41–0.60) as moderate, (0.61–0.81) as substantial, and values greater than 0.81 are considered as almost perfect (Landis and Koch 1977).

The AUC provides a single statistical parameter of the overall summary of the Receiver Operating Characteristic (ROC) curve. The ROC graph is a two dimensional graph that shows the trade-offs between true positive (TPrate=TP/P) and false positive rates (FPrate=FP/N) (Fawcett 2006) – where True Positive (TP) is the number of infested instances at time $t+1$ correctly classified as infested at time $t+1$; P is the total number of infested instances at time $t+1$; False Positive (FP) is the number of uninfested instances at time $t+1$ incorrectly classified as infested at time $t+1$; and N is the total number of infested instances at time $t+1$. The AUC value of 1 indicates an accurate model while a value of 0.5 denotes a random guess model (Bradley 1997).

3   Study Area and Data Sets

The study area of Bulkley-Nechako is a census subdivision of British Columbia, Canada (Fig. 2) that covers an extent of about 76 km x 57 km and was selected because of the large number of infested areas from 2004-2008 and 2008-2012. Geospatial data from various sources were used including: points and polygons of infested areas prior to years 2004 and for years 2004, 2008 and 2012 (British Columbia Ministry of Forests 2000); 0.75 arc second (~20 m) Canadian Digital Surface Model (CDSM) (Government of Canada 2012); land cover data for 1992 (Government of Canada 1992), and national road network data (Government of Canada 2014). The datasets were integrated in a spatial database using the ArcGIS software (Esri 2011) at a 50 m spatial resolution. The values for attributes (factors) were defined for each raster cell for elevation, slope, aspect, solar radiation, Euclidean distance from road, and land use class.

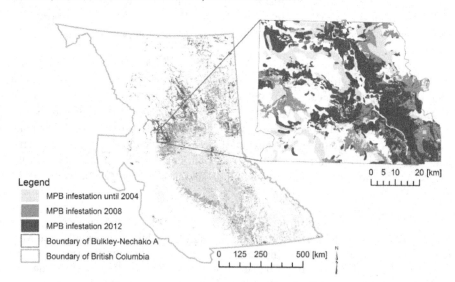

Fig. 2. Study area: Bulkley Nechako, British Columbia, Canada

There are approximately two million raster data cells in the study area and hence it was necessary to perform data sampling. The original data for the infested areas were defined as polygons: recorded locations of dead trees as polygons on 1:100,000 or 1:250,000 base maps; and as points which represent recorded locations of small areas of infestations of up to 30 trees (0.25 ha) and infestation from 31 to 50 trees (0.5 ha). The infestation for each polygons was given as level of severity by percent of infested trees in regards to area of each polygon. Given the high number of polygons that contain larger surfaces with uninfested trees than infested and that the point datasets are more indicative of infestation conditions (Chen et al. 2014), for the purpose of this study only point datasets were considered as sample data to represent locations where the infestation occurred. The total number of points (raster cells) which represent the infested area is relatively small (approximately 1000) compared to the approximately two million cells that represent the entire study area. Moreover, applying the most commonly used random data sampling method is not suitable because it implies all samples have an equal probability of being selected at any location within the study area. For the training dataset, all points which represent the infested area between 2004 and 2008 were selected as samples for infested area. Further, the same number of sample data of uninfested areas were selected randomly across the entire study area ensuring that they are located in forest but not near or inside the polygons of infested areas and on a minimum of 200m away from infested points. The same sampling procedure was used for the test datasets by selecting all points which represent the infested area between 2008 and 2012 and the same number of sample data of uninfested areas (Fig. 3). The sampled training dataset contain a total of 1029 cells while the sampled test dataset contains 1219 cells; where each cell in the training and the test datasets contains values for the attributes presented in Table 1.

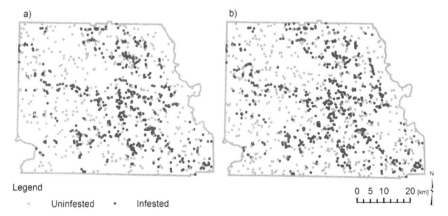

Fig. 3. Sampled (a) training datasets and (b) test datasets

652   M. Samardžić-Petrović and S. Dragićević

Table 1. List of attributes used for modeling the MPB infestation

Training dataset	Test dataset
Infested / Uninfested at 2004	Infested / Unifested at 2008
Elevation	Elevation
Slope	Slope
Aspect	Aspect
Solar radiation	Solar radiation
Euclidean distance from Road	Euclidean distance from Road
Land use class	Land use class
Infested / Unifested at 2008	Infested / Unifested at 2012

4   Results and Discussions

Using the training dataset, the CFS selection of attributes was done to find the most informative subset of attributes from Table 1. The resulting subset of the four most important attributes were: infested/uninfested at 2004, elevation, Euclidean distance from road, and land use class. The selection method indicated that other (unselected) attributes are highly correlated with some of the selected attributes or that they do not have an influence on the spread of the MPB infestation. In this research, two DT-based MPB infestations models were developed and validated. The first model, labelled as M, was developed and validated on training and test datasets that contain all attributes presented in Table 1. The second model, labelled as M^{CFS}, was developed and validated on datasets that contain only the subset of attributes selected by the CFS method. The obtained measures of validation for both models are presented in Table 2.

Table 2. Values for Area Under Receiver Operating Characteristic Curve (AUC) and kappa measures for the two models M and M^{CFS}

Model	Kappa	AUC
M	0.54	0.91
M^{CFS}	0.70	0.93

The obtained values for the AUC indicated a good performance for both models. The values for both the kappa and AUC measures indicate the model developed from a subset of selected attributes using the CFS exhibits better model performance. This may be explained by the smaller number of attributes producing a less complex model that is better able to generalize the MPB infestation process. This is can be closely attributed to the bias–variance tradeoff problem, where if the model complexity increased then the variance tends to increase, and the squared bias tends to decrease; while the opposite behavior occurs as the model complexity is decreased (Hastie et al. 2009). If the DT algorithm fits the training datasets too well (overfitting) then the variance term is large and the overall model accuracy is decreased.

The obtained results indicate that the DT algorithm can be used for forecasting spreading MPB infestation by using a relatively small number of considered attributes (factors). A map of agreements and disagreements between the real locations of the MPB infestation and those obtained by the M^{CFS} model is shown in Figure 4.

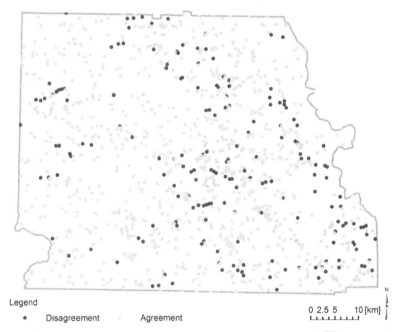

Fig. 4. Map comparing the MPB infestation locations obtained by the M^{CFS} model with the actual infestation locations for year 2012

In addition to the capability to model MPB infestation propagation, the DT method has the ability to extract relationships between factors and the infestation spread. Based on the decision trees models, it is possible to explore the influence of the obtained values of the factors. Each created DT model will derive transition rules and will split values of each considered factor in such a way to better understand relations and influence of those factors in order to complement expert knowledge. The derived decision trees contain information of importance of each considered factor as well as the importance of particular range of values for each considered factor (Fig. 5). For the study area, the elevation and information about the previous MPB infestation are the factors that have the greatest impact for the propagation of MPB infestation with respect to the other considered factors. Various studies (Perez and Dragicevic 2012; Coops et al. 2006, Kurz et al. 2008) have emphasized the importance of elevation on the propagation of MPB infestation and indicated that in the last decade MPB has started to infested forests at higher elevations. Based on the derived IF-THEN rules from the DT model it can be concluded that the probability of infestation occurrence at elevations higher than 1333 m above sea level is very small (Fig. 5). Following each branch in the tree, various information can be extracted. For example, if the forest is older than 140 years and more than 6 m high, then it is more probable that infestations will propagate at locations closer to the road (Euclidean distance to the road is less than 3.18 km). Furthermore, if the forest is younger than 140 years, and more than 6 m high, then it is more probable that infestations will propagate at locations farther to the road (Euclidean distance to the road is less than 4.29 km).

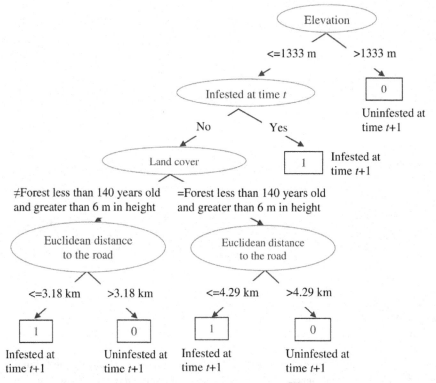

Fig. 5. Decision tree obtained from the M^{CFS} model

5   Conclusions

The results indicate that the main advantage of the decision tree (DT) ML technique is that it provides the ability to successfully derive models related to the MPB infestation and produce information on the relationships between the considered factors and the infestation propagation. The information derived from decision trees can complement expert knowledge to improve understanding of MPB infestations. Given the ease-of-use of the method as well as the visual classification tree outputs, the DT algorithm presents a useful aid to support future infestation management and decision making. Furthermore, reducing the initial set of input attributes (factors) to an informative subset by the CFS method resulted in a reduced model with better performance and that is easier to interpret. However, since the DT algorithm depends on past data it is important to use high-quality, informative and realistic datasets. Hence, a systematic workflow should be in place to carefully apply all stages of the DT-based modeling procedures including appropriate sampled training and test datasets identification, factor selection, DT based model building, and its validation. Future analysis will examine more datasets that represent different study areas in order to improve the reliability of the approach.

Acknowledgments. This study is fully supported by a Natural Sciences and Engineering Research Council (NSERC) of Canada Discovery Grant awarded to the second author.

References

1. Bone, C., Altaweel, M.: Modeling micro-scale ecological processes and emergent patterns of mountain pine beetle epidemics. Ecological Modelling **289**, 45–58 (2014)
2. Bone, C., White, J.C., Wulder, M.A., Robertson, C., Nelson, T.A.: Impact of forest fragmentation on patterns of mountain pine beetle-caused tree mortality. Forests **4**(2), 279–295 (2013)
3. Bone, C., Dragicevic, S., Roberts, A.: A fuzzy-constrained cellular automata model of forest insect infestations. Ecological Modelling **192**(1–2), 107–125 (2006)
4. Bradley, A.P.: The use of the area under the roc curve in the evaluation of machine learning algorithms. Pattern Recognition **30**(7), 1145–1159 (1997)
5. Chen, H., Ott, P., Wang, J., Ebata, T.: A positive response of mountain pine beetle to pine forest-clearcut edges at the landscape scale in British Columbia, Canada. Landscape Ecology **29**(9), 1625–1639 (2014)
6. Cohen, J.: A coefficient of agreement for nominal scales. Educational and Psychological Measurement **20**(1), 37–46 (1960)
7. Coops, N.C., Wulder, M.A., White, J.C.: Integrating remotely sensed and ancillary data sources to characterize a mountain pine beetle infestation. Remote Sensing of Envi-ronment **105**(2), 83–97 (2006)
8. Coops, N.C., Wulder, M.A., Waring, R.H.: Modeling lodgepole and jack pine vulnerability to mountain pine beetle expansion into the western Canadian boreal forest. Forest Ecology and Management **274**, 161–171 (2012)
9. Fahse, L., Heurich, M.: Simulation and analysis of outbreaks of bark beetle infestations and their management at the stand level. Ecological Modelling **222**(11), 1833–1846 (2014)
10. Fawcett, T.: An introduction to ROC analysis. Pattern Recognition Letters **27**(8), 861–874 (2006)
11. Hall, M.A., Smith, L.A.: Practical feature subset selection for machine learning. In: Proceedings of the 21st Australasian Computer Science Conference, Acsc 1998, pp. 181–191. Springer-Verlag Singapore Pte Ltd, Singapore (1998)
12. Hansen, M.C., Potapov, P.V., Moore, R., Hancher, M., Turubanova, S.A., Tyukavina, A., Thau, D., Stehman, S.V., Goetz, S.J., Loveland, T.R., Kommareddy, A., Egorov, A., Chini, L., Justice, C.O., Townshend, J.R.G.: High-Resolution Global Maps of 21st-Century Forest Cover Change. Science **342**(6160), 850–853 (2013)

13. Hastie, T., Tibshirani, R., Friedman, J., Franklin, J.: The elements of statistical learning: data mining, inference and prediction. The Mathematical Intelligencer **27**(2), 83–85 (2005)
14. Karvemo, S., Van Boeckel, T.P., Gilbert, M., Gregoire, J.C., Schroeder, M.: Large-scale risk mapping of an eruptive bark beetle - Importance of forest susceptibility and beetle pressure. Forest Ecology and Management **318**, 158–166 (2014)
15. Kim, Y., Street, W.N., Menczer, F.: Feature selection in data mining. In: Data mining: Opportunities and Challenges. Idea Group Publishing, pp. 80–105 (2003)
16. Kurz, W.A., Dymond, C.C., Stinson, G., Rampley, G.J., Neilson, E.T., Carroll, A.L., Ebata, T., Safranyik, L.: Mountain pine beetle and forest carbon feedback to climate change. Nature **452**(7190), 987–990 (2008)
17. Lamers, P., Junginger, M., Dymond, C.C., Faaij, A.: Damaged forests provide an opportunity to mitigate climate change. Global Change Biology Bioenergy **6**(1), 44–60 (2014)
18. Landis, J.R., Koch, G.G.: The Measurement of Observer Agreement for Categorical Data. Biometrics **33**(1), 159–174 (1977)
19. Latifi, H., Fassnacht, F.E., Schumann, B., Dech, S.: Object-based extraction of bark beetle (Ips typographus L.) infestations using multi-date LANDSAT and SPOT satellite imagery. Progress in Physical Geography **38**(6), 755–785 (2014)
20. Liang, L., Chen, Y.L., Hawbaker, T.J., Zhu, Z.L., Gong, P.: Mapping Mountain Pine Beetle Mortality through Growth Trend Analysis of Time-Series Landsat Data. Remote Sensing **6**(6), 5696–5716 (2014)
21. Liu, K., Liu, L., Liu, H.X., Li, X., Wang, S.G.: Exploring the effects of biophysical parameters on the spatial pattern of rare cold damage to mangrove forests. Remote Sensing of Environment **150**, 20–33 (2014)
22. Marx, A.: Detection and Classification of Bark Beetle Infestation in Pure Norway Spruce Stands with Multi-temporal RapidEye Imagery and Data Mining Techniques. Photogrametrie Fernerkundung Geoinformation **4**, 243–252 (2010)
23. N.R.C.: The threat of mountain pine beetle to Canada's boreal forest. Government of Canada, Natural Resources Canada (2015)
24. Negron, J.F., Allen, K., Cook, B., Withrow, J.R.: Susceptibility of ponderosa pine, Pinus ponderosa (Dougl. ex Laws.), to mountain pine beetle, Dendroctonus ponderosae Hopkins, attack in uneven-aged stands in the Black Hills of South Dakota and Wyoming USA. Forest Ecology and Management **254**(2), 327–334 (2008)
25. Perez, L., Dragicevic, S.: Modeling mountain pine beetle infestation with an agent-based approach at two spatial scales. Environmental Modelling & Software **25**(2), 223–236 (2010)
26. Perez, L., Dragicevic, S.: Landscape-level simulation of forest insect disturbance: Coupling swarm intelligent agents with GIS-based cellular automata model. Ecological Modelling **231**, 53–64 (2012)
27. Pouliot, D., Latifovic, R., Fernandes, R., Olthof, I.: Evaluation of annual forest dis-turbance monitoring using a static decision tree approach and 250 m MODIS data. Remote Sensing of Environment **113**(8), 1749–1759 (2009)
28. Preisler, H.K., Hicke, J.A., Ager, A.A., Hayes, J.L.: Climate and weather influences on spatial temporal patterns of mountain pine beetle populations in Washington and Oregon. Ecology **93**(11), 2421–2434 (2012)
29. Pukkala, T., Moykkynen, T., Robinet, C.: Comparison of the potential spread of pinewood nematode (Bursaphelenchus xylophilus) in Finland and Iberia simulated with a cellular automaton model. Forest Pathology **44**(5), 341–352 (2014)
30. Quinlan, J.R.: C4.5: Programs for Machine Learning. Morgan Kaufmann Publishers (1993)

31. Robertson, C., Wulder, M.A., Nelson, T.A., White, J.C.: Risk rating for mountain pine beetle infestation of lodgepole pine forests over large areas with ordinal regression modelling. Forest Ecology and Management **256**(5), 900–912 (2008)

32. Robinet, C., Van Opstal, N., Baker, R., Roques, A.: Applying a spread model to identify the entry points from which the pine wood nematode, the vector of pine wilt disease, would spread most rapidly across Europe. Biological Invasions **13**(12), 2981–2995 (2011)

33. Safranyik, L., Carroll, A.L., Régnière, J., Langor, D.W., Riel, W.G., Shore, T.L., Peter, B., Cooke, B.J., Nealis, V.G., Taylor, S.W.: Potential for range expansion of mountain pine beetle into the boreal forest of North America. The Canadian Entomologist **142**(05), 415–442 (2010)

34. British Columbia Ministry of Forests: Forest Health Aerial Overview Survey Standards for British Columbia: The BC Ministry of Forests Adaptation of the Canadian Forest Service's FHN Report 97–1 "Overview Aerial Survey Standards for British Columbia and the Yukon". Resources Inventory Committee, Victoria (2000)

35. Shannon, C.E.: A Mathematical Theory of Communication. Bell System Technical Journal **27**(3), 379–423 (1948)

Internal Areas Strategies:
From Statistical Methods to Planning Policies

Silvestro Montrone[1], Paola Perchinunno[1(✉)],
Francesco Rotondo[2], and Francesco Selicato[2]

[1] DISAG, University of Bari, Via C. Rosalba 53 70100, Bari, Italy
{silvestro.montrone,paola.perchinunno}@uniba.it
[2] DICAR, Politecnico di Bari, Via Orabona 4 70125, Bari, Italy
{francesco.rotondo,francesco.selicato}@poliba.it

Abstract. The "National Strategy for Internal Areas", made by the Italian Government for the European Union Partnership Agreement 2014-2020, defines the territory of the Italian internal areas as a set of project-areas, local inter-municipal systems each with its own territorial identity defined by social, economic, geographic, demographic and environmental characteristics. In this sense, we can define "internal" those areas significantly distant from the centers of supply of essential services (education, health, and mobility), rich in environmental and cultural resources with highly diversified natural aspects. The objective of the work is to re-elaborate the existing mapping for the identification of the internal areas, made by the Italian Government, especially taking into account the demographic, economic, morphological profiles and essential services supply, through the use of fuzzy logic. Then, trying to deep explain possible planning strategies and policies for these relevant, sometimes abandoned and extremely diffuse territories.

Keywords: Fuzzy logic · Planning policies · Territorial clusters

1   Introduction

The "National Strategy for the internal areas" interprets the territory of the Italian internal areas as a set of project-areas, or of local systems of intermunicipal, each with its own territorial identity defined by social, economic, geographic, demographic and environmental characteristics. Each project-area, selected through an investigation between Region and State is required to prepare a development strategy for the involved area or "Strategy area".

The identification of the national *Internal Areas* starts from a reading of the polycentric Italian territory, that is a nation characterized by a network of

---

The contribution is the result of joint reflections by the authors, with the following contributions attributed to S. Montrone (chapter 3), to P.Perchinunno (chapter 4), to F. Rotondo (chapter 2 and 5), to F. Selicato (chapter 1). The conclusions are the result of the common considerations of the authors.

© Springer International Publishing Switzerland 2015
O. Gervasi et al. (Eds.): ICCSA 2015, Part III, LNCS 9157, pp. 658–672, 2015.
DOI: 10.1007/978-3-319-21470-2_48

municipalities or groups of municipalities (centers offer services) around which gravitate areas with different levels of peripheral space (it has to be considered that urban Italy is made by about 8.000 municipalities and the majority of them have no more than 10.000 inhabitants). those areas significantly distant from the centers of supply of essential services (education, health, and mobility) are defined *Internal Areas* and they are rich in important environmental and cultural resources and highly diversified by nature or as a result of centuries of human processes.

The objective of this work is to verify the organization's identification of *Internal Areas* used by the Italian Government through methods of integration of all available data with fuzzy techniques able to identify the degree of belonging to the class of the *Internal Areas* compared to all the data available today and with respect to each of them considered individually.

In fact, the degree of belonging to the *Internal Areas* with respect to each criterion highlights important differences between the same areas already identified by the government and allows us to understand the spatial features on which action is needed to rehabilitate the territories and allow their valorization.

Therefore, a first point emerging from this work is related to the possibility to describe territorial phenomena through a fuzzy integrated model, which starts from the construction of indicators, with a multi-dimensional nature, and then adopt models capable of identifying "goal areas". A second point of research is related to the ability to schedule the *Internal Areas* enhancing their landscape, environmental and economic features rather than trying to change them making more similar to those that currently appear more central and dynamics areas in the globalized economy [1,2]

2   The Italian National Strategy for Internal Areas and Perspectives for European Cohesion Policies

2.1   Italian National Strategy Goals

Italy, like many other European countries, is characterized by the presence of numerous municipalities (there are more than eight thousand) often placed in areas far from major roads (highways, railways, ports and airports) and the main economic flows. It can be said that the country can be described just by these internal areas and large metropolitan cities linked by medium-sized cities in polycentric networks. The internal areas have paid their subordination in terms of depopulation, economic deficit, marginalization in national and European policies. The National Strategy for the Internal areas can contribute to the recovery of the economic and social development of this relevant part of Italy that in these two centuries has not taken advantage of the economic growth, but precisely for this reason has maintained significant environmental and landscape resources that today may become decisive factors of development. It is therefore necessary to develop a new Development Policy "Place-Based" [3] capable of touching every region and macro-region of the country, creating jobs, achieving social inclusion and reducing the costs of abandoning the territory.

As Barca, McCann, P. and Rodríguez-Pose suggest [4,5], since many core urban centres will grow without the need for significant policy interventions, but the question is to whether development goals should be shifted from promoting efficiency in the core to enhancing the potential for growth and development in the periphery. Place-based approaches offer a greater possibility of exploiting unused potential in all regions in a co-ordinated and systematic way.

This strategy provides the basis for implementing the interventions by means of a Framework Programme Agreement (FPA) between the italian state and the italian regions and the tool to communicate clearly to all citizens of the expected results and the actions taken. The document and the share path between the Region and State are required to obtain financial and organizational support (European Community Programmes and National Stability Law) for the National Strategy for the internal areas.

The strategy area is not the usual "list of actions or projects" in which several municipalities or interests "share" funds "intercepted", but the logical framework that guides the choice of actions.

A significant part of the Internal areas has gradually suffered (after World War II) a process marked by marginalization: population decline, sometimes below the critical threshold; reduction in employment and land use; a local supply waning of public and private services; social costs for the entire nation, such as the hydrogeological instability and cultural heritage and landscape degradation. But in this large part of the country there is such a strong development potential that the construction of a national strategy, robust, participatory and continuous in time may free.

This strategy has been initiated in Italy using as an occasion and a lever, the European Choesion Funds available for all regions of the country for seven years 2014-2020, combined with the provision of dedicated resources in stability law. It is a work in progress, through close cooperation with the Regions and a useful discussion with Municipalities and Provinces, aware that for a national strategy should contribute the leadership of open and innovative local communities.

2.2   Internal Areas' Classification Methods

A major part of the Italian territory is characterized by a spatial organization founded on municipalities, very often with a small demographic dimension, which in many cases are capable of providing to residents only a limited accessibility to basic services.

The analysis made has cleared that the only demographic size of the municipality does not seem sufficient to qualify the territories as poles of attraction and has therefore directed the work towards a pole declination as the center which offer specific services.

The character of "services supply center" is reserved exclusively for those municipalities, or aggregates of neighboring municipalities, offering simultaneously all the secondary school supply, hospitals with all the basic venues and Platinum, Gold or Silver railway stations. The main hypothesis is therefore that in the first instance we can identify the nature of Internal Areas on the basis of the "distance" from essential services. Note that Internal Areas, in this view, is not necessarily synonymous of "weak area". Only by examining the characteristics and dynamics of the demographic structure and socio-economic of the areas identified, we can have a complete reading of the different paths of territorial development.

In Italy there is a very different overview of Internal Areas. In some areas the remarkable capacity of local actors, together with many policy interventions that have occurred since the eighties', has helped to change the inaccessibility in an asset to be valued, triggering interesting development processes, through the involvement of local communities and succeeding to stop the population drainage.

The proposed methodology of the Italian government is substantiated in two main phases:

1) Identification of the poles, in accordance with a offer capacity of some essential services;

2) Classification of other municipalities in 4 groups: peri-urban areas; intermediate areas; peripheral areas and outermost areas, according to the distances from the poles measured in travel times.

The final mapping is therefore mainly influenced by two factors: the criteria by which to select the services supply centers and the choice of the distance thresholds to measure the degree of remoteness of the various areas. Thus classification of municipalities was obtained on the basis of an accessibility indicator calculated in terms of minute trip respect to the nearest pole. Ranges obtained are calculated using the terziles of the distribution of the index of distance in minutes from the next pole, equal to about 20 and 40 minutes. It was then inserted a third band, over 75 minutes (the 95-th percentile), to identify the ultra peripheral regions.

The approach using the dimensional threshold chooses as poles those municipalities that for various reasons, due to market conditions as well as public investments, have followed a process of agglomeration. The approach which has been reached, based on the offer of services, chooses as poles, municipalities with essential services.

According to this approach, based on the offer of services in the following map it has been shown the areas that are *Central* (first 3 categories) or *Internal* (last 3 categories).

(Source: Italian Ministry of Economic Development, Economic and Cohesion Department, http://www.dps.tesoro.it/aree_interne/, visited 16.02.2015)

Starting from this work, this research tries to deep the identification of *Internal Areas* used by the Italian Government through methods of integration of all available data with fuzzy techniques able to identify the degree of belonging to the class of the *Internal Areas* compared to all the data available today and with respect to each of them considered individually, in the case of the Apulia Region (see Figure 2).

In fact, as already said in the introduction, the degree of belonging to the *Internal Areas* with respect to each criterion may suggests important distinctions between the same areas and allows us to understand the spatial features on which action is needed to rehabilitate the territories and allowing their valorization.

Fig. 1. Maps of Italian municipalities according to the classification in Poles and areas with different degrees of remoteness from the centers of reference

3   The Construction of Indicators to Identify Internal Areas

3.1   Introduction

The starting point for this work derives from the necessity of identifying, on the basis of statistical data, what has been called by the Italian Government geographical *internal zones*, as already defined (cfr. paragraph n.2), characterized by demographic, morphological, economic and structural aspects in the specific case of the Apulia Region. From this follows the importance to define specific indicators, which are able to estimate the level of membership to the internal areas class, between municipalities. The attempt to apply a fuzzy approach to this question allows for the definition of a measurement of the degree of association to the totality, taking into account typical indicators of demographic, morphological, economics and structural aspects.

The scope of interest and the fields of application of the present work are numerous. The first consists of the possibility of quantifying and localizing the disparity of distribution of the phenomenon and therefore in the identification of phases of policy programming in those areas which may not usually be involved in development processes. Furthermore, it facilitates the possibility of organizing information with the aim of structuring and focusing intervention programs with regards to special and inherent problems, for example, the necessity to improve scholar or medical facilities.

3.2   The Fundamental Dimensions for the Identification of Indicators

According to what emerges of the Partnership Agreement 2014-20 "National Strategy for Internal areas: definition, objectives, instruments and governance" a significant part of the Internal areas suffered, since the fifties of last century, a process of marginalization that, first of all, is manifested through intense phenomena of de-human activity: a) decrease of the population below the critical threshold and aging population; b) decrease of employment c) reduction of the degree of utilization of the territorial capital. Moreover, this process was manifested in the gradual reduction of the quantity and quality of the local public services, private and collective.

The activity of prediction of possible scenarios for the Internal areas may be carried out with reference to a few dimensions. From these dimensions, analyzed for the potential development of the inland areas, it was decided instead to derive criteria for identifying the areas through the construction of the set of indicators.

1) The first dimension to be considered is the demographic aspect, mainly due to the fact that local systems of internal areas have reached a degree of aging that does not ensure a sufficient exchange of population. Many local systems are likely to suffer a demographic collapse in the medium to long term, or at least a reduction in the population of working age classes. There are some signs of demographic recovery, but are still limited and insufficient.

In Internal Areas emerges a negative rate of change of the population and a constant increase in the total population of the share of the older population (aged 65 and over), which has almost doubled between 1971 and 2011. The aging concerns both centers that Internal areas, but especially in the peripheral areas and ultra-peripheral that are recorded the highest percentages.

2) The second dimension concerns the morphological component that is associated with the non-use or improper use of the territorial capital. The settlement system will suffer an inevitable process of decay. Even the hydro-geological instability will increase, with effects of degeneration territorial that will make these territories gradually less hospitable. The demographic changes determined a change in land use and its destination, particularly in the Internal areas, creating an increase of phenomena such as the loss of active protection of the territory and the increased level

of geological risk. As for the Internal areas it is noted particularly strong presence in areas exposed to landslides, a high percentage of forest and agricultural areas.

3) The third component concerns the logic of employment that influence those territorial systems of internal areas that had a limited process of industrialization and outsourcing. In line with the physical characteristics of the internal areas, their economic structure is characterized by a strong specialization in the primary sector. In particular, we note that, while the percentage of municipalities specialized in the first sector in the centers is equal to 43%, it rises to 73% in the case of the municipalities that are located in the Internal areas. However, it is interesting to observe that within the Internal areas the regional realities present some variability: it is observed as in the internal areas of the regions of Southern Italy there is a higher agricultural specialization, compared to those of the Centre-North. Conversely, in the Internal Areas, are found lower than average percentages of specialization in Secondary and Third sector.

4) Last fundamental dimension to be considered for an exploration of individual local systems is the development of the supply of services and infrastructure. The term "pre-development conditions" refers to the production and supply services in the territory that in contemporary society qualify as rights of "citizenship": a) Health b) education c) mobility. Strictly functional and complementary to the efficient administration of services in the internal territories is the widespread dissemination of communication technologies telematics.

4   The Internal Areas of Puglia: A Methodological Comparison

4.1   The Set of Indicators

The objective of this work is to integrate mapping of internal areas with other mappings resulting from the integration of indicators of demographic (declining and aging population); morphological (exposure of the area to landslides or seismic risk); economic (reduction or increase in employment on the basis of economic sectors); supply of services and infrastructure (supply of services in transport, health, education and communication).

The case study concerns the Puglia region and, based on what was mentioned in section 3.2, for each component were constructed Set of statistical indicators:

Set 1: *DEMOGRAPHIC INDICATORS:*

      a) Changes in the population 1971/2011
      b) Share of elderly residents in the territory (2011)

Set 2: *MORPHOLOGICAL INDICATORS:*

      a) Class of population exposed to landslides (2012)
      b) Percentage of total forest area on the surface - in classes (2010)
      c) Percentage variation of utilized agricultural area (1982-2010)

Set 3: *ECONOMIC INDICATORS:*

  a) Population employed in agriculture on total employment (2001)
  b) Percentage of employees in the Manufacturing sector from 1971 to 2001
  c) Percentage of employees in the services sector from 1971 to 2001

Set 4: *INDICATORS OF THE OFFER OF SERVICES AND INFRASTRUCTURE:*

  a) Health - Number of beds in hospitals per 100,000 residents (2011)
  b) Education - Presence of school aggregated into three types: high schools, technical and professional institutes and other types (2011)
  c) Transport - Presence of a train station, at least of type "Silver" (2012)
  d) Digital divide - Proportion of population without broadband fixed network (2012)

4.2   The Fuzzy Approach

The development of *fuzzy theory* stems from the initial work of Zadeh [6] and successively of Dubois and Prade [7] who defined its methodological basis. Fuzzy theory assumes that every unit is associated contemporarily to all identified categories and not univocally to only one, on the basis of ties of differing intensity expressed by the concept of degrees of association. The use of fuzzy methodology in Italy can be traced back to only a few years ago thanks to the work of Cheli and Lemmi [8] who define their method *"Total Fuzzy and Relative"* (TFR) on the basis of the previous contribution from Cerioli and Zani [9].

Supposing the observation of k indicators of poverty for every family, the membership function of *i*-th family to the fuzzy subset of the poor may be defined thus [9]:

$$f(x_{i.}) = \frac{\sum_{j=1}^{k} g(x_{ij}).w_j}{\sum_{j=1}^{k} w_j} \qquad i = 1,.....,n \qquad (1)$$

For the definition of the function $g(x_{ij})$ please refer to other works [10,11,12].

4.3   The Thematic Maps in Apulia Region

Analyzing the situation of Puglia on the basis of the criteria specified in Section. 2, based on the offer of services, emerges as the percentage of municipalities defined as internal areas is equal to 56.2%, with lower percentages of outermost areas (2.3%) and higher intermediate areas (32.2%).

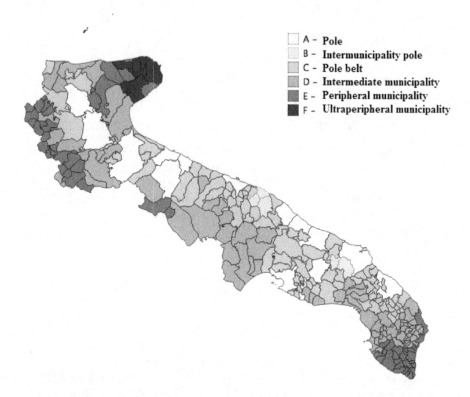

Fig. 2. Geographical classification of the municipalities on the basis of the service offer

Table 1. Percentage of the municipalities on the basis of the service offer

Classificazione	n. comuni	%
A - Polo	14	5.4%
B - Polo intercomunale	5	1.9%
C - Cintura	94	36.4%
D - Intermedio	83	32.2%
E - Periferico	56	21.7%
F - Ultraperiferico	6	2.3%
	258	**100%**

The results obtained with the fuzzy approach can be classified into 6 classes of value ranges, in which, the more the values approximate to unity, the more the ratios show a potential belonging to the set of the Internal Areas. In the following table are presented the number of municipalities that belong to the individual classes analyzed.

Table 2. Number of municipalities by classes of fuzzy values and types of indicators

Fuzzy value	Demographic aspects	%	Morphological aspects	%	Economic aspects	%	Offer of Services	%
0,00- 0,16	36	14,0%	217	84,4%	52	20,2%	89	34,6%
0,16 - 0,33	0	0,0%	5	1,9%	95	37,0%	143	55,6%
0,33 - 0,50	13	5,1%	18	7,0%	0	0,0%	6	2,3%
0,50 - 0,67	31	12,1%	12	4,7%	12	4,7%	2	0,8%
0,67 - 0,83	0	0,0%	3	1,2%	65	25,3%	0	0,0%
0,83 – 1,00	177	68,9%	2	0,8%	33	12,8%	17	6,6%
	257	**100,0%**	**257**	**100,0%**	**257**	**100,0%**	**257**	**100,0%**

In particular, as regards the application of data relating to *Demographic aspects* emerges as the most of the municipalities (about 69%) belongs to the last class or to one which has greater criticality. This is due to the fact that in many municipalities of Puglia is a reduction of the population and aging (Fig 3).

Instead, as regards the application of the data relating to morphological aspects most municipalities (84.4%) belongs to the first class, that is, the class that has not critical. This factor becomes, therefore, strongly discriminating for the identification of Internal Areas, as the municipalities that have values close to one really are areas with problems related to morphological aspects, such as exposure to landslides or percentage of forest area or surface agricultural (Fig 4).

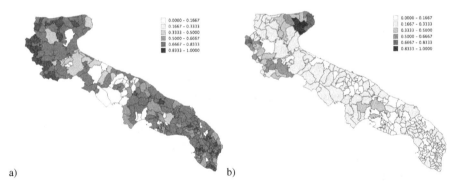

a)   b)

Fig. 3-4. Classification of the municipalities on the basis of fuzzy indicators related to aspects: *a) demographics; b) morphological*. Representations underline the discriminatory effect of the second set of indicators for the identification of possible internal areas.

On the economic aspects should be noted, however, a situation of equal distribution of the municipalities in the different classes. May be, therefore, interesting to the identification of areas inside the municipalities belonging to the last two classes next to 1, which together account for 38.1% of the municipalities of Apulia. In these municipalities emerges prevalence of employed in the agricultural sector rather than in manufacturing or services (Fig 5).

Even more discriminating than the morphological component, is the aspect of services and infrastructure. Indeed, the municipalities in which there is lack of services in the field of health, education and transport are few (6.6%). These municipalities belong to the latter class, next to 1, and then with high probability belong to the municipalities classified as Internal Areas (Fig 6).

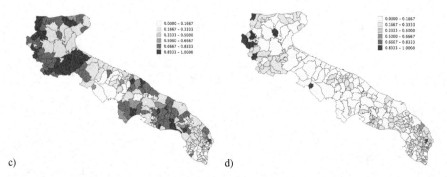

c)   d)

Fig. 5-6. Classification of the municipalities on the basis of fuzzy indicators related to aspects: c) economic; d) supply of services and infrastructure. Also in this case representations show the discriminating effect of the second set of indicators for the detection of possible internal areas.

5   The Intersection of all Sets of Indicators

Going to consider all the indicators set, a differentiated picture emerges where the municipalities belonging to the last three classes are equal to about 15% of the municipalities of Apulia Region and they are characterized by a high degree of belonging to the Internal areas (darker in Fig. 7).

This application takes into account all the components analysed: demographics, geomorphology, economy and range of services and infrastructure.

It 'should be noted that compared to the study carried out by the Ministry (see Figure 2) the areas classified as *Internal Areas*, identified considering all indicators except the distance from the services supply providers, are inferior, and in particular are not belonging to the class of Internal Areas of many municipalities in the South Salento, while confirming the belonging to the same class of Municipalities of Northern Apulia and in particular those of the "Gargano", the mountainous area of the region and the so-called "Monti Dauni", another mountainous area on the border of the region bordering the Campania Region. From this comparison it can be deduced further scale intervention priorities determined by belonging to the class of the Internal Areas even without considering the distance from the services supply centers, signal of a condition of greatest weakness of these two zones with respect to that of the south the region, which is focused by today's most important tourist flows that have increased economic flows and decreased unemployment rates.

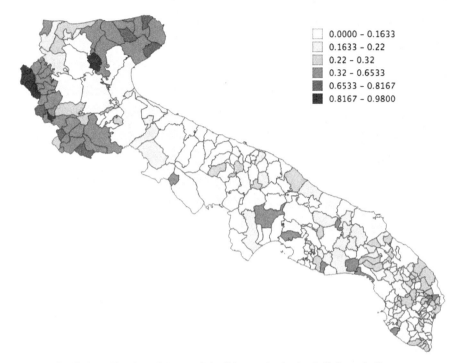

Fig. 7. Classification of the municipalities on the basis of all fuzzy indicators

The representation of the synthesis shown in Figure 7 does not consider the thresholds distance of each municipality from the service provider useful for measuring the degree of remoteness of the different areas, according to the analysis conducted by the Ministry. To understand the weight of this criterion was drafted last application considering these thresholds and, therefore, on the balance of all indicators compared to the distance from the center service provider.

Going to consider also the aspect of the distance from the center you get the classification of the municipalities of the Apulia Region in the representation shown next.

In this case remains strong the membership of the areas located in the North Region to the Internal Areas class, while it seems clearer that there is a further area of inland areas in southern Salento with a greater difference between the towns of this area and the neighboring. Specifically, having inserted, as a fifth criteria, the distance from the centers service supply, seem to have "polarized" some of the situations that have already emerged in the previous analysis, making it more clear the difference between the central areas and internal areas.

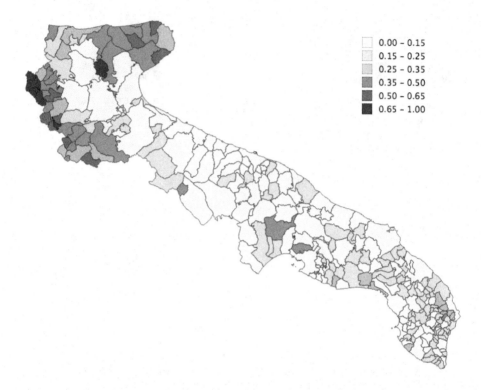

Fig. 8. Classification of the municipalities on the basis of all fuzzy indicators *weighted with the distance from the center*

6   Conclusions

The Internal Areas represent a large part of the country - about three-fifths of the area and just under a quarter of the population - very diverse within itself, far from the great agglomeration centers and service and unstable development trajectories, but endowed with resources lacking the central areas, with demographic problems but also strongly polycentric and with strong potential for attraction.

Intervene in these areas requires creative solutions capable of interpreting the territories, the traditional inhabitants and new, promote development models based on the quality of the sites. The study presented here demonstrates how to investigate the internal areas can not be limited to studying only the distance from the service supply centers, but that it is necessary to investigate all the components of the phenomenon because each area has different criteria discriminating belonging to the class of the internal areas. The case of the Apulia Region shows that, for example, in the post-industrialized nations population dynamics are not elements capable of discriminating the internal areas, as well as for the economic dynamics.

Because in this period of crisis the population may decrease even in the central areas as well as in the internal ones, as well as the growth in the number of unemployed or employment growth in areas hitherto regarded as traditional agriculture not only distinguishes the areas far from service supply centers but also important development poles, given the growing importance of its sectors such as agriculture and tourism have started to play even in countries such as Italy with a strong manufacturing development that seems to doze off over time in favor of emerging nations where the cost of labor is undoubtedly lower.

On the contrary, the geomorphological stability, unmistakable sign of of the land care and the availability of resources, and even further away from the service poles providers, mark clearly the difference between internal areas and the central event in Puglia. The case study analyzed is not a solution to generalize, because on the contrary it highlights the need to analyze in depth each component of an area to identify the criteria discriminating the degree of membership to the internal areas highlighting the aspects on which more focus the subsequent enhancement policies of these areas long neglected in the usual place-neutral policies.

In fact, as pointed out by Barca, McCann and Rodriguez-Pose [4], many of the previously accepted arguments have been called into question by the impacts of globalization and a new response to these issues has emerged in response to both these global changes and also to non-spatial development approaches. We are not yet able to assess whether the correct approach is the most radical of an happy De-growth (proposed by Latouche) or the more place-based institutional one proposed by scholars as Barca [5] who also played Institutional roles supporting technically and theoretically the work of the European Commission. The most convincing studies conducted in these areas demonstrates with certainty that to redevelop and enhance these territories through plans and programs must deeply understand the peculiarities that distinguish them and interpret them in a new way and able to self-generate a process of endogenous development from below as the same Apulia Region is trying hardly to develop through a new landscape plan that tries to bring out the multiple regional identities [13].

Acknowledgments. The research project was funded by "Azioni di Sostegno alle attività del Sistema Nazionale di Valutazione e dei Nuclei di Valutazione", PON GAT (FESR) 2007-2013 OB. I.3 "Potenziamento del Sistema Nazionale di Valutazione", implemented by FORMEZPA.

References

1. Magnaghi, A.: The Urban Village: A Charter for Democracy and Local Self-sustainable Development. Zed Books, London and New York (2005)
2. Latouche, S.: De-growth, Inequality and Poverty. In: Ventura, P., Calderon, E., Tiboni, M. (eds.) Sustainable development Policies for Minor Deprived Urban Communities, pp. 71–79. McGraw-Hill, Milano (2011)
3. McCann, P., Rodríguez-Pose, A.: Why and when development policy should be place-based. In: OECD (ed.) OECD Regional Outlook 2011: Building Resilient Regions for Stronger Economies. OECD Publishing, Paris (2011)

4. Barca, F., McCann, P., Rodríguez-Pose, A.: The case for regional development intervention: place-based versus place-neutral approaches. Journal of Regional Science **52**, 134–152 (2012)
5. Barca, F.: An agenda for a reformed cohesion policy. A place-based approach to meeting European Union challenges and expectations. Independent Report, prepared at the request of Danuta Hübner, Commissioner for Regional Policy (2009). http://www.europarl.europa.eu/meetdocs/2009_2014/documents/regi/dv/barca_report_/barca_report_en.pdf (visited February 26, 2015)
6. Zadeh, L.A.: Fuzzy sets. Information and Control **8**(3), 338–353 (1965)
7. Dubois, D., Prade, H.: Fuzzy sets and systems. Academic Press, Boston, New York, London (1980)
8. Cheli, B., Lemmi, A.: A Totally Fuzzy and Relative Approach to the Multidimensional Analysis of Poverty. Economic Notes **24**(1), 115–134 (1995)
9. Cerioli, A., Zani, S.: A fuzzy approach to the measurement of poverty. In: Dagum, C., Zenga, M. (eds.) Income and Wealth Distribution, Inequality and Poverty. Springer Verlag, Berlin (1990)
10. Montrone, S., Perchinunno, P., Rotondo, F., Torre, C.M., Di Giuro, A.: Identification of hot spots of social and housing difficulty in urban areas: scan statistic for housing market and urban planning policies. In: Murgante, B., Borruso, G., Lapucci, A. (eds.) Geocomputation and Urban Planning, Studies in Computational Intelligence, vol. 176, pp. 57–78. Springer Verlag, Heidelberg (2009)
11. Perchinunno, P., Rotondo, F., Torre, C.M.: A multivariate fuzzy analysis for the regeneration of urban poverty areas. In: Gervasi, O., Murgante, B., Laganà, A., Taniar, D., Mun, Y., Gavrilova, M.L. (eds.) ICCSA 2008, Part I. LNCS, vol. 5072, pp. 137–152. Springer, Heidelberg (2008)
12. Montrone, S., Bilancia, M., Perchinunno, P., Torre, C.M.: Scan statistics for the localization of hot spots of urban poverty. In: Conference Proceedings of the Regional Studies Association, Winter Conference, Londra, pp. 74–77 (November 28, 2008)
13. Magnaghi, A.: The apulian approach to landscape planning. Urbanistica n. 147 July September 2011, distribution by www.planum.net

Adaptive Zoning for Efficient Transport Modelling in Urban Models

Alex Hagen-Zanker[1]([✉]) and Ying Jin[2]

[1] Department of Civil and Environmental Engineering, University of Surrey, Guildford, UK
a.hagen-zanker@surrey.ac.uk
[2] Martin Centre for Architectural and Urban Studies, University of Cambridge, Cambridge, UK
ying.jin@aha.cam.ac.uk

Abstract. Transport modelling and in particular transport assignment is a well-known bottleneck in computation cost and time for urban system models. The use of Transport Analysis Zones (TAZ) implies a trade-off between computation time and accuracy: practical computational constraints can lead to concessions to zone size with severe repercussions for the quality of the transport representation in urban models. This paper investigates how a recently developed geographical topology called adaptive zoning can be used to obtain more favorable trade-offs between computational cost and accuracy than traditional TAZ. Adaptive zoning was developed specifically for representing spatial interactions; it makes use of a nested zone hierarchy to adapt the model resolution as a function of both the origin and destination location. In this paper the adaptive zoning method is tied to an approach to trip assignment that uses high spatial accuracy (small zones) at one end of the route and low spatial accuracy (large zones) at the other end of the route. Opportunistic use of either the first or second half of such routes with asymmetric accuracy profiles leads to a method of transport assignment that is more accurate than traditional TAZ based assignment at reduced computational cost. The method is tested and demonstrated on the well-known Chicago Regional test problem. Compared with an assignment using traditional zoning, an adaptive-zoning-based assignment that uses the same computation time reduces the bias in travel time by a factor 16 and link level traffic volume RMSE by a factor 6.4.

Keywords: Transport · Assignment · TAZ · Zone

1   Introduction

Traffic assignment is a long-standing problem in transport modelling: it determines the level of traffic load as well as congestion/crowding which are essential for appraising investment and regulatory policies. Owing to its importance, traffic assignment methods have undergone continuous development. The simplest form of assignment is all-or-nothing which places all traffic between an origin and destination node on the minimal cost path. More advanced models recognize that as traffic is assigned to a link, a degree of congestion takes place which affects the time required

© Springer International Publishing Switzerland 2015
O. Gervasi et al. (Eds.): ICCSA 2015, Part III, LNCS 9157, pp. 673–687, 2015.
DOI: 10.1007/978-3-319-21470-2_49

for traversing the link and hence the cost. In a static user equilibrium, traffic is assigned to the links such that all used paths from an origin to a destination carry the same cost, and no less costly path exists (Wardrop 1952). Over the years, methods of iterative weighted all-or-nothing assignment (Fukushima 1984) based on a more general quadratic programming method (Frank and Wolfe 1956) have become firmly established. In recent years, bush-based assignment methods are providing a more efficient, though perhaps not yet as well understood, alternative (Bar-Gera 2002; Dial 2006; Nie 2010). Other models of assignment include dynamics traffic assignment (Friesz et al. 1989), and stochastic user equilibrium (Lam et al. 1999). This paper is primarily concerned with all-or-nothing assignment, and its role in the Frank-Wolfe algorithm. Nevertheless, the computational load and cost of the assignment procedures is a common challenge across the above spectrum of assignment methods. The phenomenal rise in the power of personal computing has not yet overcome the issue, and the continued expansion of functional city regions and the need for urban models to cover large areas will likely exacerbate the problem.

Practically all transport models make use of Transport Analysis Zones (TAZ) for aggregate computations on groups of locations and individuals. Larger zones correspond to a greater degree of aggregation and therefore less precision and reduced computational cost. The use of zones in transport assignment as an approximation causes a bias, overestimating local traffic near the zone centroid and underestimating it elsewhere. When the zones are large, the inability to assign intra-zonal traffic can lead to significant underestimation of traffic, whereas the bundling of groups of origins and destination to representative centroids leads to a mix of over- and underestimation. Without mitigation, the effect can culminate in excessive and spurious levels of modelled congestion. The common workaround of introducing auxiliary nodes around a centroid spreads the traffic loads and devolves local overloading. However, the underlying problem of a bias in the distribution of local traffic is not overcome. A considerable body of research exists addressing the trade-offs involved in zone design and optimization of zone systems(Chang et al. 2002; Ding 1998; Martinez et al. 2005; Martinez et al. 2009; Viegas et al. 2009).

This article proposes a reformulation of the assignment problem that creates new possibilities for simultaneously improving computational efficiency and precision of local traffic assignment, using the concepts of adaptive zoning. In first instance we are concerned with are for the all-or-nothing assignment which can be a further input to the static user equilibrium assignment based on the standard Frank-Wolfe algorithm and other assignment methods. The all-or-nothing assignment is most often solved using the Dijkstra shortest path algorithm. This algorithm finds the tree that represents all shortest paths between one single node and all others. With TAZ, the nodes that are considered as origins and destinations are the zone centroids. The aggregation error occurs as the actual origin and destination of trips are not the zone centroids. It should be noted that the error in the route from an origin to a destination is mainly near the origin and destination: the approximation by zone centroids means that traffic is routed from roughly the right location to roughly the right location, and it is only for the extremes of the trip that locational accuracy is crucial. Likewise, if we would find the route between a small zone and a large zone, we can expect it to be more accurate near the small zone than near the large zone. This paper aims to exploit this last consideration: the rationale is that for a given origin and destination we find two routes: one from a large zone containing the origin to a small zone containing the destination, and another from a small zone containing the origin to a large zone containing the destination. Of both routes we will only assign traffic to the more accurate half.

Thus the assignment takes place twice, the enhanced degree of aggregation at one end of the trip provides opportunity for computational advantage. There are two arguments to increase to degree of aggregation at greater distances. Firstly, from the point of view of geometry, at greater distances the same angular error corresponds to a greater locational error. Thus if we are concerned with traffic heading off in the right direction zone size should increase with distance. Secondly, the number of trips between locations generally reduces with distance, thus for greater distances we can allow for a greater error than for smaller distances. The recently introduce geographical topology of adaptive zoning (Hagen-Zanker and Jin 2012; Hagen-Zanker and Jin 2011a) allows to model spatial interaction using zone sizes that adapt to the distance over which interactions take place.

Adaptive zoning was introduced by Hagen-Zanker & Jin (2011a) as an alternative method for representing geographical space in the context of spatial interactions. In some transport models, the zoning scheme is defined such that the strongest concentric flows of spatial interaction are represented in more detail, such as in a transport model of London and South East England where the zones become progressively larger away from central London (Jin et al, 2002). In such cases the zone plan is adaptive to the amount of traffic into and out from one group of zones (i.e. in central London), but does not account for the fact that each and every zone interacts more strongly with some zones than with others. Adaptive zoning takes the adaptation to traffic patterns one step further and adapts the size of origin and destination zones to the strength of interaction (amount of traffic) between those zones. Since the strength of interactions typically diminishes with distance, it means that trips over short distances are modelled using small zones and long distances using large zones.

Adaptive zoning is related to origin and destination sampling (Kristoffersson and Engelson 2009; Miller et al. 2007; Williams and Lindsay 2002). The idea of destination sampling is to reduce the computational load of transport models by modelling stochastically filtered destinations and apply a correction factor on the volume of traffic that reflects the sampling probabilities. Williams and Lindsay (2002) use a sampling strategy that is distance dependent, providing more precision at shorter distances. Adaptive zoning also reduces the computational load by reducing the number of destinations for each individual origin, but uses a system of zone aggregation instead.

Under adaptive zoning, each origin interacts with a smaller number of destinations when compared with a model using conventional TAZ zones. The reduction is achieved through a bespoke and adaptive aggregation of destination zones across the study area. At short distances from the origin zone there is little aggregation of destination zones, and the degree of aggregation becomes progressively higher further away from the origin. Origin-based adaptive zoning was used previously for the assignment problem (Hagen-Zanker and Jin 2011a), but it was found that the aggregation at the destination end of trips – and the associated bias – overly constrained the applicability. This article extends the methodology by using both origin- and destination-based adaptive zoning in combination with bi-partitioned assignment: the first half of the trip is assigned on the basis of an origin-based adaptive zoning system and the second half of the trip on the basis of a destination-based adaptive zoning system. This reaps the advantages of adaptive zoning, without being exposed to the negative side-effects arising from aggregating distant zones.

Earlier articles introducing and exploring adaptive zoning demonstrated its potential for traditional spatial interaction modelling (Hagen-Zanker and Jin, 2012) and city-scale mode choice modelling (Hagen-Zanker and Jin, 2013). The current paper is the first to successfully apply adaptive zoning to the computational bottleneck of transport assignment and is intended as a step towards full adaptive zoning based urban modelling. This article presents the methodology for a bi-partitioned assignment with adaptive zoning and all-or-nothing assignment. The methods are demonstrated with the well-known Chicago Regional road transport network, which is one of several commonly used networks for testing and benchmarking assignment algorithms (Bar-Gera 2010).

2   Methods

2.1   Terminology

The traffic network is represented as a weighted graph, where the links (edges) in the graph are road segments that are connected at the nodes (vertices). All trips to and from a zone are modelled as if they are starting or ending from a designated node in that zone called the *centroid*. The edge weights represent the costs of travelling typically in the form of a generalized cost that includes travel time, distance and tolls.

In this article the terms *origin* and *destination* refer to the start- and the end nodes of trips. The terms *source* and *target* on the other hand, refer to the role of nodes in the assignment algorithm. Each sub-problem assigns traffic from one source to multiple targets. When the source is an origin, the targets are destinations and the sub-problem is called *origin-based*. Conversely, a *destination-based* sub-problem assigns trips from multiple origins (targets) to a single destination (source), using a reverse assignment.

2.2   Network Assignment

The input to network assignment is an OD matrix that specifies the number of trips between each origin and destination pair. A second input is the weighted graph that represents the transport network. The purpose of all-or-nothing assignment then is to allocate the trips of each OD pair to the shortest path between the origin and destination and thus establish link intensities. The least-cost paths in the network can be found using the Dijkstra algorithm (Dijkstra 1959). The algorithm builds a shortest path tree from one source node (a zone centroid), to all nodes in the network. The algorithm starts as an empty tree and grows by successively adding nodes to the tree from the source node outwards. Once all the target zone centroids have been added, the shortest paths for the source zone are complete and the algorithm can be terminated. The Dijkstra algorithm provides the following information:

- $p_{l,t}^{D(N,s)}$,indicating whether link l is on the path from node s to t in network N
- $d_v^{D(N,s)}$,the shortest path distance from node s to v in network N

For convenience, the following intermediate variables are defined:

$$a_l^{D(N,s)} = \min\left(d_{l_1}^{D(N,s)}, d_{l_2}^{D(N,s)}\right) \tag{1}$$

$$b_l^{D(N,s)} = \max\left(d_{l_1,t}^{D(N,s)}, d_{l_2,t}^{D(N,s)}\right) \tag{2}$$

$$h_t^{D(N,s)} = \tfrac{1}{2}d_t^{D(N,s)} \tag{3}$$

where link l connects the nodes l_1 and l_2. $a_l^{D(N,s)}$ is the shortest path distance to the nearer of l_1 and l_2, and $b_l^{D(N,s)}$ is the shortest path distance to the further, and $h_t^{D(N,s)}$ is the half-way distance between nodes s and t. The $D(N,s)$ superscript is used to indicate that the associated variable are found though one application of the Dijkstra shortest path algorithm on the network N and the source node s.

The status of a link relative to the shortest path between s and t can be classified in four categories:

- not part of the shortest path,
- part of the first half of the shortest path,
- part of the second half of the shortest path, or
- partially belongs to the first half and partially to the second half.

Using these categories the proportion of a link belonging to the first half of the shortest path is:

$$p_{l,t}^{\prime D(N,s)} = \begin{cases} p_{l,t}^{D(N,s)} = 0 & : \quad 0 \\ p_{l,t}^{D(N,s)} = 1 \ \wedge \ b_l^{D(N,s)} < h_t^{D(N,s)} & : \quad 1 \\ p_{l,t}^{D(N,s)} = 1 \ \wedge \ h_t^{D(N,s)} \leq a_l^{D(N,s)} & : \quad 0 \\ p_{l,t}^{D(N,s)} = 1 \ \wedge \ a_l^{D(N,s)} < h_t^{D(N,s)} \leq b_l^{D(N,s)} & : \quad \dfrac{b_l^{D(N,s)} - h_t^{D(N,s)}}{b_l^{D(N,s)} - a_l^{D(N,s)}} \end{cases} \tag{4}$$

where, $p_{l,t}^{\prime D(N,s)}$ is the proportion of link l that is on the first half of the shortest path from s to t. The proportion for the second half of the path follows naturally:

$$p_{l,t}^{''D(N,s)} = \begin{cases} p_{l,t}^{'D(N,s)} = 0 & : \quad 0 \\ p_{l,t}^{'D(N,s)} = 1 & : \quad 1 - p_{l,t}^{'D(N,s)} \end{cases} \tag{5}$$

where, $p_{l,t}^{''D(N,s)}$ is the proportion of link l that is on the second half of the shortest path from s to t.

All-or-nothing assignment can be expressed as follows:

$$x_l^a (N,T) = \sum_i \sum_j T_{ij} p_{l,j}^{D(N,i)} \forall l \in N \tag{6}$$

where $x_l(N,T)$ is the traffic on link l, when traffic matrix T is assigned onto network N on the basis of shortest paths. Alternatively, the all-or-nothing assignment can be expressed as follows:

$$x_l^{AN} (N,T) = \sum_i \sum_j T_{ij} p_{l,j}^{'D(N,i)} + \sum_j \sum_i T_{ij} p_{l,i}^{'D(N^R,j)} \forall l \in N \tag{7}$$

$$x_l^{AN} (N,T) = \sum_i \sum_j T_{ij} p_{l,j}^{''D(N,i)} + \sum_j \sum_i T_{ij} p_{l,i}^{''D(N^R,j)} \forall l \in N \tag{8}$$

where N^R is the reverse graph of N; i.e., link $l_1 - l_2$ in N corresponds to $l_2 - l_1$ in N^R.

Equations (7) and (8) assume that there is only a single shortest path between two nodes. In reality there can be multiple paths of identical cost, therefore a path independent tie-breaking mechanism should be used.

2.3   Adaptive Zoning

Adaptive zoning (Hagen-Zanker and Jin, 2012) consists of two elements: zone hierarchy and zone neighborhood. The lowest level of the zone hierarchy consists of *atomic* zones and can be formed by a traditional zone systems. Higher levels of the hierarchy are formed by the subsequent amalgamation of zones into progressively larger *aggregated* zones. The top level of the zone hierarchy is a single zone that covers the whole study area. When each aggregated zone is formed by the amalgamation of two other zones (atomic or aggregated), then the total number of aggregated zones is one below the number of atomic zones. Thus for a system consisting of n atomic zones, there are in total *2n-1* zones in the hierarchy.

The zone neighborhood specifies for each atomic zone a specific set of zones (atomic or aggregated) that it interacts with. These zones are selected from different levels in the zone hierarchy; at short distances from the atomic zones neighboring zones are small and selected from low levels of the hierarchy and at further distances they are large and selected from high levels in the hierarchy. Each neighborhood is a zone-system in its own right that covers the whole study area. Thus, the interaction between two locations in the model area is spatially represented by a zone pair consisting of one atomic and one aggregated zone, whereby the size of the aggregated zone is a function of distance.

As atomic zones interact with small aggregated zones nearby and large aggregated zones at a large distance, it is implied that large aggregated zones interact with atomic zones at large distances and small aggregated zones interact with atomic zones at short distances. We call the set of atomic zones that interact with one aggregated zone, the inverse neighborhood of that zone.

The algorithms that form the zone hierarchy and neighborhood are based on the distance between zones as well as the size of zones in terms of traffic generated. This paper uses a point sampling approach to estimate Euclidean distances between zones (Hagen-Zanker and Jin, 2011). Upon amalgamation distances for the aggregated zone are determined as the weighted average of the constituent zones.

$$d_{i,a \cup b} = \frac{A_a d_{i,a} + A_b d_{i,b}}{A_a + A_b} , \tag{9}$$

where $d_{i,j}$ is the distance from zone i to j, A_a is the area of zone a, and $a \cup b$ is the zone that amalgamates a and b.

The zone hierarchy is created following the procedure of Hagen-Zanker and Jin (2012) and iteratively merges the pair of zones leading to the smallest increment in the estimated error for a spatial interaction model:

$$c_{a,b}^{join} = D_{a \cup b} e^{\beta d_{a \cup b, a \cup b}} - D_a e^{\beta d_{a,a}} - D_b e^{\beta d_{b,b}} , \tag{10}$$

where the algorithm joins the pair of zones a and b with lowest value for, D_a measures the size of destination zone a (here the number of trips destined for that zone), β is the distance sensitivity parameter of a best-fitting model.

The creation of the neighborhood is based on an iterative approach whereby in first instance the neighborhood consists of the top level of the zone hierarchy. Then, iteratively one zone in the hierarchy is subdivided considering both the size of the zone (in terms of distance) and the strength of the interaction. This iteration takes place until the neighborhood consists of the required number of zones.

$$c_{i,j}^{split} = T_{i,j} d_{j,j} , \tag{11}$$

where $c_{i,j}^{split}$ is the criterion for subdividing the aggregated zone j in the neighborhood of zone i. The zone j with the highest value for $c_{i,j}^{split}$ is subdivided into its constituent parts. $T_{i,j}$ is the number of trips from i to j.

Using adaptive zoning, the original interaction matrix T can be aggregated in two ways; either the origin zones are aggregated, resulting in T^O or the destination zones are aggregated, resulting in T^D. Using both these matrices the all-or-nothing assignment of equation (8) can be approximated by:

$$x_l^{AN}(N,T) \approx \sum_i \sum_j T_{ij}^O p_{l,j}''^{D(N,i)} + \sum_j \sum_i T_{ij}^D p_{l,i}''^{D(N^R,j)} \forall l \in N \tag{12}$$

Note that using this system, the source of the assignment is always an aggregated zone. Which means that the assignment takes place for one inverse neighborhood at a time: each application of the Dijkstra shortest path algorithm determines the routes between one aggregated zone and all the atomic zones that it interacts with. The relationship between zone size and distance means that the Dijkstra shortest path algorithm finds a small tree for small zones and a large tree for large zones. This relationship is the core underlying the efficiency gain of the proposed method.

2.4   Bias Correction

The adaptive-zoning based assignment introduces a new bias: when aggregating source zones, traffic that would originally go to the centers of all of its component zones will now go to only one center. The distance from a target to its aggregate source will be different from the average distance to all the original source zones. This difference is likely to be systematic, because the distribution of trips over zones is not random. It is not feasible to calculate the difference exactly, because that would undo the efficiency gain of using the adaptive zoning system. However by assuming a correlation between network distances and Euclidean distance, the following bias correction factor can be calculated:

$$f_{s,t} = \frac{\left(\sum_{i \in s} T_{it}\right) \|p_s - p_t\|}{\sum_{i \in s} \left(T_{it} \|p_i - p_t\|\right)} \tag{13}$$

where s is an aggregate source zone of which the constituent atomic are indicated by i. p_z gives the geographical coordinates of the center of zone z. The correction factor is applied by modifying the position of the half-way point, yielding the following modified form of equation (3):

$$h_t^{D(N,s)} = f_{s,t} \tfrac{1}{2} d_y^{D(N,s)} \tag{14}$$

3   Results

The model is applied to the Chicago Regional road transport network, which is one of several commonly used networks for testing and benchmarking assignment algorithms (Bar-Gera 2010). It was originally developed by the Chicago Area Transportation Study (CATS). The model area comprises 1,790 zones, 12,982 road nodes and 39,018 road links. (Figure 1-a,b). The assignment procedure that is followed is for the user equilibrium using the Frank-Wolfe algorithm that relies on an iterative evaluation of the all-or-nothing assignment presented in this paper. For details we refer to the Annex and more in particular to Van Vliet (1987).

For comparison purposes an alternative model is created that halves the number of zones (895 zones, Figure 1-c), as well as a series of adaptive zoning based models with varying degrees of aggregation (Figure 1-d). The only difference between these models and the original 1,790 zone model is due to the aggregation and approximation methods. The deviation from the 1,790 zone mode is therefore a measure of the approximation error. The approximation error is measured by means of the systematic error in total travel time in the model (travel time bias), the root mean squared error (RMSE) in link flows and travel time as well as the correlation between in link traffic intensity.

The application of the models shows that with increasing degrees of aggregation, the adaptive zoning system requires less time, but also increases approximation error, in all measures of aggregation error (Table 1). The adaptive zone model with a neighborhood size of 200 zones is nearest to the 895 zone system in terms of computation time. A comparison of the model results of these two runs shows that the adaptive zoning based model has substantially better performance: the correlation is near perfect (0.998 compared to 0.956), the bias in travel times is removed (0.64% instead of 10%) and the link level error is drastically reduced (Link volume RMSE reduces from 221 to 33 vehicles, link cost RMSE reduces from 0.33 to 0.075 seconds).

Table 1 also reports on the number of iterations that was necessary for the Frank-Wolfe algorithm to converge. It shows that the traditional aggregation method reduces the number of iterations relative to the 1790 zone system; this is consistent with the downward bias in traffic volumes, which leads to lower levels of congestion and in turn less spreading over multiple paths. Under adaptive zoning the required number of iterations remains practically the same for all runs, suggesting that the convergence of the algorithm is not materially affected by the adaptive zoning approximation.

Table 1. Comparison between traditional and adaptive zoning based models

Model	Zone system dimensions	Computation time [s]	Link volume correlation [-]	Travel time bias [-]	Link volume RMSE [veh]	Link cost RMSE [s]	Iterations [-]
Ground truth	1790 * 1790	313	1	0	0	0	32
Traditional	**895 * 895**	**193**	**0.954**	**-0.10**	**221**	**0.33**	**26**
Adaptive	1790 *50	158	0.996	-0.018	85	0.10	33
Adaptive	1790*100	168	0.997	-0.010	47	0.087	33
Adaptive	1790*150	188	0.997	-0.0083	38	0.085	33
Adaptive	**1790*200**	**195**	**0.998**	**-0.0064**	**33**	**0.075**	**31**
Adaptive	1790*300	231	0.998	-0.0035	25	0.076	32

Notes:Computation is on a 2.93 GHz Intel® Core™desktop computer using 8 processors and Windows 7, 64 Bit. Computation time excludes the time for generating the zone system, which is in the order of 15s. Link flow correlation is Pearson's correlation with ground truth of traffic volume over all links. Travel time bias is calculated as $\left(T^{model} - T^{ref}\right)/T^{ref}$, where $T^{model} = \sum_{l} x_l^{model} t_l^{model}$ and *ref* refers to the ground truth model.

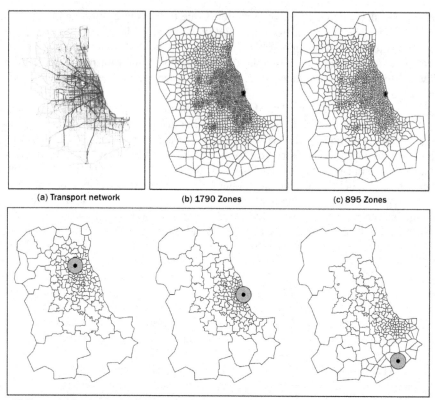

(a) Transport network   (b) 1790 Zones   (c) 895 Zones

(d) Three neighbourhoods of 150 zones each, the centre of the neighbourhood is marked with a bulls-eye

Fig. 1. Transport network and alternative zone systems for the Chicago study area

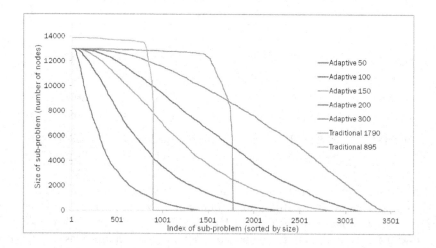

Fig. 2. Comparison of the tree sizes of the sub-problems. The distribution of sub-problem sizes explains the variations in calculation time among the alternative assignment runs.

The outcomes for a range of different neighborhood dimensions—from 50 zones to 300 zones—indicate a gradual deterioration with decreasing neighborhood size. However, even for the most severe aggregation (of 50-zone neighborhoods) the results remain reasonable and considerably better than the traditional zone system at 895 zones. Figures 2 sheds more light on the use of computation time by the algorithms. It shows how the traditional approach creates a sub-problem for each zone, whereby the size of each sub-problem, measured as the size of the shortest path algorithm tree is roughly constant and near to the total number of nodes in the network. When a coarser scale traditional zone system is used, the number of sub-problem reduces, but the size of the individual sub-problems remains roughly the same. The adaptive zoning doubles the number of sub-problems, but it drastically reduces the size of the sub-problems.

4   Discussion

4.1   Scalability

Hagen-Zanker & Jin (2011) explored the computational complexity of adaptive-zoning-based assignment and found that the complexity under adaptive zoning is O(n log n) and using traditional zoning O(n^2 log n), where n is the number of zones. That analysis depends on the assumption that the number of zones is proportional to the number of links as well as the size of the study area. Here we explore in more depth the nature of this complexity and consider separately the effect of study area size, the number of zones and the size of the network. In order to consider these independently, we define zone density as the number of zones per area, and link density as the number of links per area.

One crucial assumption is about the network size of the sub-problems. From the nested nature of the zone hierarchy that provides the aggregated zones, it follows that the size of aggregated zones is exponentially distributed. Furthermore the network size for a sub-problem is a positive function of the zone size; larger zones are found at further distances and hence the radius of an inverse neighbourhood is a positive function of zone size. Finally, there is a lower limit for zone sizes. Consequently, it is reasonable to assume that the network size for sub-problems is negatively exponentially distributed and in any case there is a negatively exponential function under which the distribution will fit. Figure 2 confirms that this is a reasonable assumption: when the level of aggregation increases, the distribution indeed takes form of an exponential decay curve.

The cost of one sub-problem of the Dijkstra algorithm is $O(m\ log\ m)$ where m is the number of nodes in the network. For n sub-problems this gives:

$$O\left(dijkstra\right) = O\left(\sum_{x=1}^{n} m_x \log m_x\right) \tag{15}$$

where m_x is the size of the sub-problem. Approximation by a continuous integral function and considering the network size constant, as under traditional zoning, gives:

$$O\left(dijkstra\middle|m_x = m\right) = O\left(\int_0^n m\log m \, dx\right) = O(nm\log m) \tag{16}$$

Under adaptive zoning m_x is assumed to be exponentially distributed:

$$O\left(dijkstra\middle|m_x = me^{-\beta x}\right) = O\left(\int_0^{2n} me^{-\beta x}\log me^{-\beta x} \, dx\right) = O(m\log m) \tag{17}$$

Substituting, m and n for the appropriate products of study area, link density, and zone density gives:

$$O\left(dijkstra\middle|traditional\right) = O\left(\rho_l\rho_z A^2\left(\log A + \log\rho_l\right)\right)$$
$$O\left(dijkstra\middle|adaptive\right) = O\left(\rho_l A^2\left(\log A + \log\rho_l\right)\right) \tag{18}$$

where A is area, ρ_z is zone density and ρ_l is link density. Table 2 summarizes the results for each individual variable, considering all others independent. There is no improvement in scalability with respect to link density, which means that if the sole objective is to increase the number of links in the model, then adaptive zoning offers no additional advantages. Also under adaptive zoning the complexity as a function of study area reduces by an order of magnitude, making it highly suited for large area applications such as city region and country level models. Finally it states that in the limiting case the computational cost is independent of the size of the atomic zones. This is a striking and perhaps counterintuitive result, it is however a direct consequence of the developed method: Each invocation of the Dijkstra shortest path algorithm uses an aggregated zone as its source and finds to path to all atomic zones. The cost of the algorithm is determined by the size of the tree, which in turn is determined by the furthest destination: the number of destination zones is inconsequential.

Table 2. Complexity by variable under traditional and adaptive zoning

Variable	Traditional Complexity	Adaptive Zoning Complexity
Area (A)	$O(A^2\log A)$	$O(A\log A)$
zone density (ρ_z)	$O(\rho_z)$	$O(1)$
link density (ρ_l)	$O(\rho_l)$	$O(\rho_l)$

5   Conclusion

This paper presents a new approach to road traffic assignment that partitions each origin-destination route into the first and the second half, which are assigned by separate sub-problems of computation. The partition opens new opportunities for adaptive trade-offs between computation time and accuracy. Specifically, the paper demonstrates that a bi-partition assignment algorithm with adaptive zoning is a significantly more efficient method. In the worked case using the benchmark Chicago regional road network, we find that when the precision of the zone system is constrained by computation time, the adaptive zoning system is far more precise than a traditional zone system. The bias in modelled travel times reduces from 10% to 0.64% when using adaptive zoning, the RMSE of link volume and link traversal cost reduce by factors of 6.4 and 4.4 respectively. Moreover, our understanding of the structure of the algorithm and an analysis of the computational complexity imply that the detail at which traffic arriving in and departing from zones can be refined without affecting the computational cost. Finally the model scales much better with the size of the study area (A) the complexity of the model is $O(A \log A)$ instead of $O(A^2 \log A)$ which makes it very attractive for modelling large study areas such as those confronted by policy analysts when assessing infrastructure investment and policy initiatives across mega-city regions and nations.

Annex

In transport systems typically the time to traverse a link (and hence the cost) is a function of the free flow time, link capacity and traffic flow volume. This article uses the following common model:

$$t_l = t_l^f \left(1 + b \left(\frac{x_l}{c_l} \right)^p \right)$$

where t_l is the link traversal time, t_l^f is the free-flow traversal time, x_l is the traffic flow volume, and c_l is the link capacity, b and p are parameters, here $b=0.15$ and $p=4$.

The co-dependence between t_l and x_l makes assignment problems more complex. The Frank-Wolfe assignment algorithm accounts for the co-dependence by means of iteration. In the first iteration, traffic is assigned on the basis of free-flow costs and the resulting link traffic loads are used to update the link costs. From then on, each iteration reassigns a fraction of the overall flows on the basis of the updated costs.

$$x_l^{FW,1}\left(N^1,T\right) = x_l^{AN,1}\left(N^1,T\right)$$
$$x_l^{FW,n}\left(N^n,T\right) = \left(1-\lambda^n\right)x_l^{FW,n-1}\left(N^{n-1},T\right) + \lambda^n x_l^{AN}\left(N^n,T\right) \forall l \in N, n > 1$$

where n indicates the iteration number and λ^n is the convergence factor. For the calculation of λ^n as well as the stopping criterion called the *relative gap*, we refer to Van Vliet (1987). The assignment algorithm iterates until a relative gap of 0.01 is achieved.

References

1. Bar-Gera, H.: Origin-based algorithm for the traffic assignment problem. Transportation Science **36**(4), 398–417 (2002)
2. Bar-Gera, H.: Transportation Test Problems (2010). http://www.bgu.ac.il/~bargera/tntp (Accessed August 13, 2012)

3. Chang, K.T., Khatib, Z., Ou, Y.M.: Effects of zoning structure and network detail on traffic demand modeling. Environment and Planning B: Planning and Design **29**(1), 37–52 (2002)
4. Dial, R.B.: A path-based user-equilibrium traffic assignment algorithm that obviates path storage and enumeration. Transportation Research Part B: Methodological **40**(10), 917–936 (2006)
5. Dijkstra, E.W.: A note on two problems in connexion with graphs. Numerische Mathematik **1**(1), 269–271 (1959)
6. Ding, C.: The GIS-based human-interactive TAZ design algorithm: examining the impacts of data aggregation on transportation-planning analysis. Environment and Planning B: Planning and Design **25**(4), 601–616 (1998)
7. Frank, M., Wolfe, P.: An algorithm for quadratic programming. Naval Research Logistics Quarterly **3**(1–2), 95–110 (1956)
8. Friesz, T.L., Luque, J., Tobin, R.L., Wie, B.-W.: Dynamic Network Traffic Assignment Considered as a Continuous Time Optimal Control Problem. Operations Research **37**(6), 893–901 (1989)
9. Fukushima, M.: A modified Frank-Wolfe algorithm for solving the traffic assignment problem. Transportation Research Part B: Methodological **18**(2), 169–177 (1984)
10. Hagen-Zanker, A., Jin, Y.: Improving geographic scalability of traffic assignment through adaptive zoning. In: 2011 Conference on Computers in Urban Planning and Urban Management, Lake Louise, Canada, p. 15.(2011a)
11. Hagen-Zanker, A., Jin, Y.: Reducing aggregation error in spatial interaction models by location sampling. In: The 11th International Conference on GeoComputation, London, p. 4. (2011b)
12. Hagen-Zanker, A., Jin, Y.: A new method of adaptive zoning for spatial interaction models. Geographical Analysis **44**(4), 281–301 (2012)
13. Hagen-Zanker, A., Jin, Y.: Adaptive zoning for transport mode choice modeling. Transactions in GIS **17**(5), 706–723 (2013)
14. Jiang, B., Claramunt, C.: A structural approach to the model generalization of an urban street network. GeoInformatica **8**(2), 157–171 (2004)
15. Jin, Y., Williams, I., Shahkarami, M.: A new land use and transport interaction model for London and its surrounding regions. In: European Transport Conference, Cambridge (2002)
16. Kristoffersson, I., Engelson, L.: A Dynamic Transportation Model for the Stockholm Area: Implementation Issues Regarding Departure Time Choice and OD-pair Reduction. Networks and Spatial Economics **9**(4), 551–573 (2009)
17. Lam, W.H.K., Gao, Z.Y., Chan, K.S., Yang, H.: A stochastic user equilibrium assignment model for congested transit networks. Transportation Research Part B: Methodological **33**(5), 351–368 (1999)
18. Martinez, L.M., Viegas, J.M., Silva, E.A.: Modifiable areal unit problem (MAUP) effects on traffic analysis zones (TAZ) delineation. Modelling and Simulation **2005**, 313–323 (2005)
19. Martinez, L.M., Viegas, J.M., Silva, E.A.: A traffic analysis zone definition: a new methodology and algorithm. Transportation **36**(5), 581–599 (2009)
20. Miller, S., Daly, A., Fox, J., Kohli, S.: Destination sampling in forecasting: application in the PRISM model for the UK West Midlands region. In: European Transport Conference, Noordwijkerhout (2007)
21. Nie, Y.: A class of bush-based algorithms for the traffic assignment problem. Transportation Research Part B: Methodological **44**(1), 73–89 (2010)

22. Van Vliet, D.: The Frank-Wolfe algorithm for equilibrium traffic assignment viewed as a variational inequality. Transportation Research Part B: Methodological **21**(1), 87–89 (1987)

23. Viegas, J.M., Martinez, L.M., Silva, E.A.: Effects of the modifiable areal unit problem on the delineation of traffic analysis zones. Environment and Planning B: Planning and Design **36**(4), 625–643 (2009)

24. Wardrop, J.C.: Some theoretical aspects of road traffic research. Proceedings, Institution of Civil Engineers Part **2**(9), 325–378 (1952)

25. Williams, I., Lindsay, C.: An efficient design for very large transport models on PCs. In: European Transport Conference, Cambridge, p. 18 (2002)

On the Problem of Clustering Spatial Big Data

Gabriella Schoier and Giuseppe Borruso[(✉)]

DEAMS – Department of Economic, Business, Mathematic and Statistical Sciences
"Bruno de Finetti", University of Trieste, Via A. Valerio, 4/1 34127, Trieste, Italy
{gabriella.schoier,giuseppe.borruso}@econ.units.it

Abstract. Different motivation are related with the analysis of Spatial Big Data (SBD). Google Earth, Google Maps, Navigation, location-based service allow to obtain a great amount of geo-referenced data. Often spatial datasets exceed the capacity of current computing systems to manage, process, or analyze the data with reasonable effort. Considering SBD history methodology as Data-intensive Computing and Data Mining techniques have been useful. In this context the problem regards the analysis of of high frequency spatial data. In this paper we present an approach to clustering of high dimensional data which allows a flexible approach to the statistical modeling of phenomena characterized by unobserved heterogeneity. We consider the MDBSCAN and compare it with the classical k-means approach. The applications concern a synthetic data set and a data set of satellite images.

Keywords: Spatial data mining · Clustering algorithms · Arbitrary shape of clusters · Efficiency on large spatial databases · Handling noise · Lagrange-Chebychev metrics · Image analysis

1   Introduction

The rapid developments in the availability and access to spatially referenced information in a variety of areas, has induced the need for better analysis techniques to understand different phenomena. In particular spatial clustering algorithms, which groups similar spatial objects into classes, can be used for the identification of areas sharing common characteristics.

Clustering is an unsupervised classification of patterns - observations, data items, or feature vectors - into groups or clusters [6]. Cluster analysis can be defined as the organization of a collection of patterns - usually represented as a vector of measurements, or a point in a multidimensional space - into clusters based on similarity.

The clustering problem has been considered in many contexts and by researchers in different disciplines. It is useful in several exploratory pattern-analysis, grouping, decision-making and machine-learning situations, including data mining (see e.g. [5]), spatial data mining (see e.g. [1], [2], [8]), document retrieval, image segmentation, and pattern classification.

Clustering techniques have been recognized as primary Data Mining methods for knowledge discovery in spatial databases, i.e. databases managing 2D or 3D points, polygons etc. or points in some d-dimensional feature space (see e.g. [8], [13, [14]]).

© Springer International Publishing Switzerland 2015
O. Gervasi et al. (Eds.): ICCSA 2015, Part III, LNCS 9157, pp. 688–697, 2015.
DOI: 10.1007/978-3-319-21470-2_50

The aim of this paper is to compare a density based algorithm (MDBSCAN) for the discovery of clusters of units in large spatial data sets with the classical k-means method. This algorithm is a modification of the DBSCAN algorithm (Ester et. al.(1996)). The modifications of this algorithm with respect to the original regard the consideration of spatial and non-spatial variables and the use of a Lagrange-Chebychev metrics instead of the usual Euclidean one. The applications concern a synthetic data set and a data set of satellite images.

2   Spatial Big Data

Spatial data mining can be used for browsing spatial databases, understanding spatial data, discovering spatial relationships, optimizing spatial queries.

Spatial data, as other kinds of data, are becoming bigger and bigger, although since the introduction of GIS and desktop GIS in particular, GIS users and experts have become facing with the issue of managing big amount of data, even though often data were much more difficult to retrieve than today. In geographical terms, the nature of data is such that an increase of dimension of the dataset is always very possible, both in terms of the number of the records to be considered, as well as in terms of the attribute of the geographical data. Both the vector and raster data formats used in GIS analysis tend to be multidimensional, i.e., containing a quantity of elements to be considered in any form of grouping and aggregation. In any case at least two fields (if not three) are needed to store the spatial information while all the attribute data contribute to increasing the dimension of the dataset. Satellite imagery in particular represents another case, in which redundant information is also considered, as very close pixels present very little differences although weighting in the processing, storage and visualization time of the data. So compression algorithms on one side and proficient clustering tools are needed in order to extract the more precise and complete set of geographic information ([15], [16], [17]).

3   The Methodology

3.1   K-means Algorithm

The K-means algorithm is very well known (see e.g. [7]). The algorithm allocates the data points (objects) into clusters, so as to minimize the sum of the squared distances between the data points and the center of the clusters.

The centers of the clusters are initialized by randomly selecting from the data or by fixing particular data points. Then the data set is clustered in the process of assigning each point to the nearest center .When the data set has been identified, the average position of the data points within each cluster is calculated and the cluster center then moved to the average position. This process is repeated until a condition of stopping is reached, in other words the algorithm has these steps:

Step 1
Place K points into the space represented by the objects that are being clustered. These points represent initial group centroids.
Step 2
Assign each object to the group that has the closest centroid.
Step 3
When all objects have been assigned, recalculate the positions of the K centroids.
Step 4
Repeat Steps 2 and 3 until the centroids no longer move or another stopping rule is achieved. This produces a separation of the objects into groups from which the metric to be minimized can be calculated.

The K-means algorithm requires three user-specified parameters: number of clusters K, cluster initialization, and distance metric.The most critical choice is K.

K-means is typically used with the Euclidean metric for computing the distance between points and cluster centers. As a result, K-means finds spherical or ball-shaped clusters in data. K-means with Mahalanobis distance metric has been used to detect hyperellipsoidal clusters (see [9]), but this comes at the expense of higher computational cost

3.2   DBSCAN and MDBSCAN Algorithms

In this section we will consider clustering methods based on the notion of density.These regard clusters as dense regions of units which are separated by regions of low density (representing noise); moreover they may be used to discover clusters of arbitrary shape (see e.g. [3], [4], [10], [11, [12]).

Among these the *DBSCAN* algorithm (Ester (1996)) is a locality-based algorithm relying on a density based notion of clustering. Density based methods can be used to filter out noise (outliers) and discover clusters of arbitrary shape. This algorithm judges the density around the neighborhood on an unit to be sufficiently dense if the number of points within a distance *EpsCoord* of an unit is greater than *MinPts* , in this case the unit is a *core point* otherwise is a *border point*. This algorithm has been generalized in different papers (e.g. Sander (1998)).

The key idea is that for each point of a cluster the neighbourhood of a given radius has to contain at least a minimum number of points, i.e. the density in the *neighborhood* has to exceed some threshold. The shape of a *neighborhood* is determined by the choice of a distance function for two points p and q, denoted by $dist(p,q)$ The *Epscoord* neighbourhood of a point q is defined by

$$N_\varepsilon(q) = \{q \in D | dist(p,q) \le \varepsilon\},$$

where D is a data set of points.

A naive approach could require for each point in a cluster that there are at least a minimum number (*MinPts*) of points in an *Eps-neighborhood* of that point. However, this approach fails because there are two kinds of points in a cluster, points inside of the cluster (*core points*) and points on the border of the cluster (*border points*).

In general, an *Eps-neighborhood* of a border point contains significantly less points than an *Eps-neighborhood* of a core point. Therefore, one would have to set the minimum number of points to a relatively low value in order to include all points belonging to the same cluster. This value, however, will not be characteristic for the respective cluster particularly in the presence of noise. Therefore, one has to require that for every point *p* in a cluster *C* there is a point *q* in *C* so that *p* is inside of the *Eps-neighborhood* of *q* and $N_e(q)$ contains at least *MinPts* points.

This definition is elaborated in the following: a point *p* is *density reachable* from a point *q* if there is a chain of points $p_1, p_2, \ldots, p_{n-1}, p_n$ where $p_1 = p$ and $p_n = q$ such that p_i is *direct density reachable* from p_{i+1}.

Moreover a point *p* is *directly density reachable* from a point *q* if *p* belongs to the neighborhood of *q* and *q* is a core point.

The clustering formed from DBSCAN follows the rules below:

1. A point can only belong to a cluster if and only if it lies within the *Epscoord-* neighborhood of some core point in the cluster.
2. A *core point o* within the *Epscoord-* neighborhood of another core point *p* must belong to the same cluster as *p*.
3. A *border point r* within the *Epscoord-* neighborhood of some core point must belong to the same cluster to at least one of the core points.
4. A border point which does not lie within the *Epscoord-* neighborhood of any core point is considered to be noise.

There are some problems regarding both the considerations of spatial and non-spatial variables and the use of different distances.

Our generalization *the Modified Density-Based Spatial Clustering of Applications with Noise* MDBSCAN ([]) considers an approach density based that takes into account at the same time spatial and non spatial variables. It has a similar structure of the DBSCAN but introduces a notion of proximity not only for spatial characteristics but also for non spatial characteristics.

The key idea is that the cardinality of the neighborhood of an unit is given not only by counting the number of units that have distance from it less than a radius *EpsCoord* (the limiting distance value for the spatial variables) but by the points that have distance less than *EpsCoord* and that are "sufficiently" similar as regards non spatial attributes. In order to have a sufficient homogeneity for the non-spatial attributes another radius *Eps*, that represent the threshold for the distances calculated on the bases of the non-spatial variables is evaluated. In so doing we want to find clusters of elements which are spatially close to each other and homogeneous as regards other observed variables. The elements of such clusters may be interpreted as elements which are similar as regards some variables and belong to the same spatial area.

The distance function determines the shape of the neighbourhood. *MinPts* is the minimum number of points that must be contained in the neighbourhood of that point in the cluster. In the following we present the main steps of our algorithm the *Modified DBSCAN (MDBSCAN)*

Step 1.

Insert the values of the parameters:

SetOfPoints representing the matrix with the values of the non-spatial variables, *Coordinates* representing the spatial variables ,

Eps the limiting distance value for the non-spatial variables ,

EpsCoord the limiting distance value for the spatial variables,

MinPts the minimum number of points to consider a point as a *core point*.

Step 2.

Chose an arbitrary point *i* in the database, if *i* is a *core point* built the cluster by choosing all the points which are density-reachable from i else if it is a *border point* pass to the point *i+1*.

Step 3.

Classify the points which are not *density-reachable* as *noise*.

The value of the parameter *MinPts* is fixed as in Ester (1996). In order to determine the parameter *EpsCoord*, regarding the spatial variables, and the parameter *Eps*, regarding the non spatial variables we consider the algorithms *SorteKdist* and respectively *SorteKdist2*. The former *SorteKdist*, used for the spatial variables, is implemented following Ester (1996), it evaluates, for every point of the data set represented by the matrix of coordinates, the distances from the nearest *k*-points (belonging to the same data set),orders them in decreasing way and gives a graphical representation of these distances (for the choice of *k* see Sander (1996) and see fig:1 for an example.

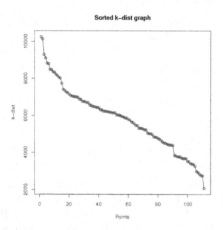

Fig. 1. Sorted k-distances graph for a sample data set

The latter *SorteKdist2* used for non spatial variables is similar, it evaluates, for every point of the data set represented by the matrix *SetOfPoints*, the distances from the nearest *k-points* (belonging to the same data set) and gives the mean value.

As written before the form of the neighborhood is determined by the choice of the distance function between two points $\underline{x}_i$ and $\underline{x}_j$. The euclidean distance generates a sphere around the point of circular shape (MDBSCAN (euc))

$$_2d_{ij} = \left[\sum_{s=1}^{p}\left|x_{is} - x_{js}\right|^2\right]^{\frac{1}{2}}$$

The Lagrange-Tchebychev distance generates a sphere around the point of square shape (MDBSCAN (lag))

$$_\infty d_{ij} = \max_{s}\left|x_{is} - x_{js}\right|$$

We have verified the validity of the MDBSCAN based on the Lagrange distance and that based on Euclidean on the base of the CPU times, as one can see from Fig. 2 the time (in seconds) with the increase of the number of points remain similar

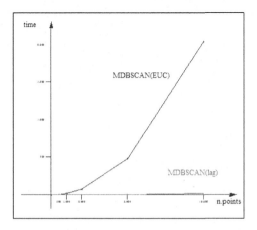

Fig. 2. MDBSCAN(lag) and MDBSCAN(euc)

4   The Application Results and Discussion

4.1   MDBSCAN versus K-means Clustering

In order to test the MDBSCAN(lag) and the K-means algorithm we consider some applications to a synthetic dataset and to a satellite images data set.

The analysis on bit-map images are a real example of raster spatial analysis, the image is formed by a regular grid of pixel and to every cell a colour has 24 bit (16 million of colours are possible).We have considered images that have 300 pixel by side (for a total of 90000 pixel) in a space of RGB colours to 24 bpp (bit by pixel, about 16 millions of colours) this colour format is known as true-colour.

Every pixel is a statistical unit, a point in the space of five dimensions: two relatively of the spatial attributes, the other to the non-spatial attributes. In order to apply the algorithm a standardization of the variables has been performed. The implemented MDBSCAN(lag) algorithm involved choosing different thresholds for the coordinates and for the other indices. As regards the spatial variables the algorithm *SorteKdist* gives the first point in the first valley. As regards the non-spatial variables the algorithm *SorteKdist2* has been implemented.

After the standardization of the variables the *R* language and a visualization language have been used for the analysis.

In the data set presented in Fig. 1 we consider a simple synthetic data set with two spherical clusters without noise.

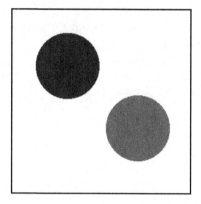

Fig. 1. (a) A synthetic data set

As one can see both algorithms find the clusters.

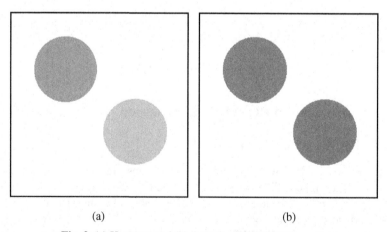

(a)   (b)

Fig. 2. (a) K-means and (b)clustering MDBSCAN results

The difficulty for K-means algorithm is handling clusters of arbitrary shape and noise. This is not the case of the MDBSCAN.

Next applications of MDBSCAN regards data set of satellite images

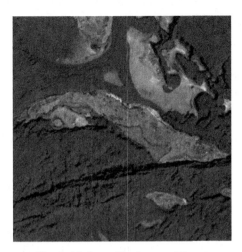

Fig. 3. (a) A satellite images data set

The algorithm was used on a satellite image of the Island of Cuba in the Gulf of Mexico in order to understand the performance of the algorithm over areas in which the land and sea borders are not so easy to detect. In Figure 3 the starting dataset is portrayed, while Figure 4 reports in false colors the different groups or clusters identified by the procedures.

The more easily detectable parts of the pictures have been assigned a common color. From Figure 4 it is therefore possible to observe in red the water component of he sea, while other colours as green and brown highlights the island and the other components as part of the Florida peninsula.

The less easily detectable parts of the pictures have been assigned a white colour representing a level of 'noise'.

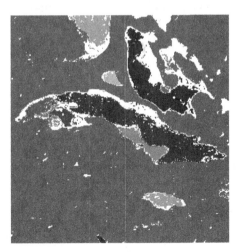

Fig. 4. MDBSCAN clustering results

5   Conclusions

In this paper two way of clustering methods applied to identify homogeneous areas are compared: K-means, and MDBSCAN.

MDBSCAN is an algorithm which is a modification of the original DBSCAN algorithm. Our algorithm takes into consideration both spatial and non spatial variables relevant for the phenomena that has to be analyzed. To improve computational aspects we proposed to use a Lagrange-Chebychev metrics instead of the Euclidean one.

The applications regards both synthetic and real datasets. The spatial clustering analysis allowed to obtained good bit-map images and good representation of satellite images.

The main advantage of K-means algorithm is its simplicity and speed which allows it to run on large datasets. One problem of the application of the K-means is the necessity of knowing *a prori* the number of clusters. Other problem regard the identification of noise (outliers) and the discovering of clusters of arbitrary shape

MDBSCAN is robust enough to identify clusters in noisy data, requires just a few parameters and is mostly insensitive to the ordering of the points in the database. This algorithm is efficient even for very large spatial databases, discovers clusters of arbitrary shape and does not need to know the number of clusters in the data a priori, as opposed to K-means.

References

1. Bailey, T.C., Gatrell, A.C.: Interactive Spatial Data Analysis. Addison Wesley Longman, Edinburgh (1996)
2. Cressie, N.A.C.: Statistics for spatial data. John Wiley & Sons, London (1993)
3. El-Sonbaty, Y., Ismail, M.A., Farouk, M.: An efficient density-based clustering algorithm for large databases In: Proceedings of the 16th IEEE International Conference on Tods with Artificial Intelligence (ICTAI) (2004)
4. Ester, M., Kriegel, H.P., Sander, J., Xiaowei, X.: A density-based algorithm for discovering clusters in large spatial databases with noise. In: Proceeding of the 2nd International Confererence on Knowledge Discovery and Data Mining, pp. 94–99 (1996)
5. Fayyad, U., Piatesky-Shapiro, G., Smyth, P.: From Data Mining to Knowledge Discovery in Databases (1996). http://www.kdnuggets.com/gpspubs/aimag-kdd-overview-1996-Fayyad.pdf
6. Han, J., Kamber, M., Tung, A.K.H.: Spatial Clutering Methods in Data Mining: A Survey (2001). ftp://fas.sfu.ca/pub/cs/han/pdf/gkdbk01.pdf
7. Jan, A.K.: Data Clustering. 50 years beyond K-means. Pattern Recognition Letters, 651-666 (2010)
8. Koperski, K., Han, J., Adhikary, J.: Mining Knowledge in Geographical Data (1998). ftp://fas.sfu.ca/pubcs/han/pdf/geo_survey98.pdf
9. Mao, J., Jain, A.K.: A self-organizing network for hyper-ellipsoidal clustering (HEC). IEEE Trans. Neural Networks, 16–29 (1996)
10. Sander, J., Ester, M., Kriegel, H.P., Xiaowei, X.: Density-Based Clustering. from in Spatial Databases: The Algorithm GDBSCAN and its applications (1999). http://www.dbs.informatik.uni-muenchen.de/Publikationen/

11. Schoier, G., Borruso, G.: A clustering method for large spatial databases. In: Laganá, A., Gavrilova, M.L., Kumar, V., Mun, Y., Tan, C., Gervasi, O. (eds.) ICCSA 2004. LNCS, vol. 3044, pp. 1089–1095. Springer, Heidelberg (2004)

12. Schoier, G., Bato, B.: A modification of the DBSCAN Algorithm in a Spatial Data Mining Approach. In: Meeting of the Classification and Data Analysis Group of the SIS: CLADAG 2007, pp. 395-398 . Macerata: EUM, Macerata, settembre, 12-14, 2007

13. Steinbach, M., Ertöz, L., Kumar, V.: The Challenges of Clustering High Dimensional Data (2003). http://www-users.cs.umn.edu/~kumar/papers/high_dim_clustering_19.pdf

14. Xu, R., Wunsch II., D.: Survey of Clustering Algorithms (2005). http://ieeexplore.ieee.org/iel5/72/30822/01427769.pdf

15. Bedard, Y.: Beyond GIS: Spatial On-Line Analytical Processing and Big Data, Univ. of Maine (2014). http://umaine.edu/scis/files//09/Beyond-GIS-Spatial-On-Line-Analytical-Processing-and-Big-Data-Yvan-Bedard.pdf

16. Chen, Y., Suel, T., Markowetz, A.: Efficient query processing in geographic web search engines. In: SIGMOD 2006, Chicago, Illinois, USA, June 27–29, 2006. http://cis.poly.edu/suel/papers/geoquery.pdf

17. Worboys, M., Duckham, M.: GIS.: A Computing Perspective. CRC Press, Boca Raton (2004)

Web-Based Geographic Information Technologies for Environmental Monitoring and Analysis

Valeri G. Gitis, Alexander B. Derendyaev$^{(\boxtimes)}$, and Arkady P. Weinstock

Institute for Information Transmission Problems (Kharkevich Institute) of the Russian Academy of Sciences, Moscow, Russia
{gitis,wintsa,wein}@iitp.ru

Abstract. We describe two technologies of web-based application of GIS in the monitoring and analysis of environmental processes. One technology is a web-based GIS platform, which offers information resources and tools for two levels of environmental process analysis. Level 1 is fast shallow express analysis and level 2 is sophisticated research implemented by specialists. This platform, available at http://dcs.isa.ru/geo/2/ and http://saltlab.emsd.ru/server2/, is used for summarization of seismic activity fields. The other technology is a web GIS incorporated into a large-scale information system for environmental monitoring. It is used for the presentation and analysis of spatial and spatio-temporal data on the World Ocean. A demo version is available at http://www.geo.iitp.ru/esimo/.

Keywords: Geoinformation analysis · Environmental monitoring · Spatio-temporal processes · Emergency oil spill

1 Introduction

Systems of monitoring of natural processes operate with spatial and spatio-temporal data. The specific character of these data and the increasing complexity of their joint analysis require special techniques and technologies, which are implemented in web-based Geographic Information Systems (GIS).

The data that involve natural processes are distinguished by large volume and dynamic character. In many cases, the tasks of analysis and forecasting require intensive computations of such data both locally and on remote servers. For this reason, GIS that monitor and analyze natural processes must be able to perform, in addition to standard operations, such functions as: (1) input and integration of large volumes of data from remote servers and local networks, (2) interactive visual exploration, including animated cartography, (3) spatial and spatio-temporal modeling, data mining and forecasting, (4) submitting job forms and starting intensive calculations on remote servers.

Web-based geographic information technologies for environment monitoring are implemented using client-server architecture with thin or thick clients. As a rule, thin client technologies are applied to allow the user to view the results of process modeling in a geographical context. Some examples of this technology can be found in the works on marine pollution monitoring and forecasting [1], [2]. Thin client technology for

© Springer International Publishing Switzerland 2015
O. Gervasi et al. (Eds.): ICCSA 2015, Part III, LNCS 9157, pp. 698–712, 2015.
DOI: 10.1007/978-3-319-21470-2_51

environmental process monitoring has two limitations. On the one hand, the user needs to access a server for any operation, including the simplest visualization, which renders interactive data analysis almost impossible. On the other hand, such solutions do not support joint analysis of the data received from the remote servers and the user's confidential data stored in the local network. Thick client technologies overcome these limitations. As a rule, the solutions are implemented as Java applications. An example of such a technology designed for environmental monitoring and natural resource management is given in [3].

In this paper we consider two technologies of GIS application in environmental monitoring and analysis: (1) a web GIS platform working directly with online resources and (2) a GIS integrated into a monitoring system. The first technology is oriented at monitoring and analysis of seismic fields, while the second is used in the Unified State System of Information on the World Ocean (ESIMO).

2 Geoinformation Analysis of Spatio-temporal Processes

2.1 GeoTime Approach

Environmental monitoring and assessment are being applied for systematic estimation, analysis and forecasting of natural processes. GIS is being used for visualization and study of space structure and time behavior, for detection of abnormal changes, prediction of natural catastrophes and decision support.

In 1991 a desktop version of GIS GeoTime has been developed [4]. It was implemented for the study of pre-earthquake processes. In nineties, detection of earthquake precursors was based on the analysis of time series referring to seismotectonic characteristics in a fixed space domain (the simplest example is the time series of density of earthquake epicenters within some area) and geo-monitoring time series, averaged over the stations in the domain. Such sort of analysis disregards spatial heterogeneity of the seismotectonic data. GIS GeoTime was the first system implementing the detection of earthquake precursors on the base of time-dependent fields, i.e. functions defined on 3D space (two spatial and one time coordinates) rather than one dimensional time series.

Web GIS GeoTime 3 (www.geo.iitp.ru/GT3/) expands the ideas of the desktop version [5], [6]. It is implemented as a Java application and loaded via Java Web Start. The system can dynamically load distributed data and plug-ins, run remote computing, and store the user output. GeoTime 3 supports parallel multithreaded data processing on multiprocessor/multicore frontend computers.

GIS GeoTime 3 processes the data presented in coordinates X, Y, Z, T, where X and Y are geographic coordinates, Z is the altitude (or depth), and T is the time coordinate. Continuous entities are represented as functions defined on regular 2D, 3D and 4D grids. The functions can take either scalar values or 2D or 3D vector values. Scalar 2D fields are represented as functions of two arguments, 3D fields as functions of three arguments, including depth and time, 4D field as a function $f(x,y,z,t)$. Vector 2D fields are vector-functions with components directed along the geographical coordinate system axes: $\mathbf{s} = s_x(x,y)\,\mathbf{x} + s_y(x,y)\,\mathbf{y}$, $\mathbf{v} = v_x(x,y,z)\,\mathbf{x} + v_y(x,y,z)\,\mathbf{y}$, $\mathbf{u} = u_x(x,y,t)\,\mathbf{x} +$

$u_y(x,y,t)$ **y**, where **x** and **y** are unit vectors along longitude and latitude axis. Discrete entities present geographical objects as lines, polygons and points. Lines and polygons can be specified in geographical coordinates X, Y, or in coordinates X,Y,T. Points present geographical objects in coordinates X,Y, sequences and series of measurements in coordinates X,Y,Z and X,Y,T, catalogs of events in coordinates X,Y,Z,T. Attributes of objects can be presented as numbers, vectors or text. Coordinates and attributes of objects are assigned to their descriptors and may be changed in space and time. In visual analysis raster and tile maps can also be used.

Since various characteristics of spatio-temporal processes are presented by data of different types, in order to operate with different properties of the processes, we need to fulfill operations of converting and transforming the data. GeoTime 3 allows the following transformations of the geoinformation layers: from vector to grid representation and vice versa, from 3D to 2D and from 4D to 3D layers. In simple cases, the system itself analyzes the types of layers and the coordinate binding and then offers the user a set of allowed operations and the default values of the processing parameters. In this sense, the tools of GeoTime 3 are universal with respect to data type.

2.2   Visual Exploration

Visualization tools are intended to simplify the visual perception of the geographical information and provide a concise and descriptive data representation in response to a user's information need. It deals with managing display parameters, constructing illumination model of the surface, plotting the cross-section of any grid-based layer along a user-defined profile, browsing through the values of any grid-based layer, retrieving subsets of vector objects via SQL-like request, reading attributes of any vector-based object, etc.

Display parameters of the thematic data properties are: filling of the layer, color, thickness and style of lines, type and size of icons and label style. The parameters depend on numeric or nominal values of the layer.

Scalar 2D field is displayed as a map in a corresponding coordinate plane. Scalar 3D and 4D field can be presented in all the 2D coordinate planes. The system supports animation to visualize the changes in the field on the third coordinate. The animated maps may be saved as GIF or AVI files.

Vector 2D and 3D fields can be visualized as arrows in the geographic coordinate plane XY. Visualization of 3D fields is realized with the help of animation varying the values of the field along Z (respectively, T) axis.

In some applications, one needs to carry out a visual analysis of a point field (catalog) of events. In general, catalog of events is a marked discrete subset of a four-dimensional space X, Y, Z, T. As a rule, a mark corresponds to the magnitude of the event. Visualization of the event field is supported by animation in the coordinate planes XY, XZ, XT, YZ, YT, ZT. Visual analysis of relation between several processes is supported by (1) synchronous animation of multiple layers and (2) interactive construction of combined plots.

The local properties of the processes specified by scalar 3D fields, catalogs of the events and numeric series could be visualized using plots. For any point (x^*, y^*) in the geographic coordinate plane, the plot shows the dependence of the field (or series) value on depth (respectively, time). For a catalog, the plot shows the time sequence of magnitudes of events in a circle of a certain radius with the center (x^*, y^*). The position of the reference point (x^*, y^*) is dynamically controlled by the cursor.

Measurements in GeoTime 3 are presented by operations of several types using an interactive graphical interface. A class of available standard operations includes computation of a distance between points, computation of a polygon area, reading the attributes of vector objects, and measurement of the values of function depicted on the plots. For scalar fields, there are measurements of two types: (1) the measurement of the layer value at any point; (2) the evaluation of the elementary layer statistics in a simply-connected polygon. For a 2D field the system just measures the value at a given point, while for 3D fields it is possible to display the dependence of the measured values on depth or time. In addition, in the geographical coordinates it is possible to visualize and measure the values of a cross section of 2D field along an arbitrary profile. The dynamism of all measurements is provided by interactive control of the position of the point, polygon shape and the choice of the profile.

2.3   Spatio-temporal Modeling Techniques

Modeling techniques make it possible to carry out a joint analysis of data presented in various formats, and allow the researcher to extract new information from thematic, spatial and spatio-temporal properties of the data.

Data conversion. Converting data of different types to a common format, usually to the format of a scalar grid field, allows the user to involve into study several properties of the investigated processes at one time. Moreover, converting extends visual analysis capabilities and makes it possible to combine several map layers in one map and to perform synchronous animation.

Four types of conversion operations are provided by GIS GeoTime 3: interpolation of the values of the layer containing irregular points to get a scalar 2D or 3D grid field, creation of vector fields out of scalar fields, construction of isolines for scalar 2D and 3D fields and construction of scalar grid fields out of the contours.

Local analysis. A typical technique for the local analysis is calculation in the moving spatio-temporal window. The dimension of the window and its parameters are set by the user within the limits defined by the dimension of the analyzed layer. By default, the result is attributed to the point whose coordinates correspond to the center of the spatial area of the window and the last slice in time.

There is a set of predefined operators, which can be applied to a scalar field in a moving window. There is also a possibility for the user to construct a new operator. The predefined operators compute the illumination model for the layer; mean, median, standard deviation, the difference between the value in the center of the window and the mean value in the window; maximum, minimum and difference between maximum and minimum; the module and the azimuth of the gradient of 2D slice, the curvature of the isolines of a 2D slice in the XY plane. The user can compute a convolution of the input field with an arbitrary spatio-temporal kernel.

For two scalar 2D, 3D or 4D (either dimension) fields the layer of values of correlation coefficients between these fields in a moving (respectively, 2D, 3D or 4D) window can be calculated. When the initial fields are of different dimension an automatic cloning of field values for the missing coordinates is performed. In addition, the fields can be adjusted to a uniform grid. By default the minimum grid spacing in each coordinate is set.

Calculations in a moving window are also used to create a scalar field from the vector layers of lines and points, for instance, the field of object density, the field of distances from objects to the grid nodes, etc. In this case the dimensionality of the output scalar field is equal to or less than the dimensionality of the input vector layer.

Applied tasks of spatio-temporal analysis require sometimes to estimate the similarity of the behavior of a process in different geographical locations. The similarity of the 3D scalar field $f(x,y,t)$ at two points, (x^*,y^*) and (x,y), can be defined as the correlation coefficient $cor(f(x^*,y^*,t), f(x,y,t))$ of the corresponding time series (similarly for a 3D $f(x,y,z)$ field). The result of the analysis is a 2D field of correlation coefficients.

The user can easily interpret the results of estimating the local similarity of geographic points made on the basis of several 2D fields. The simplest measure of similarity is the distance function. Let to each point (x, y) in the map a vector $\mathbf{f}=(f_1 f_2,...,f_I)$ of thematic field properties be assigned. Then the similarity is defined as a function of distance in the space of thematic properties: $S=1-\rho(\mathbf{f^*},\mathbf{f})/R$ if $\rho(\mathbf{f^*},\mathbf{f})<R$ and $S=0$ if $\rho(\mathbf{f^*},\mathbf{f}) \geq R$, where R is chosen by the user (typically, when calculating the similarity, the values of thematic properties are normalized by standard deviation). The 2D scalar field of similarity is a field in the geographic space XY.

Integral analysis. A typical technique of integral analysis is grid-based computing. The results of operations are scalar grid-based fields, which characterize the properties of environmental processes.

In many applications it is important to compile a scalar field which is a function of one or more input fields. GeoTime 3 contains a predefined set of functions; in addition, a function can be constructed by the user from a set of elementary functions using algebraic and logical operations. It is possible to use in one and the same operation fields with different dimensions and with different grid spacing. When computing all the input fields are converted to the same dimension and uniform coordinate grid.

An important operation of grid computing is the scalar field projection. The operation consists in calculating fields of a smaller dimension, whose values are elementary statistics of the source field along the projection axis. For example, for a scalar 3D field $f(x,y,t)$ it is possible to calculate 2D fields of mean values of $f(x,y,t)$ along the axis T, Y or X, similarly for a 4D field.

The estimation of spatial similarity of two 3D fields belongs to the same sort of operations. The result is a spatial 2D field whose values are determined by the measure of similarity of corresponding time or depth series.

2.4   Natural Hazard Prediction

Support of the research in the domain of natural hazard prediction is one of the most important applications of dynamic GIS. GIS GeoTime 3 is to a large extent focused on the analysis of earthquake precursors. The analysis is based on the assumption that the process of earthquake preparation is accompanied by spatio-temporal anomalies of characteristics of the geological environment in the vicinity of a future event. These anomalies are associated with the accumulation of potential energy and a gradual transition of the geological environment from the phase of elastic deformations to the phase of plastic deformations.

GIS GeoTime 3 provides a three-stage analysis of a seismic process: (1) evaluation of the dynamic characteristics of the analyzed process, (2) detection of spatio-temporal anomalies (change-point detection), and (3) estimation of the parameters of anomalies.

To evaluate seismotectonic properties of the geological environment, GeoTime 3 calculates spatio-temporal fields (feature fields) from earthquake catalogues and geo-monitoring time series. To this end, we use the local kernel techniques, adaptive weights smoothing [7] and interpolation methods.

It is assumed that the geological environment is heterogeneous in space, but in the normal state it has stationary dynamics, which is perturbed during the preparation of geological catastrophe. The task of finding out the anomalies is reduced to the problem of change point detection in a random process. In GIS GeoTime 3 several parametric models for the detection of spatio-temporal anomalies in the feature fields are realized. In fact, there are no substantial reasons for adopting a parametric model since the distribution of the process is unknown and cannot be obtained on the basis of the available information. Therefore nonparametric approach to the problem of change point detection is more promising.

The presence of spatio-temporal anomaly gives a qualitative idea of the presence of a catastrophe precursor. However, the forecast of a catastrophe with a given energy should formally predict the location and time of the event. One may assume that the anomaly achieves maximal value in the center of the zone of the catastrophe preparation and that the value of the anomaly exponentially decreases in space towards the boundary of the zone. For the estimation of the anomaly parameters each time slice of the 3D field of anomalies is approximated by Gaussians. It is necessary to estimate the coordinates of the Gaussian center, its height and damping. To estimate the degree of confidence in the presence of an anomaly within the given slice, the normalized value of the approximation accuracy is used, which varies from 0 to 1.

3   Environmental Monitoring and Analysis Using Web-Based GIS Platform

3.1   Basic Concept

A number of websites publish observations on the environmental processes in the form of coordinate time series, temporal sequences of events and time-dependent fields.

In particular, the global data centers and regional seismological centers publish real-time earthquake information: the coordinates of the epicenter, the depth of the hypocenter, the time and magnitude, seismograms, and other characteristics. From these data one can calculate physical fields that characterize the spatio-temporal behavior of the seismic process. We consider the field of seismic activity (seismic activity is equal to the number of earthquake epicenters in a certain magnitude interval that occurred in a spatio-temporal window of 1,000 sq. km in space and 1 year in time). This field characterizes accumulation and release of seismic energy and the changes in the geological environmental structure at different stages of tectonic process.

Considering the fields of seismic activity in the present with respect to the fields of the background seismic activity in the past one can determine statistically significant regions of seismic quiescence, respectively, seismic activation. The decrease in seismic activity indicates the process of accumulation of seismic energy, while the increase in activity indicates the activation of the seismic process. It is known that both mentioned spatio-temporal anomalies can be precursors of strong earthquakes.

Monitoring technology is realized on a web platform that consists of two GISs: SeismoMap and GeoTime 3. GIS SeismoMap is realized as a thin client based on Google Maps API. It supports express analysis of seismic process and is intended both for specialists and for a wide range of Internet users. GIS GeoTime is realized as a thick client. It is a multi-functional analytic system designed for analysis of spatio-temporal processes in geographical and geological environment and intended for specialists in the Earth sciences. Thus, the platform of seismic fields monitoring provides information resources and tools for two-level analysis: (1) express-analysis of the seismic process accessible to a wide range of users and (2) detailed analysis of the same seismic process, which is performed by specialists.

A prototype of a web-based GIS platform for monitoring of 24 regions is available at the site of the Institute for Information Transmission Problems RAS http://dcs.isa.ru/geo/2/ and at the site of Kamchatka branch of the Geophysical service of RAS http://saltlab.emsd.ru/server2/. GIS platform daily loads regional earthquake catalogs for $T_1+T_2+T_\varepsilon+T_0 \times N \approx 2.5$ years from one of the sites http://earthquake.usgs.gov/earthquakes/feed/, www.isc.ac.uk, http://geonet.org.nz, or http://www.emsd.ru/ts/alldemo.php, and calculates the corresponding spatio-temporal seismic fields. The involved parameters are as follows: $T_1=2$ years is the interval of background seismicity, $T_2=2$ months is the interval of current seismicity, $T_\varepsilon=3$ months is the interval for estimation of the first time slice of the seismic fields, $T_0=7$ days is the time interval between the consequent slices of 3D seismic fields, $N=10$ is the number of the maps of time slices. When launching the platform the user selects one of the predetermined seismically active regions. Further, he can visualize the following maps of time slices: ten maps of the time slices of the field of background seismic activity, ten maps of the time slices of the field of current seismic activity, and ten maps of the time slices of the field of changes in the current seismicity compared to the background seismicity. The last field is the field of seismic activity anomalies, which is computed using change point detection method. During the analysis, the user can depict on the map the layer of epicenters of earthquakes that occurred during the time interval of the current seismicity. The interface allows the user to read the time and magnitude of each earthquake. Now we are going to give an example of detecting an earthquake precursor using GIS SeismoMap. The earthquake with magnitude M=6.1 occurred on the 24th of September 2014 in California. Figure 1 shows the sequence of several time slices of 3D anomaly field, which follow at intervals of one week. One can see that a month and a half before the earthquake an anomaly begins to form. Blue tone of the anomalous area denotes that seismic activity is lower compared to the background one. This indicates the accumulation of power. The following slices demonstrate that the anomalous area decreases and finally converges to the place where later the epicenter of the earthquake appears, which is depicted on the slice corresponding to the date 26.08.2014 as the largest circle.

After reviewing the fields of seismic activity, the user can perform a detailed analysis of the seismic process using GIS GeoTime 3. GeoTime 3 runs with the same seismological and geographic data, which were used to calculate seismic fields in GIS SeismoMap. The tools of GeoTime 3 allow the user to supplement the GIS project with 2D, 3D, 4D grid and vector data from remote or local servers. The results of analysis include the parameters and the fields of earthquake precursors as well as the maps of seismic hazard prediction.

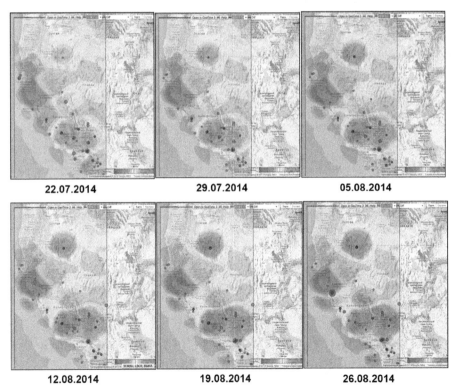

| 22.07.2014 | 29.07.2014 | 05.08.2014 |
| 12.08.2014 | 19.08.2014 | 26.08.2014 |

Fig. 1. Time slices of the 3D field of anomalies and epicenters of the earthquakes that took place in the period within two months before the date of the slice. On the slice corresponding to 26.08.2014 in the anomalous area (blue color) there is a large circle that indicates the epicenter of the Californian earthquake with magnitude M=6.1 which occurred on 24.09.2014.

We now consider the example of the application of GIS GeoTime 3 for retrospective detection of anomaly before the earthquake with the magnitude m=6.8 that took place in Japan on 11.07.2014. The forecast is based on the data from GIS SeismoMap. The earthquake catalog consists of 2975 events with magnitudes varying from 4 to 7.4 that occurred in the period from 05.01.2012 to 11.07.2014.

Figure 2 depicts 6 time slices of two fields: the field of epicenters of the earthquakes and the field of the anomalies. An anomaly appeared 5 months before the earthquake of 11.07.2014. The earthquake epicenters on the map correspond to the moving interval of 30 days. The right boundary of this interval coincides with the time slice of the anomaly field. Yellow circle on the last map is the epicenter of the earthquake under consideration.

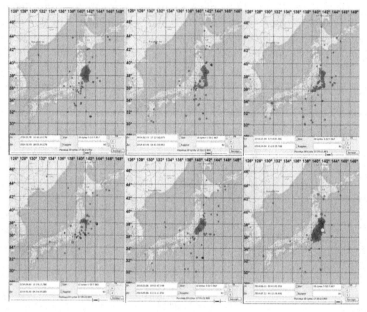

Fig. 2. Time slices of the fields of the earthquake epicenters and anomalies. The dates corresponding to the slices from left to right and from the upper row to the lower row: (1) the anomaly of 05.02.2014, epicenters from 05.01.2014 to 05.02.2014; (2) the anomaly of 06.03.2014, epicenters from 03.02.2014 to 06.03.2014; (3) the anomaly of 04.04.2014, epicenters from 04.03.2014 to 04.04.2014; (4) the anomaly of 05.05.2014, epicenters from 04.04.2014-05.05.2014; (5) the anomaly of 06.06.2014, epicenters from 06.05.2014 to 06.06.2014; (6) the anomaly of 12.07.2014, epicenters from 11.06.2014 to 12.07.2014. The yellow circle indicates the epicenter of the forecasted earthquake.

4   Data Analysis Using Web-Based GIS Incorporated into a Large-Scale Environmental Monitoring System

4.1   GIS GeoESIMO

Web GIS GeoESIMO is incorporated into the Unified State System of Information on the World Ocean (ESIMO) for the analysis of environmental monitoring data. ESIMO is supported by about 200 databases, http://portal.esimo.ru/portal/. ESIMO's distributed databases contain over 2300 diverse information resources from more than 80 data providers, of more than 5 TB total size. About 30% of the data is updated with periodicity varying from several minutes to a day. Analytical resources include more than 100 computer systems that are located on servers in different organizations. The technological basis of the ESIMO is service-oriented infrastructure and international standards, ensuring the functioning of the elements and communications of the ESIMO.

GIS GeoESIMO is developed on the base of GIS GeoTime 3. In order to integrate GIS into ESIMO, a number of additional functions has been brought into development, such as: authorized access to distributed information and computational resources, the preservation of the results of the analysis in the monitoring system for re-use or for transmission to another user, automatic registration of interaction between the GIS components (fault messages, tracking the progress of external tasks, logging GIS operations), the connection of GIS applications to the portal of monitoring systems, creating and editing by the user a problem-oriented GIS-project, etc.

GeoESIMO is connected with the ESIMO portal as a custom portlet. Access to GIS applications through http://portal.esimo.ru/portal/portal/esimo-user/services/geoesimo is regulated. Open access is available at the demo website http://www.geo.iitp.ru/esimo.

Figure 3 shows the GeoESIMO starting window with GIS-application to the region of the Barents Sea. The map shows the digital elevation model and Russian ships. The data was loaded on 04.12.2014 from the server of the enterprise "Morsvyazsputnik". For illustrative purposes the data is filtered so that the vessels with the speed greater than 1 knot are presented. The rectangle shows the attributes of the vessel "Norilsk Nickel": the speed equals 12.2 knots, the course is 264, the date and time Thu Dec 04 12:16.00 GST 2014. To the left of the map, a window of the map layers is depicted, where one can choose to load information resources from web servers and ESIMO: geodata in different formats, tile maps and maps in the WMS format. Under the layer window a fragment of the visualization attributes control window is shown. Above the layer window there is a panel of a hierarchical menu that contains the following groups: File operations, Transform operations and modeling, operations of management of the Calculation and Simulation Systems (CSS) that are performed on remote servers, and operations of selecting the Way of visualization of 3D and 4D data. Above the map window there is a panel that contains the Properties tab, which controls the selection of a map projection and attributes of the grid, and icons that start the operations of visual analysis.

Figure 4 shows a map of the ice conditions (data from the server of Arctic and Antarctic Research Institute). In the rectangular box one can see the attributes of one of the polygons, which is highlighted in blue.

Figure 5 shows a map of the time slice of the spatio-temporal field of the wind speed forecast at 12:00 4.12.2014. On the right, the plots of the time series of forecast for 4.12.2014 - 6.12.2014 at the point $32.56°$ E and $63.34°$ N are depicted. The reference point is marked by an arrow. The 3D forecast fields of air temperature, land surface temperature, precipitation and wind speed are loaded from the server of Hydrometeorological Center of Russia.

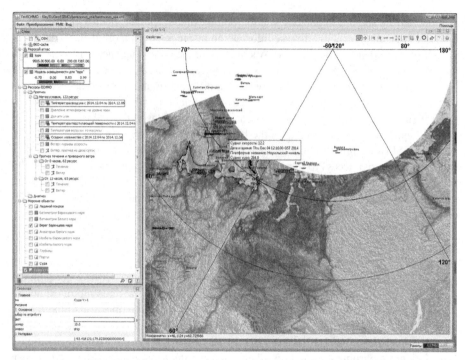

Fig. 3. Starting window of GIS GeoESIMO. The map window shows the digital elevation model and Russian ships with the speed greater than 1 knot. On the left: the window of geographic layers of GIS-project and the window that controls the visualization attributes.

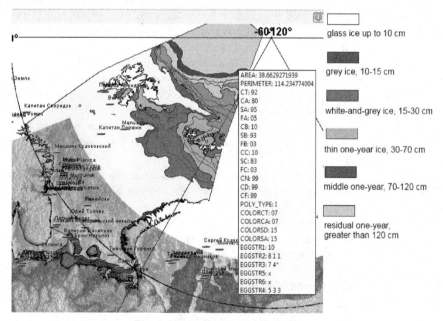

Fig. 4. The map of the ice conditions

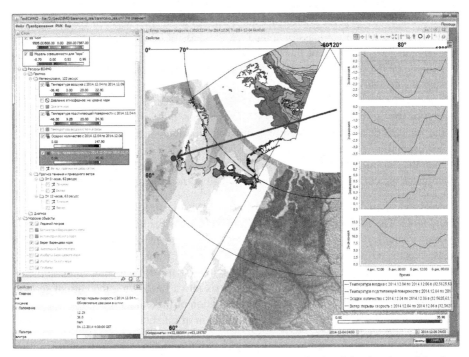

Fig. 5. A map of the wind speed forecast at 12:00 4.12.2014. On the right, the plots of the time series of meteorological parameters forecast for 4.12.2014 - 6.12.2014 at the point that is marked by an arrow.

4.2   An Example of the Environmental Monitoring

As an example of application of GIS GeoESIMO we describe the scenario of an operator work in an emergency oil spill. User tasks are as follows: (1) to estimate the volume, area and depth of contamination, to provide a forecast for the spread of the oil slick and meteorological forecast; (2) to decide on the possibility of involvement of the emergency aid in the disaster area and about the appropriate equipment delivery.

To predict the consequences of an oil spill, the server application "Express-analysis of emergency oil and oil products spill" is used. The application was developed in the State Oceanographic Institute (SOI).

The modeling process of a spill consists of three main stages.

1. *Assignment and transfer of simulation parameters from GeoESIMO to the server of SOI.* In the dialog box of GeoESIMO, the following parameters of the simulation are determined: the duration of the forecast, the iteration step in minutes, the type of the petroleum product, the date and time of the accident, the coordinates of the spill, the intensity and duration of discharge of a petroleum product. Further, the simulation parameters are sent from GIS application through web service to the server of SOI.
2. *Computer simulation.* The simulation procedure uses parameters from GIS application, as well as meteorological forecast and other data on the state of the environment. For

each simulation step, the coordinates and form of the oil slick for each gradation of oil film thickness, the total mass of spilled oil, the mass of oil on the surface, the mass of evaporated oil, the mass of the oil dispergated into the water, and the mass of oil remaining on the shore are calculated. Web service reports on the implementation of each step of calculations.

3. *Load and aggregation of the results in GIS.* Upon receipt of the notice of completion of the current iteration, GeoESIMO performs the following actions: (a) loads the forecast data (namely, the polygons for 3 possible values of oil film thickness) corresponding to the completed step of iteration, (b) displays on the interactive map the 3D polygon model of a slick corresponding to the completed step, (c) aggregates the current result with those obtained in the previous steps.

Analysis of oil spill consequences in GeoESIMO is fulfilled basically by cartography tools. The following functions of GeoESIMO are available: animation and step-by-step visualization, reading the attributes, measurement of values, distances and areas, making cross sections, designing of the composition of map layers, creating plots and tables, mapping labels and icons, saving a map with a legend and comments in a Word file, writing the history of the analysis into a file, saving the current session in the form of a GIS project for future reconstruction of the situation, etc.

Figure 6 shows several time slices of simulation of oil slick dynamics. The point of the spill is indicated by "+".

Figure 7 shows several elements of the express analysis of the situation with the oil spill. The point of the spill is indicated by a red circle near the town of Polarny. Purple color represents the oil slick 48 hours after the spill. The figure shows that the oil-spot consists of three polygons. Each polygon delineates the area with a certain thickness of the oil film. The spot area is equal to 27 sq. km. The red line indicates the line segment between the point of the spill and the center of the spot. On the left and on the right are the forecast plots of meteorological parameters in the port of Severomorsk and in the center of the spot of oil, respectively. The plots are constructed by means of GeoESIMO on the query of the server application "Meteograms" of Hydrometeorological Center.

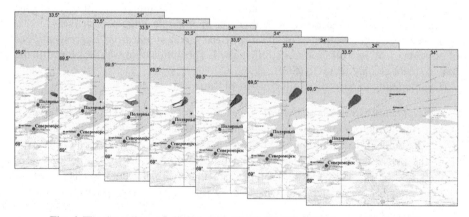

Fig. 6. The time slices of oil slick 2, 6, 12, 18, 24, 30 and 39 hours after the spill

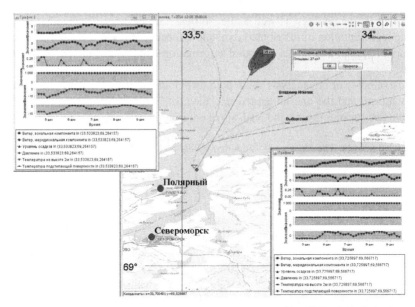

Fig. 7. Elements of express analysis of the situation with the oil spill. Red circle indicates the point of the spill; purple polygons demonstrate the forecast position and form of slick for three grades of oil film thickness after 48 hours; on the left and on the right are the forecast plots of meteorological parameters in the port of Severomorsk and in the center of the oil slick, respectively.

5   Conclusion

The key problems of environmental monitoring and assessment are associated with the study of spatial and spatio-temporal data. To process and analyze this sort of data, it is advisable to use web-based geoinformation technologies and systems.

Two technologies of web GIS applications for monitoring and analysis of environmental processes are illustrated with two case studies: 1) a GIS platform, consisting of GIS's SeismoMap and GeoTime 3, and 2) GIS GeoESIMO, integrated into the Unified State System of Information on the World (ESIMO).

A web-based GIS platform is a new technology intended for the analysis of monitoring data on spatio-temporal processes. The platform systematically sends queries to remote servers which receive in real time the results of measurement of various characteristics of the environment and information on the occurring events. Then the platform converts these data into spatio-temporal fields that describe the state of the environment, evaluate the statistical significance of changes of its properties in space and time, and generates a GIS project to run the GIS that supports comprehensive analysis of the data. Thus, the platform provides information resources and tools for two levels of the analysis: visual express analysis available to a wide range of users, and sophisticated research implemented by specialists.

GIS GeoESIMO is an example of technology of integration of GIS into a large-scale information system of environmental monitoring. GIS GeoESIMO exploitation shows that the technology as well as the software are effective for interactive carto-graphic representation of distributed dynamic information resources and for advanced analysis of spatial and spatio-temporal information about the situation in the World Ocean.

The examples of the analysis show both techniques, the one implemented as a web GIS platform, and the one integrated into the GIS GeoESIMO, can be used as basic technologies for monitoring and analysis of environmental processes.

Acknowledgments. The research is supported by the Russian Science Foundation grant (pro-ject 14-50-00150).

References

1. Hamre, T., Krasemann, H., Groom, S., Dunne, D., Breitbach, G., Hackett, B., Sandven, S.: Interoperable web GIS services for marine pollution monitoring and forecasting. Journal of Coastal Conservation **13**(1), 1–13 (2009)
2. Kulawiak, M., Prospathopoulos, A., Perivoliotis, L., Kioroglou, S., Stepnowski, A.: Interac-tive visualization of marine pollution monitoring and forecasting data via a Web-based GIS. Computers & Geosciences **36**(8), 1069–1080 (2010)
3. Tsou, M.H.: Integrating Web-based GIS and image processing tools for environmental mon-itoring and natural resource management. Journal of Geographical Systems **6**(2), 155–174 (2004)
4. Gitis, V.G., Osher, B.V., Pirogov, S.A., Ponomarev, A.V., Sobolev, G.A., Jurkov, E.F.: A System for Analysis of Geological Catastrophe Precursors. Journal of Earthquake Prediction Research **3**(4), 540–555 (1994)
5. Gitis, V., Derendyaev, A., Metrikov, P., Shogin, A.: Network geoinformation technology for seismic hazard research. Natural Hazards **62**(3), 1021–1036 (2012)
6. Metrikov, P., Derendyaev, A., Gitis, V.: Web-GIS technology for dynamic data analysis. In: ACM SIGSPATIAL GIS 2011 Conference, Chicago, pp. 31–38 (2011)
7. Goldenshluger, A., Spokoiny, V.: Recovering convex edges of an image from noisy tomo-graphic data. IEEE Transaction on Information Theory **52**(4), 1322–1334 (2006)

Efficient Combined Text and Spatial Search

Amber Han and Bradford G. Nickerson[(✉)]

Faculty of Computer Science, University of New Brunswick,
Fredericton, N.B. E3B 5A3, Canada
{amber.han,bgn}@unb.ca
http://cs.unb.ca/~bgn

Abstract. We present a search engine called `TexSpaSearch` that can
search text documents with associated locations in space. We defined
three search queries denoted as Q1(t), Q2(t,r) and Q3(p,r) for finding
documents containing text t intersecting a disc centered at position p
with radius r. Testing was performed using the UNB Connell Memorial
Herbarium database whose records normally contain the location where
plant specimens were collected along with associated textual data. The
sample herbarium database of size $N = 40,791$ records with associated
locations was indexed using a novel R*-tree and suffix tree data structure
to achieve efficient search for the defined queries. Significant preprocessing was performed to transform the database into the index data structure used by `TexSpaSearch`. Testing was performed with 20 example Q1
text only queries to compare `TexSpaSearch` to a Google Search Appliance, as well as a significant number of example Q2 and Q3 queries.
`TexSpaSearch` search results are ranked by a modified Lucene scoring
algorithm, and combined with a spatial rank for Q2 search. A theoreti-
cal analysis shows that `TexSpaSearch` requires $O(A^2\overline{|b|})$ average time for
Q1 search, where A is the number of single words in the query string
t, and $\overline{|b|}$ is the average length of a subphrase in t. Q2 and Q3 queries
require $O(A^2\overline{|b|} + Z\log_{\mathcal{M}} \mathcal{D}_N + y)$ and $O(\log_{\mathcal{M}} \mathcal{D}_N + y)$ time, respec-
tively, where Z is the number of point records in the list $\mathcal{P}$ of text search
results, $\mathcal{D}_N$ is the number of data objects indexed in the R*-tree for N
records, $\mathcal{M}$ is the maximum number of entries of an interior node in the
R*-tree, and y is the number of R*-tree leaf nodes found in range in a
Q3 query.

1 Introduction

Traditional search engines treat place names in search strings in the same way as
any other keyword. Extensive research has investigated how to efficiently answer
text queries when combined with spatial location (see e.g. [18][19][9][20] [13][8]).
Chen et al [7] present an extensive evaluation of 12 different so-called geo-textual
indices that support three types of spatial keyword search. An example text plus
spatial query might be to find all the restaurants falling with 1 km of our current
location.

© Springer International Publishing Switzerland 2015
O. Gervasi et al. (Eds.): ICCSA 2015, Part III, LNCS 9157, pp. 713–728, 2015.
DOI: 10.1007/978-3-319-21470-2_52

In this paper, we present an efficient geo-textual index and search methodology that simultaneously supports text only, text with search radius, and point location with radius queries. Like other geo-textual indices, our `TexSpaSearch` system indexes points with associated text. In addition, `TexSpaSearch` supports simultaneous efficient indexing of text having simple polygons describing their location.

To support text with spatial search, we need to add features of spatial queries to the search engine, including concepts of [22]

1. Representation of spatial data in the index,
2. Filtering by some spatial concept such as a bounding box or other shape,
3. Sorting, scoring and boosting by distance from a query point or query region.

The suffix tree is a special kind of trie, which can be used to index all suffixes in a text to carry out fast full text searches [11][10]. The first linear time algorithm for constucting suffix trees was presented by Weiner [24] in 1973, although at that time a suffix tree was called a position tree. A few years later, a more space efficient algorithm to build suffix trees in linear time was given by McCreight [21]. More recently, a conceptually different linear-time algorithm was developed by Ukkonen [23], which has all the advantages of McCreight's algorithm, but allows a much simpler explanation [14]. These classical algorithms [21, 23, 24] construct a suffix tree for a string σ of length n in $O(n \log |\sum|)$ time and $O(n)$ space, where $\sum$ is the alphabet. Given a suffix tree for σ and a pattern α of length m, an algorithm to determine whether the pattern appears in the string can be implemented to run in $O(m \log |\sum|)$ time.

In order to handle multi-dimensional point data efficiently, a number of structures have been proposed. The R-tree [15] is a height-balanced tree derived from the B-tree, and provides efficient, linear space indexing of multidimensional objects by storing their bounding rectangles. An improved version of the R-tree, the R*-tree was introduced by Beckmann et al [6]. The motivation of the R*-tree was to reduce search cost by minimizing the overlap region between sibling nodes. Arge et al [5] introduced the priority R-tree, the first linear space R-tree variant to answer rectangular range search on spatial objects in worst case optimal $O((N/B)^{1-1/d} + F/B)$ I/Os, where B is the number of objects fitting in one disk block, and F represents the number of objects in range.

2   Definitions and Objectives

We assume that text documents have associated locations in space, and we wish to search such documents with queries containing spatial components. For example, we might have a set of populated place names (e.g. cities), with associated locations (latitude, longitude) on the earth's surface. These place names can be part of larger documents or text based web pages. For the remainder of this paper, we use the word "document" to refer to an item (e.g. web page, document, database, record) indexed by the search engine.

An example query might be to find all populated places within 50 km of a specific populated place, or of a given point specified by its latitude and longitude (ϕ, λ). In this case, a query Q might be Q = ("Fredericton", 50km) or Q = ((45.95,-66.633333), 50km). Other example queries might be to find restaurants within 10 km of your current position or of a known restaurant, then we have Q = ("Golden Triangle", 10km) where "Golden Triangle" is the name of a known restaurant. In any case, the search returns a ranked list of populated places or restaurants nearby. We represent search strings (e.g. city or restaurant names) by t, spatial position (ϕ, λ) by p and radius by r to define three query forms, as follows:

1. $Q1(t)$, search returns a ranked list of items matching the search string t, along with their associated spatial information (e.g. latitude, longitude).
2. $Q2(t, r)$, search returns a ranked list of documents with at least one spatial component having its location falling within the circle of radius r centred at the position p of the ranked documents.
3. $Q3(p, r)$, search returns a ranked list of documents with at least one spatial component having its location falling within the circle of radius r centred at position p.

To our knowledge, ours is the first paper to define the $Q2(t, r)$ combined text and spatial query, and to provide a data structure for answering such queries.

As an example, assume there are six records consisting of plain text and associated spatial information in a database as shown in Figure 1. Figure 2 shows examples of the three query types and query answers using the database shown in Figure 1.

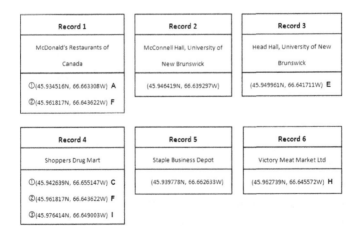

Fig. 1. An example database with records consisting of plain text and associated spatial information

Query 1: Q = ("Mc")
Search returns:
(1). McDonald's Restaurants of Canada, (45.934516N, 66.663308W).
(2). McDonald's Restaurants of Canada, (45.961817N, 66.643622W).
(3). McConnell Hall, University of New Brunswick, (45.946419N, 66.639297W).
Query 2: Q = ("McDonald", 1.5km)
Search returns:
(1). McDonald's Restaurants of Canada, (45.934516N, 66.663308W).
(2). McDonald's Restaurants of Canada, (45.961817N, 66.643622W).
(3). Shoppers Drug Mart, (45.961817N, 66.643622W), 0km from (2).
(4). Victory Meat Market Ltd, (45.962739N, 66.645572W), 0.185km from (2).
(5). Staple Business Depot, (45.939778N, 66.662633W), 0.592km from (1).
(6). Shoppers Drug Mart, (45.942639N, 66.655147W), 1.098km from (1).
Query 3: Q = (position P = (45.952567N, 66.646001W), 1.5km)
Search returns:
(1). Head Hall, University of New Brunswick, (45.949961N, 66.641711W), 0.441km from P.
(2). McConnell Hall, University of New Brunswick, (45.946419N, 66.639297W), 0.862km from P.
(3). McDonald's Restaurants of Canada, (45.961817N, 66.643622W), 1.043 km from P.
(4). Shoppers Drug Mart, (45.961817N, 66.643622W), 1.043 km from P.
(5). Victory Meat Market Ltd, (45.962739N, 66.645572W), 1.122km from P.
(6). Shoppers Drug Mart, (45.942639N, 66.655147W), 1.315km from P.

Fig. 2. Examples of answers to the three query types using the database shown in Figure 1

3 Sample Database

UNB houses the Connell Memorial Herbarium database [2], each record of which contains text describing plant sample species and spatial information (e.g. latitude and longitude) describing where the samples were collected. The electronic version of the UNB Connell Memorial Herbarium database was provided to us by the UNB Biology department as a comma separated file exported from a database. The database contains 40,791 records in total, each of which consists of 22 fields including e.g. UNB accession number, full name, latitude, longitude, and collector(s) name(s). The fields that can provide spatial information are: latitude, longitude, location, County, Prov/State. Fields are separated by commas and each field is surrounded by quotation marks. The records are uniquely recognized by their UNB accession numbers. Each database record was transformed to a web page with appropriate metadata and content, and then placed on a web server running on the UNB network. For the records having latitude and longitude (ϕ, λ), we use their (ϕ, λ) values to index them; otherwise, we use their county boundaries to represent their spatial location. Further details on data preprocessing are in [17].

4 Indexing

A suffix tree and a packed R*-tree were used to realize the geo-textual index, as shown in Figure 3. The full text of the data in the complete collection was

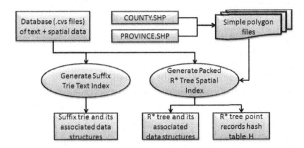

Fig. 3. The geo-textual indexing scheme using an R*-tree and a suffix trie

indexed, and we generated a suffix tree and its associated data structures. To generate an R*-tree, if the point data (ϕ, λ) exists, we index the spatial data with this point. Otherwise, we index by the polygon or the county where the object is located. While building the packed R*-tree, an R*-tree point records hash table H is generated at the same time. H is a hash table from a record's primary keys to their corresponding (ϕ, λ). The key value only exists when the record has an associated point data (ϕ, λ) pair. H is used in a Q2 query to check if a record has an associated point (ϕ, λ).

4.1   Text Indexing

We use a suffix tree [4] for indexing English language text from a database. In the text index, we first skip all the stop words such as "the", "a", and "and". A complete list of stop words is given in [3]. In this example, we then get the compact phrase "cat in hat". To speed up text search, we index all the possible sub-phrases of this compact phrase. In this example, we have 6 sub-phrases in total; they are: "cat", "cat in", "cat in hat", "in", "in hat" and "hat". There are always $\frac{A(A+1)}{2}$ sub-phrases with $A = $ no. of single words in the compact phrase.

4.2   Ranking Text Search Results

We use the Lucene ranking algorithm [16] to compute the scores of text search results. Apache Lucene [12] is an open source information retrieval (IR) software library, originally created by Doug Cutting [12]. Lucene provides a scoring algorithm to find the best matches to document queries, which ranks documents resulting from a search query based on their content. The default scoring algorithm considers such factors as the frequency of a particular query term within individual documents, and the frequency of the term in the total population of documents. The Lucene scoring algorithm considers the rarity of a matched term within the global space of all terms for a given field. In other words, if you match a term that is not very common in the data then this match is given a higher score [1].

We modified the Lucene ranking algorithm for our system. The original version can be found in [16]. To illustrate Lucene's scoring algorithm, we have the following definitions:

1. m: number of sub-phrases in the query string t.
2. m_{td}: the total number of sub-phrases of query string t found in a document d.
3. b: a possible sub-phrase in the query string t.
4. D: set of records in the index, and $n = |D|$.
5. D_b: number of documents containing the specific sub-phrase b.
6. I_b: inverse document frequency, which is computed as: $I_b = 1 + \log(\frac{n}{D_b+1})$.
7. F: term frequency, which is the number of times a sub-phrase b appears in a document d.
8. n_w: the number of words in a sub-phrase b.

Let R_{td} stand for the ranking score of query t for a document d. R_{td} is computed as follows:

$$R_{td} = \frac{m_{td}}{m} \frac{1}{\sum_{\forall b \in t}(I_b)^2} \sum_{\forall b \in t}(\sqrt{n_w}F I_b^2) \tag{1}$$

The item $\frac{m_{td}}{m}$ is a score factor based on how many of the sub-phrases in t are found in the specific document d. Typically, a document containing more query's sub-phrases will receive a higher score than another document with fewer sub-phrases. Documents that have more occurrences of a given sub-phrase b receive a higher score is guaranteed by the factor F. The item $\frac{1}{\sum_{\forall b \in t}(I_b)^2}$ is a normalizing factor used to make scores between different queries comparable. This factor does not affect document ranking (since all ranked documents are multiplied by the same factor). As $I_b \geq 1.5$, the second I_b^2 term indicates that rarer sub-phrases give a higher contribution to the total ranking score . The item $\sqrt{n_w}$ ensures that the longer sub-phrases contribute more to the score.

4.3   Spatial Index Using a Packed R*-tree

After adding locations to the records as explained in Section 4, we have two different types of records: records with (ϕ, λ) point locations describing them, and records with polygons describing their locations. For the records having point locations, we first pack B points in the smallest bounding box that encloses them, where B is the maximum number of data points contained in one leaf node. We then insert the generated bounding boxes containing B points as leaf nodes into an R*-tree. For the records having polygon descriptions (e.g. York County), we directly get the bounding box of the polygon, and insert it together with the simplified polygon data containing w_B or fewer points as a leaf node of the same R*-tree.

We use the data structure RStarDataObject to represent the data object stored on the R* leaf nodes. For each RStarDataObject, the nearest B points are chosen as shown in Algorithm 1.

Algorithm 1. PackPoints(S, B, S_p)

Input: The list of all the points S, The maximum number of data points contained in one leaf node B ;

Output: A list of `RStarDataObject` objects S_p;

1 S_p is an empty list of `RStarDataObject` ;
2 **if** S *is not* **then**
3   `RStarDataObject` r ;
4   Sort the points in S by their latitude ;
5      $temp \leftarrow S_0$; Pack $temp$ to r ;
6   Remove S_0 from S ;
7   Sort S by their distances to $temp$;
8   **if** $|S| \geq B - 1$ **then**
9   Pack the first $B - 1$ points in S to r ;
10   remove the first $B - 1$ points from S ;

11   **else**
12   Pack all points in S to r ;
13   Delete S ;

14     $S_p \leftarrow S_p + r$;
15 **return** S_p ;

A $Q3(p, r)$ query returns two lists; L_1 containing point results (records having (ϕ, λ) locations), and list L_2 containing polygon results (records with no (ϕ, λ) pairs but with county names), respectively. Both lists L_1 and L_2 are sorted based on the attribute d_t. For a point result $\ell \in L_1$, d_t is the distance from point p to ℓ's (ϕ, λ) location. For polygon result $\ell \in L_2$, d_t is the distance from p to the nearest edge of polygon ℓ's boundary.

4.4 Combined Text and Spatial Query Q2

As explained in Section 2, a Q2 query $Q2(t, r)$ returns the ranked list of records having their locations intersecting a circular disk of radius r centred at the locations of the records matching search string t. To perform a $Q2(t, r)$ query, a text query $Q1(t)$ is performed first, which returns a set $\mathcal{P}$ of search results. We then perform a set of $Q3(\mathcal{P}_i, r)$ queries for points $\mathcal{P}_i \in \mathcal{P}$, which will return all points and polygons intersecting a disc of radius r centered at $\mathcal{P}_i$.

Combined Score for Text with Spatial Search. For the text ranking score R_{td} from a text result in $\mathcal{P}$ and distance d_t obtained from the $Q3(\mathcal{P}_i, r)$ search, the combined score `finalScore` of `TexSpaSearch` is computed as follows:

$$f = W_t \frac{R_{td}}{S_m} + W_s \cos(\frac{\pi}{2} \frac{d_t}{S_m}) \tag{2}$$

where W_t is the weight for the text search and W_s is the weight for the spatial search, with the further restriction that $W_t \in [0, 1]$, $W_s \in [0, 1]$ and $W_t + W_s = 1$.

By default, W_t and W_s are both set to 0.5. S_m is the maximum text ranking score returned by the Q1(t) search. Since a Q1(t) search returns a ranked list $\mathcal{P}$ of text search results, we can obtain S_m as $S_m = \mathcal{P}_0.Rtd$ ($\mathcal{P}_0.Rtd$ stands for the text ranking score of $\mathcal{P}_0$), which means the score of the first result in the ranked list. For the text search score, we have $\frac{R_{td}}{S_m} \in [0, 1]$. For the spatial search score, since only points and polygons having $d_t \leq r$ are returned, we have $\cos(\frac{\pi}{2}\frac{d_t}{r}) \in [0, 1]$, where $d_t \in [0, r]$ is inversely proportional to the spatial score.

Performing a Q2 Query. As mentioned above, a Q1 search using Q(t) is first performed to return a ranked list $\mathcal{P}$ of text search results. If the i_{th} record $\mathcal{P}_i$ is a point record (i.e. not a polygon), then we perform a search for other point or polygon records intersecting the disk of radius r centred at $\mathcal{P}_i$. Assume R_1 and R_2 are initially empty lists for `TexSpaSearch` point results (records containing (ϕ, λ) pairs) and `TexSpaSearch` polygon results (records with no (ϕ, λ) pairs but with county names) initially. For a text search result $\mathcal{P}_i$, we first obtain the text ranking score R_{td} and corresponding primary key $textPrk$. We then check if $textPrk$ can be found in the `point record hash table` H. If the record with key $textPrk$ is a point record (can be found in H), the corresponding (ϕ, λ) pair p associated with the $textPRK$ is retrieved. List L_s stores the points that have already been used as inputs to the `Q3Querys` in the entire Q2 query process. If p is not in L_s we perform a `Q3InternalSearch(p,r)`, getting the lists L_1 for Q3 point results and L_2 for Q3 polygon results. We then compute the combined scores, generate the corresponding Q2 point results r_1 and Q2 polygon results r_2 for all the results in L_1 and L_2. The newly generated Q2 point and Q2 polygon results are added to the lists R_1 and R_2, respectively. Finally, the lists R_1 and R_2 are sorted on their combined text + spatial scores `finalScore`. Complete details of the Q2 search algorithm are given in [17].

4.5   Search Complexity

This section analyses the asymptotic search complexity of queries using the `TexSpaSearch` data structure. For a Q1 query Q1(t), where t is the query string of length τ, we perform a suffix tree look up for each possible sub-phrase b of t. We denote the length of a sub-phrase b by $|b|$, so the algorithm searches for b in $O(|b|)$ time using the suffix tree as described in Section 5.1. For the query string t containing A single words, the time complexity $\mathcal{C}1$ is computed as follows:

$$\mathcal{C}1 = \sum_{i=1}^{\frac{A(A+1)}{2}} O(|b_i|) \tag{3}$$

Since the longest sub-phrase of t has length τ, the upper bound of $\mathcal{C}1$ is $O(\frac{A(A+1)}{2}\tau)$. Assuming the average length of sub-phrases is $\overline{|b|}$, the average time complexity for a Q1 query $\overline{\mathcal{C}1}$ can be computed as:

$$\overline{\mathcal{C}1} = O(\frac{A(A+1)}{2}\overline{|b|}) = O(A^2\overline{|b|}) \tag{4}$$

For a Q3(p, r) query, assume there are $\mathcal{D}_N$ data objects indexed in the R*-tree for N records, and the maximum number of entries of an interior node is $\mathcal{M}$, the query complexity $\mathcal{C}3$ for the Q3 search is $\mathcal{C}3 = O(\log_{\mathcal{M}} \mathcal{D}_N + y)$, where y is the number of R*-tree leaf nodes found in range. In the worst case when all the bounding boxes in the R*-tree overlap the bounding square defined by p and r, the worst case query complexity is $O(\mathcal{D}_N)$.

For a Q2(t,r) query, the search engine first performs a Q1(t) query returning a list $\mathcal{P}$ of text search results, which takes $\mathcal{C}1 = O(A^2|\overline{b}|)$ time. We then perform a Q3(p, r) query for each point record in $\mathcal{P}$. Assuming there are Z point records in $\mathcal{P}$, the time complexity $\mathcal{C}2$ for Q2 query can be computed as:

$$\mathcal{C}2 = O((\sum_{i=1}^{\frac{A(A+1)}{2}} |b_i|) + Z \log_{\mathcal{M}} \mathcal{D}_N + y)$$

$$= O(A^2|\overline{b}| + Z \log_{\mathcal{M}} \mathcal{D}_N + y)$$

(5)

on average. As we can see, the Q2 query time heavily depends on the number of point results returned by the text search list $\mathcal{P}$.

5   Web Server Architecture

We wrote a Java and C++ server called `TexSpaSearch` that provides a web user interface for our search engine. The web server we used is Apache Tomcat. `HTMLHandlerServlet` is the server side Java servlet that handles HTML requests. The server architecture is shown in Figure 4. A total of 8,571 lines of C++ and Java source code was written to implement the `TexSpaSearch`.

We use JSP (Java Server Pages) to implement the client page code. JSP allows Java code to be interleaved with static web markup content, so we can define a `ResultClass` to store the search results. It is very helpful to be able to associate some data with each client in a web server. For this purpose, a session can be used in JSP. A session is an object associated with a client. Data can be put in the session and retrieved from it, and operates like a hash table. In our servlet, an object of `ResultClass` called `finalResult` is used to store the search result for each session. The session can be obtained from the HTTP request. Each client has a session with our server, so each client has their own `finalResult` object. In this way the server can process queries for each client separately without causing critical section issues. On the server side, the servlet stores the search results in the session object. The client page gets the result object of `ResultClass` from the session in the header of the `index.jsp` file.

6   Test Results

In the R*-tree construction, their are 28,435 records having (ϕ, λ) pairs, 11,909 records having county names but no (ϕ, λ) pairs, and 447 records having no related spatial information ((ϕ, λ) pair or county name). The total number of

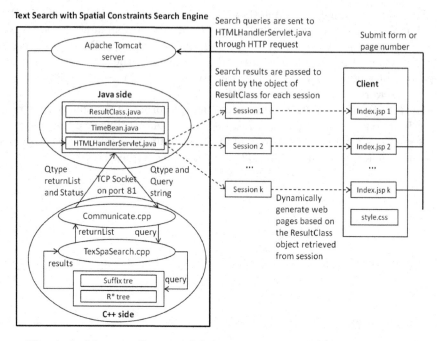

Fig. 4. Architecture diagram of the `TexSpaSearch` web application server

records in the Herbarium database we used (circa 2012) is 40,791, in which 40,344 records are indexed by the R*-tree. We chose a maximum number of entries $\mathcal{M} = 8$ for an interior node of the R*-tree. The time for constructing the R*-tree and suffix tree was 249.213 and 142.157 seconds, respectively. There are several key time epochs in the entire query process, as shown in Figure 5.

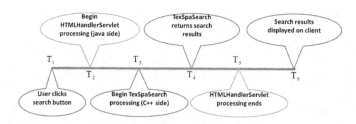

Fig. 5. Time epochs in the `TexSpaSearch` query process

The following parameters are used in timing our experiments:

1. r: radius for Q2 and Q3 query in meters.
2. N_r: is the total number of search results returned by a Q1 query.
3. N_{pt}: the total number of (ϕ, λ) pairs returned for Q2 and Q3 point results.

4. R_{pt}: the total number of records returned for Q2 and Q3 point results.
5. N_{pl}: the total number of counties returned for Q2 and Q3 polygon results.
6. R_{pl}: the total number of records returned for Q2 and Q3 polygon results.
7. $T_c = T_4 - T_3$, is the C++ side processing time.
8. $T_s = T_5 - T_2$, is the server side (including C++ and Java) processing time.
9. $T_e = T_6 - T_1$, is the total query time from the user clicking the search button to the search results displayed on the web page.

The `TexSpaSearch` testing environment had the web server and web browser running on the same workstation in the UNB ITB214 Communication and Networking Laboratory.

6.1   Q1 Test Results

We choose 20 sample query strings for Q1 tests as shown in Table 1.

Table 1. Time comparison of the `TexSpaSearch` engine and the Google Search Appliance (GSA) on the 20 sample queries. All times are shown in ms.

Query string	TexSpaSearch			GSA		T_e/T_g	T_c/T_g
	N_r	T_e	T_c	N_g	T_g		
apple	60	524	27	59	20	26.2	1.35
crab	10	516	3	10	20	25.8	0.15
crab apple	69	10	520	1	10	52	1
Red spruce	2706	1424	915	290	30	47.7	30.5
Hieracium pilosella	361	646	122	40	10	64.6	12.2
Rosa virginiana	661	725	213	72	20	36.25	10.65
seaside arrow grass	3759	1854	1361	47	20	92.7	68.05
Red pitcher plant	2149	1263	744	0	20	63.15	37.2
pitcher plant	413	659	136	3	30	22.0	4.53
Eupatorium perfoliatum	142	547	38	32	20	27.4	1.9
Bromus Inermis Leyss	151	555	44	28	20	27.8	2.2
Amelanchier laevis wieg	1244	955	449	188	20	47.8	22.45
Vesce des haies	1959	1214	698	30	10	121.4	69.8
Poison ivy	70	525	14	40	20	26.3	0.7
Lilac	15	519	13	15	20	26.0	0.65
Lady slipper	267	575	60	119	20	28.75	3
lady's slipper	200	524	16	119	20	26.2	0.8
Herbe aux ?crevisses	1054	986	481	54	20	49.3	24.05
Verge d'or des bois	2654	1450	943	34	20	72.5	47.15
vanilla	43	535	14	43	20	26.8	0.7
Average						45.5	16.95

The line graph in Figure 6 shows the changes of T_s and T_e measured in ms vs. N_r. As we can see, T_s and T_e rise proportional to the number of results

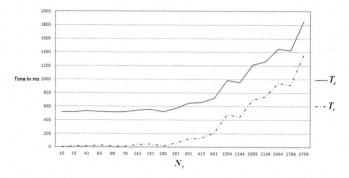

Fig. 6. Search engine server side (including C++ and Java) processing time T_s and total query time T_e plotted versus the number N_r of returned search results for Q1 queries

returned. An approximately 500 ms offset between the two curves is constant for most values of N_r, and is due to the overhead of sending the query request and response from and to the client, and dynamically generating the displayed results. Since each page can only display a limited number of records, data contained in the generated web page is approximately constant, which leads to a constant value of $T_e - T_s$, the average of which is 507.2 ms. To return a certain number of results, the time cost for the entire search process (T_e) is significantly higher than the server side (including C++ and Java) processing time T_s. The processing time T_s is slightly higher than T_c as shown in Table 1, which accounts for the extra time (5.65ms, on average) for Java to reformat the search results for web display. For the Q1 test, the average costs for T_c, T_s and T_e are 315.05ms, 323.95ms and 825.8ms, respectively. The average number of records returned is 899.35.

6.2 Comparing with GSA Test Results

Let N_g stand for the total number of search results returned by a GSA query, and T_g stand for the corresponding total query time. The comparison of the TexSpaSearch engine and the Google Search Appliance (GSA) on the 20 sample queries is shown in Table 1.

The times for GSA queries were obtained by manually recording the search time and count shown in the upper right corner of the test front end we built for testing the GSA with our sample text with spatial data collection. The GSA was residing in the UNB server facility located adjacent to the building where the workstation running the browser was located. From Table 1 and the contents of the search results, we make the following observations:

1. The search time for GSA is significantly lower (either 17 or 45 times faster, on average) than for TexSpaSearch. If we assume that the GSA reported search time T_g includes the time to display the search results on the screen,

then the GSA is 45.5 times faster (on average) than `TexSpaSearch`. If the GSA is reporting only the search engine search time, then `TexSpaSearch` is about 17 times slower than the GSA.

2. For some of the sample queries, `TexSpaSearch` returns more results than GSA. The reason is that GSA only returns records that contain all of the single words in the query string t, while `TexSpaSearch` returns records containing one or more sub-phrases of t. So we have: $N_g \leq N_r$.

3. Both `TexSpaSearch` and the GSA index French characters well.

4. For query strings like `Lady slipper` and `lady's slipper`, the GSA treats them as the same phrase, while TexSpaSearch regards them as different phrases.

5. We checked 7 of the 20 samples (crab, crab apple, pitcher plant, Eupatorium perfoliatum, Bromus Inermis Leyss, Lilac, Verge d'or des bois), and the top N_g records in the TexSpaSearch Q1 results are exactly the same as the GSA search results. For 6 of these 7 search results, the ranking within those top N_g results is different. This is because GSA only returns results containing all the single words in the query string t. In `TexSpaSearch` Q1 query results, the records containing exactly t or all the single words in t usually rank higher than other records.

6.3   Q2 Test Results

We tested Q2 text with spatial search using the 20 sample query strings with radius 2m, 20m and 200m respectively. Compared to Q1 results, Q2 test results are 22.8 times slower, on average. The main reason is that the search complexity on the C++ side is higher than for Q1. The line graph in Figure 7 shows the changes of the T_c, T_s and T_e measured in ms versus $N_{pt} + N_{pl}$. As we can see, T_c, T_s and T_e rise proportionately to the total number of point results and polygon results returned. Note that the value of T_c fluctuates more frequently

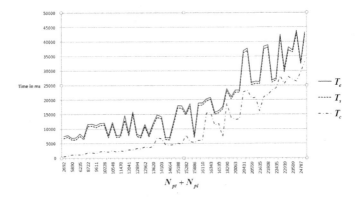

Fig. 7. Search engine C++ processing time T_c, server side (including C++ and Java) processing time T_s and total query time T_e plotted versus the number of returned search results for Q2 queries $N_{pt} + N_{pl}$

than that in Q1. A possible reason leading to the fluctuation is that the time for a Q2 search heavily depends on the number of text search results returned, which is an uncertain factor. Similar to Q1 results, the relatively constant value of $T_e - T_s$ still holds, with an average value of 732.45 ms. To return a certain number of results, the time cost for the entire search process (T_e) and the server side (including C++ and Java) processing time T_s are significantly higher than the C++ processing time T_c. The average value of $T_s - T_c$ for Q2 processing is 7,619.32 ms. The value of $T_s - T_c$ is considerably higher than that in Q1, because the format of results returned by C++ for Q2 is more complex than Q1, which leads to a higher overhead for analysing and interpreting the results to generate the ResultClass object. The total number of results returned by Q2 queries being higher than that of Q1 queries also gives rise to the larger $T_s - T_c$ values. For the Q2 tests, the average T_c, T_s and T_e are 10,488ms, 18,107.3ms and 18,839.8ms, respectively, and the average value of $N_{pt} + N_{pl}$ is 15,433.6.

6.4   Q3 Test Results

We tested Q3 point plus radius search using 15 sample query points spread in the 15 counties in New Brunswick, Canada with radius 5m, 50m, 500m and 5000m, respectively. The line graph in Figure 8 illustrates the changes of the T_c, T_s and T_e measured in ms versus $N_{pt} + N_{pl}$. T_s and T_e show an increase in proportion to the total number of point results and polygon results returned, while T_c remains relatively steady for most values of $N_{pt} + N_{pl}$ at around 200 ms. The steady performance of T_c is consistent with the average search complexity $O(\log_\mathcal{M} \mathcal{D}_N + y)$ of the R*-tree (see section 4.5). Similar to Q1 and Q2, a constant value of $T_e - T_s$ still holds, with an average of 580.68 ms. The time cost for the entire search process (T_e) and the server side (including C++ and Java) processing time T_s are significantly higher than the C++ processing time T_c. The average value of $T_s - T_c$ is 3,411.63ms. The value of $T_s - T_c$ is considerably higher than that for Q1, but lower than that for Q2. A main reason for the lower

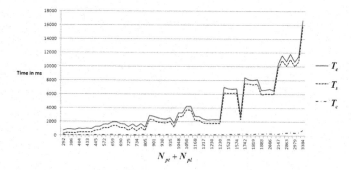

Fig. 8. Search engine C++ processing time T_c, server side (including C++ and Java) processing time T_s and total query time T_e plotted versus the number of returned search results for Q3 queries $N_{pt} + N_{pl}$

value is due to the size of the results returned by Q3, which is less than that of Q2, but generally more than that of Q1. The time spent on analysing and interpreting the `returnList` gives rise to these differences. For the Q3 test, the average T_c, T_s and T_e are 191.5ms, 3,603.2ms and 4,183.9ms, respectively, and the average value of $N_{pt} + N_{pl}$ is 1,313.4.

7 Conclusions

The `TexSpaSearch` engine can perform $Q1(t)$, $Q2(t,r)$ and $Q3(p,r)$ queries successfully, and can rank the search results reasonably. The experimental results on the 20 sample query strings for the Q1 text only query indicate an average 45.5 times slower search time compared with a Google Search Appliance, but returns a wider range of results (records containing any subphrase of the query string) than the GSA. The average query time for Q1 is 0.826 s, with on average 899.35 results returned, while the average theoretical query time for Q1 is $O(A^2 \overline{|b|})$. For a Q2 query, the average query time is 18.84 s to return an average of 15,433.6 records. The average theoretical query time for Q2 is $O(A^2 \overline{|b|} + Z \log_{\mathcal{M}} \mathcal{D}_N + y)$. Q3 tests gives an average query time of 4.184 s for an average of 1,313.4 returned records, while the average theoretical query time for Q3 is $O(\log_{\mathcal{M}} \mathcal{D}_N + y)$. A constant value of $T_e - T_s$, where T_e stands for the total query time and T_s is the server side (including C++ and Java) processing time, holds for all three query types due to the data contained in the generated web page being approximately constant (the number of results displayed on each page is limited to e.g. 5 for Q1, and 3 for Q2 and Q3). Although not currently implemented, the `TexSpaSearch` engine can be used to support a Q3 search followed by a Q1 search of the Q3 results.

References

1. Lucene as a ranking engine. http://www.wortcook.com/pdf/lucene-ranking.pdf (accessed November 10, 2013)
2. Specimen Label Data for the Connell Memorial Herbarium. http://herbarium.biology.unb.ca/fmi/iwp/res/iwp_auth.html
3. Stopwords. http://www.ranks.nl/stopwords (accessed May 5, 2014)
4. Suffix tree. http://en.wikipedia.org/wiki/Suffix_tree (accessed June 23, 2011)
5. Arge, L., de Berg, M., Haverkort, H.J., Yi, K.: The priority r-tree: A practically efficient and worst-case optimal r-tree. ACM Transactions on Algorithms 4(1) (2008)
6. Beckmann, N., Kriegel, H.P., Schneider, R., Seeger, B.: The r*-tree: an efficient and robust access method for points and rectangles. In: SIGMOD Conference, pp. 322–331 (1990)
7. Chen, L., Cong, G., Jensen, C.S., Wu, D.: Spatial keyword query processing: An experimental evaluation. PVLDB 6(3), 217–228 (2013). http://www.vldb.org/pvldb/vol6/p217-chen.pdf
8. Christoforaki, M., He, J., Dimopoulos, C., Markowetz, A., Suel, T.: Text vs. space: efficient geo-search query processing. In: Proceedings of the 20th ACM Conference on Information and Knowledge Management, CIKM 2011, Glasgow, United Kingdom, pp. 423–432, October 24–28, 2011. http://doi.acm.org/10.1145/2063576.2063641

9. Fan, J., Li, G., Zhou, L., Chen, S., Hu, J.: SEAL: spatio-textual similarity search. PVLDB **5**(9), 824–835 (2012). http://vldb.org/pvldb/vol5/p824_jufan_vldb2012.pdf
10. Farach, M.: Optimal suffix tree construction with large alphabets. In: FOCS, pp. 137–143 (1997)
11. Ferragina, P., González, R., Navarro, G., Venturini, R.: Compressed text indexes: From theory to practice. J. Exp. Algorithmics **13**, 12:1.12–12:1.31 (2009). http://doi.acm.org/10.1145/1412228.1455268
12. Foundation, A.S.: Apache lucene - scoring (2011). letzter Zugriff: 20, Oktober 2011. http://lucene.apache.org/java/3_4_0/scoring.html
13. Göbel, R., Henrich, A., Niemann, R., Blank, D.: A hybrid index structure for geo-textual searches. In: Proceedings of the 18th ACM Conference on Information and Knowledge Management, CIKM 2009, Hong Kong, China, November 2–6, 2009, pp. 1625–1628. http://doi.acm.org/10.1145/1645953.1646188
14. Gusfield, D.: Algorithms on Strings, Trees, and Sequences - Computer Science and Computational Biology. Cambridge University Press (1997)
15. Guttman, A.: R-trees: a dynamic index structure for spatial searching. In: SIGMOD Conference, pp. 47–57 (1984)
16. Han, D., Nickerson, B.G.: Comparison of text search ranking algorithms. Tech. rep., TR11-209, Faculty of Computer Science. University of New Brunswick, August, 2011
17. Han, D.A.: Efficient text search with spatial constraints. Tech. rep., TR14-233, Faculty of Computer Science. University of New Brunswick, August, 2014
18. Heuer, J.T., Dupke, S.: Towards a spatial search engine using geotags. In: Probst, F., Keßler, C. (eds.) GI-Days 2007 - Young Researchers Forum. IfGIprints (2007). http://www.gi-tage.de/downloads/acceptedPapers/heuer.pdf
19. Jones, C.B., Abdelmoty, A.I., Finch, D., Fu, G., Vaid, S.: The SPIRIT spatial search engine: architecture, ontologies and spatial indexing. In: Egenhofer, M., Freksa, C., Miller, H.J. (eds.) GIScience 2004. LNCS, vol. 3234, pp. 125–139. Springer, Heidelberg (2004)
20. Li, Z., Lee, K.C.K., Zheng, B., Lee, W., Lee, D.L., Wang, X.: Ir-tree: An efficient index for geographic document search. IEEE Trans. Knowl. Data Eng. **23**(4), 585–599 (2011). http://dx.doi.org/10.1109/TKDE.2010.149
21. McCreight, E.M.: A space-economical suffix tree construction algorithm. J. ACM **23**(2), 262–272 (1976)
22. Roussopoulos, N., Leifker, D.: Direct spatial search on pictorial databases using packed r-trees. SIGMOD Rec. **14**(4), 17–31 (1985). http://doi.acm.org.proxy.hil.unb.ca/10.1145/971699.318900
23. Ukkonen, E.: On-line construction of suffix trees. Algorithmica **14**(3), 249–260 (1995)
24. Weiner, P.: Linear pattern matching algorithms. In: Proceedings of the 14th Annual Symposium on Switching and Automata Theory, SWAT 1973, pp. 1–11. IEEE Computer Society, Washington, DC (1973) http://portal.acm.org/citation.cfm?id=1441424.1441766

Urban Development as a Continuum: A Multinomial Logistic Regression Approach

Ahmed M Mustafa$^{(\boxtimes)}$, Mario Cools, Ismail Saadi, and Jacques Teller

LEMA, University of Liège, Liège, Belgium
a.mustafa@ulg.ac.be

Abstract. Urban development is a complex process influenced by a number of driving forces, including spatial planning, topography and urban economics. Identifying these drivers is crucial for the regulation of urban development and the calibration of predictive models. Existing land-use models generally consider urban development as a binary process, through the identification of built versus non-built areas. This study considers urban development as a continuum, characterized by different level of densities, which can be related to different driving forces.

A multinomial logistic regression model was employed to investigate the effects of drivers on different urban densities during the past decade in Wallonia, Belgium. Sixteen drivers were selected from sets of driving forces including accessibility, geo-physical features, policies and socio-economic factors.

It appears that urban development in Wallonia is remarkably influenced by land-use policies and accessibility. Most importantly, our results highlight that the impact of different drivers varies along with urban density.

Keywords: Urban development · Driving forces · Multinomial logistic · Regression model · Cadastral data · Urban densities

1   Introduction

Urban development is a global issue with paramount socioeconomic and environmental implications, which may affect our well-being in terms of society, economy and/or culture. It may lead to a number of problems related to water quality degradation, air pollution, socio-economic disparities and social fragmentation [1, 2]. Several modelling approaches have been adopted in the analysis of urban growth. Those models can be aimed at predicting spatial location and/or the amount of change.

Prediction of the location of urban growth can be modeled through different approaches including statistical methods, cellular automata and agent-based models. For different modelling approaches, it is important to explore and analyze the main drivers of urban growth on space bases in order to better understand, control and model the future growth of urban settlements.

Urban growth models do usually not differentiate between high-density and low-density urban development [e.g. 3–5] . The main goal of this paper is to investigate the major drivers of different urban densities. This requires analyzing the relationship

© Springer International Publishing Switzerland 2015
O. Gervasi et al. (Eds.): ICCSA 2015, Part III, LNCS 9157, pp. 729–744, 2015.
DOI: 10.1007/978-3-319-21470-2_53

between urban growth and a number of forces related to people choices in terms of spatial location of new urban developments.

Empirical estimation models use statistical methods to model the relationship between urban growth and its drivers based on past observation. In this context, logistic regression models are commonly employed to model urban development potentials [e.g. 3, 5–8].

A multinomial logistic regression model (MNL) is a generalization of a logistic regression model for more than two discrete and unordered response categories. Thus, the MNL methodology allows for the consideration of several classes or urban densities as the dependent, responsible, variable (Y), using a set of independent, explanatory, variables (X), which were selected based on the following literature review.

1.1   Potential Driving Forces

The identification of dominant urban growth driving forces is the main objective of this paper. In literature [e.g. 3, 9–17], numerous explanatory urban development drivers are proposed, which can be grouped into four main sets: accessibility indicators, geo-physical features, land-use policies and socio-economic factors.

Accessibility indicators are often implemented in urban models by means of simple accessibility indicators, such as distance to cities, distance to the road network and distance to water bodies [10, 12, 13]. Road infrastructure consumes a high amount of urban land, around 25% of the total urban area in Europe and 30% in the USA [18]. In this paper, we considered Euclidean distances to different roads categories and to major 11 Belgian cities.

Geo-physical factors are reported as a fundamental driver of the spatial distribution and expansion of urban areas [3]. There is often a relationship between urban growth and a number of these factors, especially the topography of the study area [3, 19]. We considered elevation and slope as geo-physical factors in our study.

Zoning status (policies) is often considered as one of the major urban development drivers worldwide. It has been classified as the most pervasive driver in USA [20]. In Wallonia, land allocation is controlled by several regulations including the regional development plan, referred to as "plan de secteur (PDS)". In this paper, we consider this zoning plan, which defines the legally authorized land-use type for all the territory.

This study also selects a number of socio-economic factors. Population is one of the most active driver of urban development [15]. In this respect, the evolution of net and gross population densities was considered. Economic development could also be considered as a driver of urban development; there is a relation between economic increase and urban development [15] and furthermore economic development has an important influence on people's location choices. In this respect, employment potential, richness level, housing and land prices are considered. The number of households is another factor to be considered in this paper. This number may rely on population lifecycle, migration, societal values, gender relationships and the relationships between parents and children.

2   Methodology

2.1   Study Area

The study area is Wallonia, occupying the southern part of Belgium (Fig. 1). Wallonia is the predominantly French-speaking region of Belgium. It has a territory of 16,844 km² that makes up 55% of the territory of Belgium but with only a third of its population. The population volume in 2010 was 3,498,384 inhabitants [21]. Administratively, it comprises five provinces: Hainaut, Liège, Luxembourg, Namur, and Walloon Brabant. It has 20 administrative arrondissements and 262 municipalities. The geography. of the area goes from flat to hilly with altitude ranges from 0 to 693 m above see-level. Over the last decades, land-use change in Wallonia was related to urban growth and a subsequent loss of agricultural land.

Major cities in Wallonia are characterized by a strong center–periphery structure with well-off households located in the peripheries [22]. The main metropolitan areas are Charleroi, Liège, Mons and Namur. They are all characterized by a historical city-center around which the urban development was spread. Urban sprawl has affected Wallonia for decades leading to fragmented and isolated landscapes that were developed in space and time [23]. Wallonia is highly affected by its neighbors especially by the trans-border workers number evolution.

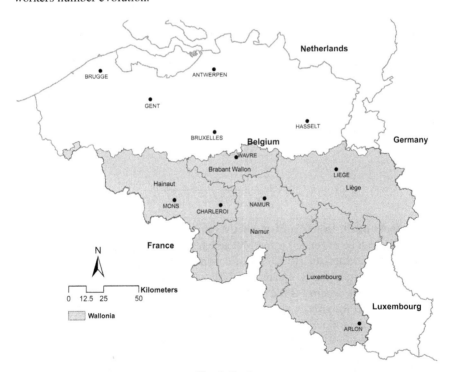

Fig. 1. Study area

2.2   Multinomial Logistic Regression Model

An MNL model was applied to investigate the contribution of the selected driving forces on the probability of urban development along different densities. MNL consists of three components: multi-temporal urban maps, a multivariate function of the hypothesized drivers of change, and the resulting prediction map of urban development potential [24].

MNL analysis yields coefficients for each X. These coefficients are then interpreted as weights in a formula that generates a map for each urban density class depicting the probability of each cell in the landscape to be converted into this class. If the Y variable is a categorical map with k classes, taking on values $0, 1,..., k-1$ and X is a set of explanatory variables $X_1, X_2,..., X_n$ then the logit for each non-reference class $k_1,..., k_n$ against the reference class k_0 model is calculated through:

$$\log(k_1) = \ln\left[\frac{P(Y = k_1 \mid X)}{P(Y = k_0 \mid X)}\right] = \alpha_{k_1} + \beta_{k_1 1}X_1 + \beta_{k_1 2}X_2 + ... + \beta_{k_1 n}X_n$$

$$...\tag{1}$$

$$\log(k_n) = \ln\left[\frac{P(Y = k_n \mid X)}{P(Y = k_0 \mid X)}\right] = \alpha_{k_n} + \beta_{k_n 1}X_1 + \beta_{k_n 2}X_2 + ... + \beta_{k_n n}X_n$$

where $\log(k)$ is a logit function of class k against the reference class, α is the intercept, β is the regression coefficients of class k.

The conditional probabilities of each class can be calculated with the following formula:

$$P(Y = k_0 \mid X) = \frac{1}{1 + e^{\log(k_1)} + ... + e^{\log(k_n)}}$$

$$...\tag{2}$$

$$P(Y = k_n \mid X) = \frac{e^{\log(k_n)}}{1 + e^{\log(k_1)} + ... + e^{\log(k_n)}}$$

The goodness-of-fit, in terms of predictive ability and the interpretability, of the MNL outcomes can be evaluated using McFadden pseudo R-square (MFR2) and Relative Operating Characteristic (ROC) statistic respectively [e.g. 5, 13, 17, 25, 26].

MFR2 tries to mimic the R-squared analysis of linear regression. An MFR2 of 1 indicates a perfect fit, where MFR2 of 0 indicates no relationship. Clark & Hosking [25] stated that a MFR2 greater than 0.2 can be considered a good fit and it is calculated according to the following formula:

$$MFR2 = 1 - \frac{\ln(L_m)}{\ln(L_0)}\tag{3}$$

where L_m is the value of the likelihood function for the full model as fitted with X and L_0 is the value of the likelihood function if all β except α are 0.

The ROC statistic compares the probability map, produced by MNL, to a map with the observed changes of the urban cells for each class between two time-steps. It first divides the probability outcomes into percentile groups from high to low probability and then calculates the proportion of true-positives and false-positives for a range of specified threshold values and relates them to each other in a graph. The ROC measures the area under the curve and its value could range between 0.5 (no relationship) and 1 (perfect fit). ROC values higher than 0.70 are considered as a reasonable fit [19, 27, 28].

Prior to performing the MNL model, we have to consider three aspects that may exist in X: disparity in units, autocorrelation and multicollinearity. These aspects potentially affect any regression analysis results. Due to disparity in units and scale of X (table.1), the logit coefficients cannot be used directly to measure the relative contribution of the X in urban development process. Consequently, all continuous X were standardized before performing MNL analysis according to the following formula:

$$z_i = \frac{x_i - \mu}{\sigma} \tag{4}$$

where z_i is the standardized score of cell i, x_i is the original value, μ is the mean value of X variable and σ is the standard deviation of the X variable. z score is negative when the raw value is below the mean and positive when above. Categorical X were not standardized to keep the meaning of the dummy variable.

Spatial autocorrelation in one or more X will bias the results of the regression analysis. Autocorrelation is the propensity for cell value to be similar to surrounding cells. Moran's I statistic was processed to detect spatial autocorrelation for each X. It is given as:

$$M\_I = n \frac{\sum_{i=1}^{n}\sum_{j=1}^{n} w_{ij}(x_i - \mu)(x_j - \mu)}{\left(\sum_{i=1}^{n}(x_i - \mu)^2\right)\left(\sum_{i \neq j}\sum w_{ij}\right)} \tag{5}$$

where M_I is the Moran's I statistic for each X, n is the number of neighbor cells to be taken into account, w spatial weights and $x_{i/j}$ cells values at location i/j. Locations depend on cell neighbors which can be considered as only shared-border neighbors to the cell under evaluation (x_i) or also diagonal neighbors (x_j). We considered only x_j neighbors. Moran's I value ranges between -1 and +1, where +1 means absolute autocorrelation and -1 none autocorrelation. All X show strong degree of spatial autocorrelation with Moran's I value between 0.746 for zoning and 0.999 for distance to cities. A number of scholars [e.g. 3, 19, 29] suggested that this problem could be addressed through a data sampling approach. A random sample of 15,675 cells, around 1.15% of the study area, distributed throughout the study area were used in the MNL model.

Multicollinearity represents a high degree of dependency among a number of X. It commonly occurs when a large number of X are introduced in a regression model. It is because some of X may measure the same phenomena. Strong collinearities cause the erroneous estimation of parameters and further affect the MNL results [26]. In this context, a number of procedures is proposed to detect multicollinearity among X such as tolerance value, variance inflation factor and Belsley diagnostics [30–33]. We used Belsley diagnostics, one of the most common procedures, to detect multicollinearity. Although Belsley diagnostics are normally applied to linear regression models, it is still valid to apply it for MNL since multicollinearity is a problem among X [34]. The outcomes of Belsley diagnostics are condition indices and variance-decomposition proportions for each X. A condition index greater than 30 represents strong multicollinearity [33]. In this case, it is highly recommended to omit all X with variance-decomposition proportions exceeding the tolerance of 0.5 [33]. All X show low degree of multicollinearity with condition indices between 1 and 9.15 for all X maps and 1 to 9.86 for the selected samples. Thus, all X will be used in the MNL model.

2.3   Data

Dependent variable.
The Y is constituted by cells whose status not changed from non-urban to urban and changed from non-urban to one of different urban classes between 2000 and 2010.

The cadastral dataset (CAD) was used to develop Y. CAD is a vector map representing buildings in two dimensions as polygons provided by the Land Registry Administration of Belgium. Each building comes with different attributes from which the construction date is the most important attribute for our study. Using construction date, two urban land-use maps were developed for 2000 and 2010 years.

Preparing the data.
CAD vector data were rasterized at a very fine cell dimension ($2x2$ m^2). The rasterized cells were aggregated with a factor of 50 by which to multiply the cell size to obtain $100x100$ m^2 raster grid. Each aggregated cell has a density value that represents the number of rasterized 2x2 cells. This value will be used to introduce dwellings density in the aggregated CAD maps.

Scholars always define a minimum map unit (MMU) to avoid overestimation of one class in land cover data [35, 36]. The CORINE Land-Cover (CLC) dataset, as an example, was provided with MMU of 25 ha. Due to the nature of the study area, it is possible to find several scattered buildings in a hectare. Consequently, to produce more accurate data we set MMU at one hectare. In order to avoid overestimation of urban lands, two procedures were applied to the aggregated data: minimum building density per cell (MBDC) and minimum building density per neighbor (MBDN).

Minimum building density per cell.
The average size of residential building in Belgium is about $10x10$ m^2 [37]. This figure somehow corresponds to the average size of households in Wallonia [38] multiplied by the single family house floor space per capita in north & west Europe, which is about 41 m^2 [39]. The multiplication gives an average size of $2.3*41= 94.3$ m^2. Thus, we set MBDC at 25 (100 m^2).

Minimum building density per neighbor.

A threshold of five dwellings per ha was fixed for considering that a cell was urbanized. Neighborhoods with such a density are indeed observable in Wallonia. We then performed an analysis using different thresholds of MBDN using a search window of 3x3 cells for each MBDN cell less than 125 (5x25). These thresholds are 125, 250, 625, 1250 and 2500. Table 1 lists a comparison between CLC data, CAD original aggregated data and different MBDN thresholds.

Table 1. Comparison of area (km²) between CLC, CAD_Org original aggregated CAD data, MBDN_125, MBDN_250, MBDN_625, MBDN_1250 and MBDN_2500

Year	CLC	CAD_Org	MBDN1 25	MBDN 250	MBDN 625	MBDN 1250	MBDN 2500
2000	2506	3229	2599	2468	2093	1744	1579
2006	2513	-	-	-	-	-	-
2010	-	3339	2716	2594	2230	1868	1693

We assumed that the number of changed cells between two time-steps would increase until a specific value of MBDN and then start declining along with the increase of MBDN. Actually, those cells that are under urban development at time-step 1 and reach the threshold of MBDN at time-step 2 are then considered as urban. If the MBDN threshold is very high, this condition will not be reached because this threshold exceeds the observed number of built cells at time-step 1 and 2. The number of changed cells calculated in two provinces of Wallonia confirmed our assumption (Fig. 2). The result showed that the most appropriate threshold for MBDN is 625.

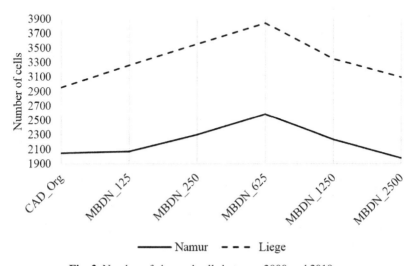

Fig. 2. Number of changed cells between 2000 and 2010

Measuring density.
Different definitions of urban density can be found in the literature, according to the application it is used for. Density can refer to either dwellings or inhabitants per unit area [40]. In this paper, we defined urban density as a number of built-up units per cell of one hectare. We performed different MNL models for 4, 6, 8 and 10 urban densities quantile classes and measured the goodness-of-fit. An MFR2 statistic is not appropriate to compare the goodness-of-fit in this case because MFR2 depends on Y which is changeable, in terms of number of classes. Consequently, we measured mis-classification rate instead. It equaled 24.23%, 23.60%, 22.70% and 26.07% respectively. As a result, for the final MNL, we used eight classes for urban densities, from class0 (non-urban) to class7 (highest density), each class has almost the same number of cells except for class0 (Fig. 3). Table 2 lists the density range for each class.

Urban class7 can be considered as urban cores and urban classes 4, 5 and 6 may be considered as urban peripheries and suburbs. Urban classes 2 and 3 may be considered as rural areas whereas urban class1 may be considered as remote locations. This can be further assessed by measuring how land-use policies control urban development within each class. Land-use policies can highly control urban developments within urban cores whereas urban development in peripheries and suburbs is not strictly following policies. Rural areas are expected to follow policies whereas very remote sites are normally not controlled by policies.

Table 2. Urban classes density ranges in number of 2x2 pixels (% of 100x100 cell area) covered by building footprints

Class	Min	Max	μ	Mode
Class0[a]	-	-	-	-
Class1	25 (1.0%)	78 (3.2%)	51.5	32
Class2	79 (3.2%)	132 (5.3%)	105.5	127
Class3	133 (5.3%)	180 (7.2%)	156.5	138
Class4	181 (7.2%)	243 (9.8%)	212.0	182
Class5	244 (9.8%)	330 (13.2%)	287.0	254
Class6	331 (13.2%)	491 (19.7%)	411.0	333
Class7	492 (19.7%)	2500 (100.0%)	1365.9	504

[a] class0 represents all non-urban cells and affected cells by MBDC and MBDN procedures

Independent variables.
Statistical data related to population volume, households, employment rate, richness index and mean land/housing price were acquired from the official Belgian statistics [38, 41] and mapped with a resolution of 100x100 m² raster grid at municipality level. Gross population density was calculated for each municipality as the number of inhabitants divided by the area of municipality in km² whereas net population density was calculated as the number of inhabitants divided by the area of built-up zones of the municipality in km².

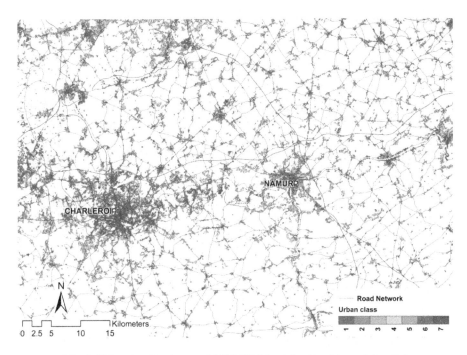

Fig. 3. Urban density classes of 2010 (7 highest density, 1 lowest density)

Digital Elevation Model (DEM) provided by the Belgian National Geographic Institute was used to calculate elevation and slope in percentage for each cell.

Accessibility was measured in terms of Euclidean distance of a cell from the nearest
road and city. Road networks for 2002 were provided by Navteq. Four functional classes of roads were introduced in MNL (R_class1: high speed and volume controlled
access roads, R_class2: quick travel between and through cities, R_class3: moderate
speed travel within cities and R_class4: moderate speed travel between neighborhoods).
Major cities of Belgium (Antwerp, Brussels, Wavre, Brugge, Gent, Charleroi, Mons,
Liege, Hasselt, Arlon and Namur) were used to develop a map of distances to cities.

According to the most recent zoning plan of Wallonia, urban development is only
allowed in those zones that are designated for residential, economic or leisure development. In other zones, such as agricultural and forest areas, urban development is
not permitted unless specific conditions. A zoning map was developed by discerning
zones where urban development is not permitted (code 0) and zones that are designated for urban development (code 1).

All maps were created as raster grids with a resolution of 100x100 m² (table 1).
The spatial resolution is defined based on the availability of data. The statistical data
are available at municipality level whereas other variables could be calculated at cell
level. That is common to coupling both resolution levels [e.g. 19, 42].

Table 3. List of the selected drivers of urban development

Driver	Name	Type[a]	Unit	Resolution[b]	μ	σ
X_1	Elevation	1	m	1	257.14	183.4
X_2	Slope	1	%	1	5.51	57.02
X_3	Dist to city	1	m	1	29028.16	15479.34
X_4	Dist to R_class1	1	m	1	7936.12	8282.57
X_5	Dist to R_class2	1	m	1	4174.5	3757.23
X_6	Dist to R_class3	1	m	1	1668.27	1425.25
X_7	Dist to R_class4	1	m	1	818.63	850.46
X_8	Dist to rail stations	1	m	1	6962.07	5710.64
X_9	Num households	1	number	2	6421.52	12040.71
X_{10}	Mean housing price	1	€	2	139487	31965.1
X_{11}	Mean land price	1	€/m²	2	51.1	99.73
X_{12}	Employment potential	1	%	2	48.39	98.42
X_{13}	Richness index	1	%	2	95.71	61.62
X_{14}	Growth population density	1	inh/km²	2	206.95	354.33
X_{15}	Net population density	1	inh/urban km²	2	819.52	522.97
X_{16}	Zoning status	2	binary	1		

[a] 1. Continuous, 2. Categorical. [b] 1. Cell level, 2. Municipality level.

3   Results and Discussions

In the last decade, Wallonia experienced urban growth distributed between different urban classes. The total area of urban land in 2000 was 2093 km², accounting for 12.4% of the total area, and in 2010, urban land increased to 2230 km², accounting for 13.2% of the total area. More than 137.5 km² of land was converted between 2000 and 2010 from non-urban to different urban densities as: 50.4 km² to urban class1, 34.7 km² to class2, 23 km² to class3, 10.5 km² to class4, 7.12 km² to class5, 5.3 km² to class6 and 6.68 km² to class7. Of course, one km² of urban class1 has not the same weight as one km² of urban class7, as densities are very different.

The goodness-of-fit for predictive ability was evaluated using MFR2 and it equals 0.244. The model reveals a very good correspondence with ROC values: 0.775, 0.819, 0.829, 0.805, 0.793, 0.813 and 0.914 for classes 1, 2, 3, 4, 5, 6 and 7 respectively. This indicates that the MNL performs well and the MNL's outcomes could effectively interpret the process of urban development in Wallonia.

The results of the MNL model are illustrated in table 4. Scholars and decision-makers are not only interested in the identification of potential driving forces of urban development process but also interested in measuring the relative contribution of these drivers to urban development process [3, 10]. To relatively measure the contribution of each X to urban development process, Odds Ratio (OR) that equals $exp(\beta)$ is calculated. It is difficult to directly interpret an MNL coefficients and they become useful when they are converted into an OR. An OR greater than 1 indicates a positive effect, the probability of urban development increases by increasing the OR of the variable, whereas less than 1 indicates a negative effect, the probability of urban development decreases by increasing the OR of the variable. An OR of 1 means neutral contribution to the urban development process [13]. Table 5 lists OR values for all X.

For interpreting continuous X, multiply a one unit increase in X variable by OR. For instance, five units increase in elevation increase the class7 portability by $exp(5x0.110) \sim 1.7$ whereas decrease the class1 portability by $exp(5x-0.078) \sim 0.7$. For categorical X, zoning status, the OR value for class1 is about 15.4. This means that it is around 15.4 $(exp(2.735))$ times more likely to find a new urban cell related to class1 in zones designated to urban development than in zones designated to other uses in zoning plan.

Generally, the impact of different drivers varies with different urban densities. These drivers can be grouped into common drivers with impacts on different urban classes and special drivers with impacts on individual classes.

Table 4. The coefficients (β) of MNL model (class0 is the refrence class)

	Class1	Class2	Class3	Class4	Class5	Class6	Class7
α	-4.158	-4.603	-4.700	-4.583	-4.409	-4.500	-5.969
Elevation	-0.078	-0.032	0.043	0.105	0.023	0.204	0.110
Slope	-0.237	-0.078	-0.210	-0.644	-0.693	-0.840	-1.185
Dist to city	0.072	0.051	-0.099	-0.036	0.130	0.008	0.070
Dist to R_class1	-0.129	0.014	-0.004	-0.165	-0.146	-0.306	-0.917
Dist to R_class2	-0.113	-0.036	-0.021	-0.084	-0.244	-0.215	-0.587
Dist to R_class3	-0.265	-0.257	-0.141	-0.283	-0.214	-0.197	-0.278
Dist to R_class4	-0.651	-0.536	-0.587	-0.552	-0.427	-0.394	-0.228
Dist to rail stations	0.002	-0.020	0.069	0.024	-0.136	-0.141	-0.289
Num households	0.001	-0.016	-0.079	-0.098	-0.131	-0.082	-0.086
Mean housing price	0.027	0.029	0.074	0.037	0.096	0.038	-0.170
Mean land price	0.079	0.127	0.083	0.191	-0.054	0.250	0.198
Employment potential	-0.174	-0.098	-0.027	0.001	0.105	0.205	0.236
Richness index	0.159	0.032	0.057	-0.076	-0.043	-0.371	-0.305
Gross population density	0.311	0.161	0.210	0.213	0.257	-0.076	0.045
Net population density	-0.362	-0.364	-0.451	-0.233	-0.078	0.120	-0.070
Zoning status	2.735	3.639	3.745	3.278	2.942	2.807	3.775

Table 5. The OR value for X (class0 is the refrence class)

	Class1	Class2	Class3	Class4	Class5	Class6	Class7
Elevation	0.925	0.969	1.043	1.111	1.023	1.226	1.117
Slope	0.789	0.925	0.811	0.525	0.500	0.432	0.306
Dist to city	1.074	1.053	0.906	0.965	1.139	1.008	1.072
Dist to R_class1	0.879	1.014	0.996	0.848	0.864	0.736	0.400
Dist to R_class2	0.893	0.964	0.980	0.920	0.783	0.807	0.556
Dist to R_class3	0.767	0.774	0.869	0.754	0.807	0.822	0.758
Dist to R_class4	0.521	0.585	0.556	0.576	0.653	0.674	0.796
Dist to rail stations	1.002	0.980	1.071	1.024	0.873	0.868	0.749
Num households	1.001	0.984	0.924	0.907	0.877	0.922	0.918
Mean housing price	1.027	1.029	1.076	1.038	1.101	1.039	0.843
Mean land price	1.082	1.136	1.087	1.210	0.947	1.284	1.219
Employment potential	0.840	0.907	0.974	1.001	1.111	1.228	1.266
Richness index	1.172	1.033	1.059	0.927	0.958	0.690	0.737
Gross population density	1.365	1.175	1.233	1.237	1.293	0.927	1.046
Net population density	0.696	0.695	0.637	0.792	0.925	1.128	0.932
Zoning status	15.405	38.050	42.317	26.523	18.952	16.555	43.598

We found that the likelihood of urban development is markedly influenced by policies (zoning status). Zoning status has the strongest impact on urban development within all urban classes. Slope, distance to R_class4, distance to R_class3, net/gross population densities and mean land price respectively also demonstrate an impact on all urban classes, but far less important than zoning status. The result shows that the impact of slope is generally increasing with built-up densities. Quite logically dense urban projects are rather developed in flat areas, in many cases floodplains in Wallonia. In contrast, the impact of distance to R_class4, namely intra-urban or inter-villages roads, is generally decreasing with built-up densities. Distances to R_class1 and R_class2 have a noticeable impact on the development of high density projects (class7) with OR of 0.40 and 0.56 respectively. That indicates a strong relationship between highest urban density class and proximity to high speed roads. It should be stressed at this respect that a number of urban cores are directly accessible via high-speed roads in Wallonia. Employment potential has a significant attraction impact on urban class7. It is generally increasing with density, which is what can be expected. Richness index and elevation have moderate impacts on urban class6. Distance to rail stations has a moderate positive influence on urban class7. Still this influence is much lower than the proximity of high-speed roads, suggesting that urban areas located nearby train stations are not yet sufficiently attractive for new dense urban developments in Wallonia. This should be a major source of concern for urban policy makers. Quite significantly, mean housing price represents a low influence on urban growth. This influence is negative for high density developments, which is another source of concern given the shortage of available housing, especially apartments, in areas characterized by a strong pressure on the real estate market.

Our assumption regarding general identification of different urban classes can be assessed by measuring the influence of land-use policies on urban development. OR values for zoning show that policy has a very strong impact on the highest density developments (class7). Those high density developments will most naturally be developed in areas where the legally-binding plan allows such developments, in order to minimize the administrative and financial risks of such operations. Further on they are often located in dense urban areas where the existing plan already allocates a significant share of the land to urban development and where non-urban zones (parks, green areas, etc) are strongly protected for environmental, social and/or heritage reasons. Policies impact is taken a downward trend with classes 4, 5 and 6 respectively. We consider those classes as suburbs. Quite understandably that urban developments in suburbs do not strictly follow policies. The impact of policy on class1 is very low compared to other classes. This class can be considered as remote developments, consisting in scattered constructions (1 to 3 buildings per ha), which can sometimes deviate from existing zoning plans especially in agricultural zones. Land-use policies also show a noticeable impact on classes 2 and 3. We considered those both classes as low density developments in rural areas. It is not surprising that new developments are mainly directed to urbanizable zones, where there is an excess supply of such land. Fig. 4 presents urban development probability maps for classes 1 and 7 as examples of MNL model outcomes.

4   Conclusions

Urban development process in Wallonia is dynamic and diverse. The prediction of spatial distribution of such development can be modelled based on a set of geo-physical and socioeconomic attributes that represent proximate causes of urban growth. Considering urban development as a continuum allows us to better understand the interactions between different drivers and different urban densities.

In this paper, we examined the driving forces of urban development process in Wallonia over a period of 10 years (2000 to 2010). Multinomial logistic regression model was employed to relatively measure the impact of different drivers on probability of urban development. Sixteen drivers were selected from four sets of driving forces including geo-physical features, land-use policies, socio-economic and accessibility. Generally, result reveals that policies and accessibility are the most important determinants of urban growth process. Most importantly, our results highlight that the impact of different drivers varies along with urban density. This is especially the case of land planning driver, whose effects are much more significant for smaller densities than for higher ones, except in the case of urban cores. This study findings can support urban growth modeling, urban planning, and decision-making process in identifying urban development likelihood for each location in the present and medium term.

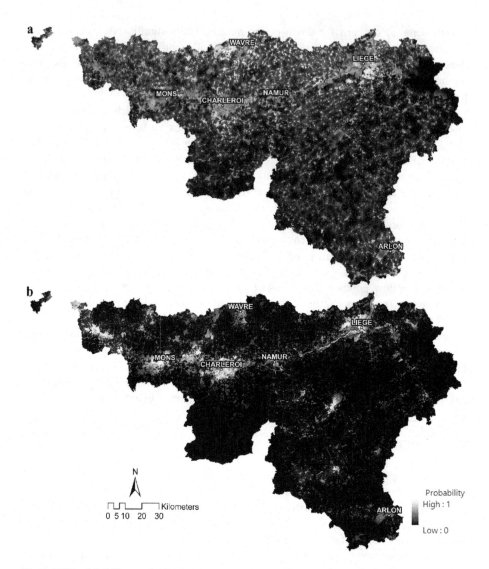

Fig. 4. MNL probability maps for: **a.** class1 (scattered), **b.** class7 (generally located in urban zones)

Acknowledgments. The research was funded through the ARC grant for Concerted Research Actions, financed by the Wallonia-Brussels Federation.

References

1. Arnfield, A.J.: Two decades of urban climate research: a review of turbulence, exchanges of energy and water, and the urban heat island. Int. J. Climatol. **23**, 1–26 (2003)
2. Xian, G., Crane, M.: Assessments of urban growth in the Tampa Bay watershed using remote sensing data. Remote Sens. Environ. **97**, 203–215 (2005)

3. Li, X., Zhou, W., Ouyang, Z.: Forty years of urban expansion in Beijing: What is the relative importance of physical, socioeconomic, and neighborhood factors? Appl. Geogr. **38**, 1–10 (2013)
4. Maimaitijiang, M., Ghulam, A., Sandoval, J.S.O., Maimaitiyiming, M.: Drivers of land cover and land use changes in St. Louis metropolitan area over the past 40 years characterized by remote sensing and census population data. Int. J. Appl. Earth Obs. Geoinformation **35**, Part B, 161–174 (2015)
5. Mustafa, A., Saadi, I., Cools, M., Teller, J.: Measuring the Effect of Stochastic Perturbation Component in Cellular Automata Urban Growth Model. Procedia Environ. Sci. **22**, 156–168 (2014)
6. Puertas, O.L., Henríquez, C., Meza, F.J.: Assessing spatial dynamics of urban growth using an integrated land use model. Application in Santiago Metropolitan Area, 2010–2045. Land Use Policy **38**, 415–425 (2014)
7. Wu, F.: Calibration of stochastic cellular automata: the application to rural-urban land conversions. Int. J. Geogr. Inf. Sci. **16**, 795–818 (2002)
8. Munshi, T., Zuidgeest, M., Brussel, M., van Maarseveen, M.: Logistic regression and cellular automata-based modelling of retail, commercial and residential development in the city of Ahmedabad. India Cities **39**, 68–86 (2014)
9. Bičík, I., Jeleček, L., Štěpánek, V.: Land-use changes and their social driving forces in Czechia in the 19th and 20th centuries. Land Use Policy **18**, 65–73 (2001)
10. Serneels, S., Lambin, E.F.: Proximate causes of land-use change in Narok District, Kenya: a spatial statistical model. Agric. Ecosyst. Environ. **85**, 65–81 (2001)
11. Verburg, P.H., Schot, P.P., Dijst, M.J., Veldkamp, A.: Land use change modelling: current practice and research priorities. Geo. Journal **61**, 309–324 (2004)
12. Quan, B., Chen, J.-F., Qiu, H.-L., Römkens, M.J.M., Yang, X.-Q., Jiang, S.-F., Li, B.-C.: Spatial-Temporal Pattern and Driving Forces of Land Use Changes in Xiamen. Pedosphere **16**, 477–488 (2006)
13. Braimoh, A.K., Onishi, T.: Spatial determinants of urban land use change in Lagos. Nigeria. Land Use Policy **24**, 502–515 (2007)
14. Poelmans, L., Van Rompaey, A.: Detecting and modelling spatial patterns of urban sprawl in highly fragmented areas: A case study in the Flanders-Brussels region. Landsc. Urban Plan. **93**, 10–19 (2009)
15. Liu, C., Ma, X.: Analysis to driving forces of land use change in Lu'an mining area. Trans. Nonferrous Met. Soc. China **21**(Supplement 3), s727–s732 (2011)
16. Hallowell, G.D., Baran, P.K.: Suburban change: A time series approach to measuring form and spatial configuration. J. Space Syntax **4**, 74–91 (2013)
17. Shu, B., Zhang, H., Li, Y., Qu, Y., Chen, L.: Spatiotemporal variation analysis of driving forces of urban land spatial expansion using logistic regression: A case study of port towns in Taicang City. China. Habitat Int. **43**, 181–190 (2014)
18. Camagni, R., Gibelli, M.C., Rigamonti, P.: Urban mobility and urban form: the social and environmental costs of different patterns of urban expansion. Ecol. Econ. **40**, 199–216 (2002)
19. Cammerer, H., Thieken, A.H., Verburg, P.H.: Spatio-temporal dynamics in the flood exposure due to land use changes in the Alpine Lech Valley in Tyrol (Austria). Nat. Hazards. **68**, 1243–1270 (2013)
20. Brueckner, J.K.: Lectures on Urban Economics. MIT Press (2011)
21. Belgian Federal Government: Population. http://statbel.fgov.be/fr/modules/publications/statistiques/population/population_-_chiffres_population_1990-2010.jsp

22. Verhetsel, A., Thomas, I., Beelen, M.: Commuting in Belgian metropolitan areas: The power of the Alonso-Muth model. J. Transp. Land Use **2** (2010)
23. Antrop, M.: Landscape change and the urbanization process in Europe. Landsc. Urban Plan. **67**, 9–26 (2004)
24. Lambin, E.F.: Modelling Deforestation Process - A review - Trees Tropical Ecosystem Environment Observations by Satellites. European Commission Luxembourg (1994)
25. Clark, W.A.V., Hosking, P.L.: Statistical Methods for Geographers. Wiley, New York (1986)
26. Lin, Y., Deng, X., Li, X., Ma, E.: Comparison of multinomial logistic regression and logistic regression: which is more efficient in allocating land use? Front. Earth Sci., 1–12 (2014)
27. Hosmer, Jr., D.W., Lemeshow, S.: Applied Logistic Regression. John Wiley & Sons (2004)
28. Poelmans, L.: Modelling urban expansion and its hydrological impacts (2010)
29. Poelmans, L., Van Rompaey, A.: Complexity and performance of urban expansion models. Comput. Environ. Urban Syst. **34**, 17–27 (2010)
30. Belsley, D.A., Kuh, E., Welsh, R.E.: Regression Diagnostics. John Wiley and Sons, New York (1980)
31. Judge, G.G., Griffiths, W.E., Hill, R.C., Lütkepohl, H., Lee, T.-C.: The Theory and Practice of Econometrics. Wiley, New York (1985)
32. Belsley, D.A.: Conditioning diagnostics. Wiley Online Library (1991)
33. Kennedy, P.: A Guide to Econometrics. MIT Press (2003)
34. Flom, P.L.: Multinomial and ordinal logistic regression using PROC LOGISTIC. NESUG. Baltimore (2010)
35. Knight, J.F., Lunetta, R.S.: An experimental assessment of minimum mapping unit size. IEEE Trans. Geosci. Remote Sens. **41**, 2132–2134 (2003)
36. Saura, S.: Effects of minimum mapping unit on land cover data spatial configuration and composition. Int. J. Remote Sens. **23**, 4853–4880 (2002)
37. Tannier, C., Thomas, I.: Defining and characterizing urban boundaries: A fractal analysis of theoretical cities and Belgian cities. Comput. Environ. Urban Syst. **41**, 234–248 (2013)
38. Belgian Federal Government: Statistics Belgium.
    http://statbel.fgov.be/fr/statistiques/chiffres/
39. Economidou, M., Atanasiu, B., Despret, C., Maio, J., Nolte, I., Rapf, O.: Europe's buildings under the microscope. Brussels, Buildings Performance Institute Europe (BPIE) (2011)
40. Jenks, M., Dempsey, N.: Future Forms and Design for Sustainable Cities. Routledge (2005)
41. Institut wallon de l'évaluation, de la prospective et de la statistique: Statistiques. http://www.iweps.be/themes-page
42. Roy Chowdhury, P.K., Maithani, S.: Modelling urban growth in the Indo-Gangetic plain using nighttime OLS data and cellular automata. Int. J. Appl. Earth Obs. Geoinformation **33**, 155–165 (2014)

Ecosystem Services Along the Urban–Rural–Natural Gradient: An Approach for a Wide Area Assessment and Mapping

Marco Vizzari[1](✉), Sara Antognelli[1], Mohamed Hilal[2],
Maurizia Sigura[3], and Daniel Joly[4]

[1] Department of Agricultural, Food, and Environmental Sciences,
University of Perugia, Perugia, Italy
{marco.vizzari,sara.antognelli}@unipg.it
[2] INRA, UMR 1041 CESAER, Dijon, France
mhilal@dijon.inra.fr
[3] Department of Agricultural and Environmental Sciences, University of Udine, Udine, Italy
maurizia.sigura@uniud.it
[4] Laboratoire ThéMA UMR 6049 Centre National de la Recherche Scientifique,
Université de Franche-Comté, Besançon, France
daniel.joly@univ-fcomte.fr

Abstract. Landscapes can be viewed as a continuum and studied using spatial gradients along which environmental modifications determine the structural and functional components of ecosystems. The analysis and quantification of Ecosystem Services, intended as the benefits people obtain from ecosystems, play a crucial role in sustainable landscape planning. In this framework we developed a novel method for the identification and characterization of the landscapes nested along the urban-rural-natural gradient and the analysis of potential ES supply and demand within said landscapes. The Kernel Density Estimation technique was applied to calculate continuous intensity indicators associated with urbanization, agriculture, and natural elements, considered as key components of the gradient. The potential ES demand and supply within each landscape area were assessed using expert–knowledge based indices associated to the LULC CORINE classes. Results showed a complex organization of "pillar" and transitional landscapes along the gradient, which match different bundles of ES demand and supply.

Keywords: Urban–rural–natural gradient · LULC gradients · Ecosystem services mapping · Kernel density estimation · Principal component analysis

1 Introduction

Landscape has usually been considered as an arrangement of relatively homogeneous patches (vegetation communities, forest types, land covers) which are repeated across the space [1] Land-use and land-cover (LULC) have largely been used as indicators of environmental condition and landscape quality and numerous studies point out land

© Springer International Publishing Switzerland 2015
O. Gervasi et al. (Eds.): ICCSA 2015, Part III, LNCS 9157, pp. 745–757, 2015.
DOI: 10.1007/978-3-319-21470-2_54

uses as determinant of the state of the natural environment. The view of landscapes as continua and spatial gradients represents a challenge to this conventional view of how the natural (and human) environment is organized [2]. Anthropogenic gradients generated by the increasing intensity of LULC were defined as the specific succession in the space of natural–managed–cultivated–suburban–urban landscapes [3]. In this view urbanization can be considered as a particular environmental gradient that produces modifications in the structures and functions of ecological systems [4,5] with a magnitude that depends on the steepness of the gradient itself [6]. Along this gradient, urban fringes represent spaces with undefined boundaries [7] inside of which transitions and changes in equilibrium and relationships can be observed [8,9]. The gradient view implies the identification of fringe regions or transitional landscapes, where different pressures due to human activities arise, causing more or less marked, unstable conditions involving both the internal configuration and the relationships with the surroundings [10]. Thus fringe areas show a particular vulnerability towards urban sprawl and structural and functional modifications occurring in rural spaces [7,8], [11,12,13,14,15], since the pressures due to urbanization affect the supply of functions and services expected from the ecosystems of these areas. As a consequence an increasing interest can be observed in understanding relationships between the socio-ecological issues and land use changes at a landscape scale [16,17,18,19,20].

GIS spatial analysis and modelling can support the definition and calculation of continuous indicators allowing better assessment and interpretation of the gradients characterizing landscape [21]. Along landscape gradients, generated by the intensity of human use of land, different mosaics of LULC and different characteristics of landscape structural elements (patches) have been identified [22,23,24,25,26]. Density analysis tools, available within the GIS environment, allow researchers to transform values measured at specific locations on continuous surfaces to obtain the general trend of the spatial distribution for the considered variable [27]. Kernel Density Estimation (KDE) has already been used to represent and analyse spatial trends generated by landscape features as well as their potential ecological interactions or influences on the surrounding landscape [21], [24], [28], and for the spatial modelling of landscape quality [23]. KDE produces smoothed surfaces by applying a moving window superimposed over a grid where the density of studied variables is estimated at each location according to a kernel function [27]. The degree of smoothing is controlled by the kernel bandwidth [29]; generally, its definition is a sensitive step because a wider radius shows a more general trend, smoothing the spatial variation of the variable, while a narrower radius highlights more localised effects in the distribution [30,31]. Two main approaches to determine bandwidth can be found in the literature: the first, more frequently employed, uses a fixed bandwidth to analyse the entire distribution, while the second implements a local adaptive bandwidth [32]. However, the examination of resultant surfaces for different values of bandwidth remains a common method supporting the definition of this parameter [27], [33].

Clustering procedures are commonly used to describe multivariate data in terms of groups (clusters) that are characterised by strong internal similarities [34,35]. Unsupervised classification methods are very powerful because they may identify a number and composition of classes that do not correspond to predetermined notions about the landscape structure [36]. The k-means procedure is among the most widely used unsupervised clustering algorithms [37]. A very popular variant of this technique is

ISODATA (Iterative Self–Organising Data Analysis Technique) [38], which is commonly used for unsupervised classification of digital images (see, e.g. [39,40,41,42]). Like k-means, this technique organises data iteratively into a number of groups, where objects of the same group are more similar than those belonging to different groups, computing the minimum square average distance of each point from its nearest centre. The main difference from the k-means algorithm is that the user is required to indicate a rough estimate of the number of clusters. The ISODATA algorithm uses different heuristic methods to optimise this number by removing small clusters, merging neighbouring clusters or splitting larger and more widespread clusters.

Since the Millennium Ecosystem Assessment [43], many scientists are increasing their efforts to studying and quantifying goods and services produced by ecosystems. These Ecosystem Services (ES) have recently become a primary objective in ecology conservation research, as a framework supporting decisions in natural resource management [44]. The concept of ES integrates the ecological perspective and the anthropological viewpoint to analyse the ecosystem functions (carried out by ecosystem components and processes) according to how useful it is perceived [45,46,47]. Different approaches to assess and map ecosystem services on the landscape scale have been developed and applied at different spatial scales by several authors (see e.g. [48,49,50,51,52,53]). However, special attention has recently been paid to approaches based on LULC and expert knowledge with the aim of introducing the ES concept to decision makers and land managers as a tool to support sustainable landscape planning at wider scales. In the framework of ES mapping methods, the use of proxy indicators to assess the potential ES supply, demand and budget offered by LULC can be viewed as a proper solution to overcome the difficulty of systematically considering a wide, complex set of ES within landscape management and land-use planning processes [25], [54,55].

In this framework, this research is aimed to develop a methodology, easily applicable to various scales of analysis and starting from EU-wide available LULC data, effective to: a) identify and characterise the types of landscape along a urban–rural–natural gradient generated by LULC intensity; b) analyse the potential ES demand, supply within the types of landscape expressed along the gradient. To this aim, KDE, multivariate spatial analysis, ES LULC-based indicators, and Principal Component Analysis (PCA) were combined to develop a spatially explicit methodology with parsimonious data. This approach was applied to the country of France, in order to test the method at a national scale, to perform a detailed classification of the landscape gradient, and to characterise the related supply and demand of ES.

2   Method

Three main steps were taken: (a) spatial modelling of gradients generated by key landscape components; (b) multivariate spatial analysis and landscape classification; (c) assessment and mapping of potential ES supply and demand.

The model for landscape gradient detection and analysis was based on the calculation of continuous density indicators, associated with urbanisation, agriculture and natural elements. These indicators were computed using KDE on LULC classes derived from a reclassification of the Corine Land Cover dataset (year 2006) (Table 1). In order to apply KDE, layers associated to these classes were extracted from Corine data in grid format (250 cell size) and converted to points. The objective of the KDE application was to transform the values at specific locations on continuous surfaces to measure the general spatial distribution trend for the considered variable. The quartic function, a simplification of the Gaussian model [30], [56] was adopted and four different band-widths (1000, 2500, 5000, and 7500 m) were tested. The value of 2500 m was finally chosen by means of a visual analysis [21], [33]. Cell size for KDE layers was set at 250 m, consistent with the scale of analysis [57] and the resolution of data source.

Table 1. Reclassification of CORINE Land Cover classes for calculation of density indicators

CLASS CODE	CLASS NAME	CORINE LULC Class codes (level 3)
U	Urban fabric	111, 112, 121, 122, 123, 124, 131, 132, 133, 141, 142
A	Arable lands	211, 212, 213
P	Permanent crops	221, 222, 223
H	Pastures and heterogeneous agricultural areas	231, 241, 242, 243, 244
F	Forests	311, 312, 313, 324
S	Semi-natural areas	321, 322, 323, 331, 332, 333, 334, 335
W	Wetlands and Water bodies	411, 412, 511, 512

The seven KDE layers were analysed by using a PCA to eliminate correlations and redundancy between the indicators. All the resulting seven PCs were subsequently processed by an ISODATA cluster analysis to obtain the final urban–rural–natural gradient classification and landscape map of France.

In order to understand the potential ES demand and supply for each landscape type, the habitat focus approach [45], [48], [52], [57,58] was adopted. This approach is based on the use of matrices, compiled by experts, which estimate the ES demand and supply for each LULC Corine class (Level 3). The supply matrix expresses the different LULC abilities to provide selected ecosystem goods and services, whereas the demand matrix expresses the requirements of ecosystem services for humans living of the various LULC classes. Both demand and supply are assessed by a weight (P) between 0 and 5, which makes all the demand and supply values directly comparable. Thus, for each landscape area, ES demands and supplies were calculated using the equation (1) where P is the weight of the j-th ES demand or supply related to i-th LULC and I is the averaged value of the normalised intensity indicator of the i-th LULC composing the landscape of each area.

$$ES_j = \sum_{i=1}^{n} P_{ij}I_i \qquad (1)$$

The weighted average supply and demand of the 31 ES suggested by [57] was then calculated for each area. A further PCA was calculated by the "prcomp" R module, subsequently applied both on the average ES supply and demand values, in order to assess and map bundles of the ES produced and demanded in each area and discover general trends along the gradient. A final Kruskal-Wallis test was performed in R by the "kruskal.test" module [59,60] to statistically verify the significance of relationship between landscape classes and ES supply and demand.

3   Results and Discussion

3.1   Landscape Gradient Analysis

Using ISODATA cluster analysis, 13 landscape types, including 7877 landscape areas, were identified and mapped. Average values of intensity indicators enabled the identification of the landscape type and the distinction between so–called "pillar" and "transitional" landscapes (Fig. 1). The former are characterized by the relative prevalence of a single LULC intensity, while in the latter two or more featuring LULC classes can be observed. Prevalence was detected by comparing the local average response of each indicator with the average indicator values of the other landscape classes observed along the gradient. The spatial distribution of landscapes within the study area is represented in Fig. 2, while the landscape codes are explained in Table 2.

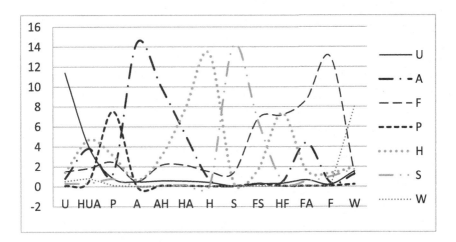

Fig. 1. LULC indicators responses in the different landscape classes. On Y axis is the density expressed in number of cells (250x250m wide) per Km^2. Pillar landscapes, differently than transitional ones, are coded by a single letter. (see Table 2 for legend)

As expected, the spatial distribution trends of pillar landscapes show the natural contexts linked mainly with the wider wooded areas of the country, the urban areas of the major towns of the country, whereas agricultural landscapes appear more extensively spread over the country. On the contrary, transitional landscapes assume a more complex, heterogeneous configuration among the more compact pillar landscapes.

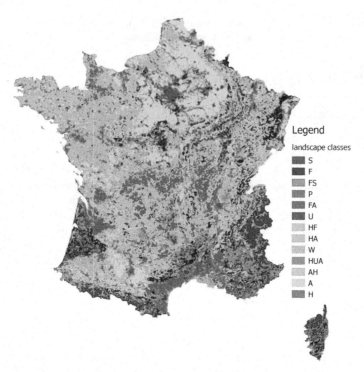

Fig. 2. Map of landscape classes

Table 2. Landscape codes and descriptions

CODE	DESCRIPTION
U	Urban dominated landscapes
HUA	Heterogeneous agricultural peri–urban transitional landscapes
P	Permanent crops dominated landscapes
A	Arable lands dominated landscapes
AH	Arable lands with heterogeneous agricultural areas transitional landscapes
HA	Heterogeneous agricultural transitional landscapes
H	Agricultural heterogeneous dominated landscapes
S	Semi–natural dominated landscapes
FS	Forest with semi–natural areas transitional landscapes
F	Forest dominated landscapes
HF	Heterogeneous agricultural with forests transitional landscapes
FA	Forests with arable lands transitional landscapes
W	Wetlands or water bodies dominated landscapes

3.2   ES Supply and Demand Analysis

PC1 and PC2 calculated on ES supply average values of each landscape area are able to explain around 90% of the total variance (77% PC1 and 13% PC2). PC1 explains primarily the supply of so-called "natural ES" - mainly regulating ones, typically produced by ecosystems with high presence of natural vegetation - , but also cultural services, typically supplied by high quality landscapes (Table 3). PC1 explains also specific provisioning services – namely timber and wood fuel – usually associated with wooded areas. PC2 explains mostly the supply of so-called "agricultural ES", i.e. services typically produced by agricultural areas such as food and fibre.

Table 3. Eigenvectors of PC1 and PC2 of ES supply. *ES type code: R=regulating, P=provisioning, C=cultural*

ES type	ES name	PC1	PC2
R	global climate regulation	0.233	0.037
R	local climate regulation	0.197	0.188
R	air quality regulation	0.239	0.188
R	water flow regulation	0.089	0.163
R	water purification	0.284	0.045
R	nutrient regulation	0.239	0.073
R	erosion regulation	0.258	-0.052
R	natural hazard regulation	0.176	0.088
R	pollination	0.173	0.111
R	pest and disease control	0.118	0.096
R	regulation of waste	0.124	0.024
P	crops	-0.205	0.430
P	biomass for energy	-0.123	0.418
P	fodder	-0.158	0.258
P	livestock	-0.046	-0.248
P	fibre	-0.128	0.484
P	timber	0.281	0.094
P	wood fuel	0.276	0.080
P	fisheries seafoods and edible algae	-0.001	-0.018
P	acquaculture	0.000	-0.018
P	wild foods and resources	0.230	0.042
P	biochemicals and medicines	0.061	0.314
P	freshwater	0.000	-0.016
P	mineral resources	-0.001	-0.006
P	abiotic energy sources	-0.119	-0.087
C	recreation and tourism	0.204	-0.069
C	landscape aesthetic and inspiration	0.218	-0.050
C	knowledge systems	0.182	0.016
C	religious and spiritual experience	0.157	-0.024
C	cultural heritage and cultural diversity	0.062	0.083
C	natural heritage and natural diversity	0.265	-0.023

PCA application on ES demand average values produced two components explaining around 89% of the total variance, (66% PC1 and 23% PC2). PC1 (Table 4) explains the demand of "natural ES", similarly to the supply PC1. PC2 explains two groups of demanded ES: agricultural services, i.e. services typically demanded by agricultural areas, negatively correlated with the PC, and the so-called "urban goods and services", positively correlated with the same PC. The agricultural ES demand is conceptually different from the agricultural ES supply, since the former is represented mainly by services able to improve crop production and farm resilience, while typical

Table 4. Eigenvalues of PC1 and PC2 of ES demand. *ES type code: R=regulating, P=provisioning, C=cultural*

ES type	ES name	PC1	PC2
R	global climate regulation	0.219	-0.116
R	local climate regulation	0.233	-0.021
R	air quality regulation	0.230	0.110
R	water flow regulation	0.242	-0.002
R	water purification	0.207	0.110
R	nutrient regulation	0.236	-0.360
R	erosion regulation	0.136	-0.287
R	natural hazard regulation	0.254	-0.065
R	pollination	0.175	-0.289
R	pest and disease control	0.321	-0.355
R	regulation of waste	0.230	-0.308
P	crops	0.135	0.186
P	biomass for energy	0.203	0.073
P	fodder	0.123	-0.041
P	livestock	0.136	0.189
P	fibre	0.121	0.168
P	timber	0.137	0.143
P	wood fuel	0.056	0.078
P	fisheries seafoods and edible algae	0.144	0.200
P	acquaculture	0.144	0.200
P	wild foods and resources	0.144	0.200
P	biochemicals and medicines	0.227	0.101
P	freshwater	0.228	0.167
P	mineral resources	0.124	0.173
P	abiotic energy sources	0.119	0.165
C	recreation and tourism	0.125	0.174
C	landscape aesthetic and inspiration	0.125	0.173
C	knowledge systems	0.164	0.013
C	religious and spiritual experience	0.093	0.128
C	cultural heritage and cultural diversity	0.135	-0.031
C	natural heritage and natural diversity	0.092	0.127

agricultural demanded ES are nutrient regulation, erosion regulation, pest and disease control, pollination etc. The "urban goods and services" are represented by services used directly by people, such as most of the provisioning services and some of the cultural services.

PC1 and PC2 of ES supply and demand were also mapped in order to highlight and investigate their spatial distribution (Fig. 3).

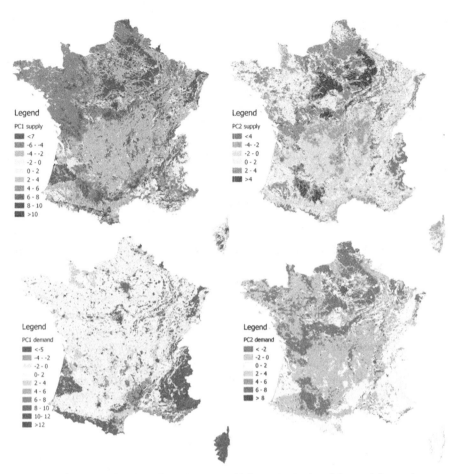

Fig. 3. PC1 *(left)* and PC2 *(right)* maps of ES supply *(top)* and demand *(bottom)*

As expected, supply of natural ES is higher in more natural or semi-natural land-scapes characterized by low urbanization and low agricultural uses, which is typical of the southern part of France. The lowest values of natural ES supply are individuated in the agricultural landscapes of Paris basin and around the other major cities. The highest demand of "natural services" is clearly localized in the urban centres, while the landscapes dominated by natural element are the less demanding. The highest supply of agricultural ES is localized in agricultural areas. Obviously, also the demanded agricultural ES is localized in agricultural areas, while the demand of "urban services" is concentrated in the most urban landscapes (main cities and southern coast). The Kruskal-Wallis rank sum test statistically confirmed the high correlation between the gradient, expressed by the 13 landscape typologies, and the bundles of ES supplied and demanded expressed by the four PC taken in consideration ($p < 0.001$).

4   Conclusions

This study can contribute to identifying the need to focus on the composition and spatial configuration of the entire urban–rural–natural gradient expressed by landscapes, in order to obtain integrated information supporting decisions regarding the arrangement, intensity and functionality of land uses. To this aim, KDE techniques and multivariate analysis to model and classify the types of landscape along such gradient demonstrated their applicability and replicability, since the suggested approach begins from a free, commonly used data source, represented by Corine Land Cover data.

The gradient characterization allows a better contextualization of ES demand and supply, which is expected to be an answer to structural dissimilarities and peculiarities especially in fringe areas or, more generally, in transitional landscapes (e.g. urban–rural landscapes, rural–natural landscapes). Thus, the approach can be effective to integrate consideration of the entire landscape gradient into planning process topics such as landscape multi–functionality and ecosystem services provided for human well-being.

The assessment of the ES demand and supply, even though based on a simplified approach, can represent an interesting attempt to overcome the difficulties related to the comparison and integrated analysis of numerous, different services and functions. Such an analysis along the gradient can be very relevant for a more effective landscape analysis and planning at regional or national level. The approach could be improved by means of a more detailed and effective ES selection, also taking into account the landscapes' structural characteristics and focusing on spatio–functional features of transitional landscapes.

Acknowledgements. This research was developed within the framework of the project TRUSTEE (Towards RUral Synergies and Trade-offs between Economic development and Ecosystem services) funded by the funding bodies partners of RURAGRI ERANET (contract FP7 235175)

References

1. Forman, R.T.T.: Some general principles of landscape and regional ecology. Landscape Ecology **10**, 133–142 (1995)
2. Bridges, L., Crompton, A., Schaffer, J.: Landscapes as gradients: The spatial structure of terrestrial ecosystem components in southern Ontario Canada. Ecological Complexity **4**, 34–41 (2007)
3. Forman, R.T.T., Godron, M.: Landscape ecology. John Wiley and Sons, New York (1986)
4. Luck, M., Wu, J.: A gradient analysis of urban landscape pattern: a case study from the Phoenix metropolitan region, Arizona USA. Landscape Ecology **17**, 327–339 (2002)
5. McDonnell, M.J., Pickett, S.T.A.: Ecosystem structure and function along urban rural gradients - an unexploited opportunity for ecology. Ecology **71**, 1232–1237 (1990)
6. McDonnell, M.J., Hahs, A.K.: The use of gradient analysis studies in advancing our understanding of the ecology of urbanizing landscapes: current status and future directions. Landscape Ecology **23**, 1143–1155 (2008)

7. Burrough, P.A., Frank, A.U.: Geographic objects with indeterminate boundaries. Taylor & Francis (1996)
8. Cavailhès, J., Peeters, D., Sékeris, E., Thisse, J.-F.: The periurban city: why to live between the suburbs and the countryside. Regional Science and Urban Economics **34**, 681–703 (2004)
9. Valentini, A.: Il senso del confine – Colloquio con Piero Zanini. Ri-Vista Ricerche per la Progettazione del Paesaggio **4**, 70–74 (2006)
10. Vejre, H., Jensen, F.S., Thorsen, B.J.: Demonstrating the importance of intangible ecosystem services from peri-urban landscapes. Ecological Complexity **7**(3), 338–348 (2010)
11. Llausàs, A., Nogué, J.: Indicators of landscape fragmentation: The case for combining ecological indices and the perceptive approach. Ecological Indicators **15**(1), 85–91 (2012)
12. Baker, W.L.: A review of models of landscape change. Landscape Ecology **2**, 111–133 (1989)
13. Pryor, R.J.: Defining the rural-urban fringe. Social Forces **47**, 202–215 (1968)
14. Thapa, R., Murayama, Y.: Land evaluation for peri-urban agriculture using analytical hierarchical process and geographic information system techniques: A case study of Hanoi. Land Use Policy **25**, 225–239 (2008)
15. Wehrwein, G.S.: The rural-urban fringe. Economic Geography **18**, 217 (1942)
16. Brook, R., Davila, J.D.: The peri-urban interface: a tale of two cities. Development Planning Unit, UCL (2000)
17. Vizzari, M., Sigura, M.: Landscape sequences along the urban–rural–natural gradient: A novel geospatial approach for identification and analysis. Landscape and Urban Planning **140**, 42–55 (2015)
18. Potschin, M.: Land use and the state of the natural environment. Land Use Policy **26**, 170–177 (2009)
19. Wiggering, H., Dalchow, C., Glemnitz, M., Helming, K., Müller, K., Schultz, A., Stachow, U., Zander, P.: Indicators for multifunctional land use - Linking socio-economic requirements with landscape potentials. Ecological Indicators **6**(1), 238–249 (2006)
20. Plieninger, T., Dijks, S., Oteros-Rozas, E., Bieling, C.: Assessing, mapping, and quantifying cultural ecosystem services at community level. Land Use Policy **33**, 118–129 (2013)
21. Vizzari, M.: Spatial modelling of potential landscape quality. Applied Geography **31**, 108–118 (2011)
22. Sigura, M., Peccol, E., Piani, L.: High Nature Value Farmland (Hnvf) and Ecological Networks: Their Role in the Sustainability of Trans Border Regions. disP – The Planning Review **46**(183), 60–68 (2010)
23. Vizzari, M.: Spatio-temporal analysis using urban-rural gradient modelling and landscape metrics. In: Murgante, B., Gervasi, O., Iglesias, A., Taniar, D., Apduhan, B.O. (eds.) ICCSA 2011, Part I. LNCS, vol. 6782, pp. 103–118. Springer, Heidelberg (2011)
24. Modica, G., Vizzari, M., Pollino, M., Fichera, C.R., Zoccali, P., Di Fazio, S.: Spatio-temporal analysis of the urban–rural gradient structure: an application in a Mediterranean mountainous landscape (Serra San Bruno, Italy). Earth System Dynamics **3**, 263–279 (2012)
25. Vizzari, M., Sigura, M.: Urban–rural gradient detection using multivariate spatial analysis and landscape metrics. Journal of Agricultural Engineering XLIV **91**(1), 453–459 (2013)
26. Gentili, S., Sigura, M., Bonesi, L.: Decreased small mammals species diversity and increased population abundance along a gradient of agricultural intensification. Hystrix, the Italian Journal of Mammalogy 25(1), (2014)
27. Bailey, T.C., Gatrell, A.C.: Interactive spatial data analysis. Longman, Harlow (1995)

28. Cai, X., Wu, Z., Cheng, J.: Using kernel density estimation to assess the spatial pattern of road density and its impact on landscape fragmentation. International Journal of Geographical Information Science 27(2), 222–230 (2013)

29. Gatrell, A.C., Bailey, T.C., Diggle, P.J., Rowlingson, B.S.: Spatial Point Pattern Analysis and Its Application in Geographical Epidemiology. Transactions of the Institute of British Geographers 21, 256–274 (1996)

30. Borruso, G.: Network Density Estimation: A GIS approach for analysing point patterns in a network space. Transactions in GIS 12, 377–402 (2008)

31. Jones, M.C., Marron, J.S., Sheather, S.J.: A Brief Survey of Bandwidth Selection for Density Estimation. Journal of the American Statistical Association 91, 401–407 (1996)

32. Danese, M., Lazzari, M., Murgante, B.: Kernel density estimation methods for a geostatistical approach in seismic risk analysis: the case study of potenza hilltop town (southern italy). In: Gervasi, O., Murgante, B., Laganà, A., Taniar, D., Mun, Y., Gavrilova, M.L. (eds.) ICCSA 2008, Part I. LNCS, vol. 5072, pp. 415–429. Springer, Heidelberg (2008)

33. Lloyd, C.D.: Local models for spatial analysis. CRC Press (2007)

34. Everitt, B.S., Landau, S., Leese, M., Stahl, D., Shewhart, W.A., Wilks, S.S.: Cluster Analysis, 5th edn. Wiley Series in Probability and Statistics (2011)

35. Irvin, B.J., Ventura, S.J., Slater, B.K.: Fuzzy and isodata classification of landform elements from digital terrain data in Pleasant Valley Wisconsin. Geoderma 77, 137–154 (1997)

36. Macqueen, J.B.: Some methods for quantification of the multivariate observations, western management science institute, University of California, working paper 96 (1967)

37. Ball, G.H., Hall, D.J.: ISODATA, a novel method of data analysis and pattern classification. Stanford Research Institute, Menlo Park (1965)

38. Ali, A., de Bie, C.A.J.M., Skidmore, A.K., Scarrott, R.G., Hamad, A., Venus, V., Lymberakis, P.: Mapping land cover gradients through analysis of hyper-temporal NDVI imagery. International Journal of Applied Earth Observation and Geoinformation 23, 301–312 (2013)

39. Lioubimtseva, E., Defourny, P.: GIS-based landscape classification and mapping of European Russia. Landscape and Urban Planning 44, 63–75 (1999)

40. Richards, J.A.: Remote Sensing Digital Image Analysis: An Introduction. Springer, Berlin (1999)

41. Usha, K., Singh, B.: Potential applications of remote sensing in horticulture: a review. Scientia Horticulturae 153, 71–83 (2013)

42. Van der Kwast, J., Uljee, I., Engelen, G., Van de Voorde, T., Canters, F., Lavalle, C.: Using remote sensing derived spatial metrics for the calibration of land-use change models. Joint Urban Remote Sensing Event, pp. 1–9 (2009)

43. Millennium Ecosystem Assessment: Ecosystems and Human Well-being: Synthesis. Island Press/World Resources Institute, Washington, DC (2005)

44. Burkhard, B., Kroll, F., Nedkov, S., Müller, F.: Mapping ecosystem services, supply and budgets. Ecological Indicators 21, 17–29 (2012)

45. de Groot, R., Alkemade, R., Braat, L., Hein, L., Willemen, L.: Challenges in integrating the concept of ecosystem services and values in landscape planning, management and decision making. Ecological Complexity 7, 260–272 (2010)

46. Haines-Young, R., Potschin, M.: Landscapes, sustainability and the place-based analysis of ecosystem services. Landscape Ecology 28, 1053–1065 (2012)

47. Costanza, R., D'Arge, R., de Groot, R.S., Farber, S., Grasso, M., Hannon, B., Limburg, K., Naeem, S., O'Neill, R.V., Paruelo, J., Raskin, R.G., Sutton, P., van den Belt, M.: The value of world's ecosystem services and natural capital. Nature **387**, 253–260 (1997)
48. Vizzari, M., Sigura, M., Antognelli, S.: Ecosystem services demand, supply and budget along the urban-rural-natural gradient. In: The 43th International Symposium on Agricultural Engineering, Actual Tasks on Agricultural Engineering, February 24-27, Opatija, Croatia, pp. 473–484 (2015)
49. Fisher, B., Turner, K.R.: Ecosystem services: classification for valuation. Biological Conservation **141**, 1167–1169 (2008)
50. Naidoo, R. Balmford, A., Costanza, R., Fisher, B., Green, R.E. Lehner, B., Malcolm, T.R., Ricketts, T.H.: Global mapping of ecosystem services and conservation priorities. In: Daily, G.C. (eds.) Proceedings of the National Academy of Sciences of United States of America, vol. 105, pp. 9495–9500 (2008)
51. Gimona, A., van der Horst, D.: Mapping hotspots of landscape functions: a case study on farmland afforestation in Scotland. Landscape Ecology **22**(8), 1255–1264 (2007)
52. Burkhard, B., Kroll, F.: Maps of ecosystem services, supply and demand. In: Cleveland, C.J. (eds.) Encyclopedia of Earth, Environmental Information Coalition. National Council for Science and the Environment, Washington, D.C. (2010)
53. Bastian, O., Haase, D., Grunewald, K.: Ecosystem properties, potentials and services – The EPPS conceptual framework and an urban application example. Ecological Indicators **21**, 7–16 (2012)
54. Burkhard, B., Kroll, F., Müller, F.: Landscapes' Capacities to Provide Ecosystem Services – a Concept for Land-Cover Based Assessments. Landscape online **15**, 1–22 (2009)
55. Levine, N.: Crime Mapping and the Crimestat Program. Geographical Analysis **38**(1), 41–56 (2006)
56. Hengl, T.: Finding the right pixel size. Computers & Geosciences **32**, 1283–1298 (2006)
57. Burkhard, B., Kandziora, M., Hou, Y., Müller, F.: Ecosystem Service Potentials, Flows and Demands – Concepts for Spatial Localisation. Indication and Quantification. Landscape Online **32**, 1–32 (2014)
58. Jacobs, S., Burkhard, B., Van Daele, T., Staes, J., Schneiders, A.: The Matrix Reloaded: A review of expert knowledge use for mapping ecosystem services. Ecological Modelling **295**, 21–30 (2015)
59. Kruskal, W.H., Wallis, W.A.: Use of ranks in one-criterion variance analysis. Journal of the American Statistical Association **47**(260), 583–621 (1952)
60. Hollander, M., Wolfe, D.A.: Nonparametric Statistical Methods. John Wiley & Sons, pp. 115–120 (1973)

Walkability Explorer: Application to a Case-Study

Ivan Blečić, Arnaldo Cecchini, Tanja Congiu, Francesco Fancello,
Giovanna Fancello[✉], and Giuseppe A. Trunfio

Department of Architecture, Design and Urban Planning, University of Sassari, Sassari, Italy
{ivan,cecchini,tancon,gfancello,trunfio}@uniss.it,
francesco-fancello@hotmail.it

Abstract. In a few recent papers we presented a methodology and the related planning and design support tool, Walkability Explorer, for the evaluation of walkability of places which are relevant for people's capabilities. The method is an attempt to move beyond the known approaches to evaluating walkability based on the analysis of proximity to urban places and on macro urban and socioeconomic factors, because it conceptualises walkability as the effective capability to walk offered by the environment thanks to micro-urban characteristics. It evaluates how the urban environment is conducive to walk by combining three elements: the destinations/opportunities reachable by foot, their walking distance and the quality of the path to these destinations. Following this approach, here we present and discuss an example assessment of walkability for the city of Alghero (Italy).

Keywords: Walkability · Multi criteria analysis · Planning and design support tool

1 Introduction

Research on walkability has proliferated in recent years. Walkability, intended as the propensity for walking fostered by the physical environment, represents an important requirement of urban quality of life [1, 2]. The idea of walkability, we want to stress in this paper, is something more than the crude pedestrian accessibility of places. Factors such as physical features, urban design qualities and individual reactions may influence the way an individual feels about the environment as a place to walk [3, 4]. Accordingly, walkability concerns the *quality* of the accessibility and how the urban environment (built environment, social practices, etc.) is *conducive* to walking.

The more comfortable, suitable, safer and easier it is to walk in a built environment to reach the majority of existent urban opportunities, the more comfortable, accessible and attractive an urban context becomes. Such quality of cities becomes much stronger if these conditions are broadly and equitably distributed, granting the opportunities to develop people's capabilities in cities. We use 'capability' here in the specific sense of the so-called *capability approach* [5], namely as the valuable states of being that a person has effective access to. Thus, a capability is the effective freedom of an individual to choose between different things to do or to be that one has reason to value.

© Springer International Publishing Switzerland 2015
O. Gervasi et al. (Eds.): ICCSA 2015, Part III, LNCS 9157, pp. 758–770, 2015.
DOI: 10.1007/978-3-319-21470-2_55

In this context, we may talk of *urban capabilities* as an important dimension of overall human capabilities: it counts how our cities "work" – the way they are built and organized, the way the social practice of their use are –, and walkability, *in the way we operationalize the concept*, may play a pivotal role in it [6, 7, 8].

In the last several years the topic of walkability has gained a growing attention of the research community, especially in the fields of urban design, transport planning and public health [9, 10]. Most interesting advancements report techniques and tools for measuring and assessing walkability and how the urban space is conducive to walk. Above all, we assist a wide production of analytical tools for planning which incorporate quantitative and qualitative factors affecting travel behaviour and people choices.

Many researches [11, 12, 13] analyse walkability exclusively through physical features at a macro, neighbourhood or a city block scale, being such features easier to measure and operationalize through common tools of spatial analysis. However, these physical features are not exhaustive to what's relevant for the experience of walking down a particular street. Quantitative measures often tend to miss qualitative factors like sense of enclosure, imageability and "liveliness" [3, 4, 14, 15].

Our work is an attempt to push beyond these methods and conceptualizes walkability as the effective capability of individuals to walk, influenced directly or indirectly by specific micro-characteristics of urban environment. To evaluate to what extent the urban environment is conducive to walk, we combine three elements: the number and type of destinations/opportunities reachable by foot, their walking distance, and the quality of the paths and their surrounding environment towards these destinations. These elements are put together into an indicator – "walkability score" – we assign to each point in space.

Our research benefits from the development and evolution of GIS software and open source tools and applications such as Google Street View and Open Street Map which allow to simplify and somewhat automate the processes of data collection [16, 17] bringing to more precise measurements, better reproducibility and easier transferability and comparability of results [18].

In this paper first briefly illustrate the method used to evaluate walkability and the related planning and design support tool. We in particular present the advancements over the previous versions: the extension of the set of analysed variables, the refinement of the evaluation method, and the improvements of the Walkability Explorer (WE) design support tool. Finally, an application to the city of Alghero allows us to show the results that may be obtained with WE, and offers suggestions for further developments.

2   The Evaluation Model of Walkability

In this section we offer a brief overview of the evaluation model of walkability. For a more detailed presentation, including the discussion of the rationale behind the modelling choices, see [6].

The approach is based upon a model of how people at different points in space can walk to destinations of interest in an urban area, using a detailed graph representation of the street network. Destinations may be divided into separate categories each representing a different type of urban opportunity (e.g. green areas, commercial and retail, services, etc.). For each category of destinations, we assume that a resident living at a point in space will walk to available destinations a certain amount of times, and will derive from that some benefit β defined by the following constant elasticity of substitution (CES) function:

$$\beta(x) = \left(\sum_{i=1}^{n} X_i^{\rho} \right)^{\frac{1}{\rho}}$$
(1)

where n is the number of available destinations, X_i is the number of times the resident visits the i-th destination and $1/(1 - \rho)$ is the elasticity of substitution among destinations.

We impose the following constraint to the pedestrian:

$$\sum_{i=1}^{n} c_i X_i \leq M$$
(2)

where c_i is the "cost" the pedestrian foregoes to reach the destination i, and M is the available budget with a conventional constant value.

In the model, the street network is represented by a graph, composed of nodes and edges connecting the nodes. A path from an origin to a destination is thus a set of interconnected edges of the graph.

The important bit of the model is that, beside mere distances, edges are described on further attributes which are relevant for their walkability. These attributes serve to model the "cost" of a path used in the constraint expression (2), since the "cost" of a path of p edges is defined as:

$$c = c_0 + \sum_{k=1}^{p} l_k \left(1 - \left(\sum_{l=1}^{r} w_l a_{k,l}^{r} \right)^{\frac{1}{r}} \right)$$
(3)

where c_0 is the fixed cost, l_k is the length of the k-th edge in the path, $a_{k,l} \in [0,1]$ is the value of that edge's l-th attribute, w_l is the weight of the attribute ($\sum w_l = 1$), and r is a parameter with $1/(1 - r)$ being the elasticity of substitution among attributes. The variable part of the expression (3) yields unit cost of 1 when all attributes are at their minimum (i.e. 0), and approaches 0 when attributes approach the maximum (i.e. 1).

In Table 1 and Table 2 we report the attributes, including their weights and scales of measurement, used in the application described in the present paper. These attributes are revised and expanded over the original set in [6].

Among many alternative paths from an origin to a destination in a street network, we plug the less costly (i.e. the one with the best combination of length and quality as channelled through equation (3)) into expression (2). Finally, the walkability score

w we attribute to a point in space is defined as the maximum benefit which, under the assumption of the above model, may be yielded by a person residing at that specific point. In other words, for each node in the graph the walkability score W is:

$$W = \max \beta(x) \tag{4}$$

Under the constraint given by Eq. (2), the maximum benefit is obtained when:

$$X_i = \frac{c_i^{\frac{1}{\rho-1}} M}{\sum_{j=1}^{n} c_j^{\frac{\rho}{\rho-1}}} \tag{5}$$

3   Implementation

A destination (i.e. urban opportunity) in space can in the new version of WE be represented in two ways: as a point-feature, for activities and services located in a specific point in space; or as an areal-feature, for facilities more extended in space such as open spaces, parks and natural areas, and so on.

The new, extended list of edges attributes (Table 1 and Table 2) has been defined with particularly attention for urban aspects relevant for the comfort and efficiency of the walk, elements that generate the sense of safety and influence the pleasantness of walking and the attractiveness of the path and its surroundings. As previously declared, our interest for the quality of urban design features originates from the need to understand and assess urban conditions and elements which subjectively (and depending on individual abilities) influence people spatial behaviours, choices, and inclination for walking [4].

For this reason we extended the analysis to micro aspects of the urban environment often neglected in several walkability studies (e.g. see [11, 12, 13]). Such aspects are determining factors of the relationship between physical features of the street environment and the walking behaviour; they concern elements and urban conditions that objectively and subjectively influence people's spatial choices and attitudes (Figure 1). Some of them are observable phenomena and features easily recognisable and measurable: cyclability (the possibility of bicycle riding on the road); width of the roadway (number of lanes on the street); car speed limit (km/h); one way street; on-street parking (authorized or not); path slope; paving quality and degree of maintenance (condition of the path in terms of materials and presence of bumps, cracks, holes); lighting (equipment of street lights).

Table 1. Physical features of the path

Attributes	Weight	Scale values
Cyclability	2/30	exclusive lane (0.8); off-road lane (0.5); on.road lane (0.3); not possible/prohibited (0.1)
Width of the roadway	1/30	pedestrian way (0.8); one car lane (0.6); 2 car lanes (0.5); 3 car lanes (0.3); >3 car lane (0.1)
Car speed limit	2/30	pedestrian way (0.8); 20 Km/h (0.7); 30 Km/h (0.5); 50Km/h (0.3); 70 Km/h (0.1)
One way street	1/30	pedestrian way (0.8); yes (0.5); no (0.1)
On-street parking	1/30	prohibited parking (0.8); permitted (0.5); illegal parking (0.1)
Width of sidewalk (accessible)	2/30	wide (0.8); comfortable (0.7); minimum (0.5); inadequate (0.3); lacking (0.1)
Separation of pedestrian route from car roadway	2/30	marked/strong (0.8); weak (0.5); lacking (0.1)
Path slope	2/30	smooth (0.8); light (0.5); rise (0.1)
Paving (quality and degree of maintenance)	2/30	fine (0.8); cheap (0.5); bumpy (0.1)

Table 2. Quality of urban design features

Attributes	Weight	Scale values
Lighting	1/16	excellent (0.8); good (0.6); inadequate (0.3); lacking (0.1)
Shelter and shade	1/16	strong (0.8); weak (0.5); lacking (0.1)
Sedibility	1/16	extended (0.8); thin (0.5); lacking (0.1)
Frequency of services and activities	1/16	abundant (0.8); somewhat (0.6); rare (0.3); no services/activities (0.1)
Attractiveness from an architectural and urban viewpoint	1/16	preponderance of pleasant elements (0.8); presence of a few pleasant elements (0.6); lack of pleasant or disturbance elements (0.4); presence of a few disturbance elements (0.2); preponderance of disturbance elements (0.1)
Attractiveness from an environmental point of view	1/16	preponderance of pleasant elements (0.8); presence of a few pleasant elements (0.6); lack of pleasant or disturbance elements (0.4); presence of a few disturbance elements (0.2); preponderance of disturbance elements (0.1)
Transparency and permeability of the public-private space	1/16	permeable (0.8); filtered (0.5); separated (0.1)
Urban texture	1/16	dense (0.8); park or green space (0.6); low density (0.4); undeveloped land (0.1)

Other attributes refer to urban conditions and combination of urban features that require evaluative judgments, so they deserve to be better explained:

— the effective accessible *width of the sidewalk* or pedestrian area: rather than the width of the sidewalk, we examine qualitatively the effective possibility to walk through the space without any detractor (physical or perceptible); possible configurations range from wide sidewalk to complete lack of pedestrian space, and have been defined based on the number of individuals able to comfortably walk side by side;

— *separation of pedestrian route from car roadway*: relates to the presence of physical or perceptual elements – horizontal or vertical – which increase pedestrians' safety and comfort (planting strips, raised planters, trees, stakes, walls,…).;

— *landscape perception* and *attractiveness*: the presence of pleasant features, with architectural, urban, environmental or historical values, affects pedestrian's use and interaction with space and especially influences the willingness to walk; for this aspect we considered the presence of enjoyable or disturbing elements along the path and in the surroundings;

— *shelter and shade*: evaluates the possibility to take shelter given the physical configuration of urban space; it considers both environmental and architectural features that allow pedestrians to shelter from the rain, wind or sun;

— *sedibility*: evaluates the possibility to seat and linger along the walk [3]; seating opportunities include benches as well as other urban elements not specifically designed for this purpose but used by people to sit on (walls, stoops, fountain borders, ledges, planters, etc.);

— *frequency of services and activities along the path*: the analysis of the frequency of active services and facilities aims to capture how much a street is perceived as a safe environment (due to the presence of individuals and storekeepers) and how attractive it is as a result of a high density of activities (one of the most stressed aspects of walkability argued by numerous researchers [2, 4, 14, 24]);

— *transparency and permeability* of urban space (public and private): aims to describe if and how the urban environment offers an engaging experience by the presence of transitional spaces [3] (front yards, entry seatbacks, porticos, …) and in-between elements – both architectural and landscaping – that provide a soft transition between public and private space; possible configurations range from permeable spaces (continuity and interchange between the public right-of-way and the surrounding) to separated ones (e.g. continuous walls or fences).

— *urban texture*: accounts for building density and typology ranging from dense continuous urban fabric with presence of urban park, plazas or green spaces, to low density referred to scattered urban fabric, and to undeveloped land, *terrain vague*, abandoned or obsolete spaces, buildings and green open spaces.

The weights of the attributes were assigned by a group of urban designers and planners. We are aware that the validity of the results ultimately relies on the possibility of modes of empirical validation of these parameters. We are undertaking a study of choice modelling based on experimental design of stated preferences [21] specifically devised for the purpose of parameterisation of the hereby presented evaluation model. We should add that WE is fairly flexible, and offers user-friendly interfaces which allow users to set and adjust the parameters of the model and to modify the attributes, according to particular normative assumptions or empirical findings.

Fig. 1. Example of the method employed for the analysis of edge's attributes

4   Walkability Explorer

In WE, all the nodes of the graph are assumed as origins of the trips. Compared with the previous version of the software presented in [6], the new one improves on how the destinations and their distribution in space are handled. Rather than having the potential destinations derived from the attributes of edges as before, now the software allows a more natural treatment of destinations, by allowing them to be imported and located at their real location in space. Then, for a point-type attraction, the respective destination node on the street network is identified by finding the node closest to the destination, while an areal-type attraction is represented by a (set of) rectangular cell in relation to the user-defined rasterization of space.

To compute a map of walkability scores, WE executes the following procedure:

1. using a suitable graph search procedure (i.e., the well-known Dijkstra's algorithm [22] in the current version of WE) we determine all the cheapest paths (i.e. the less costly in the sense of Eq. (3)), between each couple of origin and destination nodes;
2. then, for each origin node, the costs of the cheapest paths to each destination node are used for the computation of the node walkability score according to Eqs. (1) and (5);
3. finally, since the available street network does not represent all the areas accessible to pedestrians, we interpolate the walkability scores assigned to nodes to a raster of a given resolution representing the urban area; currently, we use the simple Inverse Distance Weighting (IDW) [23] method for interpolation.

This procedure allows to produce a set of raster maps representing both the geography of urban constraints and walkability scores in urban space. This approach makes possible to compare different planning and urban design scenarios by comparing their impacts on the respective spatial distribution of walkability and thus provide support for policy decision making.

5   The Case Study of Alghero

We tested the evaluation model and the WE tool by assessing the walkability in the town of Alghero, of about 40.000 inhabitants, in North-East Sardinia (Italy).

The street network was taken from the Open Street Map (OSM) project, and Free Java Open Street Map editor [19] was used to edit and enrich the baseline OSM map with our additional attributes. The analysis of the street network to assign values to the edge attributes was undertaken between August and October 2014. A team of urban planners carried out the field research by direct *in situ* observation and with the support of images provided by Google Street View (see Figure 1). Census data have been used to compare walkability score distribution with respect to population density.

To get the location and the type of destinations, Bing Maps, Google Maps and Yellow Pages services were used together with Bathcgeocoding [20], a free web service for address georeferencing.

First we generated maps (Figure 2) that categorise paths on the basis of their costs computed according to the urban attributes proposed in this paper. The parameters adopted for the assessment are reported in Table 3, while the weights assigned to edge attributes are those reported in Table 1 and Table 2.

Then we assessed the possibility of individuals to access different categories of urban opportunities. Three categories of urban opportunities have been considered (Figure 2): *commercial* (food shops, bars, butchers, bakeries, fish shop, supermarkets, tobacco shops), *services* (health services, schools excluding university, banks and public offices) and *leisure and green urban areas* (city parks, beaches, sport services). We explicitly chose to assess separately different categories of urban opportunities, instead of resorting to a synthetic indicator, in order to maintain the wealth of information that is a useful support for policy makers in urban projects. Coherently with the capability approach, it may be a mistake to compensate a deficit in one urban capability with another one; for this reason a compensatory evaluative method would be unsuitable, so it seems better to study separately the results for each category of opportunity.

Table 3. Parameters used for the assessment of Alghero's central urban area.

Parameter	Value
EOS – elasticity of substitution among destinations $1/(1 - \rho)$	2.5
Virtual budget M	1000.0
Fixed cost c_0	0.0
EOS among attributes $1/1 - r)$	0.3
Intensification distance δ [m]	100.0
Interpolation radius σ [m]	100.0
Power parameter α	2.0

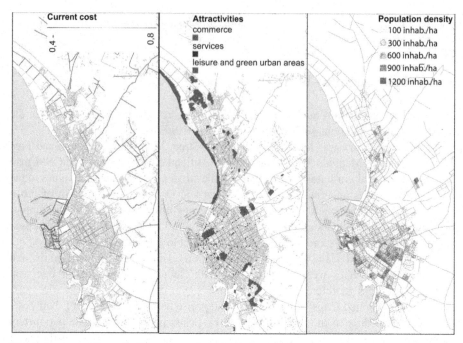

Fig. 2. Computed "costs" of the street edges (left), map of attractions (central), map of population density (right)

Figure 2 illustrates the computed costs of street edges (left) and the spatial distribution of urban attractions, distinguished by category, used for the walkability evaluation. Given the location of attractions in space, WE identifies the areas of attractions using a regular grid of cells, according to a resolution set by the user – in this study we set the resolution to 10 meters - and constructs the sets of destination nodes (i.e. the attractors are rasterized).

Once the costs of edges have been computed and given the distribution of urban opportunities, the WE software can compute walkability scores. The final output of the software are geo-referenced walkability score maps related to each of the categories of attractions. Figure 3 shows walkability score maps for the central area of Alghero, distinguished by category: commerce, services and leisure. One can observe that the walkability related to commercial attractions is higher in the central area of the city, together with the walkability for urban services. On the other hand, the map of walkability scores for leisure services reveals a stronger influence by the number of environmental attractors spread along the coast.

By definition, the theoretical minimum of the scale of walkability is 0, while the maximum is unbounded, although limited in practice by the possible configurations and number of possible destinations in urban space. To build some intuition of what walkability scores reflect on the ground, in Figure 4 we present some examples of origin-destination paths and the corresponding walkability score. Comparing the situations, one notes that the node with a walk score 26 is on average closer to reachable

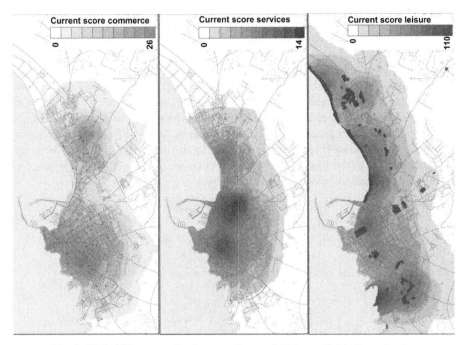

Fig. 3. Walkability scores for the central area of Alghero, divided by category

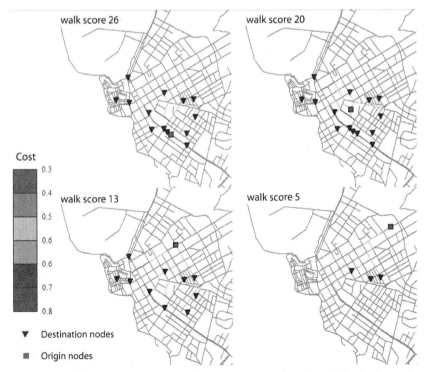

Fig. 4. Streets network computed walkability with origin and commerce destination nodes

destinations, its central location makes easier to arrive at a greater number of destinations, and the paths towards destinations have on average a comparatively low walkability costs.

WE DSS allows further more detailed explorations on the neighbourhood level. For example, one can observe that the Sant'Agostino neighborhood (Figure 5) presents relatively low walkability costs but also low walkability scores. This is above all due to the lack of services and leisure opportunities in this urban area, a detrimental condition that assumes greater significance if we consider the high population density of this part of the town (Figure 2). Conversely, in the urban area in the northern part of the city (next to the coastline) the higher costs are mitigated by the presence of several urban attractors (mostly leisure facilities).

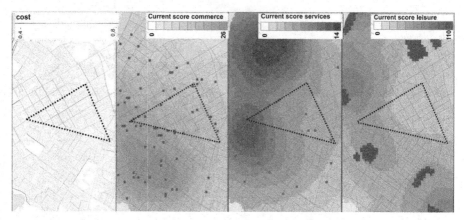

Fig. 5. Sant'Agostino neighborhood: computed costs; commerce, services, leisure scores

6   Conclusions

In this paper we presented an advancement of our research to develop of a planning support tool for walkability. The evaluation model and the design support system aim at improving knowledge, effectiveness, relevance, and inclusiveness of urban design and transportation planning, and should be considered a contribution to the effort to develop adequate methods and suitable processes for creating purposeful knowledge, i.e. analytical methods [25] to support policy design.

The application to the city of Alghero suggested different reflections: some of them stimulated by the results of the assessment, others related to the need of improving design tools and planning methods in order to support decision making.

The walkability assessment in Alghero remarked some urban circumstances which undermine the real capacity of people to access opportunities in space. For instance, areas with good pedestrian infrastructures but lacking urban opportunities, as the Sant'Agostino neighbourhood, or conversely, the Pietraia neighbourhood, in the north of the city, abundantly supplied with urban opportunities but not conducive to walk. Some other parts of Alghero present both problems (scarcity of opportunities and poor walking facilities): it is the case of areas of recent urban expansion which require a rethinking of pedestrian mobility and access to urban opportunities. All these situations need to be dealt with in future policies and planning actions according to a pedestrian-oriented design approach striving for extending the urban capabilities of individuals.

With regard to the tool, it is worth considering some possibilities to extend and improve its analytical and decision support aspect. Procedures should be devised for a faster and more automated construction of spatial datasets. This requires a deeper exploration of the possibilities of integrating different sources and of (semi)automatic data harvesting, perhaps in combination with techniques of pattern recognition and computer vision. Furthermore, the set of attributes should be enriched by considering new elements of the street such as intersections that act as 'barriers' for walking; the volume of pedestrians on the street that affect attractiveness and security of places; the opportunity of interchange with different means of transportation as a condition that enhances people ability to access. We also want to mention the importance of the time variable: it should be preferable to assess the conductivity for walking in different times of the day or in different seasons.

Another subject that deserves a deeper reflection is the definition of weights and trade-offs between attributes. For this purpose urban planning and transportation experts have to work together with other specialists to better understand people's behaviour, choices and preferences. A further development of our research concerns precisely the use of the methods of choice modelling in order to define weights of attributes in accordance to experimentally determined pedestrian preferences. This approach can also be an useful step to better consider differences among profiles of pedestrians based on age, gender, disabilities and others factors (i.e. individual abilities) which influence the development of person's capabilities in the city.

Finally, to become a more complete decision support tool for assessing urban capabilities, WE should be able to also consider car and public transportation accessibility, and the way they interact with pedestrian accessibility. Such incorporation of non-pedestrian mobility into WE is an indispensable step to go beyond the neighbourhood dimension and to extend the evaluation of the quality of accessibility to urban, metropolitan and regional level. All this is due to more comprehensively investigate overall urban capabilities of people in the city.

References

1. Talen, E.: Pedestrian access as a measure of urban quality. Planning Practice and Research **17**(3), 257–278 (2002)
2. Frank, L.D., Sallis, J.F., Saelens, B.E., Leary, L., Cain, K., Conway, T.L., Hess, P.M.: The Development of a Walkability Index: Application to the Neighborhood Quality of Life Study. British Journal of Sports Medicine **29**, 1–38 (2009)
3. Porta, S., Renne, J.L.: Linking urban design to sustainability: formal indicators of social urban sustainability field in Perth. Western Australia. Urban Design International **10**(1), 51–64 (2005)

4. Ewing, R., Handy, S.: Measuring the unmeasurable: Urban Design Qualities Related to Walkability. Journal of Urban Design **14**(1), 65–84 (2009)
5. Sen, A.: Capability and well-being. In: Nussbaum, M., Sen, A. (eds.) The Quality of Life, pp. 30–53. Clarendon Press, Oxford (1993)
6. Blečić, I., Cecchini, A., Congiu, T., Fancello, G., Trunfio, G.A.: Evaluating walkability: a capability-wise planning and design support system. International Journal of Geographical Information Science (2015). doi:10.1080/13658816.2015.1026824
7. Blečić, I., Cecchini, A., Congiu, T., Fancello, G., Trunfio, G.A.: Walkability explorer: an evaluation and design support tool for walkability. In: Murgante, B., et al. (eds.) ICCSA 2014, Part IV. LNCS, vol. 8582, pp. 511–521. Springer, Heidelberg (2014)
8. Blecic, I., Cecchini, A., Congiu, T., Pazzola, M., Trunfio, G.A.: A design and planning support system for walkability and pedestrian accessibility. In: Murgante, B., et al. (eds.) ICCSA 2013, Part IV. LNCS, vol. 7974, pp. 284–293. Springer, Heidelberg (2013)
9. Saelens, B.E., Handy, S.L.: Built Environment Correlates of Walking: A Review. Med. Sci. Sports Exerc. **40**(7), 550–566 (2008)
10. Capp, C.J., Maghelal, P.K.: Walkability: A Review of Existing Pedestrian Indices. Journal of the Urban and Regional Information Systems Association **22**(2), 5–9 (2011)
11. Krizek, K.J.: Operationalizing Neighborhood Accessibility for Land Use-Travel Behavior Research and Regional Modeling. Journal of Planning Education and Research **22**, 270–287 (2003)
12. Iacono, M., Krizek, K.J., El-Geneidy, A.: Measuring non-motorized accessibility: issues, alternatives, and execution. Journal of Transport Geography **18**, 133–140 (2010)
13. Ewing, R., Cervero, R.: Travel and the Built Environment. A meta analysis. Journal of Ametrican Planning Association **76**(3), 265–294 (2010)
14. Jacobs, J.: The Death and Life of Great American Cities. Random House, New York (1961)
15. Mehta, V.: Walkable streets: pedestrian behavior, perceptions and attitudes. Journal of Urbanism. **1**(3), 217–245 (2008)
16. Rundle, A.G., Bader, M.D., Richards, C.A., Neckerman, K.M., Teitler, J.O.: Using Google Street View to audit neighborhood environments. Am. Journal of Preventive Medicine **40**(1), 94–100 (2011)
17. Clarke, P., Ailshire, J., Melendez, R., Bader, M., Morenoff, J.: Using Google Earth to conduct a neighborhood audit: reliability of a virtual audit instrument. Health Place **16**(6), 1224–1229 (2010)
18. Graham, D.J., Hipp, J.A.: Emerging technologies to promote and evaluate physical activity: Cutting-edge research and future directions. Frontiers in Public Health **2** (2014)
19. Open Street Map. https://josm.openstreetmap.de/
20. Bathcgeocoding. http://www.findlatitudeandlongitude.com/batch-geocode/
21. Hanley, N.S., Mourato, S., Wright, R.E.: Choice Modelling Approaches: a Superior Alternative for Environmental Valuation. Journal of Economic Surveys **153**, 435–462 (2001)
22. Dijkstra, E.W.: A note on two problems in connexion with graphs. Numerische Mathematik **1**(1), 269–271 (1959)
23. Shepard, D.: A two-dimensional interpolation function for irregularly-spaced data. In: Proceedings of the 1968 23rd ACM National Conference, ACM 1968, pp. 517–524 (1968)
24. Cervero, R., Kockelman, K.: Travel demand and the 3ds: density, diversity, and design. Transportation Research D **2**(3), 199–219 (1997)
25. Tsoukias, A., Montibeller, G., Lucertini, G., Belton, V.: Policy analytics: an agenda for research and practice. EURO Journal on Decision Processes **1**, 115–134 (2013)

Economic Life Prediction Model of RC Buildings Based on Fragility Curves

Marco Vona and Benedetto Manganelli$^{(\boxtimes)}$

School of Engineering, University of Basilicata, Potenza, Italy
{marco.vona,benedetto.manganelli}@unibas.it

Abstract. Retrofitting and management strategies of existing RC buildings are actually a crucial topic. In this work, a novel approach in order to define a retrofitting and management strategies is reported. The main goal is define a framework based on multidisciplinary approach. An accurate seismic performance evaluation model is considered. This probabilistic model is combined with a probabilistic depreciation model in order to define a completely probabilistic model to economic life prediction. The proposed methodology can be used in different way and applications. It can be the basis of both methods of mitigation of seismic risk strategies and as reference tool for seismic insurance model for wide building stock.

Keywords: Fragility curves · Seismic losses evaluation · Economic life prediction

1   Introduction

In this paper, the main topic is providing some effective tools to management seismic risk. To this aim, in this section some criteria, definition, and statements are reported. Based on recent studies [1], alternative and innovative approaches must be adopted in order to define possible intervention prioritization that must be associate with expected performances.

In recent years, many studies have been carried out in order to understanding and mitigation the seismic risk [2], [3], [4], [5], [6], [7], [8]. Several important project have been founded (SYNER-G, GEM, RELUIS - DPC Project, NATO Programme: Science for Peace and Security, etc.). It must be highlighted that at present, the prediction useful in seismic risk management is actually a hard work. In fact, several methodologies to manage crisis have been developed and applied with goods results in particular on short periods (little days – some months). For instance, mitigation and prevention of seismic risk could be more efficient by setting a careful maintenance and retrofitting of the built.

About this important topic, it is the opinion of the authors that quantitative economic model should be defined. This model would be defined considering an economic point of view, which involves that the higher level of loss must be adequately captured. Moreover, in order to define mitigation strategies, the decision to perform a seismic retrofit of existing buildings should be based on rational criteria, which lead

© Springer International Publishing Switzerland 2015
O. Gervasi et al. (Eds.): ICCSA 2015, Part III, LNCS 9157, pp. 771–781, 2015.
DOI: 10.1007/978-3-319-21470-2_56

into account a number of factors that are able to define the cost-effectiveness of the investment.

In this context, seismic capacities of existing RC buildings have a key role, as shown in recent seismic events (e.g. Southern Italy 1980, Turkey 1999, L'Aquila 2009, Lorca, 2011, Emilia plan 2012). In particular, old RC buildings have been showed a poor behaviour. Low seismic performances are the main reason of significant earthquake losses. In fact, earthquake losses in terms of economic, social and political activities can be considered generally as a direct consequence of physical damages on the buildings.

As consequence, studying recent Italian earthquake the urban resilience is often insistent [1]. The main consideration is that the damage suffered by buildings (private and public) has been the main reasons of losses.

Then, the resilience of communities has been observed as inexistent or very low. On long terms (e.g. after ten or more years), the return to normal conditions (when it happened) is due to considerable economic investments. At this point, a simple consideration must be made. If these resources had been used in peace time for mitigation policies, they would have produced significant economic growth and avoided many casualties.

In order to investigate about the resilience of the cities, the problem is strongly multidisciplinary and nonlinear. Simplified models (for example [9], [10]) do not seem available or easily applicable due to practical issues.

In the last years, many efforts have been carried out in order to investigate about the seismic risk of building stock. These studies have been made considering only the priority based on the seismic risk of buildings.

This simplified way may be considered substantially insufficient in order to address correctly the communities' urban vulnerability and resilience. In fact, because the best possibility to mitigate the seismic risk is to intervene a priori on vulnerability a new approach aiming at mitigating urban seismic vulnerability, maximizing system resilience, must be constructed mainly considering strategies based on reducing the vulnerability of buildings. These strategies must consider several important questions as the social and economic convenience related to the seismic retrofit of existing building.

In [7] an interesting procedure has been applied to the building stock of Potenza, in order to obtain the total repair cost due to several seismic events. Nevertheless, significant problems have been highlighted but the main topic was simple. The fundamental question is the accurate definition of damage states for each building type and their economic quantification. This topic is even more important when old existing buildings are investigated in order to perform a seismic retrofitting based on rational criteria.

These existing RC buildings are generally characterized from a non-ductile behaviour during seismic events (for example, [11], [12]). Then, during seismic events they could be suffered significant damage and consequent consistent economic losses.

As consequence, for oldest RC concrete structures, the probability of the exceeding of a given damage level should be added to their economic losses and normal depreciation (normally based on two only main factors: age and income decay).

In this paper, these different aspects are considered in order to obtain a rational instrument to perform retrofitting investment based on the relationship between cost-effectiveness-suitability.

2 Methodological Approach

In modern earthquake, retrofitting and management strategies of existing RC buildings are actually a crucial topic. This topic requires a novel approach in order to define a retrofitting and management strategies based on multidisciplinary approach. It can be the basis of both methods of mitigation of seismic risk strategies and as reference tool for seismic insurance model for wide building stock.

In particular, the qualitative trend of total depreciation of a RC buildings (due to age and income decay and without regular maintenance) is showed in Figure 1 (continuous line). After seismic event, this curve could quickly change due to seismic losses (step in figure 1). In fact, the real impact of seismic event can be looked upon as a step due to repair cost or a drastic reduction of service life (if the building is repairable). Generally, the original service life may be modified due to repair or retrofitting (marked lines vs dashed line in Figure 1.a). On the other side, due to high seismic damage, the building may be no repairable, then there is not any residual service life (Figure 1.b).

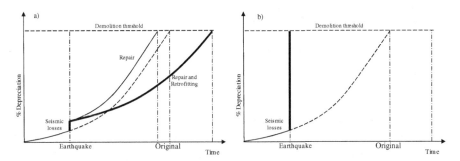

Fig. 1. Qualitative trend of total depreciation

In past years, several studies have been carried out in order to investigate and study in seismic risk. Several studies and projects have been funded to define the better way in seismic risk assessment and management. Where the problem is treated following a probabilistic approach, four different aspects should be considered:

— Seismic hazard;
— Structural performance;
— Economic losses;
— Economic life prediction.

Based on the above consideration, a novel framework may be proposed. In fact, seismic-risk strategies for existing buildings are an optimal link between strongly different aspects. A procedure based on multidisciplinary approach may be developed. The procedure should be based on probabilistic approach but in any cases, some deterministic choice could be considered. In some case, the probabilistic approach is further difficult. In fact, the several factors are affecting from by epistemic and aleatory uncertainties.

In this work, the above mentioned models are conceptually integrated. On the basis of the above methodology, the probability of damage can be defined using the total probability theorem as reported in the following:

$$P[D_r \mid s] = P_D[D_r \mid C_{r,r}] \cdot P_{Si}[C_{r,r} \mid d_{Si}] \cdot P[d_{Si} \mid I] \cdot \lambda[I \mid s] \qquad (1)$$

Where $\lambda[I \mid s]$ is the probability of the exceeding of a given seismic event; $P[d_{Si} \mid I]$ is the probability of the exceeding of a given damage; $P_{Si}[C_{r,r} \mid d_{Si}]$ is the economic losses model, generally based on the ratio between the cost of repair and the cost of replacing of considered building type. Finally, in $P_D[D_r \mid C_{r,r}]$ is the depreciation economic losses based on the buildings value (with respect to their replacement cost minus depreciation) and the land value (related with the buildings value).

The models defined from the authors in previous works may be considered: Seismic performance evaluation model [11]; Economic losses model (for example [7]); Economic life prediction model [13].

3   Seismic Performance Evaluation Model

In seismic risk assessment and consequent management strategies, a fundamental step should be a reliable procedure to define the relationship between structural performance, engineering parameters and damage. As discussed in previous studies ([14], [5], [11]) the more significant parameters is Interstorey Drift (IDR). Extremely more relevant is the identification of the performance levels as a function of the main considered parameter (IDR) on the basis of accurate NLDAs. In fact, the Non Linear Dynamic Analysis (NLDAs) is now well recognized as the best way in order to assess the demands regarding to different earthquakes.

In this paper, a new approach, as defined in [11], has been considered in order to assessment the seismic performance. In this procedure, the performance levels for investigated building types have been defined based on a well note approach (EMS 98, [15]). This procedure overcomes the expertise or deterministic assumptions that are often available in the literature. Moreover, the results are a probabilistic evaluation.

The results of NLDAs have been used to define a new relationship among Seismic Performance (SP), Damage Levels (DL) and Interstorey Drift (IDR). Then, at each condition of lateral deformation (IDR depending), it has associated a limit state reflecting a section yield level for main structural elements (columns and beams).

Following these procedure, the damage states have been defined considering the deformation capacities for structural element (columns and beams as reported for example in Figure 2). These capabilities have been compared with demand obtained from NDLAs.

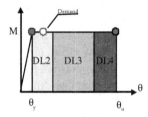

Fig. 2. Structural element perfomance and damage level

Different damage levels (described in [15]) can be considered at different IDR for each structural investigated type. Moreover, for each Damage levels have been evaluated the number of structural elements at each section yield level [11].

This approach is consistent with previous section. In a simple way, DL3 and DL2 can be considered when the structure is reparable. DL4 has been considered when the structure has suffered at least heavy damage, with low residual lateral strength and stiffness and the building is obviously not repairable (Figure 1.b).

Significant differences have been found for building types due to: different ages (e.g., old or no seismic standard design), number of storeys, infill panel distribution, material properties, and reinforcement details, among others. New Fragility Curves (FCs) have been developed. They are simply "Cumulative Distribution Functions" (CDF). Finally, for each seismic intensity value, the probability ($P\big[d_{Si}\big|I\ \big]$) of the exceeding of a given damage level can be obtained.

Based on the two-parameter log normal distribution function, the FC curve is defined in equation (1):

$$P\big[d_{Si}\big|I\ \big]=\Phi\left[\frac{1}{\beta_{d_{Si}}}\cdot\ln\left(\frac{I}{\mu_{ds}}\right)\right] \tag{2}$$

where, μ_{ds} is the median value of the Housner Intensity of the damage state (d_{Si}); $\beta_{d_{Si}}$ is the standard deviation of the natural logarithm of the seismic Intensity for the damage state, Φ is the standard normal cumulative distribution function.

In seismic risk mitigation policies, the next step is to define FCs considering the repair/retrofitting costs. In order to apply the above procedure, a model cost must be considered. An economic losses estimate should include repair cost. This step of procedure is a fundamental topic. Generally, the cost model is defined as a relationship between the repair cost and the reconstruction cost. It to be noted that it is a hard work to divide the repair cost from those due to seismic upgrading cost. In fact, also after past earthquakes the ration between repair cost and seismic upgrading cost is not clearly achievable. Moreover, the relationship between repair and seismic upgrading cost is clearly not linear. Similarly, the relationship between seismic upgrading cost and the obtained capacity-demand ratio cannot to be linear. In fact, significant different strengthening alternative are available based on strongly different alternative choice (for example [16]). In order to obtain a roughly indicating the relative costs of different alternative choice, significant effort are need. Obviously, this is another work; a complete and thorough discussion is not possible in this paper. In any case, it is important to note that the costs required by different strengthening strategies (for a fixed value of final capacity-demand ratio) could be significantly different.

For each damage level, the probable repair cost is investigated. In this way, after evaluation of the expected repair costs, it is possible to obtain the main topic of the work.

$$P_{Si} \lfloor C_{r,r} \mid d_{Si} \rfloor \qquad (3)$$

In this model, the estimation of the repair cost may be carried out using the data collected in some past Italian earthquakes [7]. Economic damage index $C_{r,r}$ was defined as relative repair cost, equal to the ratio of cost of repair to cost of replacement of the building. Then $C_{r,r}$ was defined as a random variable, ranging between 0 and 1.

Finally, consistently with the topics of this work, it to be noted that some significant improvement should be made in order to define the more realistic seismic hazard in short and medium period (some years / some decades). This is a fundamental topic in order to investigate about the depreciation economic losses due to possible seismic events.

4   Economic Life Prediction Model

The cost-effectiveness should be preliminary to any intervention of reuse of the building stock [17], [18]. The economic feasibility of different forms of intervention on the existing city is conditioned by some key economic variables: the buildings value with respect to their replacement cost minus depreciation, the land value and the relationship between it and the buildings value. These two components, taken together, make the property value. However, they have absolutely different dynamics. The land value due to the land rent and in particular the so-called absolute component, normally in time does not undergo any depreciation, indeed more readily it grows.

Over time, the reproduction cost undergoes depreciation. Given these two probable dynamics, [19], [20], [13] identified the demolition threshold at the time when the curve representing the land value intersects the curve of the total value of the property (land + building) to which is subtracted the demolition costs and charges (transport and delivery to dumping sites). The depreciation shown in Figure 3 does not take into account the possible economic losses caused by seismic actions on buildings not designed to withstand them.

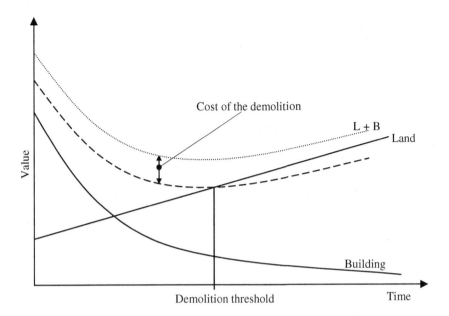

Fig. 3. Property value, sum of the aliquots related to the land and to the building

The cost-effectiveness of any maintenance on the structure of a reinforced concrete building (retrofit) located in seismic areas should bring in account the magnitude of the likely damage caused by the earthquake, estimating the cost for repair. In turn, this cost should be compared with the depreciated replacement cost of the property. In fact, usually the cost-effectiveness of the retrofit regards existing structures made 20, 30 or 40 years ago, which have a residual service life conditioned by the execution of this type of intervention. In practice, the choice depends on the cost of the retrofit but also on the damage caused by the likely seismic actions, on the building age and on the boundary conditions.

Please note that the cost-effectiveness of the retrofit should be assessed in comparison with the direct benefit generated by this intervention, also referring to other alternatives. One of these is, for example, the demolition and reconstruction of the building. In [20] have been developed several curves of value of the building replacement cost minus depreciation depending on age. These curves result from the sum of depreciation cost functions, developed for individual functional elements in which the building can be broken down.

The loss in value that characterizes each functional element over time, compared to replacement cost of it, is a function of the marginal contribution of individual depreciation factors: the age, the income decay, and the functional obsolescence.

The depreciation of the buildings is therefore the weighted sum of the depreciation of its functional elements. With the only exception of the building structure, all other functional elements, systems or finishes, are replaceable at the end of their useful life. A redevelopment or modernization that involves replacing of a portion or all of these elements (systems or finishes) could extend the economic life of the whole building. This extension of the useful life is however conditional upon the residual service life of the concrete structure. This is the only part of the building not replaceable independently. That is, the economic life of a building may never exceed the economic life of its structure.

5   Discussion

Figure 4 shows trends of total depreciation of a reinforced concrete structure due to age and income decay, without regular maintenance. This curve could be changed by implementing the economic losses (damage level) caused to the building by probable seismic action. The curve modified in this way is useful in order to check when the depreciation that leads into account also the effects of the earthquake can be considered curable. Depreciation is curable if the expenditure for retrofit is less than the increase in value due to this intervention. As the graph shows, the cost-effectiveness of the retrofit depends on the time when action is taken with respect to the building age.

Due to the damage caused by the earthquake the depreciation curve, moves upwards by an amount equal to the economic losses (cost of repair).

If the cost of the retrofit is lower than this damage level, the retrofit is obviously convenient. When, however, the cost-benefit analysis of the retrofit takes place in a time not too far from the end of the service life of the structure (point c), the retrofit is economically efficient even if its cost is at least equal to the damage level (point a). In this case, in fact, the action of retrofit is able to prevent the immediate attainment of the end of service life. From point a to point b, the cost-effectiveness decreases because there is advantage with retrofit costs gradually smaller. After step b retrofitting is no longer effective also with regard to minimum expenditure.

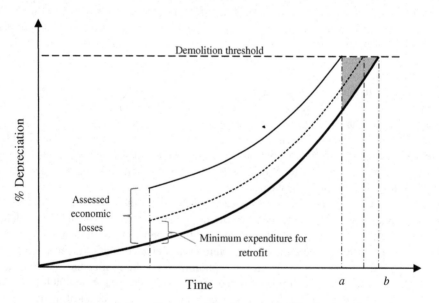

Fig. 4. Depreciation with economic losses by probable seismic action

Based on widely validated models, the proposed approach is the optimal solution in order to investigate on the depreciation due to seismic event. The approach is based on a probabilistic model. It is useful for mitigation of seismic risk strategies and as reference tool for seismic insurance model for wide building stock.

Finally, a framework can be proposed considering an optimal link between strongly different aspects.

Following the flowchart (Figure 5), a procedure based on multidisciplinary approach may be easily applied. In order to improvement the existing model and obtain a consistent result, the flowchart should be used with only probabilistic models. Consequently, the proposed methodology shall be used in different way, applications, and territorial scale. On a wide territorial scale, the approach can be the basis of both seismic risk mitigation strategies and tool for a seismic insurance model for wide building stock. On local scale the methodological approach can be used both seismic risk mitigation strategies and urban planning.

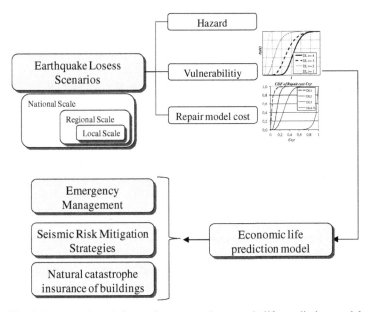

Fig. 5. Framework to define and use a novel economic life prediction model

References

1. Vona, M., Harabaglia, P., Murgante, B.: Thinking about resilience cities studynig Italian earthquake. In: Proceedings of the Institution of Civil Engineers, Urban Design And Planning (2015). http://dx.doi.org/10.1680/udap.14.00007
2. Borzi, B., Vona, M., Masi, A., Pinho, R., Pola, D.: Seismic demand estimation of RC frame buildings based on simplified and nonlinear dynamic analyses. Earthquakes and Structures **4**(2) (2013)

3. Puglia, R., Vona, M., Klin, P., Ladina, C., Masi, A., Priolo, E., Silvestri, F.: Analysis of site response and building damage distribution due to the 31 October 2002 earthquake at San Giuliano di Puglia (Italy). Earthquake Spectra **29**(2), 497–526 (2013). doi:10.1193/1.4000134

4. Masi, A., Chiauzzi, L., Samela, C., Tosco, L., Vona, M.: Survey of dwelling buildings for seismic loss assessment at urban scale: the case study of 18 villages in Val d'Agri, Italy. Environmental Engineering and Management Journal **13**(2), 471–486 (2013)

5. Masi, A., Vona, M.: Vulnerability assessment of gravity-load designed RC buildings: evaluation of seismic capacity through nonlinear dynamic analyses. Engineering Structures **45**, 257–269 (2012)

6. Chiauzzi, L., Masi, A., Mucciarelli, M., Vona, M., Pacor, F., Cultrera, G., Gallovič, F., Emolo, A.: Building damage scenarios based on exploitation of Housner intensity derived from finite faults ground motion simulations. Bulletin of Earthquake Engineering **10**(2), 517–545 (2012)

7. Dolce, M., Kappos, A.J., Masi, A., Penelis, G., Vona, M.: Vulnerability assessment and earthquake scenarios of the building stock of Potenza (Southern Italy) using the Italian and Greek methodologies. Engineering Structures **28**, 357–371 (2006)

8. Goretti, A., Bramerini, F., Di Pasquale, G., Dolce, M., Lagomarsino, S., Parodi, S., Iervolino, I., Verderame, G.M., Bernardini, A., Penna, A., Rota, M., Masi, A., Vona, M.: The Italian contribution to the USGS PAGER Project. In: Proc. of 14th World Conference on Earthquake Engineering, Beijing, China (2008)

9. Whitman, R.V., Anagnos, T., Kircher, C.A., Lagorio, H.J., Lawson, R.S., Schneider, P.: Development of a national earthquake loss estimation methodology. Earthquake Spectra **13**(4), 643–661 (1997)

10. Cornell, C.A., Krawinkler, H.: Progress and challenges in seismic performance assessment. PEER Center News **3**, 1–3 (2000)

11. Vona, M.: Fragility curves of existing RC buildings based on specific structural performance levels. Open Journal of Civil Engineering **4**(2) (2014)

12. Ditommaso, R., Vona, M., Gallipoli, M.R., Mucciarelli, M.: Evaluation and considerations about fundamental periods of damaged reinforced concrete buildings. Natural Hazard and Earth System Science **13**(7), 1903–1912 (2013)

13. Manganelli, B.: Real Estate Investing. Springer International Publishing, Switzerland (2015)

14. Masi, A., Vona, M., Mucciarelli, M.: Selection of natural and synthetic accelerograms for seismic vulnerability studies on RC frames. Journal of Structural Engineering **137**(3), 367–378 (2011)

15. Grünthal, G. (ed.): European Macroseismic Scale 1998 (EMS-98). European Seismological Commission, sub commission on Engineering Seismology, working Group Macroseismic Scales. Conseil de l'Europe, Cahiers du Centre Européen de Géodynamique et de Séismologie, Luxembourg, vol. 15 (1998)

16. Calvi, G.M.: Choices and Criteria for Seismic Strengthening. Journal of Earthquake Engineering **17**(6), 769–802 (2013). doi:10.1080/13632469.2013.781556

17. Anelli, A., De Luca Picione, M., Vona, M.: Selection of optimal seismic retrofitting strategy for existing RC building. In: 5th ECCOMAS Thematic Conference on Computational Methods in Structural Dynamics and Earthquake Engineering, COMPDYN 2015, Crete Island, Greece, May 25–27, 2015

18. Cardone, D., Flora, A., Manganelli, B.: Cost-benefit analysis of different retrofit strategies following a displacement-based loss assessment approach: a case study. In: Tenth U.S. National Conference on Earthquake Engineering, (10NCEE) July 21-25 2014 at Anchorage, Alaska (2014). DOI: 10.4231/D3CF9J706
19. Manganelli, B.: Maintenance, Building Depreciation and Land Rent. Applied Mechanics and Materials **357–360**, 2207–2214 (2013)
20. Manganelli, B.: Economic Life Prediction of Concrete Structure. Advanced Materials Research **919–921**, 1447–1450 (2014)

A Collaborative Multi-Criteria Spatial Decision Support System for Multifunctional Landscape Evaluation

Raffaele Attardi[✉], Maria Cerreta, and Giuliano Poli

Department of Architecture (DiARC), University of Naples Federico II,
via Toledo 402 80134, Naples, Italy
{raffaele.attardi,cerreta,giuliano.poli}@unina.it

Abstract. The paper aims at the development of a Collaborative Multi-Criteria Spatial Decision Support System (C-MC SDSS) for the evaluation of multifunctional landscape according to a human smart perspective. The C-MC SDSS addresses landscape transformation and preservation processes, leading to consistent choices with the principles of local self-sustainable development, considering the decision-making environment as a heterogeneous field, where conflicting interests interplay. The methodological approach, implemented in the case study of the National Park of Cilento, Vallo di Diano and Alburni (Southern Italy) demonstrates that the inclusion of web-based data and common knowledge in MC-SDSSs is a major challenge. The multidimensional nature of the landscape evaluation requires a mutual understanding, communication and collaboration, allowing an expansion of the knowledge-base to sharpen questions and improving the quality of the decision process. The paper investigates potentials of a C-MC SDSS for supporting transparent and democratic decision processes, helpful in selecting landscape enhancement policies.

Keywords: Collaborative Decision-Making · Multi-Criteria analysis · Spatial decision support system · Multifunctional landscape · Fuzzy logic

1 Introduction

Collaborative Spatial Decision Support Systems (C-SDSS) allow enhancing the effectiveness of complex social decisions analysis; they are dynamic tools that combine structured data with information related to social preferences, thus building a shared knowledge base in order to guide decision-making processes in the public domain, while increasing democracy and transparency. When dealing with the valorisation of landscape resources, the application of C-SDSS in landscape evaluation is essential to implement the approach proposed by the European Landscape Convention (ELC) (Florence, 2000). According to ELC's definition, indeed, "landscape" is a way of reading the "territory" through a multidimensional approach. From an ecological perspective, landscape contains environmental structures and functions that provide multiple tangible and intangible services that satisfy human needs.

Landscape Services (LS) are an ecological economics well-established multi-scale approach to the assessment of natural resources. Therefore, landscape can be interpreted

© Springer International Publishing Switzerland 2015
O. Gervasi et al. (Eds.): ICCSA 2015, Part III, LNCS 9157, pp. 782–797, 2015.
DOI: 10.1007/978-3-319-21470-2_57

as a supplier of multiple environmental, social and economic functions in a specific local context, taking into account the interests of multiple social actors [1]. In this context, the Geographical Information Systems (GIS) combined with web 2.0 become a decision support system able to collect spatial data, in order to implement a territorial learning process based on sharing, interaction, and integration of knowledge and, consequently, active civic participation in decision processes for territorial policies. The support of Information and Communication Technologies (ICT), indeed, ensures accessibility to a wide public and generates a common knowledge-base for the construction and analysis of spatial data in landscape preservation and enhancement processes. In this regard, ITC simplify the communication level of complex data making easier access to geographical information. The Web-GIS provides heterogeneous spatial data-set that constitutes landscape knowledge base. The integration of suitable Environmental Decision Support Management tools in the Web-GIS platform enables spatial analysis capabilities for the system. The spatial data-set is composed not only by thematic geographical data that describe the spatial characteristics in a multi-level framework (remote sensing data, digital satellite images, technical Municipal cartography, technical Regional cartography, cadastral parcels, environmental hazard and risk maps, parks and woods, coastal zones, etc.), but also by dynamic data of events uploaded by users enabling time series analysis. All this makes the system dynamic and allows a continuous enrichment of the landscape knowledge base, combining institutional data with information, non-structured or structured data deriving from the interaction of users.

In section 2, the paper makes a literature review related to Multi-Criteria approaches and GIS, focusing on multifunctional landscape and its evaluation; in section 3, the case study of National Park of "Cilento, Vallo di Diano and Alburni" has been investigated through the perspective of the human smart landscape concept, developed in the research project "Cilento Labscape: an integrated model for the activation of a Living Lab in the National Park of Cilento, Vallo di Diano and Alburni" funded by FARO Program 2012-2014 "Funding for the Start of Original Research", University of Naples Federico II; in section 4, the proposed approach is tested on the case study and the results are shown and analysed in conclusions (section 5).

2 Spatial Decision Support System: Integrating GIS and Multi-Criteria Decision Analysis

Spatial Decision Support System (SDSS) provides a computer support for decision-makers (DM), when the decision problem has a geographic or spatial dimension [2]. The contribution of GIS in decision-making processes, indeed, lies in their ability of storing and manipulating spatial data, based on their spatial location and their attributes. For those real-world applications where GIS is only one of the components of a complex decision support system, the actors of the decision-making process can be more or less interested in the geographical features of the system. In these cases, the main issue for GIS is to provide explicit problem representations based on the users' points of view: different users may have different representations of the problem.

Therefore, structuring SDSS concerns not only interface design and representation issues, but mostly database construction and problem modelling. Consequently, a key requirement of SDSS is flexibility: a GIS software, indeed, must be able to communicate with other software for the management and resolution of specific problems. A SDSS should also guarantee good interaction for the user, based on his own preferences. In the early Eighties, Group Decision Support Systems (G-DSS) arose. De Sanctis and Gallup [3] define the G-DSS as combination of computational and communication decision support technologies working simultaneously, in order to help in the identification, formulation and resolution of the decision problem involving multiple decision-makers. In 1988, Kraemer and King [4] introduced the concept of Collaborative Decision Making Support System (C-DSS), related to an interactive system enabling the participation in decision-making of a greater number of cooperating actors, in order to facilitate the synergistic solution of ill-structured problems. The decision process should profit by the inclusion of a wider knowledge contribution; while a critical success factor is the proper organization of a greater amount of information generated by the group of actors involved. Starting from approximately 1995, the World-Wide Web and Internet have provided a global technological platform to further extend potentialities and use of DSS. Indeed, web-based C-DSS overcome the "physical" limit of the G-DSS (which require the presence of decision makers group in a place simultaneously) and enable, collaboration among multiple decision-makers randomly geographically distributed. The main features of the web-based C-DSS Web-based are: no-completeness of data available; multiple agents (both human and artificial) cooperation to reach a shared acceptable decision; agents distributed in the network and possibly moving within the network.

Spatial decision problems generally concern several alternatives and multiple evaluation criteria, often in conflict with each other and not always comparable as measured with different scales. The alternatives are usually evaluated by multiple actors (decision-makers, stakeholders, interest group), which are able to express relative importance of the alternative evaluation criteria. Therefore, the combination of Multi-Criteria Decision Analysis (MCDA) and GIS meets the requirements of such a kind of decision problems. Indeed, in last twenty years, a constantly growing literature has proofed that GIS and MCDA can profitably co-work [5, 6, 7, 8] to address complex spatial decision problems involving conflictual objectives and problem descriptions. The MCDA provides a wide range of techniques to structure decision-making problems, evaluate alternatives and rank, sort or select them. The combination of MCDA and spatial analysis allows the construction of Multi-Criteria Spatial Decision Making Support Systems (MC-SDSS), which integrate geographic data with DM preferences and allow data manipulation and aggregation, on the basis of specific and appropriate decision rules [8]. The GIS-MCDA approach is used in many different fields, as highlighted in the literature review performed by Malczewski [8]; fields of application include environmental planning and management, ecology, transports, urban and regional planning, waste management and agriculture. According to Malczewski [8], the GIS-MCDA applications have a series of dual classifications. On the one hand, there are the characteristics of the spatial model (geographic data model, spatial dimension of the evaluation criteria, spatial definition of the alternatives);

on the other hand, there are the characteristics of the multi-criteria decision problem (nature of criteria, number of individuals involved in the decision-making process, typology of uncertainty considered).

With regard to the aggregation algorithms and decision rules, in GIS applications, the mostly used multi-criteria decision rules refer to linear aggregations, distance from ideal point and outranking methods. The weighted sum and related methodologies are common approaches because of their computational simplicity and they are often associated with Boolean operations, linear transformations for the standardization of the criteria and pairwise comparisons for the weights elicitation. The popularity of linear aggregation method in GIS-MCDA problems lies in relative simplicity of operations required to run them and in the easy and intuitive understanding for the DM. However, linear aggregation rules are often used without a full check of the assumptions on which they are based. Moreover, they are often applied without a valid understanding of the meaning of weights assigned to criteria in the evaluation phase (weights in linear aggregations rules does not have the meaning of importance coefficients but they are trade-off ratios among criteria and hence they are dependent on the measurement scale of criteria). Some of the difficulties of using the weighted sum or AHP can be overcome through comparison approaches, such as ideal point methods (as TOPSIS algorithm) and outranking methods (as ELECTRE and PROMETHEE). Generally, the decision-making problems can be categorized under certainty conditions or under uncertainty conditions, according to the amount and typology of knowledge available to the DM. If the knowledge available to the DM enables a complete description of the decision environment, then the approach to the decision problem is deterministic and the decision can be made under certainty conditions. However, real-world problems often involves imprecise or incomplete information, since many aspects of the problem are difficult to quantify, estimate or predict. Consequently, the problem description and analysis are under uncertainty conditions. This uncertainty can result from different sources: limited information of decision context, imprecision in semantic description of events, phenomena, facts. Therefore, the decision problems in contexts of uncertainty can be probabilistic (when the uncertainty is related to difficult or impossible predictions) or fuzzy (when the uncertainty is related to imprecise assessment).

The integration of GIS and MCDA permits the development of SDSS paradigm, in which geographical data are directly available for the DM in order to simplify evaluation of policies or development scenarios. The greatest benefit is that the DM inserts his preferences first-hand, also getting a real-time feedback on the implication of the evaluation in policy choices. This process leads to an increase of confidence in results, trying to explore and identify compromise solutions among different stakes and issues of the decision problem. GIS-MCDA combination has made a significant contribution to participatory processes, improving communication and understanding in situations of interaction among multiple DM, thus facilitating consensus building to reach political compromises. The GIS-MCDA applications have potential to improve the quality in cooperative decision-making, providing a flexible environment for solving spatial problems, in which all involved actors can explore, understand and redefine a decision problem. The main problems in the application of GIS-MCDA concern both the GIS systems and spatial analysis potentials, and the need for conceptual and operational validation of the MCDA use in real-world problems. Indeed, it is necessary an extreme attention on the fundamental assumptions on which MCDA methodologies are built: many applications actually demonstrate a lack of a proper scientific grounding, while some methodologies include a set of assumptions that probably cannot be validated in real-world situations. The evolutionary trends of GIS, from closed and expert-oriented toward open and user-oriented, allows increasing democracy in decision-making processes through public participation. The developments in this context will be crucial to determine the success of GIS-MCDA in real-world situations, mostly through web applications.

Today, this vision allows the development of exploratory map-centred approaches in GIS-MCDA applications, whose main objective is to provide an insight to the DM about the spatial nature of problems, as opposed to conventional analysis methods. The GIS-MCDA become significant for the DM when he gains confidence in these tools giving a quick feedback of his comprehension of the decision problem.

3   Multifunctional Landscape Evaluation

The Landscape Services (LS) concept, based on Ecosystem Services (ES) definition, is used to enlighten the spatial scale where these services are supplied and demanded [9], involving a dynamic variety of human and environmental forces. According to De Groot et al. [10], the stakeholders who manage very large ecosystems usually benefit only a partial set of the ES provided; while the landscape scale reduces the distance between local actors and environment, enhancing service benefits. Landscape, as opposed to ecosystem, is an action framework for not strictly ecological disciplines and it involves the life of the community in its relationships with the environment. For this reason, in recent years geographers, ecologists and planners consider landscape as a multifunctional category, in which it is possible to identify both natural and material/immaterial services provided to humans [11, 12, 13, 14, 15]. Through these services it is possible to find suitable conditions for the sustainable development of a community.

In this perspective, the landscape category can be seen as a human-ecologic system which can offer a wide range of benefits, assessed by humans through their ecological, socio-cultural and economic values [16, 17, 18]. On these theoretical bases, Termorshuizen and Opdam's [19] recommend using the concept of LS as specification (not as alternative) of the concept of ES. The concept of services aims to enlighten, at landscape scale, the interaction between physical system, defined by natural processes, and the use and non-use values recognized by the community. LS are defined in literature as essential "goods" for the existence of the community living in a specific region [20, 21]. Ecological and monetary evaluations are traditional tools to assign value to ecosystem functions. However, they only take partially into account the total economic value, because they exclude the third domain of values, namely the socio-cultural one. Since humans constantly change their environment through a multiplicity of uses, perceptions and values, landscape evaluation must focus on the complex and dynamic relationship between the community and the surrounding environment, rather than on landscape itself.

The concept of landscape service recalls the need for integrated and spatial assessment methodologies in order to involve different kind knowledge and analyse local stakeholders benefits. In recent years, many scientific studies focused on landscape knowledge and evaluation methods through integrated and participatory approaches [22, 23, 24, 25, 26, 27, 28]. Often, the contemporary policy and planning do not consider total benefits deriving from a multifunctional use of landscape, and multifunctional landscape are converted into single land-use [10]. However, several studies show that the total value of multifunctional landscapes is often much higher than the value of the same landscape converted into a single-use. The reasons why the total economic value of landscape functions benefits continue to be underestimated are the following ones: difficulties in describing the importance of landscape features in monetary terms; the lack of data about many landscape functions and values; many benefits are appreciable only at macro or global scale, or they are usually considered as market externalities. LS have assumed considerable importance since policy makers need to handle a growing specific LS demand resulting from many different stakeholders [10]. Actually, a coherent and integrated approach for practical applications of landscape functions concept in planning, management and decision-making is lacking.

4   The National Park of Cilento Vallo di Diano and Alburni: The Human Smart Landscape Perspective

The traditional approach to landscape planning is based on the prescriptive approach of traditional urban and regional planning. Nowadays the need for alternative approaches to landscape planning is rising, on the basis of new stakes for local resources enhancement, able to trigger positive effects on the quality of life of the local community. The research project "Cilento Labscape: an integrated model for the activation of a Living Lab in the National Park of Cilento, Vallo di Diano and Alburni" tries to explore new approaches for a brand-new multidimensional and multifunctional vision of landscape that is the "human smart landscape" [29]. The "smart" approach is wide diffused and applied to urban policies regarding technological, energetic and transportation issues. Recently, the "human smart city" concept has been introduced, referring to cooperation between local administration and citizens in order to support the co-design of innovative processes and services for technical and social innovation, through peer to peer relationships based on mutual trust and cooperation [30, 31]. The human smart landscape innovation is not just technology-related, but also process-related because it aims at the development of a bottom-up process for micro-actions and based on: social innovation; knowledge creation in a systemic logic; creative development model; new technology and infrastructure systems.

The human smart landscape becomes a complex space where it is possible to deal with the modern issues of local competitiveness and self-sustainable development taking into account social cohesion and quality of life. Thus an interpretive and evaluation model, that integrates the Ecological Economy approach with Regional

Science ones, can be able to describe and assess the human smart landscape dimensions, highlighting the need related to: integrating new infrastructure with reduced resource consumption; recovering cultural and environmental heritage; activating synergies between natural, rural and urban areas; creating ecological networks and enhancing the cultural diversity; increasing landscape attractiveness through the spaces quality and functions.

The smart landscape evaluation becomes relevant only if the goal is clearly defined and if the problem is well structured. Relating to present research the goal aims at identifying landscape complex values of Cilento National Park (CNP) in order to build a network among municipalities for self-sustainable development. Therefore it is necessary to focus on a multidimensional reading of the landscape, thus combining its morphological features with the cultural, environmental, social and economic systems. The proposed approach forwards the local potential development considering issues as awareness, flexibility, transformation potential, synergy, strategic behaviour. In particular, awareness seems to be an important aspect for smart local development, because of some development forms can only be activated with the participation of inhabitants, businesses, local governments through People, Public, Private Partnership (PPPP) models.

4.1   Constructing Spatial Composite Indicators for the Smart Landscape

The methodological process for building of National Park of Cilento territorial platform (GeoLab) and of Spatial Decision Making Support System (SDSS) uses an approach that combines spatial analysis through Geographic Information Systems (GIS) with Web 2.0 tools (Figure 1).

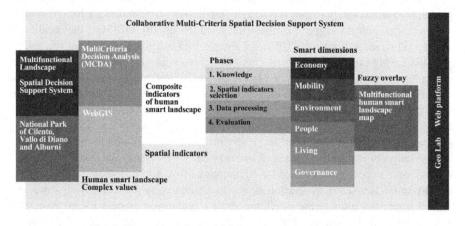

Fig. 1. The methodological framework: phases and contents

The design of GeoLab and of Web-Gis platform has highlighted the ability related to simply communicate complex data and no-conventional information for a greater public. The methodology for the construction of the SDSS consists of four basic steps:

- *Knowledge*. The analysis took place through the collection of hard and soft data from different sources, enriching the spatial data-set;
- *Spatial indicators selection*. A core set of relevant human smart landscape indicators is selected;
- *Data processing*. Spatial analysis tools allowed the construction of spatial indicators, from the spatial data-set;
- *Evaluation*. This step involves the construction of smart landscape composite indicators through a multi-criteria approach. Composite indicators depict a "big picture" of the analysed smart dimensions.

In the following, we specifically describe the four stages of the methodological process that guided the construction of the SDSS in GIS environment.

The *knowledge* phase is related to search for heterogeneous information, by type and sources, which describe the study area from different points of view: economic, social, cultural and environmental dimensions are able to outline the complexity that is typical of the historical landscapes, according to the contemporary vision oh UNESCO. In this first stage a wide and inclusive approach is required, considering a variety of factors that are not related to a specific field of knowledge, thus calling for a multi-disciplinary approach. The description of landscape cannot be characterized only by its own physical and scenic elements, but it shall include an analysis of the relationships between components and values attached by the community that creates, transforms and uses landscape over time. The smart landscape evaluation methodology in GIS environment works on 50x50 raster grid: each cell is an evaluation unit. Indeed, we set each cell a value related to each thematic layer; these values are the input data-set for the multi-criteria aggregation procedure.

The second step concerns the selection of a core-set of suitable indicators for smart landscape evaluation. The selection is based on references regarding the evaluation of the smart city dimensions [32], according to data availability. In Table 1 it is shown the list of human smart landscape indicators selected, categorized by dimension and thematic area.

Table 1. Spatial indicators of the human smart landscape

Smart Dimension	Thematic Area	Indicator	Source
Economy	Flexibility	Trend of the employment rate 1971-2011	ISTAT
Economy	Flexibility	Trend of the unemployment rate between 2001 e 2011	ISTAT
Economy	Flexibility	Trend of the number of employees in local business units between 1971 and 2011	ISTAT
Economy	Flexibility	Trend in agricultural manpower between 2000 and 2010	ISTAT
Economy	Entrepreneurship	Economic vitality index	Chamber of Commerce, Salerno
Economy	Entrepreneurship	Youth entrepreneurship rate	Chamber of Commerce, Salerno

Table 1. (*Continued*)

Economy	Local typical activities	Trend of the number of employees in local manufacturing business units between 2001 and 2011	ISTAT
Economy	Local typical activities	Trend of the number of employees in local accommodating business units between 2001 and 2011	ISTAT
Economy	Local typical activities	Trend in number of working days in the agricultural sector between 2000 and 2010	ISTAT
Economy	Local typical activities	Density index of the producers of local products	Personal analysis
Economy	Innovation	Trend in number of employees in the research and development sector between 2001 and 2011	ISTAT
Economy	Innovation	Number of firms with the label "The 100 Friends of the Park"	Cilento National Park (CNP)
Economy	Innovation	Number of innovative firms and associations	Personal analysis
Mobility	Accessibility	Accessibility index	Personal analysis on the basis of CNP data
Mobility	Technological infrastructure	Percentage of population not connected broadband internet	MISE-DPS
Environment	Ecosystem services	Ecological integrity index	European Environmental Agency
Environment	Ecosystem services	Provisioning services index	European Environmental Agency
Environment	Ecosystem services	Regulation services index	European Environmental Agency
Environment	Ecosystem services	Variation of the Agricultural Land Use	Ministry of Economic Development
Environment	Environmental preservation	Degree of preservation of natural reservoir (SPA and SIC)	Ministry of Environment, Preservation of Territory and Sea
Environment	Environmental preservation	Degree of preservation of natural reservoir (IBA)	Ministry of Environment, Preservation of Territory and Sea
Environment	Naturalistic significance	Classes of distance for the fruition of panoramic points	Personal analysis on the basis of CNP data
Environment	Naturalistic significance	Density index of naturalistic pathways	Personal analysis on the basis of CNP data
Environment	Naturalistic significance	Density index of geosites	Personal analysis on the basis of CNP data

Table 1. (*Continued*)

Environment	Naturalistic significance	Density index of natural caves	Personal analysis on the basis of CNP data
Environment	Naturalistic significance	Density index of historical built assets	Personal analysis on the basis of CNP data
Environment	Historic and cultural significance	Density index of historical built assets	Personal analysis on the basis of CNP data
Environment	Historic and cultural significance	Density index of historic routes	Personal analysis on the basis of CNP data
Environment	Historic and cultural significance	Density index of archaeological sites	Personal analysis on the basis of CNP data
Living	Touristic sector business	Density index of touristic routes	Personal analysis on the basis of CNP data
Living	Touristic sector business	Trend in the supply of touristic accommodations (number of beds) between 2008 and 2013	ISTAT
Living	Touristic sector business	Frequency of citations of the name of the municipality in touristic guides	Personal analysis on the basis of CNP data
Living	Health services	Number of beds in hospitals	Ministry of Economic Development
Living	Educational facilities	Presence of secondary education schools grouped in three typologies (liceum, technical school, professional school)	Ministry of Economic Development
Living	Educational facilities	Density index of educational farms	Campania Region
Living	Cultural vitality	Number of religious events, patronal feast, traditional events, food&wine events, cultural events	Personal analysis
Living	Demography	Demography trend between 1971 and 2011	ISTAT
Living	Demography	Variation in the average number of empty rooms with respect to the total number of rooms available over 1000 inhabitants between 1971 and 2011	ISTAT
Living	Income	Trend average income between 2010 and 2012	Ministry of Economy and Finance
Living	Income	Gini coefficient in the year 2012	Ministry of Economy and Finance
Living	Environmental risk	Seismic risk index	Ministry of Economic Development

Table 1. (*Continued*)

People	Ethnic and social plurality	Variation in the rate of foreign population between 2001 and 2011	Ministry of Economic Development
People	Ethnic and social plurality	Variation in the ratio of population over 65 years between 2001 and 2011	Ministry of Economic Development
People	Public participation	Variation in the percentage of voter turnout for the European elections from 2009 to 2014	Ministry of Interior
People	Education	Index of high school diploma	ISTAT
People	Education	Index of non-attainment of the first school cycle	ISTAT
People	Commuting	Commuting index	ISTAT
Governance	Attraction capacity for public funding	Number of scientific research programs and number of R&D participatory workshops	Personal analysis
Governance	Attraction capacity for public funding	European funds of 2007-2013 period for inhabitant (population of 2011)	Ministry of Economic Development

The third stage concerns the data processing for the construction of spatial indicators. Appropriate spatial analysis technique have been used in order to analyse and represent landscape phenomena over study area; the selection of these techniques is based on their own potential in analysing the studied phenomena and their spatial features. In particular, the Kernel density functions (or Kernel density) are geo-statistics analysis tools assessing the spatial density of the occurrence of some phenomena. All indicators of the data-set have been normalised in order to obtain a basis of comparable data for the evaluation stage. The normalisation, as shown in equation 1, uses the variation range of all data population values as a reference and scale them into a range from 0 to 100.

$$I_{i,j} = \frac{100 \times (a_{i,j} - a_{j,min})}{a_{j,max} - a_{j,min}} \tag{1}$$

Where: $I_{i,j}$ is the normalised value of the indicator j for the cell i;
$x_{i,j}$ is the value of the indicator j for the alternative i;
$x_{j,min}$ is the minimum value of the indicator;
$x_{j,max}$ is the maximum value of the indicator j.

In the next evaluation step we propose the aggregation process for the construction of spatial composite indicators through a fuzzy logic approach that will be further discussed in the next section.

4.2   The Multidimensional Evaluation of the Human Smart Landscape Through a Fuzzy Approach

In the evaluation phase a fuzzy approach has been tested in order to take into account the uncertainty inherent in smart components evaluation through the selected indicators. The uncertainty regards the spatial location of data, and the value or the significance of each indicator. In this regard, we tried to answer the following question: "What is the degree of possibility that a portion of the study area can be considered smart?". The processed spatial indicators have been transformed from crispy to fuzzy in order to establish their level of credibility through a specific linear membership function, where 0 indicates impossibility and 1 means full possibility of membership. The spatial problem, then, has been broken down into six smart dimensions describing the human smart landscape: economy, mobility, environment, living, people and governance. The fuzzy indicators have then been aggregated for each category, thus obtaining the six composite indicators of the smart landscape dimensions through a fuzzy overlay that has produced six output representing their fuzzy combination, using the Boolean function "OR". This function takes into account more optimistic possibility for which an alternative (raster image cell) belong to a specific fuzzy set; for this reason each alternative is attributed to a credibility class characterised by the value of the indicator with highest membership among all indicators considered in the overlay. After obtaining the six fuzzy smart dimensions maps, a further fuzzy overlay has been performed, in order to obtain a complexity map of smart landscape values. Finally, the output map is classified into nine classes expressing semantic judgments. Therefore on their basis it is possible to establish the degree of likelihood for which a specific area can be generally considered *smart*.

5   Results and Discussion

The human smart landscape value maps aim at identifying homogeneous areas with high medium and low values. In particular, the high value zones allow the definition of new potential territorial systems where it is necessary to implement landscape enhancement policies; moreover, the areas belonging to these territorial systems may help the ones with low smart features, in order to build a network among the municipalities (Figures 2, 3). As an example, the north-western system of National Park of Cilento, Vallo di Diano and Alburni, which could include municipality of relevant economic value (such as the most populated city of Agropoli and Capaccio, or Castellabate and Pollica), and at the same time relevant areas from an environmental point of view. The strength of these areas could synergistically be combined in order to build networks between municipalities, where some of them become driving forces to increase attractiveness and to improve quality of life the weaker municipalities. Another zone with strong smart-related character consists of municipalities at the boundaries between the Monti Alburni area and Vallo di Diano. The approach of disjunctive algorithm, namely the fuzzy "OR", shows existence of a transverse continuous band of municipalities with high smart value. This band could acquire an important role, along with coastal and internal (on Basilicata Region border) ones, generating a common belt linked into a widespread values network.

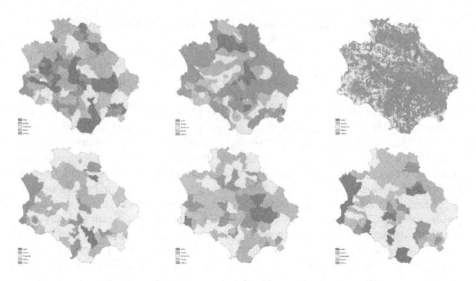

Fig. 2. Smart dimensions maps through fuzzy overlay: economy, mobility, environment, living, people, governance

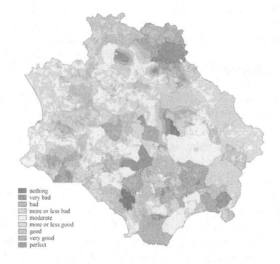

Fig. 3. Multifunctional human smart landscape map

6   Conclusions

The goal of the project Cilento Labscape is the activation of endogenous landscape enhancement processes through human smart landscape approach to the concept of landscape. From this is derived the need to build an interpretative code based on innovation, ability to organize the knowledge of the area according to the logic systemic models of creative development, but also on new technologies and infrastructure systems. The paradigm of Ecological Economics provides a multifunctional approach to landscape, paying attention to environmental and cultural issues. In order to activate a process of landscape enhancement a holistic approach to landscape analysis becomes necessary, encompassing environmental, social, cultural and economic system [19, 22, 33, 34, 35]. In this research paper we tried: to broaden the reasoning on assessment tools to support innovative landscape policies; to define and decline the paradigm of human smart landscape, through the construction of spatial indicators, starting from heterogeneous data. The methods of information collection, analysis and processing of data allow to expand and iterate the process experienced; the inclusion of information gathered through the web-GIS platform allows the direct collaborative involvement of end-users. The use of GIS in defining the SDSS has contributed not only to obtain an effective spatial representation of the selected indicators, but also to: determine the spatial distribution of data and indicators on the study; produce density indices of spatial features; provide geo-statistical analysis on the living conditions in the study area; manage a complex information base supporting the multi-criteria evaluation processes. The methodological process considers the recognition of the different landscape value components (geographical, economic, social, environmental, anthropological, cultural), and the activation of different forms of knowledge (explicit, systematized, experiential and practical, contextual, implicit) for the assessment of local complex values and their spatial representation. Indeed landscape value maps allows the exploration of the opportunities of the landscape within a broader decision context. The methodological approach used in the case study, ultimately, could be considered a basis for the improvement of the integration between expert and common knowledge in a broader participatory process involving different skills in order to improve the completeness of hard and soft data and to check consistency the overall assessment process [34]. Within a Collaborative Spatial Decision Support System local communities can actively contribute to build and update spatial data, improving the assessment of the complexity of landscape. The results of the tested methodology allow to draw a new "geography" of the National Park based on social, human, cultural and environmental values that go beyond mere municipal boundaries and that are integrated into larger landscape systems.

References

1. Lovell, S.T., Johnston, D.M.: Creating multifunctional landscapes: how can the field of ecology inform the design of the landscape? Frontiers in Ecology and the Environment **7**, 212–220 (2009)
2. Keenan, P.B.: Spatial decision support systems. In: Mora, M., Forgionne, G., Gupta, J.N.D. (eds.) Decision Making Support Systems: Achievements and challenges for the New Decade, pp. 28–39. Idea Group Publishing, Hershey (2003)
3. De Sanctis, G., Gallupe, R.B.: A Foundation for the Study of Group Decision Support Systems. Management Science **33**(5), 589–609 (1987)
4. Kraemer, K., King, J.: Computer-Based Systems for Cooperative Work and Group Decision Making. ACM Computing Surveys **20**(2) (1988)
5. Laaribi, A., Chevallier, J.J., Martel, J.M.: A spatial decision aid: a multicriterion valuation approach. Computers, Environment and Urban Systems **20**, 351–366 (1996)
6. Malczewski, J.: GIS and multi-criteria decision analysis. Wiley, New York (1999)

7. Thill, J.C.: Multi-Criteria Decision-making and Analysis: A Geographic Information Sciences Approach. Ashgate, New York (1999)
8. Malczewski, J.: GIS-based multicriteria decision analysis: a survey of the literature. International Journal of Geographical Information Science 20(7), 703–726 (2006)
9. Limburg, K.E., O'Neill, R.V., Costanza, R., Farber, S.: Complex systems and valuation. Ecological Economics 41, 409–420 (2002)
10. De Groot, R., Alkemade, R., Braat, L., Hein, L., Willemen, L.: Challenges in integrating the concept of ecosystem services and values in landscape planning, management and decision making. Ecological Complexity 7, 260–272 (2010)
11. Bastian, O.: Landscape ecology-towards a unified discipline? Landscape Ecology 16, 757–766 (2001)
12. Fry, G.L.A.: Multifunctional landscapes-towards transdisciplinary science. Landscape Urban Planning 57, 159–168 (2001)
13. Tress, B., Tress, G., De Camps, H., D'Hauteserre, A.M.: Bridging human and natural sciences in landscape research. Landscape Urban Planning 57, 137–141 (2001)
14. Stephenson, J.: The cultural values model: an integrated approach to values in landscapes. Landscape Urban Planning 84, 127–139 (2008)
15. Veeneklaas, F.: Over ecosysteemdiensten. Wettelijke Onderzoekstaken Natuur & Milieu 16, 1–6 (2012)
16. Chee, Y.E.: An ecological perspective on the valuation of ecosystem services. Biological Conservation 120, 549–565 (2004)
17. De Fries, R.S., Foley, J.A., Asner, G.P.: Land-use choices: balancing human needs and ecosystem function. Frontiers in Ecology and the Environment 2, 249–257 (2004)
18. De Groot, R.: Function-analysis and valuation as a tool to assess land use conflicts in planning for sustainable, multi-funcional landscapes. Landscape and Urban Planning 75, 175–186 (2006)
19. Termorshuizen, J.W., Opdam, P.: Landscape services as a bridge between landscape ecology and sustainable development. Landscape Ecology 24, 1037–1052 (2009)
20. De Groot, R.S., Wilson, M., Boumans, R.: A typology for the description, classification and valuation of ecosystem functions, goods and services. Ecological Economics 41, 393–408 (2002)
21. MEA: Ecosystems and Human Well-being: Multiscale Assessment, Millennium Ecosystem Assessment Series, 4. Island Press, Washington, DC (2005)
22. Frank, S., Fürst, C., Koschke, L., Makeschin, F.: A contribution towards a transfer of the ecosystem service concept to landscape planning using landscape metrics. Ecological Indicators 21, 30–38 (2012)
23. Uuemaa, E., Antorp, M., Roosaare, J., Marja, R., Mander, U.: Landscape Metrics and Indices: An Overview of Their Use in Landscape Research. Living Reviews in Landscape Research 3(1), 5–28 (2009)
24. Luesink, E.: Cultural heritage as specific landscape service Stimulus of cultural heritage in the Netherlands. Wageningen University, Netherlands (2013)
25. Fagerholm, N., Käyhkö, N., Ndumbaro, F., Khamis, M.: Community stakeholders' knowledge in landscape assessment- Mapping indicators for landscape services. Ecological Indicators 18, 421–433 (2012)
26. Hein, L., van Koppen, K., de Groot, R., Ekko, C., van Ierland, E.C.: Spatial scales, stakeholders and the valuation of ecosystem services. Ecological Economics 57, 209–228 (2006)
27. Attardi, R., Bonifazi, A., Torre, C.M.: Evaluating Sustainability and Democracy in the Development of Industrial Port Cities: Some Italian Cases. Sustainability 4, 3042–3065 (2012). doi:10.3390/su4113042

28. Attardi, R., Cerreta, M., Franciosa, A., Gravagnuolo, A.: Valuing cultural landscape services: a multidimensional and multi-group SDSS for scenario simulations. In: Murgante, B., et al. (eds.) ICCSA 2014, Part III. LNCS, vol. 8581, pp. 398–413. Springer, Heidelberg (2014)
29. Cerreta, M., Malangone, V., Panaro, S: Human smart landscape: a living lab in the national park of cilento, vallo di diano and alburni. In: Proceedings of the 18th ICOMOS General Assembly and Scientific Symposium, Florence, Italy, November 10–14, 2014
30. Concilio, G., De Bonis, L., Marsh, J., Trapani, F.: Urban Smartness: Perspectives arising from the Periphéria Project. Journal of Knowledge Economy 4(2), 205–216 (2013)
31. Oliveira, A., Campolargo, M.: Form smart cities to human smart cities. In: 2015 48th Hawaii International Conference on System Sciences (2015). Doi: 10.1109/HICSS.2015.281
32. Centre of Regional Science TU Vienna. Smart Cities. Ranking of European medium-sized cities (2007). http://www.smart-cities.eu/download/smart_cities_final_report.pdf
33. Perchinunno, P., Rotondo, F., Torre, C.M.: The evidence of links between landscape and economy in rural park. International Journal of Agricultural and Environmental Information Systems 3(2), 72–85 (2012)
34. Cerreta, M., Panaro, S., Cannatella, D.: Multidimensional spatial decision-making process: local shared values in action. In: Murgante, B., Gervasi, O., Misra, S., Nedjah, N., Rocha, A.M.A., Taniar, D., Apduhan, B.O. (eds.) ICCSA 2012, Part II. LNCS, vol. 7334, pp. 54–70. Springer, Heidelberg (2012)
35. Selicato, M., Torre, C.M., La Trofa, G.: Prospect of integrate monitoring: a multidimensional approach. In: Murgante, B., Gervasi, O., Misra, S., Nedjah, N., Rocha, A.M.A., Taniar, D., Apduhan, B.O. (eds.) ICCSA 2012, Part II. LNCS, vol. 7334, pp. 144–156. Springer, Heidelberg (2012)

A Quantitative Prediction of Soil Consumption in Southern Italy

Federico Amato[1], Federico Martellozzo[2], Beniamino Murgante[1(✉)],
and Gabriele Nolè[3]

[1] University of Basilicata, 10,Viale Dell'AteneoLucano 85100, Potenza, Italy
fdrc.amato@gmail.com, beniamino.murgante@unibas.it
[2] University of Rome "La Sapienza", Via Del Castro Laurenziano 9 00161, Rome, Italy
f.martellozzo@hotmail.com
[3] Italian National ResearchCouncil, IMAA, C.Da Santa Loja 85050, Tito Scalo, Potenza, Italy
gabriele.nole@imaa.cnr.it

Abstract. Landuse/cover evolution dynamic is a subject widely and thoroughly investigated, especially concerning consumption of natural and other lands, due to anthropogenic activities. This paper focuses on a region in southern Italy, where soil consumption is known to represent a urging matter of concern. However, although negative impacts of soil consumption are well known, to our knowledge there are no case studies presenting a precise quantitative measurement of the intensity of such phenomenon for the region of interest. This study aims at forecasting the development of urban settlements through the application of the cellular automata model SLEUTH; the first region to be investigated has been the Municipality of Altamura (Apulia region, Italy). This area has been used as a pilot case study to explore many difficulties and advantages in applying such a methodology to the whole southern Italian region. The final goal was to frame and populate an atlas of soil consumption in southern Italy, which intends to offer useful support to sustainable planning and policies.

Keywords: Land use change models · Soil consumption · Urban sprawl · Built-up areas · Sustainability

1   Soil Consumption: Definition and Measurement Problems

Soil is a non-renewable resource and, like water and air, it has to be considered a common good. Soil degradation can be caused by many factors (e.g. erosion, organic matter decline, compacting, landslides, floods, desertification).

Only in recent times the attention has been focused on the main threat of soil: soil sealing. Consequently, the realization of new residential areas, industrial zones, trade and services centres, roads and other infrastructures, is a major problem because it produces soil sealing. [1] [2].

A sealed soil loses its biological value, becoming unable to absorb and filter rainwater. Excessive urbanization, especially when it is due to ineffective urban planning, generates a strong fragmentation of natural and agricultural landscape [3] [4];natural

© Springer International Publishing Switzerland 2015
O. Gervasi et al. (Eds.): ICCSA 2015, Part III, LNCS 9157, pp. 798–812, 2015.
DOI: 10.1007/978-3-319-21470-2_58

islands constrained by urban edges are therefore too small to accommodate the life of particular species, consequently this urbanization produces a negative impact on bio-diversity, as well as on agricultural production [5].

These issues are typically included within the definition of soil consumption, a term which means a land cover transformation from natural to artificial. The Euro-pean research LUCAS allows to compare general characteristics of land cover in 27 European countries. The portion of territory with artificial cover in Italy has been estimated to be 7.8% of the total compared with 4.6% of EU average. Italy is ranked at the fifth place among the countries with the greatest artificial cover, after Malta (32.9%), Belgium (13.4%), the Netherlands (12.2), Luxembourg (11.9%), and slightly before Germany, Denmark and the UK [6].

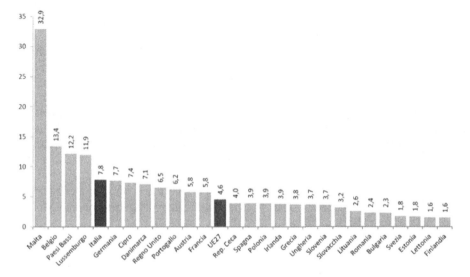

Fig. 1. Territory with artificial cover by country, LUCAS survey, 2012 (percentage incidence on total area); Source: ISTAT elaboration on Eurostat data

Analysing in detail artificial cover, distinguishing between residential areas, ser-vices and other artificial areas such as infrastructures and annexed areas, it is clear that in the 27 EU countries the value of residential areas and services only is 1.5 % of total area, approximately one third of the artificial surface, while other artificial areas covering is 3.0% of the territory. In Italy, there are 7.8% of artificial areas, with 2.7% residential areas and services and 5.1% other artificial areas.

Italian national legislation in the field of soil consumption, and more generally in territory governance, is unfortunately obsolete and unable to tackle problems of mod-ern cities and the territories. However, several legislative drafts, concerning territorial government, are currently deposited at the Chamber of Deputies, and they deal, in a more or less pronounced way, with soil consumption topic, also.

Express judgment about the quality of suchlegislative draftsis not the purpose of this paper, nor is the discussion of their potential effectiveness. It is very interesting that almost all legislative proposals identify soil consumption monitoring as a key step towards its reduction. In this scenario the application of SLEUTH model could be useful, which is based on the methodology of cellular automata and has been already successfully employed worldwide.

2   Methodology

2.1   Cellular Automata

The literature recognizes several groups of geo-computational methods applied to modelling of urban phenomena [7].

Cellular automata is a methodology developed in 1940 by Jhon von Neumann, he tried to create a machine capable to autonomously reproduce its structure, but only in recent times, due to the prodigious development of hardware technology, it was possible to develop wider applications [8].

A cellular automata is a mathematical model able to reproduce the system evolution in time and space through a set of rules, expressed in graphs or tables that describe behavioural aspects. These are very useful tools in the analysis of complex systems characterized by a behaviour with strong non-linearity. A cellular automata is defined by a n-dimensional space, generally with n = 2, divided into a grid where the single unit (i.e. the single cell) is defined by a set of characteristics that define the status, which may be subject to change according to defined functions and rules. More particularly, the change of state of a cell depends not only by the characteristics of the cell itself, but also by the state of cells in a defined neighbourhood of the analyzed cell.

Cell neighbourhood is a set of adjacent cells, disposed according to a certain configuration. There are different types of neighbourhoods; Von Neumann and Moore's ones are certainly the most known.

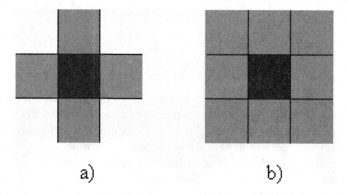

a)   b)

Fig. 2. Von Neumann (a) and Moore (b) neighbourhoods

Let us call F the function that determines changes of state of a cell. This function is defined as the set of local rules defining the relationship in space and time between states of cells. Therefore, considering k possible states and n cells included in the neighbourhood, one can obtain $(k2)*n$ possible rules.

2.2   SLEUTH Model

SLEUTH model is precisely based on cellular automata [9]. The name is an acronym of factors to be considered in its application: Slope, Land Use, Excluded (i.e. the set of soils to be excluded totally or in part by transformations provided by the model), Urban (i.e. urbanized areas), Transportation and Hill shade.

To date the usefulness of this model application is now proven not only in the identification of future development trends, but also in understanding the factors that have determined the growth of urban settlements in the past [10] [11].

SLEUTH model is the integration of Urban Growth Model, developed in 1998 by Keith Clarke, with Deltatron model. The first one simulates only urban growth, while the second allows the assessment of land use changes [12] [13].

The basic time unit adopted by SLEUTH is the cycle of growth, conventionally equal to one year. For each time unit the module "Urban Growth Model" performs four types of analysis of urban transformations: spontaneous growth, new spreading centres growth, edge growth, road-influenced growth. These steps define the simulation of urban growth.

Spontaneous Growth describes the possibility that a generic cell of the grid can spontaneously be urbanized. To each cell a probability of being urbanized at each interval time is associated. This probability depends by the state of the cell, but also by the values of some particular coefficients, Dispersion Coefficient and Slope Coefficient, defined by users that will be explained in a detailed way later on. The spontaneous growth takes place according to a stochastic process. New Growth Spreading Centre is the process that determines for each of the cells spontaneously urbanized, according to the result of the previous step, the probability that it will become an urban centre, able to spread as long as you have at least two cells developable in the neighbourhood of the cell object of analysis. Edge Growth defines growth of suburban areas, considering the expansion of centres on the edge generated by the new spreading centre growth, or already existing in reality or identified from previous application cycles of the model. Even in this case the simulation trend is determined by a parameter, the Spread Coefficient, defined by the user. This parameter identifies the probability of urbanization of a cell having at least three cells in his neighbourhood defined as developable.

Road-Influenced Growth is the most complex phases of simulation. Firstly, as a function of Breed Coefficient, some just urbanized cells are selectedfor which identifying the closest streets. The maximum radius distance in which the streets are searched is defined as a function of the parameter Road Gravity Coefficient, imposed once again by the user. If within that radius a street is found, the model poses an urban temporary cell in the point of the road closest to the cell under analysis. This temporary urban cell walks randomly along a length determined as a function of the previously mentioned Dispersion Coefficient. The cell where the road ends is then considered the new nucleus of urban expansion, and therefore in its surroundings there are developable cells, one or more of these cells will be randomly selected and urbanized.

The definition of transition rules, and consequently the execution of the four types of transformations described above, is done through the entering by the user of the values of five parameters: dispersion, breed, spread, road gravity and slope.

Dispersion Coefficient defines the number of times that a cell is randomly selected for a possible urbanization during the phase of spontaneous growth.

The dispersion value is at most equal to 50% of the diagonal of the image, and therefore it increases with increasing of image size.

Breed Coefficient defines the probability that an urbanized cell during the spontaneous growth becomes a new urban centre, able to expand. This parameter intervenes in the phase of new spreading centre growth.

It has also an important role in the phase of road-influenced growth, because it determines the number of times that it goes through a road path.

Spread Coefficient determines the probability that a cell belonging to a spreading centre, or of a unit consisting of at least two urbanized cells in a neighbourhood of 3x3 cells, generates a further urbanized cell in its surroundings.

Gravity Road Coefficient intervenes in road-influenced growth. It also determines, as a function of size of pattern input image, the search for the maximum distance of a given cell from a road.

Slope Coefficient is the only parameter that defines a restrictive condition of the expansion of the urban phenomenon; it is in fact also said Slope Resistance or resistance driven by slope. Slope resistance influences all growth stages of simulation. The relationship between urbanization and slope is not considered linear, but is a function of a multiplier. In this way, if coefficient values are high, areas with higher slope will have gradually more chances to fail in slope test. This means that even if a cell should be urbanized as a result of the application of other parameters at different stages of simulation, however it will change the state if the slope exceeds certain thresholds. When slope coefficient tends to zero, the influence of the slope on the probability of urbanization of a cell tends gradually to decrease. As a general rule, we can affirm that the tendency to build up in points with highest slope is variable and is a function of the amount of available flat areas in the study area and of presence of areas with a higher steepness in proximity of already urbanized soils. The parameters are not uniquely associated with individual growth stages that compose the simulation: single parameters are involved in most phases, and each phase involves generally more parameters.

This lack of linearity between the value of a parameter and the result, generates difficulty in identifying, within the final simulation, the effects caused by each single parameter. This "black box" effect is a typical characteristic of great part of simulation models based on cellular automata. Growth phases follow one another within each time cycle. At the end of each of them a further process is performed, said self-modification (i.e. self-modification of parameters). In this process model parameters, with the exception of slope resistance, which remains constant and unchanged over time, are modified following any trend growth, very fast or very slow, highlighted by the simulations. The user, during model calibration, can establish parameters limits beyond which increases or decreases in dispersion, breed and spread are imposed,. Threshold values are critically high and critically low. When growth within a cycle exceeds the set value for the critically high coefficient, the three growth parameters are incremented by a certain user-defined amount, said boom, generally equal to 10% of the initial value.

Table 1. Parameters and relative growth stages involved in SLEUTH model

Parameters	Phases of urban growth
Dispersion	Phases of urban growth
	Spontaneous growth
Breed	Road-influenced growth
	New spreading centre growth
Spread	Road-influenced growth
	Edge growth
	New spreading centre growth
Slope	Road-influenced growth
	Edge growth
	Spontaneous growth
Road Gravity	Road-influenced growth

Such an increase is intended to be representative of the trend of rapid growth defined by the simulation in the just completed cycle. Similarly, when the growth rate falls below the value defined as critically low, growth parameters are decremented by an amount of the bust, also user-definable and generally corresponding to 10%. In this way it simulates growth decrease of saturated and depressed systems. Self-modification phase is crucial for simulation success. In absence of it simulations characterized by constant or exponentially increase (or decrease) over time would be obtained, following developments very distant from actual urban dynamics. The main result of the application of Urban Growth Model is a map that simulates urban evolution in a period defined by the user. It also allows to evaluate the probability that each cell in the map will be urbanized.

Completed the execution of "Urban Growth Model", SLEUTH proceeds to the application of "DeltaTron" module, which deals with land use changes assessment and prediction. From a conceptual point of view, DeltaTron is based on the assumption that urbanization phenomena are the main generators of changes in coverage of non-urbanized soils.

The term DeltaTron means a cluster (i. e. a group of neighbouring pixels having a common feature) composed by at least 5 cells. This cluster is considered as a vehicle for change and follows temporal and spatial effects of territory transitions. DeltaTron is not related to a particular land use class, because it is only an identifier of the position and time when a certain change occurs. The module DELTATRON describes the phenomenon of land use changes in four sequential steps.

The first phase is the initiate change. A potential land use change is associated to each new urbanized cell identified with UGM. A DeltaTron is therefore generated in a not urbanized cell randomly selected; this can keep the class of land use unchanged or may instead induce a variation of state. The possibility of changing is defined as a probability function calculated depending on weighted average of slopes for each land use. This defines a vector which identifies the probability of transition from each land use class to any other class. The next step is cluster change. The newly generated DeltaTrons are aggregated, as well as land cover transitions associated with them. While in the previous phase each cell had the chance to change state assuming any other land use class, in this second step only two possibilities can occur: the cell maintains its unchanged state, or it changes in a DeltaTron associated class.

The third phase is the propagated change. All not DeltaTron cells adjacent to at least two DeltaTron cells from at least two cycles of growth (two years of simulation) may remain in the same state or assume the same use class DeltaTron reference.

The last phase is the aged DeltaTron. DeltaTrons are literally made aging, passing to the next step. The user defines the maximum number of cycles for which the DeltaTrons can be maintained in life. Exceeded this limit, they are no longer considered as potential DeltaTrons in the next growth cycle. In the execution of DeltaTron module only two parameters intervene.

Cluster size is used in the process of cluster change, and controls how each DeltaTron new cluster can grow in terms of size.

Minimum years between transitions define instead, in the phase of age DeltaTron, the maximum number of cycles for which the DeltaTrons can live. The implementation of the model generally is developed in two different stages. The first is the calibration, where values of coefficients to be subsequently used through the analysis of historical characters of growth of the settlement are defined, the second is the prediction, where historic features of urban growth are used to simulate future developments [14] [15].

3   The Case Study

3.1   The Study Area

Despite a growing attention in Europe and in Italy to soil consumption phenomenon, in southern Italy there is almost a total absence of studies on measuring the phenomenon extent and on analyzing its possible future developments.

In order to solve this gap, an application of SLEUTH model has been developed in Altamura municipality (Apulia, Italy). The purpose is to demonstrate the applicability of the methodology also in typical southern Italy urban environments, stimulating the implementation of further studies and researches.

Altamura municipality is interesting within the context of Apulia region, because with its 70,688 inhabitants is the eighth municipality in the region for resident population. Moreover, Altamura is the thirteenth municipality in term of territorial extension in Italy. Population dynamics, studied through national census data of Italian Institute of Statistics (ISTAT), show the vitality of the town. The population grew from 57,874 inhabitants in 1991, to 64,167 in 2001, to 69,529 in 2011 and 70,688 in 2014.

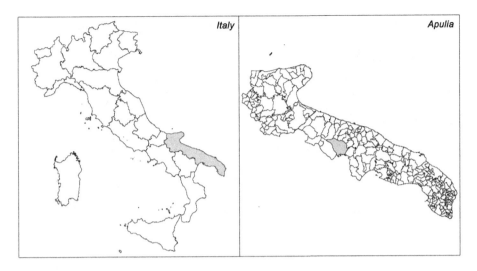

Fig. 3. Left: location of Apulia Region (highlighted in grey); Right: location of Altamura municipality (highlighted in grey) within Apulia region

Population growth is easily explained by the socio-economic dynamism generated by the proximity with the city of Bari, the regional capital and important urban centre in southern Italy. The City of Altamura, in fact, is part of the metropolitan city of Bari, the new Italian local authority that from 2015 replaces the Province of Bari.

In order to define urban areas at different dates and to use them as input data in SLEUTH model, an analysis was conducted on the historical evolution of built-up areas in the municipality of Altamura. This analysis was developed using the cartography at 1: 25.000 scale realized in 1954 by the Italian Military Geographic Institute (IGM), orthophotos at three dates (1985,1996, 2011) available in Web Map Service (WMS) format on the Italian National Geoportal (www.pcn.ambiente.it), and the cartography at municipality scale (2006) available on Apulia Region Geoportal.

Time series show a constant increase of built-up areas, with an overall increase of 84 hectares from 1954 to 2011.

Table 2. Hectares of built and unbuilt areas at different dates

	1954	1985	1996	2005	2011
Unbuilt [ha]	42466	42479	42447	42405	42382
Built [ha]	261	248	280	322	345

The increase in built volumes has continued even between 2005 and 2011: in this period 23 new hectares have been urbanized. This tendency is in contrasts with the national trend, where in the period 2005-2006 a significant reduction in building sector has been registered, mainly as a result of the effects of the global economic crisis.

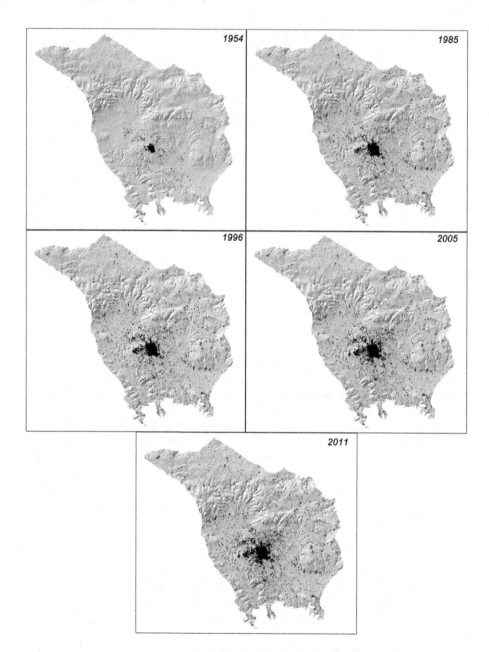

Fig. 4. Evolution of built-up areas at five different dates

3.2   Data Preparation and Calibration

As described in the previous paragraphs, the use of SLEUTH requires the use of several input information.

Table 3. Data and data sources adopted in the construction of input maps in SLEUTH model

Factor	Input data\source
Slope	Slope /Cartography
Land Use	Land Cover 2006/ Apulia region
	Land Cover 2011/ Apulia region
Excluded	Rivers, streams and waterways / Law 2004/42 Art.142
	Territories covered by forests and woodlands / Law 2004/42 Art.142
	Civic uses / Law 2004/42 Art.142
	Archaeological Areas / Law 2004/42 Art.142
	Areas with significant public interest / Law 2004/42 Article 136
	Respect Area of the network of tratturi, the historic and cultural sites and areas with archaeological interest / Regional Landscape Plan
	Wetlands / Regional Landscape Plan
	Grid connection of regional ecological network / Regional Landscape Plan
	Sinkholes, geological sites, caves / Regional Landscape Plan
	Shrub formations / Regional Landscape Plan
	Natural pastures / Regional Landscape Plan
	Buffer zone of forest areas / Regional Landscape Plan
	Hydrogeological constraint / Regional Landscape Plan
Urban	Urban area at 1954/ IGM Cartography
	Urban area at 1985/Orthophoto National Geoportal
	Urban area at 1996/ Orthophoto National Geoportal
	Urban area at 2011/Update Regional Cartography regional
Transportation	Transport infrastructures 2011/ Update Regional Cartography regional
Hillshade	Hillshade / Cartography

In order to properly apply the model, it was necessary to implement slope and hill-shade maps, starting from the use of the regional cartography of Apulia region, A map of transport infrastructure (road and rail) and two Land Cover maps, referring at 2006 and 2011 years, realized by Apulia Region have ben used. More complex was the construction of maps of excluded areas. In its construction, it was decided to take into account areas constrained by law 2004/42, which, in Articles 136 and 142, identifies landscape heritage to protect. In addition to these areas, more landscape contexts were taken into account, considering Apulia Regional Territorial Landscape Plan, recently approved. In this way it was possible to verify the impact that the new plan could have on the expansion of urban areas through the application of several constraints. After the construction of input data the calibration phase started. The purpose of calibration is to identify values of growth parameters that best describe evolution of settlement occurred during the time series used as input, in the specific case between 1954 and 2011. For this purpose SLEUTH uses a Monte Carlo method brute-force. This method allows the user to define a range of possible values for each parameter. The model repeats the analysis for each possible combination of values within the defined range. For each combination a simulation of urban growth is performed up to the most recent urban area (2011 in the case of Altamura).

The result of this simulation is compared with the actual built-up areas at the same date through a series of statistical indicators. These detailed analysis and comparison allow to identify the combination that best describes the trend of urban dynamics within the study area. Since each parameter may vary between 0 and 100, the possible combinations are extremely numerous, and test them all in advance would require a significant computational effort. In order to reduce the number of combinations to be studied, SLEUTH calibration is conventionally carried out in three successive steps: coarse calibration, fine calibration and final calibration. In the first stage the maximum range of values (0-100) is used for each parameter, using, however, a 25 long step. In this way only four values (0, 25, 50, 100) will be investigated for each parameter.After performing the first calibration, combinations that are closer to a correct description of urban development are used to determine intervals of values to be used in the subsequent calibration phase. In order to identify the best combinations we used the method of Optimum Sleuth Metric, already widely discussed and described in the literature.

4   Results

After completing calibration phase, it was possible to create a simulation of the evolution of built-up areas up to the year 2050.Values of coefficients obtained from the final calibration are reported in Table 4.

Table 4. Values of coefficients resulting from the calibration phase and used in the simulation

Diffusion	Breed	Spread	Slope	Road gravity
50	3	26	96	37

The value of diffusion coefficient, exactly equal to a half of the range of values assumed by the parameter, indicates an average chance of a dispersive growth of the settlement. This trend is further confirmed in the lower value assumed by breed coefficient, which indicates a low probability of developing new settlements separated from existing settlements.

The development of a settlement, therefore, will mainly take place through growth and densification of existing urban agglomerations, as shown also by the fairly high value of spread, which intervenes at the stage of Edge Growth, when the model simulates the expansion of suburban areas. The value identified by slope resistance is extremely high, which indicates a high dependence of urban development on territory orography. Such a result was already expected, because historically it has been observed that growth of settlements has affected only open territories of the municipality between Murgia hills, leaving substantially undeveloped part of the territory with more complex orography.

Finally, the value assumed by road gravity indicates a moderate dependence of urban growth on proximity to infrastructure for mobility, especially main arterial road mobility. Before proceeding with the discussion of simulation results, a further clarification is important. In the previous section we studied historical evolution of Altamura settlement on the basis of an analysis exclusively conducted on built-up areas. These, however, did not entirely consumed the soil, defined as the set of all sealed areas, including therefore also streets, parking areas and any other area that loses its character of permeability. The need to study trend of historical development of the town, considering only built-up areas, derives from the absence of a series of land use maps in the studied area.

However, in the simulation phase land use maps made in 2006 and 2011have been taken into account, enabling a more accurate assessment of soil consumed in recent years and expectations for next forty years (without great modification in urban development policies). Consequently, if 322 and 345 hectares of built-up areas were measured in 2006 and 2011, respectively, 1,934 total hectares of artificial surfaces in 2006, which become 2083 in 2011, are measured through a comparison with land use maps.

Through SLEUTH application artificial surfaces will increase to 882 hectares reaching a total of 2,965 hectares of consumed soil in 2050.

Table 5. Quantitative results of SLEUTH model application

	2006	2011	2020	2030	2040	2050
artificial surfaces [ha]	1934	2083	2194	2324	2566	2965

Urban development is expected mainly in the north-west part of Altamura municipality, as well as along the two main roads close to the city centre, S.S. (State Road) 99, which connects Altamura to Matera, and S.S. 96, which connects Altamura to Bari. Moreover, in terms of quantity it is possible to observe how increases in artificial surfaces provided by the simulation will not occur into a linear way, with constant growth, but they rather tend to become gradually higher.

After an initial decrease in soil consumption speed, from 29.8 hectares sealed per year between 2006 and 2011 to 12.5 per year expected by the simulation until 2020 and to 11.1 per year up to 2030, a gradual increase occurs with 24.2 hectares per year until 2040 and 29.65 hectares per year until 2050.This increase can only partly be explained by the greater difficulty of the model to accurately predict scenarios extremely distant in time.

However, it should be a clear signal of how current urban policies are inadequate at ensuring a proper long term preservation of the municipal territory. Despite the new Regional Landscape Plan can have a central role in preserving rural areas, an increase of settlement pressure on agricultural areas will occur. Measures adopted at regional scale cannot be enough at local scale.

This confirms that the complexity of the rules of territorial government will successfully achieve its objectives only through the cooperation of all administrative entities called to participate to the planning process. In the case of Altamura municipality it could be important to produce a new plan at urban scale, because current planning regulations are obsolete and inadequate to satisfy needs of the area.

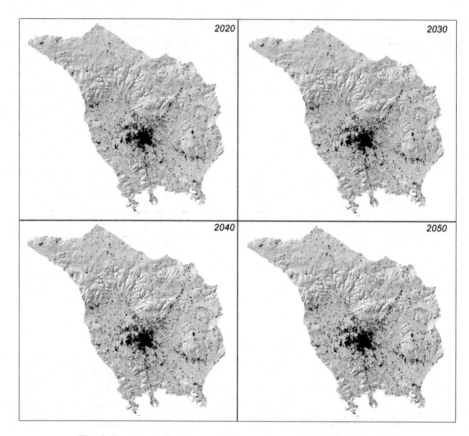

Fig. 5. Evolution of built-up areas according to the simulation results

5   Conclusions

The case study highlights how SLEUTH application provides consistent results with actual housings dynamics trend also in urban areas of southern Italy.

The analysis of coefficients obtained during the calibration phase confirms the excellent ability of the model to interpret geography of urban development based on time series of data.

The model can be useful in defining effects of alternative scenarios suitable in supporting planning choices. However, a prerequisite for its proper implementation is the continued development of technical and thematic maps from local authorities.

An important contribution, in data production, can be done by the interpretations of satellite data [16] [17].

Appling the model to a large number of municipalities in the regions of southern Italy, can be a fundamental step in monitoring soil consumption in Italy.

It is not only possible to predict the expected soil consumption according to different pressure factors on different territories, but also to compare results, especially in terms of model calibration and parameter values. Applying SLEUTH means becoming part of a large community of researchers involved in the analysis of housings dynamics. The study of Apulia Region would be significant within the Project Gigalopolis. This is the meeting point of the research community which studies the problem of modeling dynamics of urban growth at different scales. One of the purposes is to apply SLEUTH to the greatest number of territories, with the aim to analyze sprawl at global scale.

Only a complete knowledge of the phenomenon may lead to a correct definition of policies, strategies and actions to mitigate soil consumption and its negative effects [18] [19]. It is important to remember that the soil, often only as suitable for future urbanizations, plays a key role in biogeochemical cycles of water and its main vocation is agriculture.

References

1. Balena, P., Sannicandro, V., Torre, C.M.: Spatial multicrierial evaluation of soil consumption as a tool for SEA. In: Murgante, B., et al. (eds.) ICCSA 2014, Part III. LNCS, vol. 8581, pp. 446–458. Springer, Heidelberg (2014)
2. Sannicandro, V., Torre, C.M.: Monitoraggiodel land cover index e valutazionemultidimensionaledelletrasformazioniinsediativepotenziali, Atti 2015. ConvegnoRecuperiamoTerreno, IstitutoSuperiore per la Protezione e la RicercaAmbientale (ISPRA), Milano, vol. I, pp. 94–105, 6 maggio 2015. ISBN: 978-88-448-0710-8
3. Romano, B., Zullo, F.: Land urbanization in Central Italy: 50 years of evolution. J. Land Use Sci. (2012). http://dx.doi.org/10.1080/1747423X.2012.754963
4. Romano, B., Zullo, F.: Models of urban land use in Europe: assessment tools andcriticalities. Int. J. Agric. Environ. Inf. Syst. 4(3), 80–97 (2013). doi:10.4018/ijaeis.2013070105. IGI Global
5. Romano, B., Zullo, F.: The urban transformation of Italy's Adriatic Coast Strip: fifty years of unsustainability. Land Use Policy 38, 26–36 (2014)
6. ISTAT. Le problematicheconnesse al consumo del suolo (2012). http://www.istat.it/it/archivio/51331
7. Murgante, B., Borruso, G., Lapucci, A.: Geocomputation and urban planning. In: Murgante, B., Borruso, G., Lapucci, A. (eds.) Geocomputation and Urban Planning Studies in Computational Intelligence. SCI, vol. 176, pp. 1–18. Springer, Heidelberg (2009). doi:10.1007/978-3-540-89930-3_1. ISBN: 978-3-540-89929-7
8. Von Neumann, J.: Theory of Self-Producing Automata. University of Illinois Press, Urban and Chicago (1996)
9. USGS. Project gigalopolis: urban and land cover modelling. US Geological Survey (2003) http://www.ncgia.ucsb.edu/projects/gig/
10. Clarke, K.C., Hoppens, S., Gaydos, L., A self-modifying cellular automaton model of historical urbanization in the San Francisco Bay area. Environmental and Planning B: Planning and Design 24 (1997)
11. Jantz, C.A., Goetz, S.J., Shelley M.K.: Using the SLEUTH urban growth model to simualte the impacts of future policy scenarios on urban land use in the Baltimore - Washington metropolitan area. Environmental and Planning B: Planning and Design 30 (2003)

12. Clarke, K.C., Gaydos, L.J.: Loose-coupling a cellular automaton model and GIS: long-term urban growth prediction for San Francisco and Washington/Baltimore. International Journal of Geographical Information Science **12** (1998)

13. Caglioni, M., Pelizzoni, M., Rabino, G.A.: Urban sprawl: a case study for project gigalopolis using SLEUTH model. In: El Yacoubi, S., Chopard, B., Bandini, S. (eds.) ACRI 2006. LNCS, vol. 4173, pp. 436–445. Springer, Heidelberg (2006)

14. Martellozzo, F., Clarke, K.C.: Measuring urban sprawl, coalescence, and dispersal: a case study of Pordenone, Italy. Environment and Planning B: Planning and Design 2011 **38**, 1085–1104 (2011). doi:10.1068/b36090

15. Martellozzo, F.: Forecasting High Correlation Transition of Agricultural Landscapes into Urban Areas: Diachronic Case Study in North Eastern Italy. International Journal of Agricultural and Environmental Information Systems (IJAEIS) **3**(2), 22–34 (2012). doi:10.4018/jaeis.2012070102

16. Nolè, G., Murgante, B., Calamita, G., Lanorte, A., Lasaponara, R.: Evaluation of Urban Sprawl from space using open source technologies. Ecological Informatics (2014). DOI http://dx.doi.org/10.1016/j.ecoinf.2014.05.005

17. Nolè, G., Lasaponara, R., Lanorte, A., Murgante, B.: Quantifying Urban Sprawl with Spatial Autocorrelation Techniques using Multi-Temporal Satellite Data. International Journal of Agricultural and Environmental Information Systems **5**(2), 20–38 (2014). doi:10.4018/IJAEIS.2014040102. IGI Global

18. Amato, F., Pontrandolfi, P., Murgante, B.: Using spatiotemporal analysis in urban sprawl assessment and prediction. In: Murgante, B., et al. (eds.) ICCSA 2014, Part II. LNCS, vol. 8580, pp. 758–773. Springer, Heidelberg (2014). doi:10.1007/978-3-319-09129-7_55

19. Amato, F., Pontrandolfi, P., Murgante, B.: Modelli di analisi e previsionespazio-temporali per la valutazione del consumo di suoloedimplicazioninellepoliticheurbanistiche, UrbanisticaInformazioni, Anno XXXXI, Settembre-Ottobre 2014, Sessione 7, vol. 257, pp. 6–10. INU Edizioni (2014b). ISSN: 0392-5005

Discrete Spatial Assessment of Multi-parameter Phenomena in Low Density Region: The Val D'Agri Case

Giuseppe Las Casas[✉] and Francesco Scorza

School of Engineering, Laboratory of Urban and Regional Systems Engineering,
University of Basilicata, 10, Viale Dell'Ateneo Lucano 85100, Potenza, Italy
{gb.lascasas,francescoscorza}@gmail.com

Abstract. The low density regions represent the wider portion of European lagging areas. Even if several models for territorial analysis and interpretation were proposed, several issues comes when planning process start the assessment phase. The paper proposes a procedure for the spatial assessment of socio-economic phenomena based on a multiple density measures combined in a discrete model.

Keywords: Regional planning · Spatial assessment · Multidimensional kernel density estimation

1   Introduction

Principles of concentration and orientation of investments for territorial development represent new requests for the planning process [1]. In the experience of the Structural Inter-Municipal Plan of Val d'Agri, in Basilicata, these topics acquire a great importance in a situation of uncertainty which includes decisions regarding the use of territorial fossil fuels (oil and gas), the removal of constraints in public expense for development incentives, the safeguard of territorial identity through means that attracts investments.

In this point of view, considering an organic framework of priorities allows to support territorial rights on the basis of shared perspectives in which environmental sacrifices could be balanced by a revitalization of the territory that expresses growth trend and restrains a process of ageing and abandonment.

The basis for identifying the development strategy requested by the plan refers to two strategies of knowledge/analysis:

-   The first regards the identification of macroscopic issues , such as: depopulation, population ageing, risks for natural resources, the inadequate exploitation of agro-industrial and tourist resources, isolation.
-   The second is a listening and interaction strategy in political debate, for positioning in civil society, for an active participation.

Four main goals emerge in the examined area:

-   Improving the quality of life and opposing depopulation;

© Springer International Publishing Switzerland 2015
O. Gervasi et al. (Eds.): ICCSA 2015, Part III, LNCS 9157, pp. 813–824, 2015.
DOI: 10.1007/978-3-319-21470-2_59

- Fostering the employment of Val d'Agri's citizens both in traditional and innovative sectors, with reference to technologies and to means for the enhancement of environmental resources in a broad sense;
- Break of the isolation and consequent re-balancing of opportunities;
- Safeguard and promotion of environmental and cultural resources of the area.

In this sense the work presents some elaborations which aims at showing descriptive elements of suggested policies through analysis tools and the geographical representation of territorial phenomena.

The methodological innovation lies in the combination of different effects, that can be expressed through density measurements in a territorial mosaic that compares effects in pairs. This measure allows deeper understanding of functional specializations on the territory and contributes to select target areas for policies [2].

2   The Implementation Context: A Synthetic Description of Strategic Sectors

The application context of this research is represented by an inner territory of Basilicata Region. Val d'Agri is a mountainous area whose settlement system is placed along "Agri's valley floor" (S.S. 598), a road infrastructure that connects the Valley with the provinces of Potenza and Matera and Vallo di Diano in Campania Region.

This context consists of 23 municipalities, grouped into sub-areas of specialization. It is a territory characterized by very high environmental values: approximately 50% of territory is included in the National Park of Appennino Lucano and there are about 12.300 ha of SIC areas and 5.400 ha of ZPS.

The analysis of land use shows that the most of the territory is characterized by woods, pasture lands and agricultural areas.

The territory presents a lot of physical and intangible endogenous resources linked to the forms of historical villages, to agricultural techniques and products, to cultural traditions such as Lady Maria cult or arboreal rites.

The physical component of territorial heritage is widespread on the whole area. Moreover we can find unique landscapes and villages, historical and monumental riches that are not yet enhanced by a structural policy of local development.

The rural and pastoral vocation is proved by local gastronomy: Sarconi's beans, protected by IGP, Moliterno's 'Pecorino Canestrato', 'Caciocavallo podolico', Missanello's oil, cured meats, Roccanova's 'Grottino' IGT, horseradish. All of these products represent a territorial identity that connects geo-climatic features with traditional practices and knowledge. This is a structural dimension that must be considered by prospects of economic development based on innovations and on the qualification of land products.

Since the beginning of fossil energy resources exploitation, Val d'Agri has obtained an " energetic vocation" that is shown in regional planning documents (especially in PIEAR Basilicata).

This development dimension has not an endogenous foundation but represents a strategic area of interest that can attract investments, infrastructures, job, financial resources deriving from royalties and compensation forms.

It should be underlined that the identification of environmental impacts, a consequence of industrial plants and processes, is not entirely known and it depends on monitoring systems which are planned and managed at a regional level.

Today we can say that the expectations of local communities are only partially satisfied in terms of public resources invested in requalification, in support of local production activities (including agro-zootechny), in territorial promotion. Employment represents a weakest link in terms of local development since it delays a cross growth in economic sectors different from the industrial one and the migration of people in productive age is still persisting.

The "Green Economy" can be considered as a prospect for the territorial system: public and private interventions concerning the use of plants and technologies with low environmental impact are evaluated.

Renewable Energy Sources (RES) have been adopted according to the Regional Programme Framework contained in the Regional Environmental Energy Plan (PIEAR) that considers the Valley as a Regional "Energy District".

Currently this forecast is not configured in terms of systemic interventions that are able to combine production processes and technological development with installations and implementations.

As regards oil, even if the discovery of oilfields dates back to the 80s, mining activities have been started during the 90s. The Oil Center of Viggiano was built in 1996 and in subsequent years an oil pipeline was devised in order to transport crude oil to the refinery of Taranto.

The growth of mining activities and the hydrocarbon processing have brought several concerns about environmental safeguard, that is a precondition for all other forms of development.

Table 1. Stretegic development domain and specialization degree of territorial sub-areas

		Sub-area						
	Functionality	**1**	**2**	**3.1**	**3.2**	**3.3**	**4.1**	**4.2**
1	Nature tourism	5	5	5	3	5	3	1
2	Cultural Tourism	3	5	5	5	3	3	3
3	Tourist accommodation	4	4		3	3	3	
4	Health and third Age	5	5	3	1	3	1	3
5	Requalification and reuse Residences and services	3	3	5	5	3	3	5
6	District/Energy Museum	1	3	5	3	3	2	3
7	Industry	1	4	5	2	2	1	1
8	Agro-Zootechny	3	3	1	4	4	5	3

There is the need to develop a knowledge framework that allows to define an important system of choices that, despite the decline of some areas (from an environmental point of view), could reinforce the defence of other areas through the strengthening of hiking, cultural and gastronomic tourism, agriculture livestock and dairy industry, small local industries that could benefit from the low cost of energy.

Eight strategic domains can be considered as fundamentals for development prospects:

The following figure shows four sub-areas in which the implementation context has been divided.

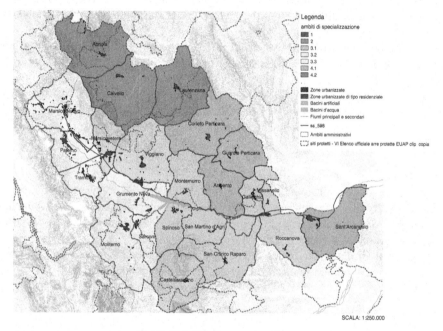

Fig. 1. Val d'Agri strategic sub-areas

3   Information Sources, Open Data and New Territorial Interpretations

The application proposes preliminary results of a territorial analysis based on several datasets concerning the following thematic areas:

- Population structure
- Investments according to regional development programs implementation
- Services supply structure.

Such basic information are characterized by the suitability for density estimation. It means that, processed by density estimation tools, such datasets express a territorial projection of each phenomena sensible for interpretation and territorial classification.

We adopted the kernel estimation density algorithm in order to asses the intensity of each phenomena: characterizing patterns of spatial distribution of event locations; determining where events are more likely to occur in space; investigating relationships between spatial "clusters" and nearby sources or other factors. Once each dimension where characterized on spatial distribution, we combine the effects in a bi-dimensional matrix highlighting the correlation of high-high (low-low) double combination of each parameter for better interpreting the local clusters and to build a geography of studied phenomena according to declared planning objectives.

It is not possible to review the enormous literature on spatial analysis here, but for relevant background material on spatial statistics it is possible to refer to Anselin and Florax [3], Basawa [4,5], Cressie [6], Ripley [7], and Tjòstheim [8].

We are in the case of a Multidimensional Kernel Density Estimation – exemplified in the figure below – where from a point pattern dataset we generate a raster estimation of spatial intensity for the phenomena.

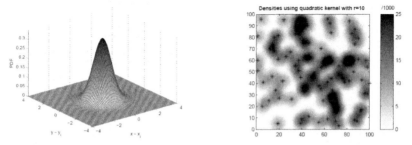

Fig. 2. Multivariate kernel density estimation schema

In the following figures we represent the point pattern distribution of datasets and its territorial density estimation according to weight models assumed for the analysis.

The population structure is represented in the map in a point-pattern dataset deriving from the local statistical census sections centroid. The kernel density estimation allows to discriminate areas with structured settlements against more unpopulated areas.

Concerning investments concentration analysis it is necessary to specify the source of information and more details concerning the procedure of territorial projections. This analysis comes from previous studies [1] concerning the exploitation of geographic dimension of open-data supporting planning process. In facts, since open data on European regional policy implementation were available through Open Coesione web-portal, it is possible to produce a spatial projection of such information. The authors described the territorialisation process and preliminary results in previous works discussing the innovative element in the assessment approach and the open procedure to obtain a punctual territorialisation of investments projects.

The detailed scale of punctual information allows to build a territorial monitoring system allowing the assessment of sectorial policy implementation and impacts.

In the figures below are represented the punctual distribution of investments delivered in the programming period 2007-2013 through Basilicata Regional Operative Programs applying EU Structural funds, and the density estimation based on the weight matrix of investments amount.

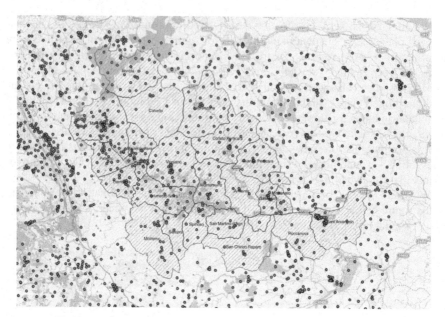

Fig. 3. Point pattern dataset of resident population (source ISTAT 2013)

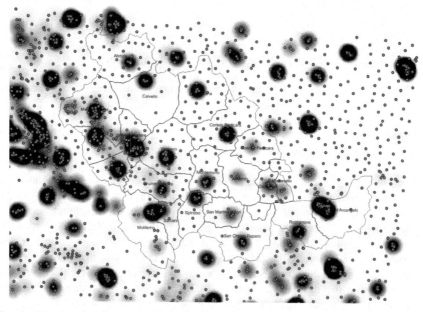

Fig. 4. Kernel estimation analysis on resident population (R=5000, I=resident_pop, cell: 30x30m)

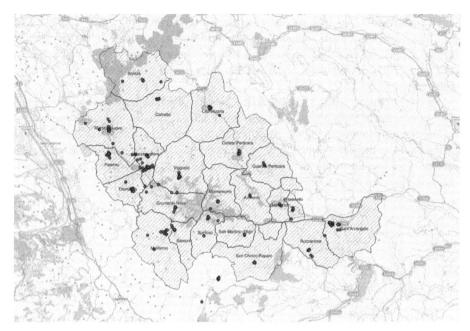

Fig. 5. Point pattern dataset of investments POR Basilicata 2007-2013

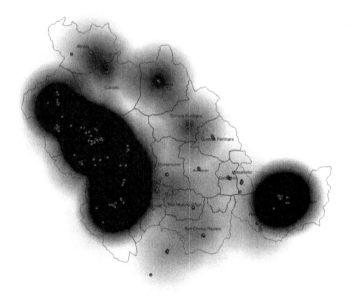

Fig. 6. Kernel estimation analysis on investments (R=5000, I=tot_€, cell: 30x30m)

The analysis of "services supply structure" in the territory of Val d'Agri is based on a data collection process based on web information services. It means that the service supply structure was generated collecting information of each service point combining: location, typology, identification information from open datasets or open web-services.

This information dataset represent the results of combining several open data sources in order to exploit spatial dimensions of information for planning support.

The figure below presents the point pattern of "services supply structure" in Val d'Agri and the density estimation based on the weight matrix that characterizes the service typology according with the invers frequency of the occurrence and the primary typology for local communities.

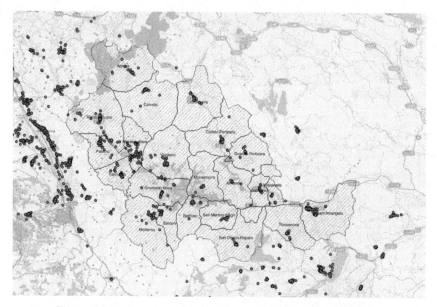

Fig. 7. Point pattern dataset of services supply structure in Val d'Agri

The obtained results allow territorial interpretations. But we combined these output in order to obtain a comprehensive multivariate classification of territorial clusters according to the contingence matrix of the phenomena.

Basically applying map algebra tools (among other authors we consider [9] [10] [11] [12]) we defined an assessment space based on raster output on which we assessed intensity of each phenomena producing distribution analysis and contingency analysis as showed in the next figures.

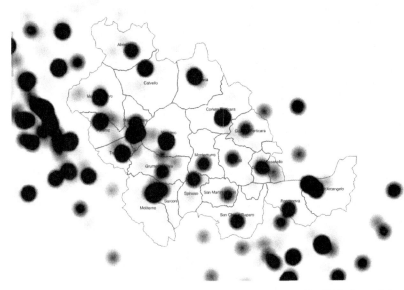

Fig. 8. Kernel estimation analysis on services supply structure (R=5000, I= weight_matrix , cell: 30x30m)

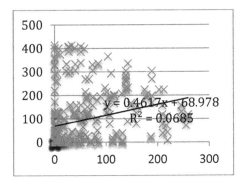

	A	B	C	D
A	163	17	18	29
B	23	14	21	17
C	10	2	7	16
D	4	10	4	2

Fig. 9. Dispersion graph Resident Population Density/ Investments Density, and contingence matrix (cell: 100x100m)

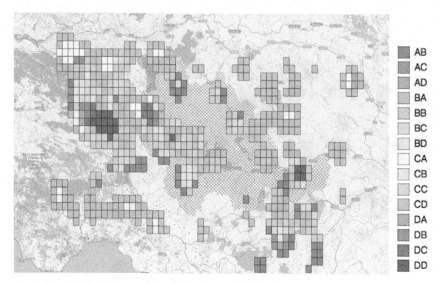

Fig. 10. Spatial assessment and clusters deriving from combining phenomena

4   Conclusions

How to express operatively the "Place based approach" (Barca [13]) in order to bring effective innovations in EU cohesion policy management? Our position starts from territorial interpretation models as operative tools for improving places understanding. Under a planning point of view it means specializations, identities, values to be integrated in a shared development strategy.

Open data phenomena represent an useful process that already driven the research from data production to exploitation of the informative value of several data sources available for everybody. In this paper we combine such new investigation domain with traditional territorial intensity estimation of discrete phenomena in order to get a comprehensive analysis.

The results represent a preliminary but promising application. Several domain need to be included in the analysis. Among others: real estate information ([14] [15] [16]), landscape and natural values, historical manufacts and cultural landscapes. The information management and exchange implies problems in interoperability [17] [18] between sources, procedures and technologies. Such investigation looks at generating specialization analysis at territorial level, as base descriptive geographies for EU cohesion programming at variable scale [19].

The perspective regards the application of such processes in developing programs involving beneficiaries and citizens looking at local chains of feasible interventions and effective projects.

References

1. Las Casas, G., Lombardo, S., Murgante, B., Pontrandolfi, P., Scorza, F.: Open Data for Territorial Specialization Assessment. Territorial Specialization in Attracting Local Development Funds: an Assessment. Procedure Based on Open Data and Open Tools, in Smart City. Planning For Energy, Transportation And Sustainability Of The Urban System Special Issue, June 2014 Editor-in-chief: Rocco Papa, TeMA Journal of Land Use, Mobility and Environment, print ISSN 1970-9889 l on line ISSN 1970-9870 (2014)
2. Las Casas, G.: Processo di piano ed esigenze informative. In: Clemente, F. (ed.) Pianificazione del Territorio e sistema informativo. F.Angeli, Milano (1984)
3. Anselin, L., Florax, R.J.G.M.: New Directions in Spatial Econometrics. Springer, Berlin (1995)
4. Basawa, I.V.: Special issue on spatial statistics, Part I. J. Statist. Plann. Inference **50**, 311–411 (1996)
5. Basawa, I.V.: Special issue on spatial statistics, Part II. J. Statist. Plann. Inference **51**, 1–97 (1996)
6. Cressie, N.A.C.: Statistics for Spatial Data. Wiley, New York (1991)
7. Ripley, B.: Spatial Statistics. Wiley, New York (1981)
8. Tjòstheim, D.: Spatial series and time series: similarities and differences. In: Droesbeke, F. (ed.) Spatial Processes and Spatial Time Series Analysis, pp. 217–228. FUSL, Brussels (1987)
9. Câmara, G., Palomo, D., De Souza, R.C.M., De Oliveira, O.R.F.: Towards a generalized map algebra: Principles and data types. In GEOINFO 2005 - 7th Brazilian Symposium on GeoInformatics (2005). http://www.scopus.com/inward/record.url?eid=2-s2.0-84870666 744&partnerID=40&md5=6c686125f42654ccb6e642b32a38b241
10. Frank, A.U.: Map algebra extended with functors for temporal data. In: Akoka, J., et al. (eds.) ER Workshops 2005. LNCS, vol. 3770, pp. 194–207. Springer, Heidelberg (2005). doi:10.1007/11568346_22
11. Takeyama, M., Couclelis, H.: Map dynamics: Integrating cellular automata and GIS through Geo-Algebra. International Journal of Geographical Information Science **11**(1), 73–91 (1997). http://www.scopus.com/inward/record.url?eid=2-s2.0-0030997404&partner ID=40&md5=fbbc1ea989765e4833ca7170a6a35dad
12. Jaakkola, O.: Multi-scale Categorical Data Bases with Automatic Generalization Transformations Based on Map Algebra. Cartography and Geographic Information Science **25**(4), 195–207 (1998). http://www.scopus.com/inward/record.url?eid=2-s2.0-0032184602 &partnerID=40&md5=4d58963356322747c432b7394536d3bc
13. Barca, F., McCann, P.: Methodological note: outcome indicators and targets-towards a performance oriented EU cohesion policy and examples of such indicators are contained in the two complementary notes on outcome indicators for EU2020 entitled meeting climate change and energy objectives and improving the conditions for innovation, research and development (2011). http://ec.europa.eu/regional_policy/sources/docgener/evaluation/performance_en.htm (accessed October 1, 2011)
14. Tajani, F., Morano, P.: Concession and lease or sale? A model for the enhancement of public properties in disuse or underutilized. WSEAS Transactions on Business and Economics **11**, 787–800 (2014)
15. Morano, P., Tajani, F.: Estimative analysis of a segment of the bare ownership market of residential property. In: Murgante, B., Misra, S., Carlini, M., Torre, C.M., Nguyen, H.-Q., Taniar, D., Apduhan, B.O., Gervasi, O. (eds.) ICCSA 2013, Part IV. LNCS, vol. 7974, pp. 433–443. Springer, Heidelberg (2013)

16. Tajani, F., Morano, P.: Evaluation of the economic sustainability of the projects in social housing. In: Murgante, B., et al. (eds.) ICCSA 2014, Part III. LNCS, vol. 8581, pp. 135–147. Springer, Heidelberg (2014)

17. Scorza, F., Casas, G.L., Murgante, B.: Overcoming interoperability weaknesses in e-government processes: organizing and sharing knowledge in regional development programs using ontologies. In: Lytras, M.D., Ordonez de Pablos, P., Ziderman, A., Roulstone, A., Maurer, H., Imber, J.B. (eds.) WSKS 2010. CCIS, vol. 112, pp. 243–253. Springer, Heidelberg (2010). doi:10.1007/978-3-642-16324-1_26. ISSN: 1865-0929

18. Laurini, R., Murgante, B.: Interoperabilità semantica e geometrica nelle basi di dati geografiche nella pianificazione urbana. In: Murgante, B. (ed.) L'informazione geografica a supporto della pianificazione territoriale, pp. 229–244. FrancoAngeli, Milano (2008)

19. Scorza, F.: Improving EU cohesion policy: the spatial distribution analysis of regional development investments funded by EU structural funds 2007/2013 in Italy. In: Murgante, B., Misra, S., Carlini, M., Torre, C.M., Nguyen, H.-Q., Taniar, D., Apduhan, B.O., Gervasi, O. (eds.) ICCSA 2013, Part III. LNCS, vol. 7973, pp. 582–593. Springer, Heidelberg (2013). doi:10.1007/978-3-642-39646-5_42

Author Index